Notes on Numerical Fluid Mechanics and Multidisciplinary Design 154

Notes on Numerical Fluid Mechanics and Multidisciplinary Design publishes state-of-art methods (including high-performance methods) for numerical fluid mechanics, numerical simulation and multidisciplinary design optimization. The series includes proceedings of specialized conferences and workshops, as well as relevant project reports and monographs.

Indexed by SCOPUS, zbMATH, SCImago.

All books published in the series are submitted for consideration in Web of Science.

Andreas Dillmann · Gerd Heller · Ewald Krämer ·
Claus Wagner · Julien Weiss
Editors

New Results in Numerical and Experimental Fluid Mechanics XIV

Contributions to the 23rd STAB/DGLR Symposium, Berlin, Germany, 2022

Springer

Editors
Andreas Dillmann
Institut für Aerodynamik und Strömungstechnik
Deutsches Zentrum für Luft- und Raumfahrt e. V.
Göttingen, Niedersachsen, Germany

Ewald Krämer
Institut für Aerodynamik und Gasdynamik
Universität Stuttgart
Stuttgart, Baden-Württemberg, Germany

Julien Weiss
ILR - Aerodynamik
Technische Universität Berlin
Berlin, Germany

Gerd Heller
Airbus Defence and Space GmbH
Manching, Bayern, Germany

Claus Wagner
Institut für Aerodynamik und Strömungstechnik
Deutsches Zentrum für Luft- und Raumfahrt e. V.
Göttingen, Niedersachsen, Germany

ISSN 1612-2909 ISSN 1860-0824 (electronic)
Notes on Numerical Fluid Mechanics and Multidisciplinary Design
ISBN 978-3-031-40481-8 ISBN 978-3-031-40482-5 (eBook)
https://doi.org/10.1007/978-3-031-40482-5

This Springer imprint is published by the registered company Springer Nature Switzerland AG
The registered company address is: Gewerbestrasse 11, 6330 Cham, Switzerland

Preface

This volume contains the papers presented at the 23rd STAB/DGLR Symposium held in Berlin, Germany (November 9–10, 2022), organized by the Chair of Aerodynamics at TU Berlin.

STAB is the German Aerodynamics/Fluid Mechanics Association (Deutsche Strömungsmechanische Arbeitsgemeinschaft) founded in the late 1970s, and DGLR is the German Society for Aeronautics and Astronautics (Deutsche Gesellschaft für Luft- und Raumfahrt – Lilienthal-Oberth e.V.). The mission of STAB is to foster aerodynamics research and its appreciation in Germany. This is accomplished by creating forums for scientific discussions and by disseminating most recent research results, thereby promoting scientific progress and avoiding unnecessary duplication in research.

STAB brings together scientists and engineers from universities, research establishments and industry. They present research and project work in numerical and experimental fluid mechanics as well as aerodynamics for a wide variety of fields, such as aeronautics, space, ground transportation, wind turbines and other applications. This format also offers an excellent opportunity for exchange on numerous joint research activities supported by different funding agencies.

Since 1986, the symposium has taken place every two years at different locations in Germany, all having an affinity for fluid mechanics and aerodynamics.

In addition, STAB workshops are regularly held at the Deutsches Zentrum für Luft- und Raumfahrt (DLR) in Göttingen in the intermediate years.

Both, STAB symposia and workshops, provide excellent forums where new research activities can be presented, often resulting in new jointly organized research and technology projects.

In this volume, symposium contributions are published for the 14th time, following a thorough peer-review process.

The review panel for the papers included in this volume consisted of *G. Babij (Göttingen), E. Bagheri (Erlangen), P. Bahavar (Göttingen), R. Barta (Göttingen), E. Batmaz (Göttingen), C. Bauer (Göttingen), J. Bell (Göttingen), L. Bertsch (Göttingen), J. Braukmann (Göttingen), C. Breitsamter (München), G. Brenner (Clausthal), C. Brückner (Göttingen), A. Buhr (Göttingen), C. Corsi (Berlin), A. Dachwitz (Göttingen), M. Danker (Neubiberg), K. Davies (Göttingen), C. Ebert (Berlin), M. Eck (Berlin), K. Ehrenfried (Göttingen), S. Ehrmanntraut (Dresden), R. Ewert (Braunschweig), M. Fehrs (Göttingen), D. Feldmann (Bremen), M. Fertig (Braunschweig), U. Fey (Göttingen), A. Fischer (Bremen), A. Gardner (Göttingen), N. Giannelis (Newcastle, Australia), A. Goerttler (Göttingen), V. Hannemann (Göttingen), U. Hartmann (München), S. Hein (Göttingen), D. Heine (Göttingen), W. Heinze (Braunschweig), G. Heller (Manching),*

S. Helm (Göttingen), S. Herzog (Göttingen), M. Hilfer (Göttingen), R. Höld (Schrobenhausen), J. Hwang (San Diego), S. Jakirlic (Darmstadt), F. Jansen (Bremen), M. Jentys (München), I. Joksimovic (Darmstadt), M. Keßler (Stuttgart), T. Kilian (Braunschweig), M. Klein (Cottbus), T. Knopp (Göttingen), A. Kohl (Göttingen), T. Kowalski (Göttingen), E. Krämer (Stuttgart), L. Krenkel (Regensburg), N. Kroll (Braunschweig), A. Krumbein (Göttingen), C. Lübon (Biberach), T. Lutz (Stuttgart), C. Mertens (Delft), O. Meyer (Neubiberg), M. Mommert (Göttingen), M. Müller (Göttingen), P. Nau (Stuttgart), G. Oblapenko (Göttingen), F. Oehme (Bremen), D. Peitsch (Berlin), M. Pfeiffer (Stuttgart), R. Radespiel (Braunschweig), M. Raffel (Göttingen), T. Raitor (Göttingen), D. Ramaswamy (Aachen), D. Reckzeh (Bremen), K. Rewitz (Aachen), U. Rist (Stuttgart), J. Ruhland (München), M. Rütten (Göttingen), C. Sabater (Braunschweig), J. Schein (Neubiberg), D. Schiepel (Göttingen), C. Schnepf (Göttingen), E. Schülein (Göttingen), A. Schütte (Braunschweig), C. Schwarz (Göttingen), B. Selent (Stuttgart), A. Shishkin (Göttingen), D. Soper (Birmingham), R. Streblow (Aachen), A. Theiß (Göttingen), M. Thieme (Berlin), J. Tschepe (Berlin), B. van der Wall (Braunschweig), F. Wachter (Erlangen), S. Wagner (Gaimersheim), A. Waldmann (Stuttgart), X. Wang (Darmstadt), F. Webner (Göttingen), K. Weinman (Göttingen), J. Weiss (Berlin), C. Westphal (Berlin), R. Wickersheim (Stuttgart), J. Wild (Braunschweig), C. Wolf (Göttingen), Y. Wu (Stuttgart) and J. Zastrow (Göttingen).

Nevertheless, the authors sign responsible for the contents of their contributions.

The editors are grateful to Prof. Dr. W. Schröder as Series Editor of the "Notes on Numerical Fluid Mechanics and Multidisciplinary Design" series and to the Springer publishing house for the opportunity to publish the results of the symposium.

February 2023

Andreas Dillmann
Gerd Heller
Ewald Krämer
Claus Wagner
Julien Weiss

Contents

Aeroelasticity and Structural Dynamics

Analysis of the Boundary Layer on a Highly Flexible Wing Based on Infrared Thermography Measurements ... 3
Christoph Mertens, Adrián Grille Guerra, Bas W. van Oudheusden, Michael Fehrs, and Markus R. Ritter

Numerical Investigation of a Spoiler Effect on the Transonic Flutter Boundary ... 14
Martin Schmalz and David Quero

Experimental Aerodynamics

Measuring the Oncoming Flow that Operational Freight-Trains Experience Using the DLR FR8-LAB ... 27
James R. Bell, Alexander Buhr, and Arne Henning

Experimental and Numerical Investigation of Passive Measures to Limit Aerodynamic Forces on Wind Turbine Rotor Blades ... 38
Marten Danker, Maximilian Maigler, Oliver Meyer, and Jochen Schein

A Miniature Flow Sensor Capable of Determining the Instantaneous High-Frequency Wall Shear Stress Fluctuations in Magnitude and Direction ... 48
Mario Eck, Dieter Peitsch, and Christian Westphal

Pressure Waves and Flow Induced by a Train in a Tunnel ... 58
Klaus Ehrenfried and Daniela Heine

Construction of a Flow Test Rig with a Predefined Shape of the Boundary Layer to Investigate Sand Absorption by Side Flaps on a High-Speed Train ... 68
Uwe Fey, Thomas Kowalski, Torsten Gries, and Manuel Gageik

Analysis of the Unsteady Loads on Train Models in Wind Tunnels ... 78
Daniela Heine, Klaus Ehrenfried, Torsten Gries, Gerrit Heine, and Max Müller

Low-Order Modeling of Bistable Side Forces on a Sphere Measured for a Transient Inflow in a Wind Tunnel ... 88
Max Müller, Klaus Ehrenfried, and Claus Wagner

Effects of Jet-Orifice Shape on the Flow-Control Effectiveness of Air-Jet Vortex Generators ... 99
Deepak Prem Ramaswamy and Anne-Marie Schreyer

Discovering Latent Physical Variables from Experimental Data in Supersonic Flow Using Physics-Informed Neural Networks (PINNs) ... 109
Lennart Rohlfs and Julien Weiss

Surrogate Based Prediction of 2D Car Wake ... 119
Mathis Thieme, Stephan Löffler, and Julien Weiss

Hypersonic Aerothermodynamics

Validation for a Polyatomic Model in a Fokker-Planck Solver Based on the Extended Master Equation Ansatz ... 133
Leo Basov, Martin Grabe, and Aaron Nagel

Uncertainty Quantification of Expanding High-Enthalpy Air Flows ... 143
Georgii Oblapenko and Volker Hannemann

Influences on Langmuir Probe Measurements by an ECR Thruster with Magnetic Nozzle ... 153
Clara Schäfer, Jana Zorn, Kristof Holste, and Peter J. Klar

Calibration of Heterodyne Dual Frequency Comb Laser Absorption Spectroscopy for NO and H_2O Detection ... 163
Leni Schmidt and Jan Martinez Schramm

Multidisciplinary Optimization

On the Feasibility of Drone Towing ... 177
Georg E. Koppenwallner

Gust and Maneuver Load Alleviation in Conceptual Aircraft Design ... 186
Markus Dino Krengel and Martin Hepperle

Digital Design of Moveables at DLR Virtual Product House ... 196
Fabian Lange-Schmuckall, Alexander Zakrzewski, Martin Rädel, and René Hollmann

Optimization of Control Surface Deflections on the High Aspect Ratio Wing to Improve Cruise Flight Performance ... 206
Tobias Franziskus Wunderlich and Florian Siebert

Computational Aerodynamics

Scalable Framework Integration of CODA for a Multidisciplinary Preconditioned Matrix-Free Newton-Krylov Method 219
Simon Ehrmanntraut, Adam Büchner, Sebastian Gottfried, and Arthur Stück

Surrogate Modeling of Hybrid Laminar Wing Aerodynamic Coefficients 231
Mauricio M. Jentys and Christian Breitsamter

Development of a Coupled Implicit Incompressible Flow Solver in CODA 241
Deepak Kunhappan and Johannes Löwe

An Analytical View on Data-Driven Turbulence Modeling and a Realization via a regularized Newton Method 252
Stefan Langer

Comparison and Unification of Finite-Volume Discretization Strategies for the Unstructured Node-Centered and Cell-Centered Grid Metric in TAU and CODA 262
Stefan Langer, Axel Schwöppe, and Tobias Leicht

Deformation of CFD Meshes with Anisotropic Cells in a Viscous Boundary Layer Using Line-Implicit Methods 273
Arne Rempke

Numerical Simulation of Vortex-Dominated Flows Using Advanced Physical Models 284
Vamshi Togiti

Preliminary CFD Validation of Unsteady Onflow Effects on Model Train Drag 294
Keith A. Weinman and Klaus Ehrenfried

HyperCODA – Towards High-Performing Time-Resolving Flow Simulations 305
Johannes Wendler, Immo Huismann, Ronny Tschüter, and Stefan Fechter

Reproducing Vortex Topology Mode Alternation of a Generic Combat Aircraft Model by Means of Hybrid RANS-LES Simulations 315
Jonas Zastrow and Axel Probst

Aeroacoustic Noise

Relevant Turbulent Time Scale for Broadband Sound and the Prediction of the Grid Cut-Off Frequency for LES Based Aeroacoustic Simulations 329
Eman Bagheri, Florian Wachter, and Stefan Becker

Numerical Study of Individuals Exposure to Road-Noise in Urban Environments 339
Amin Nabikhani, Tobias Simon Müller, and Arne Henning

Aeroacoustics of High Fidelity URANS Simulations of Distributed Electric Propellers 349
Robin Wickersheim, Manuel Keßler, and Ewald Krämer

Flow Control

A New Approach of Using Porous Bleed Boundary Conditions - Application of Local Porosity 361
Julian Giehler, Pierre Grenson, and Reynald Bur

Simulation of Suction Chambers with Variable Pressure Distribution for HLFC Applications 372
Heinrich Lüdeke

Some Specific Challenges of HLFC Design on a Long-Range Wing 382
Geza Schrauf and Thomas Kilian

Parametric Study of Vortex Generator Jets by Direct Numerical Simulation 392
Björn Selent, Ulrich Rist, and Oliver T. Schmidt

Direct Numerical Simulation of Counter-Rotating Cylindrical Roughness Pairs for Laminar Flow Control 402
Yongxiang Wu, Christoph Wenzel, and Ulrich Rist

Turbulence and Transition

Turbulent Wedge Modeling in Local Correlation-Based Transition Models 415
Michael Fehrs and Sebastian Helm

First Comparison of CFD Simulation and Wind Tunnel Test of the Forward-Swept Natural Laminar Flow Model NLF-ECOWING-FSW ... 425
Sebastian Helm, Kristopher Davies, and Michael Fehrs

Stochastic Modeling and Large-Eddy Simulation of Heated Concentric Coaxial Pipes 435
Marten Klein, Pei-Yun Tsai, and Heiko Schmidt

Large Eddy Simulations Towards Industrial Relevant Flows on Graphics Processing Units 445
Christian Lübon

Contribution of Vortices to Laminar-Turbulent Transition in a Boundary Layer Disturbed by a Roughness Element 455
Tristan M. Römer, Christoph Wenzel, and Ulrich Rist

Curvature Correction for Turbulent Diffusion Inside a Differential Reynolds-Stress Model 465
Philip Ströer and Tobias Knopp

A Surrogate-Based e^N Transition Prediction Method for Three-Dimensional Compressible Boundary Layers 475
Alexander Theiss and Stefan Hein

Overview of the Wind Tunnel Tests with Suction in the DNW-NWB 486
Richard von Soldenhoff

Flow in a Concentric Annulus with Inner- or Outer-Cylinder Rotation: A Conventional and Sensitized RANS Modelling Study 496
Xiaoyu Wang, Jeanette Hussong, and Suad Jakirlic

Technical Flows

DNS of Aerosol Particle Spreading Emitted by Coughing and Breathing in a Simplified Room 509
Ege Batmaz, Philipp Bahavar, Daniel Schmeling, and Claus Wagner

Flow Reconstruction of Urban Wind Fields for Wind-Based Path Planning 519
Carola Ebert and Julien Weiss

Towards Indirect Assessment of Surface Anomalies on Wind Turbine Rotor Blades 529
Daniel Feldmann, Felix Oehme, Lennart von Germersheim, Rubén López Parras, Andreas Fischer, and Marc Avila

Development of a Low-Cost Thermal Human Manikin with Seven Zones for Comfort Assessment in Passenger Cabins 539
Daniel Schiepel, Felix Werner, and Daniel Schmeling

Numerical Study of Aerosol Dispersion in the Aircraft Cabin 549
Andrei Shishkin, Daniel Schiepel, and Daniel Schmeling

Generation and Distribution of Surgical Smoke During High Frequency Electrocauterization 559
Vera Stelzer, Sandra Melina Tauwald, Veronika Vielsmeier, Fabian Cieplik, Arne Kandulski, Wulf Schneider-Brachert, Olaf Wünsch, Markus Rütten, and Lars Krenkel

High Spatial Resolution Tomo-PIV of the Trachea Focussing on the Physiological Breathing Cycle 569
Sandra Melina Tauwald, Maurizio Quadrio, Markus Rütten, Christian Stemmer, and Lars Krenkel

An Approach to Automated Detection of Sessile Droplets in Mixed Convection 579
Marie-Christine Volk, Konstantin Niehaus, Andreas Westhoff, and Claus Wagner

Aerosol Spread in a Generic Train Entrance: Comparison Between Experiment and Numerical Simulation 590
Florian Webner, Andreas Kohl, Daniel Schmeling, and Claus Wagner

Rotorcraft

Dynamic Stall Computations of a Double-Swept Rotor Blade with Rigid and Elastic Modelling 603
Georg Babij and Martin Michael Müller

Wind Tunnel Analysis of Helicopter Aerodynamics with Varying Rotor Head Blade Stub Lengths 613
Ulrich Hartmann, Johannes Ruhland, and Christian Breitsamter

Vortex-Induced Stall on an Actively Twisted Highly Loaded Model Rotor Blade 623
Gunther A. Wilke, Berend G. van der Wall, Yasutada Tanabe, Hideaki Sugawara, Do-Hyung Kim, Hee J. Kang, Sung N. Jung, Seong H. Hong, Joëlle Bailly, George Barakos, Rinaldo Steininger, and Joon W. Lim

High-Agility Configuration

Influence of Vertical Stabilizer on the Aerodynamic Stability of a Generic Multi Delta Wing Aircraft Configuration 637
Katharina Alt and Andreas Schütte

Invariant Analysis of Vortical Delta Wing Flow Using the Extended Optimal Triple Tensor Decomposition 647
Markus Rütten, Michael Werner, and Jonas Zastrow

Numerical Investigation of the Unsteady Behaviour of Side Jet Induced Control Concepts 658
Christian Schnepf

Airplane Aerodynamics/Propulsion Integration

Aerodynamic Design of Shock Control Bumps Considering Structural Constraints 671
Andreas Goerttler, Sven Christian Künnecke, and Christian Sabater

Increased Efficiency of Shock Control Bumps for Buffet Onset Delay Due to Variable Camber 682
Jens Müller, Andreas Waldmann, Andreas Goerttler, Sven Christian Künnecke, and Thorsten Lutz

Shock Control Bumps for Buffet Onset Delay on Sections of a Wing with Hybrid Laminar Flow Control 693
Andreas Waldmann, Jens Müller, Andreas Goerttler, and Thorsten Lutz

Author Index 705

Aeroelasticity and Structural Dynamics

Analysis of the Boundary Layer on a Highly Flexible Wing Based on Infrared Thermography Measurements

Christoph Mertens[1(✉)], Adrián Grille Guerra[1], Bas W. van Oudheusden[1], Michael Fehrs[2], and Markus R. Ritter[2]

[1] Faculty of Aerospace Engineering, Delft University of Technology, Kluyverweg 1, 2629 HS Delft, The Netherlands
C.Mertens@tudelft.nl

[2] Institute of Aeroelasticity, German Aerospace Center (DLR), Bunsenstraße 10, 37073 Göttingen, Germany

Abstract. The effects of the wing skin distortion on the boundary layer of a highly flexible wing are analyzed in a wind tunnel experiment using infrared thermography measurements. Considerable differences in the boundary layer flow are observed when comparing the sections of the wing near the ribs, where the design shape of the wing is preserved, and in between the ribs. At the spanwise locations between the ribs, the sectional wing shape distorts and triggers boundary layer transition close to the leading edge. The differences between the design behavior of the wing and the experimental results of the boundary layer analysis demonstrate the need for considering the skin deformation and its effects on the boundary layer flow when designing highly flexible wings.

Keywords: Pazy wing · laminar separation bubble · differential infrared thermography · wind tunnel test

1 Introduction

The Pazy wing is a recently introduced experimental benchmark geometry for highly flexible wing studies [1]. The purpose of the benchmark model design is to sustain very large deformations in wind tunnel experiments, with wingtip deflections above 50% of the span, to provide experimental validation data for the activities of the Large Deflection Working Group within the framework of the third Aeroelastic Prediction Workshop[1]. Since the publication of the first measurement data of the Pazy wing geometry, the data have already been used as a reference in several studies that focus on the development of novel non-linear aeroelastic prediction models [2,3]. The wing model, shown in Fig. 1, is constructed from a 3D-printed chassis that forms the leading and trailing edges as well as the ribs, which have a NACA 0018 cross-sectional shape. The stiffness

[1] https://nescacademy.nasa.gov/workshops/AePW3/public/wg/largedeflection

A. Dillmann et al. (Eds.): STAB/DGLR Symposium 2022, NNFM 154, pp. 3–13, 2024.
https://doi.org/10.1007/978-3-031-40482-5_1

of the wing is mainly provided by an aluminum spar plate that is glued inside the chassis. A thin Oralight iron-on film is used as the skin and is applied to the chassis by thermal shrinking. This construction process does not preserve the design airfoil shape between the ribs due to the flexibility of the foil and therefore locally alters the aerodynamic shape of the wing from the design shape. The first experimental studies of the Pazy wing at the Technion were focused on measurements of the structural dynamics [1], and did not include any dedicated aerodynamic characterization that would permit the analysis of this shape alteration effect.

Fig. 1. Pazy wing installed vertically in the wind tunnel test section at the Technion [1]

A further experimental study with an adapted version of the Pazy wing geometry, the Delft-Pazy wing, has been conducted at TU Delft. This study was focused on combined structural and aerodynamic measurements in the wind tunnel using a particle tracking velocimetry approach [4]. Furthermore, surface measurements with an infrared camera were conducted to analyze the boundary layer on the wing. However, while the interpretation of the infrared thermography measurement data is straightforward for steady attached flows, the analysis of flows with separation is more challenging [5]. Additionally, the high degree of flexibility of the wing exposes it to vibrations and thus unsteady flow conditions, further complicating the analysis [6].

In this paper, the boundary layer on the suction side of the Delft-Pazy wing is analyzed based on infrared thermography measurements. The purpose of this study is to demonstrate and assess an experimental approach to describe the effect of surface deformation on the aerodynamics of highly flexible wing structures. To support the analysis of the experimental data, a computational fluid dynamics (CFD) analysis of the Delft-Pazy wing is performed to provide insight into the expected flow patterns. Subsequently, the infrared measurements are analyzed using state-of-the-art data analysis methods to infer the boundary layer state at different angles of attack in steady as well as oscillating inflow conditions, produced by a gust generator.

2 Boundary Layer on the Delft-Pazy Wing

A CFD analysis using the DLR TAU-code [7] is performed to assess the aerodynamic differences that are expected to occur between the design shape of the Pazy wing and the actual Delft-Pazy wing geometry that was used in the wind tunnel tests. The first step in this investigation is to generate an accurate model of the shape of the Delft-Pazy wing to be used in the simulations. This is achieved by performing a 3D-scan of the wing in the laboratory, in unloaded conditions, using a FARO Quantum Max scanner arm (maximum permissible error below 0.1 mm)[2]. The result of the shape scan is visualized in Fig. 2. Two cross-sectional slices of the scanned geometry are shown, section I between two ribs and section II on a rib, as well as one slice along the span at $x/c = 0.25$. The differences between the outer shape of the Delft-Pazy wing to the design shape are clearly visible, the maximum thickness of the wing section between the ribs is reduced by more than 10% compared to the reference NACA 0018 airfoil.

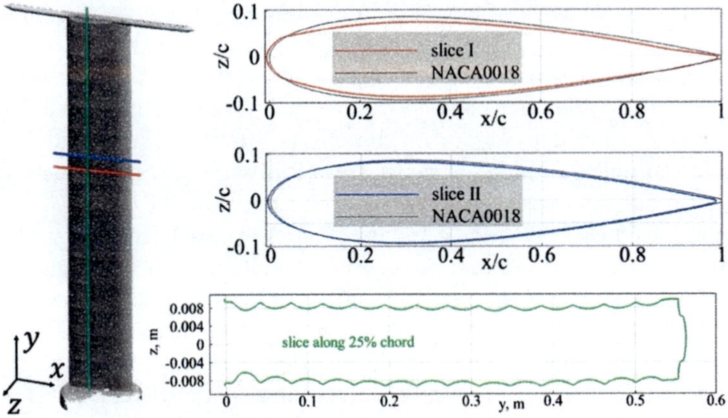

Fig. 2. Shape scan of the outer geometry of the Delft-Pazy wing

The shape scan data is further processed by generating a smooth geometry, correcting for imperfections of the shape scan due to gaps in the data, and afterward generating a grid for the CFD simulations. The CFD simulations are performed using the Reynolds-averaged formulation of the Navier-Stokes equations with the γ-model for boundary layer transition prediction [8]. Further details on the implementation of the CFD simulations can be found in [9].

[2] https://www.faro.com/en/Products/Hardware/Quantum-FaroArms

In Fig. 3, the results from the CFD simulations in terms of the skin friction coefficient distribution on the suction side at a chord-based Reynolds number of Re = 340 000 and an angle of attack of $\alpha = 4°$ are shown. On the left are the results for the design shape of the Pazy wing, and on the right, the results obtained for the scanned geometry of the Delft-Pazy wing are shown.

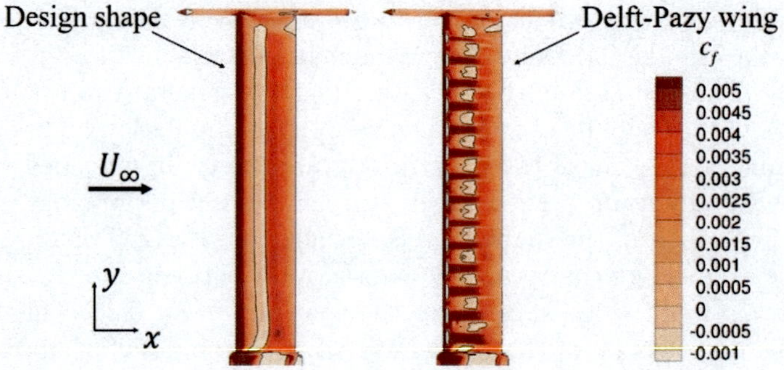

Fig. 3. **Comparison of skin friction coefficient distributions at $\alpha = 4°$** angle of attack

For the design shape, the flow is initially laminar, then separates at around $x/c = 0.3$ and reattaches at around $x/c = 0.45$, as indicated by the black lines. The increased skin friction level after the reattachment point indicates that transition to turbulence has occurred in the separated shear layer, thus forming a spanwise uniform laminar separation bubble (LSB). For the actual Delft-Pazy wing, a more complex boundary layer pattern is observed, with strong differences depending on the spanwise location. Near the locations of the ribs, the design airfoil is preserved well and the boundary layer flow closely resembles that of the design shape, forming a laminar separation bubble with a reattachment point around mid-chord. In between the ribs, the flow immediately separates downstream of the leading edge, at the kink in the sectional airfoil shape (originating from the chassis design of the Pazy wing, see Fig. 2). The flow then reattaches in a turbulent state at around $x/c = 0.1$ in the central part between the ribs. Closer to the ribs, a spanwise flow component from the rib to the sagged areas occurs, resulting in a pattern of consecutive separations in the chordwise direction. The occurrence of this complex behavior is verified experimentally in the following.

3 Experimental Setup

The infrared thermography measurements were performed at the Open Jet Facility wind tunnel at Delft University of Technology, which has a 2.85 m×2.85 m test section and a freestream turbulence intensity of around $Tu = 0.5\%$. The Delft-Pazy wing (chord $c = 0.1$ m, span $s = 0.55$ m) was mounted vertically in the wind tunnel test section and subjected to an inflow velocity of $U_\infty = 18.3\,\mathrm{m\,s^{-1}}$, corresponding to a Reynolds number of Re = 122 000. The measurements were performed with a FLIR SC7300 infrared camera, which was acquiring thermographic images of the suction side of the wing with a resolution of 320×256 pixels at a frame rate of 230 Hz, using an integration time of 250 µs. The infrared camera measured the infrared radiation emitted from the surface of the wing, which is proportional to its surface temperature. To facilitate the analysis of the boundary layer state on the wing based on the thermographic measurements, the wing surface was heated above the temperature of the freestream with a halogen lamp.

The experimental setup and its components in the wind tunnel are shown in Fig. 4. The experimental conditions for the infrared measurements include steady inflow at angles of attack in the range $0° < \alpha < 14°$, as well as one test case with unsteady inflow conditions at $\alpha = 5°$. With steady inflow, the wingtip reached a maximum tip displacement of 16% of the span. To generate the unsteady inflow, a gust generator was mounted upstream of the wing and operated continuously at a frequency of 5.7 Hz, corresponding to a reduced frequency of $k = f\pi c/U_\infty = 0.1$, and the amplitude of the sinusoidal inflow variation due to the gust was $\alpha_g = \pm 2.5°$. The wing responded to this gust inflow with a sinusoidal oscillation with a small dynamic tip deflection amplitude of 1.5% of the span width, which was in anti-phase with the gust excitation phase angle ϕ [4].

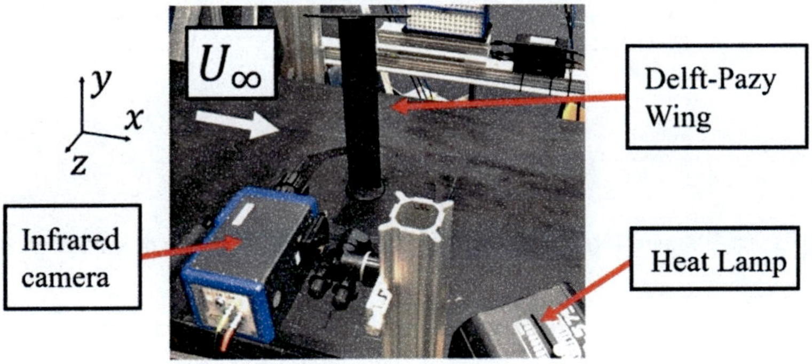

Fig. 4. Wind tunnel setup for the infrared thermography measurements

4 Infrared Thermography Data Analysis

The acquired infrared images are processed for the purpose of detecting of the boundary layer state. The physical values of the surface temperature are not required for this, hence no temperature calibration is performed and the data is analyzed in terms of the infrared radiation in camera counts. The infrared images are mapped onto the wing's reference system to account for the (dynamic) deflection of the wing by detecting the intersections of the ribs with the edges of the wing. For the steady inflow case, the field of view of the infrared camera captures the full span of the wing. For each α, 100 images are taken for time-averaging. In the gust response situation, the spatial resolution is increased by capturing only a central region of the wing, covering 20% of the span.

The main features of a laminar separation bubble are the laminar separation, transition within the separated shear layer, and turbulent reattachment. In steady inflow, these features can be extracted from thermographic measurements by analyzing the chordwise temperature gradient [5]. Two examples of this approach are shown in Fig. 5, using data measured in the wind tunnel at $\alpha = 4°$ for the two sections shown in Fig. 2.

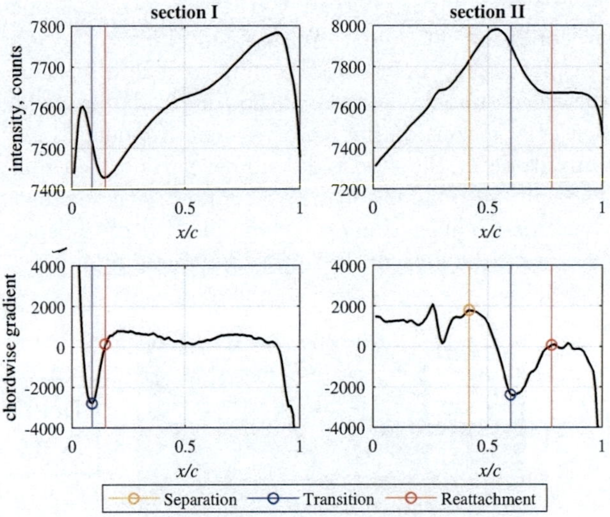

Fig. 5. Examples for the boundary layer state detection, for steady inflow at $\alpha = 4°$

As the boundary layer thickness increases along the chord, the velocity gradient at the wall decreases. The skin friction is proportional to the velocity gradient and linked to the convective heat transfer through the Reynolds analogy. Therefore, the surface temperature is generally increasing with x for a wing that is heated above the temperature of the freestream. However, a turbulent boundary layer produces significantly larger skin friction values compared to a laminar boundary layer at the same Re_x. When transition occurs over a laminar separation bubble, the surface temperature is maximal inside the bubble, where the flow speed is lowest. The laminar-turbulent transition location is therefore considered to be the position of the minimum of the chordwise temperature gradient. The separation location can be determined as the position of the local maximum of the temperature gradient upstream of the transition location. When the flow reattaches, the chordwise temperature trend reverses from decreasing to increasing again. The reattachment location is therefore the position where the chordwise gradient becomes positive again. As observed in Fig. 5, the determination of the separation location with this approach fails when transition occurs very close to the leading edge (section I), as there is no local maximum of the gradient. The detection of the reattachment is problematic when the laminar separation bubble is located further downstream (section II) because the non-uniformity of the external heating causes a reduction in the temperature increase near the trailing edge. Due to the challenges with determining the separation and reattachment locations, the quantitative results of this study are limited to the transition location x_{tr}, which is determined both on and between the ribs.

The aforementioned procedure can generally not be used in the case of unsteady inflow conditions. Here, the instantaneous surface temperature distribution may not directly correspond to the boundary layer state, as it will also be influenced by the thermal response of the surface material. To overcome this, **differential infrared thermography** (DIT, [6]) is applied for the identification of the unsteady transition location occurring on and next to the ribs of the wing. This technique consists of the subtraction of two thermograms obtained closely in time and the identification of unsteady features from the differential image. The approach is illustrated in Fig. 6. The small time separation between consecutive frames makes the two intensity distributions almost identical, but the application of DIT reveals a negative DIT peak linked to the transition process in the separated shear layer of an unsteady laminar separation bubble [10].

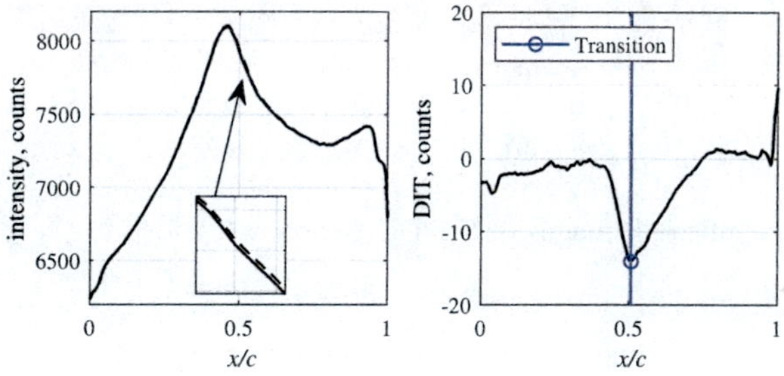

Fig. 6. Example of the identification of unsteady transition location using DIT

5 Results

The mapped and time-averaged thermographic images measured at eight angles of attack α in steady inflow conditions are shown in Fig. 7. For $\alpha = 0°$, a region of increased surface temperature is present between $0.55 < x/c < 0.75$, which is relatively uniform along the span (the decrease in overall temperature for $y/s > 0.75$ can be attributed to the non-uniform external heating). The reduced convective heat transfer in this region can be explained by the presence of a laminar separation bubble. At $\alpha = 0°$, the spanwise variation of the sectional shape of the wing has little influence on the boundary layer flow. However, as α increases, these variations cause differences in the spanwise flow pattern. The main effect is a reduced surface temperature downstream of the kink near the leading edge in the airfoil shape between the rib locations. The kink causes the boundary layer to transition to turbulence, increasing the skin friction and thus the convective heat transfer when compared to a laminar boundary layer state. This effect sets in locally at individual spanwise locations at $\alpha = 2°$, it is established along all spanwise sections at $\alpha = 4°$, and further increases with increasing α. For very high values of α, the flow over the wing appears as fully turbulent at $\alpha = 12°$, before at $\alpha = 14°$ the wing stalls, causing an increased surface temperature at some sections of the span, in comparison to the turbulent attached flow. The high infrared count regions in the bottom part of the wing for the larger angles are measurement artifacts (reflections of the heat lamp).

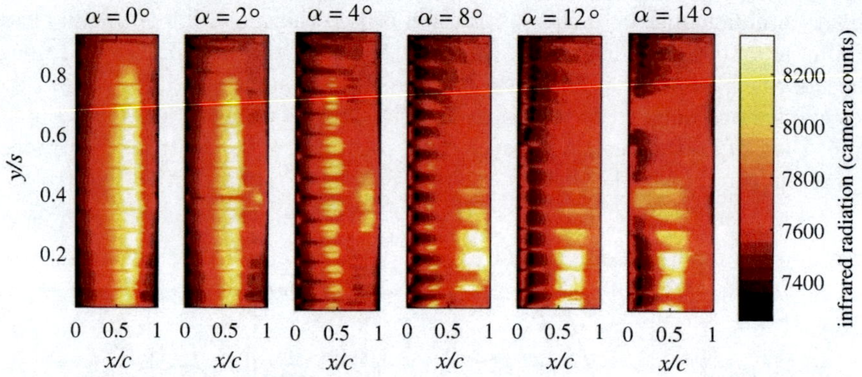

Fig. 7. Infrared images of the Delft-Pazy wing for $0° < \alpha < 14°$

The quantitative values of the extracted boundary layer transition location along the span are shown in Fig. 8. The locations of the ribs of the Delft-Pazy wing are indicated as well, showing that the boundary layer transition is triggered near the leading edge between the ribs for $\alpha \geq 4°$. With increasing α and thus increasing adverse pressure gradients on the suction side of the wing, the transition location on the ribs moves gradually upstream and the region of turbulent flow covers a larger fraction of the span.

For analyzing the unsteady test case, the mapped infrared images are ordered in terms of the phase angle of the gust ϕ. Consecutive images are subtracted to create DIT images, and the unsteady transition location is identified from the mean DIT curve for 50 phase bins. This procedure is applied to every pixel row along the span that is close to the ribs of the wing. Between the ribs, transition remains approximately unaltered and close to the leading edge, as explained for the static measurements. The DIT peaks measured along the cycle next to a rib are shown in Fig. 9. Their sign changes depending on whether the transition location is moving upstream or downstream. This causes the appearance of two regions where no clear peak could be detected. To provide an estimation of the

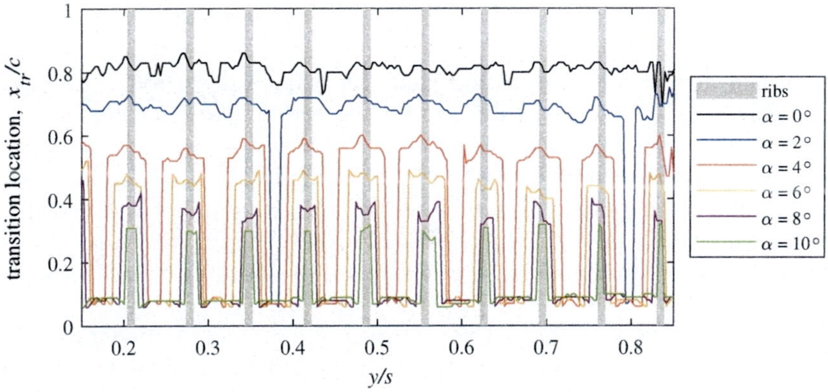

Fig. 8. Boundary layer transition location along the span for $0^\circ < \alpha < 10^\circ$

full cycle, a sinusoidal fit is constructed using the experimental data. From there, it can be observed that there is approximately a 10% lag between the motion and the unsteady transition process. Extending the approach along the span gives the unsteady transition fronts shown in Fig. 10, at five different phase angles along the cycle. The results indicate that the unsteady conditions have no effect in the region between ribs, while they induce variations in the transition location of more than 20% of the chord on and next to the ribs of the wing.

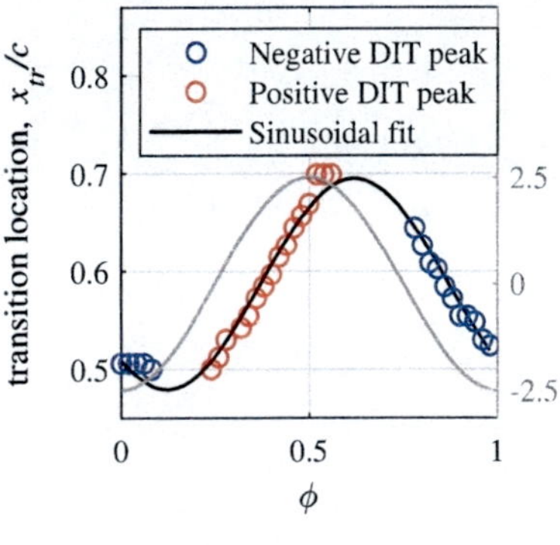

Fig. 9. Sinusoidal fit applied to the transition location measurements

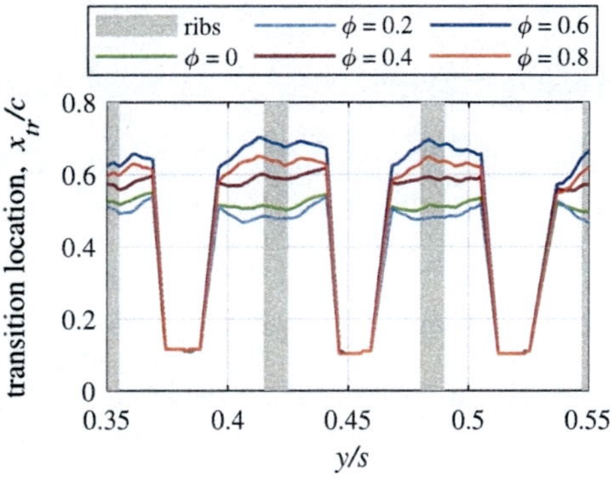

Fig. 10. Unsteady transition location along the span for the sinusoidal gust response

6 Conclusion

The boundary layer that develops on the suction side of the highly flexible Pazy wing geometry was analyzed in this study based on infrared thermography measurements obtained in a wind tunnel experiment. The complex behavior of the boundary layer, where laminar-turbulent transition occurs over a laminar separation bubble, complicates the data analysis of these measurements with respect to attached flow conditions, in particular when unsteady aerodynamic effects have to be considered. Insights into the flow physics from a CFD analysis of the wing geometry were therefore used to support the analysis and interpretation of the experimental data. The laminar-turbulent boundary layer transition location was determined for a range of angles of attack in steady inflow and for one test case with unsteady inflow based on the infrared thermography measurements, whereas the determination of the separation and reattachment locations of the laminar separation bubble based on the experimental data was found to be more challenging. The deformation of the wing shape that occurs between the ribs of the investigated wing due to the flexibility of the wing skin causes considerable differences in the boundary layer transition behavior across the span for positive angles of attack. Near the locations of the ribs, the behavior of the design shape is preserved and a relatively long laminar separation bubble forms around mid-chord, moving upstream with increasing angle of attack. Between the ribs, the distorted airfoil shape produces a small region of separated flow close to the leading edge, downstream of which the flow reattaches in a turbulent state. The spanwise extent of this effect increases with increasing angle of attack. Because this complex boundary layer behavior deviates strongly from the aerodynamic behavior of the design shape of the wing, it is recommended to dedicate more attention to preserving the sectional wing shape on highly flexible wing designs in the future, with a particular emphasis on maintaining a smooth shape in the region of the leading edge.

References

1. Avin, O., Raveh, D.E., Drachinsky, A., Ben-Shmuel, Y., Tur, M.: Experimental aeroelastic benchmark of a very flexible wing. AIAA J. **60**, 1745–68 (2022). https://doi.org/10.2514/1.J060621
2. Riso, C., Cesnik, C.E.: Correlations between UM/NAST nonlinear aeroelastic simulations and the pre-Pazy wing experiment. In: AIAA Scitech 2021 Forum, Virtual Event, Paper No. 1712 (2021). https://doi.org/10.2514/6.2021-1712
3. Goizueta, N., Wynn, A., Palacios, R., Drachinsky, A., Raveh, D.E.: Flutter predictions for very flexible wing wind tunnel test. J. Aircr. **59**, 1082–97 (2022). https://doi.org/10.2514/1.C036710
4. Mertens, C., Costa Fernández, J.L., Sodja, J., Sciacchitano, A., van Oudheusden, B.W.: Integrated aeroelastic measurements of the periodic gust response of a highly flexible wing. In: International Forum on Aeroelasticity and Structural Dynamics IFASD, Madrid, Spain (2022)
5. Carlomagno, G.M., Cardone, G.: Infrared thermography for convective heat transfer measurements. Exp. Fluids **49**, 1187–218 (2010). https://doi.org/10.1007/s00348-010-0912-2
6. Wolf, C.C., Gardner, A.D., Raffel, M.: Infrared thermography for boundary layer transition measurements. Meas. Sci. Technol. **31**(112002) (2020). https://doi.org/10.1088/1361-6501/aba070

7. Schwamborn, D., Gerhold, T., Heinrich, R.: The DLR TAU-code: recent applications in research and industry. European Conference on Computational Fluid Dynamics ECCOMAS, Delft, The Netherlands (2006)
8. Fehrs, M.: Boundary layer transition in external aerodynamics and dynamic aeroelastic stability. Dissertation, TU Braunschweig, ISSN 1434-8454, ISRN DLR-FB-2018-11, also NFL-FB 2017-27, Braunschweig (2018)
9. Fehrs, M., Ritter, M., Helm, S., Mertens, C.: CFD simulations of the Pazy wing in support of the third aeroelastic prediction workshop. In: International Forum on Aeroelasticity and Structural Dynamics IFASD, Madrid, Spain (2022)
10. Grille Guerra, A., Mertens, C., Little, J.C., van Oudheusden, B.W.: Investigation of the unsteady behaviour of a laminar separation bubble using infrared thermography. In: 20th International Symposium on Application of Laser and Imaging Techniques to Fluid Mechanics, Lisbon, Portugal (2022)

Numerical Investigation of a Spoiler Effect on the Transonic Flutter Boundary

Martin Schmalz(✉) and David Quero

DLR, Institute of Aeroelasticity, 37073 Göttingen, Germany
{martin.schmalz,david.queromartin}@dlr.de

Abstract. In this paper, the effect of a statically deflected spoiler on the transonic flutter boundary is investigated by means of numerical airfoil models. For this purpose, a spoiler geometry is added to the NACA64A010 airfoil and a wing cross-section of the NASA's Common Research Model. The deflection angle and the hinge-line position of the spoiler are varied. Unsteady aerodynamic loads are obtained with a linearized CFD method in the frequency domain. The flutter results show an increase in flutter velocity in the transonic dip region due to the spoiler deflection, which seems to be connected to a change of the aerodynamic loading of the airfoil. However, deterioration in the flutter stability boundary can be found for some specific spoiler deflection angles, especially at high Mach numbers. A careful use of the spoiler and a dedicated scheduling scheme dependent on the Mach number are therefore suggested if the spoiler effect is exploited as part of future flutter suppression technologies.

Keywords: flutter stability boundary · flutter suppression · transonic dip · shock wave · static spoiler deflection

1 Introduction

With the advent of new types of aircraft with high aspect ratio wings, the demand for increasing performance and extension of the stability margin is high, as these aircraft feature a great flexibility and are therefore more susceptible to aeroelastic instabilities such as flutter [1]. Especially in the transonic flow regime, the risk of flutter is increased due to the occurrence of nonlinear aerodynamic effects such as shock waves. In this context, the transonic dip describes a sudden drop in the flutter boundary [2]. Due to its risk of severe damage, flutter has to be avoided over the entire flight envelope by all means. In fact, in order to certificate a new aircraft, it is required to ensure a flutter speed which is at least 15% greater than the highest allowable airspeed, i.e., the design dive speed V_D [3].

As a result, active means for flutter suppression [4], e.g., active control systems consisting of ailerons or morphing devices, are gaining more importance in the design process of new aircraft. However, one disadvantage of active flutter

A. Dillmann et al. (Eds.): STAB/DGLR Symposium 2022, NNFM 154, pp. 14–23, 2024.
https://doi.org/10.1007/978-3-031-40482-5_2

suppression devices is their need for maintenance and their risk of malfunction. Therefore, auxiliary flutter suppression devices that act as fail-safe systems have to be considered as well if active flutter suppression shall be part of flutter control systems in future aircraft. One possible concept could be the exploitation of a static spoiler deflection, as it offers a rather simple and directly available technology. In the literature, only a few references can be found on using spoilers for such a purpose. Flutter experiments with statically deflected spoilers on a low aspect ratio, rectangular wing were conducted by Doggett [5] and Cole [6]. By choosing an appropriate design of the spoilers in terms of size and deflection angle, the authors could achieve an increase in flutter dynamic pressure.

This paper is intended to present a first numerical assessment of the effect of a static spoiler deflection on the flutter stability boundary of airfoils in the transonic dip region. To this aim, numerical models of the NACA64A010 airfoil and of a wing cross-section of the NASA's Common Research Model (CRM) [7] with an added spoiler are created. The geometrical parameters of the spoiler, like spoiler hinge-line position and spoiler deflection angle, are varied. By using CFD methods, the impact of the spoiler on transonic flow characteristics like shock waves can be investigated. Further nonlinear effects, which may be relevant, including transitional flow and buffet phenomena, are beyond the scope of this work and are proposed for future work.

2 Spoiler Actuation Concept

The content of this paper follows the ideas found in a patent by Seeler and Zimmermann [8], who propose a concept of a flutter brake for high-speed aircraft, including a spoiler for flutter suppression. The general idea is to provide sufficient damping of a flutter motion of a wing by deflecting a spoiler into a static position. Figure 1 depicts three distinct phases that describe how this flutter suppression device works in theory. Above a threshold of a critical amplitude or frequency of the wing motion, a sensor signal is triggered in phase 1 in order to actuate the spoiler within a short period of time of about only 200 to 400 ms. As a result, the spoiler is rapidly deflected into its static position in phase 2, which typically takes less than 150 ms (considering for example a spoiler deflection of 15° with a deflection rate of 100 °/s). The effect of the spoiler on the flutter boundary becomes apparent in phase 3. The spoiler is retracted immediately after the instability has been effectively damped.

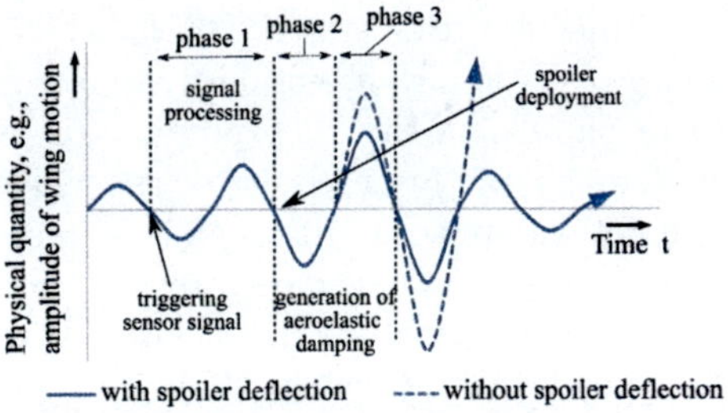

Fig. 1. Principle of flutter suppression with spoiler deflection over time according to Seeler and Zimmermann [8].

In general, deflecting a spoiler reduces aerodynamic efficiency because aerodynamic drag increases and lift decreases. However, the device is thought to be used at speeds beyond V_D for flutter clearance certification, a region in which aircraft performance deterioration is of no practical concern. In addition, it is assumed that the time scale for the spoiler to be deflected and the corresponding steady flow state around the spoiler to be reached is at least one order of magnitude less than the time required to attain a new trimmed condition for the entire aircraft due to the spoiler deflection. Numerical investigations of the transient aerodynamic response of 2D airfoil models due to a spoiler deflection are conducted in [9]. It is shown that the new final state of the aerodynamic response is almost immediately attained after the spoiler deflection process is completed (e.g. deflecting the spoiler from 0° to 30° in 125 ms). With this prompt aerodynamic response of the flow field around the spoiler, simultaneous damping of a flutter motion is expected to be feasible before a catastrophic failure of the airframe occurs.

It should be noted that the present paper only focuses on the numerical computation of the generated aeroelastic damping in phase 3, when the spoiler has already reached its static position and the final aerodynamic response of the flow around the spoiler has been achieved. The dimensioning of the actuators and sensors required for the spoiler deployment process is not covered.

3 Numerical Methods

3.1 Aerodynamic Model

The NACA64A010 airfoil and a wing cross-section of the CRM are used as the objects of this study. The latter is used in order to additionally address a more realistic airfoil section flying at high speeds and is extracted at 65% half span parallel to the longitudinal axis of the full CRM model. It is scaled with $\cos(\phi_{25\%} = 35°)$ to account for the wing sweep $\phi_{25\%}$ at quarter cord. Both airfoil geometries are scaled in x- and z-direction to obtain a chord length of $c = 1\,\text{m}$. A spoiler geometry is integrated on the upper surface of the airfoils. The spoiler geometry has a generic shape with a blunt leading edge (the spoiler length equals 14% the length of the chord c) and is connected to the rest of the airfoil leaving no hinge gap in between. Three different locations of the spoiler hinge-line are studied. The selected positions are defined with $x/c = 0.4$, 0.5 and 0.6. Values of 5°, 7.5° and 10° are employed for the spoiler deflection angle (SDA).

All required CFD meshes are created with the commercial software ANSYS ICEM CFD (V 19.1). The clean mesh of the NACA64A010 airfoil and of the CRM airfoil comprises 238904 nodes and 420950 nodes, respectively. Unstructured quadrilateral cells are used for meshing the circular far field (radius equals 100 chord lengths), and a blockstructured O-grid is used around the airfoil geometry in order to guarantee an adequate refinement of the boundary layer. A value of below 1 for the nondimensional wall distance y^+ is achieved for all simulated cases. The spatial grid convergence of these meshes is verified with a grid convergence study.

The flow field is computed by solving the Reynolds-averaged Navier-Stokes (RANS) equations with the DLR-TAU code [10]. All convective fluxes are spatially discretized with a second order central scheme. The first order implicit backward-Euler method is chosen as a local time stepping scheme for the steady simulations. In order to close the RANS-equations, the one equation turbulence model according to Spalart and Allmaras [11] is used.

The linear frequency domain (LFD) solver [12] is employed for all unsteady calculations. For very small-amplitude harmonic perturbations, this solver linearizes the unsteady RANS (URANS) equations around the steady state, transforms them into the frequency domain and then solves for the first harmonic of the flow response. Complex nonlinear unsteady flow features such as transonic buffeting or buzz cannot be predicted by the LFD. However, since the flutter analysis involves a linear stability analysis of the aeroelastic system, the use of LFD is appropriate. The first harmonic is in good agreement with the first harmonic produced by time marching solvers as long as the underlying steady solution is well converged. This holds true even for transonic separated flows [12,13].

The CFD computations consider Mach numbers M_∞ in a range between 0.58 and 0.79. The Reynolds number Re of the simulations with the NACA64A010 and the CRM airfoil is $12.5 \cdot 10^6$ and $40 \cdot 10^6$, respectively. The angle of attack is automatically adjusted by TAU at each Mach number to a targeted lift coefficient of $c_l = 0.5$ for the clean configuration.

3.2 Linear Flutter Analysis

The airfoil configurations are considered as three-degree-of-freedom (3 dof) models characterized by motions for heave, pitch and an additional flap denoted by h, α and β, respectively, as shown in Fig. 2. The flutter boundary limit of these models is calculated as a function of the Mach number M_∞ by solving the equations of motion in terms of the following nonlinear eigenvalue problem in nondimensional form

$$\left[\mathbf{M}p^2\left(\frac{U_\infty}{b}\right)^2 + \mathbf{K}\right]\begin{Bmatrix} h(p)/b \\ \alpha(p) \\ \beta(p) \end{Bmatrix} = \frac{1}{\pi\mu}\left(\frac{U_\infty}{b}\right)^2 \begin{Bmatrix} -c_l(p, M_\infty) \\ 2c_m(p, M_\infty) \\ 2c_n(p, M_\infty) \end{Bmatrix} \tag{1}$$

with the p-k-method [14] for the dimensionless frequency $p = g + ik$. Here, g denotes the nondimensional damping and $k = \omega c/U_\infty$ the reduced frequency with ω being the eigenfrequency of the corresponding mode and U_∞ the freestream velocity. The mass matrix $\mathbf{M}$ and the stiffness matrix $\mathbf{K}$ can be written as

$$\mathbf{M} = \begin{bmatrix} 1 & x_\alpha & x_\beta \\ x_\alpha & r_\alpha^2 & (c_\beta - e)x_\beta + r_\beta^2 \\ x_\beta & (c_\beta - e)x_\beta + r_\beta^2 & r_\beta^2 \end{bmatrix}, \quad \mathbf{K} = \begin{bmatrix} \omega_h^2 & 0 & 0 \\ 0 & r_\alpha^2\omega_\alpha^2 & 0 \\ 0 & 0 & r_\beta^2\omega_\beta^2 \end{bmatrix}, \tag{2}$$

where e defines the position of the elastic axis, x_α the airfoil's center of gravity, x_β the flap's center of gravity, r_α and r_β the gyration radii, c_β the position of the hinge-line of the flap and ω_h, ω_α and ω_β the eigenfrequencies of the respective mode. The actual values for these structural parameters are taken from Yang and Batina [15]. The aerodynamic load coefficients c_l, c_m and c_n are calculated in two steps. First, a steady CFD solution is calculated using a specific combination of Mach number and Reynolds number as mentioned in Sect. 3.1. Second, based on the steady solution, the LFD method is employed, in which the required harmonic perturbation is realized by deforming the CFD mesh with respect to the three mode motions of the airfoil. The resulting generalized aerodynamic forces in the form of frequency response functions are used to derive the aerodynamic load coefficients.

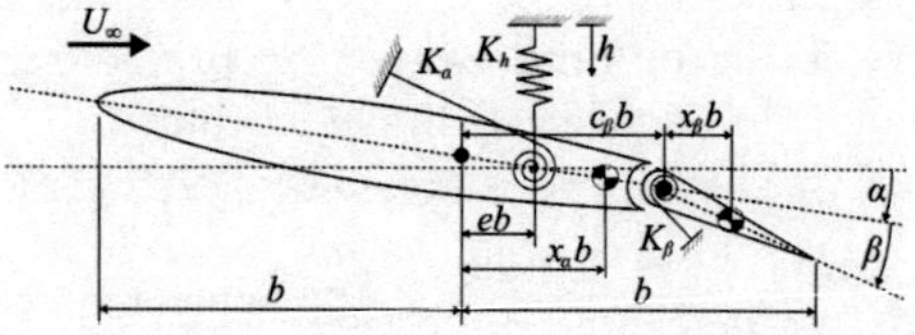

Fig. 2. Representation of the 3 dof aeroelastic airfoil model.

By solving Eq. 1, a sweep over the freestream velocity U_∞ is performed for each considered Mach number in order to calculate the flutter velocity U_F at which the damping g becomes zero. Actually, sweeping the freestream velocity would correspond to a variation of the Reynolds number that would change the aerodynamic load coefficients. However, as the Reynolds number only varies by less than one order of magnitude, its influence is neglected and the aerodynamic coefficients depend only on the Mach number and the frequency. The values for the semi-chord $b = c/2$ and the mass ratio $\mu = 23.48$ are fixed, where the latter is set according to Schulze [16].

4 Results and Discussion

4.1 CFD Solutions

A large change in the steady flow characteristics is caused by a deflection of the spoiler on the upper surface of the airfoil. The flow separation behind the spoiler is the most noticeable feature. Figure 3 depicts the distribution of the local Mach number for the NACA64A010 clean and spoiler configurations with an SDA of 5° and 10° exemplarily at $M_\infty = 0.74$. It can be observed that increasing the SDA leads to a larger displacement of the streamlines aft of the spoiler resulting in a greater spatial extension of the separated flow region. Both SDAs achieve no reattachment of the flow in front of the trailing edge. It can be noted that the spoiler deflection changes the circulation around the airfoil, which is consequently associated with a variation in the surrounding pressure field. Figure 4 shows the distribution of the local pressure coefficient c_p along the surface of the NACA64A010 airfoil for the clean and spoiler configurations at $M_\infty = 0.74$. An increase in pressure can be observed on the upper surface in front of the spoiler. The larger the SDA the larger the increase in pressure. A marginal increase in pressure is also achieved in this area by moving the spoiler towards the leading edge. In contrast, the spoiler deflection leads to a decrease in pressure in the separated flow region aft of the spoiler. A general increase in pressure results from the spoiler deflection on the lower surface. As depicted in Fig. 3, the change of the circulation around the airfoil due to the spoiler deflection is associated with changes in the velocity field, as the supersonic flow region which is formed on the upper surface of the clean airfoil is clearly reduced in size. Figure 4 shows that the spoiler accounts for a mitigation of the strong pressure increase at the shock location at $x/c \approx 0.39$. With an SDA of 10°, the flow barely accelerates to supersonic velocities on the upper surface, realizing a full suppression of the shock.

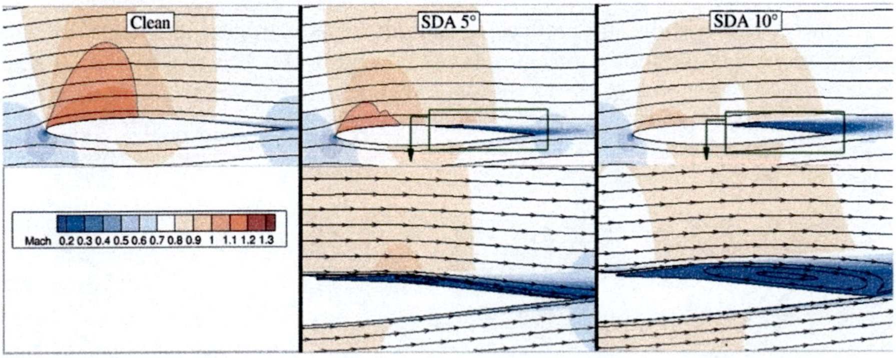

Fig. 3. Local Mach number distribution around the NACA64A010 clean and spoiler configurations (spoiler hinge-line at x/c = 0.5) at $M_\infty = 0.74$.

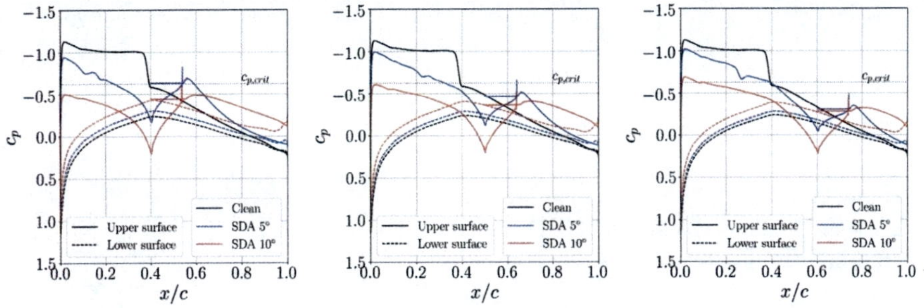

(a) $x/c = 0.4$ (b) $x/c = 0.5$ (c) $x/c = 0.6$

Fig. 4. Distribution of the local pressure coefficient of the NACA64A010 airfoil configurations at Mach 0.74 for three different spoiler hinge-line locations.

As a consequence of the change in pressure distribution due to the spoiler deflection, there is a different load distribution and, thus, a different composition of the applied aerodynamic lift force and pitching moment compared to the clean configuration. The results of the unsteady simulations clearly show these differences. To show an example, the frequency response of the dynamic lift derivative $c_{l,\alpha}$ is depicted in Fig. 5 at $M_\infty = 0.72$ for the NACA configurations. A general decrease in magnitude with increasing the reduced frequency can be observed for all airfoil configurations. In addition, a decrease in magnitude is achieved with increasing the spoiler deflection angle. A drop of about 25% in magnitude is determined with an SDA of 10° at a reduced frequency of $k = 0$. This difference becomes less with an increase in the reduced frequency. At the highest frequency of $k = 1.05$, all spoiler configurations show a larger magnitude than the clean configuration. There is also a phase difference between clean and spoiler configurations that increases with increasing the frequency. It is to mention that similar differences in magnitude and phase also exist for the derivative of the dynamic pitching moment $c_{my,\alpha}$. Finally, it can be stated that the spoiler causes a change in the dynamic response behavior of the airfoil, which can be attributed to the previously mentioned flow displacement and separation behind the spoiler.

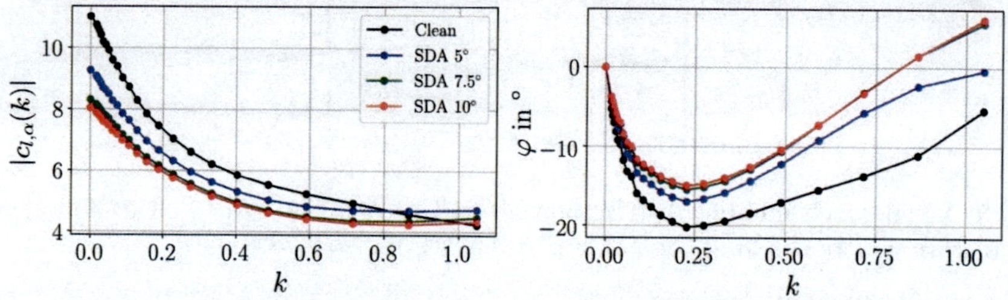

Fig. 5. Dynamic lift response due to pitch motion of the NACA64A010 clean and spoiler configurations (spoiler hinge-line at $x/c = 0.5$) at $M_\infty = 0.72$.

4.2 Flutter Solutions

In order to validate the presented flutter analysis method and the utilized CFD method in particular, the flutter results are compared to results from the literature as given in Schulze [16], who used Euler aerodynamics and the structural model of [15]. For this purpose, a flutter analysis of the 3 dof NACA64A010 airfoil is conducted for an angle of attack of $\alpha = 0°$. Figure 6 shows the normalized flutter velocity (ω_α as the wind-off pitching frequency) as a function of the Mach number regarding the own results and the results by Schulze. Both methods are in good agreement,

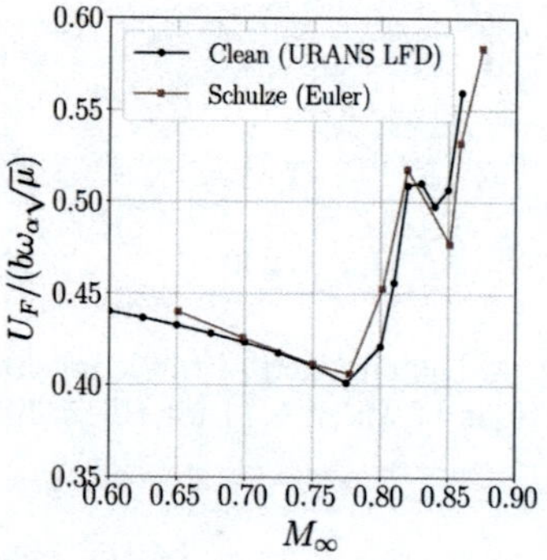

Fig. 6. Flutter boundary of the 3 dof NACA64A010 airfoil ($\alpha = 0°$).

as the two transonic dips are similarly predicted at around $M_\infty = 0.76$ and $M_\infty = 0.85$, respectively.

The flutter boundaries of the 3 dof NACA configurations are depicted in Fig. 7 **regarding the three different spoiler hinge-line locations. The transonic** dip of the clean configuration can be found between Mach 0.72 and Mach 0.73. Up to a Mach number of 0.74, a spoiler deflection leads to a robust increase in flutter stability. The flutter velocity increases with increasing the SDA or by moving the spoiler hinge-line towards the leading edge. For example, at Mach 0.72, the flutter velocity for the spoiler configurations with the spoiler hinge-line located at $x/c = 0.4$ has increased by 18% and by 27% with an SDA of 5° and 10°, respectively. Although some configurations show an increase in flutter boundary in a Mach number range between 0.75 and 0.79, e.g., with an SDA of 5° at $x/c = 0.6$, still no improvements in flutter stability are achieved here.

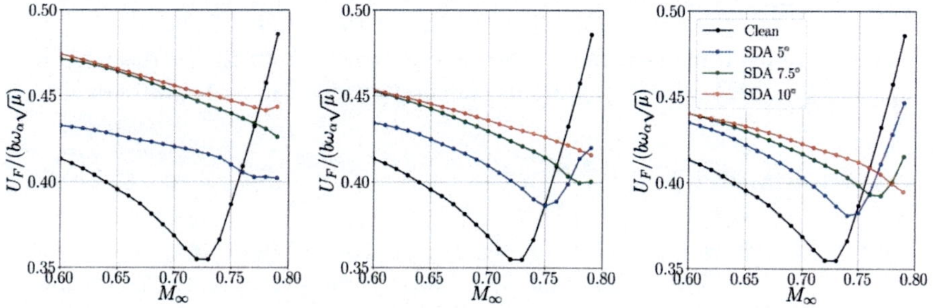

(a) $x/c = 0.4$ (b) $x/c = 0.5$ (c) $x/c = 0.6$

Fig. 7. Flutter boundaries of the 3 dof NACA64A010 airfoil configurations for different spoiler hinge-line locations.

Figure 8 shows the flutter boundaries of the CRM clean and spoiler configurations with the spoiler hinge-line located at $x/c = 0.5$. The transonic dip of the clean airfoil is located at around Mach 0.72. The flutter boundary increases with a spoiler deflection. This holds true up to Mach 0.73. In the region of the transonic dip, the lowest SDA achieves the highest increase in the flutter boundary as opposed to the NACA64A010 configurations. The increase in flutter velocity at Mach 0.72 is at around 8% for an SDA of 5°. In general, the flutter boundary decreases for the spoiler configurations by increasing the Mach number leading eventually to lower flutter velocities than achieved with the clean airfoil.

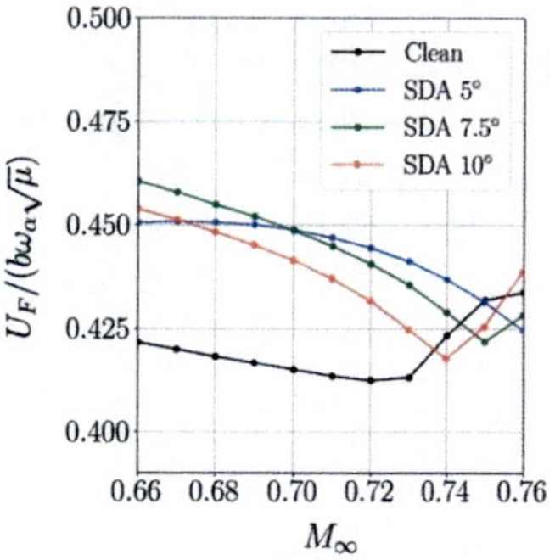

Fig. 8. Flutter boundaries of the 3 dof CRM airfoil configurations.

The changes in the flutter boundary due to the spoiler deflection can only be attributed to the changes in the flow field around the airfoil, as the structural parameters remain unchanged. In Sect. 4.1, a different aerodynamic loading on the airfoil due to the spoiler deflection is pointed out affecting the transonic flow characteristics in terms of a suppression of the supersonic area and the shock on the upper surface. It is obvious that the lift force applied in the area close to the leading edge of the airfoil is reduced since the spoiler causes a pressure increase in this area. Therefore, the resulting aerodynamic pitching moment is changed. Considering the flutter plots, it can be assumed that the spoiler accounts for a more favorable flow condition around the airfoil which causes an increase in the flutter boundary. The phase lag of the aerodynamic lift force and pitching moment can also be considered very decisive for influencing the flutter stability boundary. Of course, an increase in flutter velocity is not achieved in every case with a spoiler deflection, especially at high Mach numbers. With increasing Mach numbers, the shock moves downstream on the upper surface of the two clean airfoil configurations. At these Mach numbers, the aerodynamic loading of the clean airfoil offers a more favorable flow condition for increasing flutter velocity compared to the aerodynamic loading attained with a spoiler deflection.

5 Conclusions

A numerical assessment of the impact of a statically deflected spoiler on the transonic flutter stability boundary was carried out using 3 dof airfoil models with an added spoiler. The spoiler deflection angle and the spoiler hinge-line position were varied.

A robust behavior of the spoiler effect within the transonic dip can be stated, as an increase in flutter velocity is achieved by the spoiler deflection for every spoiler configuration considered. An attempt was made to explain the increase in flutter stability. It is assumed that the spoiler accounts for a more favorable aerodynamic loading of the airfoil in terms of the applied aerodynamic lift force and pitching moment, leading to an increase in flutter velocity. However, a decrease in flutter stability can be observed with specific spoiler deflection angles at very high Mach numbers. This indicates a careful use of the spoiler and a dedicated scheduling scheme, depending on the Mach number. Still, the increase in flutter stability achieved with the supercritical CRM airfoil section clearly shows a good potential of using a spoiler as a semi-active flutter suppression device for a backup solution of active devices. For a more profound assessment, future work has to include flutter investigations performed on 3D wing models with added spoilers in order to ensure that a spoiler can generally be exploited to dampen a flutter motion of an aircraft wing.

Acknowledgements. The authors would like to thank Sebastian Helm from DLR Institute of Aerodynamics and Flow Technology for laying the groundwork for the meshing process.

References

1. Hermanutz, A., Hornung, M.: Influence on the flutter behavior of pre-stressed wing structures under aerodynamic loading. In: 2019 Proceedings of the Forum of Aeroelasticity and Structural Dynamics IFASD, 9–13 June 2019 (2019)
2. Bendiksen, O.O.: Review of unsteady transonic aerodynamics: theory and applications. Prog. Aerosp. Sci. **47**, 135–167 (2011). https://doi.org/10.1016/j.paerosci.2010.07.001
3. Wright, J.R., Cooper, J.E.: Introduction to Aircraft Aeroelasticity and Loads. Wiley, Hoboken (2007)
4. Livne, E.: Aircraft active flutter suppression: state of the art and technology maturation needs. J. Aircr. **55**, 410–450 (2018). https://doi.org/10.2514/1.C034442
5. Doggett, R.V.: Some effects of aerodynamic spoilers on wing flutter. NASA Tech. Memorandum 101632 (1989)
6. Cole, S.R.: Aeroelastic effects of spoiler surfaces on a low-aspect-ratio rectangular wing. J. Aircr. **29**, 768–773 (1992). https://doi.org/10.2514/3.46244
7. Vassberg, J.C., DeHaan, M.A., Rivers, S.M., Wahls, R.A.: Development of a common research model for applied CFD validation studies. In: Proceedings of the 26th AIAA Applied Aerodynamics Conference, 18–21 August 2008 (2008). https://doi.org/10.2514/6.2008-6919
8. Seeler, H., Zimmermann, H.: Method and flutter brake for an aircraft. US 4,615,497 A (1986)
9. Geisbauer, S.: Numerical simulation and validation of aerodynamics of static and dynamic spoilers. J. Aircr. **58**, 1187–1203 (2021). https://doi.org/10.2514/1.C036145
10. Schwamborn, D., Gerhold, T., Heinrich, R.: The DLR TAU-code: recent applications in research and industry. In: Proceedings of the European Conference on Computational Fluid Dynamics (ECCOMAS), 5–8 September 2006, Egmond aan Zee, The Netherlands (2006)
11. Spalart, P., Allmaras, S.: A one-equation turbulence model for aerodynamic flows. In: Proceedings of the 30th Aerospace Sciences Meeting and Exhibit, 6–9 January 1992, Reno, NV, U.S.A. (1992). https://doi.org/10.2514/6.1992-439
12. Thormann, R., Widhalm, M.: Linear-frequency-domain predictions of dynamic-response data for viscous transonic flows. AIAA J. **51**, 2540–2557 (2013). https://doi.org/10.2514/1.J051896
13. Widhalm, M.: Flight dynamic stability prediction with a linear frequency domain method. Ph.D. thesis, Technische Universität Carolo-Wilhelmina zu Braunschweig, Braunschweig, Germany (2020)
14. Hassig, H.J.: An approximate true damping solution of the flutter equation by determinant iteration. J. Aircr. **8**, 885–889 (1971). https://doi.org/10.2514/3.44311
15. Yang, T.Y., Batina, J.T.: Transonic time-response analysis of 3-degree-of-freedom conventional and supercritical airfoils. J. Aircr. **20**, 703–710 (1983). https://doi.org/10.2514/3.44932
16. Schulze, S.: Transonic aeroelastic simulation of a flexible wing section. In: AGARD Report 822 Numerical Unsteady Aerodynamic and Aeroelastic Simulation; AGARD, pp. 10-1–10-18 (1997)

Experimental Aerodynamics

Measuring the Oncoming Flow that Operational Freight-Trains Experience Using the DLR FR8-LAB

James R. Bell(✉), Alexander Buhr, and Arne Henning

German Aerospace Center (DLR), Institute of Aerodynamics and Flow Technology, 37073 Göttingen, Germany
james.bell@dlr.de

Abstract. The DLR FR8-LAB is a self-contained swap-body container, equipped with an on-board power supply, data acquisition system and remote-access communication. The oncoming flow is derived from the surface pressure measurements on the FR8-LAB container's front/rear/side surfaces and a calibration developed in a 1:15 scaled closed-return wind-tunnel experiment. The wind-tunnel model of the container was statically yawed over a range of −90:90°, mapping the relationship between changing surface-pressure and yaw angles. This relationship is used to infer the oncoming flow characteristics – until now, uncharacterized – that the FR8-LAB experiences in real-world operation, through application of the calibration to the measured transient surface-pressure. Oncoming flow conditions can be used to discover and characterize the real-world operating conditions at specific locations where aerodynamics is important, as well as input into aerodynamic/multi-body models to assess the risk of overturning, and modelled in simulations/experiments. Measurements on operational freight trains - validated against ultrasonic anemometers - have demonstrated that the FR8-LAB can be used to not only determine peak pressure loads and their resulting forces and moments, but also as a multi-hole dynamic-pressure probe to infer the oncoming flow conditions that freight trains experience.

Keywords: freight train · aerodynamics · crosswind

1 Introduction

Freight trains operate in a complex variety of conditions in the real-world: interaction with infrastructure, local topography and exposure to the environment. This can result in a wide range of oncoming flow characteristics (wind velocity, direction, turbulence and transient features such as gusts) that an operational freight train and its freight are exposed to in real-world operation [1, 2], that until now, are not well characterized in literature.

The DLR FR8-LAB is a 'swap-body' shipping-container (Fig. 1), that can be loaded in different configurations. The FR8-LAB is self-contained, with an on-board power supply (rechargeable by roof solar panels), data acquisition system and remote-access communication capability that has been certified for transport on normal operating freight-trains in Europe. The FR8-LAB was developed to investigate the global aerodynamic

A. Dillmann et al. (Eds.): STAB/DGLR Symposium 2022, NNFM 154, pp. 27–37, 2024.
https://doi.org/10.1007/978-3-031-40482-5_3

pressures and the resulting aerodynamic forces and moments (that are important for a train's energy efficiency and safety of operation) that a shipping container is exposed to when loaded on freight trains in real-world operation.

In this work, the FR8-LAB was utilized as a multi-hole dynamic-pressure probe to measure the wide variety of oncoming flow characteristics that freight trains and containers are exposed to in real-world operation. The identified oncoming flow conditions can be used to characterize the operating conditions at specific locations where aerodynamics is important, as well as input into aerodynamic/multi-body models to assess the risk of overturning (side-wind stability) as well as modelled in numerical simulations and replicated in wind-tunnel experiments. The inclusion of realistic oncoming flow can improve the accuracy of risk-assessment models, and enable any aerodynamic optimizations identified in simulations/experiments to be more transferable to real-world application.

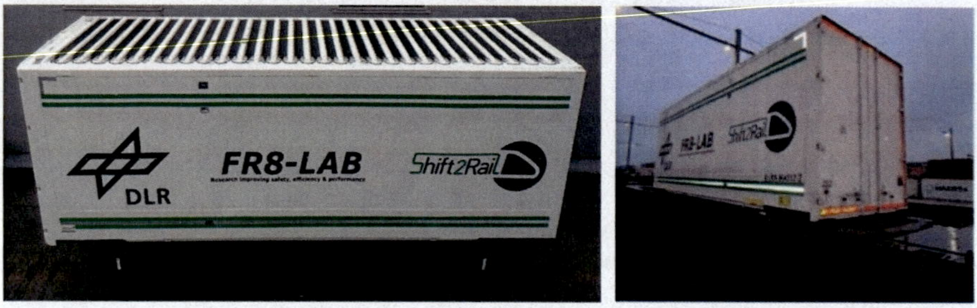

Fig. 1. The DLR FR8-LAB measurement container

2 Methodology

2.1 Full-Scale Experimental Setup

The DLR FR8-LAB. The FR8-LAB is a modified 'Jumbo' swap-body container, WB-type: WK7.7STG with length 7.82 m, height 2.75 m and width of 2.55 m, that is certified/codified to be transported on normal operating freight-trains (and trucks). The FR8-LAB is useful as a probe to measure velocity, as it can be loaded and transported like a normal container (unlike specially mounted sensors). Further, it can take measurements in locations unable to install stationary probes, and measures the flow that a moving train experiences.

Loading Configuration. The FR8-LAB was loaded onto a test-wagon consist, directly after the locomotive(s), as illustrated in Fig. 2. The FR8-LAB was positioned on a 6-axle articulated - Sggmrs 714 wagon, coupled with an additional 4-axle Sgns 691 wagon. This configuration resulted in a minimum gap in front and behind of the measurement container of ~66 m & ~10 m respectively. This ensured the FR8-LAB was ideally positioned in order to take robust cross-wind measurements – being as exposed as possible to crosswind, rather than being affected by the locomotive/other containers in front of it, within the train consist. Two ultrasonic anemometers (USAs) were located centrally in the second Sgns 691 wagon from the locomotive with 9m spacing between them, resulting in distances of ~45 m and 36m from the USAs respectively to the closest end of the FR8-LAB.

Fig. 2. The loading configuration of the DLR FR8-LAB on the train consist during measurements

Surface-Pressure. Time-resolved (200–1000 Hz) surface-pressure measurements at up to 330 positions on the container can be obtained. Integration of pressure on the front and rear surface determines pressure drag, where longitudinal rows and rings along the side and roof determines side force, yaw and roll moments. In this work, 9 pressure measurement positions (highlighted in Fig. 3) each on the sides, front and rear are used as calibration surfaces for determining the oncoming flow.

Individual ±4 kPa *Honeywell* piezoresistive-silicon differential-pressure, temperature compensated, digital (I2C), 3.3 V, 12-bit resolution sensors were used. The reference side of each sensor is open to the interior pressure of the container which has shielded vents to the outside, allowing the inside of the container to experience local atmospheric pressure. Each sensor is connected to a machined Acetal pressure manifold mounted through the container surface; having an external 40 mm diameter, 1 mm thick disc with a 0.6 mm diameter hole through it, which expands after 5 mm to a 1 mm diameter hole of 25 mm length, to which 1 mm inner-diameter, 70 mm silicone tubing connects to the sensor inside. This short distance from measurement point to sensor enables a high frequency response. The pressure sensors are mounted to manifolds on the inside of the container surface with vibration isolating elements; to reduce the effect physical vibrations have on the transient pressure measurements.

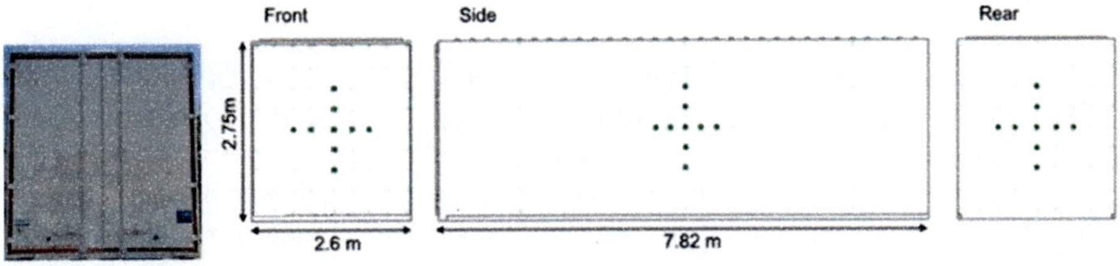

Fig. 3. Surface pressure-measurement on the FR8-LAB external surface.

Environmental Monitoring. A global navigation satellite system (GNSS) determines the location and the velocity over ground (VoG). Seven single-point LIDAR distance sensors with ~40 m range are located on the container's sides, roof and front/rear surface to quantitatively characterize the physical environment. Two wide-angle 75° field-of-view (FOV) thermal cameras (able to operate at night and in poor weather conditions) provide additional qualitative information on the local topography. Accelerometers and temperature sensors also measure the conditions inside the container.

Data Acquisition System. A bespoke Internet-of-things 'IoT' data-acquisition system (DAQ) – consisting of both software and electronic component architecture – was developed for the FR8-LAB. This system was required due to the novel experimental requirements: high number of synchronized sensors, relatively high sampling rate, operating in a challenging environment (vibrations, heat/cold, moisture) with low overall power consumption.

Clusters of up to 16 sensors are connected to multiple *Nucleo* F767Zi microcontroller nodes, that trigger and log data within the C/C++ based *RIOT* operating system. Data is synchronized with a temporal precision of ~1 μs using I2C digital communications protocol in parallel ('bit-banging') between the sensors and nodes, User Datagram Protocol (UDP) communication from the node to the server, and precision time protocol (PTP) for the node clocks.

Remote-access is achieved through a 4G/LTE connection which shares an external antenna mounted on the roof, with the GNSS. This enabled data management, sanity checking, trouble-shooting, and direct manual control of systems if necessary. In addition, the embedded system contained multiple 'intelligent' software sub-systems, such as automatic system idle and data-acquisition triggers based on speed, geofencing.

Power. The core of the power supply system is 16 individual Sun Power VRM 6V 200 Ah absorbent glass matt (AGM) batteries. AGM batteries were specifically used to be able to best withstand vibration, shocks and tipping over. Combined, the batteries provide up to 20 kWh, which at a conservative total system running requirement of 100 W, corresponds to approximately 8 days of possible system duration. The roof of the FR8-LAB is covered in 30 individual 50 W solar panels (Fig. 1) that recharge the batteries and extend the system duration. An external power socket on the door enables manual charging by cable if necessary.

2.2 Calibration Wind-Tunnel Experimental Setup

The oncoming flow was derived from the surface-pressure measurements on the FR8-LAB container's front/rear/side surfaces through a calibration developed in a wind-tunnel experiment (Fig. 4). The wind-tunnel model of the container was statically yawed over a range of −90:90°, mapping the relationship between changing surface-pressure characteristics and different relative yaw-angles (i.e. the oncoming flow the train sees as result of its motion, and atmospheric wind). This pressure-yaw angle relationship is then used to infer the oncoming flow characteristics the FR8-LAB experiences in real-world operation, through application to the transient surface-pressure measured by the FR8-LAB.

Wind-Tunnel Setup. The wind-tunnel calibration experiment was performed in the Crosswind Simulation Facility (SWG) at DLR Göttingen, a closed-circuit wind tunnel (Göttingen-type) with a test-section of 2.4 m × 1.6 m × 9 m. The 0.5 MW fan can achieve flow speeds within the test section of 2 m/s to 67 m/s.

A 2380 mm long, 1300 mm wide rounded-corner splitter-plate was raised 195mm from the wind-tunnel floor, reducing the influence of the artificial ground-boundary layer that exists on the floor of the wind-tunnel. The 1:15 scale test-model consists of a generic wagon with wheelsets with an adjustable length, in addition to dummy containers and a generic locomotive – used to vary the upstream gap configuration. The test-model was fixed above a 1:15 single-track ballast and rail, modelling a generic ground configuration that the freight train would typically operate above.

The calibration experiment was performed over a velocity and Reynolds-number range of 10–60 m/s, Re = 1–6 × 10^5 respectively, different freestream turbulence intensities of ~1,3 and 5% (generated with 40/80 mm wide vertical bars with 200/400 mm spacing) and with different loading configurations; upstream locomotives/containers resulting in full-scale equivalent gaps of 0.23–17.8 m. These variables provide insight into the sensitivity of the calibration to conditions, as well as possible range of calibrations depending on the loading configuration.

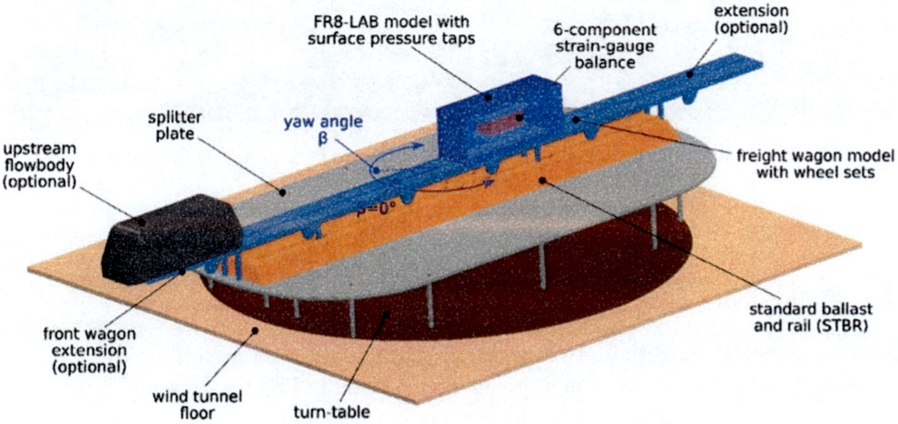

Fig. 4. The calibration wind-tunnel experimental setup, with 1:15 scale freight train model.

Surface-Pressure Measurement. Surface-pressure measurements in the wind-tunnel experiment were performed using two 64 channel, PSI ESP64HD ±5 kPa differential-pressure modules connected to a DTC Initium data acquisition system. Pressure was measured at the same scale-relative positions on the scaled FR8-LAB as the full-scale. Each measurement had a duration of 30 s, and a sampling rate of 300 Hz.

Flow Visualisation. The way in which air moves around the FR8-LAB was visualized using smoke in the wind-tunnel experiment. At 0° yaw (Fig. 5a, b), the flow stagnates on the front surface resulting in high pressure, then separates at the leading edges of the side and roof corresponding to lower pressure Ahead of the front surface, above the wagon, the flow recirculates on itself. The flow separates and recirculates behind container, corresponding to low pressure.

At crosswind angles of 10–60° (Fig. 5c, d), the flow around the FR8-LAB exhibits different features and pressure. At the windward-side (facing towards the wind), the flow also stagnates, causing high surface pressure. Flow separates from at the leading windward edge, forming a 3D longitudinal vortex on the roof that has low pressure. At the upper edge of the leeward side (facing away from the wind) the flow separates and recirculates behind container causing low pressure. At even relatively small angles of crosswind $>15°$, with enough upstream gap, the flow that the FR8-LAB experiences is not affected by upstream locomotive/container (Fig. 5d).

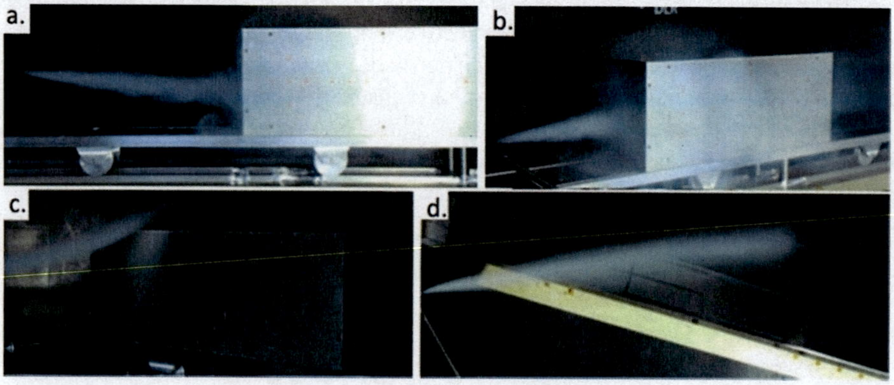

Fig. 5. Visualisation of flow around the model FR8-LAB container in the wind-tunnel

2.3 Calibration

Surface-Pressure Sensitivity to Yaw Angle. The variation of pressure at different locations on the container with changing yaw is presented in Fig. 6a. Pressure at the front, and left side (wind-ward) exhibit high sensitivity to yaw angle, with almost linear change in pressure with changing yaw angle. In contrast, the right (lee-ward) and rear show less sensitivity to yaw angle. These characteristics fit with the insight and explanations provided from the flow visualizations; high, low pressure being caused by stagnated flow and separation/recirculation respectively. These varying sensitivities of pressure to yaw angle result in a defined relationship, that is used as the calibration.

On each surface (front, rear, sides), 9 pressure positions in a cross formation (Fig. 6a) were averaged, to generate a calibration surface-pressure for each yaw angle. Multiple pressure taps were averaged, rather than using a single pressure sensor, to increase robustness/reliability in the full-scale FR8-LAB measurements.

Magnitude. A combination of the calibration surface pressures is used to estimate the dynamic pressure, Q_R, which infers the magnitude of the wind velocity in the wind tunnel, or the FR8-LAB is exposed to. As dynamic pressure is the total pressure, minus the static pressure in the flow-field, an analogous combination of the calibration pressures was used to estimate the dynamic pressure:

$$Q\_R = \surd((p\_front\ -\ p\_rear)^2 + (p\_left\ --p\_right)^2) \quad (1)$$

where p_front, p_rear, p_left, p_right are the average pressures from the 9-sensor cross at each respective calibration surface. This 'resultant' dynamic pressure, Q_R, compares well to the dynamic pressure, Q, measured directly in the wind-tunnel experiment with a Pitot-static tube in the test-section (Fig. 6a).

Direction. The yaw angle is determined by assessing the pressure differential between either side of the FR8-LAB, and using the dynamic pressure to normalize, to develop a calibration coefficient that is a function of yaw angle:

$$\textit{Calibration Coefficient} = (p\_left - p\_right)/Q\_R \tag{2}$$

This concept of using surface pressures at different locations on the FR8-LAB is a similar concept as 'train/car-as-probe' in vehicle aerodynamics, and dynamic-pressure multi-hole probes (used as wind-tunnel measurement devices) utilized previously in applied aerodynamic research [3–5].

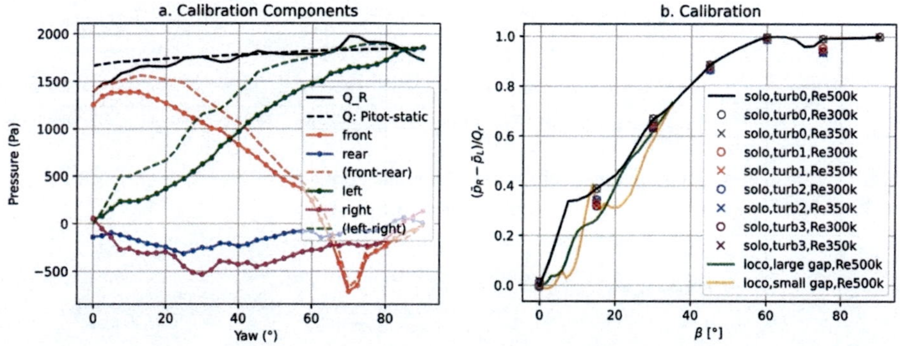

Fig. 6. a. Pressure at each calibration surface for different yaw angles. b. Calibration sensitivity

Robustness. The robustness of the calibration coefficient profile was investigated through repeatability tests, and sensitivity to Reynolds Number, loading configuration and freestream turbulence intensity. In Fig. 6b, the calibration is demonstrated to be repeatable, and largely insensitive to differences in the Reynolds number and turbulence intensities tested. Minor sensitivity is visible at ~10° yaw angle, which corresponds to relatively low crosswinds. There is greater sensitivity to loading configuration; as expected, the closer an upstream object (locomotive), the more it shields the FR8-LAB from flow at low yaw angles, affecting the magnitude of pressure acting on the front surface, affecting the accuracy of the calibration coefficient. However, even at the smallest gap tested in full-scale of 9.3 m (corresponding to the 'small gap' labelled in Fig. 6b), beyond 30° degrees yaw, the calibration is consistent across all loading configurations, and for a gap of 17.8 m (corresponding to the 'large gap' labelled in Fig. 6b), consistent from 20° yaw. These results demonstrate the calibration is useful to determine crosswind oncoming flow, with large enough upstream gaps. For future measurements, without the large-gap test-wagon configuration tested, a calibration profile that is dependent on the upstream gap could be developed.

Application. An example of the transient pressure at each calibration surface during a single 90 s measurement with no crosswind and crosswind exposure is presented in Fig. 7, exhibiting the same crosswind dependent pressure characteristics described above. The resultant dynamic pressure, Q_R and calibration coefficient (Fig. 8) and yaw angle (Fig. 9) are also presented for both examples. The dynamic pressure, Q_R is then converted to the resultant velocity, V_R, and with vector addition: $\overrightarrow{v_R} = \overrightarrow{v_T} + \overrightarrow{v_W}$, , the train-speed induced wind V_T (determined from the GNSS), and atmospheric wind V_W are derived and presented for the two examples in Fig. 9.

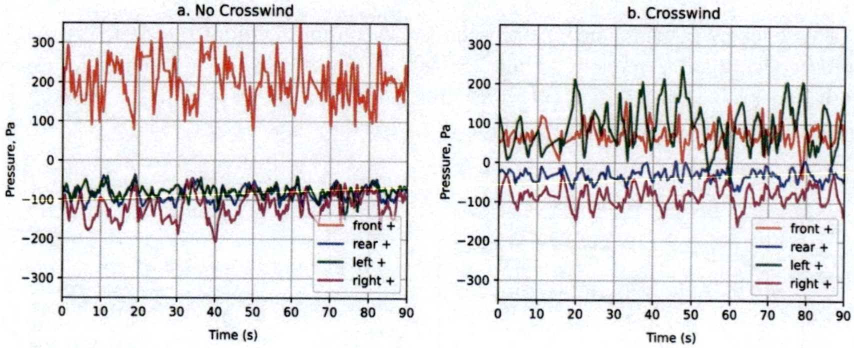

Fig. 7. Transient pressure – with a 0.5 s moving average applied – measured on the FR8-LAB at each calibration surface during a. no crosswind, b. crosswind.

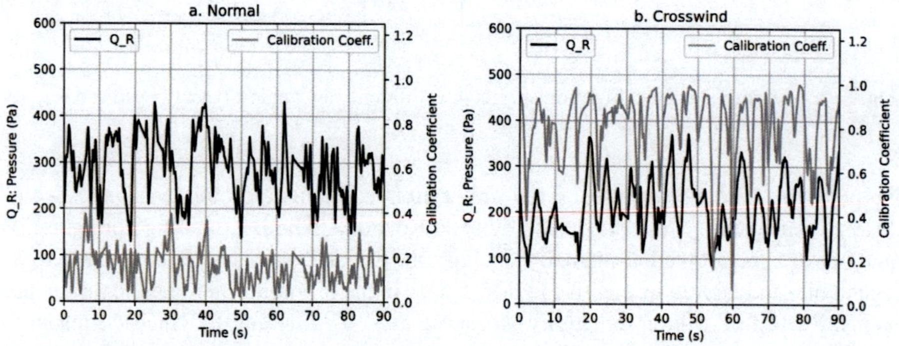

Fig. 8. Estimated dynamic pressure Q_R, and Calibration coefficient – with a 0.5 s moving average applied – from FR8-LAB measured data during a. no crosswind, b. crosswind

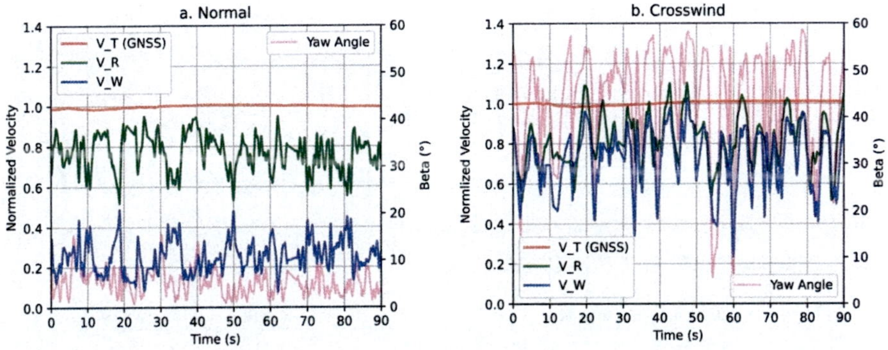

Fig. 9. Derived resultant wind V_R and atmospheric wind, V_W, normalized by mean train speed V_T and oncoming angle β – with a 0.5 s moving average applied – from FR8-LAB measured data during a. no crosswind, b. crosswind.

3 Results

The derived resultant velocity, V_R (the total velocity the probes would measure on the moving train) and the atmospheric wind component, V_W, of the velocity (i.e. the velocity measured by the equipment, with the train speed component removed) are presented for a specific example of a high-crosswind measurement (wind > 5 m/s, angle > 10°) in Fig. 10 upper. The derived velocities from the FR8-LAB measurements are presented both with a 0.5 s moving-average applied, as well as a 60 s moving-average and compared to measurements made by the two ultrasonic anemometers: USAs 1 & 2 – provided by DB Systemtechnik GmbH. Comparison of the different measurement shows strong agreement in both the long-term trend visible in the 60 s profiles, as well as the temporal fluctuations visible in the 1s profiles. Similarly, in Fig. 10 (lower), two additional crosswind examples are presented showing strong agreement between the FR8-LAB derived velocity and the USAs. The different examples all have crosswind present, but occurred at different days, with different magnitudes, and different time-varying characteristics.

Some differences exist between the FR8-LAB derived velocity and the USA measured velocities (there are visible differences even between the two USAs) but some variation is to be expected due to the different measurement positions of the equipment, the turbulent nature of the wind that has temporal and spatial fluctuations of different scales, as well as the different manner in which the velocity is determined between the USAs and the FR8-LAB.

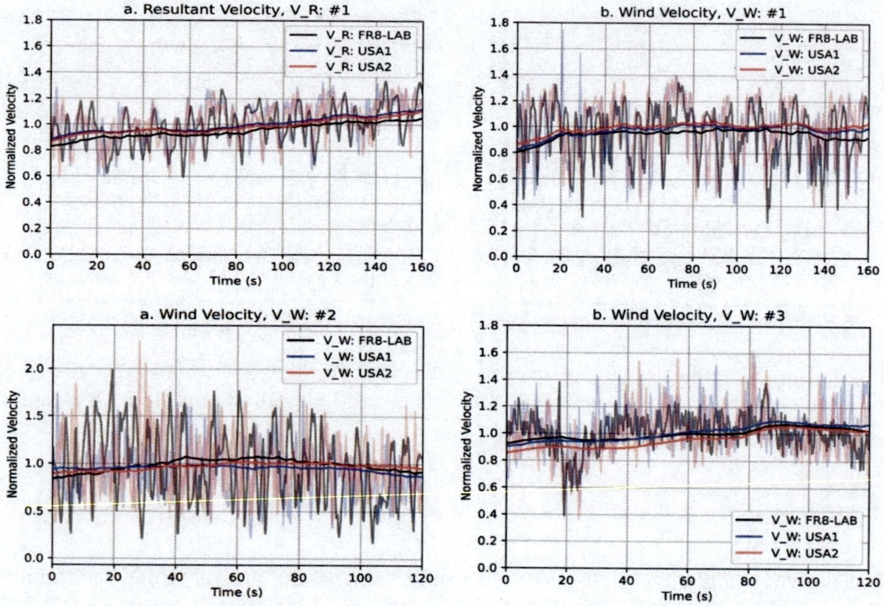

Fig. 10. Upper: a. Resultant velocity V_R and b. atmospheric wind V_W - normalized by local 160 s mean, with 0.5 s and 60 s moving average applied - derived from FR8-LAB measurements during crosswind measurement, and ultrasonic anemometers USA1 and USA2, Example #1 Lower: V_W, Example #2 and b. #3.

4 Conclusions

The oncoming flow a freight train/container experiences in real-world operation has been measured using the DLR FR8-LAB, a self-contained swap-body container equipped with sensors, data acquisition, power supply and communication system. The combination of pressure on different surfaces has successfully been utilized to generate a useable calibration function, similar to that of a multi-hole dynamic pressure probes, that can derive the oncoming crosswind flow the FR8-LAB measurement container is exposed to. This method was validated against ultra-sonic anemometer measurements and showed strong agreement.

Acknowledgements. The shipment of the container for measurements was facilitated by Christoph Lackhove from DB Cargo, and the ultrasonic anemometer measurements used for validation were provided by Michael Meister at DB Systemtechnik GmbH. The DLR FR8-LAB was developed as part of FR8Rail IV project that has received funding from the ER-JU under the European Union's Horizon 2020 research and innovation programme GA 101004051.

References

1. Quazi, A., Crouch, T., Bell, J., McGreevy, T., Thompson, M.C., Burton, D.: A field study on the aerodynamics of freight trains. J. Wind Eng. Ind. Aerod. **209**, 104463 (2021)
2. Quazi, A., Crouch, T., Bell, J., McGreevy, T., Thompson, M.C., Burton, D.: A field study on the aerodynamics of freight trains with different stacking configurations. J. Wind Eng. Ind. Aerod. **232**, 105245 (2023)
3. Gallagher, M., et al.: Trains in crosswinds – Comparison of full-scale on-train measurements, physical model tests and CFD calculations. J. Wind Eng. Ind. Aerod. **175**, 428–444 (2018)
4. Jessing, C., Wilhelmi, H., Wittmeier, F., Dillman, A., Wiedemann, J.: Characterization of the transient airflow around a vehicle on public highways. 12th FKFS Conference, 1–2 October 2019, Stuttgart, Germany (2019)
5. Cogotti, F., Pfadenhauer, M., Wiegand, T.: Potential of porsche reference cars for aerodynamic development. In: 12th FKFS Conference, 1–2 October 2017, Stuttgart, Germany (2017)

Experimental and Numerical Investigation of Passive Measures to Limit Aerodynamic Forces on Wind Turbine Rotor Blades

Marten Danker[1](✉), Maximilian Maigler[2], Oliver Meyer[1], and Jochen Schein[2]

[1] Institut für Aerodynamik und Flugsimulation, Universität der Bundeswehr München, Werner-Heisenberg-Weg 39, 85579 Neubiberg, Germany
marten.danker@unibw.de

[2] Institut für Plasmatechnik und Mathematik, Universität der Bundeswehr München, Werner-Heisenberg-Weg 39, 85579 Neubiberg, Germany
maximilian.maigler@unibw.de

Abstract. The approach and results of a study on the optimization of airfoils for wind turbines through geometrical modification of the leading edge are presented. The objective is to produce a constant lift-coefficient after the airfoil stalls and to remain at this constant lift for as long as possible. An experimental and numerical methodology that yields optimized geometries that influence lift this certain manner has been devised. The procedure is performed by experimentally testing a base airfoil (TEG2618) for a wide range of α at $\mathrm{Re} = 8 \times 10^5$ ($u_\infty = 38\,\mathrm{m\,s^{-1}}$) and using the produced experimental data as a reference for validating computational fluid dynamic (CFD) models. A software tool is developed to automatically carry out analyses of all predefined geometrical variations between $\alpha = -10$ and $20°$. The leading edge variants yielding the best results in the CFD analyses according to the criteria are selected, 3D printed and experimentally analyzed. The results demonstrate that the maximum lift coefficient at stalling can be maintained for at least another $\Delta\alpha$ of $6°$ with reasonable loss of the total maximum lift compared to the unmodified base airfoil.

Keywords: airfoil design · wind turbine overload protection

1 Introduction

As the demand of renewable energy increases, extensive efforts are spent on optimizing wind turbines and their safety mechanisms. The present approach refers to an aerodynamic method in overload control by modifying the leading edge of rotor blades and is based on a principle used for stall indicators installed on aircraft wings in the 1940s. In Fig. 1, the forces acting on a rotor blade airfoil are outlined. According to Betz's law, Horizontal Axis Wind Turbine (HAWT) blades are approached by 2/3 of the initial windspeed V_0. The rotation of the

A. Dillmann et al. (Eds.): STAB/DGLR Symposium 2022, NNFM 154, pp. 38–47, 2024.
https://doi.org/10.1007/978-3-031-40482-5_4

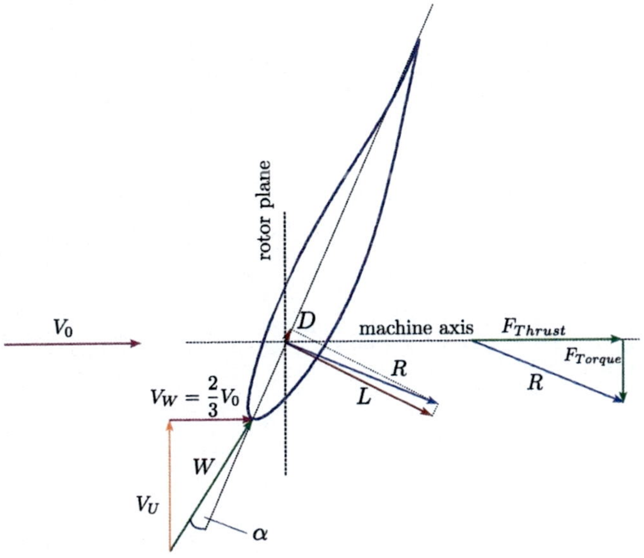

Fig. 1. Schematic of forces acting on a rotor blade airfoil

blades causes an additional flow V_U to act on the rotor blade. These two wind vectors add to form a resulting incident flow W, which acts locally on the rotor blade. Any change in these flow velocities causes a change in the local incident flow. In particular, a change in the ratio to each other causes a change in the direction of the airflow and thus the local angle of attack α. The change in α results in a change in lift L and drag D acting on the rotor blade. Lift and drag can be combined to a resulting force vector R. Decomposing R again, one part F_{torque} points in the rotary plane and drives the wind turbine while the other one (F_{thrust}) points in axial direction to the wind turbine, loading the turbine blades and acting as thrust. Figure 1 shows that especially an influence of the lift at the rotor blade is relevant, because in normal operation L roughly equals R and thus also the power and total load on the turbine. The efficiency of the system is proportional to the lift-to-drag ratio L/D, therefore it is desirable to keep the rotor blade configuration in a regime, where the glide ratio remains close to its maximum.

In order to protect the turbine from overload, wind turbines can adjust lift by changing the pitch, i.e. α. However, the response rate compared to the characteristic time scale in gusty wind conditions may be too slow to act as an overload protection while keeping the turbine running in gusty wind circumstances. In those ambient conditions, most wind turbines are shut down and brought to the so-called vane position. For optimized efficiency, the turbine should be able to operate for a larger range of α even in gusty and high wind speed scenarios without rapid overloading. Accordingly, the outlined approach modifies the rotor blade airfoil geometry edge to limit the maximum lift and influence the post-stall behavior. Safety in wind turbine aerodynamics has been subject to extensive research, especially on the trailing edge of its blade airfoils. Various

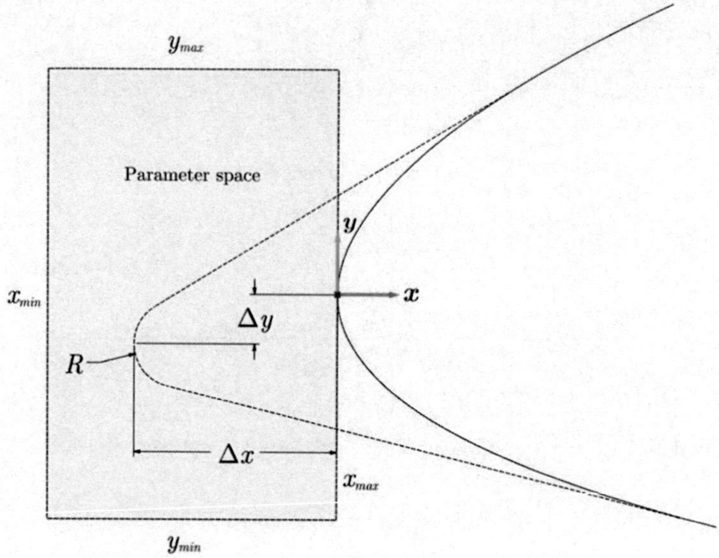

Fig. 2. Schematic of leading edge modification parameter space

passive and active measures for load control on wind turbines were put forward by [1] , including traditional and nontraditional trailing-edge flaps and microtabs with leading edge measures, such as Leading Edge Vortex Generators (LEVoGs by DLR) and stall strips also being investigated. A measure of active stall strips at the leading edge of rotor blades were implemented by [2], outlining microspoilers, which can be extended on the suction side near the leading edge in order to quickly minimize lift in case of overload to reduce load from the blades. The present study is based on the work of [3] with the focus of leading edge modifications of a different airfoil (TEG2618, *base airfoil*) which features a larger leading edge radius and less critical stall behavior. Additionally, methods of numerical preselection and approaches in the fabrication of experimental models for wind tunnel tests are highlighted.

Standard wind turbines usually feature higher Reynolds numbers ($Re > 1 \times 10^6$) compared to the currently investigated one of $Re = 8 \times 10^5$ which is attributed to wind tunnel limitations. The effect of higher Reynolds numbers for different wind turbine airfoils was analyzed in detail by [4]. The present investigations are focused on smaller wind turbines and essential basic principles of these kind of airfoil manipulations are studied.

2 Methodology

2.1 Leading Edge Modification Approach

The airfoil modification analysis features 3 geometrical degrees of freedom (DOF) including the leading edge radius R, horizontal shift Δx and vertical shift Δy.

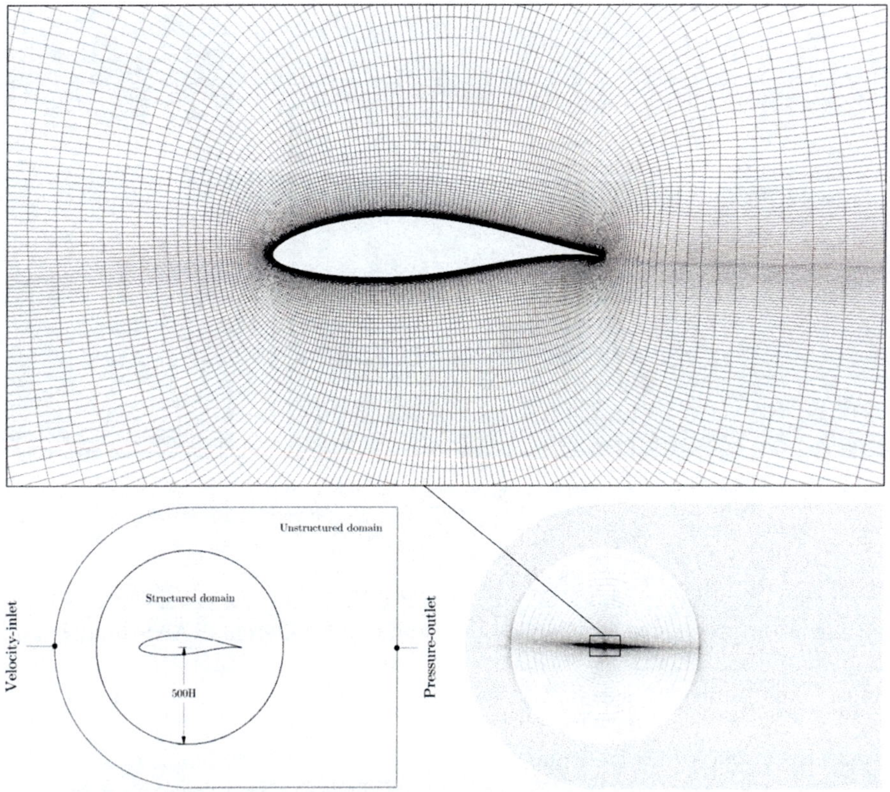

Fig. 3. Meshing approach and outer domain with numerical boundary conditions

The entire parameter space (grey area) along each respective DOF is shown in Fig. 2. The base airfoil leading edge tip at x, $y = 0$ serves as the reference point for the parametric study from where all airfoil variations are derived from. A tool written in *Mathematica* was implemented to automatically create all possible combinations of the discretized parameter space and to tangentially connect the modification to the base airfoil.

2.2 Numerical Approach

The CFD simulations are based on a 2D approach, where a structured grid is extruded from the airfoil for at least 500 chord-lengths and connected to an unstructured C-Grid in the farfield. In Fig. 3, a schematic of the general meshing approach is highlighted (left) along the numerical boundary conditions and actual mesh (top and right, respectively).

The CFD solver Fluent 2022R1 was employed using an incompressible, steady-state approach with the density and dynamic viscosity of air set to $\rho = 1.1\,\mathrm{kg\,m^{-3}}$ and $\mu = 1.793 \times 10^{-5}\mathrm{Pa\,s}$, respectively. In order to maintain

Fig. 4. Glued leading edge retrofit and 3D printing process

a constant Reynolds number of 8×10^5, all airfoil variants are normalized. An optimization procedure was carried out employing different RANS models and numerical discretizations in order to determine the appropriate numerical settings. Wind tunnel measurements of the base airfoil as well as a pre-defined demo variant ($\Delta x/\mathrm{L} = -0.012$, $\Delta y/\mathrm{L} = -0.018$ and $\Delta R/\mathrm{L} = -0.005$) were used as reference, in particular the polar lift ($\alpha - c_l$) and Lilienthal plots ($c_d - c_l$). The widely used Spalart-Allmaras, Shear-Stress Transport (SST) and recently introduced generalized $k - \omega$ (GEKO) [5] model were analyzed. The first layer was meshed with a y^+ of 1 and the semi-implicit, pressure-based, coupled solver was used with pseudo-transient time-stepping as the residuals and force coefficients tend to oscillate around a mean value with this approach.

An open source Python package [6] was developed to automatically prepare, run and analyze all simulations based on the chord length L, minimum and maximum α, freestream conditions and number of increments for each airfoil. The range of α has been discretized by 15 steps from $\alpha = -10$ to $20°$ for 45 airfoil variations yielding a total number of 675 simulations, which serve as the database for obtaining the optimum airfoil variation.

2.3 Experimental Approach

The wind tunnel is of Göttinger type and has an open test section ($0.9 \cdot 1.3 \cdot 2\,\mathrm{m}$, turbulence intensity $I = 0.35\%$) and Reynolds number driven fan control. TEG2618 airfoil test model features a chord length $\mathrm{L} = 0.35\,\mathrm{m}$ and the suitability of the wind tunnel for this flow scenario was demonstrated by [7], including the correction factors and formulas applied to the measured raw data. The lift force F_l is measured with a scale while the drag force F_d is determined by a pitot pressure rake located in the wake of the airfoil.

2.4 3D Printed Leading Edges

Additive manufacturing process is employed to 3D-print the optimized leading edges, after successful determination of the optimized variants. They are subsequently sanded and retrofitted to the base airfoil using glue and adhesive tape for ensuring tangent transition. Experimental results from the wind tunnel facility demonstrated that the 3D printed leading edges agree well with the results produced by a CNC milled geometry of the same variant [8]. An image of a glued leading edge part before applying adhesive tape is shown in Fig. 4 (left), while the central support structure necessary for ensuring stability during the printing process is shown on the right.

3 Results

3.1 Base Airfoil Analysis and Validation of Numerical Setup

The results of the base airfoil (*base*) and pre-defined airfoil variant (*demo*, y–position below *base*) are illustrated in Fig. 5. It is clear that the Spalart-Allmaras performs poorly at very high α due to its one-equation approach. **Interestingly enough, the SST model performs worse if transition effects are** not taken into account. The results improve drastically when enabling the γ-algebraic model, in particular for the *demo* version when determining the critical stall angle α_c.

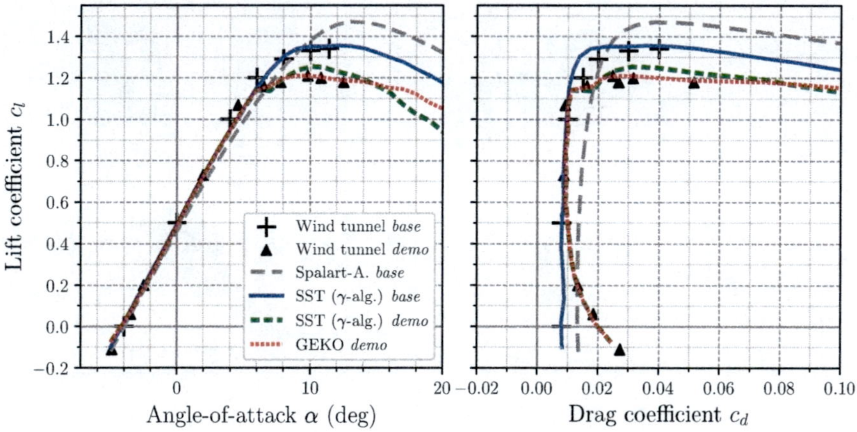

Fig. 5. Base profile lift and drag predictions with different turbulence models

However, for the modified variants which introduce small recirculation and flow separation bubbles, the SST model failed to consistently produce accurate results, in particular for higher Δy positions due to overproduction of specific dissipation rate ω. For this reason, the GEKO model features a freely adjustable parameter (C_{Sep}), that allows to change the production rate of ω. For low Δy positions, the results produced with the default C_{Sep} parameter in the GEKO model agree very well with experimental data. However, with increasing Δy shift, the C_{Sep} parameter were linearly reduced to counteract the overprediction of the ω production rate. The minimum C_{Sep} value was determined based on wind tunnel data of another variant where $\max(\Delta y) = 0.04\,\mathrm{L}$.

3.2 Analysis of Airfoil Variants

In this section the results for the investigated airfoil variants are presented. The previously outlined 3 DOF (leading edge radius R, x-shift Δx and y-shift Δy) were analyzed for a wide range of α spanning from -5 up to 25°, yielding a 4-dimensional parameter results space. Primarily, the aim was to determine the DOF exerting the biggest influence on the progression of the lift coefficient while maintaining high glide-ratios. For more convenient visualization, the amount of airfoil variants displayed in the respective plots has been reduced to the minimum and maximum for each parameter. A 2D scatter plot of the parametric results is shown in Fig. 6, including the base profile reference results. The $y-$position is denoted by the coloring, reduced to 3 major positions (y_{min}, y_0 and y_{max}) with each region colored from its respective *min* to *max* values, and the radius of the leading edge is indicated by the type of marker. Since the parameter space for Δx remained relatively small for realistic airfoil shapes, the effects of this DOF were found to be negligible and are omitted in the plots for simplicity.

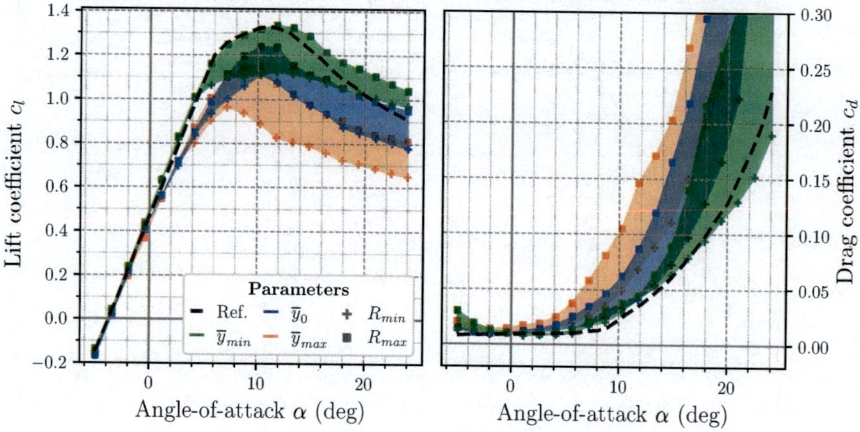

Fig. 6. Lift and drag polars of CFD analyzed airfoil variants

According to Fig. 6 it is evident, how the leading edge $y-$position exerts the largest influence on both the lift and drag coefficient. The maximum lift is produced by the variants where y is the lowest and vice versa. Furthermore, with increasing Δy, the airfoil suffers stall at lower α, including a generally

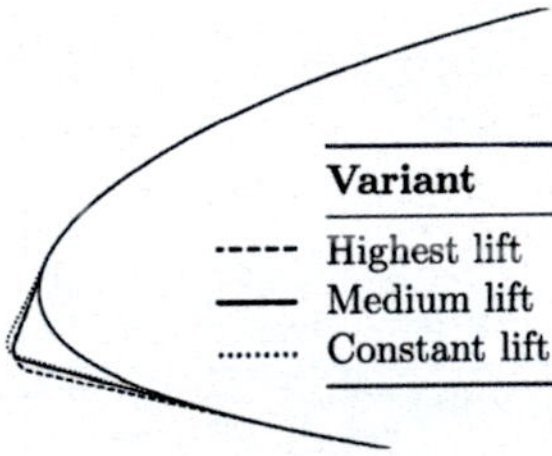

	Variant	x/L–shift	y/L–shift	R/L Radius
-----	Highest lift	-0.01	-0.02	0.008
——	Medium lift	-0.01	-0.02	0.004
....	Constant lift	-0.0125	-0.018	0.0045

Fig. 7. Optimized variants leading edge properties

reduced slope of c_l in the area of non-separated flow (-5 to $6°$). At large positive Δy, a recirculation region is produced on the upper airfoil surface, resulting in flow separation that diminishes the negative static pressure in this region. Interestingly enough, the reduced leading edge $y-$position results in better normal alignment with the stagnation point, thus maintaining a low drag coefficient which is reflected in the right plot of Fig. 6. Another effect seems to be prevalent at high α in this regime, where a small local laminar separation bubble (LSB) occurs at the leading edge. This effect was investigated by a Detached-eddy simulation carried out for the optimized variant (*constant lift*, see Sect. 3.3) and the results of the velocity magnitude are illustrated in Fig. 9. During stall and post-stall, the LSB produces a local negative static pressure region that causes the flow to stay better attached to the anterior upper airfoil surface compared to the base airfoil (see Fig. 9, right). The unmodified and upper leading edge $y-$positions cannot maintain this small LSB since the adverse pressure gradient and flow separation become too large. It should also be mentioned that by shifting the leading edge y-position, the camber line of the airfoil becomes slightly more curved (and vice versa). Apart from these local effects, another combination of global effects take place in the post-stall regime. As $\alpha > 6°$ and is further increased, still, the lower airfoil surface produces more lift while the upper airfoil surface produces increasingly less lift, canceling each others effects out. For the stall and post-stall regime, it is assume that the interaction of these local and global effects are responsible for maintaining a constant c_l.

The leading edge radius R plays a minor role in the progression of the post-stall behavior for both c_l and c_d. By reducing R, the lift remains more slightly more constant but is lower including increased drag, while variants with larger R tend to feature a higher maximum lift, followed by a faster decline in c_l and increase in c_d. This effect can be attributed to larger and more smooth flow curvature, until abrupt flow separation occurs close to the leading edge.

3.3 Determination of Optimized Variants And Testing

The generated and validated aifoil variants database allowed the determination of multiple optimized variants. It is clear, that the blue and orange regions in Fig. 6 do not satisfy the constant lift criteria nor do they yield acceptable c_l and glide ratios. Therefore, 3 variants with different negative $y-$shifts, leading edge

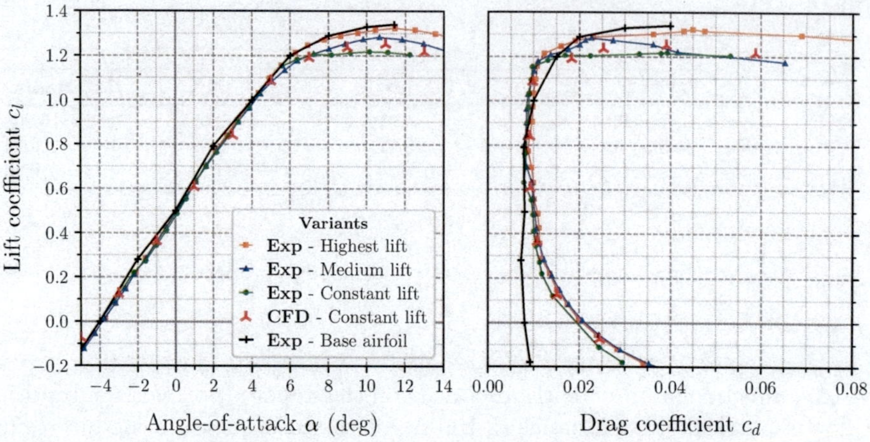

Fig. 8. Experimental results of different optimized variants

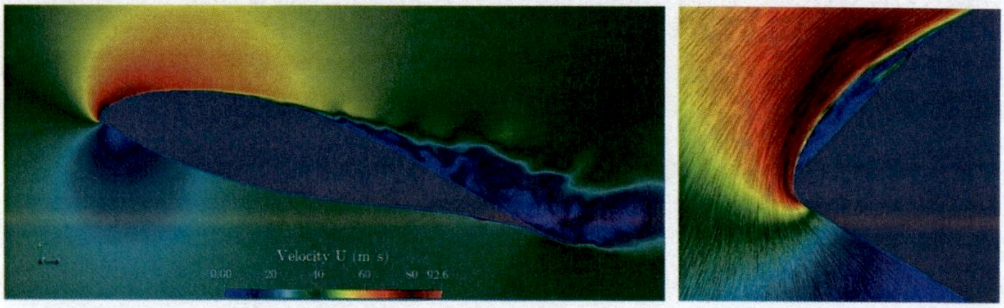

Fig. 9. DES of *constant lift* variant at $\alpha = 15°$ showcasing LSB

radius R and $x-$positions were tested. Details including a schematic for each variant compared to the base airfoil and their naming are outlined in Fig. 7.

Results of the wind tunnel analyses are outlined in Fig. 8. As expected, the variant with larger leading edge radius produces the highest lift and closely resembles the behavior of the base airfoil. It is assumed, that the combination of wind tunnel corrections at very high α and the three effects described in Sect. 3.2 are responsible for the *highest lift* variant remaining close to the base airfoil plot. However, c_l and c_d are majorly effected in the regime where $\alpha < 4°$, which is attributed to the same issue (see Sect. 3.2) that occur for variants with negatively shifted Δy. We determined that the *constant lift* variant exhibits the most desirable lift and drag coefficient progression, featuring an abrupt gradient of c_l at $\alpha = 6°$, followed by a constant c_l and c_d up to $\alpha = 14°$.

4 Conclusion

An extensive parametric study of modified leading edges with 3 DOF of a wind turbine airfoil (TEG2618) was carried out both experimentally and numerically.

The numerical models provided a general overview of how the variants behave for a wide range of α, although some expected deviations at post-stall angles were encountered using the RANS approach. With a suitable measure, the airfoil investigated by [3] was able to achieve a lift limitation of about 15% compared to 8% of the base airfoil in the present study, due to its superior stall characteristics. The importance of the nose radius and y-position became particularly apparent. Three variants were obtained from the numerical study and tested in the wind tunnel facility, revealing the final optimized variant featuring the most constant lift after stalling takes place as well as reasonable glide-ratios in the operating regime. Future research will focus on developing active measures that are based on the same operating principles but produce a broader range of functions by allowing the shape of the airfoil to be changed. From a numerical perspective, scale-resolving methods will be applied for improved predictions of the complex flow structures occuring at the leading edge and upper airfoil surface that significantly influence the lift and drag coefficient.

References

1. van Dam, C.P., Berg, D.E., Johnson, S.J.: Active load control techniques for wind turbines (2008). https://doi.org/10.2172/943932
2. Mertz, B., Bortolotti, P., Abbas, N.J., Johnson, N., Paquette, J., Jordan, W.: Leading edge microspoilers for load dumping in wind turbine rotors. J. Phys.: Conf. Ser. **2265**, 032024 (2022). https://doi.org/10.1088/1742-6596/2265/3/032024
3. Meyer, O.: On the effectiveness of retrofitted airfoil leading edge devices to limit lift. In: Dillmann, A., Heller, G., Krämer, E., Wagner, C. (eds.) STAB/DGLR Symposium 2020. Notes on Numerical Fluid Mechanics and Multidisciplinary Design, vol. 151, pp. 729–738. Springer, Cham (2021). https://doi.org/10.1007/978-3-030-79561-0_69
4. Ge, M., Tian, D., Deng, Y.: Reynolds number effect on the optimization of a wind turbine blade for maximum aerodynamic efficiency. J. Energy Eng. **142** (2016). https://doi.org/10.1061/(ASCE)EY.1943-7897.0000254
5. Menter, F.R., Matyushenko, A., Lechner, R.: Best practice: generalized k-omega two-equation turbulence model in ANSYS CFD (GEKO). ANSYS Inc. (2019)
6. Maigler, M.: pyaoa - automatic angle of attack analysis (2022). https://github.com/inflowencer/pyaoa
7. Meyer, O.: 2D Airfoil-testing in an open wind tunnel test section: test setup, instrumentation, interference-corrections, validation. In: Dillmann, A., Heller, G., Krämer, E., Wagner, C. (eds.) STAB/DGLR Symposium 2020. Notes on Numerical Fluid Mechanics and Multidisciplinary Design, vol. 151, pp. 719–728. Springer, Cham (2021). https://doi.org/10.1007/978-3-030-79561-0_68
8. Meichsner, C.: Validierung von Untersuchungen zur Wirksamkeit von Vorderkantenmaßnahmen zur Beeinflussung des Hochauftriebsverhaltens bei Flügelprofilen. Universität der Bundeswehr München (2022)

A Miniature Flow Sensor Capable of Determining the Instantaneous High-Frequency Wall Shear Stress Fluctuations in Magnitude and Direction

Mario Eck[1(✉)], Dieter Peitsch[1], and Christian Westphal[2]

[1] Institute of Aeronautics and Astronautics, Technische Universität Berlin, Marchstraße 12-14, 10587 Berlin, Germany
{mario.eck,dieter.peitsch}@tu-berlin.de

[2] Institute of Fluid Dynamics and Technical Acoustics, Technische Universität Berlin, Müller-Breslau-Straße 15, 10623 Berlin, Germany
christian.westphal@tnt.tu-berlin.de

Abstract. The present paper deals with a wall shear stress sensor that is an improved modification of a conventional sublayer fence probe. Using a novel probe geometry - a so-called star probe - the measurement technique is capable of deternining the shear stress or the velocity close to the wall in magnitude and direction instantaneously at high temporal resolution. Hence, the novel probe enables significant gain in sensor capabilities compared to the conventional probe configuration. However, the directional resolution becomes more and more biased and damped with increasing frequency.

Keywords: Flow measurement technique · wall shear stress sensor · sublayer fence probe · Computational Preston Tube Method (CPM3)

1 Introduction

Many direct and indirect shear stress measurement devices have been developed over the years all of which differ to some extent in their applicability, robustness, spatial/temporal resolution and limitations [1–4].

The measurement technique addressed in this paper is a surface fence. Its underlying principle is based on intentionally disturbing the viscous sublayer by means of a small obstacle in the form of a microscopic surface fence. The pressure difference developed across the fence can be considered proportional to the local wall shear stress τ_W. The proportionality implies the probe to be an indirect measurement technique, which must be calibrated using a reference.

The conventional probe (single fence) is only suitable to be applied in straight 2D flows, where the probe must always be aligned perpendicular to the flow. If the probe is instrumented with piezoresistive pressure transducers, the probe is capable of determining fluctuations of the local wall shear stress $\tau_W(t)$. Another

A. Dillmann et al. (Eds.): STAB/DGLR Symposium 2022, NNFM 154, pp. 48–57, 2024.
https://doi.org/10.1007/978-3-031-40482-5_5

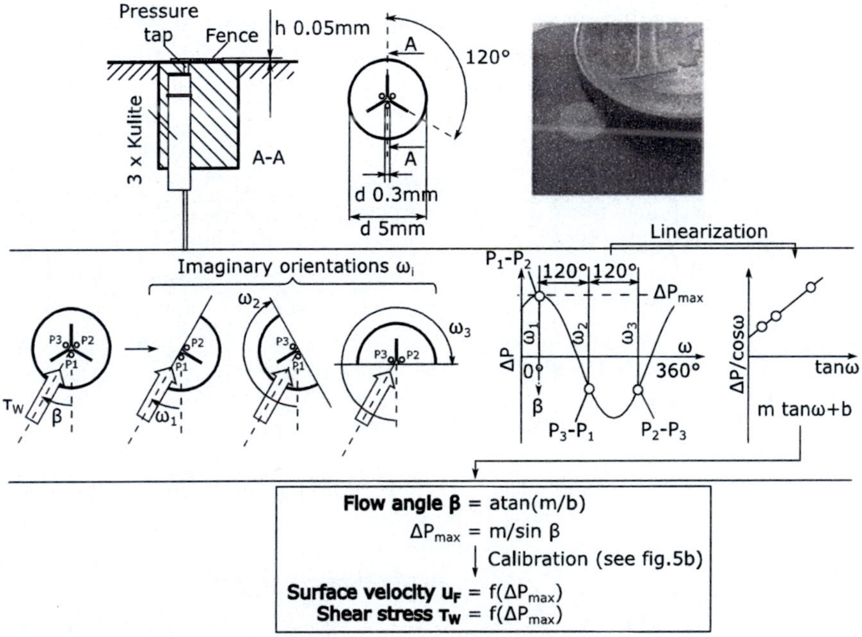

Fig. 1. Star probe, working principle and flow angle calculation; resonance frequency of the pressure tap - sensor cavity volume $f_{res} = 12.5\,\text{kHz}$

unique feature of the conventional probe allows to determine the direction β of τ_W. If the probe is rotated around its vertical axis, the pressure difference recorded describes an almost perfect cosine, with the maximum corresponding to the direction β. By applying a linearization procedure from [6], the measured pressure differences and orientation angles of the probe are transformed in such a way that they describe a straight line. A linear regression yields the interception and the slope, allowing the magnitude and the direction to be finally calculated. At least two orientations are required for determining the flow direction, with an error decreasing as more orientations are considered [5].

The new probe geometry presented in this paper exploits all the previously listed properties of the single fence. The probe has three fences arranged in a star shape, each spanning a sector of 120°. This specific geometry allows determining the direction of the shear stress, firstly without the need of rotating the probe and secondly providing the magnitude and direction of the shear stress instantaneously which was unattainable until now.

2 Experimental Setup

2.1 Probe

Figure 1 shows a cross-section and the top view of the new probe including all main dimensions. The probe is equipped with three XCS-062 Kulite sensors. For further data processing, the probe can be cut imaginarily in such a way that three individual fences with different orientations evolve. The probe figuratively contains three single fences with imaginary orientations $\omega_{1,2,3}$ each having an offset of 120°. The pairs of ΔP and imaginary omegas $\omega_{1,2,3}$ are located on a harmonic distribution having a global maximum (ΔP_{max}) corresponding to the magnitude and direction of the wall shear stress. Linearization of the data triplet finally yields the linear regression and computing β is based upon the slope m and interception b.

To validate the linearization routine applied to the star probe, the latter has been rotated from $\omega = 0$–360° in 10° increments. Figure 2a) shows the normalized $\Delta P_{i,j}/\Delta P_{i,j,max}$ for various inflow velocities set during τ_W-calibration presented in Sect. 2.2. After redistributing the ΔP-triplet to the sensor-fixed orientations ω_S (Fig. 2b) and applying the linearization, the calculated flow angles β were compared with the probe orientations ω (Fig. 2c). The deviation between the data points from the line through the origin is represented by the Root Mean Square Error (RMSE) which amounts to $RMSE = 4.14°$.

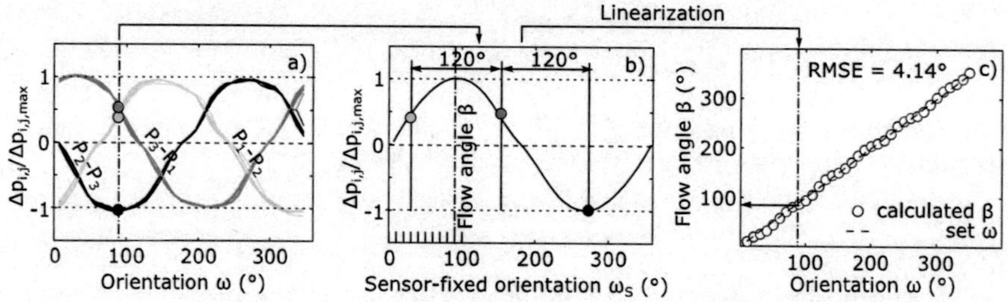

Fig. 2. Error in processing the flow angle by means of the linerization procedure applied to the star probe data

The relation between ΔP_{max} and u_F or τ_W still needs to be provided by carrying out an additional calibration procedure, which is explained in the following section.

2.2 Probe Calibration

While β can be determined analytically, the relationship between ΔP_{max} and u_F or τ_W must be established by a calibration procedure. For this purpose, a reference measurement technique is necessary, which provides the desired target quantity. In this work the Preston tube is used as CPM3 (Fig. 3).

If the star probe is not to be used to determine the local shear stress but the near-wall velocity u_F at half the fence height $h/2$, the τ_W determined by the CPM3 probe can also be converted into u_F. If τ_W is substituted into the law of

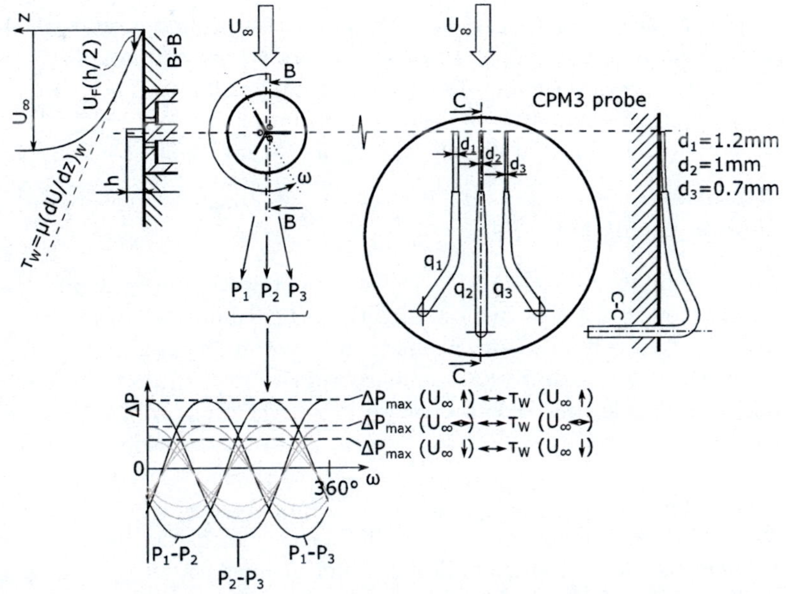

Fig. 3. Calibration of the star probe using CPM3 as reference

the wall, a boundary layer velocity profile plotted against the wall distance z is obtained. The required u_F can now be read out from the profile at $z = h/2$.

During calibration, the star probe and the CPM3 probe are positioned simultaneously and at the same longitudinal position on a flat plate. Using a conventional surface fence probe, the single fence would be oriented perpendicular to the flow in order to ascertain ΔP_{max}. In contrast, determining a representative ΔP_{max} for the star probe is possible only if the probe is rotated by 360° at constant inflow conditions (cf. Fig. 2a).

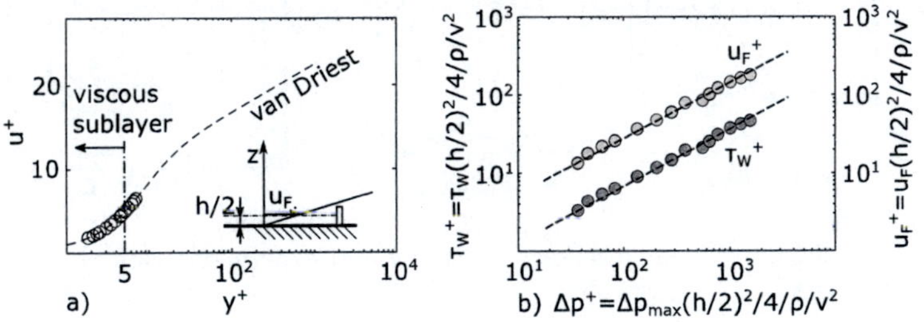

Fig. 4. Law of the wall and calibration points a); calibration of the star probe b)

The pairs $[\Delta P_{max}, \tau_W]$ or $[\Delta P_{max}, u_F]$ respectively have been converted to their corresponding non-dimensional values ΔP^+_{max}, τ^+_W, u^+_F and plotted in the calibration diagram (see Fig. 4b). The calibration function is derived from an exponential regression applied to the data points (dashed line). To assess the boundary layer region, the calibration points are to be allocated, u^+ and y^+ have been processed by inserting u_F and τ_W into the law of the wall. Figure 4a) shows the data points to be located within and slightly above the viscous sublayer.

2.3 Fluidic Oscillator

The fluidic oscillator provides an excellent environment for generating specified high-frequency changes in flow direction. The fluidic element from Park [7] was chosen, since it has been geometrically parameterized and comprehensively investigated regarding the frequency ranges covered and the maximum sweeping angles achievable. For the validation of the star probe, the narrowest cross-section of the outlet nozzle was chosen as the installation location inside the fluidic element, since the center of rotation of the sweeping jet is presumed to be located here.

A total of three fluidics is used, which, starting from the base line configuration (geometry from Park; fluidic B), differ geometrically by scaling in size (Fig. 5). This allows different frequency ranges and jet velocities to be achieved at various inlet pressures. Following results show the maximum sweeping angle of all fluidics to be almost constant at about 25° in a frequency range of $300\,\text{Hz} < f < 600\,\text{Hz}$. Below 300 Hz the actuators do not activate and operation above $600\,Hz$ is not possible due to the pressure supply system choking.

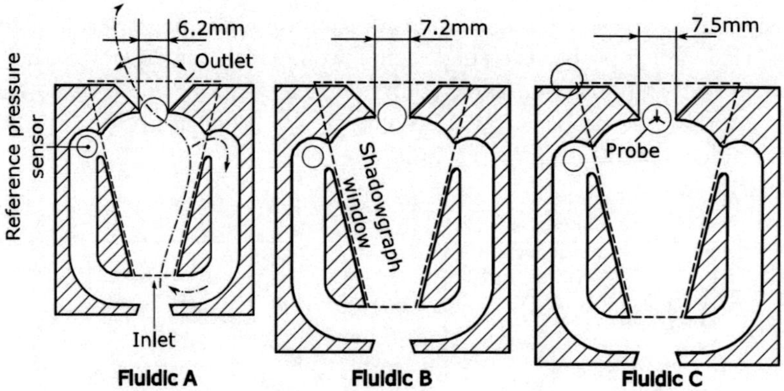

Fig. 5. Fluidic oscillators, scaling and instrumentation

Before experimentally validating the probe's ability to resolve high-frequency fluctuations in flow angle, the actual time-resolved sweeping angle must be acquired first. For this purpose, high-speed shadowgraph visualizations are performed (see Sect. 2.4). Using this measurement technique requires the fluidics to be visually transparent. By means of a special data reduction method applied to the image data, the time-resolved sweeping angle can be extracted automatically (see Sect. 3). Once the actual sweeping angles have been established, the parametric investigations are repeated with the star probe installed at the nozzle throat.

2.4 Shadowgraph Visualization

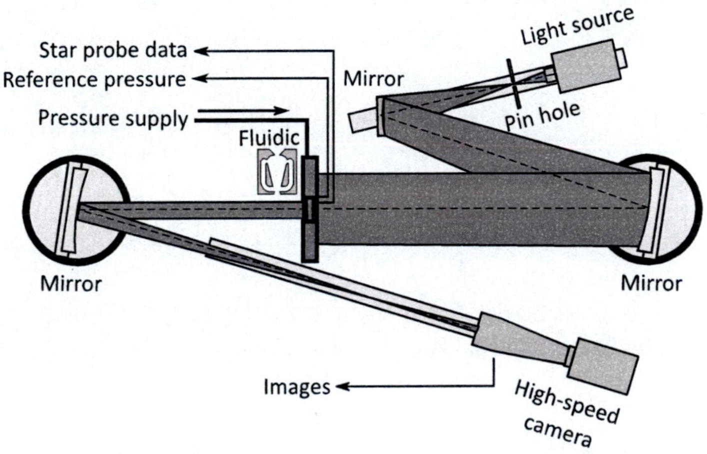

Fig. 6. High-speed shadowgraph visualization

To visualize the flow in the nozzle area of the fluidic element, shadowgraph visualization in a conventional Z-type setup is employed [10]. The reflection angle of the two concave mirrors is 12.5° and the experimental setup is sketched in Fig. 6. The high-speed camera (Phantom v2512, Vision Research) spatially resolves the translucent part of the fluidic with 170×300 pixels. The recordings are made with 25.000 frames per second and an exposure time of 1.5 µs. In order to minimize the influence of measurement noise and optical artefacts, recordings without flow were also taken as a reference for post-processing.

3 Image Post Processing

The sweeping angle is computed in four successive steps (Fig. 7a)). The image data is being post-processed (brightness, contrast, noise reduction). Then the image is cropped to a specific target window covering the area around the nozzle including part of the free stream area (step 1). The choice of the window size is based on investigations of Hirsch et al. (2018) [9]. Afterwards, the so-called canny edge method [8], identifying gradients in the grayscale, is applied to the cropped image (step 2). The center of mass of the canny edge pattern is determined in step 3 and by connecting this center of mass with the center of the nozzle throat, the sweeping angle can be calculated for each image sample (step 4). Figure 7b) shows the time-resolved sweeping angles as an approximate rectangle function covering approximately $-25° < \Theta_{sg} < 25°$.

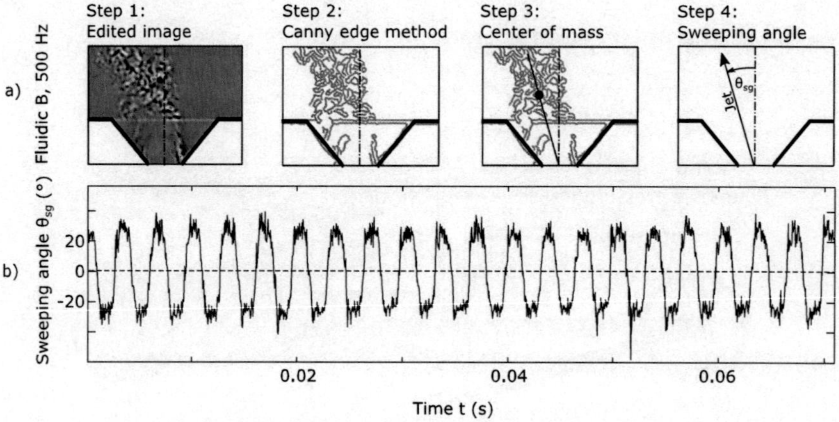

Fig. 7. Shadowgraph image post processing

4 Results

4.1 Sweeping Jet Angle

First, we show the results of the unfiltered and non-phase-averaged sweeping angles of the probe Θ_{Probe} recorded in fluidic B at $f = 500$ Hz (Fig. 8). The distribution exhibits a clear harmonic variation of Θ_{Probe} in the expected range between $-25° < \Theta_{Probe} < 25°$. **However, there is a slight offset around the x-**axis, i.e. the jet appears to be oscillating asymmetrically around the symmetry axis of the fluidic. This peculiarity occurs in all fluidics equally and as can be proved in the further course of this work, this is not a measurement error.

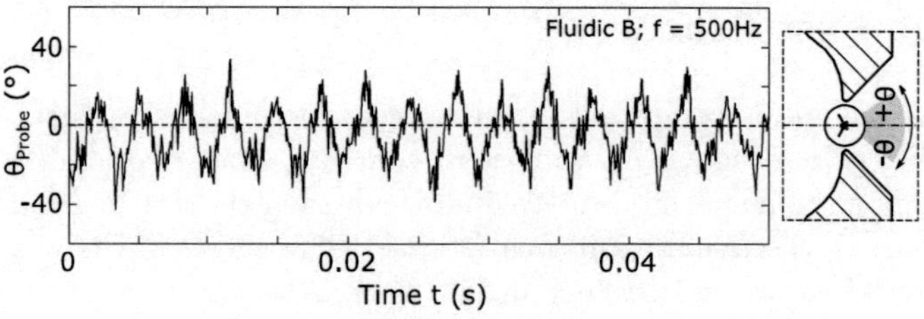

Fig. 8. Time-resolved sweeping angle calculated from the star probe data

Figure 9 left shows the phase-averaged sweeping angles Θ_{sg} and Θ_{Probe} (fluidic B, $f = 300\,\text{Hz}$). The low-pass filtered reference pressure signal has been used for phase-averaging. In general, there is good agreement between the two distributions with a maximum sweeping angle of the probe

$$\bar{\Theta}_{max,Probe} = (\Theta_{max,Probe} + |\Theta_{min,Probe}|)/2 \tag{1}$$

being slightly smaller with $\bar{\Theta}_{max,Probe} = 23.5°$ than $\bar{\Theta}_{max,sg} = 27.3°$. Nevertheless, both distributions show the asymmetry already identified in Fig. 8, as well as a pronounced local minimum at $\varphi = 230°$ and the subsequent plateau around $\varphi = 300°$. At $\varphi = 40°$, the Θ_{Probe} distribution exhibits a local minimum, indicating a briefly swing back of the jet when sweeping over to the left. While this feature is not evident in Θ_{sg}, sporadic swingback of the jet was already visually evident when viewing the shadowgraph image data.

The probability distributions in Fig. 9 right confirm the asymmetric sweeping angle with each distribution shifted to negative Θ. In addition, the probability distribution of the probe is broader, which is attributed to the higher fluctuation level. The unintentional asymmetry is a remarkably purposeful feature of the fluidic to provide evidence that the probe is even capable of capturing this unique fluctuation. Presumably, the asymmetry is due to the installation of the reference pressure sensor. Although the sensor was installed flush with the surface, the cavity between the sensing screen and the piezo chip creates a slightly larger volume in the corresponding feedback channel.

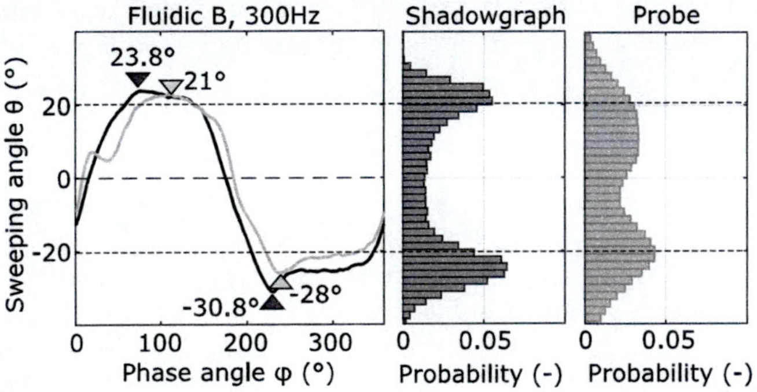

Fig. 9. Phase-averaged sweeping angles Θ_{sg} (dark grey), Θ_{Probe} (light grey) and probability distributions

4.2 Error and Damping

Before presenting the results of the frequency-dependent deviation or damping of the phase-averaged sweeping angles, another variable, the jet velocity, must be considered. Regardless of which fluidic is used, an increase in frequency is achieved by increasing the inlet pressure. This also inevitably causes the jet velocity to be increased. Figure 10a) clearly mirrors this relationship and the star probe velocities u_F of the larger fluidics are seen significantly higher compared to configuration A at similar frequencies.

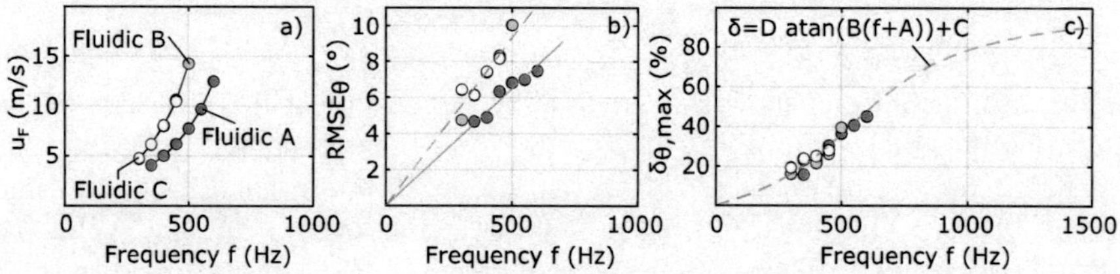

Fig. 10. Jet velocities, root mean square error and damping of the probe angles

The Root Mean Square Error from the phase-averaged sweeping angles Θ_{sg} and Θ_{Probe} are summarized in Fig. 10b).

$$RMSE_{\Theta} = \frac{1}{n}\sum_{i=1}^{n}\left(\Theta_{Probe}(\phi) - \Theta_{sg}(\phi)\right)^2 \tag{2}$$

It becomes evident that the error of the probe data increases with increased probe velocity. These results indicate that for constant maximum sweeping angles but larger magnitudes in probe velocity, the probe error increases, with the error generally decreasing at lower frequencies. Figure 10c) summarizes the damping of maximum sweeping angles, where the damping $\delta_{\Theta,max}$ has been processed as follows:

$$\delta_{\Theta,max} = \left(1 - \frac{(\Theta_{max,Probe} + |\Theta_{min,Probe}|)/2}{(\Theta_{max,sg} + |\Theta_{min,sg}|)/2}\right) \cdot 100\% \tag{3}$$

The maximum amplitudes of the probe angles are increasingly damped with increasing frequency, although unlike the RMSE, the different velocities do not appear to have any influence. Assuming the damping increases with increasing frequency but asymptotically approaching 100% damping, a regression can be applied to the data points. The underlying *atan*-function can be used as a transfer function to correct the flow angles in future experiments.

5 Conclusions

A novel probe is presented that is capable of detecting high-frequency fluctuations of the local wall shear stress in magnitude and direction while installed fixed. The probe represents a further development of a conventional surface fence probe and is thus a pressure-based and indirect measurement technique. Validation has revealed the probe to be generally capable of determining unique changes in τ_W direction. However, the time-resolved direction of the local wall shear stress becomes even more inaccurate the higher the average magnitude of the shear stress or the higher the frequency is. Also, the results indicate the fluctuation amplitude of the direction being damped as the frequency increases. However, the question remains whether there is actually a damping of the probe's data or whether the probe represents the prevailing flow conditions in the fluidic nozzle rather accurately. Especially at higher frequencies, fundamental deviations are to be expected between the boundary layer flow being captured by the star probe and the main flow made visible by the shadowgraph visualization and which served to compute the sweeping angles. Further investigations in this direction are still pending.

Acknowledgment. The authors are deeply grateful to our collegues from "research contracts, patents and corporate investments" for the trust and the financial support within the framework "ProTUTec" initiated by the Technical University of Berlin.

References

1. Nitsche, W., Haberland, C., Thünker, R.: Comparative investigations on friction drag measuring techniques in experimental aerodynamics. In: ICAS-84-2.4.1, 14th Congress of the international council of the aeronautical sciences, 9–14 September 1984, Toulouse, France, pp. 391–403 (1984)
2. Fernholz, H., Janke, G., Schober, M., Wagner, P., Warnack, D.: New developments and applications of skin-friction measuring techniques. Meas. Sci. Technol. **7**(10), 1396 (1996)
3. Hanratty, T., Campbell, J.: Measurement of wall shear stress. In: Goldstein, R. (ed.) Fluid Mechanics Measurements. Hemisphere Publishing Corporation, Washington/New York/London (1983)
4. Staats, M., Weiss, J.: Measuring wall shear stress in direction and magnitude. In: Aerospace Europe Conference 2020, Bordeaux, France (2020)
5. Eck, M., Rückert, R., Peitsch, D.: Advanced application of a sublayer fence probe in highly instationary turbomachinery flows: observations on prestall instabilities. Exp. Fluids **60**(3), 1–17 (2019). https://doi.org/10.1007/s00348-019-2686-5
6. Dambach, R., Hodson, H.P.: A new method of data reduction for single sensor pressure probes. ASME Paper No. 99-GT-304 (1999)
7. Park, S., Ko, H.C., Kang, M., Lee, Y.: An experimental study of the characteristics of a supersonic fluidic oscillator utilizing the design of experiment. In: Conference: AIAA Aviation 2019 Forum (2019)
8. Canny, J.: A computational approach to edge detection. IEEE Trans. Pattern Anal. Mach. Intell. **8**(6), 679–698 (1986)
9. Hirsch, D., Gharib, M.: Schlieren visualization and analysis of sweeping jet actuator dynamics. AIAA J. **58**(8), 2947–2960 (2018)
10. Settles, G.S.: Schlieren and Shadowgraph Techniques: Visualizing Phenomena in Transparent Media (2001). ISBN 978-3-54-066155-9

Pressure Waves and Flow Induced by a Train in a Tunnel

Klaus Ehrenfried(✉) and Daniela Heine

German Aerospace Center (DLR), Institute of Aerodynamics and Flow Technology, Bunsenstraße 10, 37073 Göttingen, Germany
Klaus.Ehrenfried@dlr.de

Abstract. The flow around a train in a tunnel was experimentally investigated at the Tunnel Simulation Facility of the DLR in Göttingen. A 1:25 scaled moving train model equipped with flow sensors passed through a 10 m long model tunnel. Parallel to the experiments simulations based on a quasi-1D formulation of the flow were performed. It could be shown that with appropriate adjustment of empirical parameters, the simulation can reproduce the measured pressure and velocity. All deviations between the measured and calculated data can be plausibly explained.

Keywords: pressure waves · train · tunnel

1 Introduction

The numerical simulation of pressure waves generated by a train passing through a tunnel has a long tradition. The results are used, for example, to estimate the loads on the infrastructure in the tunnel or on oncoming trains. To calculate the wave propagation in the entire tunnel – including the reflection at the exits and the complex superposition of waves running back and forth – usually quasi-1D formulations are used (see, e.g., [1,2]). In this context, the pressure, density and velocity in a cross section are represented by their mean values. Variations of these flow quantities within a cross section are not considered. Apart from the complex flow fields at the head and tail of the train, the pressure can be assumed to be constant in a cross section. However, the velocity field around the train with the boundary layers at the train surface and the tunnel wall is very inhomogeneous. So the question is: How well does a quasi-1D formulation represent the flow around the train? This is particularly interesting with regard to the estimation of the drag of a train in a tunnel on the basis of quasi-1D calculations, as investigated by Vardy and Reinke [3] (see also [4]).

Viscous effects are described in the quasi-1D formulations by means of empirical parameters. These include skin friction coefficients for the train surface and tunnel wall, as well as coefficients that determine the pressure loss due to 3D flow effects at the head and tail of the train. These parameters are typically determined in driving tests where the pressure in the tunnel is measured at a certain

A. Dillmann et al. (Eds.): STAB/DGLR Symposium 2022, NNFM 154, pp. 58–67, 2024.
https://doi.org/10.1007/978-3-031-40482-5_6

distance from the entrance. The parameters are then adjusted so that the calculated pressure curves agree with the measured curves. However, measurements of the flow velocity inside tunnels often show larger deviations from the calculation, see for example Vardy and Reinke [3]. One problem is the positioning of the probe. Under certain circumstances, it can be reached by the boundary layer of the train or of the tunnel wall. In addition, the velocity can be easily disturbed by meteorological effects and remaining flows from previous trains. Thus, the interpretation of the measured velocities is often difficult.

In order to determine how well quasi-1D computations can predict not only the pressure waves but also the flow velocity, model tests were carried out under laboratory conditions. First, it had to be shown that the quasi-1D formulation can be adapted to the model tests, where the train and tunnel length is relative short.

2 Quasi-1D Formulation

The calculation of the pressure waves and the flow around the train in the tunnel is based on a quasi-1D formulation. A stream tube is assumed whose cross-sectional area $A(x,t) = A_{\text{tu}}(x) - A_{\text{train}}(x,t)$ can vary spatially and temporally. The situation is illustrated in Fig. 1. In the area without train, the stream tube simply corresponds to the tunnel tube $A_{\text{tu}}(x)$. Along the train, the stream tube is defined by the free area between train surface and tunnel wall.

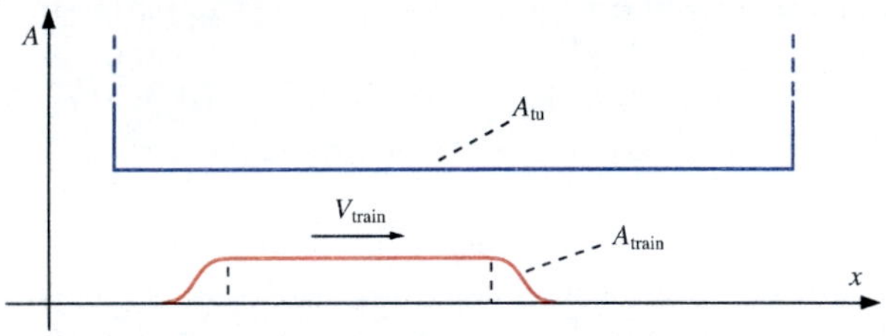

Fig. 1. Illustration of the cross-sectional areas of tunnel A_{tu} and train A_{train}.

When the train moves, the cross-sectional area $A(x,t)$ changes in the head and tail section of the train not only spatially but also temporally. For numerical reasons the area outside of the tunnel is realised by a finite cross section, which is huge compared to the actual tunnel cross section. All values, such as pressure p, density ρ or flow velocity u are regarded as constant over a cross section of the stream tube. Hence, in the quasi-1D formulation they depend spatially only on the x-coordinate. The conservation of mass, momentum and energy is supposed to apply for the stream tube. Accordingly, conservation equations for the three quantities are solved numerically using a finite volume method, where the conservation equations for a 1D volume $\Phi = [x_1, x_2]$ are indicated as follows:

$$\frac{\mathrm{d}}{\mathrm{d}t} \int_{\Phi} A\,\mathbf{q}\,\mathrm{d}x + \Big[A\,\mathbf{f} \Big]_{x_1}^{x_2} = \int_{\Phi} \mathbf{H}\,\mathrm{d}x. \tag{1}$$

$\mathbf{q}$ is the vector of conservative variables and $\mathbf{f}$ the vector of convective flux for mass, momentum and energy:

$$\mathbf{q} = \begin{pmatrix} \rho \\ \rho u \\ e_{\mathrm{tot}} \end{pmatrix} \quad \text{und} \quad \mathbf{f} = \begin{pmatrix} \rho u \\ \rho u^2 + p \\ (e_{\mathrm{tot}} + p)\,u \end{pmatrix}, \tag{2}$$

with $e_{\mathrm{tot}} = \rho e + \frac{1}{2}\rho u^2$ being the total specific energy per unit volume and e the internal energy per unit mass. A thermally and calorically perfect gas is assumed, where $p = \rho R T$ and $e = c_v T$. T is the temperature, R the specific gas constant and c_v the specific heat at constant volume. The source vector $\mathbf{H}$ on the right-hand side of (1) is given by

$$\mathbf{H} = \begin{pmatrix} 0 \\ p\,\frac{\partial A}{\partial x} - f_{\mathrm{tu}}\frac{1}{2}\rho u|u|\Psi_{\mathrm{tu}} - f_{\mathrm{train}}\frac{1}{2}\rho(u - V_{\mathrm{train}})|u - V_{\mathrm{train}}|\Psi_{\mathrm{train}} \\ -p\,\frac{\partial A}{\partial t} - f_{\mathrm{train}}\frac{1}{2}\rho(u - V_{\mathrm{train}})|u - V_{\mathrm{train}}|\Psi_{\mathrm{train}}\,V_{\mathrm{train}} + \dot{Q} \end{pmatrix} \tag{3}$$

It describes the transfer of momentum and energy from the surfaces of train and tunnel to the flow. $f_{\mathrm{tu}}(x)$ and $f_{\mathrm{train}}(x,t)$ are friction coefficients for the tunnel and the train, and $\Psi_{\mathrm{tu}}(x)$ and $\Psi_{\mathrm{train}}(x,t)$ represent the perimeter of tunnel and train respectively. f_{train} and Ψ_{train} are temporally constant in the train-fixed reference frame. Their temporal changes are only caused by the train movement. f_{tu} is the skin friction coefficient of the tunnel, which is constant inside the tunnel and set to zero outside the tunnel. Ψ_{tu} is also constant and corresponds to the circumference of the tunnel cross-section. The coefficient f_{train} is assumed to be piecewise constant. In the head and tail sections its value equals f_{head} and f_{tail} respectively. These coefficients describe the local frictional losses in the 3D flow around the train head and tail. In the middle section between head and tail f_{train} is equal to f_{middle}, which is the skin friction coefficient along the train. In the middle section, Ψ_{train} corresponds to the circumference of the train. In the head and tail section $\Psi_{\mathrm{train}} = |\partial A_{\mathrm{train}}/\partial x|$ is set.

The quantity $\dot{Q}$ in the third component of $\mathbf{H}$ describes the heat transfer between the flow and the surfaces of the train and tunnel. For reasons of simplicity, the train is considered adiabatic and only the heat transfer to the tunnel wall is taken into account using $\dot{Q} = \alpha\Psi_{\mathrm{tu}}(T - T_0)$. α is a heat transfer coefficient and T_0 is the temperature of the tunnel wall, which is set equal to the initial temperature. It should be noted, that by the chosen approach for $\dot{Q}$ linear damping of the waves is introduced in the computational model. All other friction terms in $\mathbf{H}$ result in nonlinear damping.

The parameters Ψ_{tu} and Ψ_{train} are given geometrically. The four friction coefficients f_{tu}, f_{head}, f_{tail} and f_{middle} are empirical parameters which are usually

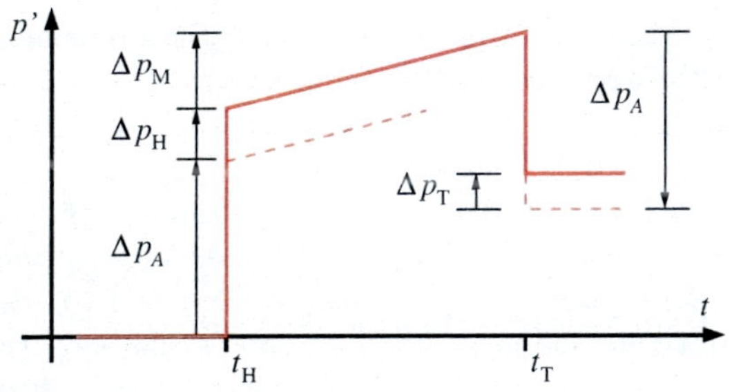

Fig. 2. Schematic representation of the pressure curve caused by train entry far inside a long tunnel.

determined on the basis of pressure measurements inside the tunnel. In Fig. 2 a corresponding pressure curve is shown schematically. The train generates a compression wave at the entrance, which reaches the observer at time t_H. The pressure increase in this so-called head wave consists of two parts: Δp_A and Δp_H. Δp_A results from the displacement of the air in front of the train. The value depends on the train speed and the blockage ratio A_{train}/A_{tu}. The flow losses, which occur in the three-dimensional flow around the head, cause the additional pressure increase Δp_H. In the quasi-1D formulation, these losses are realised by additional friction in the head section described by the coefficient f_{head}. Hence, the coefficient f_{head} is adjusted accordingly so that in the simulation the pressure increase Δp_H matches the measured value. At time t_T the tail wave – generated by the entry of the train tail into the tunnel – reaches the observer. In the meantime, the pressure has increased continuously by Δp_M due to surface friction along the train. The coefficient f_{middle} is chosen in such way, that Δp_M in the simulation corresponds to the observed increase. Without extra losses in the tail region, the tail wave would decrease the pressure by Δp_A. The pressure drop observed at t_T is reduced by the pressure losses in the flow around the tail Δp_T. By adjusting f_{tail}, the simulated pressure drop is adapted to the measurement at this point.

In principle, the skin friction at the tunnel wall contributes to Δp_M as well so that this pressure increase also depends on the coefficient f_{tu}. In most cases f_{tu} is simply set to a literature value suitable for the tunnel in question. Occasionally, the damping of the waves and consequently the coefficient f_{tu} is determined by measuring the decay of the head wave between two points in the tunnel.

3 Experiment

The model experiments were performed in the Tunnel Simulation Facility of the DLR in Göttingen. A 1:25 scaled model was accelerated by a hydraulic catapult and travelled on rails through a 10 m long model tunnel. Figure 3 shows a sketch of the train model. The model has a generic head and an end car with a pointed tail. Pressure sensors and a mobile data acquisition system are integrated in the middle part. One of the pressure sensors records the pressure difference Δp_{Pitot} between a Pitot tube and a reference point on the surface. The Pitot tube has a diameter of 2 mm and ends exactly 28 mm above the reference pressure tap. This distance from the surface is sufficient, to ensure that the tube is always outside the boundary layer. A sensor connected to a second pressure tap measures the static pressure at the model. The Pitot probe and the pressure taps all have the same distance (0.9 m) from the train nose. In addition to the pressure values, the acceleration of the model is recorded and an optical sensor at the underbody records position markings generated by LEDs in the model track. By integrating the acceleration value twice a mapping function between the train position x_{train} and the measurement time t can be obtained. A position marker at the end of the tunnel is used to verify this mapping. With the assignment $x_{\text{train}}(t)$ a conversion of the measuring time into probe position is possible, which is used for the presentation of the results further below.

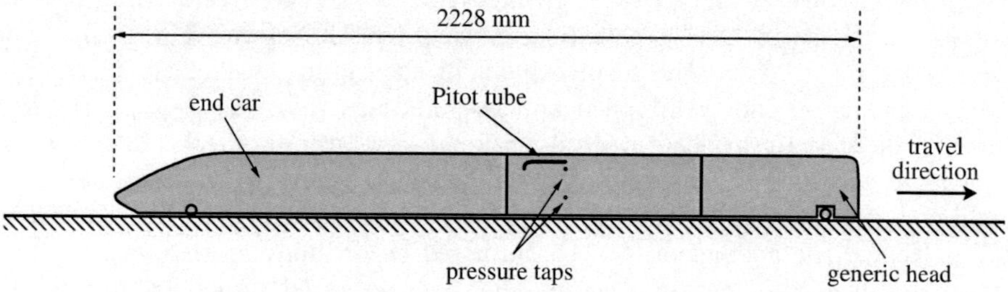

Fig. 3. Sketch of the train model with Pitot tube and pressure taps.

The model with the generic head was chosen, because it was particularly easy to integrate the instrumentation. All signals were sampled at 5 kHz and stored directly on a flash device. The recording starts when a predefined acceleration value is exceeded, whereby 0.2 s history are also stored ("pre-trigger"). After a measurement run, the values can be read out via Bluetooth connection. The model tunnel has a cross-sectional area of approximately 0.07 m^2, which corresponds to about 44 m^2 in full scale. The cross-sectional area of the train model is 0.02 m^2, which results in a blocking ratio of $A_{\text{train}}/A_{\text{Tu}} \approx 0.29$. Pressure sensors were installed at three positions in the model tunnel wall. Their signals were recorded using a stationary data acquisition system. The speed of the train model was measured using a light barrier located 0.65 m behind the tunnel entrance. The light barrier additionally serves to synchronise the stationary data acquisition with the mobile data aquisition, so that time data from the tunnel can also be assigned to the position of the train model. For details on the pressure measurements in the tunnel see Heine and Ehrenfried [5].

4 Results

4.1 Pressure at the Tunnel Wall

In the following, an example is described where the model train enters the tunnel at approximately 31.6 m/s. Figure 4 shows the pressure fluctuation, which is observed in the middle of the tunnel, 5 m away from the entrance. The red curve shows the measured signal and the blue curve the result of the simulation with adjusted empirical parameters. The time axis was chosen so that the nose of the train reaches the tunnel exactly at $t = 0$. Since the computational program is optimised for full-scale computations, all model dimensions were scaled up by a factor of 25 for the simulations resulting in a tunnel length of 250 m. The 1D volumes Φ had an extension of $\Delta x = 0.5$ m. The time step was $\Delta t = 1/4000$ s. The numerical results were then transferred back to the model scale for comparison. In this scale the numerical step sizes correspond to $\Delta x = 2$ cm and $\Delta t = 0.01$ ms.

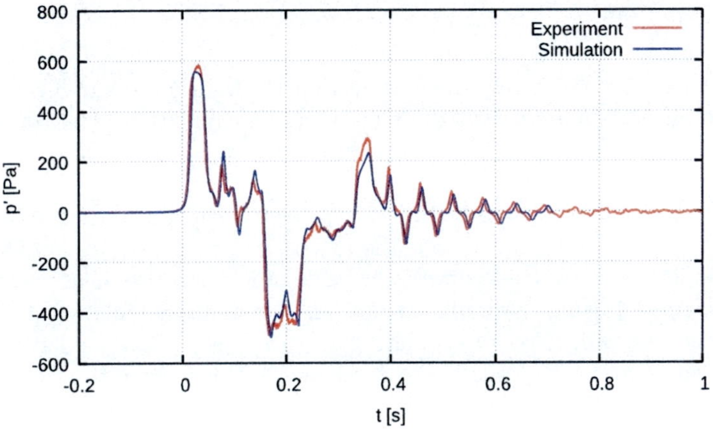

Fig. 4. Pressure history in the middle of the tunnel; red: experiment; blue: simulation.

In the short tunnel the measured pressure fluctuations deviate considerably from the ideal curve in long tunnels, see Fig. 2. Before the model has completely entered the tunnel, reflected waves have already passed through the entire tunnel. The gradual pressure increase Δp_{M} and the pressure drop in the tail wave are superimposed by reflected waves. Hence, the friction coefficients have to be adjusted in a slightly different way. At least the first pressure rise of nearly 600 Pa is not yet disturbed. Therefore, the parameter f_{head} could be adjusted as described above. Furthermore, the skin friction coefficient of the tunnel was set to $f_{\mathrm{tu}} = 0.006$, which is a typical literature value for smooth tunnels (see, e.g., [3]). The train model leaves the tunnel at $t \approx 0.3$ s and generates strong exit waves, which also run back and forth in the tunnel subsiding slowly. The linear damping was adjusted via the parameter α so that there is a good agreement of the post exit fluctuations between simulation and measurement. The two remaining parameters $f_{\rm tail}$ and $f_{\rm middle}$ were adjusted by trial and error so that overall an optimal match of the curves was achieved. It has to be noted that the agreement of the curves could certainly be improved by an iterative optimisation with respect to a mathematical norm. In order to keep the numerical effort low, this was not done here. Nevertheless, the comparison in Fig. 4 shows that by the quasi-1D formulation the pressure fluctuations at the tunnel wall can also be described quite well in a short model tunnel.

The decay of the pressure fluctuations after the train model has left the tunnel fits to a linear damping of the waves quite well. This is realised in the computational model using the term $\dot{Q}$. However, the question arises whether in the experiment actually the heat transfer to the tunnel wall is responsible for the observed damping. Compared to real tunnel structures, the model tunnel is relatively flexible and linear damping due to the elasticity of tunnel wall is also conceivable. It is likely that the observed damping is dominated by a physical process which is not included in the computational model.

4.2 Pressure and Flow Velocity at the Train

Using the example from the previous section the values measured on the train model are now compared with the results of the simulation. The top graph in Fig. 5 shows the inflow velocity at the Pitot tube $u_{\rm Pitot}$ and the bottom graph shows the fluctuations of the static pressure at the train, $p'_{\rm train}$. Note that the curves are not plotted over time but over measurement position. $x = 0\,{\rm m}$ corresponds to the tunnel entrance, and the exit is located at $x = 10\,{\rm m}$. The experimental inflow velocity (red curve) was calculated from the measured pressure difference $\Delta p_{\rm Pitot}$ using Bernoulli's equation. The simulated inflow velocity (blue curve) was determined by interpolating the calculated flow velocity u at the respective measurement position and then subtracting it from the train velocity: $u_{\rm pitot} = V_{\rm train} - u$. It should be noted that along the train inside the tunnel $u < 0$ holds. The air flows against the travel direction. Thus, $u_{\rm Pitot}$ is greater than the train speed. It must also be pointed out that in the experiment the train model slows down slightly due to aerodynamic and mechanical friction. In contrast, the train speed is constant in the simulation. It is identical to the experimental train speed at the beginning of the tunnel, where the light barrier is located.

The two experimental curves depicted in Fig. 5 indicate noisier behaviour compared to the calculated ones. This can be explained by the pressure fluctuations in the turbulent boundary layer at the train. In front of the tunnel, the measured pressure $p'_{\rm train}$ is slightly lower than the calculated pressure. For the velocity $u_{\rm Pitot}$, the opposite is true. Although the train model is a slender body, it still displaces the air to some extend, also in free space. This leads to an increase of the relative flow velocity next to the train and a corresponding decrease in pressure. In the computational model these effects of three-dimensional flow around the train in free space are not included. The flow velocity outside the tunnel is practically zero in the simulation. This explains the discrepancies between the measurement and simulation in front of the tunnel.

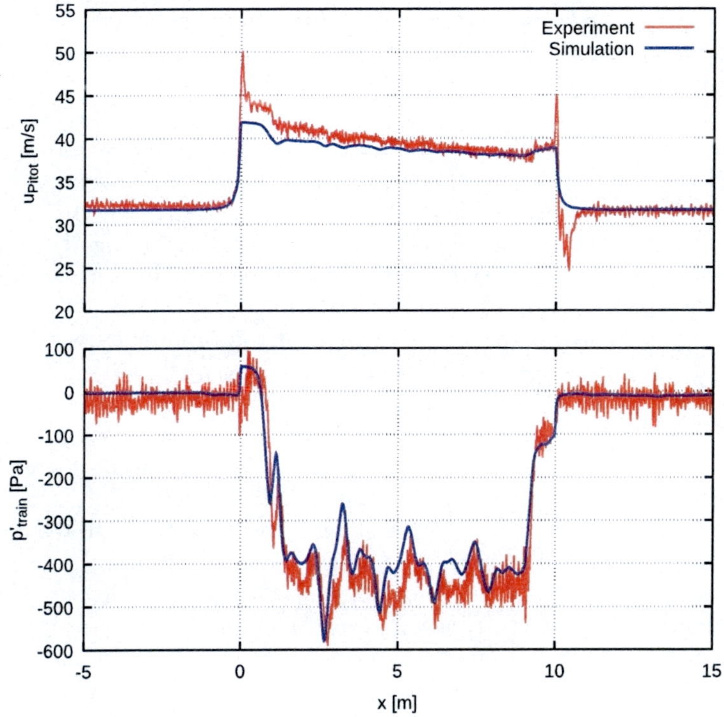

Fig. 5. Inflow velocity at the Pitot tube (top) and pressure fluctuation at the train (bottom) as a function of the measurement position; Red: Experiment; Blue: Simulation.

In the tunnel, as expected, the flow around the train results in a significant increase of u_{Pitot} and the pressure p'_{train} becomes negative. u_{Pitot} decreases continuously inside the tunnel both in the experiment and in the simulation. The back and forth travelling pressure waves gradually generate a flow in driving direction which reduces u_{Pitot}. Since the velocity in the simulation represents the average value over the cross section, and the Pitot probe in the experiment measures the velocity in the core flow outside the boundary layers, a higher u_{Pitot} value is expected in the experiment. Indeed, such a deviation is observed in the first section of the tunnel. In the further course the u_{Pitot} curves approach each other more and more. This can be explained by the deceleration of the train model due to friction in the experiment.

The deviations of the u_{Pitot} curves in the area of the tunnel ends are striking. At $x = 0\,\mathrm{m}$ and $x = 10\,\mathrm{m}$ positive peaks can be observed in the measured curve which are not present in the simulation. The calculation of u_{Pitot} from $\Delta p_{\mathrm{Pitot}}$ assumes that the static pressure at the reference point on the surface is more or less equal to that at the Pitot tube. But the flow field at the entrance and exit of the tunnel is certainly more complex than the flow around the train in the tunnel. Here, a 3D flow effect can easily cause a pressure gradient normal to the train surface and spoil the velocity calculation. Thus, in these regions the measured u_{Pitot} values have to be interpreted with caution. Behind the tunnel, the measured curve shows a drop of u_{Pitot} below the train speed. The air is accelerated in the tunnel by the pressure waves and flows out of the tunnel. The flow separates at the tunnel exit and forms a jet like flow structure behind the tunnel. When the train leaves the tunnel it is exposed to a lower inflow velocity within this flow structure. This explains the drop of u_{Pitot} over a distance of more than half a meter behind the tunnel. Because the quasi-1D formulation does not include the separation of the flow at the tunnel exit the simulation does not capture this drop of u_{Pitot}.

Both p'_{train} curves in the bottom graph of Fig. 5 show fluctuations in the tunnel caused by pressure waves travelling back and forth. Certainly the height of the fluctuations and their timing does not match exactly, but all major details of the measured fluctuations are described by the simulation as well. The timing deviation might be caused by a slight difference in the speed of sound, which could be related to elasticity of the tunnel walls. Furthermore, in the tunnel the measured pressure seems to be systematically lower than the value in the simulation. This fits the fact that the flow velocity u in the simulation is lower than the velocity measured in the core flow. Since the flow velocity in the core determines the pressure distribution along the train – not the velocity value averaged over the cross section, the observed difference is reasonable. The higher core velocity results in a lower value in the experimental pressure curve.

5 Conclusion

The investigations have shown that a quasi-1D formulation of the tunnel flow can describe the model experiments despite the relatively short model tunnel. The necessary empirical parameters can be determined based on pressure fluctuation at the tunnel wall analog to full-scale tests so that the calculated wall pressure agrees well with the measured data. The quasi-1D calculation also reflects the pressure and flow velocity measured at the train model quite well. Deviations of the flow velocity occur only upon entry and exit of the train and can be explained by 3D flow effects.

In order to describe the decay of the waves in the tunnel after the train model has left the tunnel, a linear damping of the waves is required in the computational model. In the present approach this has been formally realised by the heat transfer between the flow and tunnel wall. It is, however, questionable whether it is actually this thermal or another physical effect, which causes the observed damping of the waves. In terms of fitting the simulation to the measured results, it does not matter how the linear damping is physically motivated. It is only important that a linear damping is present in the formulation. However, for a better understanding it should be clarified by further investigations, which physical process is responsible for the observed damping.

Future work will concentrate on using the quasi-1d calculation to estimate the drag of a train in a tunnel following the idea of Vardy [4]. The aerodynamic calculation can be coupled with a driving dynamics simulation to determine the energy consumption for a tunnel passage. This option is already implemented in the code and can be used for an energetic optimisation of the tunnel infrastructure or operational parameters. One challenge will be the validation of the drag calculation using model experiments or full-scale testing. The present study delivers the promising result that the flow around the train is reasonably described by the quasi-1d model. Hence there is chance that the drag can also be predicted with sufficient accuracy. However, this must be proven in further experiments.

References

1. Woods, W.A., Pope, C.W.: A generalised flow prediction method for the unsteady flow generated by a train in a single track tunnel. In: Proceedings of International Symposium on Aerodynamics of Transportation, Niagara, USA, pp. 137–151. American Society of Mechanical Engineers, New York (1979)
2. Schultz, M., Sockel, H.: Pressure transients in short tunnels. In: Haerter, A. (ed.) Proceedings of 7th International Symposium on Aerodynamics and Ventilation of Vehicle Tunnels, Brighton, 27–29 November 1991, pp. 221–237. BHR Group, Cranfield (1991)
3. Vardy, A.E., Reinke, P.: Estimation of train resistance coefficients in tunnels from measurements during routine operation. Proc. Inst. Mech. Eng. Part F: J. Rail Rapid Transit **213**(2), 71–87 (1999)
4. Vardy, A.E.: Aerodynamic drag on trains in tunnels. Proc. Inst. Mech. Eng. Part F: J. Rail Rapid Transit **210**(FI), 29–49 (1996). (2 papers)
5. Heine, D., Ehrenfried, K.: Experimental study of the entry of a high-speed train into a railway tunnel. Int. J. Railway Technol. **1**, 1–18 (2012). https://doi.org/10.4203/ijrt.1.4.1

Construction of a Flow Test Rig with a Predefined Shape of the Boundary Layer to Investigate Sand Absorption by Side Flaps on a High-Speed Train

Uwe Fey[1(✉)], Thomas Kowalski[1], Torsten Gries[1], and Manuel Gageik[2]

[1] German Aerospace Center DLR, 37073 Göttingen, Germany
Uwe.Fey@dlr.de
[2] Siemens Mobility GmbH, 47829 Krefeld, Germany
http://www.dlr.de/as,
https://www.mobility.siemens.com/global/en/portfolio/rail.html

Abstract. For a high-speed train operating in the desert, a certain proportion of sand may be lifted up from the ground in the vicinity of the underfloor caused by the turbulent slipstream. The train under investigation here has a number of lattice doors (side flaps) near the ground where air is sucked in with a certain amount of volume flow. To experimentally address the situation in the desert, a test rig was designed and constructed which establishes a flow on the surface of the side flap in full scale, whereas in addition the flow can be enriched with a constant, predetermined mass fraction of sand. The setup finally allows for comparison of different kind of filters and sealing attached to the side flap. Here, the design and construction of the experimental rig is presented.

Keywords: flow control · baffles · trip wires · train operation in the desert

1 Definition of the Problem

When the high-speed train under investigation runs along the track, air is sucked in through lattice doors (called "side flaps" in the following) used for conditioning and cooling of internal devices (see Fig. 1). If the train operates in the desert, it is of interest if a proportion of sand (which may be lifted up caused by the slipstream) could also be sucked in during operation. To address this question, a custom-built rig should be designed and built up for experimental determination of the possible sand flow through the side flap, when a certain suction rate is applied. Since enriching the flow of a conventional wind tunnel with sand may finally damage it, the test rig was based on an existing air blower which enables about $4\,\mathrm{m}^3/\mathrm{s}$ of volume flow, as the drive. If this blower were destroyed on contact with sand, the damage would be minor.

The flow through the exit opening of that air blower will be guided by a custom-made nozzle such that the side flap surface is completely blown over

A. Dillmann et al. (Eds.): STAB/DGLR Symposium 2022, NNFM 154, pp. 68–77, 2024.
https://doi.org/10.1007/978-3-031-40482-5_7

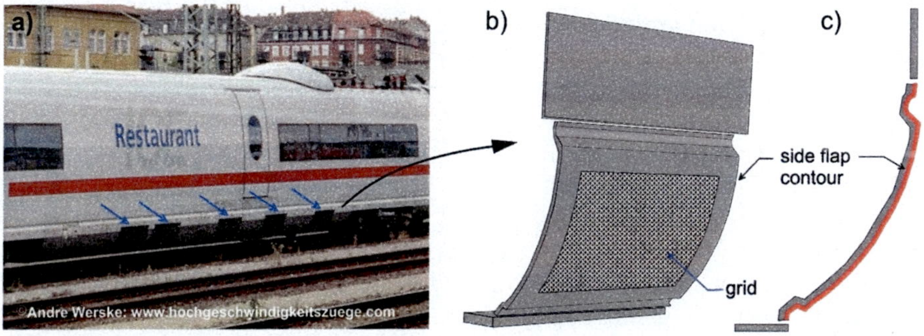

Fig. 1. The part to be examined: side flap (lattice door) of the high speed train. a) as seen on the real train, b) in perspective view and c) as seen in cross-section view with the given contour highlighted by a red line

with air. The problem here is, that the surface of the side flap is neither flat nor continuously curved, but convexly curved with indented channels. A second (suction) air blower is used to simulate the intake of air through the side flap.

1.1 Requirements and Objectives

Clear requirements were defined in order to set up a test rig that can be used for the planned experiments.

Support Box for the Side Flap. The side flap must be fitted to a suitable box in such a way that it connects seamlessly to the outer wall of this box, which reflects the shape of the real train. The support box should be of a closed volume so that a vent connection can be made to apply the required suction. There had to be a gap (some mm in width) to the surrounding frame around the side flap, as given for the real train because of technical reasons.

Shear Flow Alongside the Side Flap. Air, which has to be enriched with a certain amount of sand, should flow parallel to the side flap. The boundary-layer velocity profile near the surface of the flap should somehow resemble the boundary-layer profile built up on the real train face. The mentioned boundary-layer flow is to be established for a train speed of 230 km/h and three different positions in the train set: near the head at 10 m, in the middle of the train set at about 100 m and near the rear end at about 190 m. In addition, three different train speeds are to be simulated for the 100 m position: 80, 160 and 230 km/h.

Homogeneous Insertion of Sand. The shear flow along the side flap surface is said to be enriched with sand with a certain concentration C_{sand} (in g/m^3). The sand is to be distributed homogeneously in the flow so that the entire area of the flap and the surrounding gaps are exposed to the same concentration of sand during testing. Particular care should be taken to fill the lower and upper indented channel with sand.

Design of a Customized Nozzle. A nozzle should be designed which is mounted on the existing air blower and generates the required boundary layer flow on the side flap. Based on the available volume flow of the air blower of approx. $4\,\mathrm{m}^3/\mathrm{s}$, the contraction and shape of the nozzle outlet must be designed in such a way that the required boundary layer shapes for the different cases can be achieved.

Suction Through Side Flaps. The box holding the side flaps must be fitted with a suction port to allow controlled suction through the side flap and surrounding gaps. A standard volume flow of the suction Q_{st} (in m^3/s) is to be built up by a second blower behind the box (adjustable to lower and higher suction rates for comparison tests). The volume flow must be kept as constant as possible.

Examination of Absorbed Sand. The sand that is absorbed by the grid of the side flap and the surrounding gaps should be collected in full. Its mass fraction based on the total sand used in the test runs is to be measured and its consistency in terms of grain size distribution is to be determined.

Collection of Sand Which is Blown by. Since only a small portion of the sand (out of the total sand used for a test run) is expected to be absorbed by the side flap, most of the injected sand flies past the flap and is to be collected in an appropriate manner.

2 Realization of the Requirements

The finally created setup based on the requirements is shown in Fig. 2. A support box was built with the described, discontinuously-curved shape at the front

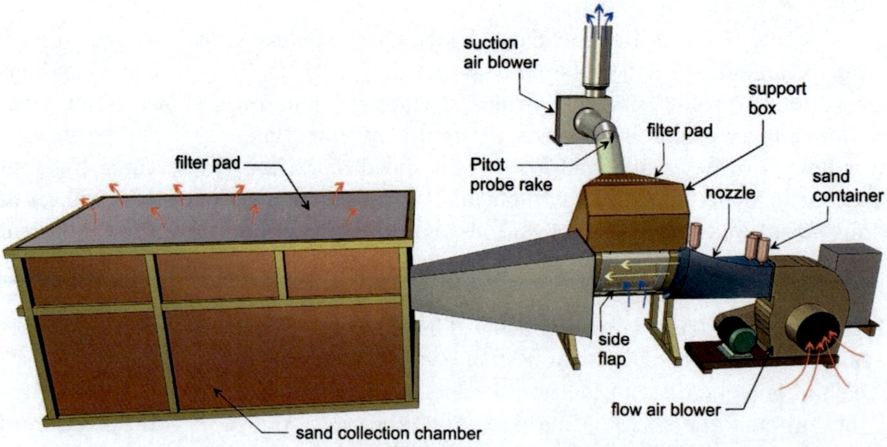

Fig. 2. Final design of the complete test rig as built up in the test hall

holding the side flap. Ambient air is sucked from the test hall (red arrows, right-hand side) and fed to the nozzle. Inside the nozzle, the flow is enriched with sand by injection tubes connected to four bottles serving as sand containers.

The air-sand mixture with high flow speed passes the side flap (yellow arrows) where the suction flow is superimposed (indicated by the blue arrows). A large funnel collects the by-blown sand and feds it to the sand collection chamber. Cleaned air passes through the large filter pad at the ceiling of this chamber (small, red arrows) and the sand is held back by the filter pad (smaller particles) or collects at the bottom of the chamber (larger particles). The amount of sand sucked through the side flap's grid falls down inside the support box. The fine-dust content here is caught by a filter pad mounted inside the box and clean suction air is blown out to the test hall (blue arrows on top). A Pitot probe rake at the end of the linear part of the tube connection on the support box allows for measuring the instantaneous velocity profile of the pipe flow and thus, the instantaneous volume flow of the suction. Both flow velocities (side-flap flow and suction flow) can each be set precisely with their own frequency converter.

It is of importance to built up the test rig in a way that the side flap is oriented in its natural position, i.e. about 45° inclined with respect to the bottom. Reliable measurements of the amount of sand which is sucked in through the side flap's grid is only possible, when the gravitational force acts as it does on the real train, i.e. downwards.

2.1 Construction of the Nozzle

The cross-section of the nozzle outlet is restricted with a specified, given maximum volume flow of the air blower of $4\,\mathrm{m}^3/\mathrm{s}$ and a maximum speed to be simulated atop the side flap of 230 km/h. Since the height results from the size of the side flap plus the surrounding channels, the associated width of the nozzle outlet was calculated accordingly. This resulted in a nozzle-exit width of 9 cm which led to the final shape of the nozzle outlet as shown in Fig. 3a. The adaptation of the nozzle from a rectangular cross section to this contour was designed based on the contraction shape of efficient wind tunnel nozzles (see Fig. 3b). Particular efforts were caused as the upper and lower channels had to be integrated appropriately (Fig. 3c).

Based on the designed geometric shape, a positive mold was built, consisting of 50 mm styrofoam sheets made on the milling machine. The actual nozzle was then created by laminating the positive mold with carbon fiber fabric. The finished setup showed that the transition from the nozzle outlet to the support box with the side flap fitted seamlessly.

2.2 Establishment of Boundary Layer Flow

The requirements for the tests to be conducted demand for specific velocity profiles normal to the side flap surface. They roughly should resemble the boundary layer (BL) as given on the real train for the different cases. Since the real boundary layer velocity distribution on the train side face is not known, it was decided (also based on numerical results) to use an empirical formula from literature.

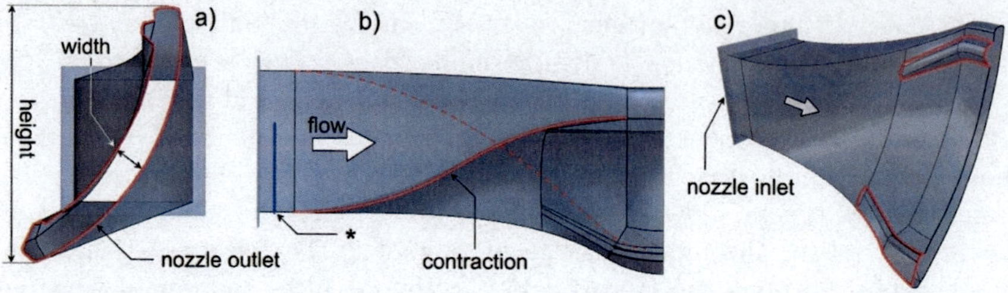

Fig. 3. Custom-designed nozzle for the test rig. a) front view with the contour of the nozzle exit highlighted, b) top view with the flow contraction highlighted and c) perspective view with the extensions highlighted which were needed to address the upper and lower channels (the meaning of the * indication will be explained later)

One common, empirical approximation for the time-averaged velocity profile of a turbulent flat-plate boundary layer is the one-seventh-power law (see, for example [3], p. 529):

$$\frac{u}{U} = \left(\frac{y}{\delta}\right)^{1/7} \quad \text{for} \quad y \leq \delta, \tag{1}$$

where U represents the flow velocity (here: the train speed) and u denotes the velocity component in x-direction in the boundary layer as a function of wall-normal distance y. The boundary layer thickness δ is *not* the 99 % thickness as given for laminar flow but rather the actual edge of the boundary layer. It can be calculated based on empirical formulas, for example [3], p. 530:

$$\frac{\delta}{x} = \frac{0.38}{(\mathrm{Re}_x)^{1/5}}, \tag{2}$$

with x being the distance to the flat plate's upstream edge (here: the position in the train set to be simulated) and Re_x being the corresponding Reynolds number: $\mathrm{Re}_x = Ux/\nu$, with ν representing the kinematic viscosity of air.

In the following Fig. 4, the calculated boundary-layer velocity profiles $u(y)$ for the five simulation cases (shown in the left-hand table of the figure) are presented. The boundary layer thickness δ shown in the table was calculated based on Eq. 2. Except for the first case ($U = 230$ km/h, $x = 10$ m), the boundary layer thickness is of the order of 1 m and can not completely be experimentally created using a nozzle of 9 cm in width (but that is not absolutely necessary).

To create the requested BL velocity profiles by some sort of flow manipulation, it is necessary to measure the velocity profiles $u(y)$ at different positions, with respect to streamwise direction x and height (see Fig. 3a). Here, only the profiles measured in the middle of the side flap, with respect to height, will be shown. Furthermore, in the following only results for the measured BL shapes for fixed train speed $U = 230$ km/h are presented (the established boundary layers for other train speeds showed similar results). All boundary-layer velocity profiles were obtained by means of a Pitot probe rake and without sand being

U [km/h]	Position x [m]	U [m/s]	Re_x	δ [m]
230	10	63.89	4.17E+07	0.11
230	100	63.89	4.17E+08	0.72
230	190	63.89	7.92E+08	1.20
80	100	22.22	1.45E+08	0.89
160	100	44.44	2.90E+08	0.77

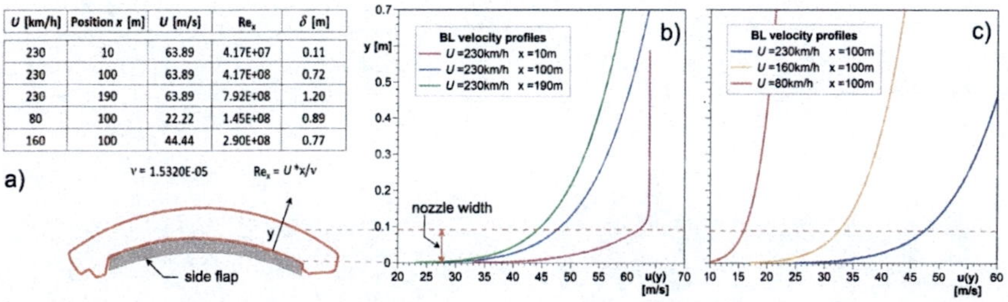

Fig. 4. Turbulent boundary layer velocity profiles calculated by the one-seventh-power law. a) used values (seen in the table), b) calculated velocity profiles for fixed train speed of $U = 230\,\text{km/h}$ and c) velocity profiles for fixed position $x = 100\,\text{m}$

injected (Fig. 5a). The *undisturbed* velocity profiles near the nozzle outlet are presented in Fig. 5b (clean) in comparison to the requested BL shapes (requ).

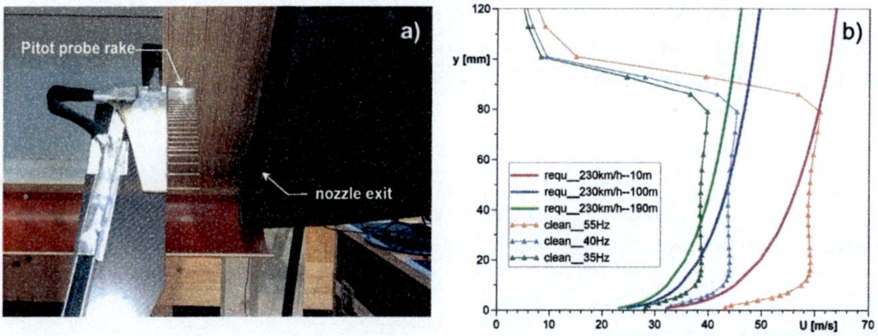

Fig. 5. Results of velocity profile measurements 100 mm downstream of the nozzle exit. a) used setup and b) measured profiles. "clean" denotes the undisturbed flow, "requ" denotes the requested (i.e. calculated) shape. The value in "Hz" is a measure for the flow speed of the air blower (in the left-hand image, the side flap is not yet mounted)

The velocity profiles shown in the following where gathered above the center of the side flap, after it has been mounted to the support box. At first, it was planned to find the desired BL shape of the flow by the use of baffle plates at the nozzle inlet only. Therefore, baffle plates of different shapes and sizes were applied at a location as indicated by the asterisk in Fig. 3b. Finally, a spire-like shape [1] was used as seen in Fig. 6a, leading to BL shapes as shown in Fig. 6b. None of the baffles (triangles or rectangles of different shapes and sizes) nor a simple plate (blocking about 1/3rd of the inlet area completely) did change the outcome in a way, that the bump in the range $0 \leq y \leq 30\,\text{mm}$ becomes closer to the requested shape.

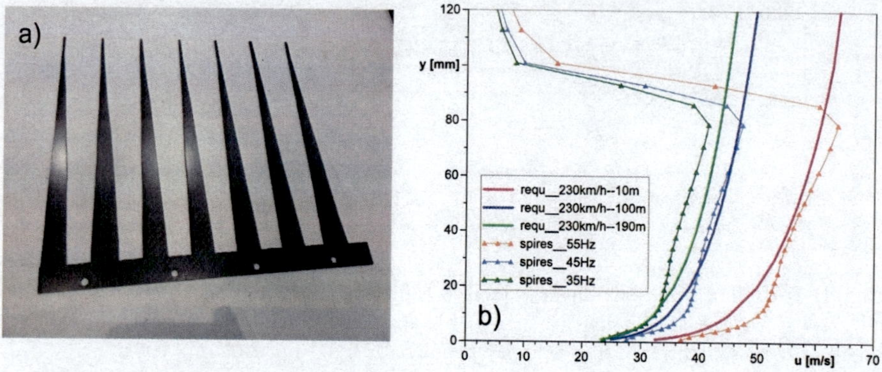

Fig. 6. a) baffle-plate based on spires and b) resulting velocity profiles when using the spires. The backcurvature of the curves for $y > 80\,\mathrm{mm}$ is caused by the shear layer developing at the outer edge of the nozzle where the flow mixes with the ambient air

The final solution was found when adding trip wires (see, for example, [2]) of different diameter to the inner surface near the nozzle exit, in addition to the baffle plate near the inlet. Extensive pre-tests were performed to find the suited number of wires, their diameters and positions. Finally, two wires with $d = 5$ and $8\,\mathrm{mm}$, located 200 and 400 mm (respectively) upstream of the nozzle exit worked very well, as can be seen in Fig. 7. Concluding, the approximation to the requested BL shape (atop the center of the side flap, up to $y = 80\,\mathrm{mm}$ wall distance) was deemed good enough to be used for the planned test runs.

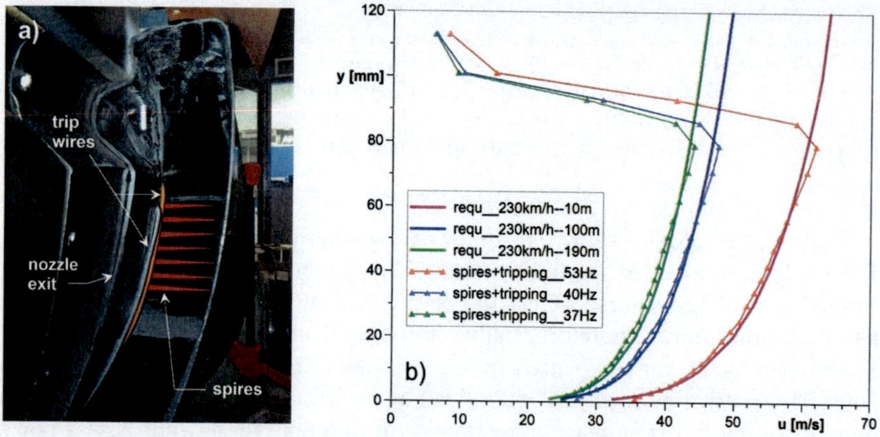

Fig. 7. a) trip wires located near the nozzle exit (the spires mounted near the nozzle inlet in the background are highlighted in red) and b) resulting velocity profiles.

2.3 Implementation of Sand Injection

As a sand which represents a grain-size distribution similar to the desert sand, a so called "standard sand" was used, provided by KSL-Staubtechnik. In order to enrich the flow along the side flap with sand, injection pipes were let into the nozzle at suitable positions, which are connected to bottles at the upper end, which serve as sand containers (see Fig. 2). The connection is made via silicone hoses in such a way, that the sand inflow can be started or stopped at any time. Positions near the nozzle inlet were chosen as the preferred locations for the sand injections, a few cm downstream from the baffle plate, so that the inserted sand is distributed as evenly as possible by the turbulent flow taking place here.

The problem is, that there is overpressure at this position, which ensures that the sand does not get into the nozzle by itself, but is pushed up and out of the bottles. This was remedied by sealing the sand containers airtight at the top so that compressed air could be applied to the sand surface via a hose connection. The pre-pressure can be controlled individually for each bottle in order to adjust the sand flow within a certain range. An aperture on the bottle neck with a previously experimentally determined diameter defines the interval in which regulation can take place.

The positions for the three injection pipes were optimized in extensive preliminary tests with the aim of enriching the flow over the side flap with sand as homogeneously as possible. The important thing here was precise positioning in the mm-range and the final positions could be found just by "trial-and-error". Nevertheless, it was not possible to enrich the upper quarter of the flow volume and in particular the upper channel with sand in this way, since of course the sand falls down on its way through the nozzle caused by gravity.

Therefore, an additional injection tube was placed in the upper part of the nozzle outlet, directly in the bulge that opens into the upper channel (#1 seen in Fig. 8 a). Because of the accelerated flow at the nozzle outlet there is negative pressure (compared to the ambient pressure) and it was not necessary to apply pre-pressure to the bottle at this position. The front sand container is, so to speak, sucked empty over the course of the test run and the sand flow rate could be adjusted by choosing a suitable diameter for the aperture on the bottle neck.

The amount of sand in the front bottle was then chosen to harmonize with the emptying of the three (fully filled) rear bottles. Uneven emptying was immediately counteracted by readjusting the respective pre-pressures. In this way, the given volumes of sand can be emptied in approximately the same time each (302 ± 12 s) for all test runs. The test runs then resulted in a constant sand mass flow $\dot{M}$ (in g/s), with the sand being homogeneously distributed in the flow volume. The areal distribution of sand was assessed on the basis of video recordings of the sand grains flying by in front of the side flap, using an image enhancement method (Fig. 8b). The homogeneity of the wall-normal sand distribution was tested by installing 6 paint-coated cylinders normal to the side flap surface at three different X-positions (in front-, in the middle- and at the rear of the side flap) and evaluating the distribution of paint abrasion over the length of the cylinders, after approximately 30 s of operation of the test stand (Fig. 9).

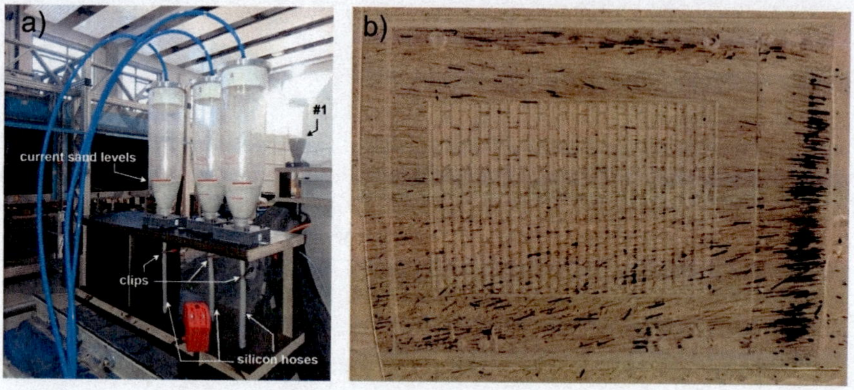

Fig. 8. a) pressurized bottles used as the sand containers and b) video recording of sand grains in fly-by afront the side flap (all four bottles are operated)

The sand concentration C_{sand} can be determined by knowing the flow volume's cross-section area and the effective mean velocity. For this purpose, the flow cross-section on the middle of the side flap was used, which results from the shape of the nozzle-exit cross-section plus the adjacent shear layer. Velocity profiles $u(y)$ were measured at five different height positions, and the mean velocity for the entire cross-section was thus determined. This results in the volume flow Q (in m^3/s) for the given flow speed and position. With the known mass flow of sand $\dot{M}$, the sand concentration then calculates to $C_{\mathsf{sand}} = \dot{M}/Q$.

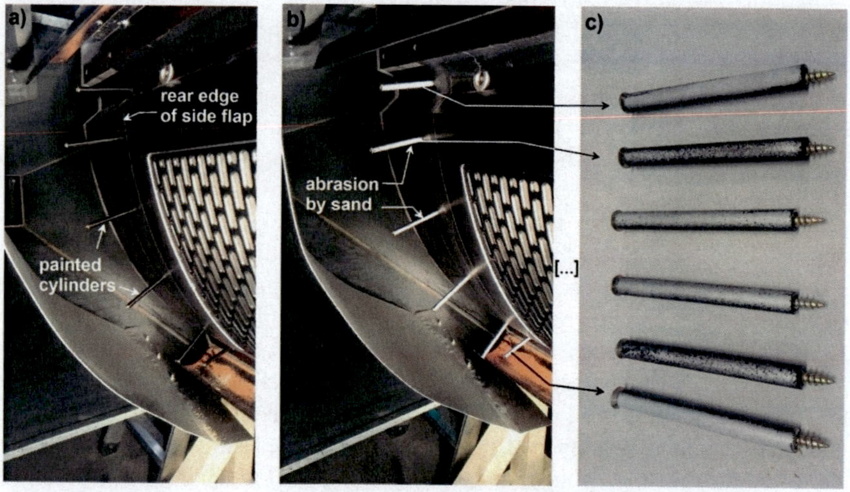

Fig. 9. a) attachment of painted cylinders (in this case at the rear of the side flap), b) cylinders after the test run and c) abrasion of the paint caused by the sand flow

2.4 Adjustment of the Flow Rate of the Suction

The exact setting of the volume flow of the suction air blower was carried out by instantaneous measurement of the velocity profile at a suitable point in a tube in front of the air blower with the help of a second Pitot probe rake (its position is indicated in Fig. 2). Assuming rotational symmetry, the current flow rate can be determined by integrating the measured cross-sectional profile over the 360° full circle. This is far more accurate than using the fan *rpm*, for example (after a previous calibration), especially since the required fan speed changes depending on the pressure drop. Different pressure drops apply to different variants of the side flap. Furthermore, the pressure drop in the suction part changes during the test runs due to progressive clogging of the filter pad in the support box caused by accumulated fine dust. With the selected procedure, the volume flow of the suction could be kept constant within 1 % deviation from the specified value.

2.5 Collection of Absorbed Sand

In order to increase the amount of sand sucked up behind the side flap, three consecutive test runs of about 300 s each were performed, using completely filled bottles at the nozzle inlet and the accordingly adjusted amount of sand in the bottle at nozzle exit. After the three test runs the side flap was opened and all the sand that had gotten caught in its structure was picked up using an industrial vacuum cleaner with an upstream cyclone separator. The sand that was found on the bottom of the support box was added, and finally the filter pad inside the box was dismounted and vacuumed. The sand obtained in this way (which all corresponds to the portion sucked in through the side flap) could then be weighed with a laboratory balance and analyzed with laboratory sieves of different mesh sizes with regard to its grain size distribution.

3 Conclusions

The test setup described here was well suited to compare different variants of side flaps in terms of their effectiveness in sand separation. Thus, a special type of side flap and sealing could be determined to give best protection against sand absorption when the train operates in the desert.

References

1. Irvin, H.P.A.H.: The design of spires for wind simulation (short communication). J. Wind Eng. Ind. Aerodyn. **7**, 361–366 (1981)
2. Gibbings, J.C.: On boundary layer transition wires. Aeronautical Research Committee Technical Report, C.P. No. 462 (1959)
3. Cengel, Y., Cimbala, J.M.: Fluid Mechanics: Fundamentals and Applications, 1st edn. McGraw-Hill, New York (2006)

Analysis of the Unsteady Loads on Train Models in Wind Tunnels

Daniela Heine(✉), Klaus Ehrenfried, Torsten Gries, Gerrit Heine, and Max Müller

German Aerospace Center (DLR), Institute of Aerodynamics and Flow Technology, Bunsenstraße 10, 37073 Göttingen, Germany
Daniela.Heine@dlr.de

Abstract. Wind tunnel tests were performed to measure the loads on a train model under disturbed incoming flow conditions. For this purpose, a wing system with movable flaps was used to generate harmonic fluctuations in the flow upstream of the train model. The model was mounted above a moving belt to generate a more realistic underbody flow and the loads were measured using an internal balance. The tests show that the mean drag and the lateral force response can be analyzed although vibrations in the setup influence the measured force signals. Evidence was found that the drag coefficient increases when incoming flow disturbances are present. Phase-averaging was used to filter out the part of the lateral force fluctuation related to the incoming flow disturbance. The resulting lateral force response depends strongly on the frequency of the incoming flow disturbances.

Keywords: unsteady loads · wind tunnel · train model · crosswind force · drag

1 Introduction

Knowing the forces acting on trains is crucial for their safe and efficient operation. Recent studies show that the wind speed has increased over the last decades due to climate change. This does not only hold true for the mean wind speed but also for the speed of gusts, as for example Zeng et al. showed in [1]. Furthermore, new train concepts are more vulnerable to crosswind forces as the trains are often built in lightweight design and as double-deck trains with a larger surface area being exposed to the wind.

In terms of cars is well known that the loads under real-life conditions are higher than in the ideal undisturbed case (see e.g. [2,3] or [4]). However, for trains, most studies of the loads are performed under steady wind conditions (see [5]). New trains have to meet several requirements concerning aerodynamic forces as specified in the EN 14067 norms [6]. These requirements comprise an approach to analyse the unsteady forces. However, this approach is a simplification of the real-life conditions. As the crosswind forces are crucial for safe train operation,

A. Dillmann et al. (Eds.): STAB/DGLR Symposium 2022, NNFM 154, pp. 78–87, 2024.
https://doi.org/10.1007/978-3-031-40482-5_8

they should be further studied – also for unsteady wind conditions. Another important point is the influence of the disturbed incoming flow on the drag of a train and its energy demand. An attempt to study the loads on a train under real-life conditions was made by Baker et al. [7], who compared wind tunnel tests to tests in full-scale.

The purpose of the present study is to show the feasibility to measure the unsteady loads on a train model in a wind tunnel under disturbed incoming flow conditions. In Sect. 2 the experimental setup is described. In Sect. 3.1 first results are presented showing the influence of the disturbed incoming flow on the mean drag. Then, in Sect. 3.2 lateral forces are considered and it is demonstrated that by phase-averaging the lateral force response can be analysed.

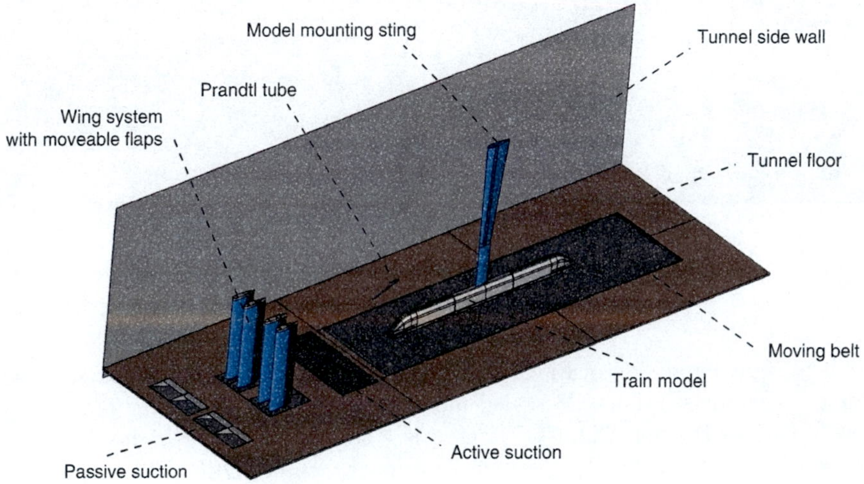

Fig. 1. Illustration of the setup in the test section of the wind tunnel.

2 Experimental Setup

The tests were performed at the Crosswind Test Facility Göttingen (SWG). This facility is a closed-circuit low-speed wind tunnel with a maximum velocity of 65 m/s. It has a closed test section of 1.6 m in height, 2.4 m in width and 9 m in length. The setup for the present study is shown in Fig. 1. The model is mounted from above over a moving belt with a clearance of only a few millimeters. The belt has a width of 1 m and a length of 4 m. Its maximum speed is 30 m/s. To remove the boundary layer from the wind tunnel floor, a passive and active suction in front of the moving belt is used. Upstream of the belt – between the area of passive and active suction – a wing system with movable flaps is installed. The system consists of four wings with a NACA0018 profile and a cord length of

30 cm. The wings have a height of 0.8 m, which corresponds to half the height of the test section. The spacing of the inner wings is 34 cm and of the outer wings 18 cm. The left-hand side of Fig. 2 shows a photo of the wing system. The flaps extend over 30% of the cord. They are directly driven by servo motors and can swing up to ±10°. The flaps can move independently or synchronously. Both an irregular or a harmonic motion are possible. The maximum oscillation frequency is 50 Hz. The idea is that the flap motion generates flow disturbances with a lateral wind component. Further information about the facility and the wing system can be found in [8] and [9].

Fig. 2. Photos of the experimental setup; left: wing system with moveable flaps; right: train model above moving belt.

The right-hand side of Fig. 2 shows the train model above the moving belt. For the present study a 1:25 scaled model of the Next Generation Train Cargo (NGT Cargo) is used. The NGT Cargo is a train concept for high-speed cargo transport developed by the German Aerospace Center (DLR). The model consists of two end cars and a middle car. It has an overall length of 2.58 m, and is aligned in mean flow direction. The distance between the wing system and the model is 1.64 m. When the flaps stand still in neutral position the model experiences a straight incoming flow and the wakes of the wings pass the model on both sides. An internal balance is used to measure the total force on the model. In Fig. 3 the inner structure of the model can be seen. The internal balance is based on four piezoelectric force transducers manufactured by Kistler (type 9317B). They are arranged between two steel plates to form a six-component balance as described by Psolla-Bress et al. [10]. The upper steel plate is mounted on the sting using an extra plate with the possibility of fine alignment. An aluminum square tube is attached to the lower steel plate. The outer shells of the train model are fixed at this aluminum tube thus ensuring the stability of the model. The shells are 3D printed in PLA (polylactide) using low-cost 3D printers.

Figure 1 – and also the right-hand side of Fig. 2 – show the Prandtl tube which is located close to the train head. There is a second Prandtl tube 30 cm below the ceiling of the test section in front of the mounting sting. While the upper Prandtl tube sees undisturbed flow, the idea of the lower Prandtl tube is to obtain information about the dynamic pressure behind the wing system at a

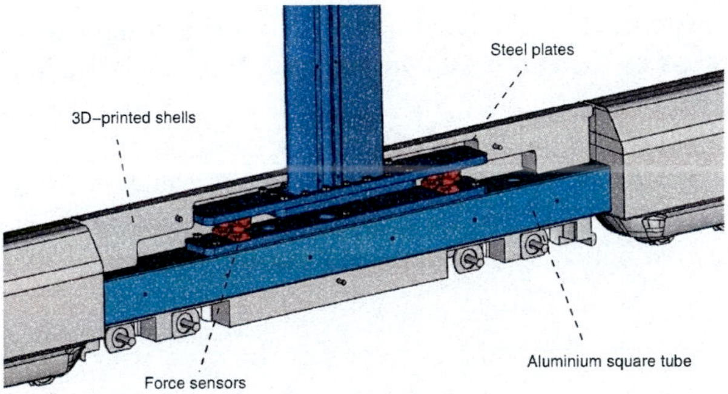

Fig. 3. Inner design of the train model with arrangement of force transducers (red).

point close to the model. The position of the lower Prandtl tube relative to the train model is illustrated in Fig. 4. The tip of the tube is 210 mm downstream of the train nose. In lateral direction, the position of the lower Prandtl tube corresponds to the center between the outer wings. It is located 0.4 m above the belt, which corresponds to half the height of the wings.

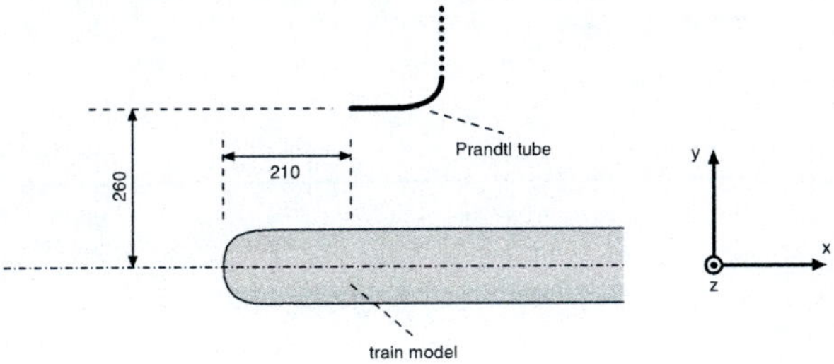

Fig. 4. Position of the lower Prandtl tube relative to the train model (top view); dimensions in millimeters.

3 Results

3.1 Drag

First the drag force in the main flow direction (x-direction) is investigated. In the selected test case the belt is moving with a velocity of 30 m/s. The flaps are oscillating synchronously and harmonically at a frequency of 30 Hz. Their angle amplitude is $\pm 3°$. In Fig. 5 the time course of several quantities is depicted. Graph a) shows the filtered drag force $F_{x,\text{filter}}$, where a moving average filter with a time constant of 10 s is applied. In graph b) the unfiltered drag force F_x can be seen. In graph c) the dynamic pressure at the upper (blue) and lower (red)

Prandtl tube is depicted. Finally, graph d) shows the difference in static pressure between two taps in the side wall, which are 2.35 m apart in flow direction. The taps are located close to the x-position of the head and the tail of the train model. Hence, this difference represents the pressure gradient in the test section along the model. Analog to the force $F_{x,\mathbf{filter}}$, the curves in graph c) and d) are smoothed using a 10 s moving average filter. The arrows in graph a) mark different moments, which are explained in Table 1.

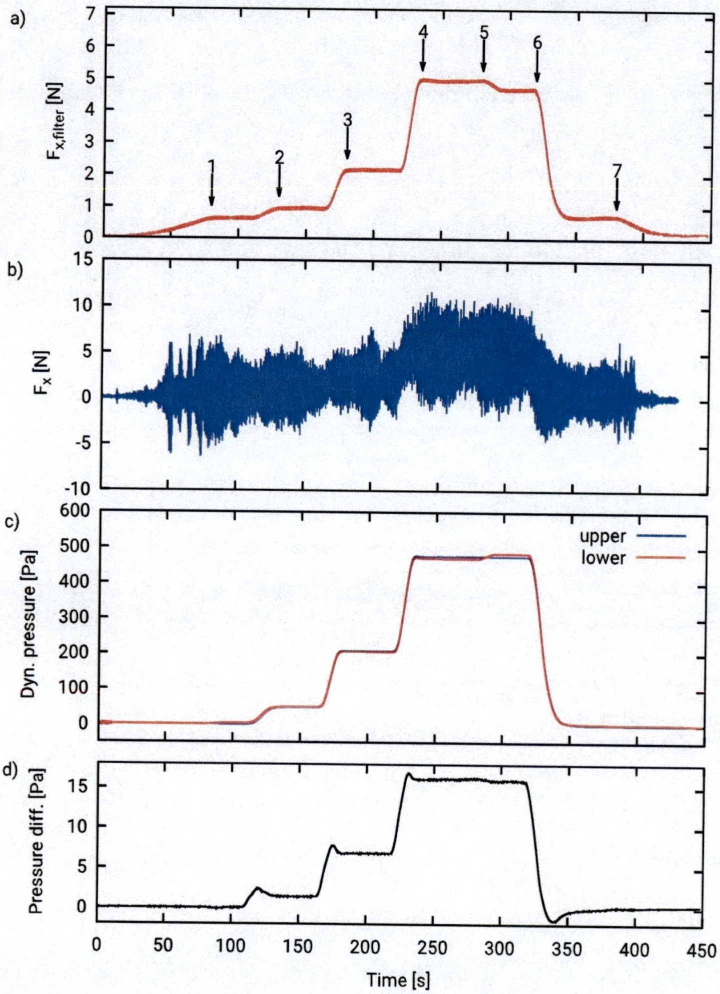

Fig. 5. Time history of force and pressure: a) filtered force $F_{x,\mathrm{filter}}$; b) unfiltered force F_x; c) dynamic pressure at upper (blue) and lower (red) Prandtl tube; d) static pressure difference between train head and tail.

Table 1. Notes on the arrows in Fig. 5a).

Arrow	Status
1	Moving belt has reached 30 m/s
2	Wind speed has reached 10 m/s
3	Wind speed has reached 20 m/s
4	Wind speed has reached 30 m/s
5	Flaps are switched off
6	Wind is switched off
7	Moving belt is switched off

The curves in Fig. 5 capture the complete test run. At the beginning the moving belt, the active suction and the flaps are started. The wind is still switched off. Without wind the moving belt generates a mean drag force on the model of approximately 0.6 N (arrow 1), and the F_x curve in graph b) shows already strong fluctuations. Next, the wind is switched on and increased step-wise up to a speed of 30 m/s. Note that this flow velocity is measured using the pressure difference at the nozzle of the wind tunnel. Due to the blockage in the tunnel caused by the wing system and the model, it does not perfectly match the velocity at the model. For safety reasons, the wind speed is gradually increased in order to detect problems regarding vibrations of the mounting or the moving belt at an early stage. The wind leads to an increase in the mean drag force $F_{x,\text{filter}}$, and a dynamic pressure at both Prandtl tubes as well as a pressure gradient becomes visible. Interestingly, the wind has no significant influence on the strength of the F_x fluctuations. The initial fluctuations without wind are in the same order of magnitude.

Stopping the flaps (arrow 5) leads to a drop in the mean drag force of approx. 0.3 N, a visible increase in the dynamic pressure at the lower Prandtl tube and a slight decrease in the pressure gradient. The dynamic pressure at the upper Prandtl tube shows no change. It appears that the unsteady incoming flow causes an increase in the drag force. This cannot be explained by the change in the pressure gradient along the train. The pressure difference changes less than 1 Pa. With the cross-sectional area of $A_{\text{train}} = 0.02\,\text{m}^2$ of the model, this results in a change of drag one order of magnitude lower than the observed drop of 0.3 N. The dynamic pressure at the lower Prandtl tube is reduced when the flaps are oscillating. This can be explained by the wakes of the wings. The oscillation of the flaps generates periodically lateral wind components which shift the wakes in y-direction so that they regularly reach the lower Prandtl tube. Without crosswind the wakes do not reach the tube which is centered between two wings. It can be assumed that the wakes of the inner wings also hit the train model. Thus, the disturbed incoming flow leads to an increase in drag and at the same time to a reduction of dynamic pressure. It can be concluded that the drag coefficient is increased by the fluctuating crosswind.

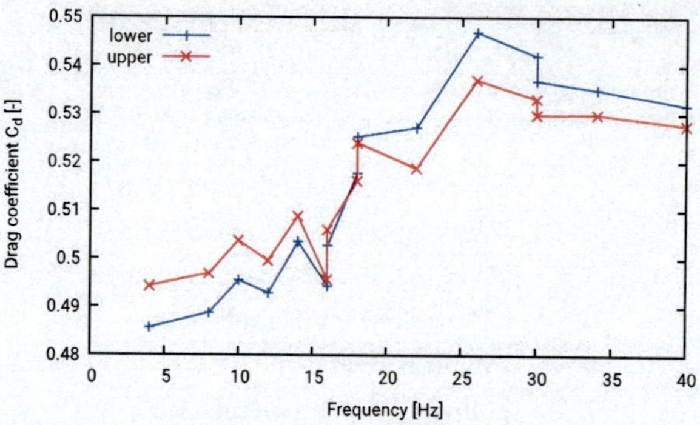

Fig. 6. Drag coefficient as a function of the excitation frequency; calculated using the dynamic the pressure from the upper (red) and lower (blue) Prandtl tube.

In Fig. 6 the drag coefficient with the disturbed incoming flow is shown as a function of the frequency at which the flaps oscillate and the disturbances are excited. In all cases the angle amplitude was $\pm 3°$ and the wind speed 30 m/s. The drag coefficient $C_\mathrm{d} = F_{x,\mathbf{filter}}/(q \cdot A_\mathrm{train})$ was calculated using the cross-sectional area $A_\mathrm{train} = 0.02\,\mathrm{m}^2$ and the dynamic pressure q of the Prandtl tubes. Although the lower tube is more representative in terms of the conditions at the train, the resulting C_d for both Prandtl tubes are depicted in Fig. 6. A frequency dependency of the drag is quite evident. The drag appears to reach a maximum at an oscillation frequency around 26 Hz. Note that so far no correction of the drag force for the pressure gradient in the wind tunnel or for a possible drift of the force transducers due to thermal effects is applied. In Fig. 5a) the offset of $F_{x,\mathbf{filter}}$ at the end of the measurement indicates such a drift.

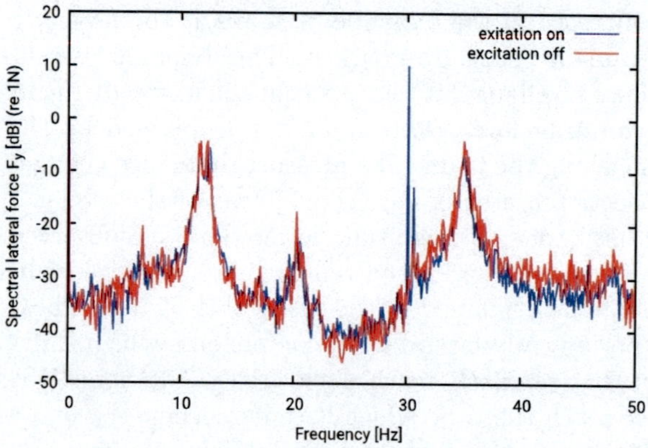

Fig. 7. Spectrum of lateral force F_y with and without excitation.

3.2 Lateral Force

Next, the lateral force F_y acting on the train model is studied. Since the unsteady force response is of special interest, the force fluctuations introduced by vibrations in the setup are important. In Fig. 7 the spectrum of the lateral force is plotted for the same test case as in the previous section. The flaps are oscillating at 30 Hz with an amplitude of $\pm 3°$, and the wind speed is 30 m/s. For comparison purposes the spectrum without flap motion is included in the plot (red curve). For both curves the spectral resolution is 0.1 Hz. The dB values are calculated using 1 N as reference value. The peak at the excitation frequency of 30 Hz is most prominent in the spectrum (blue curve). This peak is missing in the reference spectrum (red curve). Otherwise the spectra are very similar. Near the excitation frequency a sharp additional peak occurs in both spectra at approx. 30.6 Hz. Further tests, which cannot be discussed here in detail, showed that this peak is introduced by the moving belt. Additionally, broader maxima occur at frequencies of 12 Hz, 20 Hz and 35 Hz. It is assumed that these maxima belong to mechanical resonances in the model mounting which are exited by the vibrations introduced by the moving belt, the active suction and the wind tunnel itself. It has to be noted that the mounting sting is fixed at an outer frame, which is not directly connected to the wind tunnel but is supported on the floor of the hall. Nevertheless, vibration can propagate through the floor into the mounting.

A reference signal of the servo motors is used to perform phase-averaging of the lateral force response. By doing so the portions of the F_y fluctuations which are not related to the flap motion are averaged out. Figure 8 depicts the resulting $F_{y,\mathrm{pa}}$ fluctuations as a function of the phase angle for three different excitation frequencies. Since the disturbances have to travel from the wing system to the model, a different phase shift is expected for the three frequencies. However, the results also show different amplitudes of the phase-averaged fluctuations $F_{y,\mathrm{pa}}$. While the curves for 18 Hz and 30 Hz show a proper sinusoidal form, the 10 Hz curve obviously contains higher harmonics. It has to be noted that a fix average time of 50 s was used to calculate the phase average. This means that the number of averaged values is smaller for the lower frequency, which might also account for the deviation from the sinusoidal form.

Figure 9 shows the peak-to-peak amplitude of $F_{y,\mathrm{pa}}$ as a function of the excitation frequency in order to understand how the swing of $F_{y,\mathrm{pa}}$ depends on the frequency. The resulting curve reveals a strong dependency of the amplitude on the frequency. The maximum at 34 Hz is the most prominent and a second one occurs at 14 Hz. It is noticeable that the maxima appear close to those in the spectra in Fig. 7. Hence, the reason for the increased amplitudes might be a mechanical resonance of the mounting rather than an aerodynamic effect.

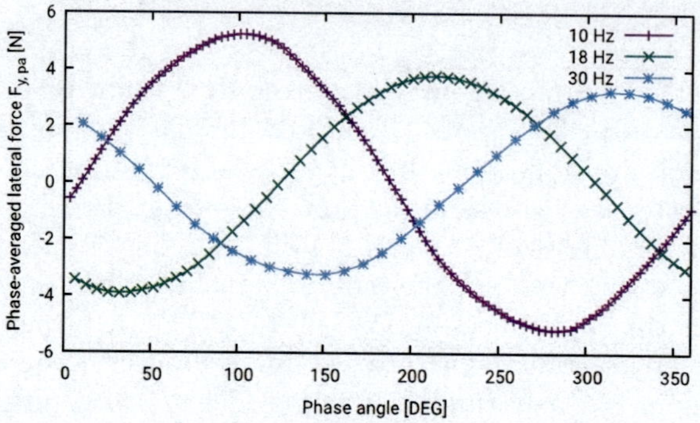

Fig. 8. Phase-averaged lateral force $F_{y,\mathrm{pa}}$ as a function of the phase angle relative to the flap motion.

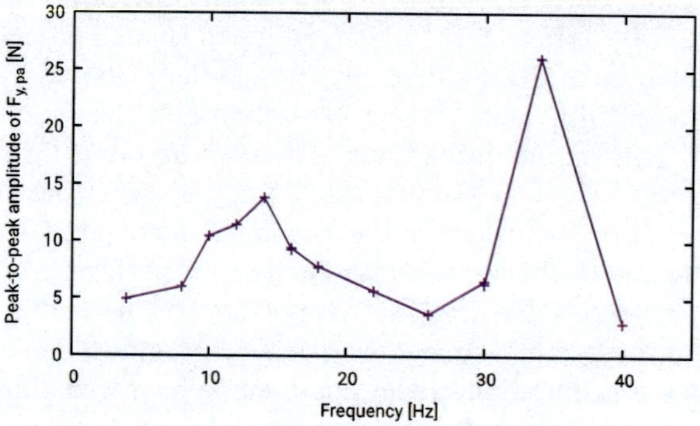

Fig. 9. Peak-to-peak amplitude of the phase-averaged lateral force $F_{y,\mathrm{pa}}$ as a function of excitation frequency.

4 Conclusion

First wind tunnel tests were performed with a train model mounted above a moving belt and an upstream wing system to disturb the incoming flow at the model. Strong evidence was found that a harmonic disturbance of the incoming flow can increase the drag coefficient of the train model. The drag coefficient increases with the oscillation frequency and reaches a maximum at a certain point. It is clear that the presented results are preliminary. The drag force has to be corrected for the pressure gradient in the wind tunnel and also for a possible drift of the force transducers due to thermal effects. One point that needs to be addressed in future studies is the characterization of the incoming flow conditions at the position of the model for the various flap oscillations. Therefore, the wind speed and its fluctuations have to be measured at the position of the model without the model. Like this, the correct dynamic pressure at the model position can be determined. The incoming flow conditions are also important for the analysis of the lateral forces. These are for example required to calculate the aerodynamic admittance as described by Baker et al. [7]. The calculation of the phase-averaged $F_{y,\mathrm{pa}}$ variations showed that with a measurement time of several minutes a statistically valid result can be obtained – at least for the higher frequencies. However, the curve in Fig. 9 indicates that a form of dynamic calibration of the balance is required so that the values are corrected in terms of the mechanical resonances of the mounting. Overall, more studies are needed to understand the unsteady response of the model to the incoming flow disturbances.

References

1. Zeng, Z., Ziegler, A.D., Searchinger, T., et al.: A reversal in global terrestrial stilling and its implications for wind energy production. Nat. Clim. Change **9**, 979–985 (2019). https://doi.org/10.1038/s41558-019-0622-6
2. Wilhelmi, H., Jessing, C., Bell, J., Heine, D., Wagner, C., Wiedemann, J.: Aerodynamic characterisation of a compact car overtaking a heavy vehicle. In: Dillmann, A., Heller, G., Krämer, E., Wagner, C., Tropea, C., Jakirlić, S. (eds.) DGLR 2018. NNFMMD, vol. 142, pp. 794–804. Springer, Cham (2020). https://doi.org/10.1007/978-3-030-25253-3_75. ISBN 978-3-030-25252-6:794-804
3. Jessing, C., Wilhelmi, H., Wittmeier, F., Wiedemann, J., et al.: Characterization of the transient airflow around a vehicle on public highways. In: 12th FKFS-Conference: Progress in Vehicle Aerodynamics and Thermal Management, Stuttgart (2019)
4. Schröck, D., Widdecke, N., Wiedemann, J.: On-road wind conditions experienced by a moving vehicle. In: Wiedemann, J. (ed.) Progress in Vehicle Aerodynamics and Thermal Management V, Expert, Renningen (2008). ISBN 978-3-8169-2771-6
5. **Baker, C.: A framework for the consideration of the effects of crosswinds on trains.** J. Wind Eng. Ind. Aerodyn. **123**, 130–142 (2013)
6. Europäische Kommission. EN 14067-6 Bahnanwendungen - Aerodynamik - Teil 6: Anforderungen und Prüfverfahren zur Bewertung von Seitenwind (2020)
7. Baker, C., Jones, J., Lopez-Calleja, F., Munday, J.: Measurements of the crosswind forces on trains. J. Wind Eng. Ind. Aerodyn. **92**, 547–563 (2004)
8. https://www.dlr.de/content/en/research-facilities/tunnel-simulation-facility-goettingen-tsg.html
9. Müller, M., Ehrenfried, K., Bell, J., Wagner, C.: Analysis of model mount configurations with regard to force measurements with transient inflow. In: Dillmann, A., Heller, G., Krämer, E., Wagner, C., Tropea, C., Jakirlić, S. (eds.) DGLR 2018. NNFMMD, vol. 142, pp. 684–694. Springer, Cham (2020). https://doi.org/10.1007/978-3-030-25253-3_65. ISBN 978-3-030-25252-6
10. Psolla-Bress, H., Haselmeyer, H., Hedergott, A., Höhler, G., Holst, H.: High roll experiments on a delta wing in transonic flow. In: Proceedings of the 19th ICIASF 2001 Record, Cleveland, Ohio, 27–30 August 2001, p. 369. IEEE, New York (2001)

Low-Order Modeling of Bistable Side Forces on a Sphere Measured for a Transient Inflow in a Wind Tunnel

Max Müller[1](✉), Klaus Ehrenfried[1], and Claus Wagner[1,2]

[1] German Aerospace Center, Institute of Aerodynamics and Flow Technology, Bunsenstr. 10, 37073 Göttingen, Germany
max.mueller@dlr.de

[2] TU Ilmenau, Institute of Thermodynamics and Fluid Mechanics, Am Helmholtzring 1, 98693 Ilmenau, Germany

Abstract. The unsteady forces acting on a sphere laterally mounted in a wind tunnel with transient sinusoidal inflow are investigated. The study shows that the low-frequency dynamic is characterized by a spontaneous switching between two partially stable states, also known as bistability. Additionally, oscillatory forces exist, which, on the one hand, are induced by the sinusoidal transient inflow and, on the other hand, originate from natural vortex shedding. By separating the bistability and the oscillatory forces using proper orthogonal decomposition, a low-order model based on the Duffing equation is developed both for the bistability and the oscillatory multi-scale induced forces. The bistability is represented by a chaotic bistable Duffing equation, whose parameters are determined by manual adjustment. In contrast, the parameters of the oscillatory multi-scale forcing are determined by expanding the forcing of the Duffing equation to three terms and preserving the phase angles between the three characteristic frequencies, the flapping frequency, its multiples and the natural vortex shedding.

Keywords: Reduced-order Model · Chaos · Bistability

1 Introduction

Bluff bodies can experience non-zero side forces with spontaneous switching between two distinct partially stable states, known as bistability. The latter was observed by Müller et al. [6] in wind tunnel measurements with a sphere installed on a vertical lateral model mount. They found partially stable states lasting up to several seconds coinciding with low frequencies in the spectrum below 5 Hz. Similar results are reported by Bauer et al. [2], who performed wind tunnel tests and numerical simulations to analyze the side forces as well as the deflection of the sphere's wake. Comparable bistabilities have also been observed for other bluff bodies such as the Ahmed body. Studying the latter, Varon et al. [7] used the position of the wake's barycenter to describe its bistability. They were able to prove that – while the overall behavior of the wake is stochastic – the low-frequency dynamic reflecting the bistability is a chaotic process with two distinct

A. Dillmann et al. (Eds.): STAB/DGLR Symposium 2022, NNFM 154, pp. 88–98, 2024.
https://doi.org/10.1007/978-3-031-40482-5_9

attractors. They also proposed to reflect the dynamics of the bistability with a low-order model. Based on the latter, a control strategy – as for example used by Wagner et al. [8] – for a non-linear oscillator can be adapted to the control of the sphere's wake flow.

A low-order differential equation suitable to describe a bistable chaotic process is the Duffing equation proposed by Georg Duffing [3], who added a cubic term to the equation of a forced oscillator. Nowadays, this equation is often used in the form $\ddot{x} + \alpha x + \beta x^3 + \delta \dot{x} = \gamma \cos(\omega t)$ as presented by Jordan and Smith [4] where α reflects the linear stiffness, β the Duffing parameter representing the nonlinear stiffness, δ the damping coefficient and γ the amplitude of the forcing. Jordan and Smith [4] showed that enhancing the forcing amplitude γ can lead to frequency doubling and a chaotic bistability.

Furthermore, the Duffing equation is used in fluid dynamics to provide a low-order representation of oscillatory forces acting on a body. In this sense, Armin et al. [1] were able to model the forces on a cylinder caused by the interaction with the flow around other cylinders in its vicinity. Such oscillatory forces occur for example as vortex-induced vibration on the legs of tension-leg platforms such as oil drilling rigs. Another application of the Duffing equation was reported by Yang et al. [9], who expanded the forcing part to the sum of three different forcing terms in order to analyze the impact of multiple scales on the resonance, and therefore, fatigue of structures.

The aim of this work is to find a low-order model demonstrating the behavior of the side forces acting on a sphere laterally mounted downstream of a transient sinusoidally varying inflow. The experimental set-up was also used in Bauer et al. [2]. In the present work, the signal reflecting the bistability is separated from the complete force signal by means of Proper Orthogonal Decomposition (POD) to define parameters of the Duffing equation serving as a low-order model for the bistable dynamics. Additionally, the multi-scale forcing is modeled by a separate Duffing equation using the expanded forcing terms analogous to the work of Yang et al. [9].

2 Measured Data

Even though the measurement itself is not part of this paper, it is described briefly to explain the origin of the data. Figure 1 shows the test set-up including the sphere (diameter: 100 mm) in the lateral vertical model mount (diameter: 25 mm) in the Crosswind Test Facility (SWG), which is a closed-circuit wind tunnel in Göttingen. The set-up includes a flap system, which is installed 1.5 m upstream of the sphere and consists of four wings with movable flaps. These flaps impose a sinusoidally varying cross-flow component on the undisturbed inflow, thus modeling a gusty inflow. The forces acting on the sphere were measured for 25 s with an internal six-component balance at a sample rate of 1 kHz. The undisturbed inflow velocity in the wind tunnel was $u_0 = 15$ m/s, and the flap system was operated at a flapping frequency of $f = 40$ Hz with a flap amplitude of 5°. To simplify the derivation of the parameters for the reduced-order model, the side force coefficient is normalized such that in the two partially stable states the dimensionless coefficient takes the values $c = \pm 1$. Further, a time period of 20 s of the complete time series is extracted for the POD, see Sect. 3.1.

In Fig. 2 the normalized side force coefficient s_{raw} is plotted in a) and the corresponding frequency spectrum in b). The time series in Fig. 2a) reflects a noisy signal varying

Fig. 1. Set-up of the wind tunnel measurements of the forces acting on a sphere mounted on a lateral model mount in transient inflow

around two different states. The force signal switches spontaneously between these two states, resulting in different time periods for the partially stable states. The absolute values of the amplitude of the fluctuations around these states are sometimes higher than the absolute mean value of the state itself leading to zero crossings in the force signal. The frequency spectrum depicted in Fig. 2b) is obtained using the Fast Fourier algorithm with Dirichlet windows of 2 s and an overlap of 50%. For a better identification of the single peaks, the amplitude spectrum corresponding to the absolute values of the complex Fourier transform is displayed in Decibel. Three narrow peaks occur at frequencies of f = 40 Hz, f = 80 Hz and f = 120 Hz. They are attributed to the flapping frequency and its multiples. The peak at f = 40 Hz corresponds to the sinusoidal inflow, whereas the multiples are induced by the wakes of the flap system. The additional peak at f = 90 Hz is attributed to natural vortex shedding. The other minor peaks reflected in the amplitude spectrum stem from mechanical eigenfrequencies of the set-up. Furthermore, a decline in the amplitude can be observed for low frequencies between f = 0 Hz and f = 5 Hz. This is due to the aforementioned low-frequency dynamic, reflecting the bistability in the time series.

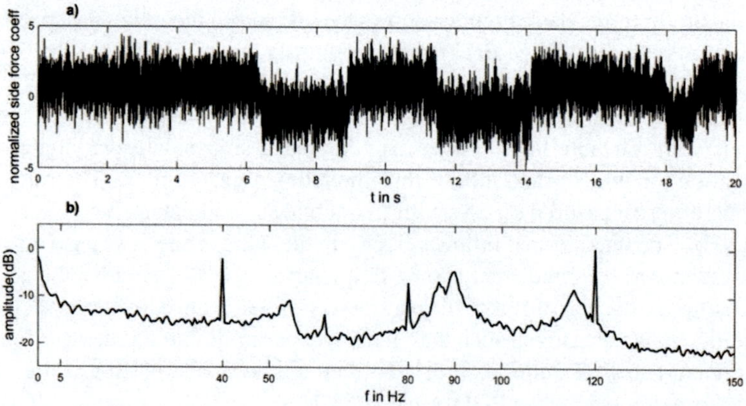

Fig. 2. a) Time series of the normalized side force signal, b) amplitude spectrum of the signal indicated in Decibel

3 Modeling of the Data with Duffing Equation

3.1 Mode Decomposition

The dynamics of the bistability and the multi-scale forcing are hidden in the complex measured force signal together with other responses of the set-up. To extract the former, a POD - as presented by Lassila et al. [5] - is performed for which modes are computed following the singular spectrum analysis discussed in Zhigljavsky [11]. The latter is based on a Singular Value Decomposition (SVD) algorithm using a snapshot matrix M, which results in a matrix U comprising the left eigenvectors, a square diagonal matrix Σ, consisting of the eigenvalues in declining order, and V^T reflecting the right eigenvectors. The snapshot matrix M of the signal s_{raw} shown in Eq. (2) consists of column vectors with the length of 100 timesteps, while the complete time series is reflected in the first line of the matrix.

$$\mathrm{svd}(M) = U\Sigma V^T \tag{1}$$

$$M = \begin{bmatrix} s(t_1)\ s(t_2) & \dots & s(t_{20000}) \\ s(t_2)\ s(t_3) & & s(t_{20001}) \\ \vdots & \ddots & \vdots \\ s(t_{100})\ s(t_{101}) & \cdots & s(t_{20100}) \end{bmatrix} \tag{2}$$

3.2 Modeling the Bistable Low-Frequency Dynamic

During the analysis of the modes, it was found that the bistability is connected to the first mode and can hence be reconstructed by multiplying the first eigenvalue with its respective left and right eigenvectors. In Fig. 3 the reconstruction s_{m1} is plotted in a), the low-frequency part of the amplitude spectrum in b) and the resulting phase plot in c). Figure 3a) shows the time series of the complete signal s_{raw} in black in the background and the time series of the reconstruction s_{m1} in light blue. The reconstructed signal reflects the two states of the bistability with less noise than the raw signal s_{raw}, i.e. the fluctuations are not crossing the zero line. Since the bistability is a low-frequency process, as discussed by Varon et al. [7] and others, the observation in Fig. 3b) – with amplitudes in the spectrum of the two signals agreeing in the low frequency range $0 < f < 5$ Hz – underlines that s_{m1} still reflects the bistable behavior and is therefore well-suited for the construction of a low-order model. Further, the corresponding phase plot of the time derivate ds_{m1}/dt over s_{m1} (Fig. 3c)) reveals the existence of the two attractors at $s_{m1} = \pm 1$. The shape of the phase lines circling around the attractors reflects Cassini ovals, described for example by Yates [10]. This is also characteristic for a bistable Duffing oscillator and is therefore a further motivation to model the bistability with the Duffing equation.

The Duffing Eq. (3) proposed by Jordan and Smith [4] to describe mechanical oscillations contains parameters with a physical meaning. In this context α denotes the linear stiffness, β (called Duffing parameter) the cubic stiffness, and δ the damping coefficient. Additionally, the forcing term on the right-hand side of Eq. (3) contains the forcing amplitude γ and the forcing frequency f.

$$\ddot{x} + \alpha x + \beta x^3 + \delta \dot{x} = \gamma \cos(2\pi f t) \tag{3}$$

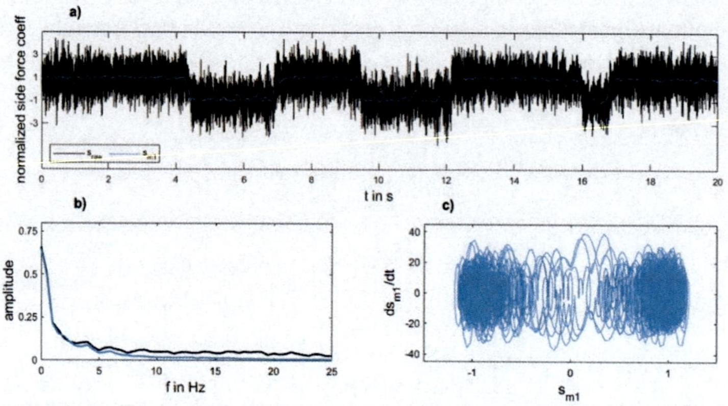

Fig. 3. a) Time series of the normalized raw signal s_{raw} (black) and the reconstruction s_{m1} (light blue), b) amplitude spectrum of s_{raw} (black) and s_{m1} (light blue), c) phase plot of s_{m1}

Since chaotic systems are known to be very sensitive to small disturbances, the parameters resulting in a bistable duffing oscillator stem from manual adjustment based on an example described by Jordan and Smith [4]. Here, the forcing frequency is f = 5 Hz, since the evenness for f > 5 Hz in the amplitude spectrum in Fig. 3b) indicates that the frequencies associated with the bistability are in the low-frequency band f < 5 Hz. Jordan and Smith [4] also reported that a ratio between the forcing frequency and the eigenfrequency of the Duffing equation f0 of approx. 1.2 can lead to a chaotic bistability. The initial value of the stiffness α was calculated from the eigenfrequency f_0, $\alpha = -\omega_0{}^2 = -(2\pi f_0)^2$, and the Duffing parameter β was α = -β. Additionally, a fairly low damping was prescribed by setting δ = 0.0001.

Further, it is also known from Jordan and Smith [4] that the forcing amplitude γ determines the chaotic bistability. This means that for a certain combination of parameters the behavior of a Duffing oscillator can be controlled by changing γ. For low γ a period-one oscillator is obtained characterized by a single orbit in the phase plane around one stable state, while for higher γ multiple period oscillators can develop, reflecting periodicity in time and multiple orbits in the phase plot. For a certain γ a time signal with two partially stable states is observed, reflecting oscillations around two attractors in the phase plot. This γ is then adjusted manually until x_{bi} stays in the partially stable states for time periods depicted in Fig. 3a). Moreover, values of its time derivate are obtained, which are similar to those shown in Fig. 3b). Therefore, the Duffing equation was numerically integrated using Eq. (4).

$$x_{bi}(t_{n+1}) = dt^2\left(\gamma\cos(2\pi f t_n) - \alpha x_{bi}(t_n) - \beta(x_{bi}(t_n))^3\right) - dt\delta(x_{bi}(t_n) - x_{bi}(t_{n-1})) + 2x_{bi}(t_n) - x_{bi}(t_{n-1}) \quad (4)$$

The best agreement of the Duffing oscillator with the measured force signal was obtained for $\gamma = 65$ and $\alpha = -675$, resulting in the time series shown in Fig. 4(left). In order to demonstrate how sensitive the Duffing oscillator responds to a change in γ, the time series obtained for a slightly lower value, $\gamma = 64.99$, is shown on the right. In contrast to the time series of x_{bi} obtained for $\gamma = 65$ reflecting two partially stable states with spontaneous switching, the time series obtained for $y = 64.99$ loses its bistable character at $t = 6$ s.

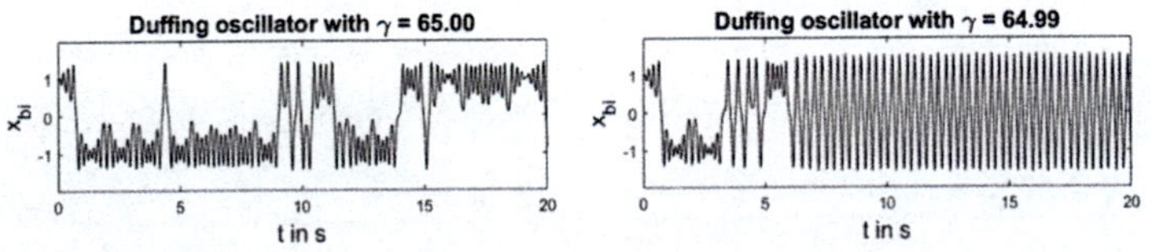

Fig. 4. Duffing oscillation predicted with $\gamma = 65$ (left) and $\gamma = 64.99$ (right)

The comparison of the 20 s long signal predicted by solving the dimensionless Eq. (5) of the Duffing oscillator using the above-discussed parameters with the reconstructed normalized force signal of the measurement is presented in Fig. 5.

$$\ddot{x}_{bi} + 0.00011/s\,\dot{x}_{bi} - 675\left(x_{bi} - x_{bi}^3\right) = 65\cos(2\pi 5Hz \cdot t) \quad (5)$$

In addition to the two time series plotted in Fig. 5a) and b), the amplitude spectrum is shown in (c) and the phase plot in d). The time series of x_{bi} (red) and s_{m1} (light blue) are partially different. On the one hand, x_{bi} depicted in Fig. 5a) contains very high single peaks (example: $t \approx 4.4$ s) including values belonging to the other stability state. For the here studied bistability, this is not physical. On the other hand, the amplitudes x_{bi} of the oscillation of higher frequencies are larger than those observed in the reconstructed signal s_{m1} of the measurement. However, the number of switches between the stability states is similar. The switching behavior depends on the interference of two close frequencies – also known as beat –reflected as recurring patterns in x_{bi}, but not in s_{m1}. Despite these obvious differences, the main characteristic of the bistability – which is the time period the signal stays in the partially stable states – is similar. Since there are only few switches from one state to the other in the 20 s long signal, the maximal partially stable duration was chosen as criterion. The time between switches from one stability state to the other is determined by assigning the zero crossing of the signal to a switching event. It can be seen that the maximal stable time for s_{m1} is $t_{s,max} = 4.346$ s and for the Duffing oscillator x_{bi} it is $t_{s,max} = 4.729$ s. In the amplitude spectrum shown in Fig. 5c), the beat between the eigenfrequency f_0 and the forcing frequency f is also visible, where the amplitudes of x_{bi} are higher for frequencies around $f = 5$ Hz than those of s_{m1}. The phase plot in

Fig. 5d) reveals that the time derivate of x_{bi} is in the same order of magnitude as s_{m1}. In addition, the elliptical shape is also similar, even though x_{bi} is smoother and shows a slightly larger orbit than the orbit of the measured signal s_{m1}.

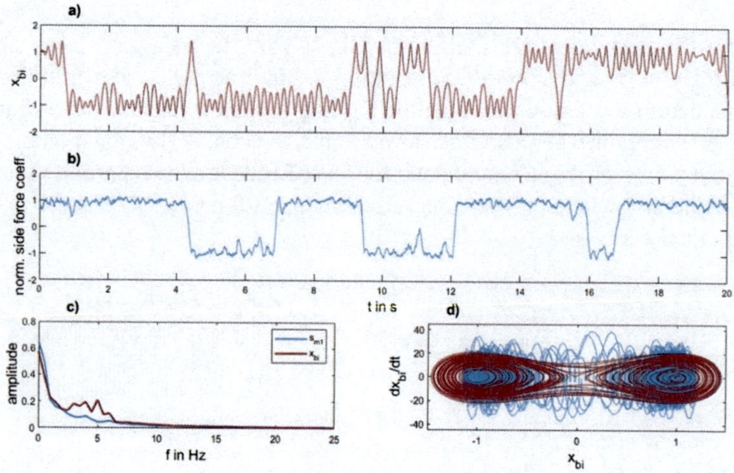

Fig. 5. Time series of the manually adjusted simulated time line x_{bi} (a), course of the first mode s_{m1} (b), their respective low-frequency amplitude spectrum (c) and their phase plot (d)

3.3 Modeling of the Multi-modal Oscillatory Forcing

In addition to the bistability, the multi-scale forcing due to the oscillatory forces can also be modeled with a separate Duffing oscillator. In a first step, further POD modes were identified which are associated with the first three dominant frequencies, i.e. the flapping frequency ($f_1 = 40$ Hz), the natural vortex shedding frequency ($f_2 = 90$ Hz) and the frequency related to the flap's wakes ($f_3 = 120$ Hz). These frequencies are connected to the modes 2 to 7. Hence, a reconstruction of these modes led to the low-order representation s_{m27}, see Fig. 6. Its time series is shown in Fig. 6a) and its amplitude spectrum in Fig. 6b). Even though the time series appears to be dominated by noise, it only contains oscillation of three high frequencies, as can be seen in the amplitude spectrum. It is also observed that the amplitudes for the frequencies f_1, f_2 and f_3 are depicted qualitatively well, indicating that all modes containing the multi-scale forcing are represented, see Fig. 6a).

To reproduce the multi-scale forcing with one Duffing equation, the right-hand side of Eq. (3) has to be replaced by a sum of three forcing terms as discussed in Yang et al. [9]. The resulting Eq. (6) depends on three frequencies and the associated phase angles ϕ_i. A suitable combination of ϕ_i inducing oscillations in the simulated x_{f3} signal and reflecting the phase angle differences between the oscillations with the frequencies f_1 and f_2, i.e. $\Delta\phi_{12}$, and f_1 and f_3, i.e. $\Delta\phi_{13}$ is required. The parameters of ϕ_i were varied to deduce the phase angle differences from the simulated signal compared to s_{m27}, which are $\Delta\phi_{12,m} = -2.9729$ and $\Delta\phi_{13,m} = 2.724$, respectively.

$$\ddot{x}_{f3} + \delta\dot{x}_{f3} + \alpha x_{f3} + \beta x_{f3}{}^{3} = \sum_{i=1}^{3} \gamma_i \cos(2\pi f_i t + \phi_i) \quad (6)$$

In detail, to solve Eq. (6), the phases angles ϕ_i were varied between 0 and 2π in steps of $1/50\pi$. For each ϕ_i the parameters α, β, δ, γ_i were estimated with a least squares algorithm to minimize the differences in relation to s_{m27} and its time derivatives for an overdetermined equation system for the six unknown variables α, β, δ and γ_i.

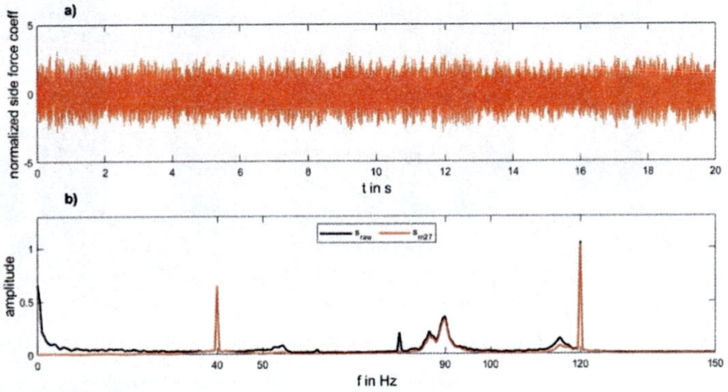

Fig. 6. a) Time series of the modes 2 to 7 and b) amplitude spectrum with the spectrum of the raw measured signal (black) and the spectrum of the modes 2 to 7

$$\begin{bmatrix} \ddot{s}_{m27}(t_1) \\ \vdots \\ \ddot{s}_{m27}(t_{20000}) \end{bmatrix} + \delta \begin{bmatrix} \dot{s}_{m27}(t_1) \\ \vdots \\ \dot{s}_{m27}(t_{20000}) \end{bmatrix} + \alpha \begin{bmatrix} s_{m27}(t_1) \\ \vdots \\ s_{m27}(t_{20000}) \end{bmatrix} + \beta \begin{bmatrix} s_{m27}(t_1)^3 \\ \vdots \\ s_{m27}(t_{20000})^3 \end{bmatrix} = \sum_{i=1}^{3} \gamma_i \cos\left(2\pi f_i \begin{bmatrix} t_1 \\ \vdots \\ t_{20000} \end{bmatrix} + \phi_i \right) \quad (7)$$

With the parameters of the Duffing oscillator, x_{f3} was determined, and the resulting phase angle differences were obtained using a Fast Fourier transform. This procedure led to a 4D solution space with the ϕ_i on the axes and the phase angle differences $\Delta\phi_{12}$ and $\Delta\phi_{13}$ as the solution. Considering slices of the solution space for fixed values of ϕ_1, it was found that for $\phi_1/\pi = 0.96$ there is a parameter ϕ_1, ϕ_2 and ϕ_3, for which the phase angle differences $\Delta\phi_{12}$ and $\Delta\phi_{13}$ are preserved.

The 3D subspaces of the color-coded phase angle differences at $\phi_1/\pi = 0.96$ are illustrated in Fig. 7, on the left for $\Delta\phi_{13}$ and on the right for $\Delta\phi_{12}$: Green represents values close to the desired values of $\Delta\phi_{12,m}$ and $\Delta\phi_{13,m}$ obtained from s_{m27}. White areas reflect non-physical parameter settings found by the least squares algorithm and are therefore excluded. In Fig. 7(left), the color pattern is vertical indicating that $\Delta\phi_{13}$ is only a function of ϕ_1 and ϕ_3. Hence, ϕ_3 was determined to be $\phi_3/\pi = 0.8$. Figure 7(right) shows diagonal patterns so that ϕ_2 was found to be $\phi_2/\pi = 0.06$. The obtained phase angle differences for this combination of ϕ_i are $\Delta\phi_{12} = -2.9518$ ($\Delta\phi_{12,m} = -2.9729$) and $\Delta\phi_{13} = 2.7248$ ($\Delta\phi_{13,m} = 2.724$) and the Duffing parameters from the least squares estimation are summarized in Table 1.

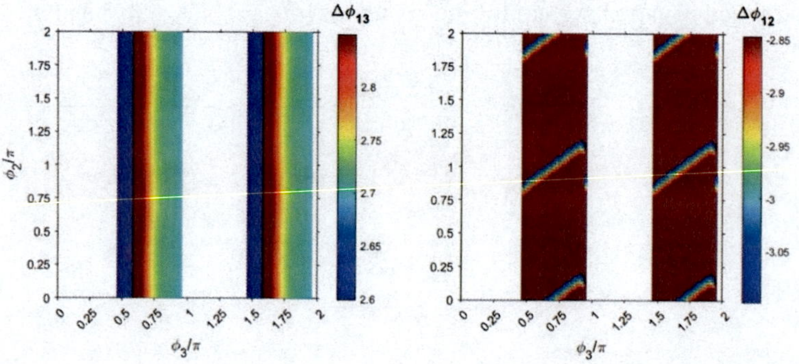

Fig. 7. Color-coded phase angle between the gust frequency and the higher harmonic (left) and between the gust frequency and the natural shedding frequency (right) at $\phi_1/\pi = 0.96$

Table 1. Parameters of the Duffing oscillator providing x_{f3}

α [1]	β [1]	δ [1/s]	γ_1 [1]	ϕ_1/π	γ_2 [1]	ϕ_2/π	γ_3 [1]	ϕ_3/π
3.4 × 105	−1.2 × 103	91	-1.8×10^5	0.96	−6.7 × 103	0.06	−2.2 × 105	0.8

The time series of x_{f3} (blue) and s_{m27} (orange) are plotted in Fig. 8a). Apparently, s_{m27} contains more noise and x_{f3} appears more regular. The corresponding amplitude spectrum shown in Fig. 8b) reveals that the peak attributed to the flap frequency $f = 40$ Hz is represented quite well. However, the amplitude associated with the vortex shedding is underestimated and the amplitude of the higher harmonic is overestimated. The reason might be that the damping δ of the Duffing equation does not depend on the frequency, whereas the width of the peaks for s_{m27} indicates different damping effects in terms of the vortex shedding and the forces induced by the sinusoidal transient inflow. The phase plot in Fig. 8c) further reveals that the time derivate of x_{f3} and s_{m27} is of the same order of magnitude. Based on the three different orbits of x_{f3}, it can be concluded that the signal predicted with the proposed low-order model reflects a period-three oscillator.

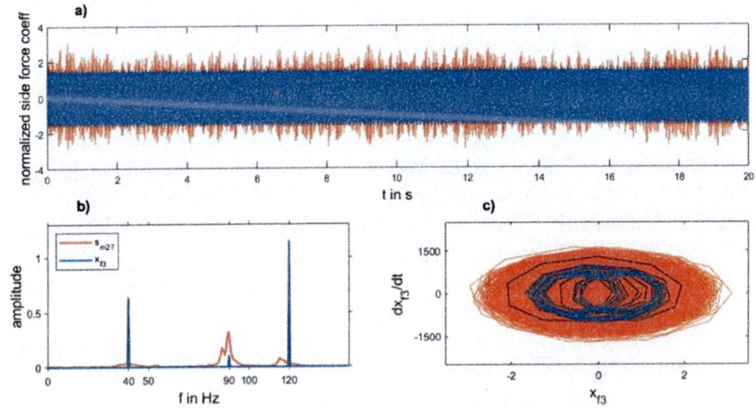

Fig. 8. a) Time series of the modes 2 to 7 (orange) and the Duffing equation with three forcing terms (blue), b) respective amplitude spectrum, c) respective phase plot

4 Conclusion

It was shown that the side forces acting on a sphere in sinusoidal inflow with underlying bistability can be modeled with the Duffing equation after separation of the bistable mode and the multi-scale oscillatory forcing, respectively.

For this purpose, the bistable mode was extracted from the raw measured signal performing a POD. The parameters of a chaotic bistable Duffing equation were then determined by manual adjustment. Solving the obtained equation, it was shown that the resulting bistable signal resembles the duration of partial stability.

In addition, the low-frequency part matches the spectrum of the measure signal in the amplitude spectrum. However, a major difference between the simulation and the signal measured in the wind tunnel tests is the beat between the eigenfrequency of the equation and the forcing frequency. In a second step, the oscillatory forces induced by the transient sinusoidal inflow and natural vortex shedding were also extracted via a POD and the forcing in the Duffing equation was extended to three forcing terms. The right set of phase angles was found by determining the correct phase angles between the forces induced by the external forcing as well as between the force connected to the flap frequency and the forces stemming from vortex shedding, obtained from the Duffing oscillator. Finally, the remaining parameters of the Duffing equation were determined applying a least squares approximation.

In the future, the developed reduced-order model (ROM) can be used to generate a deeper understanding of the bistability, the response force to transient inflow and their interaction, by performing parameter studies for different inflow conditions. Additionally, control strategies for bistable wake flows can be developed using the ROMs.

Acknowledgment. I would like to offer my special thanks to Annika Köhne for her careful and valuable proof-reading of this manuscript.

References

1. Armin, M., Day, S., Karimirad, M., Khorasanchi, M.: On the development of a nonlinear time-domain numerical method for describing vortex-induced vibration and wake interference of two cylinders using experimental results. Nonlinear Dyn. **104**, 3517–3531 (2021)
2. Bauer, C., Müller, M., Ehrenfried, K., Wagner, C.: Bi-stability detection in the flow around a sphere by means of experiments and lattice Boltzmann simulations. In: Dillmann, A., Heller, G., Krämer, E., Wagner, C. (eds.) STAB/DGLR Symposium 2020. NNFMMD, vol. 151, pp. 441–450. Springer, Cham (2021). https://doi.org/10.1007/978-3-030-79561-0_42
3. Duffing, G.: Erzwungene Schwingungen bei veränderlicher Eigenfrequenz und ihre technische Bedeutung, Sammlung Vieweg. Heft 41/42, Braunschweig (1918)
4. Jordan, D.W., Smith, P.: Nonlinear Ordinary Differential Equations, 4th edn. Oxford University Press (2007). ISBN 978-0-19-920824-1
5. Lassila, T., Manzoni, A., Quarteroni, A., Rozza, G.: Model order reduction in fluid dynamics: challenges and perspectives. In: Quarteroni, A., Rozza, G. (eds.) Reduced Order Methods for Modeling and Computational Reduction, vol. 9, pp. 235–273. Springer, Cham (2014). https://doi.org/10.1007/978-3-319-02090-7_9
6. Müller, M., Ehrenfried, K., Bell, J., Wagner, C.: Analysis of model mount configurations with regard to force measurements with transient inflow. In: Dillmann, A., Heller, G., Krämer, E., Wagner, C., Tropea, C., Jakirlić, S. (eds.) DGLR 2018. NNFMMD, vol. 142, pp. 684–694. Springer, Cham (2020). https://doi.org/10.1007/978-3-030-25253-3_65
7. Varon, E., Eulalie, Y., Edwige, S., Gilotte, P., Aider, J.L.: The chaotic dynamics of large-scale structures in a turbulent wake. Phys. Rev. Fluids **2**, 034604 (2018)
8. Wagner, C., Stelzel, W., Hübler, A., Lüscher, E., Altmann, W.: Resonante Steuerung nichtlinearer Schwinger. Helv. Phys. Acta **61**(1–2), 228–231 (1988)
9. Yang, S., Nayfeh, A.H., Mook, D.T.: Combination resonances in the response of the Duffing oscillator to a three-frequency excitation. Acta Mech. **131**, 235–245 (1998)
10. Yates, R.C.: A Handbook on Curves and Their Properties, Ann Arbor (1952)
11. Zhigljavsky, A.: Singular spectrum analysis for time series: introduction to this special issue. Stat. Interference **3**, 255–258 (2010)

Effects of Jet-Orifice Shape on the Flow-Control Effectiveness of Air-Jet Vortex Generators

Deepak Prem Ramaswamy(✉) and Anne-Marie Schreyer

Institute of Aerodynamics and Chair of Fluid Mechanics, RWTH Aachen University, 52062 Aachen, Germany
d.ramaswamy@aia.rwth-aachen.de

Abstract. An experimental investigation was carried out to assess the effectiveness of non-circular air-jet vortex generators (AJVGs) in controlling shock-induced separations. Single rows of spanwise-inclined AJVGs with circular, rectangular, and four triangular orifice shapes were tested in an incoming supersonic crossflow at Mach 2.52 and $Re_\theta = 8225$. All cases were characterised by the same hydraulic diameter for comparability. Surface oil-flow visualisations and PIV in a wall-parallel plane are used for flow diagnostics. The results reveal an improved performance by all non-circular AJVGs in controlling the shock-induced separation, in comparison to the traditional circular AJVG. The improved effectiveness is expected to be a result of increased turbulent mixing and favourable jet/jet interactions between neighbouring jets in the array.

Keywords: non-circular jets · air-jet vortex generators · separation-control · compression-ramp · jet-array in crossflow

1 Introduction

In many air and space transportation application, shock-wave/boundary-layer interactions (SWBLIs) are prevalent and can be detrimental to the operation of the vehicle [1]. A strong shock wave can induce large-scale flow separation, which impacts the regular operation of the vehicle due to enhanced pressure and thermal loads. A peculiar aspect of SWBLIs is the low-frequency oscillation of the shock/separation bubble system [2]. This unsteadiness can induce strong structural vibrations, which may lead to an overload failure in the short term or a fatigue failure in the long term. A suitable technique to deter flow separation and prevent its adverse effects is the application of air-jet vortex generators (AJVGs) [3–8,10]. Jets of air are injected into the boundary layer, and generate streamwise vortices downstream of their installation when interacting with the crossflow [4,9]. The resulting momentum redistribution makes the boundary layer more resistant to separation. A number of studies have subsequently identified geometrical and flow parameters that yield favourable flow-control

A. Dillmann et al. (Eds.): STAB/DGLR Symposium 2022, NNFM 154, pp. 99–108, 2024.
https://doi.org/10.1007/978-3-031-40482-5_10

effectiveness. Some of the identified major AJVG parameters are, amongst others, the (a) air-jet orifice diameter [5,6], (b) jet-injection orientation [3,13], (c) jet/jet spacing in the array [8,10,11], and (d) jet-injection pressure [10–12].

For effective flow control, a spanwise-oriented configuration of AJVGs [3,13] with an injection pressure equal to or slightly larger than the stagnation pressure of the crossflow [10–12] is desirable. For such a configuration, a jet-diameter of $hd_{jet} \leq 0.25\delta$ (where $\delta = 10.4$ mm is the boundary layer thickness; see Sect. 2.2) [5,6] and a jet/jet spacing of $D = 7 - 13hd_{jet}$ [8,10,11] is most advantageous. The favourable control effect under these conditions is related to strengthened jet-induced vorticity that sustains for a longer downstream distance, thereby enhancing the momentum transfer across the boundary layer [9,14].

So far, mostly air-jet vortex generators with circular cross-sections have been implemented to control shock-induced flow separations, presumably due to their relative simplicity and ease of manufacturing. However, investigations on non-circular jets in crossflow (JICF) have identified flow structures that are potentially favourable for flow-control applications. Most notably, improved turbulent mixing was observed for jets from elliptical [18], rectangular [15], and wedge- or diamond-shaped [16,17,19] orifices (compared with circular orifices).

Taking advantage of the desirable topology downstream of jet injection, Rao [20] applied rectangular air-jet vortex generators to suppress shock-induced separation and showed their effectiveness. Furthermore, recent PIV measurements [21] have demonstrated an improved control effectiveness of elliptical AJVGs compared with their circular counterparts. This improvement, however, comes at the cost of increased jet-induced separation and presumably a stronger jet-induced shock, due to the blunt-edged elliptical injectors [18]. This effect can potentially be alleviated by using sharp-edged AJVGs instead. Such devices are expected to induce weaker separation upstream of jet injection and a decreased jet-induced pressure loss [17,19], while maintaining better mixing characteristics than circular jets. Furthermore, the sharp-edges are expected to generate small-scale turbulent structures which can further improve turbulent mixing [20].

Motivated by these findings, we study five arrays of non-circular AJVGs with sharp-edged orifice shapes. We characterise their effect on the topology of the boundary layer and SWBLI, as well as their respective control effectiveness. All control cases are applied to a 24° compression-ramp-induced shock-wave/boundary-layer interaction and are investigated using particle image velocimetry (PIV) and oil-flow visualisations. The configurations include four isosceles triangles with their respective vertices oriented at 0°, 90°, 180° and 270° to the incoming cross-flow direction and a rectangular AJVG with an aspect ratio of $AR = 2$. The results are compared with traditional circular AJVGs.

2 Experimental Methodology

2.1 Flow Facility

The experiments for this investigation were conducted in the in-draft type trisonic wind-tunnel facility located at the Institute of Aerodynamics, RWTH

Aachen University. Details on the wind-tunnel operation is available in Ref. [7,8] For this study, experiments were carried out at Mach 2.52 and $\mathrm{Re}_\infty = 9.62 \times 10^6\,\mathrm{m}^{-1}$. Optical access was provided by three *quartz* windows; one each on either side of the wind-tunnel side walls and one along the top-wall. The experimental conditions are summarised in Table 1, where P_o, T_o and U_∞ are the wind tunnel total pressure, temperature and the freestream velocity, respectively.

Table 1. Experimental conditions

M_∞	U_∞	Re_∞	Po	To
$2.52 \pm 0.6\%$	581.26 m/s	$9.6 \times 10^6\,\mathrm{m}^{-1} \pm 1.5\%$	99.49 kPa ±1.5%	297 K ±1.3%

2.2 Wind Tunnel Model

The wind tunnel model (see Fig. 1) consists of a 0.9 m × 0.4 m flat plate, onto which a 24° ramp is installed ~69.5δ downstream of the flat-plate leading edge (See Ref. [7] for details). A zig-zag trip is placed ~1δ downstream of the leading edge to ensure a uniform turbulent boundary layer of $\delta = 10.4\,\mathrm{mm}$ at 4.5δ upstream of the compression-ramp corner. The maximum length of the inclined ramp was limited to 3.78δ to prevent wind-tunnel blockage.

The AJVGs are installed as modular inlets at 7.69δ upstream of the ramp corner. A pressure plenum underneath the inlet and within the plate ensures a uniform air supply at the required injection pressure (P_{oj}). The pressure and temperature in the plenum are monitored using a WIKA S-20 pressure transducer and a J-type thermocouple. Each array consists of a single row of 23 jet orifices that are spanwise-oriented with a pitch angle of $\phi = -90°$. The spacing between adjacent orifices is D = 8 mm (i.e. 0.77δ), and the jets are injected at a pressure equal to the stagnation pressure of the crossflow, i.e., APR = 1. Previously parametric studies carried out under the same conditions have identified this configuration of AJVGs as most effective [7,8,12].

With all other geometrical and flow parameters remaining constant, we tested five different non-circular AJVGs in addition to a traditional circular AJVG array: an array of rectangular AJVGs with an aspect ratio of 2 and four arrays of triangular AJVGs at four different orientations (0°, 90°, 180° and 270°) with respect to the crossflow direction. A conventional drilling process was used to drill circular orifices with a diameter of $hd_{jet} = 0.1\delta$. The non-circular orifices were manufactured using a laser fine-cutting process at the Fraunhofer Institute of Laser Technology, Aachen. The dimensions of the non-circular orifices were chosen to maintain a hydraulic diameter equal to the diameter of the circular orifice to ensure comparability. The orifice parameters and orientations with respect to the crossflow and jet-injection directions are summarised in Table 2. Note that the schematics represent the cross-sectional geometry.

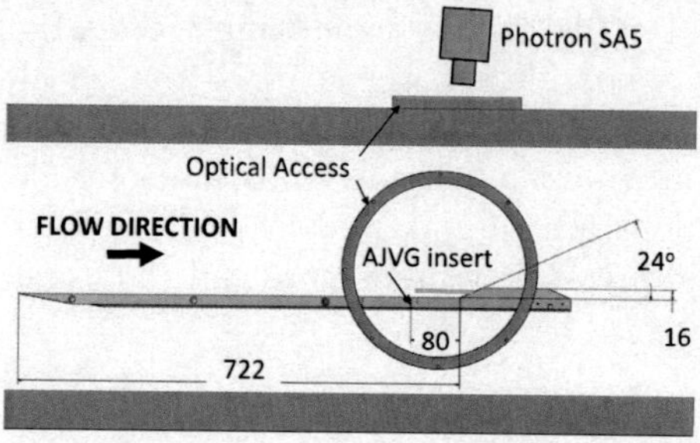

Fig. 1. 24° compression-ramp model with the AJVG insert. All dimentions are in mm

Table 2. Geometric and flow parameters of the orifices. The black down-arrow represents the crossflow and jet directions

Case	Circle	Rectangle	Tri-0	Tri-90	Tri-180	Tri-270
Orifice	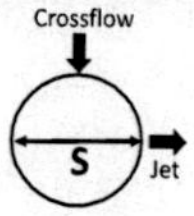	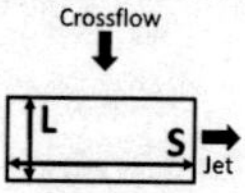	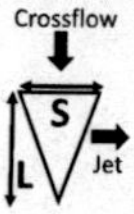	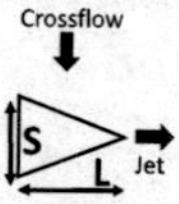		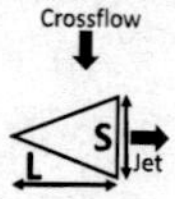
S	1 mm	1.5 mm	1.3 mm	1.3 mm	1.3 mm	1.3 mm
L	-	0.75 mm	2.43 mm	2.43 mm	2.43 mm	2.43 mm

2.3 Data Acquisition and Post-processing

We studied the baseline 24° compression-ramp interaction and the effects of flow control using surface oil-flow visualisations and particle-image velocimetry.

The surface flow features were visualised by tracking the oil-flow pattern under wind-on conditions using a PCO-Pixelfly camera at ~0.2 mm/px. For these visualisations, we applied a thin layer of oil made up of a mixture of TiO_2 particles, oelic acid, and hydraulic oil. For a quantitative assessment of the jet-induced vortex structures and their effects on the compression-ramp interaction, we used a wall-parallel PIV setup. Laser light from a Quantronix Darwin-Duo 100 Nd:YLF system was shaped into a sheet of ~1.5 mm thickness, which illuminated a streamwise/spanwise plane at $y = 0.24\delta$. To seed the flow, DEHS particles were generated and filtered using an in-house generator and cyclone setup. A Photron SA.5 high-speed camera recorded the particle images via the top-wall optical access and covered a field of view (FoV) of $8.8\delta \times 8.8\delta$. The FoV extends between $-4.4\delta \le z = 0 \le 4.4\delta$ and $-8.3\delta \le x = 0 \le 0.5\delta$, with the origin at the ramp-corner at the model centerplane. The system was synchronised and operated at 1000 Hz for 1.5 s using an ILA high-speed synchroniser.

The acquired images were processed with Dantec DynamicStudio 6.4 using an adaptive PIV algorithm. A final interrogation window size of 16 px × 16 px at 50% overlap was used. Furthermore, an universal outlier filter and a constrain on the correlation peak ratio (1.15) was used to replace spurious vectors. Details on the experimental set-up and the post-processing can be found in Ref. [8].

3 Results and Discussion

In this section, we introduce the flow topology of the 24° compression-ramp interaction with and without control and assess the control effectiveness of the studied AJVG-orifice shapes. We measured an undisturbed boundary-layer thickness of $\delta = 10.4\,\text{mm}$, a displacement thickness of $\delta^*{}_c = 2.66\,\text{mm}$, and a momentum thickness of $\theta_c = 0.85\,\text{mm}$. This results in $Re_{\theta_c} = 8225$. A detailed characterisation of the incoming turbulent boundary-layer is available in Refs. [7,8].

3.1 Baseline Shock-Wave/Boundary-Layer Interaction

To analyse the flow across the SWBLI, we present contours of the normalised mean streamwise velocity (U/U_∞), mean spanwise velocity (W/U_∞), and rms of streamwise velocity fluctuations ($\sqrt{u'^2}/U_\infty$) for the baseline 24° compression-ramp interaction along the streamwise/spanwise plane at $y = 0.24\delta$ in Fig. 2. The flow is from left to right and a clean insert with no holes was used to measure the baseline velocity fields. The mean *surrogate* separation line, as identified using the technique of Ganapathisubramani et al. [22], is added as dashed lines.

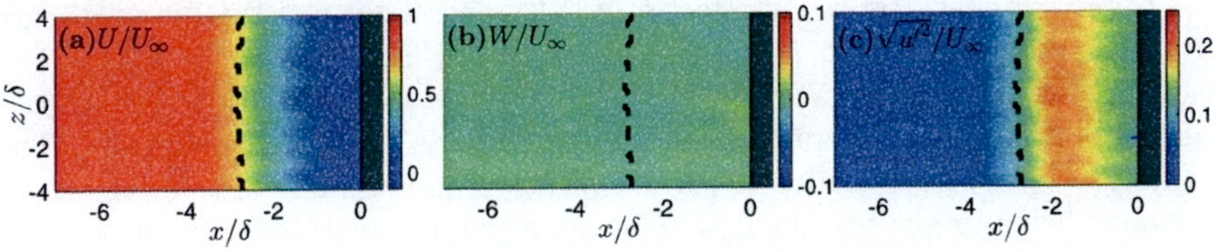

Fig. 2. PIV (a,b) mean and (c) *rms* statistics of the 24° interaction at $y = 0.24\delta$.

The results show a nearly 2D SWBLI (Fig. 2a) with a negligible mean spanwise velocity component (Fig. 2b). The streamwise velocity component strongly decelerates downstream of the *surrogate* separation line due to the adverse pressure gradient imposed by the separation shock. Consequently, the flow separates and a large region of very low velocity is captured by PIV beyond $y = -2\delta$. The streamwise turbulent intensities are also greatly amplified across the interaction, as visible downstream of the *surrogate* separation line in Fig. 2c. For more details on the structure and organisation of this baseline case, see Refs. [7,8].

3.2 Effect of AJVG Control

The effect of jet injection via an AJVG array can be observed in the contours of the mean streamwise velocity component and the streamwise turbulent intensities in Fig. 3. We exemplarily show the influence of the rectangular AJVG array;

the other control cases induce similar modifications of the boundary layer. Similar to Fig. 2, the flow is from left to right, with the AJVG insert located upstream at $x = -7.69\delta$ (outside of the field-of-view in the figure).

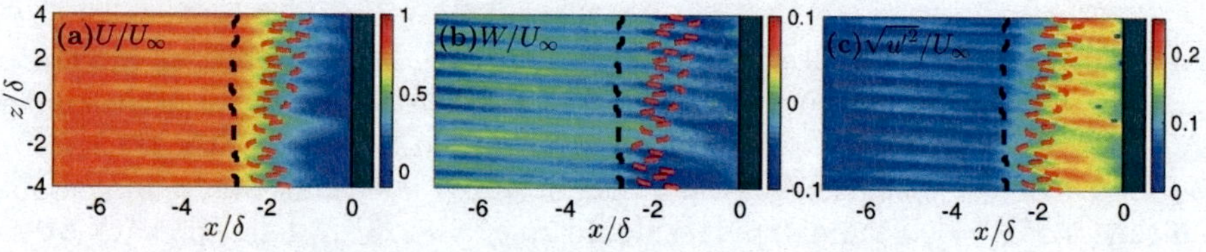

Fig. 3. **Effect of rectangular AJVGs**

The injected jets introduce a system of streamwise vortices [9], which modulate the boundary layer into a spanwise series of high- and low-speed regions (see Fig. 3a) and a periodically altering spanwise velocity component (see Fig. 3b). Consequently, the previously 2D *surrogate* separation line (black dashed lines) is corrugated in the control case (red dashed lines). Furthermore, there is an overall downstream movement of the separation line along the entire spanwise extent of the measurement domain under the influence of control. The jet-induced structures increase the turbulence level in the boundary layer as seen in the high-intensity streaks downstream of jet injection in Fig. 3c.

To further characterise the effect of AJVG control on the SWBLI, we discuss the surface flow features, as visualised in the oil-flow pattern in Fig. 4. The baseline case is shown for comparison. The figure shows the top view of the central ~30% of the interaction region; the flow is from top to bottom. The oil-film movement is sensitive to surface pressure and shear forces, and hence its topology helps to identify the flow separation (S) and reattachment lines (R).

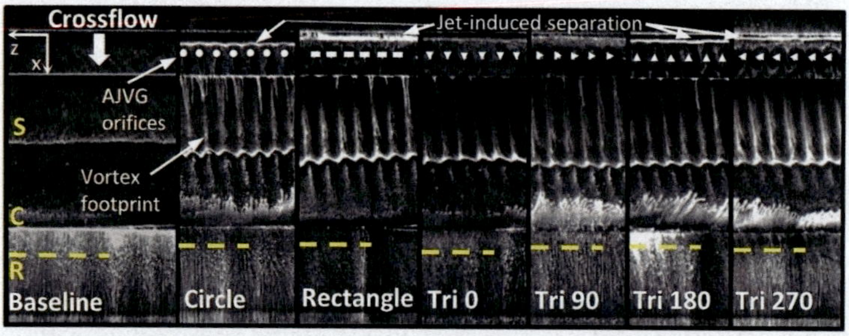

Fig. 4. Surface flow visualisation of the baseline and control cases. S - Separation line; C - Ramp corner; R - Reattachment line.

Oil accumulates upstream of the jet orifices for all control cases due to the jet-induced separation and the associated obstruction to the incoming crossflow. Downstream of the orifices, the streamwise vortices manifest themselves as streaky lines of oil film. The separation line is corrugated; the respective locally higher and lower separation lengths (as measured from the ramp corner (C)) correspond to the spanwise locations of high-speed (absence of oil-film) and low-speed (accumulation of oil-film into a streak) flow. A well-defined horizontal line of oil-accumulation is also observed upstream and downstream of the AJVG orifices; these lines signify the interface between the modular AJVG insert and the flat-plate, and were verified to have minimal impact on the turbulent boundary layer evolution [8].

Downstream of the separation line, the boundary layer first detaches from the surface and then impinges on the ramp surface. As a consequence, an interface forms on the ramp surface, from which the surface flow travels in either direction. This interface indicates the flow-reattachment line. All control configurations achieve a downstream movement of the mean separation line and an upstream movement of the mean reattachment line.

To quantify the control effectiveness, we extracted the mean separation and reattachment lengths from the oil-flow images along one jet pitch in the model mid plane ($-0.5D \leq z = 0 \leq 0.5D$). The ensuing total separation lengths for the control cases ($L_{sep;C}$), normalised with the baseline separation length ($L_{sep;B}$) are presented as black circles in Fig. 5. The extent of corrugation – as indication by the local maximum and minimum separation lengths – is also included as error bars in the figure. To corroborate the observation from the oil-flow images and to characterise the spanwise variations in the separation length behaviour, we also include the normalised spanwise-averaged separation length as estimated from PIV as blue markers and error bars in Fig. 5. The separation length estimates from both techniques show good agreement.

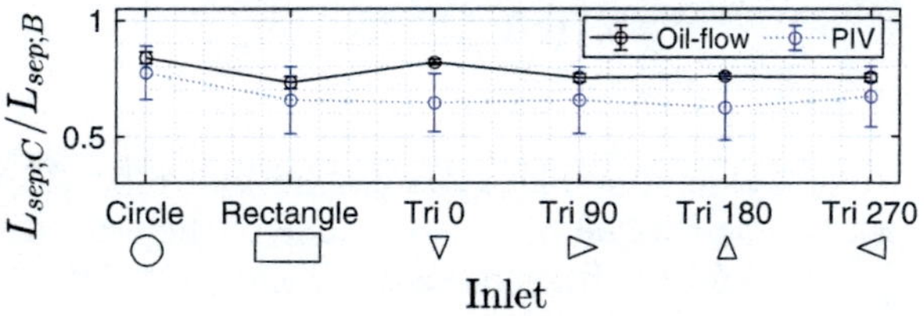

Fig. 5. **Control effectiveness of the AJVG arrays**

All cases exhibit values below one, i.e., all control arrays exhibit a favourable control effect on the SWBLI. The standard circular AJVGs achieve a 17% reduction in normalised separation length compared with the baseline case. The rectangular AJVGs improve the control effectiveness and achieve an overall reduction of 25% in total separation length. All triangular AJVGs, except for the Tri-0 case, show a similar effectiveness to that of the rectangular AJVG array, with a reduction in normalised separation length of approximately 25% over the baseline case. For the Tri-0 case, the upstream separation location is similar (see Fig. 4), and the reduction in reattachment length is weaker. This behaviour translates into a slightly weaker overall control effectiveness of ∼19%. The reason for this behaviour is unclear at this time and is topic of further investigation.

3.3 Characteristics of Jet-Array-Induced Structures

An explanation for the observed favourable control effect of the non-circular AJVGs can be obtained by inspecting the jet-array-induced structures downstream of jet injection. Figure 6(a) shows the spanwise variation of the *rms* of streamwise velocity fluctuations at $x = -5\delta$ upstream of the ramp-corner (∼2.7δ downstream of jet-injection location) extracted from PIV. The uncontrolled case (black dashed lines) and the circular AJVG case (black line with circular markers) are plotted in bold to isolate them from the non-circular cases.

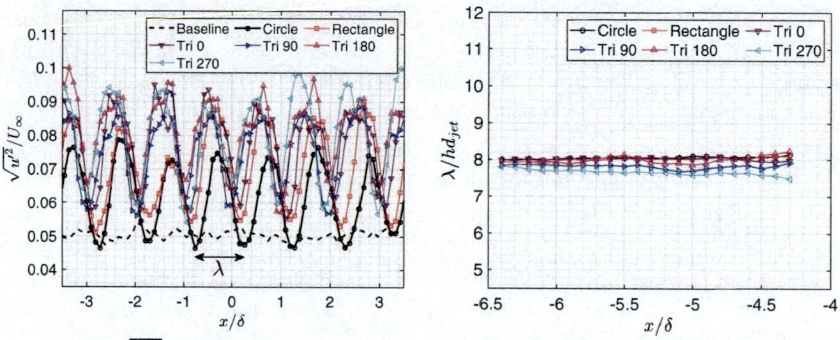

(a) $\sqrt{u'^2}/U_\infty$ at $x = -5\delta$ (b) Wake-width evolution

Fig. 6. Wake characteristics

The profiles show the expected sinusoidal behaviour related to the spanwise series of streamwise vortices. The peak magnitude of the circular AJVGs is, on average, 52% larger than for the baseline case. Hence, the AJVGs improve turbulent mixing in the boundary layer [7,8]. The corresponding peak magnitudes of the other non-circular AJVGs are, on average, ∼80% stronger than for the baseline case (∼22% stronger than that of the circular case). This increase is most likely due to the presence of sharp corners in the orifice shape that induce small-scale turbulence. This small-scale turbulence appears to considerably increase turbulent mixing, as indicated by the increased turbulence level as well as the improved control effectiveness.

The control effectiveness of AJVG arrays cannot only be improved by increasing the addition to improving turbulent mixing effected by each individual jet, but also effectiveness can also be brought about by enabling favourable jet/jet interactions between the jets in an array [8,14]. The extent of such jet/jet interactions can be estimated from the streamwise evolution of the width of the wake (λ) from a single jet in the array. We applied a first order Fourier model onto the sinusoidal *rms* of streamwise turbulent intensity from Fig. 6(a) to extract the wavelength of the best-fit curve from the model (see Ref. [21] for details on the implementation). The streamwise evolution of this wavelength, which represents the width of the jet wake, is shown in Fig. 6(b).

For all tested cases, the wake-width (λ) normalised with the hydraulic diameter (hd_{jet}) equals the spacing between the jets in the array, i.e., 8 mm. This establishes that the width of the wake converges to the jet-to-jet spacing and hence denotes the onset of jet/jet interactions as early as 1.3δ downstream of jet injection. Jet/jet interactions improve the momentum transfer [14]. This increase, in addition to the higher turbulent mixing downstream of jet injection, explains the improved control effectiveness of the non-circular AJVGs.

4 Conclusions

An experimental analysis was carried out to assess the effects of non-circular air-jet vortex generators (AJVGs) applied to control a strongly separated shock-wave/boundary-layer interaction induced by a 24° compression ramp. The incoming freestream Mach number was 2.52 and the momentum-thickness-based Reynolds number was 8225. The control setup included a single row of AJVGs installed 7.69δ upstream of the ramp corner. The jets were spanwise-inclined and pitched at 45° with respect to the flat plate. In addition to classical circular AJVGs, five non-circular AJVGs were tested - one array of rectangular AJVGs with an aspect ratio of 2 and four cases of triangular AJVGs oriented at 0°, 90°, 180° and 270°. Oil-flow visualisations and PIV were used for flow diagnostics.

The qualitative effects of non-circular and circular AJVGs on the shock-induced separation are similar. The 2D separation line observed in the uncontrolled SWBLI is corrugated due to jet injection. For all control cases, the separation line moves downstream, and the reattachment line moves upstream. The circular AJVG reduced the overall separation length by 17%, while the non-circular AJVGs achieved a larger reduction of 25%. Turbulence measurements downstream of the non-circular AJVGs revealed ~22% higher peak intensities than for circular AJVGs. Furthermore, measurements of the width of the jet wakes downstream of the AJVG array indicated favourable jet/jet interactions. These interactions, combined with the increased turbulent mixing, result in the observed increase in control effectiveness of the non-circular AJVGs.

Acknowledgements. This research was funded by the German Research Foundation in the framework of the Emmy Noether Programme (grant SCHR 1566/1-1; project number 326485414).

References

1. Delery, J., Marvin, J.G.: Shock-wave boundary layer interactions. AGARDograph AGARD-AG-280 (1986)

2. Clemens, N.T., Narayanaswamy, V.: Low-frequency unsteadiness of shock wave/turbulent boundary layer interactions. Annu. Rev. Fluid Mech. **46**(1), 469–492 (2014)
3. Wallis, R.A., Stuart, C.M.: On the control of shock-induced boundary-layer separation using discrete air jets. Aeronautical Research Council CP 595 (1958)
4. Souverein, L.J., Debiève, J.F.: Effect of air jet vortex generators on a shock wave boundary layer interaction. Exp. Fluids **49**(5), 1053–1064 (2010)
5. Szwaba, R.: Comparison of the influence of different air-jet vortex generators on the separation region. Aerosp. Sci. Technol. **15**(1), 45–52 (2011)
6. Szwaba, R.: Influence of air-jet vortex generator diameter on separation region. J. Therm. Sci. **22**(4), 294–303 (2013)
7. Ramaswamy, D.P., Schreyer, A.M.: Control of shock-induced separation of a turbulent boundary layer using air-jet vortex generators. AIAA J. **59**(3), 927–939 (2021)
8. Ramaswamy, D.P., Schreyer, A.M.: Effects of jet-to-jet spacing of air-jet vortex generators in shock-induced flow-separation control. Flow Turbul. Combust. **109**(1), 35–64 (2022)
9. Sebastian, R., Lürkens, T., Schreyer, A.M.: Flow field around a spanwise-inclined jet in supersonic crossflow. Aerosp. Sci. Technol. **106**, 106209 (2020)
10. Verma, S.B., Manisankar, C.: Control of compression-ramp-induced interaction with steady microjets. AIAA J. **57**(7), 2892–2904 (2019)
11. Ramaswamy, D.P., Hinke, R., Schreyer, A.-M.: Influence of jet spacing and injection pressure on separation control with air-jet vortex generators. In: Dillmann, A., Heller, G., Krämer, E., Wagner, C., Tropea, C., Jakirlić, S. (eds.) DGLR 2018. NNFMMD, vol. 142, pp. 234–243. Springer, Cham (2020). https://doi.org/10.1007/978-3-030-25253-3_23
12. Sebastian, R., Ramaswamy, D.P., Schreyer, A.M.: Spanwise-inclined jets in supersonic crossflow: effects of injection pressure and separation-control effectiveness. AIAA J. **61**(9), 3833–3847 (2023)
13. Johnston, J.P., Nishi, M.: Vortex generator jets-means for flow separation control. AIAA J. **28**(6), 989–994 (1990)
14. Sebastian, R., Schreyer, A.M.: Influence of jet spacing in spanwise-inclined jet injection in supersonic crossflow. J. Fluid Mech. **946**, A39 (2022)
15. Rizzetta, D.P.: Numerical simulation of slot injection into a turbulent supersonic stream. AIAA J. **30**(10), 2434–2439 (1992)
16. Tomioka, S., Jacobsen, L.S., Schetz, J.A.: Sonic injection from diamond-shaped orifices into a supersonic crossflow. J. Prop. Power **19**(1), 104–114 (2003)
17. Bowersox, R.D., Fan, H., Lee, D.: Sonic injection into a Mach 5.0 freestream through diamond orifices. J. Prop. Power **20**(2), 280–287 (2004)
18. Gruber, M.R., Nejad, A.S., Chen, T.H., Dutton, J.C.: Mixing and penetration studies of sonic jets in a Mach 2 freestream. J. Prop. Power **11**(2), 315–323 (1995)
19. Barber, M.J., Schetz, J.A., Roe, L.A.: Normal, sonic helium injection through a wedge-shaped orifice into supersonic flow. J. Prop. Power **13**(2), 257–263 (1997)
20. Rao, M.K.: An experimental investigation of the use of air jet vortex generators to control shock induced boundary layer separation. Ph.D. thesis. City University London (1988)
21. Ramaswamy, D.P., Schreyer, A.M.: Separation control with elliptical air-jet vortex generators. Exp. Fluids **64**(105) (2023)
22. Ganapathisubramani, B., Clemens, N.T., Dolling, D.S.: Low-frequency dynamics of shock-induced separation in a compression ramp interaction. J. Fluid Mech. **636**, 397–425 (2009)

Discovering Latent Physical Variables from Experimental Data in Supersonic Flow Using Physics-Informed Neural Networks (PINNs)

Lennart Rohlfs(✉) and Julien Weiss

Fachgebiet Aerodynamik, Technische Universität Berlin, Berlin, Germany
l.rohlfs@tu-berlin.de

Abstract. In recent years, advancements in the machine learning and applied mathematics community have led to the development of Physics-Informed Neural Networks (PINNs) which implement a set of governing partial differential equations (PDEs) into the framework of a classical Neural Network. Through these equations it is possible to discover latent physical quantities from an incomplete dataset such as pressure and density fields from velocity measurements. While the potential has been shown in several studies with synthetic data, there are very few publications with experimental data and even fewer for compressible, high-speed applications. In this contribution, a PIV dataset of a turbulent shockwave-boundary layer interaction at Ma = 2 is used as a base to test the PINNs ability to work with experimental data. Multiple cases are constructed to enable comparisons with analytical results and show the influence of different domain sizes and boundary conditions. The results are able do match the theoretical predictions and appear to be quite robust towards measurement noise. This suggests that PINNs can be a powerful tool to extract additional information from a given set of experimental data and future improvements in the computational framework will only increase their versatility.

Keywords: PINN · SBLI · Data Assimilation

1 Introduction

When performing wind-tunnel experiments on any object, it is of high interest to aquire as much information about the flow field as possible. However, due to spatial-, time- and financial constraints, it is usually only possible to perform a limited amount of measurements with a few different methods.

On the other side of the spectrum, numerical simulations can provide all physical quantities simultaneously, but they require carefully selected boundary conditions, schemes and high quality meshes to replicate the physics of interest.

In between pure experimental or numerical approaches, recent advancements in machine learning and especially the development of so called Physics-Informed

A. Dillmann et al. (Eds.): STAB/DGLR Symposium 2022, NNFM 154, pp. 109–118, 2024.
https://doi.org/10.1007/978-3-031-40482-5_11

Neural Networks (PINNs) [8] have opened up new possibilities to process and combine data with computational methods. By implementing the underlying differential equations such as the Euler equations [6] or even the full Navier Stokes equations [5] into the framework of a classical NN, it is possible to, for example, discover pressure and density from velocity fields or even purely qualitative data [9].

While these new methods have been proven to work in many different publications using synthetic or CFD data [1,3], there have been very few attempts to actually use them with experimental data [2] and even fewer in compressible, high-speed environments [4]. For this reason the objective of our contribution is to use PINNs to infer latent variables such as pressure and density from PIV data of supersonic flow features. The original dataset features time averaged velocity data of a turbulent shockwave-boundary layer interaction (SBLI) at Ma = 2 in the high-speed wind tunnel at TU Berlin [10]. From this snapshot multiple subsets of the inviscid flow areas are extracted and used to train the neural network. In this contribution we investigate the networks ability to accurately predict shock- and expansion waves and compare the results with theoretical values. Additionally, we investigate extrapolation capabilities and demonstrate the effect of an arbitrary inviscid wall boundary condition on the predictions. This represents to our best knowledge the first application of PINNs to experimental data in supersonic flow.

2 Setup

In this section the setup and methodology for the presented work is described. First, the experimental setup and the flow configuration is introduced. In the second part, the PINN architecture is introduced together with the imposed boundary conditions and the loss function for the training process.

2.1 Experiments

Wind Tunnel Setup. The dataset for the present investigation was acquired in the high-speed wind tunnel at the Chair of Aerodynamics of *Technische Universität Berlin.* Figure 1 shows a sketch of the indraft low-turbulence wind tunnel and the installed test section with a cross section of 150 mm × 150 mm and a length of 544 mm. A fixed Laval nozzle is used to create a nominal inflow Mach number of 2.0 in the test section. Stagnation conditions are atmospheric and constant throughout a test, with a typical unit Reynolds number of around $Re \approx 12.5 \times 10^6 m^{-1}$.

At the top of the test section a wedge is mounted to deflect the incoming flow which leads to the formation of a shockwave that impinges on the turbulent boundary layer at the bottom wall creating a shockwave-boundary layer interaction. In the present study the deflection angle is set to a fixed $\theta = 10°$ which leads to an incident shock angle of $\beta \approx 40°$ and a separation bubble at the wind tunnel floor. Details on the experimental setup and the characteristics of the SBLI can be found in [10].

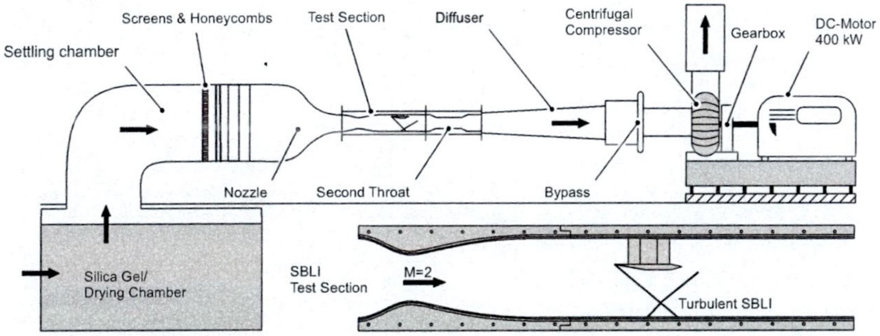

Fig. 1. Schematic diagram of the high-speed wind tunnel and the test section with a shock generator in place.

PIV Setup. Planar two-component particle image velocimetry (PIV) was performed in the x-y plane along the centerline of the test section. A dual pulsed Nd:YAG laser (Quantel EverGreen, 532 nm, 120 mJ at 15 Hz repetition rate) was used to create a light sheet through an insert in the bottom wall and a high resolution sCMOS camera (pco.edge, $5.5MPx$) was mounted in front of a window in the side wall panel. Tracer particles (DEHS) were inserted into the drying chamber far upstream of the test section to create a homogeneous seeding distribution. 500 images were acquired with a sample rate 10 Hz and processed using PIVTEC PIVview 3.9 with a multiple grid algorithm resulting in a single, time-averaged velocity field.

2.2 Physics-Informed Neural Network

Physics-informed neural networks are deep neural networks capable of mapping spatial and temporal coordinates to flow field variables such as velocity, pressure and density while being constrained by a set of partial differential equations (PDEs) and boundary conditions. In our case we have 2D velocity data ($u(x, y)$ and $v(x, y)$) and want to infer pressure ($p(x, y)$) and density ($\rho(x, y)$) fields while obeying the stationary, conservative Euler equations. A schematic illustration of the PINN used in this work is shown in Fig. 2. The left side is a fully-connected neural network (FNN) with two inputs (x, y) and four outputs (u, v, p, ρ) and L hidden layers with N neurons per layer in between. The number of hidden layers, neurons, as well as the activation function for each neuron varies greatly over the available publications with no consensus on the ideal combination. For the present investigation the setup detailed in Jagtap et al. [4] was chosen, since their work also features supersonic flow and uses the Euler equations. The resulting FNN has 8 hidden layers with 50 neurons each and uses a tanh actvation function. The computational framework is Tensorflow 1.15 in a Python 3.7 environment.

The right part of Fig. 2 shows the physics-informed part of the network architecture. By implementing a secondary structure which takes the outputs of the FNN, computes their derivatives via automatic differentiation (AD) and finally

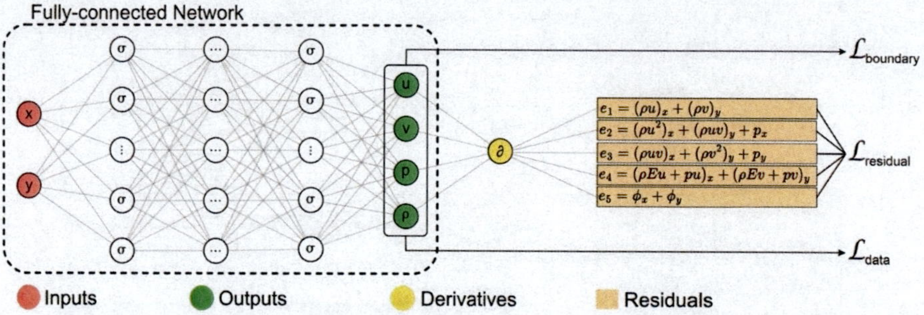

Fig. 2. Schematic diagram of the physics-informed neural network architecture. The fully-connected network estimates velocity, pressure and density fields based on the 2D coordinates, while the implemented equations in the residual network represent the physical constraints for the solutions. The loss term is composed of a data term for the experimental data, the residuals and a set of boundary conditions.

evaluates a set of differential equations, the underlying physics can be taken into account by the PINN. The equations for this work are the 2D, stationary, compressible Euler equations in their conservative form which are defined as:

$$\frac{\partial(\rho u_i)}{\partial x_i} = 0 \tag{1}$$

$$\frac{\partial(\rho u_i u_j)}{\partial x_j} + \frac{\partial p}{\partial x_i} = 0 \tag{2}$$

$$\frac{\partial(u_i \rho E)}{\partial x_i} + \frac{\partial(u_i p)}{\partial x_i} = 0 \tag{3}$$

Each of these equations creates a residual that can be minimized during training. To close the problem, a polytropic equation of state is used:

$$\rho E = \frac{p}{(\kappa - 1)} + \frac{\rho U^2}{2} \tag{4}$$

Additionally using an entropy-flux pair has been shown to increase the stability and accuracy of the training process [7] and is thus implemented as another conservation law in the form of an inequality:

$$\phi_x + \phi_y \leq 0 \text{ with } \phi = \frac{-\rho U s}{\kappa - 1} \text{ and } s = log(\frac{p}{\rho^\kappa}) \tag{5}$$

All residual terms are integrated over the domain and the mean squared error (MSE) is calculated to yield the residual loss term $L_{residual}$.:

$$L_{residual} = \frac{1}{N} \sum_{n=1}^{N} |e_1^n + e_2^n + e_3^n + e_4^n + e_5^n|^2 \tag{6}$$

This loss term is supplemented by a data term and a boundary condition term. The data loss L_{data} represents the error between the predicted velocity fields and the experimental data from the PIV measurements in a given area. It can be written as:

$$L_{data} = \frac{1}{N}\sum_{n=1}^{N}|(u(x^n, y^n) - u^n_{data}) + (v(x^n, y^n) - v^n_{data})|^2 \tag{7}$$

It is worth noting that the number of points at which the PDEs are evaluated can be much higher than the number of points used for computing the data loss. Finally, the freestream static pressure and density at the inlet of the domain are enforced to provide additional boundary conditions to select the correct solution. The corresponding loss $L_{boundary}$ is defined as:

$$L_{boundary} = \frac{1}{N}\sum_{n=1}^{N}|(p(x^n, y^n) - p^n_{boundary}) + (\rho(x^n, y^n) - \rho^n_{boundary})|^2 \tag{8}$$

The static pressure can be easily captured with a pressure tap in the wind tunnel wall. By additionally measuring the total pressure and temperature of the air in the settling chamber and assuming isentropic flow through the nozzle, the freestream density can also be calculated.

These three loss terms are summed up and each weighed by a factor ω_i, that can either be set manually or automatically adjusted during the training process as detailed in [11]). Consequently the total loss is defined as:

$$L_{Total} = \omega_1 L_{residual} + \omega_2 L_{data} + \omega_3 L_{boundary} \tag{9}$$

During the training process, this total loss is minimized by the Adam optimizer, which seeks the optimal parameter combination for the networks weights and biases through gradient descent. If the training converges, the result is a neural network that on the one hand obeys all boundary and data conditions, but also follows the underlying PDEs and thus the flow physics. For the present investigation an exponential decay learning rate schedule was used with a starting learning rate of 0.001 and a decay rate of 0.9.

3 Results

In this section the results of three different cases with increasing complexity are presented. Each case is based on a small part of the SBLI dataset that is shown in Fig. 3 with the streamwise velocity component u on the left and the wall-normal velocity v on the right side. At the bottom, the separation bubble that forms because of the strong adverse pressure gradient of the incident shock can be seen. This region is dominated by viscous effects and cannot be described by the Euler equations. Above the interaction region, and especially above the intersection between the incident and reflected shock, the assumption of inviscid flow is valid.

The boxes drawn in the figure show the regions for the three prediction cases. First, the oblique incident shockwave is isolated to verify the predictions with

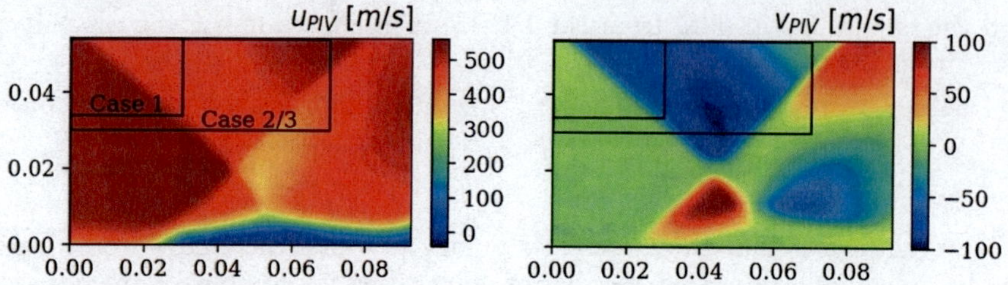

Fig. 3. Full PIV dataset with streamwise (left) and wall-normal (right) velocity components. The subsets for the prediction cases are marked by their bounding boxes.

the shock theory and the experimental dataset. Second, a larger area including a shock-expansion interaction is considered to demonstrate a case that cannot be solved analytically. Finally, the oblique shockwave training set of case 1 is combined with the increased prediction area of case 2 and an arbitrary inviscid wall boundary condition to create an extrapolation case that can be compared with theoretical results. For the entire flow field a total of 25725 data points are available with a streamwise spatial resolution of ≈0.88 mm. In the area around the shock this resolution is considerably lower than the locally refined point distributions in the synthetic examples of [6] and [4], but for an experimental dataset such a refinement cannot be achieved.

3.1 Case 1 - Oblique Shockwave

For the first case of the single oblique shock there are two things to be verified. The first is to simply check if the network can represent the experimental data used during the training while being constrained by the other loss terms. If this is not the case, the network could be too small, or the weight of the training data may be too small. Figure 4 with the prediction data on the right and the experimental data on the left confirms that the chosen parameter combination is good. The relative error between the data fields is ≈3% for the streamwise and ≈1.6% for the wall-normal velocity component. The main difference is that the predicted velocity fields look a lot smoother than the PIV fields due to the regulating PDEs in the background. This has been also observed by Cai et al. [2] who overlayed their experimental BOS data with random noise and found the PINN algorithms to be fairly insensitive to noisy data.

The second step is to validate the predicted latent variables that are discovered through the implemented equations. This is a bit more complicated without having a reference CFD available, but for the single shock case the pressure, density and Mach number ratios over the shock can be compared with the analytic solution. In the shock theory the ratios are only a function of the freestream Mach number and the deflection angle of the flow. The calculated values for the presented case are listed in the top row of Table 1. For the predicted case, the ratios at each wall-normal position are calculated and then integrated over the height of the domain. The resulting averages match up very well against the theoretical values with relative errors of –0.2% to –2.5% which indicates that the PINNs are able to correctly capture the discontinuous flow fields associated with supersonic flow even without the local refinement of the training points around the shock that has been mentioned earlier.

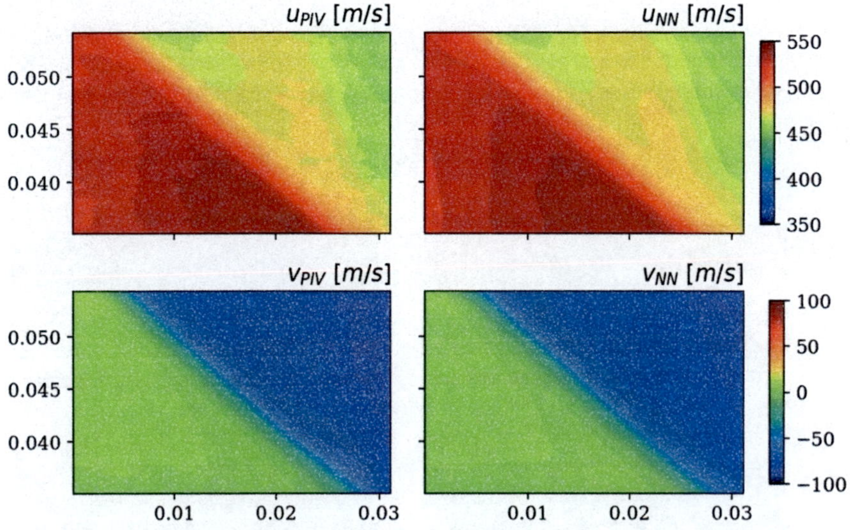

Fig. 4. Comparison of the predicted velocity fields (right) with the experimental data (left) for the single oblique shock case. Top row: Streamwise velocity, bottom row: Wall-normal velocity.

Table 1. Comparison of predicted pressure, density and Ma ratios with shock theory

	p_2/p_1	ρ_2/ρ_1	Ma_1/Ma_2
Theory	1.77	1.49	1.25
Prediction	1.71	1.46	1.22
Rel. Error	−2.5%	−0.2%	−1.5%

3.2 Case 2 - Shock-Expansion Interaction

For the second case the considered area is increased to include the expansion fans originating from the shock generator and the top of the separation bubble as well as the reflected shock that interacts with both expansions fans. This creates a fairly complex flow configuration that cannot be described analytically. Figure 5 shows the outputs of the trained PINN as well as the temperature and Mach number fields that are calculated from the network outputs. First, it can be seen that the prediction in the top left area is visually identical to the results of the first case (rel. error below 3%). This confirms that the PINNs are mostly insensitive to the investigation area, if the network is sufficiently deep. Second, the effect of the expansion fans on the latent variables is predicted as one would expect. With increasing velocity the pressure and density decrease before increasing again over the reflected shock. The highest values are located in the bottom right corner where the velocity is the smallest and the effect of the expansion fan is the weakest. Third, the derived temperature and Mach number fields also meet the expectations from isentropic flow calculations with a freestream temperature of $T \approx 164K$ and Mach number $Ma = 1.96$. Unfortunately, at this stage of the investigation, the experimental database does not allow for an independent quantitative validation of these results.

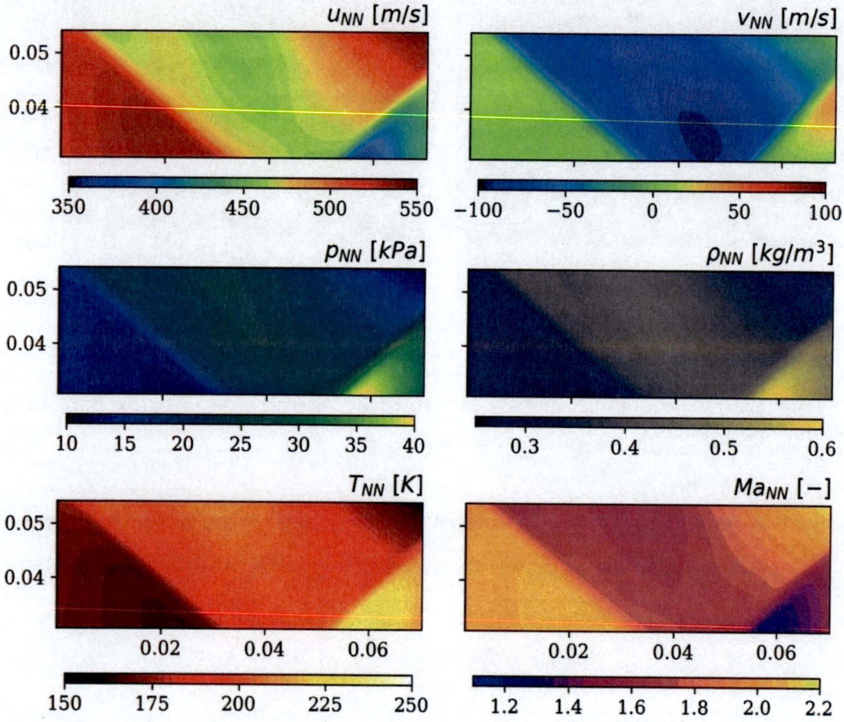

Fig. 5. Predictions and derived quantities for the shock-expansion case. The four fields at the top (velocities, pressure, density) were directly predicted by the PINN while the temperature and Mach number at the bottom were derived later.

3.3 Case 3 - Inviscid Wall Reflection

Up to this point the size of the predicted area was always the same as the available datafield. For the third case the training dataset is reduced to the area of case 1 while keeping the prediction area at the size of case 2. This creates an extrapolation case where a majority of the evaluated coordinates have no validation datapoint nearby and only the residual loss term is responsible for the solution. In addition, an arbitrary inviscid wall condition is added to the bottom of the domain by enforcing zero massflux through the patch as an additional boundary condition. In our case this is equal to setting the wall-normal velocity to zero. The expected result is an inviscid shock reflection where the reflected shock angle β_2 is larger than the incident shock angle β_1 and the flow is parallel to the wall before and after the reflection.

Figure 6 shows the predicted velocity fields and the expectations are confirmed. The incident shock is correctly extrapolated until it hits the wall and is reflected. While the reflected shock is not perfectly resolved as a sharp discontinuity, it is clearly recognizable. By increasing the resolution of the residual data field this could be improved, but for the present investigation it was kept constant for all cases. The angle of the reflected shock is $\approx 47.2°$ (solid line) which is in good agreement with the angle derived from invicid theory (49.3°, dashed line). The wall-normal velocity behind the reflection is again close to zero, which means the flow is parallel to the wall. This result confirms that the integrated equations in the PINN are solved with sufficient accuracy to find a good solution outside the areas where training data is available.

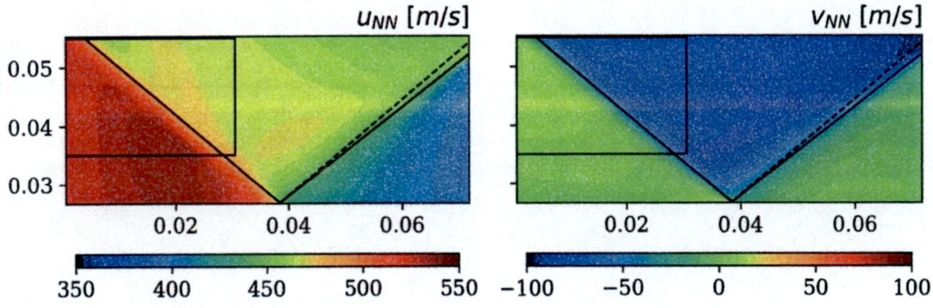

Fig. 6. Predicted velocity fields for the inviscid wall reflection case. The square marks the area where the velocity was provided during the training. Solid lines: Predicted shock angle, dashed line: Theoretical angle

4 Conclusion

A set of supersonic PIV data with a complex shockwave-boundary layer interaction has been analyzed with Physics-Informed Neural Networks with the goal to extract pressure and density fields from experimental velocity data. While this has been shown to work well with synthetic data, there have been no attempts, to our best knowledge, to use PINNs with experimental data in a high-speed flow environment. The results suggest that the noise and inaccuracies present in the experimental data do not limit the networks ability to accurately predict the flow fields and the latent variables.

Three subsets of the initial dataset have been considered. First, the incident shockwave of the SBLI was isolated to enable analytical comparisons and the extracted pressure, density and Mach number ratios match the theory very well. Second, a larger area with a complex shock-expansion interaction was considered. The predicted ratios over the incident shock again fit the theory very well and the relationship between velocity, pressure and density is correctly predicted. Second order variables such as temperature and Mach number also show expected trends and values. Finally, an arbitrary inviscid wall condition was implemented to simulate an inviscid shock reflection. The predicted flow field correctly resolves the reflected shock with the angle of the reflection matching the theory and highlighting the excellent extrapolation capability of the PINNs.

The results suggest that it is feasible to use experimental data together with PINNs and obtain additional information about a flow field without performing additional measurements. In future studies the implementation of viscous terms into the equations and improvements in the computational framework will further increase the flexibility and enable new investigations.

References

1. Cai, S., Mao, Z., Wang, Z., Yin, M., Karniadakis, G.E.: Physics-informed neural networks (PINNs) for fluid mechanics: a review. Acta Mech. Sin. **37**, 1–12 (2022)
2. Cai, S., Wang, Z., Fuest, F., Jeon, Y.J., Gray, C., Karniadakis, G.E.: Flow over an espresso cup: inferring 3-D velocity and pressure fields from tomographic background oriented Schlieren via physics-informed neural networks. J. Fluid Mech. **915**, 1–17 (2021)
3. Cuomo, S., Di Cola, V.S., Giampaolo, F., Rozza, G., Raissi, M., Piccialli, F.: Scientific machine learning through physics-informed neural networks: where we are and what's next. J. Sci. Comput. **92**(3), 88 (2022)
4. Jagtap, A.D., Mao, Z., Adams, N., Karniadakis, G.E.: Physics-informed neural networks for inverse problems in supersonic flows. J. Comput. Phys. **466**, 111402 (2022)
5. Jin, X., Cai, S., Li, H., Karniadakis, G.E.: NSFnets (Navier-Stokes flow nets): physics-informed neural networks for the incompressible Navier-Stokes equations. J. Comput. Phys. **426**, 109951 (2021)
6. Mao, Z., Jagtap, A.D., Karniadakis, G.E.: Physics-informed neural networks for high-speed flows. Comput. Methods Appl. Mech. Eng. **360**(April), 112789 (2020)
7. Patel, R.G., et al.: Thermodynamically consistent physics-informed neural networks for hyperbolic systems. J. Comput. Phys. **449**, 110754 (2022)
8. Raissi, M., Perdikaris, P., Karniadakis, G.E.: Physics-informed neural networks: a deep learning framework for solving forward and inverse problems involving nonlinear partial differential equations. J. Comput. Phys. **378**(October), 686–707 (2019)
9. Raissi, M., Yazdani, A., Karniadakis, G.E.: Hidden fluid mechanics: learning velocity and pressure fields from flow visualizations. Science **367**(6481), 1026–1030 (2020)
10. Rohlfs, L., Stab, I., Weiss, J.: Experimental investigations of incident shockwave boundary layer interactions in a continuously operating supersonic wind tunnel. In: AIAA AVIATION 2022 Forum. American Institute of Aeronautics and Astronautics, Reston, Virginia, June 2022
11. Wang, S., Teng, Y., Perdikaris, P.: Understanding and mitigating gradient flow pathologies in physics-informed neural networks. SIAM J. Sci. Comput. **43**(5), 3055–3081 (2021)

Surrogate Based Prediction of 2D Car Wake

Mathis Thieme([✉]), Stephan Löffler, and Julien Weiss

Fachgebiet Aerodynamik, Technische Universität Berlin, Berlin, Germany
mathis.thieme@tu-berlin.de

Abstract. In this study, different surrogate models for flow field prediction are applied to a size-limited experimental dataset of a 2D car wake. The dataset consists of DrivAer model wakes influenced by different rear wing positions. The objective is to identify the most robust and accurate method for small training datasets in order to exploit the time-reducing benefits of surrogate models in the experimental field. Three methods (random forest regression, Gaussian process regression and feedforward neural network (FNN) with different activation functions) trained with different dataset sizes are compared. It is shown that Gaussian process regression is the most accurate but most time-consuming method. Interestingly, the random forest regression predicts the flow fields as good as the FNN, but its training time is two orders of magnitude faster. In terms of FNN, rectified linear unit 6 (ReLU6) stood out from the other activation functions as it achieved the best wake predictions with only five training samples.

1 Introduction

The investigation and visualization of a flow field dependent on different parameters has always been of interest in aerodynamics for a better understanding of flow phenomena. For example, Lohse et al. [15] reconstructed a three-dimensional boundary layer with a single traversed three-hole pressure probe to observe the influence of active flow control on cross-flow instabilities. This is a common method in experimental aerodynamics, where a probe is mounted on a traversable system to measure discrete points on a plane or a volume. The flow field is reconstructed by interpolation between the measured points. Traversing a single probe is inexpensive but, depending on the spatial resolution or size of the field, can be time-consuming. Using a probe rake to survey a plane or volume can save time but increases measurement setup effort and costs. In the field of active flow control, Ciobaca et al. [3] used a traversing five-hole pressure probe rake consisting of eighteen equidistantly spaced probes to visualize the time-averaged impact of different active flow control setups on the wake flow of an outer-wing model. A potential drawback of this technique is that probes inserted into the flow may negatively affect the flow field, as several researchers have shown (e.g. [1] and [12]). Another experimental method to obtain information on a flow field is particle image velocimetry (PIV). Since PIV is an optical technique, it is mostly non-intrusive. Small tracer particles are added to the flow,

A. Dillmann et al. (Eds.): STAB/DGLR Symposium 2022, NNFM 154, pp. 119–129, 2024.
https://doi.org/10.1007/978-3-031-40482-5_12

which are illuminated in a plane of interest twice within a short time interval [23]. Using this technique, Steinfurth [25] acquired sequentially multiple planar three-component velocity fields to reconstruct phase-averaged 3D vortex structures for several active flow control setups. This is a time-consuming process and could technically only be accelerated by an expensive tomographic PIV system.

In computational fluid dynamics (CFD), physics-based models are used to compute flow fields. Depending on the complexity of the physical model and the computational grid, single simulations can take several hours or days. Again, a parameter study that includes a flow field is a time-intensive process. Thus, there is a strong focus on drastically reducing the simulation time to so-called real-time simulations. While various methods may be used, most approaches have one commonality. They first require a high-fidelity dataset to be constructed, because they are data-driven. One approach is to apply proper orthogonal decomposition (POD) to the dataset and thereby create a reduced order model (ROM). The flow field is then reconstructed using only a few data points obtained outside of the training set. Ebert et al. [6] applied the GappyPOD method to reconstruct the flow field around a building, while Patrikar et al. [19] and Xiao et al. [28] utilized Gaussian process regression (GPR) and Yu et al. [29] trained a multilayer feedforward neural network (FNN) for the reconstruction.

Another approach is to train one or more surrogate models (SM) that predict the flow field based only on the quantification of the boundary conditions. Guo et al. [9] and Ribeiro et al. [4] transferred their computational domain into a signed distance function and predicted the flow field with a convolutional neural network. White et al. [27] compared the predictive performance of a feedforward neural network (FNN) with a cluster network with a tiny dataset. The size of the dataset is a crucial constraint for the quality of the prediction. Especially in experimental aerodynamics it is often unfeasible to generate datasets with more than 1000 samples just for training, as Guo et al. [9] and Ribeiro et al. [4] did. The amount of data is either untractable in terms of time or the experimental setup does not have enough influencing parameters. Consequently, it is essential for experimental aerodynamics to utilize robust methods which predict the flow field accurately with limited amounts of training data.

In this contribution, different SMs for flow field prediction are applied to a size-limited experimental dataset. The objective is to identify the most robust and accurate method for small training datasets in order to exploit the time-reducing benefits in the experimental field. Here, two-dimensional wakes of a DrivAer car model [11] were acquired using an absolute pressure probe rake. The wake was influenced by the traversable rear wing (spoiler) of the car. Different examples of FNN architectures, GPR and random forest regression (RFR) are trained with several training dataset sizes and compared in terms of their predictive quality.

2 Methodology

The following section explains the methodology used in this work. First, the experimental setup to acquire the dataset is described. Then, the three methods

(FNN, GPR, and RFR) applied for the prediction of the wake are introduced. For GPR and RFR the Python library Scikit-learn [20] was utilized, while for the FNN the Python library PyTorch [18] was chosen.

2.1 Experimental Setup

The experiments were conducted on a 15% scaled DrivAer model in a low-speed blow-down open-circuit wind tunnel at a freestream velocity of 13 m/s. For the DrivAer model, the fastback configuration was chosen for the rear end and the smooth configuration for the underbody [11]. The closed test section is 1300 mm mm in height and 1000 mm mm in width. An additional splitter plate ensures that a new bottom boundary layer is formed in front of the model. Figure 1 shows the experimental setup. 34 pressure taps are located in the center section of the DrivAer to obtain the static pressure distribution on the model surface. The DrivAer model is equipped with a traversable rear wing, which is adjustable horizontally (x-position) and vertically (z-position), and with its angle of attack α. Here α can be varied from $-10°$ to $30°$ and the rear wing by a total of 16 mm in both the x- and y-directions. Three stepper motors integrated in the car model are used to move the rear wing. They enable a spatial resolution of 0.005 mm/step. The wake influenced by the rear wing is obtained by an absolute pressure probe rake, which is located 130 mm behind the model. A total of eleven equidistantly spaced total pressure probes are used. They are aligned along the y-direction with a spacing of 20 mm. By using an automatic system, the wake rake was traversed in the z-direction between 24 mm and 264 mm from the ground with a step size of 5 mm. One reconstructed wake thus consists of 539 measuring points or 11 sensor positions of the rake with 49 different heights. The wake was only obtained from one half of the car, based on the assumption that the wake of a symmetrical car model is also symmetrical.

A total of 70 different wakes that correspond to different rear wing positions were acquired and form the full dataset. Here, the Halton sequence [21] was chosen as sampling strategy to fill the parameter space efficiently. The dataset was subsequently divided into a training dataset (first 30 samples) and an evaluation

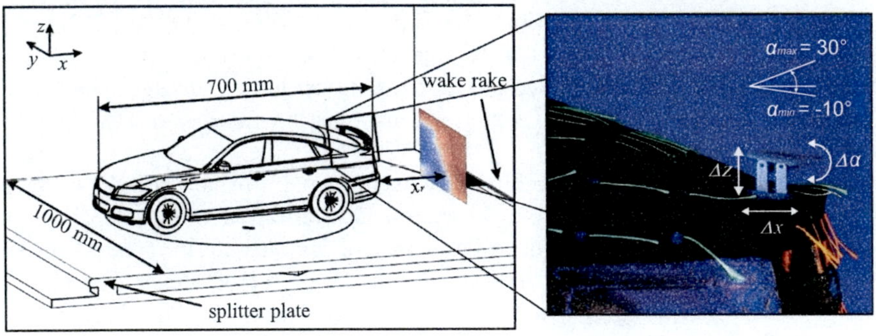

Fig. 1. Experimental setup of the DrivAer model with a traversable rear wing.

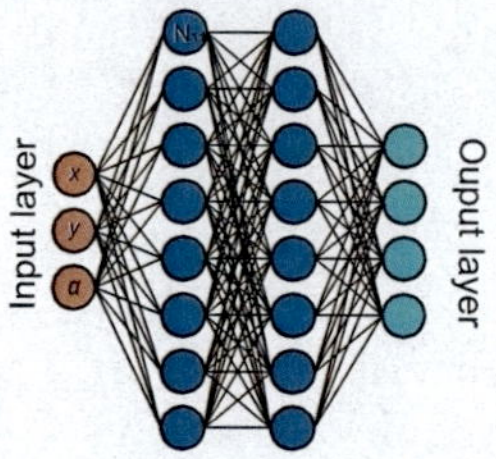

Fig. 2. Symbolic representation of a fully-connected FNN.

Table 1. Investigated non-linear functions and the corresponding plot.

Plot	Function
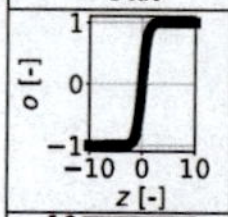	$\text{Tanh} = \frac{e^{z} - e^{-z}}{e^{z} + e^{-z}}$
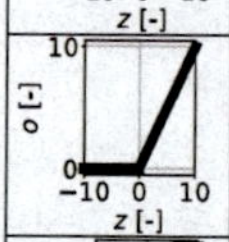	ReLU = max(0, z)
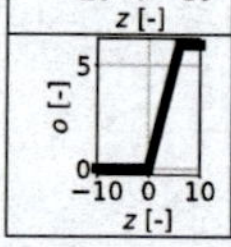	ReLU6 = min(max(0,z),6)

dataset (remaining 40 samples) in order to implement and evaluate the different surrogate models.

2.2 Multilayer Feedforward Neural Network

FNN can be applied as an inherently multi-output SM. In Fig. 2, the common symbolic representation of FNN is shown to explain the approach. An FNN consists of several layers and each layer is composed of multiple neurons. In this contribution, only fully-connected FNN are considered, thus each neuron of one layer is connected to all neurons of the following layer. Each connection represents a weight w_{ln}. At the beginning stands the input layer, where the input parameters are specified. In this work, the input parameters are the adjustable rear wing parameters (x, y and α), which means that our input layer consists of a three-dimensional array. Accordingly, the amount of output layer neurons in this work is determined by the 539 measurement points of the 2D wake which are to be predicted. The values of the neurons of the first hidden layers result from the bias b and the sum of the weighted input parameters [14]. Then a non-linear activation function (AF) is applied to get the output of the neurons. The FNN is trained by modifying the weights using backpropagation [26] to minimize a predefined error or loss function. According to Guo et al. [9] and Ribeiro et al. [4], the mean squared error (MSE) is a valid loss function to assess the flow prediction quality. For the choice of the training optimization algorithm, small preliminary tests were made in which stochastic gradient descent (SGD) with momentum performed best.

Since choosing the FNN architecture is not an intuitive process, different FNNs were trained and evaluated on the evaluation dataset. The depth of the network and the amount of neurons was varied. Hyperbolic tangent (Tanh), a typical sigmoid AF [14], the rectified linear unit (ReLU) [10], which is commonly used nowadays ([17] and [8]), and ReLU6 [13] were examined. The three investigated non-linear functions are listed with their respective plots in Table 1. The training of the FNN was influenced in two ways. On one hand by modifying the learning rate (lr) and the momentum of the SGD optimization and on the other hand by varying the number of training samples before the weights are updated (batch size) and the amount of times the training dataset is used for training (epochs). All the parameters that were varied are listed in Table 2.

Table 2. Parameters of the FNN architecture variation.

hidden layer	neurons per layer	AF	lr	momentum	batch size	epochs
0–6	300, 500, 1000,	Tanh, ReLU,	0.1, 0.5	0.3, 0.5, 0.7	1, 2, 4,	10, 30,
	2000, 3000 or 5000	ReLU6			5	50, 70

2.3 Gaussian Process Regression

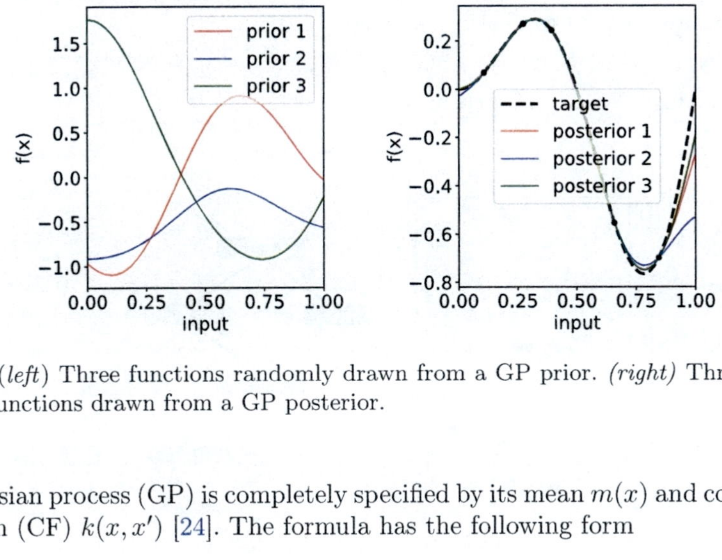

Fig. 3. (*left*) Three functions randomly drawn from a GP prior. (*right*) Three conditioned functions drawn from a GP posterior.

A Gaussian process (GP) is completely specified by its mean $m(x)$ and covariance function (CF) $k(x, x')$ [24]. The formula has the following form

$$f(x) \sim \mathcal{GP}(m(z), k(x, x')). \tag{1}$$

$$k_{\mathrm{RBF}}(x, x') = \exp\Big(-\frac{(x - x')^2}{2l^2}\Big) \tag{2}$$

The CF determines how the model generalizes to new data [5]. In this work, the radial basis function (Eq. 2) was chosen as CF, since it achieved the best results in preliminary tests with the given dataset. This CF is infinitely differentiable, which indicates that GPs with this CF have mean square derivatives of all orders and are consequently very smooth [24]. During training, the hyperparameters l of the covariance function are optimized by numerically maximizing the log-marginal-likelihood (LML) [16]. To optimize the LML, the L-BFGS-B [30] algorithm is used. Figure 3 shows a one-dimensional example of GPR. On the left of Fig. 3 Gaussian priors are generated with

$$y_* \sim \mathcal{N}(0, k(x_*, x_*)) \ [7]. \tag{3}$$

x_* is a vector with any number of test inputs. The prior has to be conditioned on the training data x and the corresponding output y. The predictive distribution of the output y_* containing a noise term with the variance σ_n^2 for x_* has the form [7]:

$$\begin{aligned} y_* \mid x_*, x, y \sim \mathcal{N}(&k(x_*, x)[k(x,x) + \sigma_n^2 I]^{-1} y, \\ &k(x_*, x_*) - k(x_*, x)[k(x,x) + \sigma_n^2 I]^{-1} k(x, x_*)) \end{aligned} \tag{4}$$

Three conditioned posteriors are shown on the right side of Fig. 3. In this work, a separate GPR is trained for each wake measurement point, which means that 539 GPRs are trained to obtain a surrogate wake.

2.4 Random Forest Regression

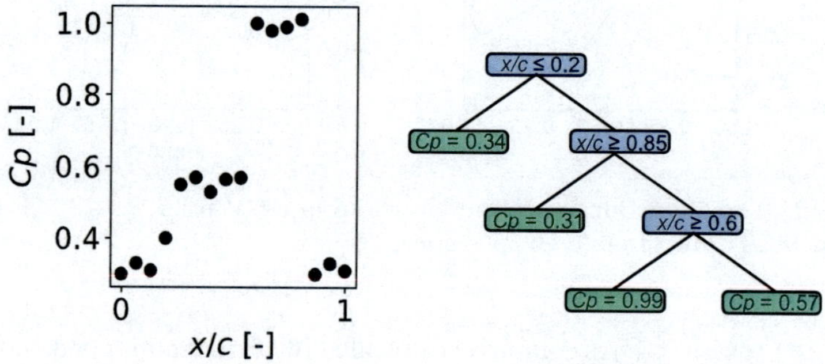

Fig. 4. (*left*) Exemplary graph of the pressure coefficient over a defined length at discrete points. (*right*) The corresponding DT to the left graph.

RFR consists of multiple randomly generated decision trees (DT) [2]. The given dataset is partitioned by the DT into smaller and smaller subsets [22]. Figure 4 shows a decision tree for a regression as example. The accuracy of the RFR can be increased by the amount of DT since the most numerous DT output is determined as the RFR output. The RFR will not be addressed in more detail here. As further information is presented in [22] and [2]. RFR was chosen as the benchmark model because the methodology is simple and represents an inherently multi-output regression.

3 Results

Firstly, the SM prediction quality of the DrivAer wake is compared. Subsequently, the best FNN architecture depending on the AF and the training dataset size will be described. Thus, a large parameter variation of the FNN architecture was performed (see Table 2).

3.1 Wake Prediction

Each SM was trained with different sample sizes to find not only the most accurate SM for car wake prediction, but also the most robust. A maximum dataset size of 30 different wakes was available for training. Therefore, the SMs were trained with dataset sizes of 5, 10, 15, 20, and 30 samples, respectively. The prediction quality of the trained SMs was tested using the evaluation dataset (in total 40 samples) by calculating for each rear wing position of the evaluation dataset the MSE of the SM prediction to the measured wake and averaging over the 40 evaluation samples. Figure 5 shows the averaged MSE of each method trained with different sample sizes. Presented are the results of the best performing FNNs of the respective training dataset size to show which performance could be maximally achieved within the FNN parameter variation. Accordingly, the FNN architectures are not uniform. Regardless of the training sample size, the GPR performs best. Among the three AFs, overall the FNN with ReLU6 generates the best wake predictions and is performing as the second-best SM. The FNNs with ReLU produce equally good results as the FNN with ReLU6 and a training sample size between 10–30, but the MSE is almost three times higher with only five training samples. It is noticeable that the best performing FNN architectures with the AF Tanh have worse prediction qualities than those with the ReLU and RelU6 AF. It is only with 5 training samples that the FNN with Tanh achieves a lower error than with ReLU. Interestingly, the RFR generates equivalent predictions as the FNN architectures with ReLU and ReLU6 above 10 training samples. Only for five training samples is the averaged MSE significantly higher than the best FNN with ReLU6.

The required training time of all surrogate models is listed in Table 3. What stands out here is the training time of the RFR. Even with 30 training samples, the training took less than a thirtieth of a second. In contrast, the training of the GPR always lasted almost a minute. The high training time of the GPR is especially important when new measurements are added to the training dataset during experiments. Since the GPR (as well as the RFR) must be completely retrained with each new sample. The FNN have moderate training times, but they are best suited for a continuous expansion of the training dataset because they can be continuously trained. The difference in the training times of the FNN with different AF is revealed in Sect. 3.2 below by reviewing the optimal FNN-architectures.

The difference in the quality of the SM car wake predictions becomes clear at rear wing positions with high α and relatively high y-position. In this case, the

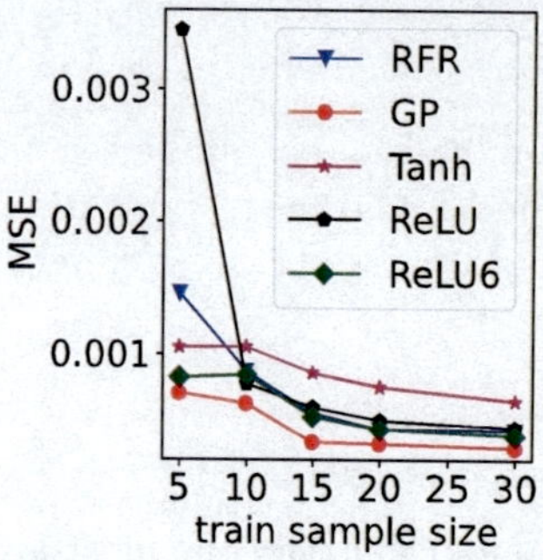

Fig. 5. Averaged MSE of the respective methods on the validation dataset.

Table 3. Training time required depending on the training dataset size.

train size	RFR	GPR	Tanh	ReLU	ReLU6
5	0.013 s	51.88 s	2.16 s	0.85 s	1.55 s
10	0.015 s	54.24 s	2.8 s	3.2 s	0.90 s
15	0.017 s	54.85 s	15.17 s	3.26 s	7.31 s
20	0.020 s	57.37 s	20.6 s	3.83 s	17.36 s
30	0.028 s	59.38 s	15.8 s	7.31 s	21.71 s

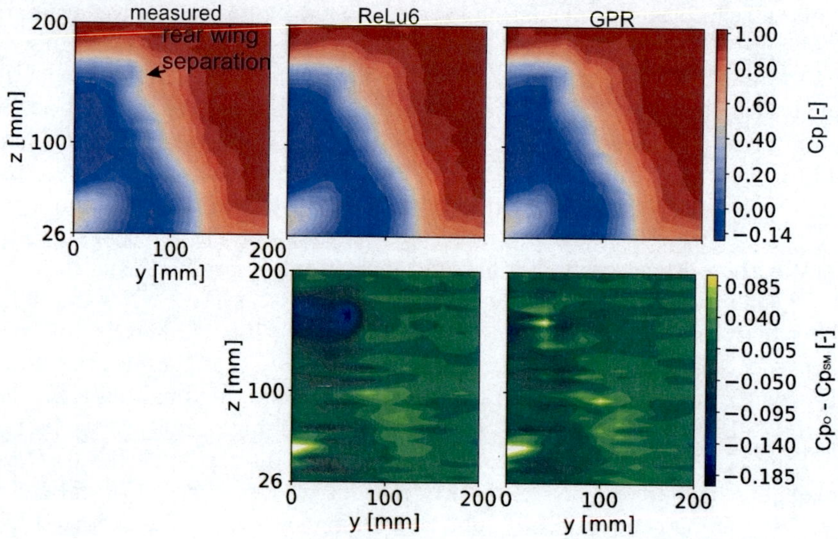

Fig. 6. C_p distribution of a measured evaluation wake and the corresponding prediction of two SMs trained with 30 sample points for a rear wing position of $x = 8.65$ mm, $y = 9.66$ mm, and $\alpha = 28.8°$.

rear wing generates its own separated zone. Figure 6 shows such a rear wing configuration with $\alpha = 28.8°$ and $y = 9.66$ mm. Compared here is the C_p distribution of the measured wake with the GPR and the FNN ReLU6. Both SMs were trained with 30 samples. Below the respective difference distributions $C_{p,o} - C_{p,SM}$ are shown. The FNN ReLU6 prediction has larger errors in the rear wing separation area compared to the GPR forecast. Compared to all other SMs, the GPR manages to reproduce this particularity of the flow field at high α rear wing positions best.

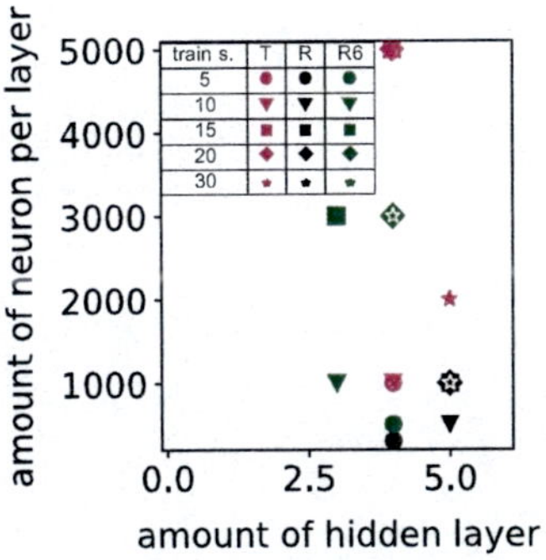

Fig. 7. Best performing FNN-architectures depending on the dataset size and the AF.

3.2 FNN Architecture

Varying the training sample size resulted in different optimal FNN architectures. Figure 7 shows the best performing FNN architectures depending on the training dataset size and the AF. These different architectures cause the differences between the FNN training times, see Table 3. Interestingly, the best performing FNN Tanh architectures always require more neurons for the same number of training samples and nonetheless does not deliver better prediction results than the FNN with ReLU and ReLU6. Especially the FNNs with ReLU achieve good results with very compact architectures, except for the case with a training of only five samples.

The other varied parameters listed in Table 2 were unambiguous for an optimal FNN and independent of AFs and training sample size. Not surprisingly, a batch size of 1 and 70 epochs always yielded the best results, since this is where the FNNs get the most information and have the most iteration steps to optimize the weights. For the SGD it was found that a learning rate of 0.5 and a momentum of 0.7 is best for the prediction of these car wakes.

4 Conclusion

The focus of this work was to find the most accurate und robust surrogate model for flow field prediction with a size-limited experimental dataset of a 2D car wake. Three methods (random forest regression, Gaussian process regression and feedforward neural network (FNN) with the Tanh, ReLU, and ReLU6 activation functions) trained with different dataset sizes were compared. The GPR stood out in this work by providing the most accurate surrogate model over the complete training dataset size variation. On the other hand, the main disadvantage of GPR was found to be its relatively long training time. The FNN with ReLU6 was the second-best performer, particularly with a limited number of training samples. With a training dataset size of 10 or more, the prediction quality of RFR, FNN ReLU and FNN ReLU6 was almost identical. Interestingly, RFR distinguished itself by its simplicity and its minimal training time.

References

1. Boutilier, M., Yarusevych, S.: Influence of hot-wire probe and traverse on low-Reynolds-number airfoil experiments. AIAA J. **52**, 2618–2623 (2014)
2. Breiman, L.: Random forests. Mach. Learn. **45** (2001)
3. Ciobaca, V., et al.: Wind tunnel experiments with active flow control for an outer wing model. In: 33rd AIAA Applied Aerodynamics Conference, pp. AIAA 2015–2728 (2015)
4. Dias Ribeiro, M., Rehman, A., Ahmed, S., Dengel, A.: DeepCFD: efficient steady-state laminar flow approximation with deep convolutional neural networks. arXiv:2004.08826 (2020)
5. Duvenaud, D.: Automatic model construction with gaussian processes. PhD thesis (2014)
6. Ebert, C., Ruwisch, C., Weiss, J., Haag, M., Silvestre, F.: Trajectory planning in windy urban environment - a Gappy POD approach for wind field estimates with sparse sensors. In: AIAA AVIATION 2022 Forum, pp. AIAA 2022–3757 (2022)
7. Forrester, A.I.J., Sobester, A., Keane, A.J.: Engineering Design Via Surrogate Modelling - A Practical Guide. Wiley, Hoboken (2008)
8. Glorot, X., Bordes, A., Bengio, Y.: Deep sparse rectifier neural networks. AISTATS **15**, 315–323 (2010)
9. Guo, X., Li, W., Iorio, F.: Convolutional neural networks for steady flow approximation. In: 22nd ACM SIGKDD International Conference, pp. 481–490 (2016)
10. Hahnloser, R.H.R., Sarpeshkar, R., Mahowald, M.A., Douglas, R.J., Seung, H.S.: Digital selection and analogue amplification coexist in a cortex-inspired silicon circuit. Nature **405**, 947–951 (2000)
11. Heft, A., Indinger, T., Adams, N.: Investigation of unsteady flow structures in the wake of a realistic generic car model. In: 29th AIAA Applied Aerodynamics Conference 2011, pp. AIAA 2011–3669 (2011)
12. Hussain, F., Zaman, K.: Free shear-layer tone phenomenon and probe interference. J. Fluid Mech. **87**, 349–383 (1978)
13. Krizhevsky, A.: Convolutional deep belief networks on CIFAR-10 (2012)
14. LeCun, Y., Bengio, Y., Hinton, G.: Deep learning. Nature **521**, 436–444 (2015)
15. Lohse, J., Berthold, A., Nitsche, W.: Experimental studies on the receptivity of stationary crossflow modes to pneumatic actuation. In: Dillmann, A., et al. (eds.) New Results in Numerical and Experimental Fluid Mechanics XI. NNFMMD, vol. 136, pp. 263–273. Springer, Cham (2018). https://doi.org/10.1007/978-3-319-64519-3_24
16. Mackay, D.J.C.: Bayesian methods for adaptive models. PhD thesis (1992)
17. Nair, V., Hinton, G.: Rectified linear units improve restricted Boltzmann machines. ICML **27**, 807–814 (2010)
18. Paszke, A., et al.: PyTorch: an imperative style, high-performance deep learning library, pp. 8024–8035 (2019)
19. Patrikar, J., Moon, B., Scherer, S.: Wind and the city: utilizing UAV-based in-situ measurements for estimating urban wind fields. In: 2020 IEEE/RSJ International Conference on Intelligent Robots and Systems (IROS), pp. 1254–1260 (2020)

20. Pedregosa, F., et al.: Scikit-learn: machine learning in Python. J. Mach. Learn. Res. **12**, 2825–2830 (2011)
21. Press, W.H., Teukolsky, S.A., Vetterling, W.T., Flannery, B.P.: Numerical Recipes 3rd Edition: The Art of Scientific Computing. Cambridge University Press, Cambridge (2007)
22. Quinlan, J.R.: Induction of decision trees. Mach. Learn. **1**, 81–106 (1986)
23. Raffel, M., Willert, C., Scarano, F., Kähler, C.J., Wereley, S.T., Kompenhans, J.: Particle Image Velocimetry - A Practical Guide (3rd Edition), vol. 1. Springer Verlag, Cham (2018). https://doi.org/10.1007/978-3-319-68852-7
24. Rasmussen, C.E.: Gaussian Processes in Machine Learning. The MIT Press, Cambridge (2006)
25. Steinfurth, B., Weiss, J.: Velocity ratio effect on flow structures of non-parallel planar starting jets in cross-flow. J. Fluid Mech. **915**, A11 (2021)
26. Werbos, P., John, P.: Beyond regression: new tools for prediction and analysis in the behavioral sciences. PhD thesis (1974)
27. White, T., Ushizima, D., Farhat, C.: Fast neural network predictions from constrained aerodynamics datasets. In: AIAA Scitech Forum, pp. AIAA 2020–0364 (2020)
28. Xiao, D., et al.: A reduced order model for turbulent flows in the urban environment using machine learning. Build. Environ. **148**, 323–337 (2018)
29. Yu, J., Hesthaven, J.S.: Flowfield reconstruction method using artificial neural network. AIAA J. **57**(2), 482–498 (2019)
30. Zhu, C., Byrd, R.H., Lu, P., Nocedal, J.: Algorithm 778: L-BFGS-B: fortran subroutines for large-scale bound-constrained optimization. ACM Trans. Math. Softw. **23**(4), 550–560 (1997)

Hypersonic Aerothermodynamics

Validation for a Polyatomic Model in a Fokker-Planck Solver Based on the Extended Master Equation Ansatz

Leo Basov(✉), Martin Grabe, and Aaron Nagel

German Aerospace Center (DLR), Bunsenstraße 10, 37073 Göttingen, Germany
leo.basov@dlr.de

Abstract. A validation of a polyatomic model for a kinetic Fokker-Planck solver based on a master equation ansatz is presented. The reference data are polynomial fits of experimentally measured heat capacity. An analytical solution based on the master equation ansatz is derived and the convergence of the Fokker-Planck solution towards it is shown. Deviations between the analytical and the reference solution are pointed out and explained.

Keywords: Fokker-Planck · rarefied gas · numerical simulations

1 Introduction

Flows encountered in space applications, like atmospheric reentry or thruster plume expansion and interactions, can be described by the well-known Boltzmann equation. A common approach to numerically solve it is the Direct Simulation Monte-Carlo (DSMC) method pioneered by Bird [1]. The method is very efficient for high Knudsen numbers but becomes increasingly computationally intensive when approaching the continuum limit. In this regime the Boltzmann equation can be numerically solved using the recently proposed kinetic Fokker-Planck (FP) method [2] which, like DSMC, relies on simulated particles to transport mass, momentum and energy through the flow domain, but does not resolve individual collisions. Due to the similarity in their formulation a hybrid method based on DSMC and FP can be derived which allows a computationally efficient simulation of flows with a broad range of Knudsen numbers [3].

Modeling flows of gas with internal degrees of freedom, such as for example CO_2, is essential for space engineering applications like studying entry missions into the atmosphere of Mars [4]. In the last years, the FP model has been extended to describe diatomic gases for single species [5,6] as well as for mixtures [7,8]. One method to treat polyatomic gas has been proposed [9]. However, this approach uses a single scalar value to model the internal energy of the gas and thus does not capture the distribution of internal energy between the individual degrees of freedom. An approach to resolve these internal degrees of freedom is described in this paper and a validation is shown.

A. Dillmann et al. (Eds.): STAB/DGLR Symposium 2022, NNFM 154, pp. 133–142, 2024.
https://doi.org/10.1007/978-3-031-40482-5_13

2 Method

2.1 Cubic Fokker-Planck

We base our further discussion on the cubic formulation of the Fokker-Planck method, which is covered in detail in the literature [6,10,11] and thus will be only briefly summarized here.

The well-known Boltzmann equation describes the evolution of a scalar distribution function f due to intermolecular collisions in rarefied flows:

$$\frac{\mathrm{D}f}{\mathrm{D}t} = S_{\mathrm{Boltz}} , \tag{1}$$

where t is time and S_{Boltz} is the Boltzmann collision integral. A recently introduced approach [2] to make the integro-differential Eq. (1) more manageable is to approximate S_{Boltz} by a Fokker-Planck collision operator S_{FP}:

$$S_{\mathrm{Boltz}} \approx S_{\mathrm{FP}} = -\frac{\partial}{\partial V_i}(A_i f) + \frac{\partial^2}{\partial V_j \partial V_j}(D^2 f) , \tag{2}$$

where the drift coefficient A_i and the diffusion coefficient D of Eq. (2) are model parameters chosen in such a way that production terms calculated using the Boltzmann collision operator are reproduced in the continuum limit. According to Itô's lemma, Eq. (2) can be written as a stochastic differential equation

$$\frac{\mathrm{d}\mathbf{X}}{\mathrm{d}t} = \mathbf{V}, \tag{3}$$

$$\frac{\mathrm{d}\mathbf{V}}{\mathrm{d}t} = \mathbf{A} + D\frac{\mathrm{d}\mathbf{W}}{\mathrm{d}t} + \mathbf{F} , \tag{4}$$

with $\mathbf{X}$ denoting particle position, $\mathbf{V}$ the particle velocity and $\mathrm{d}\mathbf{W}$ the Wiener process with zero expectation. Thus, the resulting Fokker-Planck equation can be solved through stochastic motion which in turn can be modelled using a particle method.

For a single species gas using the cubic model the drift A_i and the diffusion coefficient D are chosen as

$$A_i = -\frac{1}{\tau}c_i + \xi_{ij}c_j + \gamma_i\left(c_j c_j - \frac{3k_{\mathrm{B}}T}{m}\right) + \Lambda\left(c_i c_j c_j - \frac{2q_i}{\varrho}\right), \tag{5}$$

$$D = \sqrt{\frac{2k_{\mathrm{B}}T}{\tau m}} . \tag{6}$$

where τ is a relaxation time, γ and ξ are the coefficient matrices, $\mathbf{c}$ the velocity vector, k_{B} is the Boltzmann constant, T the temperature, m is mass, $\mathbf{q}$ the heat flux vector and ϱ the (mass) density. The parameter Λ is chosen to assure stability of the model. Equation (4) can be integrated leading to

$$V^{n+1} = \frac{1}{\alpha}\left[C_i^n \exp\left(-\frac{\Delta t}{\tau}\right) + \left(1 - \exp\left(-\frac{\Delta t}{\tau}\right)\right)\tau N_i^n + \psi_i\sqrt{\frac{D^2}{2}\tau\left(1 - \exp\left(-2\frac{\Delta t}{\tau}\right)\right)}\right] + V_i^n - C_i^n , \tag{7}$$

where $\mathbf{C}$ denotes the thermal particle velocity. The nonlinear term $\mathbf{N}$ can be expressed as

$$N_i^n = \xi_{ij} C_j^n + \gamma_i \left(C_j^n C_j^n - \frac{3k_\mathrm{B}T}{m} \right) + \Lambda \left(C_i^n C_j^n C_j^n - \frac{2q_i}{\varrho} \right) . \tag{8}$$

Here the factor α is a parameter used to scale particle velocities to ensure energy conservation for arbitrary time steps. To be compatible with DSMC a simple integration scheme for particle positions is chosen as

$$\mathbf{X}^{n+1} = \mathbf{X}^n + \mathbf{V}^n \Delta t . \tag{9}$$

2.2 Master Equation Anastz

The basic premise of the master equation ansatz [6] is the fact that the Boltzmann collision operator can be split into an elastic $\left(S_\mathrm{Boltz}^\mathrm{el}\right)$ and an inelastic part $S_\mathrm{Boltz}^\mathrm{inel}$:

$$S_\mathrm{Boltz} = S_\mathrm{Boltz}^\mathrm{el} + S_\mathrm{Boltz}^\mathrm{inel} . \tag{10}$$

The goal is to approximate both parts in such a way as to stay compatible with the Fokker-Planck formulation. The elastic operator is formulated as before, but only a fraction of the particles undergo the elastic collisions. The inelastic part may again be split into two parts:

$$S_\mathrm{Boltz}^\mathrm{inel} = S_\mathrm{v} + S_\mathrm{e} , \tag{11}$$

where S_v describes the change of the number of particles in a phase space element due to a change of particle velocities and S_e refers to the change in the number of particles in a phase space element due to a change of internal energies as a result of inelastic collisions. The first part can be modeled via the Fokker-Planck ansatz described above, where the diffusion coefficient is modified to account for energy transfer between translational and internal degrees of freedom. The second part is modeled using the master equation ansatz as

$$S_\mathrm{e} = \sum_j \left(R_{j,m} f_j - R_{n,j} f_n \right) . \tag{12}$$

Here R_{qk} are the rate coefficients modeling the transition rates between the internal energy states q and k. When combined with the original FP equation the new FP-Master Equation can be written as

$$\frac{\mathrm{D} f_n}{\mathrm{D}t} = \tilde{S}(f_n) + \sum_j \left(R_{j,m} f_j - R_{n,j} f_n \right) . \tag{13}$$

where $\tilde{S}(f_n)$ is the FP operator with the modified diffusion coefficient. Assuming that the evolution of translational velocities and internal energies can be modelled as independent stochastic processes, it can be written:

$$f_n(\mathbf{r}, \mathbf{v}, t) = f(\mathbf{r}, \mathbf{v}, t)\, g_n(\mathbf{r}, t) . \tag{14}$$

Equation (13) can be separated into a set of equations

$$\frac{\mathrm{d}f}{\mathrm{d}t} = \frac{\mathrm{d}f}{\mathrm{d}t} + \frac{\mathrm{d}f}{\mathrm{d}x_i} v_i = \tilde{S}(f_n) \tag{15}$$

$$\frac{\mathrm{d}g_n}{\mathrm{d}t} = \frac{\mathrm{d}g_n}{\mathrm{d}t} + \frac{\mathrm{d}g_n}{\mathrm{d}x_i} v_i = \sum_j (R_{j,m} f_j - R_{n,j} f_n) \ , \tag{16}$$

which are solved independently each time step [6].

The transfer of energy between the internal degrees of freedom is implicitly modeled in the rate coefficients. The coefficients can be divided into two groups: discrete and continuous. Hepp *et al.* [6] proposed two types of coefficients: *Landau-Teller relaxation* based coefficients using the continuous (CLT) and discrete (DLT) model for internal energies and *Larsen-Borgnakke relaxation* based coefficient using a discrete energy model (DLB).

2.3 Extension of the Master Equation Anastz

For the extension of the master equation ansatz to polyatomic molecules the model must be able to handle an arbitrary number of degrees of freedom. To achieve this goal two assumptions are made:

1. No direct energy exchange between the individual modes of the rotational and vibrational degrees of freedom.
2. No direct energy exchange between the rotational and vibrational energy.

An indirect energy transfer between all modes is still possible through the kinetic energy of the particles. The possible transfer paths are shown in Fig. 1.

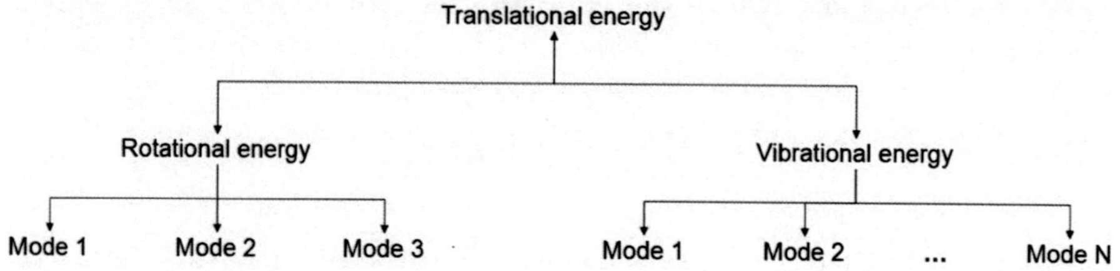

Fig. 1. Possible energy transfer paths.

In this first approximation the calculation of the coefficients ξ_{ij} and γ_i of the Fokker-Planck equation remains unchanged compared to the diatomic case. The extension of the model lies in the rate coefficients R_{qk} as they indirectly model the internal energy structure of the molecules as well as the relaxation process. The coefficients need to be extended to account for an arbitrary number of degrees of freedom. The exact changes are depend on the model of the rate coefficient which are discussed below.

Rotational Energy Relaxation – For the relaxation of rotational energy a continuous energy structure is assumed and the extended CLT method employed. The difference between the diatomic and polyatomic case in the algorithm is the inclusion of rotational degrees of freedom d_r to accommodate for non-linear molecules.

Vibrational Energy Relaxation – Compared to the original implementation the difference in the new approach is the consideration of multiple degrees of freedom. It is assumed that the total vibrational energy can be formulated as the sum of the energy of each vibrational mode multiplied by its degeneracy. Assuming no direct energy transfer between the vibrational modes the relaxation is now performed for each mode individually. Additionally for the DLB algorithm the look-up-table holding the DLB parameters must be extended to account for different vibrational modes.

3 Validation Case: Specific Heat Capacity

To validate the model, a comparison to numerical results of molar heat capacity c_v is made. The calculations are performed using the *Landau-Teller* based rate coefficients. As a reference, interpolation polynomials of experimental data provided in [12] are used.

3.1 Analytical Solution for the Heat Capacity of Polyatomic Molecules

To provide reasonable comparison the analytical solution of the heat capacity is based on the partition functions used in the derivation of the rate coefficients in the master equation ansatz. For the rotational energy the partition function is modeled as being continuous while a discrete partition function is employed for the energies of each vibrational mode. The equipartition theorem is assumed. Neglecting the total electronic energy of the molecules the total internal energy can be written as

$$e = e_\mathrm{t} + e_\mathrm{r} + \sum_{i=0}^{N} g_i\, e_{\mathrm{v},i} = \mathcal{R}T\left[\frac{3}{2} + \frac{d_\mathrm{r}}{2} + \sum_{i=0}^{N} g_i \frac{\theta_i/T}{\exp(\theta_i/T)-1}\right], \tag{17}$$

where e_t is the translational and e_r the rotational energies. The vibrational energy can be defined as the sum of the energy of each vibrational mode $e_{\mathrm{v},i}$ times the modes degeneracy g_i. Here $\mathcal{R}$ is the universal gas constant, T the temperature, d_r is the number of rotational degrees of freedom, which for this work are assumed to be fully excited, and θ_i the characteristic vibrational temperature of the i-th vibrational mode.

Using the definition $c_v \equiv \left(\frac{\partial e}{\partial T}\right)_v$, the molar heat capacity at constant volume can be calculated as

$$\frac{c_v}{\mathcal{R}} = \frac{3+d_\mathrm{r}}{2} + \sum_{i=0}^{N} g_i \left(\frac{\theta_i/(2T)}{\sinh \theta_i/(2T)}\right)^2, \tag{18}$$

which is related to the heat capacity at constant pressure via $c_p = c_v + \mathcal{R}$. For the validation process several gases were considered. The values required for evaluation of Eq. (18) for each gas are listed in Table 1.

Table 1. Gas kinetic values for used gases.

gas	Ref.	d_r	θ_1/K	θ_2/K	θ_3/K	θ_4/K	g_1	g_2	g_3	g_4
N_2	[1]	2	3371.0	–	–	–	1	–	–	–
O_2	[1]	2	2256.0	–	–	–	1	–	–	–
CO_2	[13]	2	1918.6	3382.0	959.7	–	1	1	2	–
CH_4	[13]	3	4194.9	2206.0	4341.6	1878.1	1	2	3	3

3.2 Numerical Simulation

For the numerical simulation starting at temperature T^{n}, the energy is modified by adding translational energy to each particle p in form of a velocity delta Δv_p. This energy can be translated into a molar energy increase as

$$\Delta e_t = \frac{1}{2} m_p M (\Delta v_p)^2 , \tag{19}$$

where m_p is the molecule mass and M the molar mass of the species. The system is then let to relax until a new equilibrium temperature $T^{\mathrm{n}+1}$ is reached. Using the forward differential quotient the molar heat capacity at constant volume c_v can be calculated as

$$c_v\left(T^{\mathrm{n}+\frac{1}{2}}\right) \approx \frac{\Delta e_t}{\left(T^{\mathrm{n}+1} - T^{\mathrm{n}}\right)/2} . \tag{20}$$

Values used in the numerical simulations are listed in Table 2. Here Δt is the chosen time step size, T^0 the initial temperature of the simulation, n the number density, w the particle weight and N_{iter} is the number of iterations used for every relaxation phase.

Table 2. Gas kinetic values for used gases.

gas	Δt/s	T^0/K	n/m^{-3}	Δv_p/m s^{-1}	w	N_{iter}
CO_2	1×10^{-7}	300	2×10^{22}	340	1×10^7	10 000
CH_4	1×10^{-7}	300	2×10^{22}	610	1×10^7	10 000

3.3 Results and Discussion

Good agreement is seen between the analytical solution (18) and the numerical solution produces using the Fokker-Planck solver for both CO_2 and CH_4 as seen in Figs. 2 and 3. This is expected as the analytical model used to calculate c_v is based on the partition functions and energies used in the derivation of the rate coefficients for the master equation ansatz.

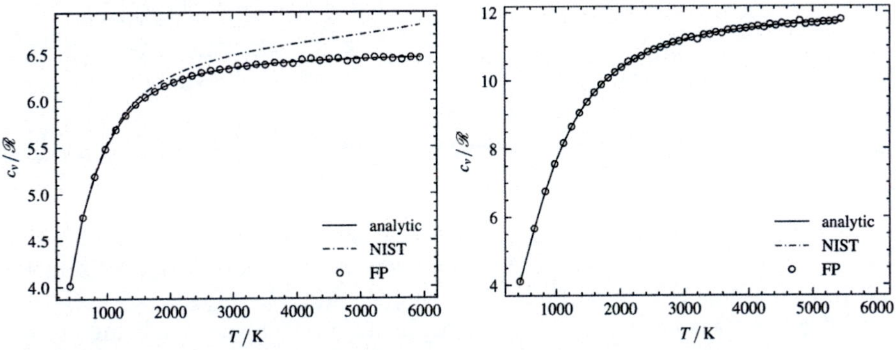

Fig. 2. Dimensionless heat capacity. Left CO_2 right CH_4. Here *analytic* represents the analytical solution, *NIST* the NIST-polynomial and *FP* the numerical solution produced by the Fokker-Planck solver.

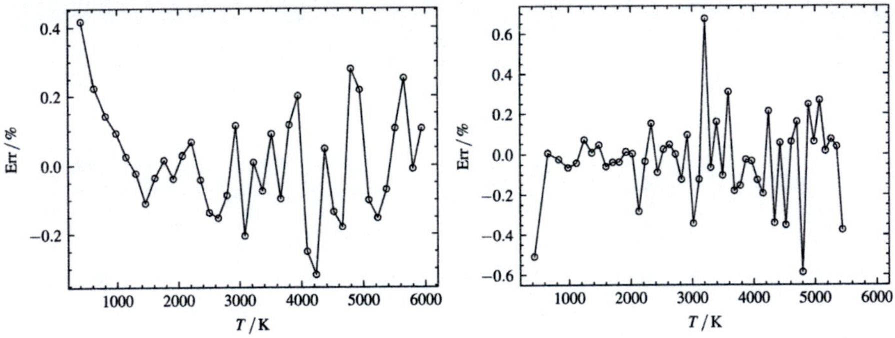

Fig. 3. Relative error between c_v calculation using Fokker-Planck and Eq. (18). Left CO_2 right CH_4.

Looking at the comparison between the analytical and the reference solution provided via the NIST-polynomials the results agree very well for CH_4. However, there is a gap in the results for the case of CO_2. For this case the relative error between the analytical and the reference solution is proportional to the temperature which can clearly be seen in Fig. 4.

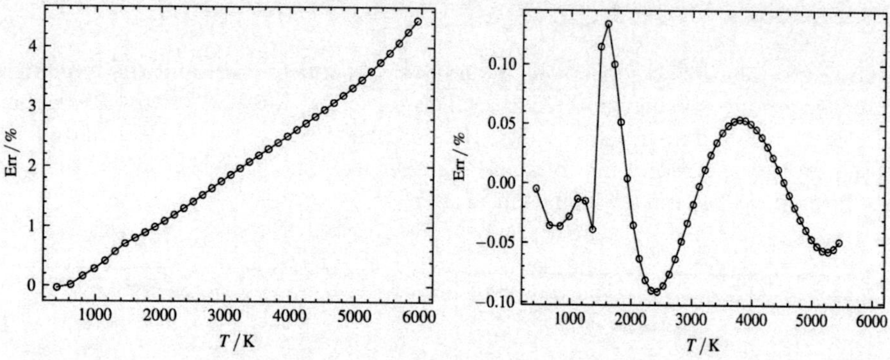

Fig. 4. Relative error between NIST-polynomials and Eq. (18). Left CH_4 right CO_2.

Assuming that the numerical solution using the Fokker-Planck operator converges towards Eq. (18) two more comparison between the analytical solution and NIST-polynomials were conducted using O_2 and N_2. As seen in Fig. 5 there are differences between the solutions with the most prominent one being present for O_2.

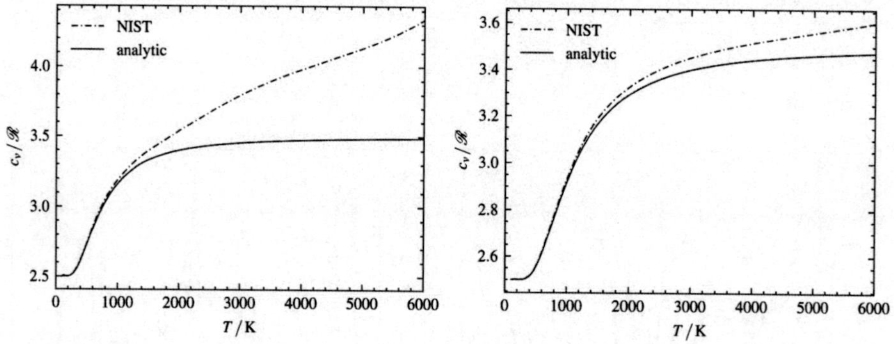

Fig. 5. Dimensionless heat capacity. Left O_2 right N_2. Here *analytic* represents the analytical solution and *NIST* the NIST-polynomial.

The gap can be explained by the neglect of the electronic energy in the formulation of the internal energy used in the derivation of the FP rate coefficients and Eq. (18). As each excited state contributes to the total internal energy it is expected that as more states are excited the value for c_v will increase. This process has been analytically investigated in [14]. Using this data the process is visualized in Fig. 6 where the cases (a), (b), (c) and (d) represent electronically increasingly excited states. The detailed description of the model and the used excited states can be found in section 3.5.1 of [14]. Starting at ground state, as it is the case for the analytical solution, more and more excited states are added which move the solution to higher values of c_v which provides an increasingly better fit to the reference solution. The influence of the electronic states is highly dependant on the molecule in question and the considered temperature range as the number of excited states is dependant on the excitation energy. Using CH_4 as an example it can be seen in Figs. 2, 4 that almost no difference is present between the analytical and the reference solution in the temperature range of 300 K–6000 K.

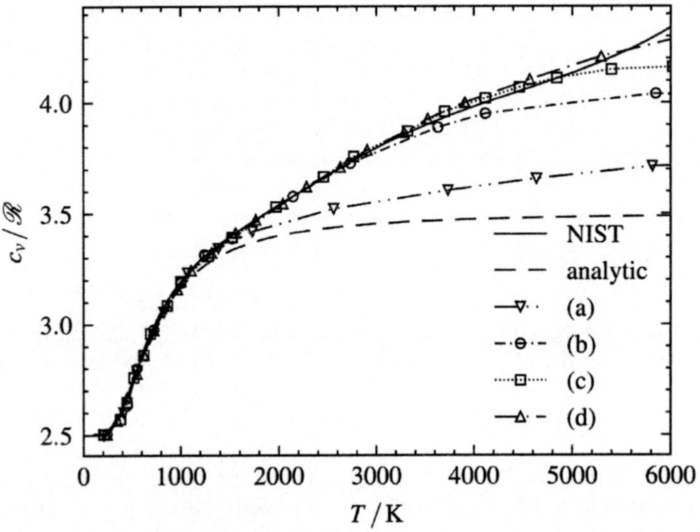

Fig. 6. Influence of electronically excited states on the dimensionless heat capacity of O_2.

4 Conclusion and Outlook

A validation of the polyatiomic model for a kinetic Fokker-Planck solver based on a master equation ansatz was carried out. The reference data were polynomial fits of experimentally measured heat capacity. An analytical solution based on the master equation ansatz was derived and the convergence of the Fokker-Planck solver towards it was shown. Deviations present between the analytical solution and the reference can be attributed to the neglect of electronic energy in our polyatomic modelling. The magnitude of the deviation is thus dependant on the excitation energy of the gas. It is open for investigation whether similar temperature dependant deviations can be observed in more complex test cases where the modeling of relaxation times plays a more important role. A comparison with experimental data as well as DSMC simulations is necessary and we will address this in future work.

References

1. Bird, G.A.: Molecular Gas Dynamics and the Direct Simulation of Gas Flows. Oxford University Press, New York (1994)
2. Jenny, P., Torrilhon, M., Heinz, S.: A solution algorithm for the fluid dynamic equations based on a stochastic model for molecular motion. J. Comput. Phys. **229**(4), 1077–1098 (2010)
3. Gorji, M.H., Jenny, P.: Fokker-Planck-DSMC algorithm for simulations of rarefied gas flows. J. Comput. Phys. **287**, 110–129 (2015)
4. Wright, M., Edquist, K., Tang, C., Hollis, B., Krasa, P., Campbell, C.: A Review of aerothermal modeling for mars entry missions. In: 48th AIAA Aerospace Sciences Meeting Including the New Horizons Forum and Aerospace Exposition, Orlando, Florida, January 2010. American Institute of Aeronautics and Astronautics (2010)
5. Gorji, M.H., Jenny, P.: A Fokker-Planck based kinetic model for diatomic rarefied gas flows. Phys. Fluids **25**(6), 062002 (2013)
6. Hepp, C., Grabe, M., Hannemann, K.: Master equation approach for modeling diatomic gas flows with a kinetic Fokker-Planck algorithm. J. Comput. Phys. **418**, 109638 (2020)
7. Hepp, C., Grabe, M., Hannemann, K.: A kinetic Fokker-Planck approach to model hard-sphere gas mixtures. Phys. Fluids **32**(2), 027103 (2020)
8. Hepp, C., Grabe, M., Hannemann, K.: A kinetic Fokker-Planck approach for modeling variable hard-sphere gas mixtures. AIP Adv. **10**(8), 085219 (2020)
9. Mathiaud, J., Mieussens, L.: A Fokker-Planck model of the Boltzmann equation with correct Prandtl number for polyatomic gases. J. Stat. Phys. **168**(5), 1031–1055 (2017)
10. Gorji, M.H., Torrilhon, M., Jenny, P.: Fokker-Planck model for computational studies of monatomic rarefied gas flows. J. Fluid Mech. **680**, 574–601 (2011)
11. Jun, E., Gorji, M.H., Grabe, M., Hannemann, K.: Assessment of the cubic Fokker-Planck-DSMC hybrid method for hypersonic rarefied flows past a cylinder. Comput. Fluids **168**, 1–13 (2018)
12. Linstrom, P.: NIST Chemistry WebBook, NIST Standard Reference Database, no. 69 (1997). Type: dataset
13. Pfeiffer, M., Nizenkov, P., Fasoulas, S.: Extension of particle-based BGK models to polyatomic species in hypersonic flow around a flat-faced cylinder. In: AIP Conference Proceedings, vol. 2132, p. 100001, Glasgow, UK, 2019. AIP Publishing LLC (2019)
14. Capitelli, M., et al.: Tables of internal partition functions and thermodynamic properties of high-temperature mars-atmosphere species from 50 K to 50000 K, p. 279 (2005)

Uncertainty Quantification of Expanding High-Enthalpy Air Flows

Georgii Oblapenko(✉) and Volker Hannemann

German Aerospace Center (DLR), Bunsenstraße 10, 37073 Göttingen, Germany
georgii.oblapenko@dlr.de

Abstract. In the present work, we study the sensitivity of expanding hypersonic air flows in the HEG facility with respect to the thermochemical relaxation rates and reservoir conditions, along with the impact of uncertainties in the rates and reservoir conditions on uncertainties of the free-stream parameters. A simplified 1-D formulation of the flow equations is utilized to allow for simultaneous variation of a large number (10) of uncertain parameters. The simulation results show large uncertainties in the vibrational temperature of oxygen and molar fractions of oxygen-containing species, mainly due to uncertainties in the reservoir parameters and data on thermochemical relaxation rates of molecular oxygen.

Keywords: shock tunnel · thermochemical relaxation · numerical simulation · uncertainty quantification · hypersonic flow

1 Introduction

Hypersonic ground test facilities use flow expansion to achieve conditions similar to those of high-velocity hypersonic re-entry in order to provide experimental data for spacecraft design. The simulation of the operation of such facilities is needed for estimation of the free-stream conditions, interpretation of experimental data, as well as verification of computational models. In shock tunnel facilities, such as the High Enthalpy Shock Tunnel Göttingen (HEG) [7], the flow is stagnated in order to achieve the required enthalpies; as it then expands through a nozzle section, thermochemical relaxation and subsequent freezing of the flow takes place, potentially leading to a chemically and vibrationally non-equilibrium flow at the nozzle exit [2]. The modeling of hypersonic flows is complicated by significant uncertainties in the data on the thermochemical relaxation rates, as well as the often significant impact on the choice of models to describe the relaxation processes [5,6,10,18].

In recent years, more attention and effort has been devoted to consistent analysis of the impact of the uncertainties in the thermochemical relaxation rates on various quantities of interest (such as surface heat fluxes) in high-enthalpy flows; however most of the effort has been focused on flows behind strong shock waves [1,21,22]. Expanding flows have received less attention, even though their analysis is required for accurate estimation of free-stream uncertainties that

A. Dillmann et al. (Eds.): STAB/DGLR Symposium 2022, NNFM 154, pp. 143–152, 2024.
https://doi.org/10.1007/978-3-031-40482-5_14

are in turn used in the aforementioned post-shock simulations of shock tunnel experiments; and the computed free-stream conditions may be quite sensitive to the choice of thermochemical relaxation rates [8]. Several recent studies have focused on improvements of modeling expanding flows in experimental facilities: in Gimelshein & Wysong [4], the impact of chemical reaction rates on nitric oxide production was studied, along with the role of variations in the reservoir density and temperature. It was shown that vibration-dissociation coupling has a large effect on O_2 and NO concentrations for a high-enthalpy HEG condition. In Gu et al. [6], the applicability of two-temperature models (as compared to higher-fidelity state-specific models) to simulations of nozzle flows was investigated, along with the possibility of simplification of vibrational relaxation modeling. It was shown that the presence of non-Boltzmann vibrational energy distributions at the nozzle exit has little impact on the flow further downstream behind a shock around the test body; as such, simplified models that assume a Boltzmann vibrational distribution may be sufficient for the computation of free-stream conditions in shock tunnels. In [15], we performed a sensitivity analysis and uncertainty quantification study of expanding binary flows of N_2/N and O_2/O, with the aim of verifying the developed framework and providing preliminary results to reduce the parameter space in uncertainty quantification/sensitivity analysis studies of air flows.

The objective of the present work is to expand our uncertainty quantification and sensitivity analysis framework to air flows, and assess the role that the uncertainties in reservoir conditions and thermochemical relaxation rates play in high-enthalpy air flows characteristic of HEG conditions.

2 Governing Equations

In the present work, we assume that the nozzle flow is steady, inviscid, and quasi-one-dimensional. This allows to use a simplified system of equations, and greatly reduces the required computational time. As numerous simulations with varying parameters are required for the uncertainty quantification and sensitivity analysis studies, the ability to rapidly perform multiple simulations is crucial.

The system of equations describing a chemically reacting vibrationally non-equilibrium flow takes on the following form [13]:

$$\frac{d\left(\rho_c v S\right)}{dx} = S R_c^{chem}, \; c = 1, \ldots, N_s, \tag{1}$$

$$\rho v \frac{dv}{dx} = -\frac{dp}{dx}, \tag{2}$$

$$\frac{dh}{dx} + v\frac{dv}{dx} = 0, \tag{3}$$

$$\frac{d\left(\rho_c S v E_v^c\right)}{dx} = S R_c^{vibr}, \; c = 1, \ldots, N_m. \tag{4}$$

Here x is the coordinate along the nozzle axis, $S = S(x)$ is the cross-sectional area of the nozzle, ρ_c is the density of species c, v is the flow velocity, R_c^{chem} is the chemical relaxation term, ρ is the flow density, p is the pressure, h is the enthalpy, E_v^c is the vibrational energy (per unit mass) of molecular species c, R_c^{vibr} is the vibrational relaxation term, N_s is the number of chemical species in the flow, N_m is the number of molecular species in the flow. Despite the inability of the utilized approach to model the viscous boundary layer, a corresponding correction can be incorporated by performing a viscous 2-D axisymmetric simulation, extracting the boundary layer thickness and using an effective nozzle profile $\hat{S}(x)$ [17].

In the present work, we consider a five-component air flow: $N_2/O_2/NO/N/O$. Vibrational relaxation due to VT exchanges was computed using the Landau–Teller formula with the Millikan–White expression for VT relaxation times [12]. For VV exchanges of vibrational energy between O_2 and N_2, the empirical correlation proposed by Klomfass was used [9]:

$$\tau_{cd}^{VV} = \frac{1}{n k_{cd}^{VV}}, \quad k_{cd}^{VV} = \hat{K}_{cd}^{VV} \sqrt{k_{cc}^{VT} k_{dd}^{VT} \exp\left(\left(\theta_c - \theta_d\right)/T\right)}. \tag{5}$$

Here τ_{cd}^{VV} is the VV relaxation times for molecules of species c colliding with molecules of species d; k_{cd}^{VV} is the corresponding VV rate coefficient, k_{cc}^{VT} is the VT rate coefficient, θ_c is the characteristic vibrational temperature of molecular species c, and $\hat{K}_{cd}^{VV}$ is a constant factor.

The exchange reactions were computed assuming no influence of the reactant and product vibrational temperatures. For dissociation reactions, the Treanor–Marrone model [11] was used to account for the preferential dissociation from higher-lying vibrational levels and compute the vibrational energy loss/gain due to dissociation and recombination reactions.

3 Uncertainty Quantification

3.1 Generalized Polynomial Chaos

In order to compute the impact of the uncertainties in the reservoir parameters and thermochemical relaxation rates on the flow, we utilise the generalized Polynomial Chaos (gPC) [19] approach. Compared to traditional Monte–Carlo methods, for moderate parameter space dimensions, far fewer model evaluations are needed. In the present work, we use the non-intrusive approach, which allows us to treat the solver of the system of governing equations Eq. (1)–(4) as a "black box" and thus does not require solving additional equations.

Denoting the vector of uncertain model parameters as $\mathbf{w}$, and assuming that all these uncertain parameters are independent random variables, a quantity of interest Y (such as pressure, temperature, etc.) can be represented in the following form:

$$Y(x, \mathbf{w}) \approx \sum_{i=0}^{N} c_i(x) P_i(\mathbf{w}). \tag{6}$$

Here $c_i(x)$ are the unknown expansion coefficients (which are dependent on the spatial location x), and P_i are the expansion polynomials (of order i up to N), which are chosen to be orthogonal with respect to the parameter distributions and are known beforehand. Due to the orthogonality of the polynomials, the coefficients of the expansion are given through the following expression:

$$c_i = \frac{\langle Y, P_i \rangle}{\langle P_i, P_i \rangle}, \quad \langle a, b \rangle = \int ab d\mathbf{w}. \tag{7}$$

Once the set of uncertain model parameters $\mathbf{w}$ and their distributions have been chosen, along with the desired order of the polynomial expansion N the quadrature appearing in Eq. (7) can be evaluated numerically. For a chosen quadrature rule, a set of sample points of $\mathbf{w}$ is generated, along with the associated weights, the model is evaluated using the sampled values of the uncertain parameters, and the quadrature is calculated in order to obtain the coefficients c_i. Smolyak sparse grids [3] are used for evaluation of the quadrature in order to reduce the number of model evaluations required. In the present work, 2nd order polynomial expansions are used, as in our previous work [15], they were found to provide sufficient convergence of the statistical properties being investigated (mean, standard deviation, sensitivity indices). Once the expansion coefficients c_i have been computed, relevant statistics can be calculated, such as the mean and standard deviation of the flow quantities of interest (for example, temperature at the nozzle exit), as well as the sensitivity of these flow quantities with respect to the uncertain model parameters (via Sobol sensitivity indices). The total Sobol sensitivity index of quantity of interest Y with respect to uncertain model variable w_i is defined as

$$S_{tot,i} = 1 - \frac{\mathrm{Var}_{\mathbf{w_{\sim i}}} \left(E_{w_i} \left(Y | \mathbf{w_{\sim i}} \right), \right)}{\mathrm{Var}(Y)} \tag{8}$$

where Var is the variance, E is the expectation, and $\mathbf{w_{\sim i}}$ is the vector of all uncertain model variable except variable w_i. The values of the Sobol indices are proportional to the influence the corresponding parameter has on the flow uncertainty: values close to 1 mean a strong influence of the parameter, and small values mean a weak influence.

3.2 Parametrization of Uncertainty Sources

In order to perform the uncertainty quantification and sensitivity analysis study, we need to choose the uncertain model parameters w which we want to incorporate into our analysis. Unlike our previous studies in binary N_2/N and O_2/O mixtures, the much larger number of relaxation pathways in air prohibits the simultaneous consideration of all possible chemical and vibrational relaxation rates. Thus, additional reduction of the possible parameter space is required. The uncertainties in the reservoir pressure p_r and temperature T_r where taken to be $\pm 5\%$ of the reference values p_r^0, T_r^0 for the conditions considered, as given in [7].

In the present work, we assume the nitric oxide in the flow to be in vibrational equilibrium, due to its very short vibrational relaxation times [14]. Our previous study found that the role of the high-temperature Park correction [16] to VT relaxation times was found to be small due to the rapid freezing of the flow [15]. Thus the value of it was kept fixed in the present work.

In the current analysis, we consider uncertainties in the following VT relaxation times: N_2 in collisions with N, and O_2 in collisions with O atoms. This is motivated by several factors. First of all, available experimental and quasi-classical trajectory computational data on these times exhibit the largest degree of discrepancy. Secondly, these VT relaxation times are much shorter than relaxation times of molecule-molecule collisions. Due to the presence of oxygen (which has a low dissociation energy), a significant fraction of oxygen atoms are present in the flow; at the high enthalpy condition considered in the present work, nitrogen also experiences a noticeable degree of dissociation in the critical cross-section. Thus, a sufficient fraction of atoms is present in the flow to have a noticeable impact on the vibrational relaxation.

We took the VT relaxation times τ_{cd}^{VT} for collisions of molecules of species c with particles of species d to be given by the Millikan–White [12] expression $\tau_{cd,MW}^{VT}$ and the Park high-temperature correction $\tau_{cd,P}^{VT}$:

$$\tau_{cd}^{VT} = \tau_{cd,MW}^{VT} + \tau_{cd,P}^{VT}. \tag{9}$$

The Millikan–White formula gives the pressure-corrected VT relaxation times in the following form:

$$p\tau_{cd}^{VT,MW} = \exp\left(A_{cd}\left(T^{-1/3} - B_{cd}\right) - 18.42\right). \tag{10}$$

While the A_{cd}, B_{cd} coefficients can be varied independently to introduce uncertainty in the vibrational relaxation data, doing so may lead to uncharacteristically low or high values of the VT relaxation times. Thus, we took these coefficients to be linearly dependent on a single model parameter α_{cd}^{VT}. The single uncertain parameter $\alpha_{VT,cd}$ is taken to be uniformly distributed between 0 and 1, and the coefficients of the linear dependence were chosen based on analysis of existing experimental and QCT data for VT relaxation times. The use of a single parameter (instead of two) also allows to reduce the dimension of the space of uncertain parameters, lowering the number of model evaluations required.

Analysis of the VV relaxation times computed using the empirical expression of Klomfass (5) showed excellent agreement with recent experimental data on O_2-N_2 VV relaxation times [20] if the factor $\hat{K}_{cd}^{VV}$ was taken to be equal to 4 (in the original work of Klomfass it is equal to 1). Thus, in the present study, $\hat{K}_{cd}^{VV}$ was assumed to be uniformly distributed between 3 and 5, which corresponds to a 25% uncertainty in the VV relaxation times, this is comparable to the 30%–40% experimental uncertainty reported by Streicher et al. [20].

In the present case of the five-component air flow, 17 chemical reactions are included in the simulation: 15 dissociation reactions (3 molecular species, each with 5 possible collision partner species) and 2 exchange reactions. Of course, considering the impact of uncertainties in all the associated reaction rates simultaneously would require an exorbitant amount of computations, therefore, some simplification have been made. We use a single parameter to vary the dissociation rates of a specific molecule, and thus we have to consider only 5 uncertainties in the reaction rates: 3 dissociation rates (of N_2, O_2, and NO), and the two Zeldovich exchange reactions $N_2+O \rightarrow NO+N$ and $NO+O \rightarrow O_2+N$. As in our previous work [15], we parametrize the dissociation rate coefficients in the following fashion:

$$k_{cd}^{diss}(T) = (a_{cd}^{diss} + b_{cd}^{diss} T) k_{cd,Park}^{diss}(T), \tag{11}$$

where $k_{cd,Park}^{diss}$ are the values computed using the Arrhenius expression and data from Park [16], and a_{cd}^{diss}, b_{cd}^{diss} are linear functions of a parameter α_c^{diss}. The coefficients for the linear dependence were chosen so that a wide rang of possible values is covered, based on existing experimental and QCT data. As already mentioned, a single α_c^{diss} parameter governs all the uncertainties in the dissociation of a molecule of species c simultaneously. Whilst a simplification, such a parametrization still allows us to determine uncertainty in which dissociation/recombination rate data of which particular species has the most significant impact on the flow, and in future studies, a detailed investigation of the sensitivity to the partner-specific dissociation rates is planned. The 2 Zeldovich exchange reactions were parametrized in the same fashion as the dissociation reactions. Our previous studies in binary flows of N_2/N and O_2/O showed the relatively small impact of uncertainties in the vibration-dissociation coupling parameters (the U_{TM} parameter in case of the Treanor–Marrone model) due to recombination playing a larger role in expanding flows. Thus, the high-temperature Park correction as kept fixed, along with the U_{TM} parameter, which was taken to be equal to $3T$.

Thus, to summarize, we consider in total 10 uncertain parameters and their associated distributions: 2 reservoir parameters, 2 parameters governing VT relaxation, 1 parameter governing VV exchanges, 3 parameters governing dissociation rates, and 2 parameters governing exchange reaction rates. All the distributions were taken to be uniform.

4 Results

We consider an air flow characteristic of condition H22R0.20 [7] in the HEG shock tunnel. The condition has a reservoir pressure p_r of 35 MPa, reservoir temperature T_r of 9100 K, and total enthalpy of 22 MJ/kg. The contoured Mach 10 Nozzle V profile was used.

Figure 1 shows the computed flow properties along the nozzle axis, with a range of one standard deviation around the mean value shown by the shaded region. The annotated text gives the values at the nozzle exit and the associated relative uncertainty in percent. We observe a strong influence of the considered model uncertainties on the vibrational relaxation of O_2 and the molar fraction of oxygen-containing species (O/O_2/NO). A significant relative uncertainty of the molar fraction of atomic nitrogen at the nozzle exit is also present, however,

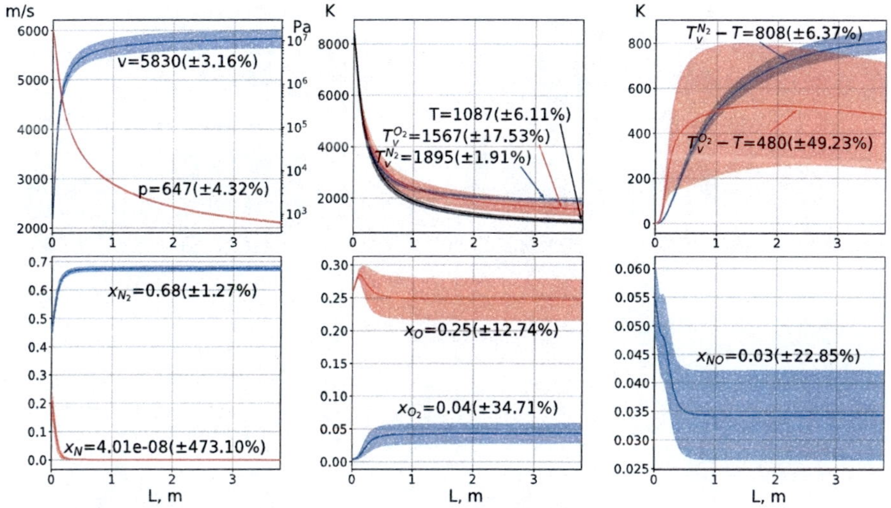

Fig. 1. Flow properties computed along the nozzle axis; shaded regions denote a range of one standard deviation, solid lines show the mean values.

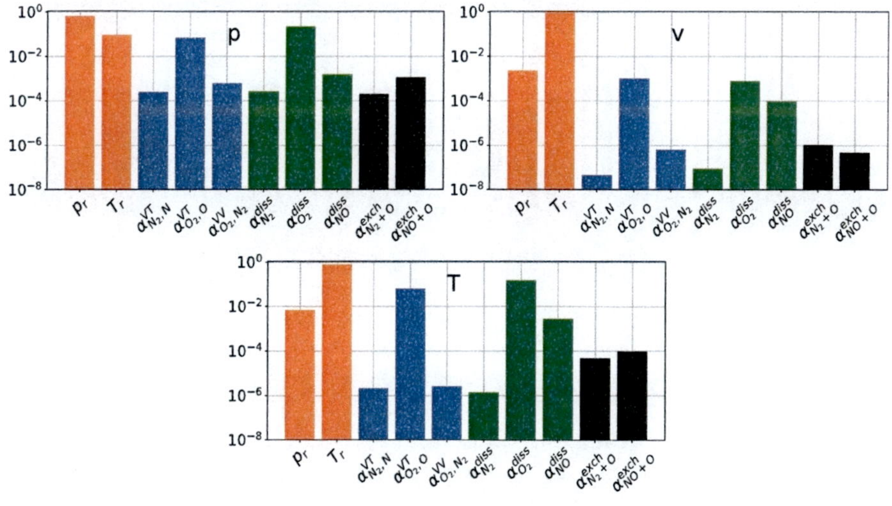

Fig. 2. Total Sobol sensitivity indices of pressure, velocity, and temperature at the nozzle exit with respect to the uncertain model parameters.

the absolute values of the atomic nitrogen concentration are extremely low, and as such, we do not investigate this quantity further.

Next we consider the sensitivity of the different flow parameters at the nozzle exit to uncertainties in the reservoir parameters and thermochemical relaxation rates. Figures 2,3 and 4 show the total Sobol sensitivity indices S_{tot} with respect to uncertainties in the reservoir parameters (shown by orange bars), parameters

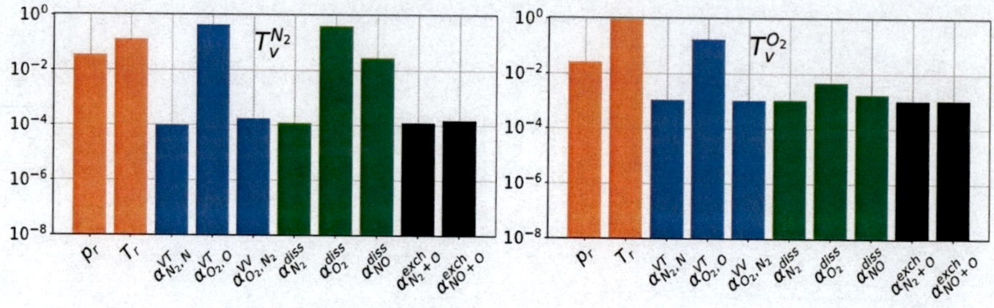

Fig. 3. Total Sobol sensitivity indices of vibrational temperatures at the nozzle exit with respect to the uncertain model parameters.

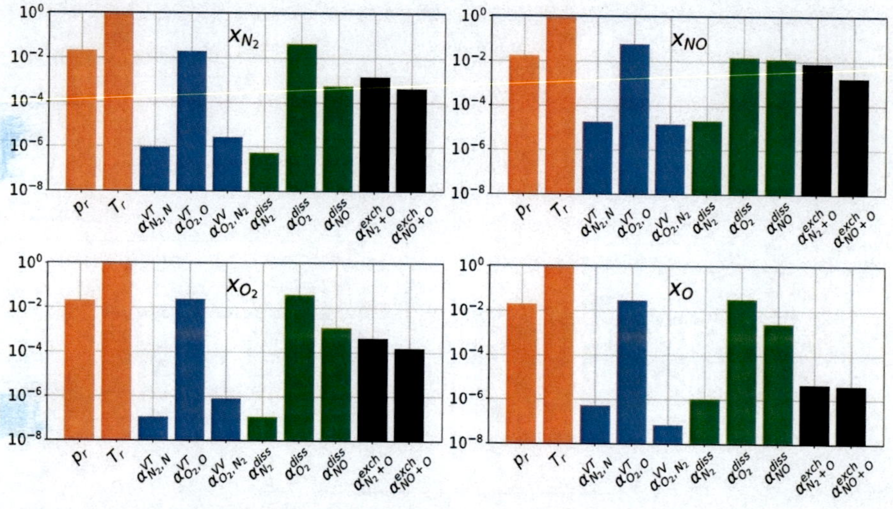

Fig. 4. Total Sobol sensitivity indices of species' molar fractions at the nozzle exit with respect to the uncertain model parameters.

governing VT and VV relaxation (blue bars), parameters governing dissociation reaction rates (green bars), and parameters governing exchange reaction rates (black bars). As expected, and seen on Fig. 2, only the reservoir temperature has any noticeable impact on the flow velocity. The flow temperature is strongly influenced not only by the reservoir temperature, but also by the VT relaxation of oxygen (due to both the large uncertainties in the O_2-O VT relaxation times and their low values) and the dissociation–recombination of O_2. As such, uncertainties in these quantities, along with uncertainty in the reservoir pressure, have the strongest effect on the uncertainty in the pressure at the nozzle exit.

For the vibrational temperatures at the nozzle exit, the VV relaxation time uncertainty has little impact on the flow, as seen on Fig. 3. This is due to the VV relaxation times being significantly larger than VT relaxation times, coupled with the rapid freezing of the flow. From Fig. 3 we also see a surprisingly large influence of the uncertainty in the VT relaxation rate and the dissociation rate of oxygen on the vibrational temperature of N_2. Analyzing the sensitivities of the molar fractions presented on Fig. 4, we see a significant sensitivity of all of the constituent species' molar fractions to the uncertainties in the VT relaxation of O_2 and dissociation–recombination of O_2. This suggests the following explanation for the strong sensitivity of $T_v^{N_2}$ to uncertainties in thermochemical relaxation rates of O_2: due to the preferential dissociation model of Treanor–Marrone, the vibrational relaxation of O_2 has a strong impact on its dissociation, which in turn affects the fraction of atomic oxygen available for the $N_2+O \rightarrow NO+N$ Zeldovich reaction. This reaction leads to a change in the vibrational temperature of N_2; thus, it is through this three-step process that the VT relaxation of O_2 affects vibrational relaxation of N_2.

5 Conclusions

We performed an uncertainty quantification and sensitivity analysis study of an expanding hypersonic air flow in the HEG nozzle. For the condition considered, apart from uncertainties in the reservoir conditions, the most significant source of uncertainty in the flow parameters at the nozzle exit were uncertainties in the data on O_2-O VT relaxation rate and uncertainties in the data on dissociation rate coefficients of O_2 and NO. The vibrational temperature of O_2, and molar fraction of O_2, O, and NO all exhibited significant (at least 10%) uncertainties at the nozzle exit. Further planned extensions to the present work include accounting for vibrational non-equilibrium of NO, incorporation of vibration-chemical coupling into the exchange reaction modeling, and a more detailed analysis of uncertainties in specific dissociation–recombination pathways on the flow.

Acknowledgments. Georgii Oblapenko acknowledges the funding provided by the Alexander von Humboldt foundation for his stay as a guest researcher at the German Aerospace Center (DLR).

References

1. Cortesi, A.F., Constantine, P.G., Magin, T.E., Congedo, P.M.: Forward and backward uncertainty quantification with active subspaces: application to hypersonic flows around a cylinder. J. Comput. Phys. **407**, 109,079 (2020)
2. Danehy, P., et al.: Non-intrusive measurement techniques for flow characterization of hypersonic wind tunnels. Flow characterization and modeling of hypersonic wind tunnels (NATO Science and Technology Organization Lecture Series STO-AVT 325), NF1676L-31725-Von Karman Institute, Brussels, Belgium (2018)
3. Gerstner, T., Griebel, M.: Numerical integration using sparse grids. Numer. Algorithms **18**(3), 209–232 (1998)
4. Gimelshein, S.F., Wysong, I.J.: Nonequilibrium effects in high enthalpy gas flows expanding through nozzles. Phys. Fluids **33**(10), 106,104 (2021)
5. Gimelshein, S.F., et al.: Kinetic and continuum modeling of high-temperature air relaxation. J. Thermophys. Heat Transf. **36**(4), 1–23 (2022)

6. Gu, S., Hao, J., Wen, C.y.: State-specific study of air in the expansion tunnel nozzle and test section. AIAA J. **60**(7), 1–15 (2022)
7. Hannemann, K., Martinez Schramm, J., Wagner, A., Ponchio Camillo, G.: The high enthalpy shock tunnel Göttingen of the German aerospace center (DLR). J. Large-Scale Res. Facil. **4**(A133), 1–14 (2018)
8. Karl, S., Martinez Schramm, J., Hannemann, K.: High enthalpy cylinder flow in HEG: a basis for CFD validation. In: 33rd AIAA Fluid Dynamics Conference and Exhibit, p. 4252 (2003)
9. Klomfass, A.: Hyperschallströmungen im thermodynamischen Nichtgleichgewicht. Ph.D. thesis, RWTH Aachen (1995)
10. Kustova, E., Nagnibeda, E., Oblapenko, G., Savelev, A., Sharafutdinov, I.: Advanced models for vibrational-chemical coupling in multi-temperature flows. Chem. Phys. **464**, 1–13 (2016)
11. Marrone, P.V., Treanor, C.E.: Chemical relaxation with preferential dissociation from excited vibrational levels. Phys. Fluids **6**(9), 1215–1221 (1963)
12. Millikan, R.C., White, D.R.: Systematics of vibrational relaxation. J. Chem. Phys. **39**(12), 3209–3213 (1963)
13. Nagnibeda, E., Kustova, E.: Non-Equilibrium Reacting Gas Flows. HMT, Springer, Heidelberg (2009). https://doi.org/10.1007/978-3-642-01390-4
14. Oblapenko, G.: Calculation of vibrational relaxation times using a kinetic theory approach. J. Phys. Chem. A **122**(50), 9615–9625 (2018)
15. Oblapenko, G., Volker, H.: Sensitivity analysis study of expanding hypersonic flows of nitrogen and oxygen. In: 9th European Conference for Aeronautics and Space Sciences (EUCASS). EUCASS (2022). https://doi.org/10.13009/EUCASS2022-6032
16. Park, C.: Review of chemical-kinetic problems of future NASA missions. I-Earth Entries. J. Thermophys. Heat Transf. **7**(3), 385–398 (1993)
17. Sharma, S.P., Ruffin, S.M., Gillespie, W.D., Meyer, S.A.: Vibrational relaxation measurements in an expanding flow using spontaneous Raman scattering. J. Thermophys. Heat Transf. **7**(4), 697–703 (1993)
18. Shoev, G., Oblapenko, G., Kunova, O., Mekhonoshina, M., Kustova, E.: Validation of vibration-dissociation coupling models in hypersonic non-equilibrium separated flows. Acta Astronaut. **144**, 147–159 (2018)
19. Smith, R.C.: Uncertainty Quantification: Theory, Implementation, and Applications, vol. 12. SIAM, Philadelphia (2013)
20. Streicher, J.W., Krish, A., Hanson, R.K.: Vibrational relaxation time measurements in shock-heated oxygen and air from 2000 K to 9000 K using ultraviolet laser absorption. Phys. Fluids **32**(8), 086,101 (2020)
21. Weaver, A.B., Alexeenko, A.A., Greendyke, R.B., Camberos, J.A.: Flowfield uncertainty analysis for hypersonic computational fluid dynamics simulations. J. Thermophys. Heat Transf. **25**(1), 10–20 (2011)
22. West, T.K., IV., Hosder, S.: Uncertainty quantification of hypersonic reentry flows with sparse sampling and stochastic expansions. J. Spacecr. Rockets **52**(1), 120–133 (2015)

Influences on Langmuir Probe Measurements by an ECR Thruster with Magnetic Nozzle

Clara Schäfer[1,2(✉)], Jana Zorn[2], Kristof Holste[2], and Peter J. Klar[2]

[1] Deutsches Zentrum für Luft- und Raumfahrt e.V.,
Institut für Aerodynamik und Strömungstechnik, Raumfahrzeuge Göttingen,
Bunsenstraße 10, 37073 Göttingen, Germany
clara.schaefer@dlr.de

[2] Justus-Liebig-Universität Gießen, I. Physikalisches Institut, Heinrich-Buff-Ring 16,
35392 Gießen, Germany

Abstract. Ensuring reliability in the measurement and interpretation of plasma diagnostics is essential for characterizing thruster concepts in the growing field of electric propulsion. Langmuir probes are often used to determine the properties of low temperature plasmas. However, not much research is found on Langmuir probe measurements on electron cyclotron resonance thrusters with magnetic nozzle, especially no detailed studies on variable probe orientations with respect to magnetic field lines. The question arises which differences can be shown in the electron energy distribution regarding the probes orientation and in what way it is possible to interpret probe alignment consequences on determined plasma characteristics. We perform our study as follows: We measure in two orientations, parallel and orthogonal to magnetic field lines, at different operating states of the thruster (variable power and volume flow settings). The Langmuir probe data shows clearly a dependency on the probes orientation. We use the measured ion energy distribution of the retarding potential analyzer as a comparison tool to the determined electron energy distribution of the single Langmuir probe measurements. Due to the determined anisotropic non-Maxwellian distribution mainly visible in parallel orientation we conclude a parallel orientation of Langmuir probes with respect to magnetic field lines is preferable for ECRT with MN.

Keywords: ECRT · magnetic nozzle · Langmuir probe · EEDF

1 Introduction

Since the beginning of the 20th century, electric space propulsion has been considered as an alternative technology to conventional chemical applications [1]. Typical challenges with existing ion thruster technologies are erosion of electrodes and grids and the need for reliable neutralisers [1]. Plasma generation

A. Dillmann et al. (Eds.): STAB/DGLR Symposium 2022, NNFM 154, pp. 153–162, 2024.
https://doi.org/10.1007/978-3-031-40482-5_15

and acceleration through an electrode- and neutralizer-less concept is therefore motivated. This includes plasma generation by electron cyclotron resonance (ECR). The electron cyclotron resonance thruster (ECRT) under investigation, developed within the project MINOTOR (Magnetic Nozzle Electron Cyclotron Resonance Thruster) under the lead of ONERA (fr. Office national d'études et de recherches aérospatiales), realises plasma generation by ECR and an electrodeless acceleration by a diverging magnetic field, the magnetic nozzle (MN) [2].

Ensuring reliability in the measurement and interpretation of plasma diagnostics is essential for characterizing plasma-based thruster concepts. In order to avoid possible influences on diagnostics as well as misinterpretation of the data, a precise understanding of the selected diagnostics is necessary. Also, a possible change of the operating state of the thruster or beam by the measuring instrument itself should not be neglected, which mainly concerns invasive diagnostics such as Langmuir probes. A distinction is made between probes with one, two or three conductive pins in various shapes, directly exposed the plasma. By measuring a current-voltage characteristic it is possible to capture properties of the plasma, such as temperatures, potentials, densities etc [3]. The effect of the magnetic field on Langmuir probe measurement is considered in Ref. [4] and recommendations for the orientation of several probes are given. In Ref. [4] it is stated, that the effective collection area for magnetized electrons is reduced to the cross-sectional area of the probe that crosses the magnetic field lines. Therefore they state that a cylindrical probe is to be oriented perpendicular to the magnetic field lines. However, they further note that the orientation of the probe is negligible if the Larmor radius of the particles is much bigger than the radius of the probe, which is in many cases given [4]. Since we are dealing with a plasma which is accelerated, there is a preferred spatial direction of the energy present [5]. Therefore, we are dealing with an anisotropic and possibly non-Maxwellian plasma. In this case Lobbia et al. recommend to align the probe parallel to this preferred direction [4]. First of all, in our case both orientation recommendations contradict each other (the direction of the anisotropic drift and the magnetic field direction are parallel to each other, compare Fig. 1). Furthermore, the excitation by the ECR additionally accelerates the electrons perpendicular to the magnetic field lines. For this reason, it is impossible to get a clear recommendation from the literature for the orientation of a Langmuir probe in an ECRT with MN. In the experimental study of Lafleur et al. [6], the effect of magnetic fields on the relationship between ion energy and electron temperature is investigated with Langmuir probe measurements on the ECRT with MN. They performed the experimental study with the probe oriented orthogonal to the magnetic field lines [6]. As a result the ratio shows to be a function of both the magnetic field strength, as well as the electron energy distribution function (EEDF) [6].

However, no detailed studies on variable orientations with respect to magnetic field lines have been performed yet to underlay the validity of the statement of Lobbia et al. [4] for ECRT with MN. The possible influences on the captured probe data by [6] due to the magnetic field is therefore not covered. The question arises what differences can be shown in the electron energy distribution regarding

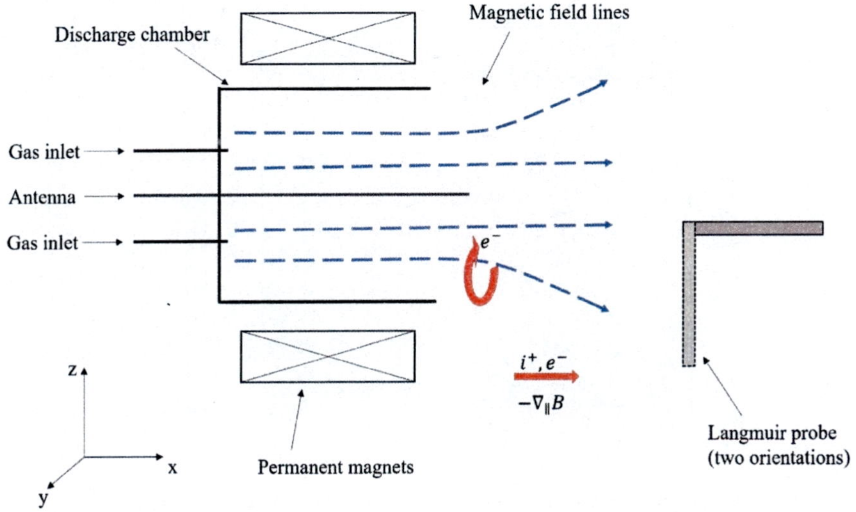

Fig. 1. Schematic of the ECRT with MN plus Langmuir probe orientation. The thruster is pictured in black. The magnetic field lines are pictured in blue, and the particle motions are pictured in red. Here e^- are the electrons and i^+ the ions. Langmuir probe orientation, parallel and orthogonal to magnetic field lines is pictured (dark grey=parallel orientation, light grey=orthogonal orientation). The schematic is not to scale. (Color figure online)

the probes orientation and in what way it is possible to interpret probe alignment consequences on determined plasma characteristics. This would enable us to specify a preferred orientation for the Langmuir probes for future measurements on an ECRT with MN. To answer these questions we perform our study as follows: We map the magnetic field of the chosen thruster in two dimensions with a Gaussmeter. This allows us to orientate the probe in a chosen manner. We measure parallel and orthogonal to magnetic field lines with a single Langmuir probe. We measure at different operating states of the thruster (variable power an volume flow settings). By mapping the beam with a Faraday cup (FC) array we can align and perform measurements with a retarding potential analyzer (RPA).

2 Experimental Setup

The reported experiments are conducted at the Justus Liebig University of Giessen in the vacuum facility *Jumbo* [7]. With a length of 6 m and a diameter of 2.6 m [7] it allows the simultaneous measurements with multiple diagnostics. This vacuum chamber is equipped with cryo- and turbopumps facilitating a pumping speed of 150 000 l/s. During experiments the cryopumps are not used, leading to a base pressure between 10^{-6} and 10^{-5} mbar. We choose the ECRT from the MINOTOR project to serve as test object since the thruster characteristics are

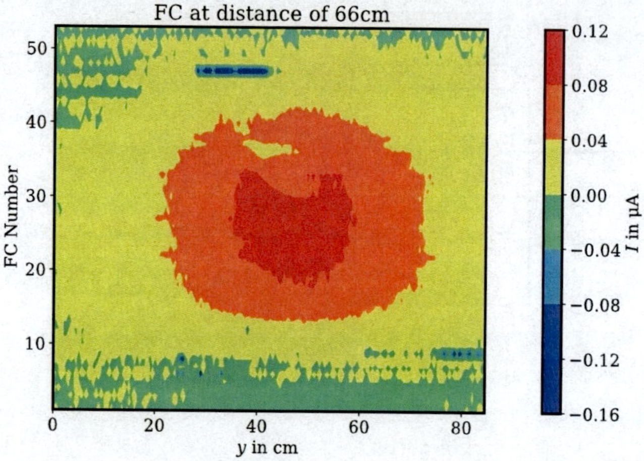

Fig. 2. FC current measurements of ECRT beam. The colorbar gives the actual measured current, not a current density of the beam. The thruster is located near cup 28 at 50 cm. Cup 48 malfunctioned in region 30–40 cm. The asymmetry in the upper area of the beam (Cup 30 to 40 at 40–50 cm) is most likely due to the Langmuir probe, which is located at this point at the time of the measurement.

investigated in multiple publications [2,6,8]. The working principle of the ECR thruster with MN is the ionization of the propellant via ECR and the acceleration by a divergent magnetic field [9]. In presence of a magnetic field, charged particles (electrons and ions) are trapped along the magnetic field lines in such a way that they are circulating (gyrating) around them, due to the Lorentz force [3]. If an electromagnetic wave with the cyclotron frequency is applied the electron is resonantly accelerated by the electric field of the wave. It absorbs the energy of the electromagnetic wave, gains kinetic energy and increases impact ionization process rate [3,10]. The acceleration of the produced ions is assumed to be described by two processes: First the faster reaction of the electrons to density disturbances or density gradients, due to their lower mass [3]. As a result of the charge separation an ambipolar field forms, leading to the acceleration of the ions towards the negative space charge [3]. The second driving mechanism is the gradient in the magnetic field. Due to the conservation of the magnetic moment of the charged particles an acceleration opposite to the magnetic field gradient direction $\nabla_{||}B$ can be observed [3,11,12]. The plasma detachment, the electromagnetically exiting of the ions and electrons, is assumed to be explained either by frozen-in magnetic flux, according to ideal magnetohydrodynamic theory for sufficient high density plasma [13] or by plasma detaching as soon as the ion gyro radius is equal to the radius of the plasma cross-section and the ions are assumed to demagnetize [14]. To create a large basis for comparison we change the thruster operating conditions around its optimal operation point. This is found in literature to be 25 W microwave input power at 2.45 GHz and 1 sccm volume flow of Xenon [2,8]. The Langmuir probe is a technically simple diagnostics tool, yet the measurements are difficult to interpret [3]. If a conductor is introduced into the plasma and the voltage U varies with respect to a reference potential, the current-voltage-characteristic can be divided into three ranges according to Ref. [15]: For sufficiently negative probe voltage the probe collects mainly positive ions, called the ion saturation current $I_{\mathrm{i,sat}}$. In the transition part of the characteristic the probe collects ions and electrons [15]. If no current is measured the ion and electron currents are equal, this is the floating potential Φ_{fl} [15]. For the set voltage equal to the plasma potential Φ_{P} the characteristic may show a kink ('knee') because the potential changes character from attracting ions and repelling electrons to repelling ions and attracting electrons [15]. For higher positive voltage the probe only collects electrons, called the electron saturation current $I_{\mathrm{e,sat}}$ [15]. By measuring this current-voltage characteristic it is possible to capture properties of the plasma, such as temperatures, potentials, densities etc [3]. This requires an appropriate theory. With the Druyvesteyn method we determine directly the EEDF $f(E)$ with the second derivative of the measured current voltage characteristic $\frac{\mathrm{d}^2 I_{\mathrm{e}}}{\mathrm{d}U^2}$, the probe surface A_{P}, the electron mass m_{e} and charge e and the energy of the electron E hitting the probe (Eq. 1) [4,16]:

$$f(E) = \frac{2}{e^2 A_{\mathrm{P}}} \sqrt{(2 m_{\mathrm{e}} E)} \frac{\mathrm{d}^2 I_{\mathrm{e}}}{\mathrm{d}E^2} \tag{1}$$

We determine the electron current I_{e} by subtracting the ion current I_{i} from the measured current I. By taking the moments of the distribution function we come to electron density (Eq. 2) and temperature (Eq. 3) of the plasma:

$$n_{\mathrm{e}} = \int_0^\infty f(E)\mathrm{d}E \tag{2}$$

$$T_{\mathrm{e}} = \frac{2}{3 n_e} \int_0^\infty f(E) E \mathrm{d}E \tag{3}$$

This model assumes an isotropic plasma which is most likely not present. Beam-like anisotropy would retain the proportionality to the second derivative, however contain additional anisotropy coefficients [4]. It is this anisotropic effect on the electron current collection stated to be mitigated when the anisotropic drifting beam component is parallel to the electrode surface according to [4]. To evaluate certain recommendations both ways, parallel and orthogonal, are tested (compare Fig. 1). We use a Hall probe and a 3D positioning system to systematically map the magnetic field strength and lines. This allows us to orientate the probe to the measured magnetic field lines of the thruster. The Langmuir probe tips are in both orientations positioned $x = 6\,\mathrm{cm}$ from the thruster exit plane and $z = 5\,\mathrm{mm}$ from the antenna position. We determine the surface of the probe to be $A_P = 1.59 \cdot 10^{-5}\,\mathrm{m}^2$ with a radius of 0.5 mm. We scan with a source and measurement unit (SMU, Keysight B2901A) over a voltage range of -100 V to +200 V in 0.1 V steps with a resolution 1 µV and measure the current I with a resolution of 100 fA coming from the probe. The vacuum chamber potential/ground is the reference potential. In order to have a comparison to the determined electron energy distribution of the single Langmuir probe measurements we want to measure the ion energy distribution with an RPA. RPAs belong to the family of electrostatic analyzers. They are often used in electric propulsion to measure electron or ion distribution functions from which one can deduce the charged particle densities and temperatures [17]. The basic operating principle of an RPA is as follows: A collector consisting of a conductive material is inserted into the plasma and shielded by a plate with a hole/grid. If a positive potential is now applied between the plate and the plasma, only ions whose energy exceeds this potential can reach the collector, thus allowing current measurement [17]. By sweeping the potential, the energy spectrum of the ions (or optionally electrons, if a negative potential is applied) can be recorded. Additional grids can be inserted to prevent secondary electrons or space charging effects [17]. The ion energy distribution function (IEDF) is then proportional to the first derivative of the measured current [17]. In order to align the RPA properly we perform a beam scan with an array of 53 separate Faraday cups (Fig. 2). The scanning of the ion beam allows the alignment of the RPA in a distance of $x = 66$ cm to the thruster exit plane. The used RPA consists of four grids. The grids have a hole diameter of 1.1 mm, a width of 1 mm and have distance of 2 mm to each other. The distance between the last grid and the collector is 1 mm. In addition, we use a collimator with a length of 20 cm. The first grid is set to chamber potential/ground. The second grid is used as a repeller for the electrons and is set at −310 V. For the third grid a positive voltage sweep is performed from 0 V to 400 V. The fourth grid is used as a suppressor to avoid secondary electrons hitting the collector and is set at −40 V. As power supply for grid 1 and 3 two MCP14-2000 Fug power supplies were used and a GwInstek power supply for grid number 4. The current of the collector is measured with a Keithley 6485 Picoamperemeter.

3 Results

To allow a discussion regarding the role of Langmuir probe orientations regarding the magnetic field of ECRTs with MN we want to systematically compare raw data and determined energy distributions as well as plasma parameters.

We investigate multiple data sets and demonstrate here just a fraction. We start with the standard case of operating settings 25 W and 1 sccm. By comparing the raw current voltage characteristic of single Langmuir probe in two orientations major differences can already be detected. At the same thruster operating conditions at the same position to the thruster exit the parallel orientation leads to a higher overall signal than the orthogonal orientation. The determined plasma potentials and floating potentials differ, however they are in the same range of $\Phi_{\mathrm{fl}} = 10 \pm 10$ V and $\Phi_{\mathrm{P}} = 90 \pm 10$ V. For both characteristics no actual electron saturation is visible. For the evaluation of the measured current we use a Savitzky-Golay filter to get a filtered signal as well as derivations. We check for the standard deviation of the residuals until an acceptable window size is found. The evaluation is then performed as follows: First the floating potential is determined. Followed by that the ion saturation current is determined, for

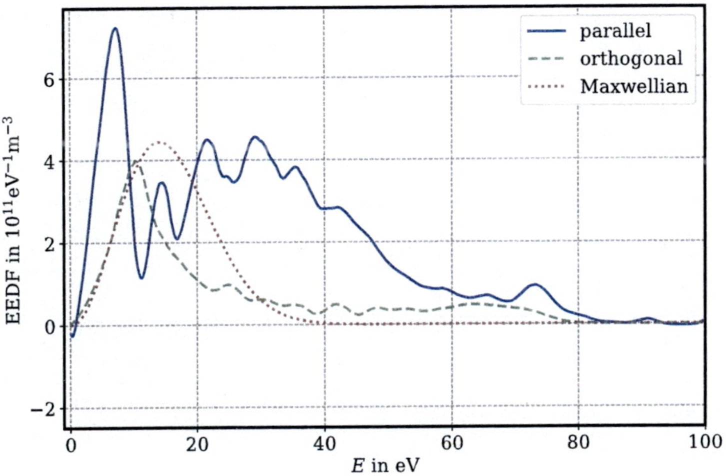

Fig. 3. Comparison of EEDF for parallel and orthogonal orientation. A Maxwellian distribution is pictured exemplarily. Thruster operating point is set at 22 W and 1 sccm.

the standard case in parallel orientation as $I_{\mathrm{i,sat}} = -2.2 \cdot 10^{-5}$ A. This is done by a linear fit for the ion saturation part of the curve until floating potential and check here for the intersection with the measured current axis. This determined ion current is then subtracted to get the electron current. The plasma potential is determined by checking for a maximum value in the first derivative. The set potential difference can now be transformed to energy scale, showing the energy an electron has regarding the bias voltage of the probe. With Eq. 1 we then determine the EEDF and compare the two orientations (see Fig. 3). We observe similar curve characteristics for all the performed evaluations for the different thruster operating conditions. In case of an orthogonal orientation the overall signal is smaller, but more important, it almost shows no deviations from a Maxwellian energy distribution. In contrary to the parallel orientation: Here we see always a high peak in the low energy region, around 8 eV, and a broadened peak in the higher energy region, with its maximum at around 30 eV. Additionally we want to point out that the position of the peak in orthogonal case, lays more in between the other two, leading to an determined energy of around 12 eV. By the integration over the EEDF one can determine the electron density (Eq. 2), as well as the electron temperature (Eq. 3). A difference in the distribution function leads also to different determined temperatures and densities. The determined electron temperature lays for both orientations for variable power and volume flow settings in the order of 18 ± 3 eV. The electron density is determined as in the order of $2.2 \pm 1 \cdot 10^{13}\,\mathrm{m}^{-3}$. As an example we want to present the raw data of the RPA measurements under the thruster conditions at 20 W and 1 sccm. In case of an Maxwellian distribution of the ions a sudden drop should be expected, leading to a sharp peak in the first derivative, hence the

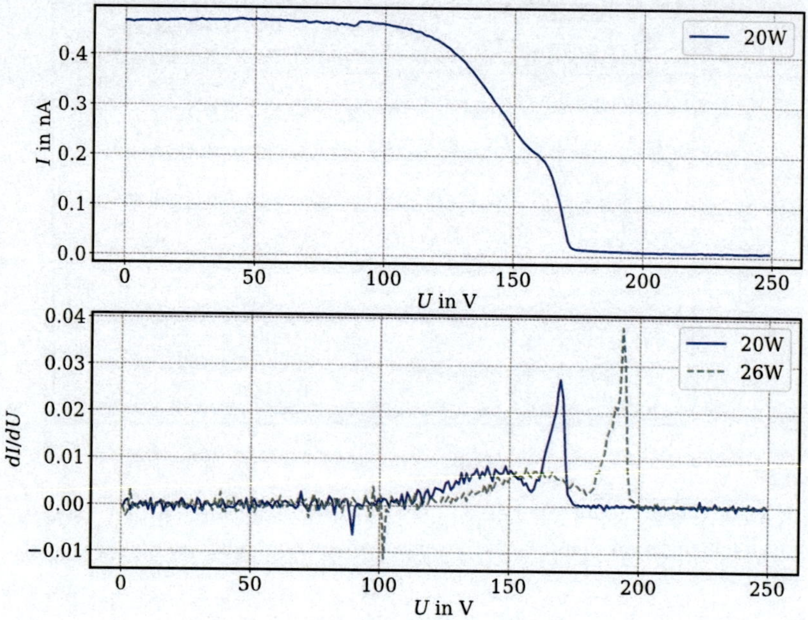

Fig. 4. RPA raw data (top) at 20 W and 1 sccm and comparison of the first derivative of RPA data (bottom) for thruster settings 20 W and 26 W at 1 sccm.

single energy distribution of the ions. Instead we see a slow decay in the range of 100 V to 1 sccm. Additionally we want to point out the small 'jump' around 80 V which is seen in many data sets. Taking the first derivative of the data set without smoothing leads us to the qualitative ion distribution, seen in Fig. 4. We plot two cases with 20 W and 26 W at 1 sccm for comparison reasons. Independent on the power settings we observe one sharp peak and a broadened. The sharp peak in the higher voltage region and the broadened one in the smaller voltage region. We see that with changing input power and volume flow rate the peaks shift position and height, but in general the shapes keep being the same, leading to a non-Maxwellian in the ion energy distribution but two energy species.

4 Discussion

The captured electron energy distribution functions (EEDF) show a highly anisotropic, non-Maxwellian character. We observe for several operating points of the thruster that the determined EEDF in parallel orientation show a stronger non-Maxwellian character than in orthogonal orientation. The resulting differences between obtained plasma parameters for the two different orientations are therefore explained. The RPA measurements show (minimum) two energy species, which could be an argument for a parallel orientation of the Langmuir probe, since the orthogonal orientation lead mainly to one species distributions.

The interpretation of the single Langmuir probe data shows a dependency on the probes orientation in the energy distribution of the electrons. It is therefore not possible to state, that the orientation of the probe with respect to the magnetic field lines does not play a role in ECRT with MN, even if the Larmor radius and probe size should assure that (according to Lobbia et al. [4]). We determined a Larmor radius around $4 \cdot 10^{-3}$ m vs a probe radius of $0.5 \cdot 10^{-3}$ m. The differences in the determined electron temperatures and densities lay, however, within typical theoretical probe accuracies, in the order of around 50% [4]. Overall we determine in every parallel orientation non-Maxwellian, minimum bi-energy distribution. Apparently anisotropy leads to just detecting the non-Maxwellian distribution with the probe orientated parallel to the magnetic field lines. We use the measured ion energy distribution of the RPA as a comparison tool to the determined EEDF of the single Langmuir probe measurements. With the RPA we detect independent of power or volume flow settings a non-Maxwellian ion energy distribution. We want to give some general remarks: Although the thruster settings and positions of the probes are kept the same, the background pressure is reported to play an important role [8]. Depending on daytime and sequence of the experiments, the background pressure varied. Furthermore the probe interferes with the plasma, which can already be seen in the Faraday cup measurements, seen in Fig. 2. A non invasive diagnostic method is much more preferable, however, not as simple in application. Inaccuracies in alignment and positioning of the probe cannot be out-ruled, as well as contamination of the probe surface by dirt or sputtered materials. In this work we did not yet taken the influence of the direction and strength of the input microwave into account, as well as plasma instabilities (plasma oscillations, drift waves etc.). Future work will include detection of prominent frequencies as well as power dependencies of energy distributions and plasma parameters.

If the acceleration process is explained by the ambipolar process, the energy distributions of electrons and ions are connected. If that is the case, information gets lost if in the EEDF just one energy species is detected while for the ions two energy species are determined. We therefore state, that due to the non-Maxwellian distribution mainly visible in parallel orientation, a parallel orientation of Langmuir probes with respect to magnetic field lines is preferable for ECRT with MN.

References

1. Holste, K., et al.: Ion thrusters for electric propulsion: scientific issues developing a niche technology into a game changer. Rev. Sci. Instrum. **91**(6), 061101 (2020)
2. Packan, D., et al.: The MINOTOR H2020 project for ECR thruster development. In: 36th International Electric Propulsion Conference (2017)
3. Stroth, U.: Plasmaphysik. Springer, Wiesbaden (2011). https://doi.org/10.1007/978-3-8348-8326-1
4. Lobbia, R.B., Beal, B.E.: Recommended practice for use of Langmuir probes in electric propulsion testing. J. Propuls. Power **33**(3), 566–581 (2017)

5. Hoegy, W.R., Brace, L.H.: Use of Langmuir probes in non-Maxwellian space plasmas. Rev. Sci. Instrum. **70**(7), 3015–3024 (1999)
6. Lafleur, T., Cannat, F., Jarrige, J., Elias, P.Q., Packan, D.: Electron dynamics and ion acceleration in expanding-plasma thrusters. Plasma Sources Sci. Technol. **24**(6), 065013 (2015)
7. Justus-Liebig Universität. Testanlagen Übersicht. https://www.uni-giessen.de/fbz/fb07/fachgebiete/physik/institute/ipi/raumfahrtphysik/Ionentriebwerke
8. Peterschmitt, S.: Development of a stable and efficient electron cyclotron resonance thruster with magnetic nozzle. PhD thesis, Institut polytechnique de Paris (2020)
9. Packan, D., et al.: H2020 MINOTOR: magnetic nozzle electron cyclotron resonance thruster. In: 36th International Electric Propulsion Conference (2019)
10. Fridman, A., Kennedy, L.A.: Plasma Physics and Engineering. CRC Press, Boca Raton (2004)
11. Longmier, B.W., et al.: Ambipolar ion acceleration in an expanding magnetic nozzle. Plasma Sources Sci. Technol. **20**(1), 015007 (2011)
12. Kosmahl, H.G., Miller, D.B., Bethke, G.W.: Plasma acceleration with microwaves near cyclotron resonance. J. Appl. Phys. **38**(12), 4576–4582 (1967)
13. Deline, C.A., et al.: Plume detachment from a magnetic nozzle. Phys. Plasmas **16**(3), 033502 (2009)
14. Cannat, F., Lafleur, T., Jarrige, J., Chabert, P., Elias, P.-Q., Packan, D.: Optimization of a coaxial electron cyclotron resonance plasma thruster with an analytical model. Phys. Plasmas **22**(5), 053503 (2015)
15. Demidov, V.I., Ratynskaia, S.V., Rypdal, K.: Electric probes for plasmas: the link between theory and instrument. Rev. Sci. Instrum. **73**(10), 3409–3439 (2002)
16. Druyvesteyn, M.J.: Der niedervoltbogen. Zeitschrift für Physik **64**(11), 781–798 (1930)
17. Lai, S.T., Miller, C.: Retarding potential analyzer: principles, designs, and space applications. AIP Adv. **10**(9), 095324 (2020)

Calibration of Heterodyne Dual Frequency Comb Laser Absorption Spectroscopy for NO and H_2O Detection

Leni Schmidt(✉) and Jan Martinez Schramm

Spacecraft Department, German Aerospace Center, Institute of Aerodynamics and Flow Technology, Bunsenstraße. 10, 37073 Göttingen, Germany
{Leni.Schmidt,Jan.Martinez}@dlr.de

Abstract. The hydrogen combustion process within the propulsion unit of wind tunnel scramjet models produces NO and H_2O. A precise determination of the concentration of both species applying laser absorption spectroscopy is necessary to validate numerical simulations. A novel approach to obtain absorption spectra of both species in the infrared region within μ-second range is based on the beating analysis of two combined frequency combs generated with quantum cascade lasers. This technique allows multi-species measurements of absorption spectra. The concentrations of species can be inferred from characteristic absorption features. The generated frequency combs are within the $1740 \pm 20\,\mathrm{cm}^{-1}$ wavenumber region which includes strong NO and H_2O spectral lines, thus enabling an identification of these species. The method offers a current-tuning based sweeping measurement mode for an increased spectral resolution. An extended calibration of all possible measurement modes under different conditions is reported.

Keywords: Laser absorption spectroscopy · Hydrogen combustion · Frequency combs · Sweeping · HEG

1 Introduction

Sustainability is one of the main driving factors in current research on civil aviation. While the environmental dimension requires new aviation concepts to be less polluting and without CO_2 emissions, there is also the desire for performance improvement by reducing the transfer time of long-range flights from an economic point of view. In this context, various concepts for high-speed vehicles are developed. The use of hydrogen combustion based scramjets as propulsion units is one promising possibility to drive hypersonic passenger transport. Although hydrogen-fueled vehicles prevent CO_2 emission, the formation of NO and H_2O as emission gases during the combustion process may have other environmental effects.

Therefore, the experimental determination of NO and H_2O production in hydrogen combustion is of interest for the study of scramjet models in the High Enthalpy Shock Tunnel Göttingen (HEG), one of the major European hypersonic test facilities. The test

A. Dillmann et al. (Eds.): STAB/DGLR Symposium 2022, NNFM 154, pp. 163–173, 2024.
https://doi.org/10.1007/978-3-031-40482-5_16

time for experiments in HEG lies typically in the ms range, which requires measurements techniques with high acquisition rates and high precision. The applicability of the heterodyne dual frequency comb based absorption spectroscopy with quantum cascade lasers (QCLs) in the infrared region within μ-second range has been demonstrated at HEG [1]. Optical fibers and couplers are necessary to guide the laser beam in and out of the spectrometer towards the HEG test section. Since the first application at HEG, the spectrometer's manufacturer developed new *sweeping* measurement modes that enable higher resolution measurements by modulating the laser current and temperature to change the spectral point spacing. The ultimate goal is to identify whether the optical fibers lead to major signal losses and how the measurement results are affected by that. But before employing the fibers, first calibration measurements are performed in a laboratory set-up to evaluate the accuracy and quality of the different modes. The paper reports on this preliminary approach, describes the experimental prerequisites for the measurements, and demonstrates the resulting comparison for the measurement modes.

2 Dual Comb Laser Absorption Spectroscopy

High-resolution molecular spectroscopy is important for many different fields of applications, including combustion diagnostics where the challenge is to observe multiple species on short time scales. Especially the mid-IR spectral domain is of interest since fundamental vibrational excitations are located in this range. The simultaneous accomplishment of crucial parameters such as good spectral resolution and wide coverage or high signal-to-noise ratio with short measurement times was often not feasible for different spectroscopic techniques. Single wavelength spectrometers only allow one narrow band measurement with high resolution, while Fourier transform spectroscopy (FTIR) measures broadband with rather low resolution and often FTIR systems include moving parts [2]. This was revolutionized by the advent of frequency combs whose generation is enabled by high-power quantum cascade lasers (QCLs). Frequency combs were first introduced to improve the precision of frequency measurements [3]. A single frequency comb is characterized by sharp spectral lines that are distanced by a repetition frequency f_{rep} at an offset frequency f_{ceo} [4] (Fig. 1):

$$f_n = n \cdot f_{rep} + f_{ceo}. \tag{1}$$

Importantly, the line-spacing (or f_{rep}) of frequency combs is extremely well defined, making them useful for spectroscopy. If two frequency combs with slightly different repetition frequencies $f_{rep,1}$ and $f_{rep,2}$ are combined, a heterodyne beating with a difference frequency of $\Delta f_{rep} = f_{rep,2} - f_{rep,1}$ is created [5]. Since this set-up effectively directly links the frequency comb from optical domain to the RF frequencies, it is possible to take measurements using conventional frequency counters [4]. The difference frequencies lie in the radio frequency (RF) range that falls within the detector's limits ($<$10 GHz). This use of two frequency combs with the multi-heterodyne detection, so-called dual-comb spectroscopy, enables Fourier transform spectroscopy with potentially high resolution, high sensitivity, and no moving parts. By tuning a frequency comb via current or temperature, the comb's repetition frequency is changed, resulting in heterodyne with slightly changed difference frequency which allows measurements at diverse spectral points.

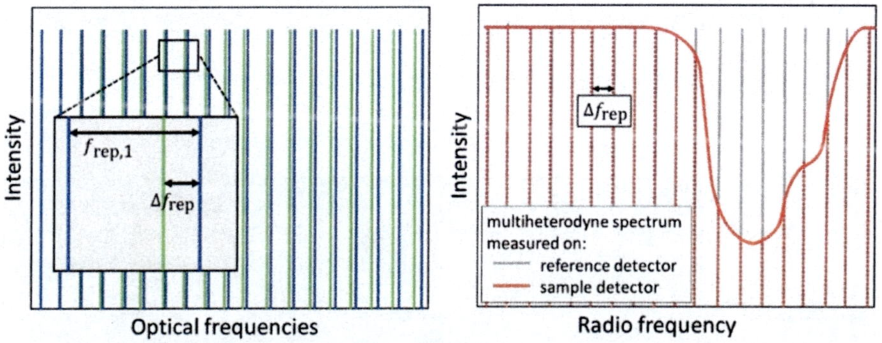

Fig. 1. Left: Illustration of two frequency combs with slightly different repetition frequencies, generated by the QCLs. Right: Resulting heterodyne beating. The sample beam is affected by an absorption feature, while the reference beam reaches the detector without attenuation.

As this sweeping procedure reduces the spectral point spacing, a spectrum has a higher resolution. More details about the sweeping technique are given in Sect. 3.1.

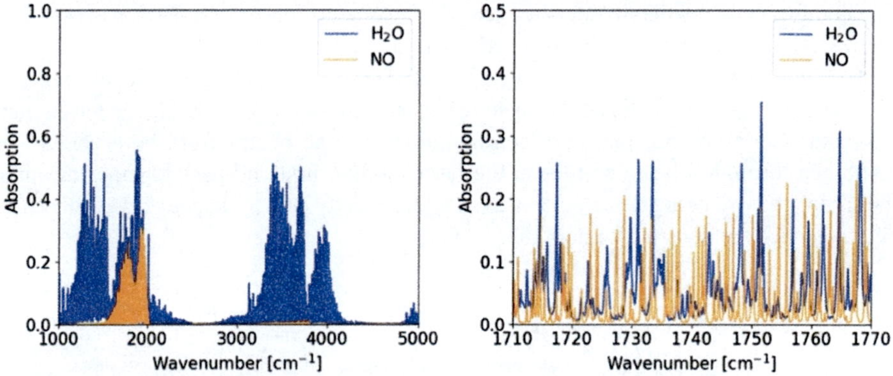

Fig. 2. Broad (left) and narrow (right) absorption spectra of NO and H_2O for a hydrogen combustion flow case in HEG with parameters: $p = 4200$ mbar, $T = 2300$ K, $l = 6.6$ cm, $n_{H2O} = 0.08$, $n_{NO} = 0.03$. Simulated with HITRAN database [7].

In the High Enthalpy Shock Tunnel in Göttingen (HEG), combustion experiments are conducted; during hydrogen combustion, NO and H_2O are produced as main components. Figure 2 (left) presents a broad NO and H_2O absorption spectra for a typical hydrogen combustion flow case ($p = 4.2$ bar, $T = 2300$ K) that once was studied in a scramjet small-scale experiment [1]. Figure 2 (right) shows features of the absorption spectra in the QCL emission region. It is concluded that the absorption of H_2O and NO can be separated and the reconstruction of the state of both species can be performed within one measurement.

3 Experimental Apparatus

3.1 Spectrometer Principle

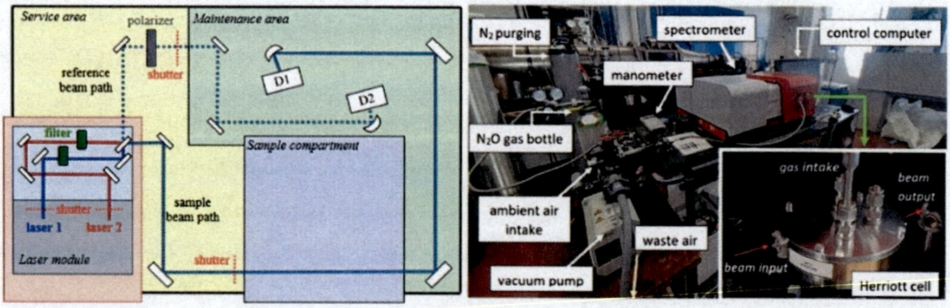

Fig. 3. Left: Beam path inside the spectrometer. The QCL generate two frequency combs that are combined with a beam splitter. Sample (solid) and reference (dotted) are aligned on detectors D1 and D2, respectively. Right: Experimental set-up with internal Herriott cell.

The commercially available dual-frequency-comb spectrometer IRis-F1 (IRsweep AG, Switzerland) uses two closely matched QCL frequency comb sources emitting at a center frequency of 1740 cm^{-1} with a span of 40 cm^{-1} and a standard repetition rate of 0.25 cm^{-1}. A schematic of the spectrometer is shown in Fig. 3. The spectrometer is operated in amplitude sensitive configuration, i.e. the beams from both QCLs are combined through a beam splitter in the laser module first and pass together through the sample before being detected on a thermoelectrically cooled detector. The reference beam reaches the detector without disturbances.

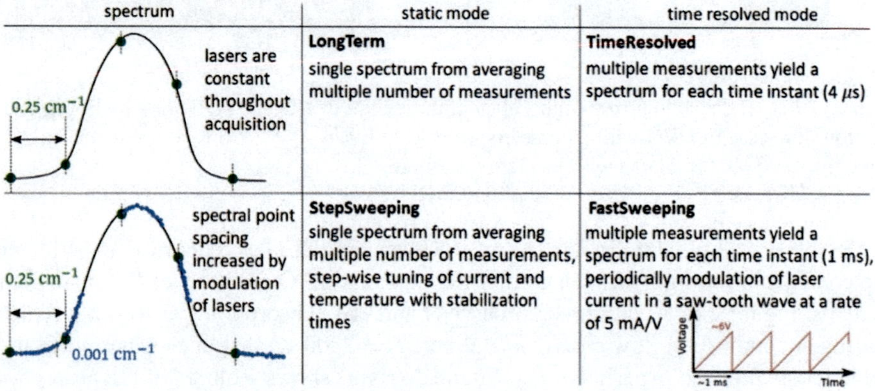

Fig. 4. Overview of possible measurement modes.

The spectrometer allows four different measurement modes (Fig. 4): *LongTerm, Time Resolved, StepSweeping* and *FastSweeping.* In *LongTerm* mode, a single spectrum is obtained from averaging multiple number of measurements; in *TimeResolved* mode, multiple measurements are taken, yielding a spectrum for each time instant. Both these modes leave the lasers current and temperature constant throughout the acquisition allowing time resolutions at 4 μs and a spectral point spacing of 0.25 cm^{-1} [5]. But this point spacing between adjacent lines is generally too large to sample all narrow absorption features of molecular gases sufficiently. The tuning of the QCL frequency combs' emission frequencies via temperature and current reduces the gaps between adjacent lines. By alternately and stepwise increasing the current of both lasers in *StepSweeping* mode, the gaps between points are closed without moving the multiheterodyne beat note to frequencies outside the detection bandwidth (Fig. 5) [6]. To prevent possible QCL mode switches during operation or the generation of subcombs due to a too large current tuning range, it is necessary to define a suitable current ramp consisting of stepwise tuning of the currents. By sweeping through this ramp at different temperatures, the gap is filled. After every current or temperature step, the lasers wait for a given stabilization time Δt_{stab} before continuing the acquisition. The standard stabilization times for current and temperature steps are Δt_{stab}= 0.7 s and 60 s, respectively. This method decreases the spectral point spacing to 2 10^{-3} cm^{-1} (Fig. 5). In *FastSweeping* mode, only the current of the lasers is periodically modulated in a saw-tooth wave at a rate of 5 mA/V, by a wave generator. The temperature is constant and the current modulation does not include stabilization times. Each ramp is set to increase from 0 V to approximately 6 V and takes about 1 ms to sweep through. The ramps are recorded and thus, the frequency changes can be precisely coordinated in the post processing. This method decreases the time resolution of the measurements, but increases the spectral resolution by a factor of almost 250. This is a theoretical value when sampling with 4 μs during 1 ms that is checked experimentally as well.

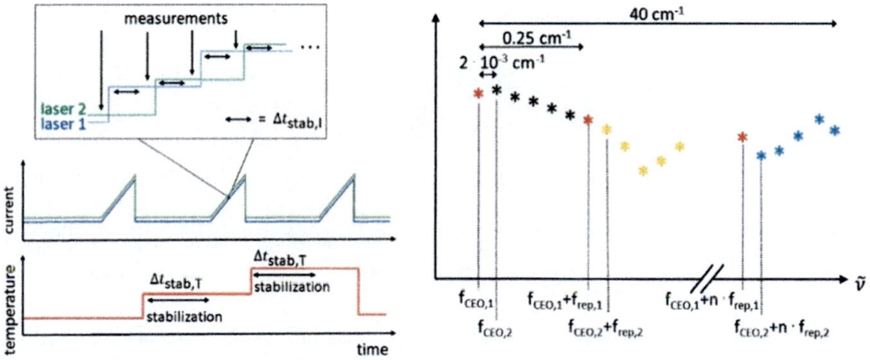

Fig. 5. Illustration of *StepSweeping* technique. Left: Laser current and temperature modulation during sweeping. Before the measurement is taken, the lasers stabilize for a given time Δt_{stab}. Right: Schematic transmission spectrum. Different colors indicate different comb lines that are tuned. Red points indicate comb lines in *LongTerm* mode.

During operation, the spectrometer is continuously purged with N_2. Before starting a measurement, transfer functions are taken with the lasers light blocked. These functions eliminate the electronic background signals of the system itself.

3.2 Experimental Details

To investigate the accuracy of the different measurement modes, experiments with exactly known conditions are carried out and compared with the reference spectra simulated with the HITRAN database [7].

By reflecting a beam multiple times in a circular shaped cell with a series of curved Au coated mirror segments on the wall, a Herriott cell achieves a long path length (here: 4 m) within a compact volume und hence is practical for precise absorption measurements. The cell is placed inside the spectrometer's sample compartment and filled with either N_2O or ambient air containing water (8500 ppm). The water content was determined by measuring temperature and relative humidity of the surrounding air. NO is not suitable as test gas since it is only weakly absorbing at room temperature and thus not optimal for the purpose of this work. Prior to the sample measurements, the cell is in vacuum and a background measurement is taken. The comparison between spectra before and after gas inlet allows observing the gas content. This procedure is done in *LongTerm, StepSweeping* and *FastSweeping* mode and at different pressures (100, 300, 500, 700 and 900 mbar). As the *FastSweeping* mode modulated the laser current on a millisecond scale without any laser stabilization time, while for *StepSweeping* it is recommended to let the lasers stabilize, the current stabilization time is varied in the *StepSweeping* measurements between $\Delta t_{\text{stab, I}} = 0.7$ and 0.00001 s to understand the accuracy of the measurement mode.

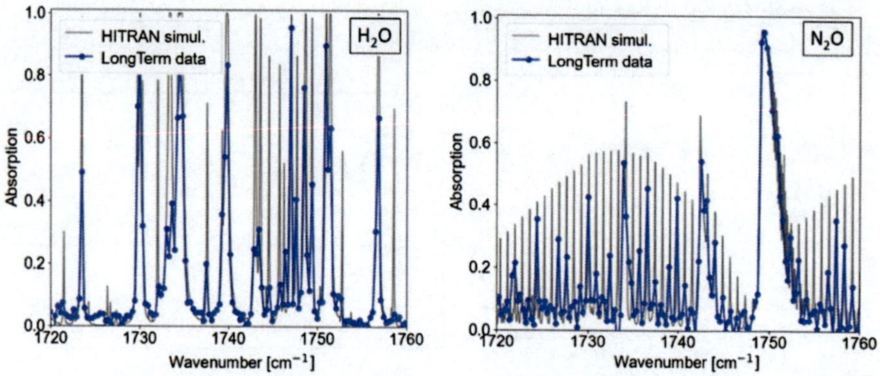

Fig. 6. Absorption spectra obtained from *LongTerm* measurements for ambient H_2O (left) and N_2O (right) with simulated HITRAN reference spectra ($p = 500$ mbar).

4 Results

First, measurements in normal *LongTerm* mode were performed. As shown in Fig. 6 for H_2O and N_2O (p = 500 mbar), the data at the measured spectral positions align well with the simulated HITRAN reference spectrum. But it has to be noted that not all absorption features are fully resolved. As inferred from a detailed closer view on specific features (Fig. 7), it is obvious that the spectral distance is too large between adjacent spectral points. This issue is overcome by using the *StepSweeping* mode. For a current stabilization time $\Delta t_{\mathrm{stab,\,I}} = 0.1$ s, the obtained spectra for $p = 500$ mbar are compared with the corresponding reference spectra in Fig. 8. In general, the increased spectral resolution results in a complete registration of all absorption features (Fig. 9).

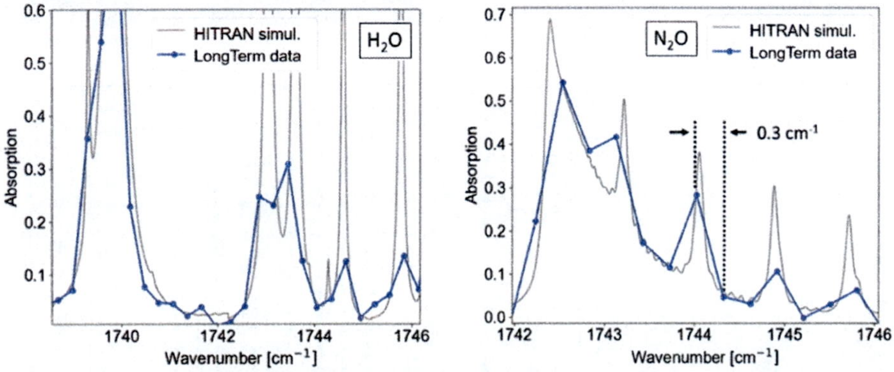

Fig. 7. A zoom on the spectra from Fig. 6 reveals that the large spectral point distance leads to missed absorption features for H_2O (left) and N_2O (right).

But despite the purging and an extended evacuation time of the Herriott cell, a few individual water molecules remain in the measurement area, thus resulting in parasitic absorption features in the measured data (marked red in Fig. 8). Based on measurements without gas insertion to the cell (p = 0 mbar), the affected wavenumber regions are masked out for the following evaluation. A cleaned spectrum corresponds to only the blue measurement points in Fig. 8 without the red marked regions.

As the current stabilization time was varied for different measurements, the absorption spectra are compared with regard to possible changes in the accuracy or shape of peaks. In the example for $\Delta t_{\mathrm{stab,\,I}} = 0.1$ s and $\Delta t_{\mathrm{stab,\,I}} = 0.00001$ s (Fig. 10), the spectra are quite similar by eye. But is has to be noted that a shorter stabilization time results in gaps in the absorption spectrum as exemplary highlighted with the red arrows in Fig. 10. This can be explained by the fact that since the lasers do not have much time to stabilize completely, they don't completely adapt to the pre-defined currents and thus the defined current range is not fully exploited as for measurements with longer stabilization time.

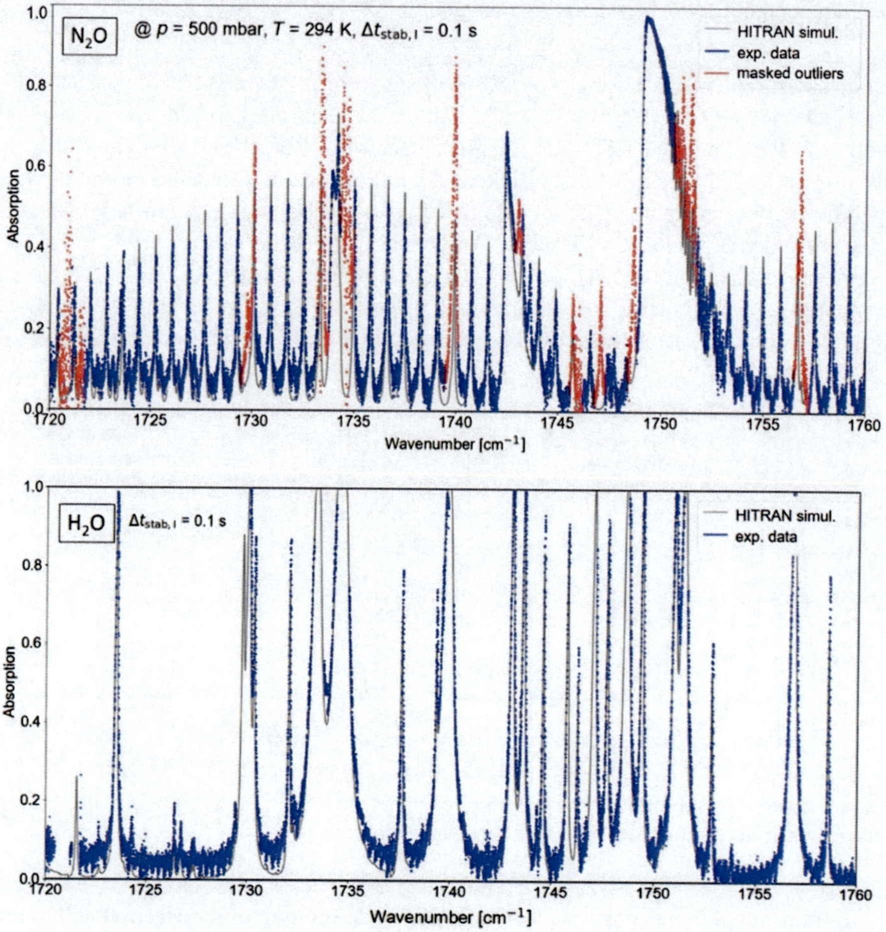

Fig. 8. Absorption spectra obtained in *StepSweeping* mode for N_2O (top) and H_2O (bottom) for $p = 500$ mbar.

A qualitative comparison of all the measured data for the different pressures and stabilization times is done based on the RMS. The RMS was calculated for deviations of experimentally measured amplitudes from the reference HITRAN amplitudes for all spectral measurement points. The results for an average of two measurement series with same conditions in each case are presented in Fig. 11. For a decreasing stabilization time, the RMS tends to slightly rise. Over the entire stabilization time range, the observed maximum increase is plus 3% (for H_2O at $p = 100$ mbar), while the RMS on average over all measurements increases by around 1% between $\Delta t_{\mathrm{stab,I}} = 0.7$ s and $\Delta t_{\mathrm{stab,I}} = 0.00001$ s, so it does not become significantly worse. Considering that the current stabilization time is reduced over almost five orders of magnitude, the deviation seems negligible.

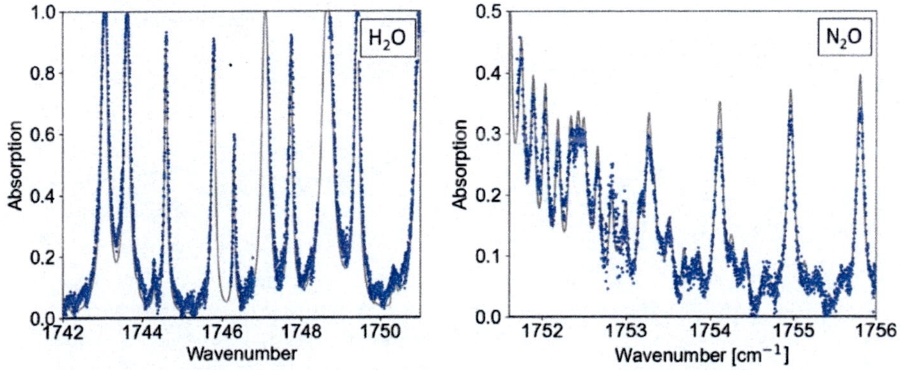

Fig. 9. A zoom on the spectra from Fig. 8 confirms that the *StepSweeping* mode reduces the large spectral point distance successfully and thus all absorption features are resolved better.

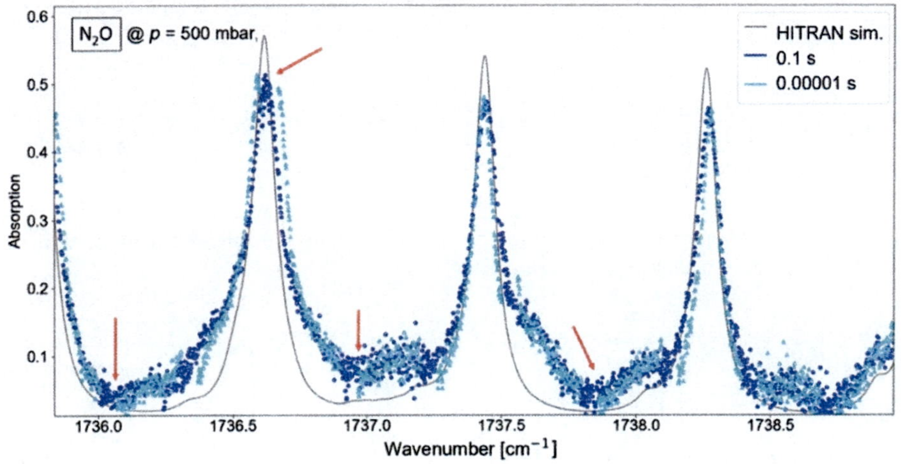

Fig. 10. Measurements with different current stabilization times align. Red arrows indicate that gaps are not completely filled for shorter stabilization time.

Another observation are saturation effects on the detector that influence the heterodyne detection for higher pressure. An example is shown in Fig. 12 for a N_2O measurement at $p = 900$ mbar. Although the features are all resolved and the upper part of the peaks align with the reference spectrum, the baseline is shifted upwards and the absolute values do not match with the reference.

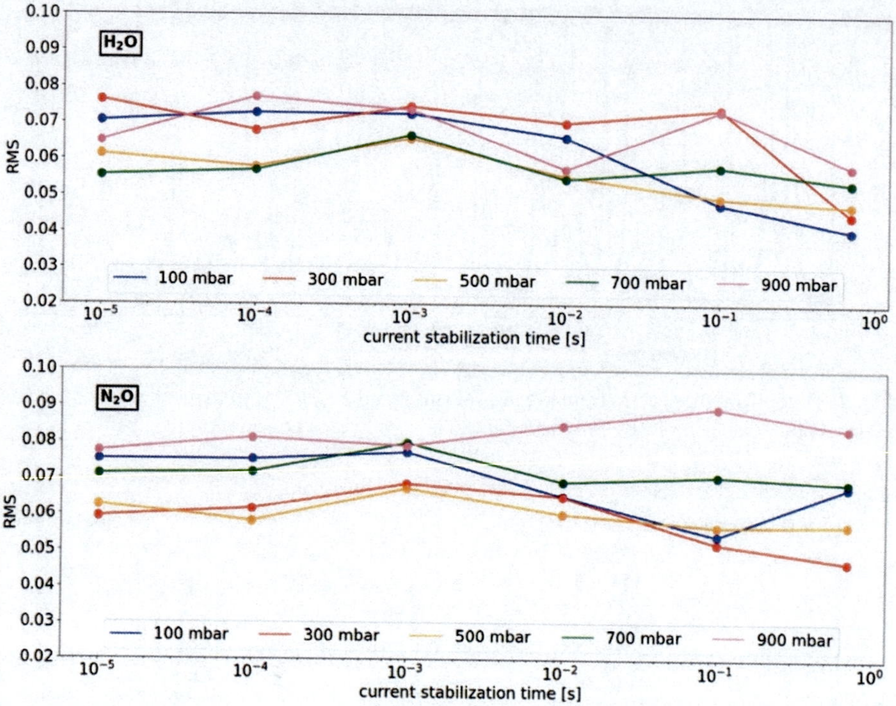

Fig. 11. The calculated RMS for H_2O (top) and N_2O (bottom) for different stabilization times and pressures, averaged over two measurements.

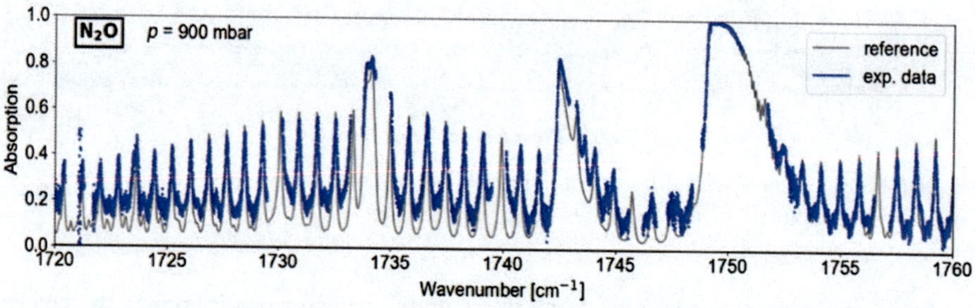

Fig. 12. Due to saturation effects on the detector caused by high absorption signals, the baseline of the spectrum seems is shifted.

5 Conclusion and Outlook

A first step in the experimental calibration of the measurement modes of dual-comb spectrometer was performed. The *StepSweeping* mode allows high spectral resolution and hence enables high precision spectroscopic measurement in good consistency with theoretical data. A variation of the stabilization time showed a negligible effect on the accuracy of the measurement, thus promising good results for the use of *FastSweeping* mode. The calibration work is continued with the Herriott cell being placed externally to investigate the influence of optical fibers that guide the sample-beam out and back to the spectrometer. This setup demands additional optics for the in- and out-coupling of the beam. It is relevant to understand how the signal is affected by this procedure. With the external Herriott cell, measurements with the same previous parameters are repeated and then compared to measurements performed with the *FastSweeping* mode. Results will be reported. In addition, future work should include tests with a small shock tube to confirm the applicability of the calibrated system before employing the technique in a real combustion case at HEG.

References

1. Martinez Schramm, J., Luís, D.: Experimental approach on concentration measurements of NO in hydrogen combustion based on heterodyne laser absorption spectroscopy using quantum cascade lasers. In: Dillmann, A., Heller, G., Krämer, E., Wagner, C. (eds.) New Results in Numerical and Experimental Fluid Mechanics XIII. Notes on Numerical Fluid Mechanics and Multidisciplinary Design, vol. 151, pp. 110–120. Springer, Cham (2021). https://doi.org/10.1007/978-3-030-79561-0_11
2. Griffiths, P.R., De Haseth, J.A., Winefordner, J.D.: Fourier Transform Infrared Spectrometry. Wiley, Hoboken (2007)
3. Udem, T., Holzwarth, R., Hänsch, T.W.: Optical frequency metrology. Nature **416**(6877), 233–237 (2002)
4. Faist, J., et al.: Quantum cascade laser frequency combs. Nanophotonics **5**(2), 272–291 (2016)
5. Villares, G., Hugi, A., Blaser, S., Faist, J.: Dual-comb spectroscopy based on quantum-cascade-laser frequency combs. Nat. Commun. **5**, 5192 (2014)
6. Lepère, M., et al.: A mid-infrared dual-comb spectrometer in step-sweep mode for high-resolution molecular spectroscopy. J. Quant. Spectrosc. Radiat. Transfer **287**, 108239 (2022)
7. Gordon, I.E., et al.: The HITRAN2020 molecular spectroscopic database. J. Quant. Spectrosc. Radiat. Transfer **277**, 107949 (2022)

Multidisciplinary Optimization

On the Feasibility of Drone Towing

Georg E. Koppenwallner(✉)

DLR Göttingen, AS-HEL, Bunsenstr. 10, 37073 Göttingen, Germany
Georg.koppenwallner@dlr.de

Abstract. The concept of aerotow for gliders with a tow-drone instead of a tow-plane is presented based on the historical development of aerotow and compared to the current state of the art for towing gliders using the measurements from the thrust flap = "Schubklappen-Antrieb" project SKA.

Keywords: Aerotow · tow-drone · tow-plane · drone towing · OTW = over the wing propeller · electric flight · approximation of polars · eco-efficient flying · regional aircraft · thrust flap

Abbreviations

a_1, b_1	Coefficients of the polar approximation
c_L	Lift coefficent
g	Earth acceleration due to gravity
k, $k_{towplane}$, $k_{tow\text{-}drone}$	$k =: F_{Eopt} / F_0$,k For tow-plane, k for tow-drone
m, $m_{towplane}$,m_{glider}	Masse, mass tow-plane, mass glider
$m_{tow\text{-}drone}$	Mass tow-drone
t	Time
v, v_{Eopt}	Flight velocity, velocity of best glide
w	Vertical velocity or climb rate
E, E_{opt}	Lift-to-drag ratio, optimal lift-to-drag ratio
$En_{tow\text{-}drone}$	Energy for a tow-launch with tow-drone
F, F_0, $F_{tow\text{-}drone}$	Thrust, static thrust, thrust tow-drone
F_{Eopt}	Thrust at the velocity of best glide
P,P_{climb}, P_{thrust}	Power, power of climb, thrust power
N_{PS}	Power in PS (HP), designation in Fig. 2
W, W_{tot}	Drag, total drag of tow-train
$W_{towplane}$, W_{glider}	Drag of tow-plane, drag glider
γ	Glide angle

1 Introduction

This paper describes a way to eco-efficient (glider) flying by proposing aerotow using a light electric tow-drone with fixed propulsion units over the wing.

A. Dillmann et al. (Eds.): STAB/DGLR Symposium 2022, NNFM 154, pp. 177–185, 2024.
https://doi.org/10.1007/978-3-031-40482-5_17

This is an approach that can greatly reduce energy consumption for takeoff, and can also be used as an approach for range extension of battery-electric regional aircraft. The starting point of this article is the idea to make the propulsion units retractable based on “Schempp-Hirth” airbrakes as so called thrust flaps. In July 2021 extensive wind tunnel measurements have been performed with different configurations of electrical powered propellers which have been mounted over the wingsection taken from a real airplane. This project is called the SKA (“Schubklappen-Antrieb”) project (see Fig. 1).

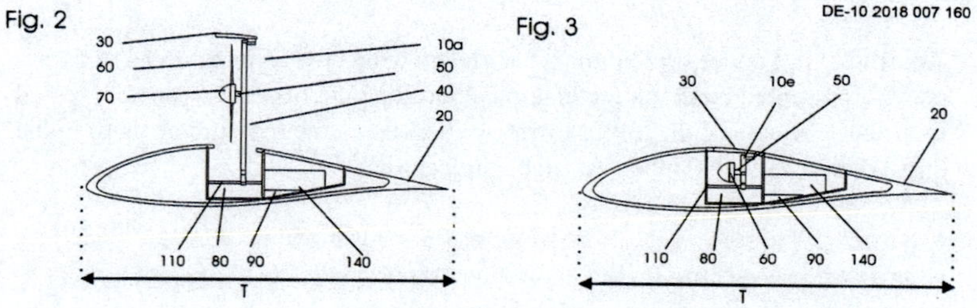

Fig. 1. Propeller thrust flap, (Propeller-Schubklappe germ.) [1], extended on the left, retracted on the right.

This propulsion configuration is an “over the wing” = OTW propeller. In the context of distributed electric propulsion, this configuration has received increased attention because the lift coefficient and the propulsion efficiency of the propeller can be increased depending on its position with respect to the wing [2] (Fig. 7.1 p.64). A simple thought experiment makes this clear. A wing with a row of propulsion propellers above the airfoil already experiences lift at static thrust. This lift could be called “static lift”. Historical investigations at the AVA-Göttingen [3], transferred to a hypothetical 1000 kg airplane with 37 kg/m^2 wing loading and an aspect ratio of 1: 4.5, also show that a single propeller of 2.61 m diameter over the wing shows a performance equal or superior to the propeller below the wing or at the height of the wing, in particular in slow and in fast flight (see Fig. 2, power is denoted by N_{PS}).

The wind tunnel tests at that time were conducted using a high-speed electric motor with gear reduction and a 30 cm or 12 inch diameter propeller. Lacking at this time powerful batteries and motors the practical step to distributed electric propulsion was blocked.

Anthony Fokker describes in [11] the principle of aerotow. The first aerotows were performed in the early 1922 by Glenn Curtiss in the USA - Manhasset Bay - with boats as well with a tow-plane [10]. In the 1920s, the aerotow was developed to practical maturity in Kassel Waldau. Antonius Raab and Kurt Katzenstein [12] describe a “connecting device for aircraft tows”. Since the 1930s aerotow has become a common takeoff-method for gliders. The very detailed history and also flight reports can be found in [4, 5, 10].

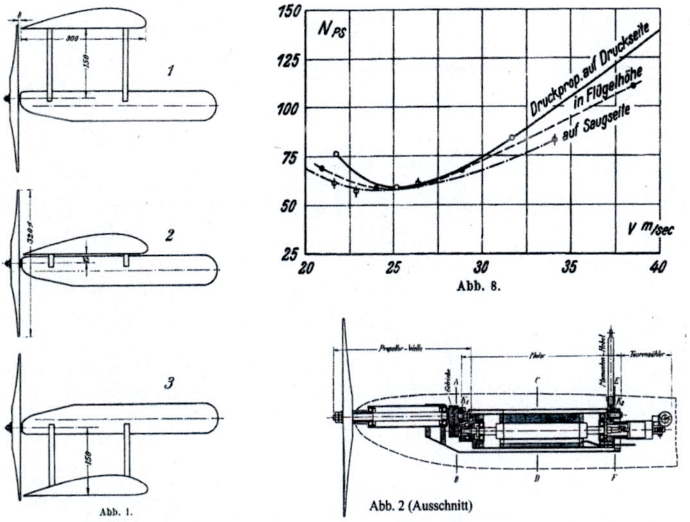

Fig. 2. Reinhold Seiferth [3] (1926) Propeller-configurations at the trailing edge. Top right: power for level flight calculated for a fictitious aircraft ("auf Saugseite" is OTW configuration)

2 Estimation of the Energy Consumption of Aerotow

2.1 Typical Tow-Planes

Table 1 lists typical tow-planes. For the DR 400 and the Rans S7, data from flight manuals regarding glide ratio and speed of best glide are available in [8, 9].

Table 1. Typical tow-planes and proposed parameters of the tow-drone

	Mass MTOW	Span	wing area	wing loading	power	V_{Max}	V_{Min}	$V_{(EMax)}$	E_{Max}	static thrust	year	aspect ratio
	kg	m	m²	kg/m²	PS	km/h	km/h	km/h		N		
Heinkel-72	820	9	20,7	39,61	168	194	77	-	-	-	1933	3,9
Morane	770	9,75	12,3	62,60	150	250	73	-	-	-	1959	7,7
DR 400 - Remoqueur	1000	8,72	14,2	70,42	180	260	87	145	9,3	-	1972	5,4
Rans-S7	559	8,92	13,67	40,89	100	190	66	102	8,4	1800	1985	5,8
SF-25 (Rotax)	485	16,6	17,5	27,71	100	180	60	85	20	-	1999	15,7
tow-drone	120	5,5	3	40,00	60			100	8,4	1000	-	10,1

It is noticeable that the typical E-class tow-plane such as "Morane" and "Remoqueur" have significantly higher wing loads than gliders. For the tow-drone, a wing loading like that of the Heinkel-72 and the Rans S7 is proposed, which is more in line with gliders. The parameters of the Rans-S7 are used for the following estimation.

2.2 Approximation of Polars

The approximation of an aircraft speed polar curve can be found in [6, 7]. In deriving formula, it is assumed that the glide angle γ is very small, so that $cos(\gamma) \approx 1$ can be assumed. This is certainly true for a lift-to-drag ratio E > 8. Thus, the parameters a_1, b_1 described by Eq. (2) can be used in Eq. (1) by means of the given aerodynamic data of an aircraft, especially the best lift-to-drag ratio E_{opt} and the velocity v_{Eopt} of the optimum lift-to-drag ratio.

$$\mathrm{w} = \mathrm{a}_1 \cdot \frac{1}{\mathrm{v}} + \mathrm{b}_1 \cdot \mathrm{v}^3 \tag{1}$$

$$\mathrm{a}_1 = \frac{\mathrm{v}_{\mathrm{E_{opt}}}^2}{2 \cdot \mathrm{E}_{\mathrm{opt}}^2}, \tag{2}$$

$$\mathrm{b}_1 = \frac{1}{2 \cdot \mathrm{v}_{\mathrm{E_{opt}}}^2 \cdot \mathrm{E}_{\mathrm{opt}}} \tag{3}$$

2.3 Climb of a Towed Airplane

The power to climb P_{climb} is the power expended in climbing with a plane or a tow-train with mass m and vertical velocity w against gravity g (4). During level flight (w = 0 m/s) the thrust F_{hover} is exactly equal to the drag W and the thrust-power P_{thrust} is equal to the hover-power P_{hover} as described in (5). Only when the thrust F is greater than the hover thrust F_{hover} climbing becomes possible. The excess power can be converted into power to climb P_{climb} (see Eq. (6)). It is assumed in a simplified way that the drag of the tow-train W_{tot} results from the drags of the tow-plane $W_{towplane}$ and the glider W_{glider} by simple summation as denoted in Eq. (7) and that the drag W is equal during climb and level flight. The subscripts *towplane*, *glider* and *tow-drone* refer to the physical quantities as mass m, drag W, thrust F of these aircrafts.

$$P_{\mathrm{climb}} = m \cdot g \cdot w \tag{4}$$

$$P_{thrust} = F \cdot v, \; P_{hover} = W \cdot v \Rightarrow F_{hover} \cdot v = W \cdot v \tag{5}$$

$$\begin{aligned} \Rightarrow P_{climb} &= P_{thrust} - W_{tot} \cdot v = m \cdot g \cdot w \\ &= P_{thrust} - \left(W_{towplane} + W_{glider}\right) \cdot v = m \cdot g \cdot w \end{aligned} \tag{6}$$

$$W_{\mathrm{tot.}} =: W_{\mathrm{towplane}} + W_{\mathrm{glider}} \tag{7}$$

$$\Rightarrow w = \frac{(F - W_{\mathrm{tot}}) \cdot v}{(m_{\mathrm{towplane}} + m_{\mathrm{glider}}) \cdot g} \tag{8}$$

To calculate the climb rate w of the tow-train by formula (8) an assumption about the thrust as a function of velocity F(v) is described in formula (9). Based on thrust and power measurements of the SKA-project with electric model aircraft motors with 24 and 26 inch propellers, one can assume an approximately linear thrust decrease with velocity, as can be seen in Fig. 3. The factor $k_{towplane}$ was chosen so that the climb rate of the Rans S7 during aerotow corresponds to experience. The factor $k_{tow\text{-}drone}$ fits to the values shown in Fig. 3.

$$F_{towplane}(v) = F_0 - \frac{(F_0 - k \cdot F_0)}{v_{E_{opt}}} \cdot v, \quad k =: \frac{F_{E_{opt}}}{F_0} \\ k_{tow-drone} =: 0.55, \; k_{towplane} =: 0.6 \tag{9}$$

The thrusts achievable with electric motors in the weight class of 1–2.5 kg and a power of around 10 kW were measured during the SKA project. The notation of the props is diameter and pitch in inch, e.g. 24/14 marks a prop with 24 inch diameter and 14 inch pitch. Roughly, it can be said that 250 N static thrust can be achieved easily with powers around 10 kW and propeller diameters between 24 and 28 inches. The thrusts and powers as shown in Fig. 3 were achieved with two 24 inch propellers at 6500 rot/min and with two 26 inch propellers at 6000 rot/min, respectively.

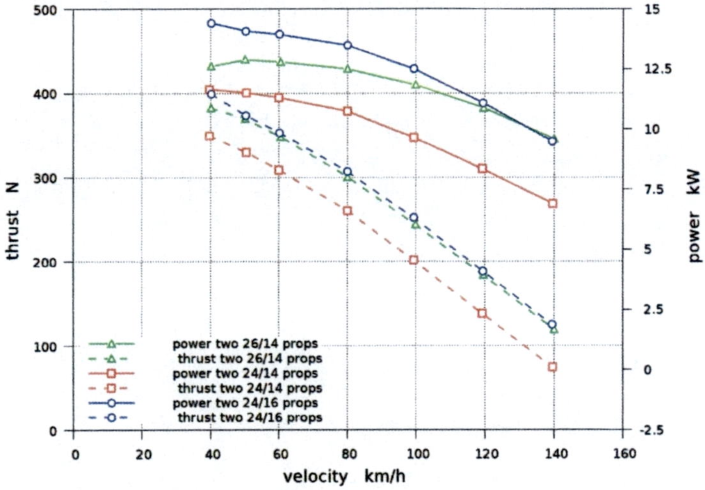

Fig. 3. Thrust and power as function of velocity, AoA = 0°. Measurements with 24 and 26 inch propellers, with 14 or 16 inch pitch, two propellers were used in total

When using 4 to 6 motors, 1000 N–1500 N static thrust is then easily achievable with series motors of e.g. Fa. Hacker and e.g. YGE controllers.

When evaluating Eq. (8) for the comparison between Rans-S7 with 1800 N static thrust and the tow-drone with 1000 N static thrust, the values of Table 1 were used. Thus the tow-plane and tow-drone have almost the same polar curve. An ASW-15 with 300 kg takeoff weight is assumed to be the towed glider (see Fig. 4).

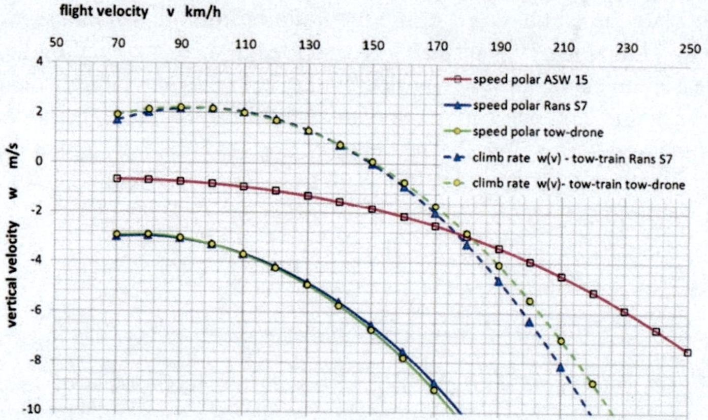

Fig. 4. Climb rates of aerotow with Rans S7 and aerotow with tow-drone with 1000 N static thrust and associated speed polars of the used aircrafts

The climb rates of aerotow with tow-plane and tow-drone are plotted in addition to the speed polars of these planes as presented in Fig. 4. There is almost no difference between the climb rates of the tow-train with tow-plane and the tow-train with the tow-drone.

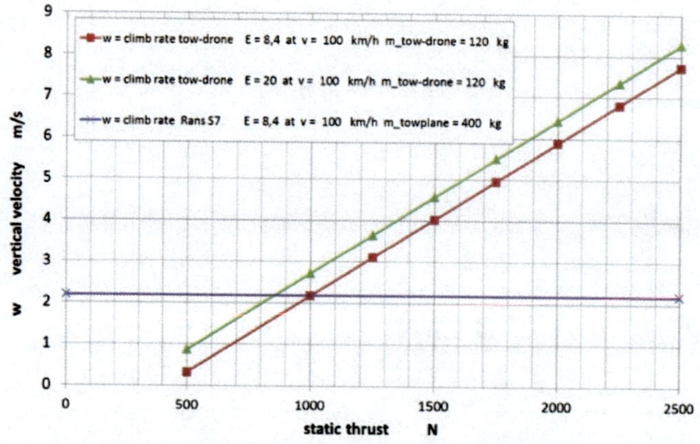

Fig. 5. Climb rates of the tow-train with ASW 15 (300 kg) as a function of static thrust with two variants of the tow-drone concerning E_{opt} at towing speed 100 km/h

This becomes different with increasing thrust of the tow-drone. Already at 1500 N thrust the climb rate of the tow-train with tow-drone is 4 m/s (compare Fig. 5).

If the thrust of the tow- drone is increased by using larger propellers than the measured 24 and 26 inch propellers or by using 6–8 motors, then thrusts between 2000 N and 2500 N and climb rates of 6–7 m/s can be achieved as shown in Fig. 5. Increasing the aerodynamic quality of the tow-drone to a lift-to-drag ratio of 20 has a minor influence on the climb rate of about 0.5 m/s (line with triangles in Fig. 5).

2.4 Characteristics of a Tow-Drone

Not considered in the estimation of the climb rate w is the lift increase due to the propeller jet, which may well cause an increase in the lift coefficient of 50%. The lift coefficient of the wing on which the motors were mounted is plotted without the mounted drive as shown at the bottom of Fig. 6.

From the readings in Fig. 6, it can be seen that it is possible to increase the lift coefficient c_L of the airfoil used at AoA = 0° by up to 50% for the tow-drone. A tow-drone around 150 kg should have a wing area of 3–4 m^2. The energy consumption can be estimated directly from the values of Fig. 3 for the case of 100 km/h towing speed and 1000 N thrust.

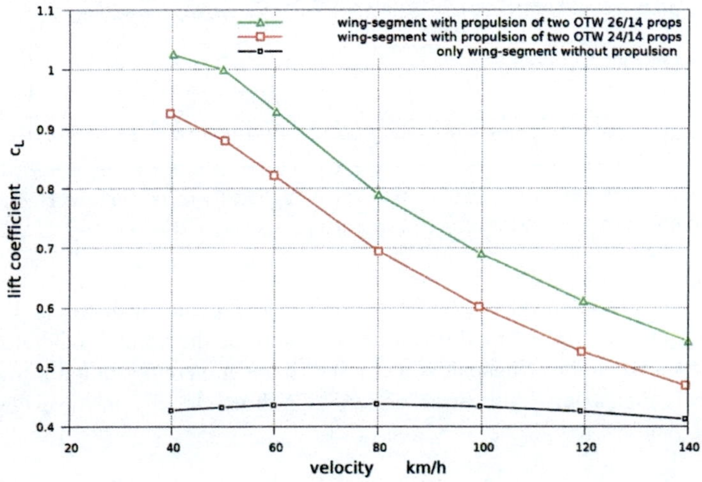

Fig. 6. Lift coefficient c_L and velocity v –SKA measurement - with 26/14 propellers and 24/14 propellers above a wing segment

Based on a power of 13 kW for 2 motors with a combined thrust of 500 N, or 26 kW for 4 motors with a thrust of 1000 N, a climb time of 4 min can be derived for a climb speed of 2 m/s as plotted in Fig. 4 (without take off and rope release). In addition, 2 min of full power for take off and descent can be "energetically" added. If we assume extra 2 min for descent without energy consumption the towing time will actually be 8 min in total. The energy consumption En_{SD} for this aerotow is then:

$$\text{En}_{\text{SD}} = L \cdot t = 26\text{kW} \cdot 6/60h = 2.6\text{kWh} \approx 3\text{kWh} \tag{10}$$

The measured aerodynamic data of the SKA project (see Fig. 3 and Fig. 6) allow the quantitative design of a tow-drone. Increasing the aerodynamic quality of the design compared to the state of the art of e.g. a taildragger like Rans S7 with only E = 8.4 can reduce the energy consumption even further, compare Fig. 5.

The proposal for a tow-drone with fixed OTW propellers looks schematically as shown in Fig. 7. The use of OTW propellers has beside the aerodynamic benefits the practical advantage of avoiding the problem of propeller ground clearance and therewith high landing gears.

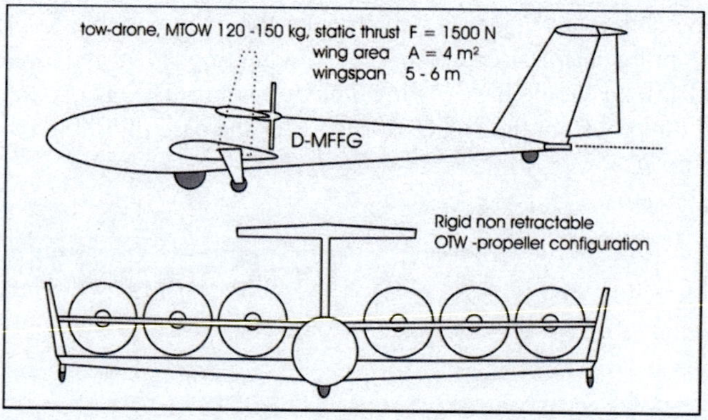

Fig. 7. Schematic sketch of a 6-engine tow-drone, wing ideally rectangular with constant airfoil thickness to minimize construction costs, Wing span 5–6 m.

In contrary to the typical single engine type of a tow-plane the distributed propulsion OTW configuration of the tow-drone allows to increase the number of motors and the propeller area when the wingspan is increased. In such a configuration a tow-drone is nothing more than a very light carrier of very high thrust. Ideally, the thrust exceeds the weight force on the tow-drone.

$$F_{\text{tow-drone}} > m_{\text{tow-drone}} \cdot g \tag{11}$$

3 Summary

This estimation of energy consumption of aerotow shows that aerotowing with a tow-drone can contribute a lot to eco-efficient flying and electric flight. Important data for the design have been measured in the SKA project. The energy content $En_{tow\text{-}drone}$ of 3 kWh for one aerotow is approximately equivalent to the energy of 300 g gasoline. Currently, about 4–5 L of gasoline are consumed per aerotow. The chronological listing in Table 1 shows that a new way in glider towing is long overdue. In a next step drone towing can increase the range and MTOW (maximum take off mass) of future electrical aircraft.

References

1. Koppenwallner, G.E.: Propeller-Schubklappe, patent pend., DE-10 2018 007 160 (2018)
2. Marcus, E.A.P.: Aerodynamic modelling and performance analysis of over-the-wing propellers, Master Thesis, TU Delft (2018). http://resolver.tudelft.nl/uuid:f37f53ef-2419-4d58-8783-3d73c1f7fdb6. Accessed 06 Oct 2022

3. Seiferth, R.: Die gegenseitige Beeinflussung zwischen Tragflügel und Propeller, S. 108ff., Heft 14/ Dez.1926 Jahrbuch der Wissenschaftlichen Gesellschaft für Luftfahrt (1926). reprint durch: Verlag: Oldenbourg Wissenschaftsverlag, eBook Veröffentlicht: 5 (2019). ISBN: 9783486752595
4. Peter, E.: Der Flugzeugschlepp von den Anfängen bis Heute, Motorbuch-Verlag, Stuttgart, erlag, 1. Auflage (1981)
5. Peter, E.: schleppte und flog Giganten, 1st edn. Motorbuch-Verlag, Stuttgart (1976)
6. Reichmann, H.: Zum Problem der Fahrtoptimierung im Streckenflug. Diss. Univ, Karlsruhe (1976)
7. Wandinger, J: Ein einfacher Approximationsansatz für die Geschwindigkeitspolare eines Segelflugzeuges (2011). https://wandinger.userweb.mwn.de/Person/Publikationen/polare.pdf. Accessed 21 Oct 2022
8. https://www.rans.com/files/ugd/6e941e2c491deb13e344149997ac93e5c5c78d.pdf - S.196. Accessed 25 Oct 2022
9. https://hvfl.de/wp-content/uploads/2019/07/D-EWAMFlughandbuch.pdf - p. 39. Accessed 25 Oct 2022
10. Short, S.: In the Wake of the Albatross and Seagull, a Look Back to 1922. AAHS J. Am. Aviat. Hist. Soc. **62**(1), 6–12 (2017). Germ. translation: Short, S.: Der erste Flugzeugschlepp, FliegerRevueX 65, p. 10–17 (2017)
11. Fokker, A.H.G.: Towing Connection for flying machines, patent, US 1,418,783 (1921)
12. Raab-Katzenstein Flugzeugwerke G.m.b.H.: Anschlußvorrichtung für Flugzeugzüge, patent, DE - 491 456 (1927)

Gust and Maneuver Load Alleviation in Conceptual Aircraft Design

Markus Dino Krengel(✉) and Martin Hepperle

German Aerospace Center (DLR e.V.), Institute for Aerodynamics and Flow Technology, Brunswick, Germany

markus.krengel@dlr.de

Abstract. Load alleviation technologies are a promising approach to increase the fuel burn efficiency. Ideally, they should be taken into account already in the early design phase, where the geometry shape is defined. Implementing the capability for a correlated evaluation requires a multi-disciplinary simulation approach. The conceptual aircraft design framework within this paper includes coupled and physics-based models of the disciplines aerodynamics, structures and flight dynamics. The software ASWING from the Massachusetts Institute of Technology is included and provides an unsteady lifting-line calculation combined with non-linear Euler beam theory. The wing shape is optimized within a nine-dimensional design space via a surrogate model approach. The objective function is a multi-point mission block fuel. The difference between an optimized aircraft design without and with maneuver load alleviation shows a potential of 5.3% combined block fuel savings. The effects of gust load alleviation on loads and the overall aircraft design are discussed.

Keywords: aircraft design · load alleviation · aeroelasticity · MDO

1 Introduction

The lift distribution of a wing is usually a compromise between the aerodynamically driven cruise performance and the capability to withstand the maximum loads. While the latter are single events, the cruise performance influences the fuel burn during the entire flight. Load alleviation control laws for maneuver and gust loads are able to alter the load distribution temporarily. This reduces the maximum structural loads and as a consequence the compromise between aerodynamic and structural requirements can potentially be shifted towards a higher cruise performance. A multidisciplinary wing and overall aircraft design with load alleviation technologies therefore requires the combination of three disciplines: aerodynamics, structures and flight dynamics in a coupled approach.

The initial idea of active control technologies in aviation came up in the 1960s [1]. Noll et al. showed 1993 the general proof of concept for maneuver (MLA) and gust load alleviation (GLA) [2]. Handojo et al. investigated the reduction potential for MLA and GLA resulting in 26.5% lower wing mass for a

A. Dillmann et al. (Eds.): STAB/DGLR Symposium 2022, NNFM 154, pp. 186–195, 2024.
https://doi.org/10.1007/978-3-031-40482-5_18

long-range aircraft design [3]. Fournier et al. showed a wing root bending moment reduction of 50% with a GLA utilizing 7° aileron deployment [4]. They further highlighted the difficulty of reducing wing bending and torsion loads simultaneously. Binder et al. showed the combined effects of MLA and GLA, as well as passive structural tailoring (PST) for a long-range aircraft. MLA alone reduced the loads already to 72% of the maximum load reduction potential achieved with all three mechanisms combined [5]. A promising approach for the structural optimization of the wing bending-torsion coupling and therefore PST is presented in Thel et al. [6]. Besides the application of load alleviation for existing wing shapes there were some investigations on wing shape optimization. Liem et al. presented a multi-point block fuel optimization of a long-range aircraft wing resulting in a higher aspect ratio and a fuel burn reduction of 6.6% [7]. Marchetti et al. presented an aspect ratio study of the wing resulting in an optimum at a large value of 15 with a block fuel saving of 9.9% for a short and medium range aircraft [8]. It is an obvious step forward to combine load alleviation and wing shape optimization. An approach applying the DLR TAU software is presented in Wunderlich et al. for a long-range aircraft [9]. The optimization resulted in a combined block fuel reduction of 12.9% [9]. Xu and Kroo assessed a short-range aircraft including a flexible wing, based on conceptual methods [10]. They showed a fuel burn reduction of 11% for the turbulent wing.

Results shown in this paper are for a generic long-range transport aircraft. The planform shape is optimized with and without applied load alleviation via deployed flaps. The design process is based on earlier activities [11].

2 Methodology of the Process

This paper utilizes a framework for conceptual aircraft design with a physics-based, flexible wing structure sizing. The framework is based on the overall aircraft design environment OpenAD [12]. The CPACS file scheme [13] is the exchange database structure within the framework. ASWING from Drela [14,15] is included as a physics-engine, integrating the flexible aircraft in an overall design framework. Figure 1 shows the general structure and data flow of the framework. The input for the entire process is a set of Top Level Aircraft Requirements, constraints, calibration factors and other assumptions that are not covered by the level of physical representation. The global output is a CPACS file of a converged aircraft design, including a more detailed wing description due to a physics-based sizing. Within the process two major loops cascade into each other.

The outer loop (**1**) feeds the wing mass and a better estimation for the aerodynamic efficiency into the overall aircraft design. The aerodynamic update is from the ASWING lifting-line approach while the wing mass is a result of the sizing process. The output of ASWING is modified. Instead of the integrated value for induced drag directly on the vortices the more precise Trefftz-Plane analysis is used. Furthermore, the wave drag is approximated locally by the Korn equation for swept wings as shown in Mason [16]. Aircraft components like

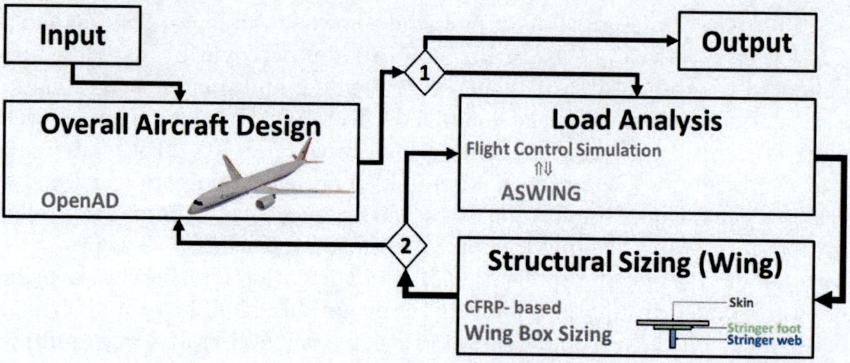

Fig. 1. Overview of the entire process flow for dynamic aeroelastic wing sizing.

the landing gears or horizontal and vertical stabilizers change in accordance with the handbook methods implemented in OpenAD [12]. The stability margin and the wing loading are constant. The inner loop (**2**) computes the stiffness and mass distributions of the main wing in an iterative process.

The design space is multi-dimensional. There are four planform parameter: aspect ratio AR, taper ratio TR, leading edge sweep angle φ_{LE} and kink position η_K. Furthermore, the twist distribution is defined via three parameters at the kink, mid ($\eta = 0.6$) and tip position. The relative thickness ratio is varied at the root and kink position. In-between these stations the distributions are linear. Figure 2 shows the design space.

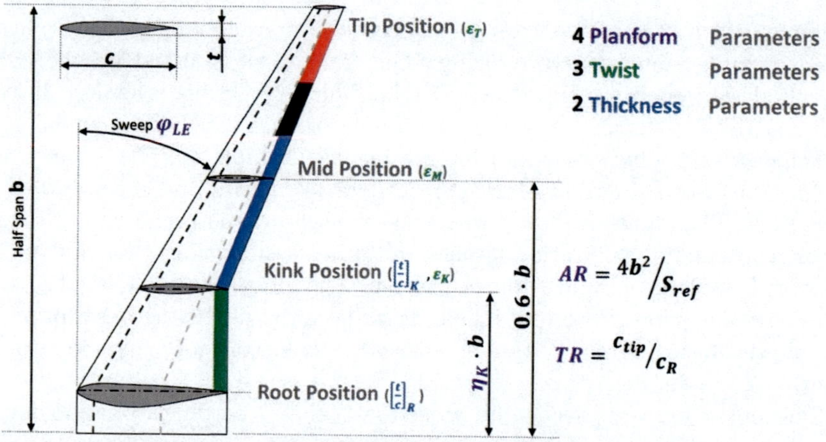

Fig. 2. Selected design space for the wing shape optimization.

Adler and Martins showed the benefits of a multi-point target function for an optimization compared to a single-point optimization [19]. Therefore, the target function here is a combined fuel consumption of three flight missions with a cruise Mach number of 0.83. The fuel consumption for each mission is weighted by a factor w_i. Besides the design mission ($w_i = 0.3$), a mission with 75% design payload and a range of 4000 NM ($w_i = 0.55$) and a mission with maximum payload for a range of 2000 NM ($w_i = 0.15$) are considered. The fuel consumption for these two missions are derived via an approximation based on the mass ratio for the take-off, climb, descend and landing segment in Jenkins [17] as well as the Breguet equation [18] for the cruise segment. Later block fuel values will refer to this combination of missions as "combined block fuel". Constraints for the optimization are:

- the mission loaded fuel fits inside the wing tank volume,
- the aircraft span $2b$ is smaller than a box limit of 65 m,
- the main landing gear tire diameter fits behind the rear spar.

For the structural sizing of the wing a set of aeroelastic maneuver and gust load cases, including a dive case, are calculated. The loads are based on four flight points presented in Table 1. EAS represents the Equivalent Air Speed, flaps UP referrers to a clean wing configuration and flaps DOWN represents a simplified high-lift point. Pull-up, pull-down and roll maneuver load cases are calculated for each flight point. Roll maneuvers are trimmed with a bank angle of 30° and a load factor of 1.67. For the push and pull load cases the altitude is reduced to 1000 m at constant EAS. This approach avoids maximum lift issues and is similar to Wunderlich's et al. [9]. Figure 3 shows the considered maneuver load cases within the flight envelope according to 14 CFR Part 25 [20]. All flaps in this paper are modeled as plain flaps. Their effect regarding local pitching moment and additional lift are estimated via the formula of Glauert [21]. The maximum movable rate for Ailerons and the Elevator is limited to 35°/s.

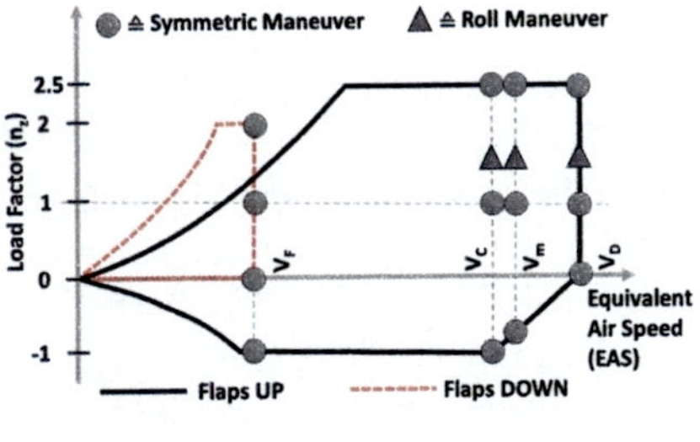

Fig. 3. Load cases for the analysis.

Table 1. Flight points for the analysis

	EAS [kts]	Altitude [m]	FLAPs
v_C	266	10 668	UP
v_m	272	5 462	UP
v_D	289	10 668	UP
v_F	131	100	DOWN

Gust load cases are calculated as dynamic 1-cos-gust encounters for each flight point in Table 1. According to CS25 the gusts are shaped for a wave length of 9, 58 and 107 m [22]. For each of those wavelengths a positive and negative gust velocity is taken into account, resulting in 24 gust load cases.

The deflection for MLA relates linearly to the equivalent airspeed and the commanded load factor n_z. The maximum deflection aligns with a load factor of

−1.0 (maximum delta to 1.0) at the dive velocity. A global factor k_{MLA} scales the MLA deflections from 0.0 ("no LA") to 1.0 for 100% MLA. The maximum deflection angles for MLA in the case of $k_{MLA} = 1.0$ are: −16° for the inner, 19° for the outer flap, 20° for the inner and 15° for the outer aileron.

A pitch damper is included to ensure equal handling qualities, and therefore equal behavior under dynamically controlled gust encounters, for all aircraft designs. Via two feedback loops to the elevator ($k_{\eta q}$ and $k_{\eta\alpha}$) the short period behavior is adjusted in damping and frequency as in Brockhaus [23]. The feedback is shown in Fig. 4.

The gust load alleviation is a feedback control based on the total aircraft lift coefficient. Due to the wing sweep and inertia effects this signal has enough lead time regarding the maximum loads to ensure a basic effectiveness. The measurement of total lift coefficient is not easily implemented for an elastic aircraft due to inertia effects. For the rigid case this signal can be derived from the vertical acceleration. In this case it is assumed though that a sufficient estimation of the total lift coefficient is available as an input for the GLA. The entire control chain is shown in Fig. 4.

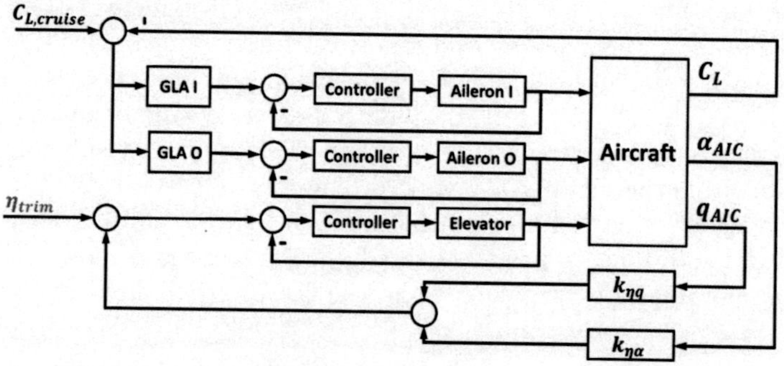

Fig. 4. Controller feedback loops for pitch damping and GLA.

A wing box sizing model for carbon-fiber-reinforced polymers sizes the primary mass and the wing stiffness distribution. 2D wing box representations, as shown in Fig. 5, are created at 10 equally distributed cuts along the wing span. The rib spacing is part of the structural stability calculation. The topology is kept constant. The sizing is done by multiplying each local laminate stack sequences by a natural number until all structural reserve factors are greater than 1.5.

The original geometry of stringer and spars is condensed in points containing the original behavior of the stringer connection. Reduced lumped areas are connected by webs with a shear stiffness of the original skin. The model allows for the consideration of local skin and stringer buckling as well as global buckling and damage tolerance. The calculation is based on the laminate composite plate theory [24]. The secondary wing mass is based on average densities for movables, leading and trailing edges.

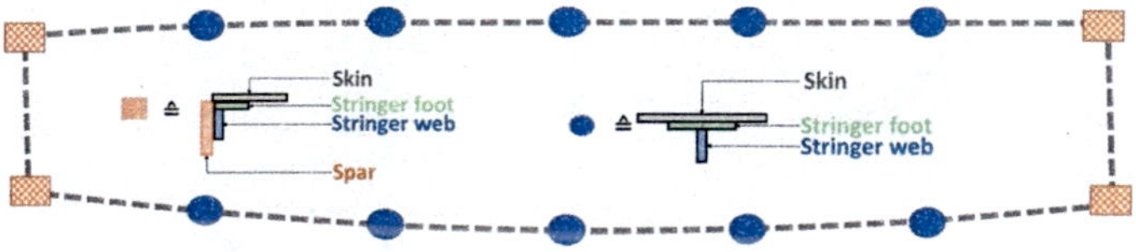

Fig. 5. Wing box sizing representation.

The design of experiments (DoE) and the optimization is performed with the SMARTy Toolbox [25]. Here, the thickness distribution is optimized manually first, followed by the optimization of the other seven twist and planform parameters in a surrogate based approach. With the SMARTy Toolbox an initial Halton sequence is created and the corresponding aircraft designs are sized with the shown process. Halton points are a deterministic sequence of low discrepancy and therefore suitable for building the initial surrogate model [26]. The surrogate model is a Kriging interpolation, as included in the SMARTy Toolbox [25]. Universal Kriging with a quadratic trend function showed the best results in a Leave-One-Out cross-validation. The boundary conditions are met with an additional, artificial penalty on the surrogate model evaluations once the design is outside of the limits. After at least 150 valid aircraft designs are available the first surrogate based optimization (SBO) is performed with the SMARTy Toolbox. The result of this optimization is a new design point of the supposed minimum and a design point, where the expected improvement is maximal. These design points are sized by the process and the surrogate model is updated, leading to a new estimated minimum. After a maximum of 15 iterations the expected minimum does not change significantly, indicating an optimum.

The generic reference aircraft (REF) is designed to transport 270 passengers over a distance of 6000 nautical miles with a Mach number of 0.83. The aircraft has a technology level of the year 2020 and a wing loading of 672.3 kg/m^2. Further details are shown in Krengel [11].

3 Results

The aspect ratio is altered, starting from the reference, to investigate the trade-off between mass and aerodynamics. This one-dimensional study is executed for the case without load alleviation and three different levels of MLA, referring to k_{MLA} of 0.5, 1.0 and 1.5. The difference in combined block fuel to the reference over the aspect ratio reveals potentially beneficial spots. The aerodynamic efficiency L/D over the relative wing mass shows the trade-off between the disciplines. Figure 6 depicts both plots.

There is a plateau for the change in block fuel within a certain aspect ratio range, depending on the level of load alleviation. For the case without LA two less pronounced local minima can be identified at aspect ratios of 11.0 and 13.2. For the cases with 100 or 150% LA equally barely significant minima can be identified at $AR = 12$ and 12.3 respectively. This implies a range where the influences of

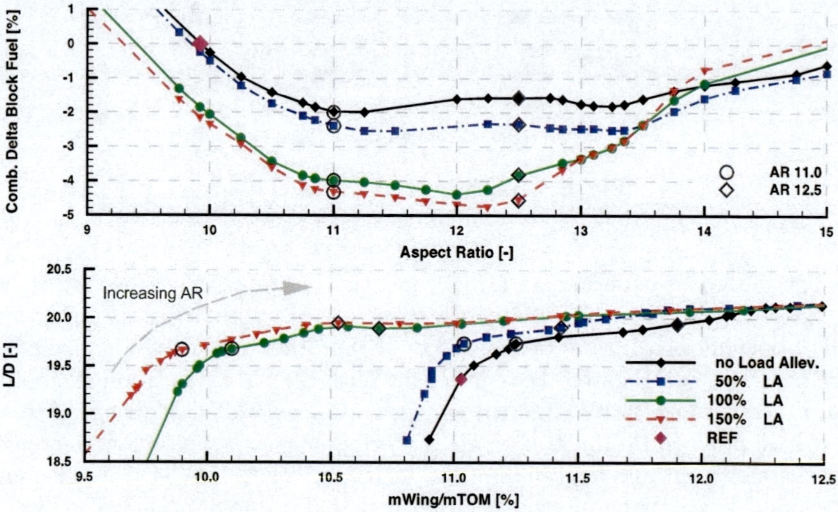

Fig. 6. Results of the aspect ratio parameter pre-study.

aerodynamics and wing mass on the overall aircraft are in balance. Towards lower aspect ratios the aerodynamic efficiency is decreasing fast, while to higher aspect ratios the aerodynamic efficiency slope is limited due to flexibility effects while the wing mass is increasing significantly. MLA reduces the wing weight on one hand and leads to a more distinct local minimum. For an aspect ratio over 14 the increase of wing mass is dominating. In the area of high aspect ratio load alleviation has no significant impact on the overall aircraft design. The optimization without LA results in an aspect ratio of 13.1. Compared to the reference aircraft the combined block fuel saving is 9.6%. Further investigations show that within these 9.6 percentage points, 2.5 are a result of the root thickness reduction, an additional 3.0 points result from a twist optimization and the final 4.1 points can be attributed to the planform optimization. The optimum aircraft with MLA resulted in a fuel burn reduction of 13.8% compared to the reference. Related to the optimum without LA this leads to a potential fuel benefit for MLA of 5.26%. Figure 7 shows the resulting planforms for these two optimizations. Some data of the aircraft designs are given in Table 2.

Figure 8 shows the lift coefficient and the total lift distribution represented by the product of local chord and lift coefficient. The lift coefficient for the case with MLA clearly shows a higher outboard loading. The higher outboard load of the airfoil is not increasing much during maneuver when MLA is shifting the load inboard. The reduced taper ratio leads to similar lift distribution as for the optimum without LA. The latter is a result of the much higher aspect ratio. Compared to the reference, the lift distribution is shifted outboard for cruise flight on both optima. Presumably the swept wings flexibility leads to a load alleviation effect already in both cases, allowing a more aerodynamically efficient distribution, while limiting the maximum bending moments of the wing. During a gust

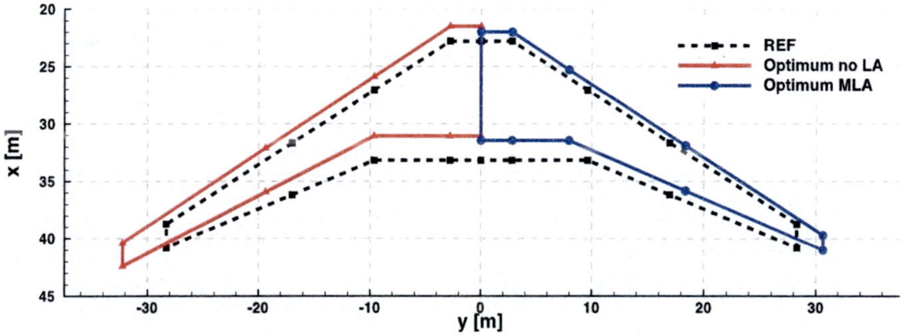

Fig. 7. Planform of the optimized aircraft designs with and without load alleviation.

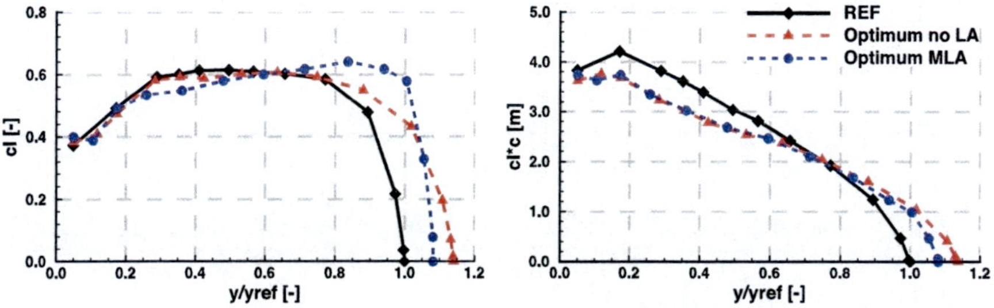

Fig. 8. Lift and wing load distribution for the optimized aircraft designs.

Table 2. Details of the optimized aircraft designs

AIC	AR	TR	φ_{LE} [deg]	t/c_R	t/c_K	m_{Wing} $[kg]$	m_{TOM} $[t]$
REF	9.92	0.20	32.0	0.12	0.12	23929	216.9
No LA	13.14	0.21	32.7	0.10	0.11	26585	212.6
MLA	12.41	0.14	32.4	0.10	0.11	22190	203.5

encounter the aerodynamics has a lead on the structural load factor of the wing due to inertia. Here the pitch damping feedback and full GLA are investigated. Figure 9 shows the aerodynamic and structural load factor as well as the aileron deflection for GLA of the inner and outer aileron over time for the reference design (see Fig. 2). The impact on the loads and on the overall aircraft design is dependent on the actual wing geometry and cruise altitude. GLA reduces the maximum bending loads (s. Fig. 9) by 14.13% while the maximum aerodynamic load factor is reduced only by 4.26%. For the reference the gust loads are not significantly relevant because they are on the same level as the reduced maneuver loads. Therefore, the impact on the overall aircraft design is neglectable. The gust loads are more relevant for the optimum MLA configuration. GLA has a small impact on the fuel saving and wing mass. The design mission block fuel saving is 0.28% and the wing mass is reduced by an additional 405 kg including all snowball effects.

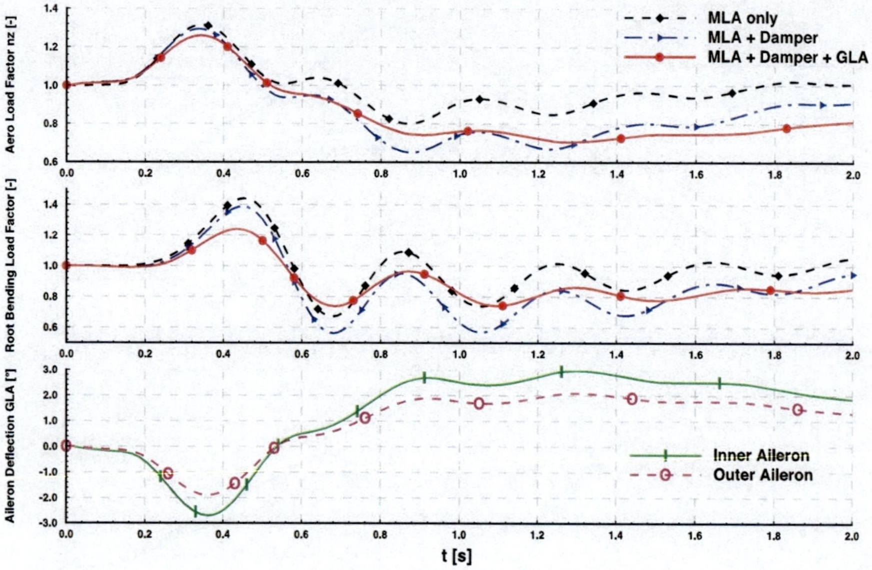

Fig. 9. Gust encounter of positive 107 m 1-cos-gust with and without GLA.

4 Conclusion and Outlook

A long-range aircraft configuration has been optimized with and without maneuver load alleviation via a surrogate based optimization of an aircraft design process, including a more detailed wing sizing. The comparison showed a fuel saving potential for maneuver load alleviation of 5.3%. The effectiveness of a simplified gust load alleviation could be demonstrated. The additional impact of gust load elevation depends on the degree of MLA, the wing shape and the wing loading. The next step will be an optimization with MLA and GLA activated at the same time to evaluate the combined potential. In addition, a more realistic approach for gust load alleviation will be provided, utilizing the vertical acceleration and sensors for the angle of attack.

Acknowledgments. The authors would like to thank Professor Mark Drela and the MIT for providing the ASWING research license.

References

1. Hönlinger, H., Zimmermann, H., Sensburg, O., Becker, J.: Structural aspects of active control technology. In: AGARD Conference Proceedings (1995)
2. Noll, T.E., et al.: Impact of active controls technology on structural integrity. J. Aircr. **30**(6), 985–992 (1993)
3. Handojo, V., et al.: Alleviation and future technologies in reducing aircraft structural mass. Aerospace **9**(8), 412 (2022)
4. Fournier, H., et al.: Robust gust load alleviation at different flight points and mass configurations. In: AIAA SCITECH 2022 Forum (2022)

5. Binder, S., Wildschek, A., De Breuker, R.: The interaction between active aeroelastic control and structural tailoring in aeroservoelastic wing design. Aerosp. Sci. Technol. **110**, 1–12 (2021)
6. Thel, S., Hahn, D., Haupt, M., Heimbs, S.: A passive load alleviation aircraft wing: topology optimization for maximizing nonlinear bending-torsion coupling. Struct. Multidisc. Optim. **65**(5), 1–17 (2022). https://doi.org/10.1007/s00158-022-03248-3
7. Liem, R.P., Kenway, G.K., Martins, J.R.: Multimission aircraft fuel-burn minimization via multipoint aerostructural optimization. AIAA J. **53**(1), 104–122 (2015)
8. Ricci, S., et al.: Ratio wings for environmentally friendly aircraft. In: AIAA Scitech 2022 Forum, p. 0166 (2022)
9. Wunderlich, T.F., Dähne, S., Reimer, L., Schuster, A.: Global aerostructural design optimization of more flexible wings for commercial aircraft. J. Aircr. **58**(6), 1254–1271 (2021)
10. Xu, J., Kroo, I.: Aircraft design with active load alleviation and natural laminar flow. J. Aircr. **51**(5), 1532–1545 (2014)
11. Krengel, M.D., Hepperle, M.: Effects of wing elasticity and basic load alleviation on conceptual aircraft designs. In: AIAA SCITECH 2022 Forum, p. 0126 (2022)
12. Woehler, S., Atanasov, G., Silberhorn, D., Fröhler, B., Zill, T.: Preliminary aircraft design within a multidisciplinary and multifidelity design environment. In: Aerospace Europe Conference 2020, Bordeaux, Frankreich (2020)
13. Alder, M., Moerland, E., Jepsen, J., Nagel, B.: Recent advances in establishing a common language for aircraft design with CPACS. In: Aerospace Europe Conference 2020, Bordeaux, Frankreich (2020)
14. Drela, M.: ASWING 5.99 Technical Description - Steady Formulation. Massachusetts Institute of Technology, Cambridge (2015)
15. Drela, M.: ASWING 5.99 Technical Description - Unsteady Extension. Massachusetts Institute of Technology, Cambridge (2015)
16. Mason, W.H.: Transonic aerodynamics of airfoils and wings. Notes in Configuration Aerodynamics, Virginia Tech 4.4:02 (2006)
17. Jenkinson, L.R., Simpkin, P., Rhodes, D., Jenkison, L.R., Royce, R.: Civil Jet Aircraft Design, vol. 338. Arnold, London (1999). ISBN: 0 340 74152 X
18. Anderson, J.D., Bowden, M.L.: Introduction to Flight. McGraw-Hill Education, New York (2005)
19. Adler, E., Martins, J.R.: Aerostructural wing design optimization considering full mission analysis. In: AIAA SCITECH 2022 Forum (2022)
20. Code of Federal Regulations, Title 14 Part 25 Airworthiness Standards - revised and re-issued annually (1988). Accessed 27 Oct 2021
21. Glauert, H.: Theoretical relationship for an airfoil with hinged flap. ARC Rep. Mem. 1095 (1927)
22. Certification Specifications and Acceptable Means of Compliance for Large Aeroplanes CS-25 (Amdt. 17). EASA, Cologne (2015)
23. Brockhaus, R.: Flugregelung. Springer, Heidelberg (2001). https://doi.org/10.1007/978-3-642-01443-7
24. Roylance, D.: Laminated composite plates. Massachusetts Institute of Technology (MIT), Cambridge (2000)
25. Bekemeyer, P., et al.: Data-driven aerodynamic modeling using the DLR SMARTy toolbox. In: AIAA Aviation 2022 Forum, p. 3899 (2022)
26. Halton, J.H.: On the efficiency of certain quasi-random sequences of points in evaluating multi-dimensional integrals. Numer. Math. **2**(1), 84–90 (1960). https://doi.org/10.1007/BF01386213

Digital Design of Moveables at DLR Virtual Product House

Fabian Lange-Schmuckall[1], Alexander Zakrzewski[1], Martin Rädel[2](✉), and René Hollmann[3]

[1] German Aerospace Center (DLR), Institute of Aerodynamics and Flow Technology, Lilienthalplatz 7, 38108 Brunswick, Germany
{Fabian.Lange,Alexander.Zakrzewski}@dlr.de

[2] German Aerospace Center (DLR), Institute of Composite Structures and Adaptive Systems, Lilienthalplatz 7, 38108 Brunswick, Germany
Martin.Raedel@dlr.de

[3] German Aerospace Center (DLR), Institute of Flight Systems, Lilienthalplatz 7, 38108 Brunswick, Germany
Rene.Hollmann@dlr.de

Abstract. This work presents the virtual design activities performed within the Virtual Product House (VPH) start-up project. In this project a multi-disciplinary process for virtual design, manufacturing and testing is developed. The VPH acts as integration plateau where the assessment capabilities of multiple partners from research, industry and certification authority are combined and applied to use cases.

The multi-disciplinary analysis setup enabled the investigation and improvement of aircraft designs and allows the impact assessment of modifications on a existing configurations.

As starting point, three disciplines are considered: aerodynamics, structural and system design. This paper presents results for selected modifications on trailing edge moveable devices of a representative research aircraft, including variations of the applied structural constituent and design, actuation system as well as number of flap tracks.

Keywords: VPH · Digital Design · Moveables · Wing · Aerodynamics

1 The Virtual Product House

The development of innovative and complex technologies requires a collaborative effort of various stakeholders to solve multi-disciplinary problems. The DLR project VicToria [1] is to be emphasized with respect to the latest developments of multi-disciplinary process chains for transport aircraft configurations. Additional disciplines, like engine sizing or systems design, have been added to the already established coupling between aerodynamic and structural design. Therefore, new methods were implemented and existing tools extended to improve the interfaces and collaboration between disciplines. The biggest challenge within multi-disciplinary design analysis (MDA) and optimization (MDO) projects is

A. Dillmann et al. (Eds.): STAB/DGLR Symposium 2022, NNFM 154, pp. 196–205, 2024.
https://doi.org/10.1007/978-3-031-40482-5_19

the coordination and implementation of interfaces between disciplines and the availability of a common research environment. In consequence, the DLR Virtual Product House (VPH) was established at the Bremen Center for Eco-efficient Materials and Technologies (ECOMAT) to improve the exchange and collaboration of research facilities, aircraft manufacturer, suppliers and certification authority. In the course of the VPH start-up project the initial use case "Multi-functional moveables" was selected together with the industry partners (Airbus, FFT, IABG, Liebherr-Aerospace). This publication presents the results of the VPH start-up project. Using the jointly created multi-disciplinary design process, different wing structural designs and wing flap designs were compared to each other. Further current activities of the Virtual Product House are the development of a design process for hydrogen tanks as well as the collaboration with the certification authorities with respect to a virtual certification approach.

2 Multidisciplinary Process

Up to this date, the main focus of most multi-disciplinary design processes was on clean wing configurations [1,2]. Therefore, the application and especially extension of available methods in regards to take off and landing configurations was required to enable the evaluation of the essential parts of a state of the art high lift system and associated requirements. Finally, using the common source approach [9] a digital continuous end-to-end process chain was established ranging from virtual design [3] over virtual manufacturing [4] up to virtual testing [5]. Advantages of this end-to-end process chain are the earlier feedback of virtual testing and manufacturing results back into the digital design process as well as a digital continuity and consistency of the process that is required for virtual certification. The development of this end-to-end process chain is an ongoing effort. For this study, it is assumed that the applied calculation methods have already been validated individually, as demonstrated for the DLR flow solver TAU for instance in several AIAA drag and high-lift prediction workshops. In this first investigation, the MDO process was set up to generate a converged wing geometry with respect to kinematic, structural and aerodynamic design. These results are presented here. In the currently ongoing follow up project of the VPH, the findings of the manufacturing

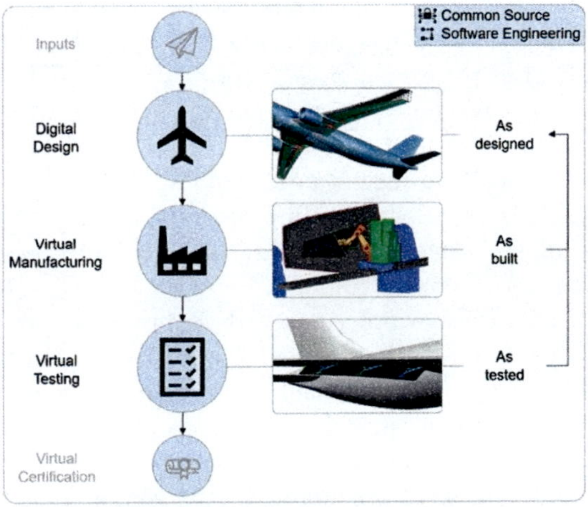

Fig. 1. VPH start-up project End-to-End process

and testing stages are fed back to the design stage in order to enhance the resulting design. A schematic of this approach can be found in Fig. 1.

As previously presented in [3], the digital design process consists of an overall aircraft design (OAD) assessment, actuation system sizing, structural model generation and a fluid-structure-interaction (FSI) based sizing loop for the sizing of the elastic wing. The FSI loop is sketched in Fig. 2. The FSI process is based on iterative compute-intensive high-fidelity simulations for the evaluation of the aerodynamic and structural behaviour of the virtual product in combination with a sizing-optimization of the primary wing structure. The flow simulation is based on the FlowSimulator, a software infrastructure that enables coupled multi-physics analyses with focus on large-scale HPC process using in this particular case the TAU code in combination with the FE solver MSC Nastran. In the course of an iterative process aerodynamic loads are provided by the RANS simulation and forwarded to the FE model, which in turn returns displacement vectors along the wing. The displacements are mapped back on the CFD grid to apply a corresponding wing deformation and thus represents a flexible wing. Then, the aerodynamic simulation is restarted to assess the flow conditions and aerodynamic loads for the adapted wing shape. This iterative coupling process is converged, as soon as the displacement alteration between the last two iterations does not exceed a defined threshold. The structural model, provided by DELiS, and aerodynamic loads are then forwarded to the structural sizing process. Here the in-house sizing tool VErSO is used.

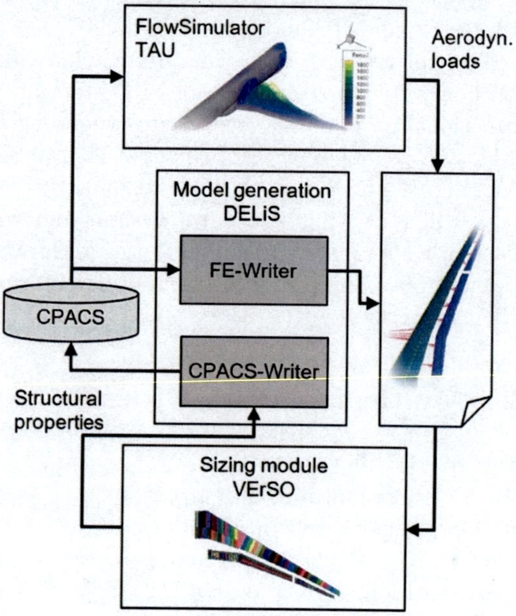

Fig. 2. FSI loop

Using this presented toolchain, different input parameters for the wing and flap geometry are selected and tested for comparison (see Fig. 4). Starting from a parametric aircraft definition in the preliminary aircraft design data exchange format CPACS [12], the design process is executed. For an existing clean wing outer shape that remains the same for all designs, the structural design as well as the kinematic mechanism designs are alternated. Starting from an aluminum (metal) wing design with a track-rear-link kinematic as a baseline, a carbon fibre reinforced plastic (CFRP) wing in combination with a composite multi-spar flap with integral load introduction (CMFIL) is introduced. Subsequently, starting from this new CFRP baseline, variations in the flap mechanisms and actuation system were performed: On the one hand, the number of flap tracks that are supporting the outer flap is changed from three to two tracks, on the other hand the track-rear-link mechanism was exchanged with an adaptive dropped hinge flap (ADHF), a trailing edge technology which implementation in the Airbus A350 aircraft [6]. A comparison of all designs in regards to overall mass, aircraft performance and estimated block fuel was performed.

3 Use Case Airbus XRF1

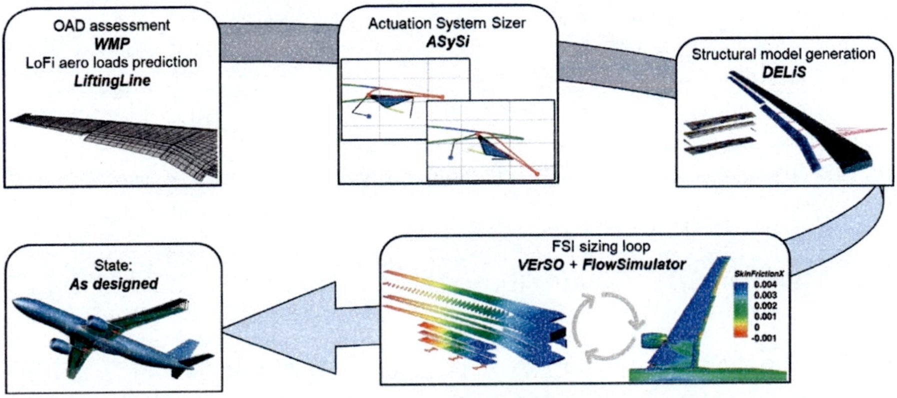

Fig. 3. VPH digital design process

The present study is based on the Airbus XRF1 configuration, a wide body research aircraft. In the course of several research projects, both the clean [2] as well as the high lift configuration [10] is characterized in detail and thus a well defined starting point to perform systematically design studies identifying potential incremental improvements. Based on the associated flight envelope a number of prevailing load cases were selected to run the design process. In total three load cases for structural and five for system sizing were chosen from the borders of the V-n-diagram. The first structural sizing load case considers a take off configuration with flaps at 28° deflection and slats extended by 22° at a load factor of $n = 2$ and Mach number (M) of 0.31. In addition, two high speed maneuver load cases are take into account with a load factor of $n = 2.5$ at $M = 0.55$ and a load factor of $n = -1$ at $M = 0.78$. For system sizing, the load cases were chosen for five different moveable settings ranging from retracted to fully extended moveables (slats, flaps, ailerons). For each load case with extended moveables, a load factor of $n = 2$ was selected in combination with the corresponding maximum allowed velocity v_F for the individual moveables setting. The multidisciplinary analysis and optimization of system, aerodynamic and structural design is performed with the help of the VPH digital design process shown in Fig. 3. The process is described in more detail in [3].

4 Results

By using the VPH design process selected parameters that define the high lift trailing edge devices are varied. An overview of the variations is shown in Fig. 4. The workflow executor can choose between metal or composite structure, differential or integral structural design, track-rear-link kinematic or adaptive dropped hinge flap and two or three flap tracks. It should be noted, that during the execution a common source approach [9] was applied including data continuity and traceability throughout the whole process chain. In the following, a selection of results will be presented by comparing different configurations.

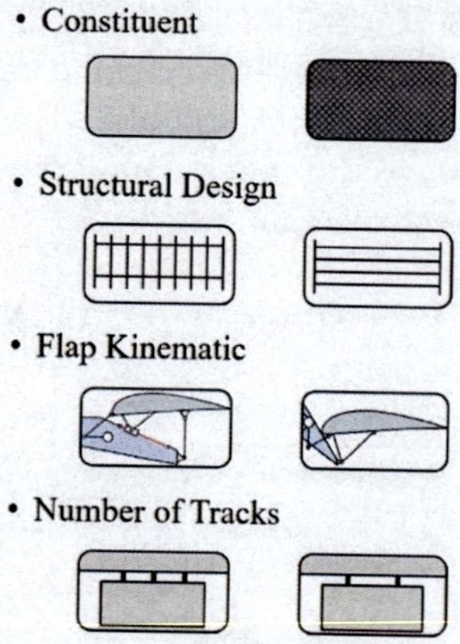

Fig. 4. Parameters varied within the VPH start-up project

4.1 Aluminum vs. Composite

During the design steps depicted in Fig. 3 a sizing for all actuation system components for the given architecture is performed. The initial architecture of the actuation system is chosen to resemble a classical mechanical shaft transmission system with a centralized power control unit (PCU) located within the aircraft fuselage. A shaft system transmits the mechanical power into each wing's trailing edge components. The inboard flap is held and actuated by two support stations, the outboard flap by three support stations. A geared rotary actuator at each station actuates the moveables. The mechanisms, which transmit the actuation torque to the desired Fowler movement is realized with a track-rear link kinematic. A more detailed description of the mechanism is given in [3] and of the actuation system architecture in [5]. For the design of the actuation system a sizing of each component is performed on the basis of the required control surface loads, which are generated with the vortex lattice method based tool LIFTING LINE (see Fig. 3) at the required deflection speeds. The sizing loads are obtained from the load cases depicted in Sect. 3. Within this first variation (Aluminum vs. CMFIL), the architecture, the wing geometry and the load cases do not change, which hence results in no mass change of the mechanisms and actuation system.

The structural architecture of the reference aircraft moveables is a differential metallic structure consisting of front and a rear spar as well as ribs along the span. To achieve a more high-rate capable approach the structural design is changed to an integral multi-spar structure made of CFRP. The structural sizing is performed based on strength, stability as well as damage tolerance criteria. The changes in the critical sizing criteria can be seen in Fig. 5. The results are given for the suction (SS) and pressure (PS) side of the wing box.

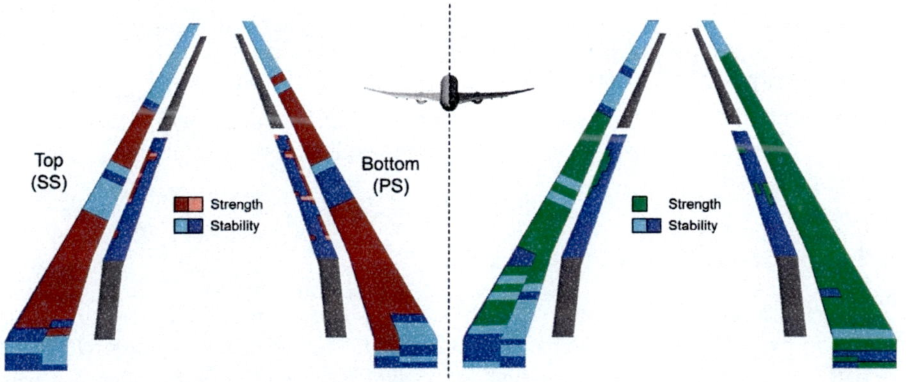

Fig. 5. Critical criteria differential metallic (left) vs. integral composite (right) design

While seeing some changes in the main wing, the critical criteria for the moveable remain almost identical between the two designs. The sizing near the flap attachments and the maximum bending stiffness around the front spar is dominated by strength criteria, both in skin and spars. The moveable domains between the track attachments is dominated by stability criteria. These findings are supported by the resulting ply shares for the CMFIL approach as shown in Fig. 6.

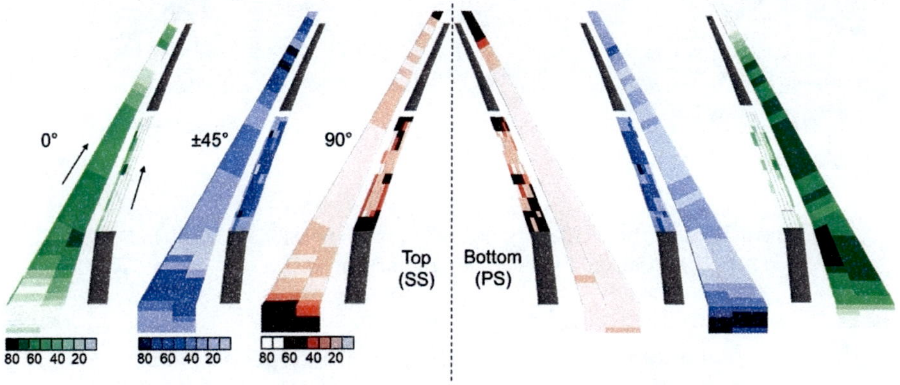

Fig. 6. Ply shares for CMF variant

The domain around the front spar with the highest effectiveness of span-wise ply orientation due to the maximum height shows the largest amount of 0° plies. The load introduction regions show the maximum of 90° plies to distribute the loads from the kinematics in chord direction. The domains between the tracks and behind the front spar are dominated by 45° plies.

Resulting from the modification of the applied material, the primary wing mass is reduced by about 23%, which in turn decreases the operating weight empty (OWE) by nearly 6%. Since the actuation and high lift mechanism mass remain constant and the aerodynamic performance is unaffected by the modification, an assessment using the Breguet range equation reveals block fuel savings of about 4%.

4.2 Three vs. Two Flap Tracks

The reduction of the number of flap tracks at the outboard flap results in an alteration of the given systems architecture of the actuation system. Figure 7 depicts the actuation system of the 2-track variant (left) and the reference configuration with three tracks (right). Due to the change, the position of the tracks in spanwise direction is altered as well. This results in newly designed mechanisms and a reduction of parts in the actuation system. Since the number of support stations of the outboard flap is reduced, the loads per station increase. This results in larger masses per stations, however the overall mechanism mass for the outboard flap is reduced by 19%. A similar trend is visible for the actuation system, where the actuation system weight is reduced by 11%.

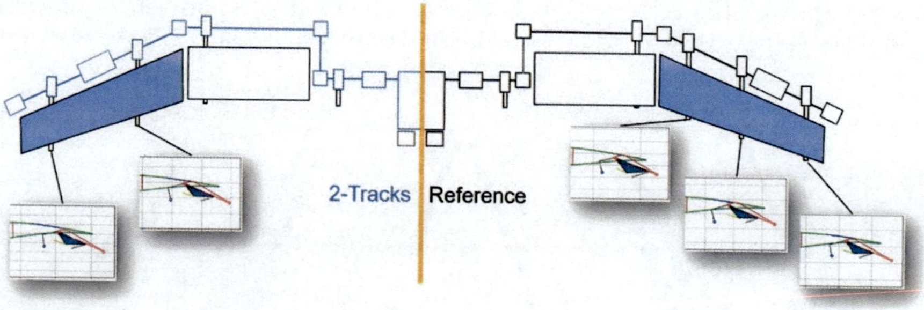

Fig. 7. Two tracks at outboard flap vs. reference configuration (illustration of actuation system based on [7] and altered for the design variation)

High lift configurations are very sensitive with respect to gap and overlap between wing and trailing edge devices. The flap position is chosen such that an optimal high lift performance is guaranteed by defining precisely the distance between wing trailing edge and flap leading edge. However, the gap might vary under the influence of aerodynamic loads. In the course of the design process, a fluid structure coupling is applied, to assess the wing and flap deformations in regards to the different load cases. This allows an analysis of the gap and overlap between wing trailing edge and flap leading edge for the take off high lift load case with $n = 2$.

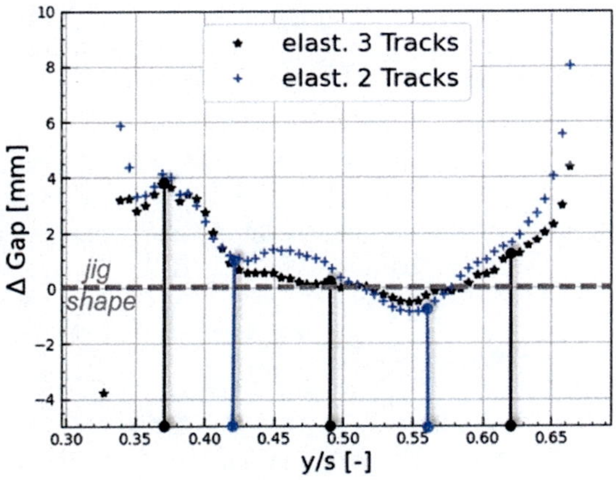

Fig. 8. Change of gap due to wing deformation

A spanwise analysis for the outer flap gap change is exemplarily shown in Fig. 8. A value of 0 represents the desired gap between wing and flap for the non-deformed high lift wing. The black data points describe the gap deviation for the three track configuration, the blue data points show the deviation for the two track configuration under the aerodynamic load conditions. The track locations are indicated by vertical solid lines. The magnitude of deviation is similar for both cases in the order of $4mm$. The maximum deviation can be found at the flap edges. Also, the impact of the track attachment points can be identified. In summary, the gap deviation due to the $n = 2$ load case is within an acceptable range. It should not significantly affect the high lift performance of the trailing edge device for both, the two and three track configurations. However, it should be noted, that the coupling between aerodynamics and structure with respect to high lift use case is not validated yet and needs additional investigations, especially with focus on high aspect ratio wings.

From an overall aircraft perspective, the modification of the number of flap tracks results in a reduction of the overall mass of kinematic mechanisms and actuation system by about 9%. In addition, the drag during cruise flight is reduced, respectively the lift-to-drag ratio (L/D) increased, due to the reduced number of flap track fairings. However, due to the flap load introduction via two instead of three tracks, the primary wing mass increases by more than 3%, resulting in an overall increase of the OWE, which compensates the aerodynamic benefits, such that the block fuel is almost identical for the three and two track configuration.

4.3 Track-Rear-Link vs. Adaptive Dropped Hinge Flap

Finally, the trailing edge high lift system was extensively modified by switching from a TRL to an ADHF mechanism. A detailed description of the modification, including a multi-fidelity dimensioning with respect to aerodynamic and system design can be found in [11]. In the course of this optimization, the inboard flap chord is increased by about 0.5% of the local clean wing chord, while the outer flap chord is increased by up 2.5% at the outer edge to reach the desired high lift performance during take off and landing. The modification from the reference to the ADHF configuration, furthermore, is accompanied by a actuation system architecture change in order to take the functional flexibility into account, which can be utilized with an ADHF trailing edge as performed in the Airbus A350 [8].

Different from the approach there, a decentralized actuation system is chosen with one PCU and a smaller shaft transmission system per flap. Within the sizing process, the choice of this mechanism type reduces the necessary actuation torque in comparison with the reference configuration. This results in a weight reduction of the shaft transmission system. The overall actuation system mass, however, remains mainly the same due to the need of one PCU per flap. Furthermore, as a result of the change to the ADHF, the combined structural mass of the wing and the moveable increases by 5%.

4.4 Summary of Results

It is visible, that the changes of constituent, structural design, flap mechanism design and number of tracks display their distinctive changes to the baseline design. These changes influence block fuel shown in Table 1 for a give cruise study mission of 4000 nautical miles. Since only minimal changes of drag (<2 drag counts) for a fixed lift coefficient are observed for the -1 g load case due to elastic effects of the wing, no recalculation of the cruise performance was performed and the cruise performance assumed to remain similar. In general higher amounts of absolute wing displacement lead to an increase in drag.

Table 1. Changes with respect to CMFIL, 3 Tracks, Track-Rear-Link

	Block Fuel	**L/D cruise**	**OWE**	**Wing Mass**
Aluminum	4.50%	0.00%	6.18%	30.53%
2 Tracks	-0.05%	0.54%	0.70%	3.47%
ADHF	0.66%	0.00%	0.91%	4.50%

5 Conclusion and Outlook

This report shows, that a continuous and collaborative multi-disciplinary design process that allows for a design of multiple wing designs using the same overall design approach was developed as described in Sect. 3. This setup allows a quick analysis of designs and serves as a basis for further design activities, where a comparability of both result and design approach is achieved. Coupled with a single source of data input and traceability of data between the individual components, this work serves as a basis for further activities that have the goal of leading towards a virtual certification capability. Means towards this goal include the implementation of virtual manufacturing and virtual testing simulation processes, where their results can be fed back into the virtual design process. Furthermore, by having defined interfaces between the individual disciplines and their components, a modular approach with a possible interchangeability is introduced, that allows the use of alternative tools. This setup allows a quick swap of tools in the future, as well as addition of further tools into the already existing workflow. This interchangeability enables the introduction of tools from external partners as well as a high flexibility in regards to the use case.

Acknowledgments. The Virtual Product House start-up project is funded by the German federal state of Bremen and the European Regional Development Fund (ERDF).

European Union
Investing in Bremen's Future
European Regional
Development Fund

The authors would like to thank J. Wild, R. Seidler and S. Melber-Wilkending for providing the geometry of the high lift configuration, A. Schuster and S. Dähne for the support during the creation of the structural models with DELiS and the sizing with VErSO as well as Oliver Bertram, Michael Schäfer and Andreas Schäfer for the development and contributions to the system sizing methods within ASySi. Further, the authors would like to thank the Norddeutscher Verbund für Hoch- und Höchleistungsrechnungen (HLRN) for providing computational resources.

References

1. Görtz, S., et al.: Overview of collaborative multi-fidelity multidisciplinary design optimization activities in the DLR project VicToria. In: AIAA Aviation Forum 2020, Virtual Event (2020)
2. Wunderlich, T., Dähne, S., Reimer, L., Schuster, A.: Global aero-structural design optimization of composite wings with active manoeuvre load alleviation. In: Deutscher Luft- und Raumfahrtkongress 2020 (2020)
3. Lange, F., Zakrzewski, A.S., Rädel, M., Hollmann, R.W., Risse, K.: Digital multidisciplinary design process for moveables at virtual product house. In: Deutscher Luft- und Raumfahrtkongress (2021)
4. Rädel, M., Delisle, D., Krombholz, C., Bertling, D., Hein, R., Wille, T.: Towards robustness assessment in virtual testing. In: Deutscher Luft- und Raumfahrtkongress (2021)
5. Hollmann, R.W., Schäfer, A., Bertram, O., Rädel, M.: Virtual testing of multifunctional moveable actuation systems. CEAS Aeronaut. J. **13**, 979–988 (2022). https://doi.org/10.1007/s13272-022-00602-5
6. Reckzeh, D.: Multifunctional wing moveables: design of the A350XWB and the way to future concepts. In: 29th Congress of the International Council of the Aeronautical Sciences, St. Petersburg (2014)
7. Recksiek, M.: Advanced high lift system architecture with distributed electrical flap actuation. In: Proceedings of the 2nd International Workshop on Aircraft System Technologies (2009)
8. Lulla, C.: Functional flexibility of the A350XWB high lift system. In: Deutscher Luft- und Raumfahrtkongress, Bremen (2011)
9. Dressel, F., Doko, A.: Common source & provenance at virtual product house. In: Deutscher Luft- und Raumfahrtkongress (2021)
10. Pülm, S., Handojo, V., Seidler, R.B.: XRF1 high-lift load analysis of wing and high-lift system. In: 33rd Congress of the International Council of the Aeronautical Sciences, Stockholm, Sweden (2022)
11. Zakrzewski, A., Lange, F., Hollmann, R.: Multi-fidelity aerodynamic design process for moveables at DLR virtual product house. In: AIAA AVIATION 2022 Forum. AIAA 2022, p. 3938 (2022)
12. Nagel, B., et al.: Communication in aircraft design can we establish a common language. In: 28th International Congress of the Aeronautical Sciences (2012)

Optimization of Control Surface Deflections on the High Aspect Ratio Wing to Improve Cruise Flight Performance

Tobias Franziskus Wunderlich(✉) and Florian Siebert

German Aerospace Center (DLR), Institute of Aerodynamics and Flow Technology, Lilienthalplatz 7, 38108 Brunswick, Germany
tobias.wunderlich@dlr.de

Abstract. Future highly efficient long-haul aircraft take advantage of increased cruise performance due to wings of higher aspect ratio and geared turbofan engines. Further improvements can be achieved by using the control surfaces to adapt the wing shape to different flight conditions. The objective of the present work is to optimize the control surface deflections of a long-haul aircraft to reduce block fuel and thus the CO_2 emissions.

An optimization of the deflection angles of six control surfaces along the trailing edge of the high aspect ratio wing has been performed. In this optimization process, a flow solver based on the Reynolds-averaged Navier-Stokes equation is used to determine the cruise performance considering the static aeroelastic deformations for three different flight conditions. Depending on the flight mission a fuel burn reduction between 0.5% and 1.7% is predicted.

Keywords: flight performance · adaptive wing · variable camber

1 Introduction

The environmental impact and resource requirements of commercial aviation are increasing with the growth of global mobility and transportation. To protect the environment and conserve resources, aviation is undergoing a transformation process toward more energy-efficient air transport. DLR's Aviation Research Strategy [1] describes and specifies the contribution of aeronautics research to achieving the goals of the mobility strategy of the European Green Deal [2]. The corresponding target for the vehicle is a 25% reduction in specific engine fuel consumption and a 50% reduction in the aircraft's energy requirements.

The efficiency of commercial aircraft is determined by aerodynamic performance in terms of the lift-to-drag ratio, the empty mass of the aircraft and the thrust-specific fuel consumption of the engine. An aerodynamically optimized shape can further reduce drag in transonic cruise flight. This is achieved through advances in wing and airfoil design. Composite materials such as carbon fiber-reinforced polymers (CFRP) allow adapting the structural design the loads by

A. Dillmann et al. (Eds.): STAB/DGLR Symposium 2022, NNFM 154, pp. 206–215, 2024.
https://doi.org/10.1007/978-3-031-40482-5_20

exploiting the additional degree of freedom of fiber orientation. Therefore, the proportion of composites in aircraft construction is increasing and contributing to the reduction of structural mass. Geared turbofan technology promises to further reduce thrust-specific fuel consumption by increasing propulsion efficiency.

In the context of aero-structural wing optimization, the optimal trade-off between cruise performance and wing mass is achieved by combining high-fidelity methods for numerical flow simulation of the aircraft outer shape and the structural sizing of the wing box with an appropriate optimization algorithm [3]. Thereby, the interaction of aerodynamic forces and wing deformations are considered to enable accurate prediction of flight performance and static maneuver loads using fluid-structure coupling.

Further improvements can be achieved by adapting the wing shape to the present flight condition. Adaptive wing technology has long been researched by industry and academia and has been summarized, for example, by Martins [4] under the synonym of morphing wing. For practical implementation, variable camber using trailing edge devices is the most promising type of wing morphing. The basic phenomena of variable camber, the corresponding wing design philosophy and system requirements, and the advantages of the new wing concept are described by Szodruch and Hilbig [5]. Aero-structural design optimization published by Burdette, Kenway, and Martins [6] shows a potential 1.7% reduction in fuel consumption by using a 10% deep continuous morphing trailing edge. Reckzeh [7] describes the wing movables concept of the Airbus A350 in service. This concept provides a functional integration of high-lift with load and cruise performance control.

2 Process Description

A process for the optimization of control surface deflections based on high fidelity simulation methods has been derived from an integrated process chain for aero-structural wing optimization. The original process for aero-structural wing optimization and their application was published by Wunderlich et al. [3,8].

The process chain applied in the present work is shown in Fig. 1 and Fig. 2 in the form of XDSM-diagrams (Extended Design Structure Matrix) [9]. In each optimization step, the aircraft description of the baseline configuration is updated according to the current values of the design parameters. The resulting aircraft description is transferred to the subsequent simulation programs using the Common Parametric Aircraft Configuration Schema (CPACS) [10].

In the CFD volume mesh deformation process, the mesh representing the baseline configuration is deformed in parallel for all flight conditions. According to the control surface deflection to be generated, the displacement field of the surface mesh is computed for each flight condition. It is then transferred to the CFD volume mesh using the Elasticity Analogy (EA) mesh deformation method [11] available in the FlowSimulator [12] environment.

In the next step, the static aeroelastic analysis is executed and the objective function value is passed to the global optimizer. After the optimization run is

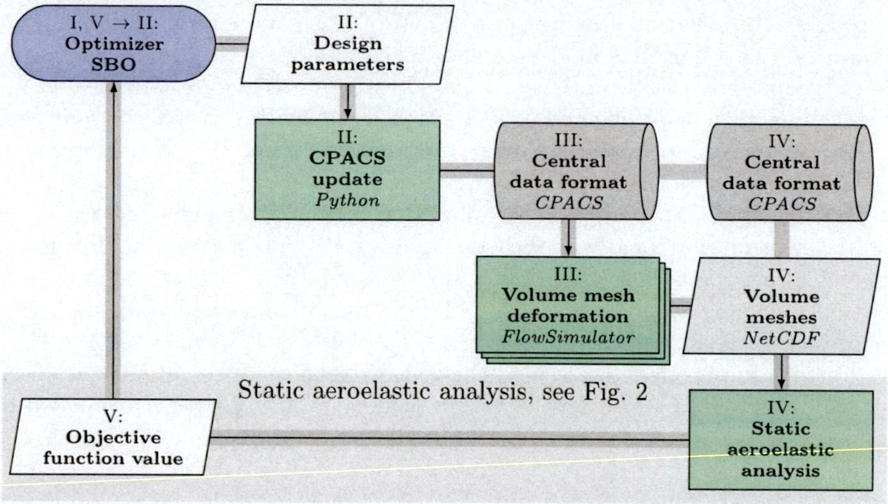

Fig. 1. Flow chart of the process chain.

completed, the optimized vector of design parameters represents the main result of the process chain for the corresponding optimization problem.

A surrogate-based optimization (SBO) method implemented by Wilke [13] has been selected for the wing optimizations in this work. This global optimization strategy represents an adequate compromise between exploring the design space and locating the optimum. The selected optimization method is an implementation of the EGO (Efficient Global Optimizer) optimization method introduced by Jones et al. [14].

Figure 2 shows the flow chart of the static aeroelastic analysis. The fluid-structure coupling loop is marked with a rounded yellow box and the lift-to-drag ratios and the objective function value are evaluated for the associated convergence check.

For all flight conditions, the aerodynamic forces and coefficients are computed with RANS-based CFD simulations. The flow simulations are performed using the DLR TAU-Code [15], which is integrated into the HPC framework FlowSimulator [12].

Based on the aerodynamic loads computed for the considered flight conditions, the corresponding wing deformations are calculated in the structural analysis. The structural analysis process uses the linear analysis module of the commercial software MSC Nastran™ to compute the deformations.

The structural deformations form the input for the CFD volume mesh deformation. The mesh deformation method available in FlowSimulator based on radial basis functions (RBF) [16] is used. The objective function is then determined and the convergence criteria of the static aeroelastic analysis is evaluated. Once the convergence of the fluid-structure coupling loop is achieved, the value of the objective function is passed to the global optimizer.

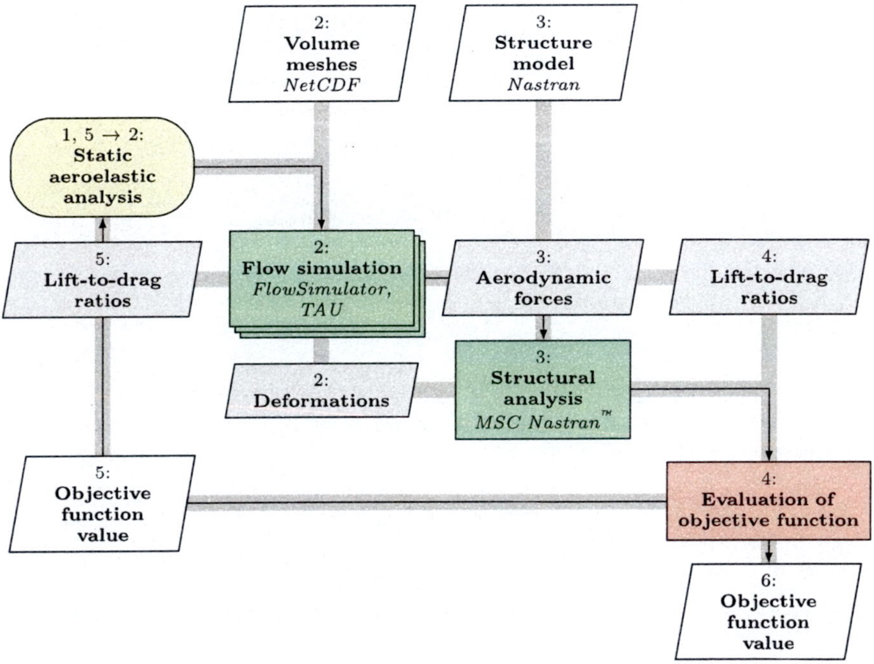

Fig. 2. Flow chart of the static aeroelastic analysis.

3 Optimization of Control Surface Deflections

To investigate adaptive and intelligent wing technologies, a baseline configuration of a long-haul aircraft with a high aspect ratio ultra-efficient wing has been designed. The aircraft represents a conventional twin-engine wide-body configuration in the ICAO Code E [17] category. A new-generation geared turbofan with a bypass ratio of 12.6 at cruise has been selected for propulsion. The large engines are positioned under the wing and the main landing gear is attached to the wing.

In a first step, the wing planform has been developed to increase the aspect ratio to a value of 12.4 under consideration of the geometric integration of multi-functional control surfaces, engine and landing gear. Based on this wing planform aero-structural wing optimization has been performed to size the wing box and optimize the twist distribution. The process chain used, the parametric simulation models, the objective function, and the constraints are the same as described in the previous chapter, except that a structural sizing method and additional load case analysis were included.

In the second step, the wing profiles have been designed for the flight shape using an inverse design method considering maximum lift at low speeds and buffet onset constraints at maximum operating Mach number. The resulting aircraft forms the baseline configuration for this work. This means that the

Table 1. Flight missions.

		Study mission	High-speed mission	Design mission
Weighting factor	w_i	0.6	0.1	0.3
Cruise Mach number	Ma	0.83	0.85	0.83
Range	R	4000 nm	4000 nm	6000 nm
Payload	m_P	40 800 kg	40 800 kg	results from m_F

twisting distribution of the reference aircraft represents the optimal compromise between cruise performance for the flight missions in Table 1 and wing mass.

Multi-functional control surfaces have been integrated into the baseline configuration to introduce variable camber technology. This section presents the results of optimizing control surface deflections to adapt the wing shape to different flight conditions to improve cruise performance.

3.1 Design Task

The objective function of the control surface deflection optimization is the combined fuel consumption per range and payload of three selected flight missions. Thus, the combined fuel consumption is the weighted sum of the fuel consumption of the corresponding missions as given in Eq. (1).

$$\frac{m_F}{R\, m_P} = \sum_{i=1}^{3} w_i \left(\frac{m_F}{R\, m_P} \right)_i \tag{1}$$

Table 1 provides an overview of the selected flight missions and weighting factors. With the selected weighting factors, the expected relative frequency of the missions in airline operation has been taken into account.

A conceptual design model has been used to calculate the fuel consumption of the individual flight missions. In this model, the cruise segment of the mission is described by the well-known Breguet range equation. The thrust-specific fuel consumption has been derived from the engine map of a geared turbofan provided by the DLR-Institute of Propulsion Technology. The lift-to-drag ratio in cruise is obtained from the aerodynamic coefficients of the flow simulation for the wing-fuselage-engine configuration, the estimated aerodynamic coefficients of the tailplane and the specified residual drag and residual thrust coefficients. The longitudinal trim of the aircraft for the given center of gravity position is taken into account. Further details on the models and equations used are described by Wunderlich et al. [3,8].

The deflection angles of six control surfaces at the trailing edge of the wing form the design parameters, as shown in Fig. 3. The wing planform, twist and thickness distribution, and fuselage shape have been kept constant during the

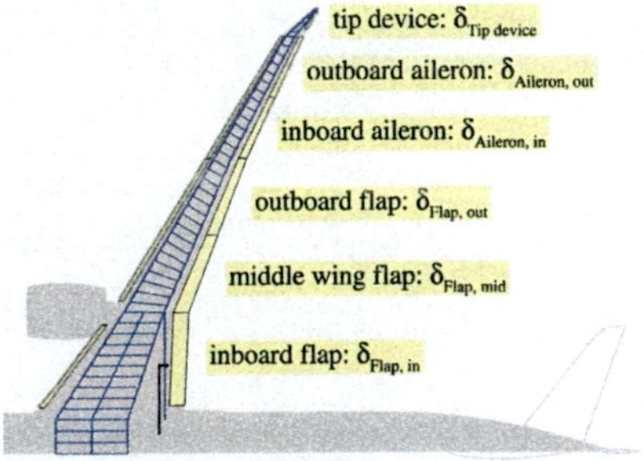

Maximum take-off mass	220 000 kg
Operating empty mass	118 620 kg
Wing mass	27 900 kg
Maximum payload	54 000 kg
Fuel tank volume	$V_F \geq V_{F,req}$
Residual drag coefficient	0.0018
Residual thrust coefficient (through-flow nacelle)	0.0030
Center of gravity position	36 % c_{MAC}

Fig. 3. Design parameters and constraints.

wing optimizations. In addition, the operating empty mass, the maximum take-off mass, the maximum payload, and specific masses of the leading and trailing edges are constant. During optimization, the required fuel tank volume is calculated for all selected flight missions and compared to the usable fuel tank volume. The aerodynamic coefficients in the aerodynamic simulation are corrected with a constant residual drag coefficient to account for drag from components that are neglected in the simulation, such as the engine pylon and flap track fairings. An additional residual thrust has been introduced to correct the coefficients in the simulation with a flow-through nacelle. Figure 3 summarizes the constraints that have been taken into account.

3.2 Optimization Results

Optimization of control surface deflections results in improved cruise performance in terms of lift-to-drag ratio for all flight missions studied. Table 2 summarizes the optimized control surface deflections and the resulting flight performance for the flight missions investigated.

A fuel burn reduction of 0.5% is achieved for the study mission. The 1.7% reduction in fuel consumption for the high-speed mission is significantly higher under these off-design flight conditions because transonic wave drag can be reduced by appropriate control surface deflections. For the design mission with the longest range, the fuel consumption is about 25% higher compared to the study mission. This is due to the limited payload for longer range missions because of the fuel required. For the design mission, a 1.2% reduction in fuel consumption is achieved. Furthermore, for this mission, the increased lift-to-drag ratio results in a higher payload and thus an additional reduction in fuel consumption compared to the study mission at a given payload. For the combined fuel consumption and the corresponding CO_2 emissions, a reduction in the order of 1% is achieved.

Figure 4 presents the lift force and lift coefficient distributions for the baseline configuration and the configuration with optimized control surface deflections. For each lift distribution, the related elliptical lift distribution is shown by a dashed dotted line, and the corresponding center of lift is indicated by a white

Table 2. Results overview for the optimization of control surface deflections.

			Baseline	Optimized control surface deflections	Diff.
Study mission with Ma = 0.83	Control surface deflections	$\delta_{Flap,\,in}$	0.0°	−1.5°	
		$\delta_{Flap,\,mid}$	0.0°	+0.0°	
		$\delta_{Flap,\,out}$	0.0°	+0.2°	
		$\delta_{Aileron,\,in}$	0.0°	+1.3°	
		$\delta_{Aileron,\,out}$	0.0°	+2.1°	
		$\delta_{Tip\ device}$	0.0°	+5.9°	
	Lift-to-drag ratio	L/D	18.52	18.62	+0.5%
	Fuel consumption	$m_F/(R\,m_P)$ in $10^{-4}\,\mathrm{km}^{-1}$	1.435	1.428	−0.5%
High-speed mission Ma = 0.85	Control surface deflections	$\delta_{Flap,\,in}$	0.0°	−2.4°	
		$\delta_{Flap,\,mid}$	0.0°	−0.7°	
		$\delta_{Flap,\,out}$	0.0°	−0.9°	
		$\delta_{Aileron,\,in}$	0.0°	+0.8°	
		$\delta_{Aileron,\,out}$	0.0°	+2.6°	
		$\delta_{Tip\ device}$	0.0°	+3.8°	
	Lift-to-drag ratio	L/D	16.76	17.04	+1.6%
	Fuel consumption	$m_F/(R\,m_P)$ in $10^{-4}\,\mathrm{km}^{-1}$	1.577	1.550	−1.7%
Design mission	Control surface deflections	$\delta_{Flap,\,in}$	0.0°	−1.5°	
		$\delta_{Flap,\,mid}$	0.0°	+0.2°	
		$\delta_{Flap,\,out}$	0.0°	+0.5°	
		$\delta_{Aileron,\,in}$	0.0°	+1.4°	
		$\delta_{Aileron,\,out}$	0.0°	+1.8°	
		$\delta_{Tip\ device}$	0.0°	+5.6°	
	Payload	m_P in kg	31519	31779	+0.8%
	Lift-to-drag ratio	L/D	18.66	18.75	+0.5%
	Fuel consumption	$m_F/(R\,m_P)$ in $10^{-4}\,\mathrm{km}^{-1}$	1.813	1.792	−1.2%
Objective	Combined fuel consumption	$m_F/(R\,m_P)$ in $10^{-4}\,\mathrm{km}^{-1}$	1.563	1.549	−0.9%
	CO_2 emissions per passenger kilometer[a]	$m_{CO2}/(R\,m_P)$ in $\mathrm{g_{CO2}/pkm}$	51.7	51.2	−0.9%

[a] Values of 3.15 $\mathrm{kg_{CO2}/kg_{Fuel}}$ for a turbofan engine that burns kerosene [18] and 105 kg for a passenger with baggage are assumed.

rectangle. The elliptical lift distribution is the optimum for planar wings in terms of lift induced drag. In both flight missions, the optimized control surface deflections cause the lift to shift outboard to reduce the induced drag. The center of lift is marked with an orange diamond for the baseline configuration and with a gray circle for the configuration with optimized control surface deflections. This outboard shift of lift is limited by an increase in the local lift coefficient at the outer wing and the corresponding non-linear increase in transonic wave drag.

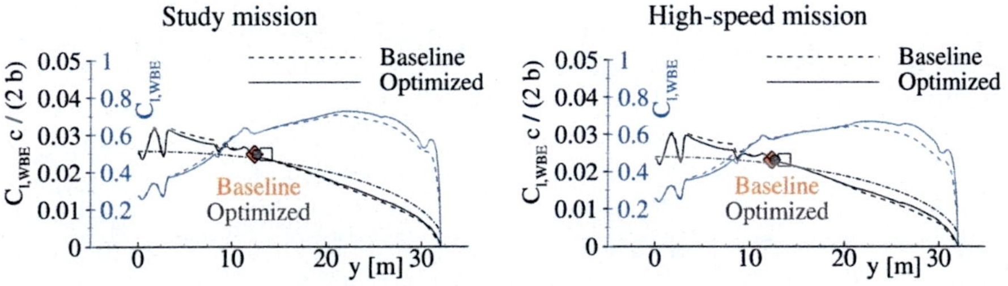

Fig. 4. Lift and lift coefficient distributions for study mission and high-speed mission cruise flight with and without optimized control surface deflections.

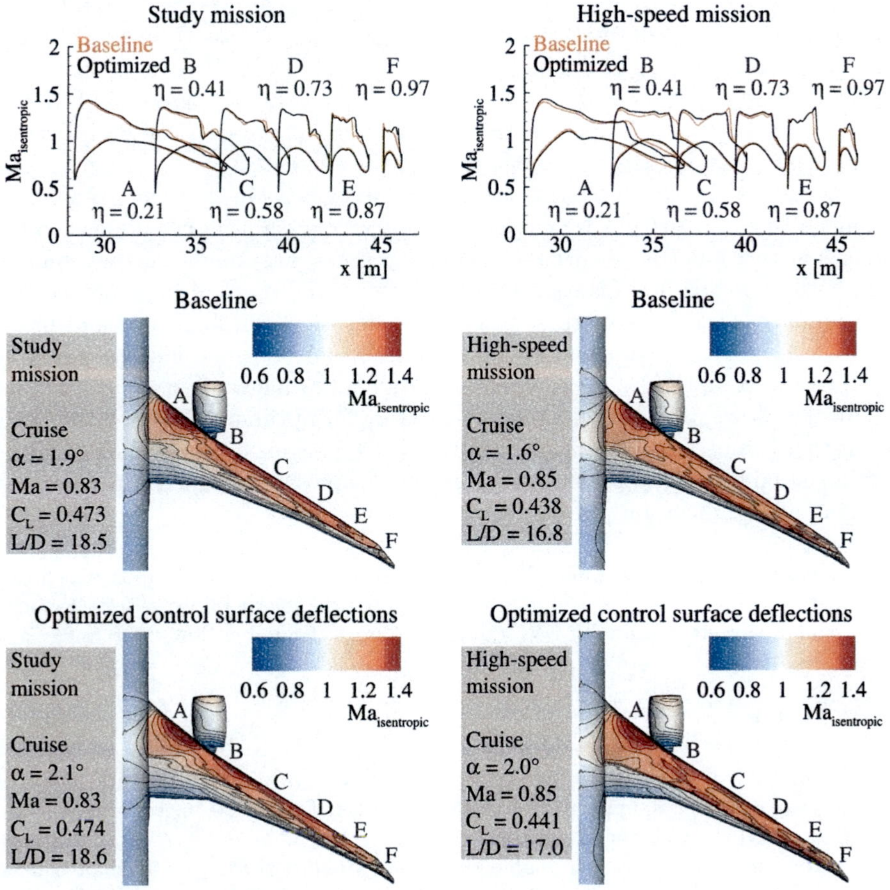

Fig. 5. Isentropic Mach number distributions for study mission and high-speed mission cruise flight with and without optimized control surface deflections.

Figure 5 presents the isentropic Mach number distributions of the baseline configuration and the configuration with optimized control surface deflections for the study mission and the high-speed mission. This figure shows the effect of the outboard shift of lift as a result of the control surface deflections. Here, the outboard shift of lift leads to an increase in compression shock strength in the middle and outboard wing regions, resulting in an increase in wave drag. The improvements in cruise performance in terms of lift-to-drag ratio result from the optimum compromise between induced and transonic wave drag.

4 Conclusion and Outlook

In this work, the use of static aeroelastic wing analysis based on high fidelity simulation methods to optimize the control surface deflections of a highly efficient long-haul aircraft has been successfully demonstrated. Based on lift distributions designed for an optimal trade-off between lift-to-drag ratio and structural mass, adapting the lift distribution to the current flight condition enables improvements in cruise performance without drawbacks in structural mass. It is assumed that the gust loads occurring during cruise flight do not dimension the wing structure and that additional gust load alleviation systems may be required. The results of this optimization show the expected reduction in combined fuel consumption due to the increase in lift-to-drag ratio in all flight missions investigated. This increase in aerodynamic performance has been achieved by finding the optimal compromise between induced and transonic wave drag.

In the future, adaptation of lift distribution for optimum cruise performance have to be considered in aero-structural wing optimization. In addition, the optimization of the control surface deflections has to be combined with an airfoil design and optimization process.

Acknowledgments. The LuFo VI joint research project INTELWI upon which this publication is based has been funded by the German Federal Ministry for Economic Affairs and Climate Action (BMWK) under the project number 20A1903L.

References

1. German Aerospace Center: Towards zero-emission Aviation - How DLR's Aviation Research Strategy supports the European Green Deal 2050. German Aerospace Center (DLR), Cologne (2022)
2. European Commision: Sustainable and Smart Mobility Strategy. Office for Official Publications of the European Communities, Luxembourg, Belgium (2020)
3. Wunderlich, T.F., et al.: Global aerostructural design optimization of more flexible wings for commercial aircraft. J. Aircr. **58**(6), 1254–1271 (2021)
4. Martins, J.R.R.A.: Fuel burn reduction through wing morphing. In: Encyclopedia of Aerospace Engineering, pp. 1–7. Wiley (2016)
5. Szodruch, J., Hilbig, R.: Variable wing camber for transport aircraft. Prog. Aerosp. Sci. **25**(3), 297–328 (1988)

6. Burdette, D.A., et al.: Aerostructural design optimization of a continuous morphing trailing edge aircraft for improved mission performance. In: 17th AIAA/ISSMO Multidisciplinary Analysis and Optimization Conference (2016)
7. Reckzeh, D.: Multifunctional wing moveables: design of the A350XWB and the way to future concepts. In: 29th Congress of the International Council of the Aeronautical Sciences, ICAS 2014 (2014)
8. Wunderlich, T., Reimer, L.: Integrated process chain for aerostructural wing optimization and application to an NLF forward swept composite wing. In: Heinrich, R. (ed.) AeroStruct 2015. NNFMMD, vol. 138, pp. 3–33. Springer, Cham (2018). https://doi.org/10.1007/978-3-319-72020-3_1
9. Lambe, A.B., Martins, J.R.R.A.: Extensions to the design structure matrix for the description of multidisciplinary design, analysis, and optimization processes. Struct. Multidisc. Optim. **46**, 273–284 (2012). https://doi.org/10.1007/s00158-012-0763-y
10. Liersch, C.M., Hepperle, M.: A distributed toolbox for multidisciplinary preliminary aircraft design. CEAS Aeronaut. J. **2**, 57–68 (2011). https://doi.org/10.1007/s13272-011-0024-6
11. Rempke, A.: Netzdeformation mit Elastizitätsanalogie in multidisziplinärer FlowSimulator-Umgebung. In: 20. DGLR - Fach - Symposium der STAB 2016, vol. 2016, pp. 128–129 (2016)
12. Reimer, L., Heinrich, R., Ritter, M.: Towards higher-precision maneuver and gust loads computations of aircraft: status of related features in the CFD-based multidisciplinary simulation environment FlowSimulator. In: Dillmann, A., Heller, G., Krämer, E., Wagner, C., Tropea, C., Jakirlić, S. (eds.) DGLR 2018. NNFMMD, vol. 142, pp. 597–607. Springer, Cham (2020). https://doi.org/10.1007/978-3-030-25253-3_57
13. Wilke, G.A.: Variable-fidelity methodology for the aerodynamic optimization of helicopter rotors. AIAA J. **57**(8), 3145–3158 (2019)
14. Jones, D.R., et al.: Efficient global optimization of expensive black-box functions. J. Glob. Optim. **13**(4), 455–492 (1998). https://doi.org/10.1023/A:1008306431147
15. Gerhold, T.: Overview of the hybrid RANS code TAU. In: Kroll, N., Fassbender, J.K. (eds.) MEGAFLOW - Numerical Flow Simulation for Aircraft Design. NNFM, vol. 89, pp. 81–92. Springer, Heidelberg (2005). https://doi.org/10.1007/3-540-32382-1_5
16. Barnewitz, H., Stickan, B.: Improved mesh deformation. In: Eisfeld, B., Barnewitz, H., Fritz, W., Thiele, F. (eds.) Management and Minimisation of Uncertainties and Errors in Numerical Aerodynamics. NNFM, vol. 122, pp. 219–243. Springer, Heidelberg (2013). https://doi.org/10.1007/978-3-642-36185-2_9
17. International Civil Aviation Organization: Annex 14 to the Convention on International Civil Aviation - Aerodromes, Volume I, Aerodrome Design and Operations, 8th edn. International Civil Aviation Organization, Montreal (2018)
18. Antoine, N.E., Kroo, I.M.: Aircraft optimization for minimal environmental impact. J. Aircr. **41**(4), 790–797 (2004)

Computational Aerodynamics

Scalable Framework Integration of CODA for a Multidisciplinary Preconditioned Matrix-Free Newton-Krylov Method

Simon Ehrmanntraut(✉), Adam Büchner, Sebastian Gottfried, and Arthur Stück

German Aerospace Center (DLR), Institute of Software Methods for Product Virtualization, Dresden, Germany
simon.ehrmanntraut@dlr.de

Abstract. Solving multidisciplinary problems by coupled black-box solvers using splitting methods, such as the nonlinear block-Gauß-Seidel method, has severe limitations on convergence. We implemented a MPI-parallel full Newton-Krylov method involving the next-generation CFD solver *CODA*, combining the HPC platform *FlowSimulator* with the open-source multidisciplinary analysis and optimization *Python* framework *OpenMDAO*. The framework implementation uses the *Python* entry points from *CODA* for its residual vector, its Jacobian and its linear solution routine as matrix-free operators. For the ONERA M6 wing ($1.3{\cdot}10^6$ DoF) with torsional spring attachment, we recovered the quadratic convergence of Newton's method and showed the reduction in computational work compared to the baseline method. The implementation is capable of handling the computationally large NASA DPW5-CRM half-plane ($144{\cdot}10^6$ DoF) distributed across a HPC cluster.

Keywords: Multidisciplinary · Newton-Krylov · Partitioned · Matrix-free · CFD · HPC Framework

1 Motivation

Scalable solution algorithms for multidisciplinary problems, such as aeroelasticity, are a key enabler towards large-scale design analyses and optimizations of future aircraft configurations. Usually, instead of discretizing and solving the multidisciplinary system of partial differential equations in a single monolithic solver, multiple existing high-performance single-disciplinary software tools are coupled in a partitioned approach. One wishes to retain the high computational efficiency of the single-disciplinary solvers involved. In library approaches, such as in *preCISE* [1], the disciplinary tools are treated as black-boxes with generally no derivative information exposed. Such approaches tend to converge slowly and only if the coupling between the disciplines is not too tight. In framework approaches, such as in *FlowSimulator* [see 2,3] and *OpenMDAO* [4], the internal

A. Dillmann et al. (Eds.): STAB/DGLR Symposium 2022, NNFM 154, pp. 219–230, 2024.
https://doi.org/10.1007/978-3-031-40482-5_21

routines of the disciplinary tools are used, making derivative information available to the framework. Such approaches, using the derivatives, can achieve more efficient and robust convergence for tightly coupled problems. Using a framework approach, we implemented a scalable MPI-parallel matrix-free globalized full Newton-Krylov method operating on the complete multidisciplinary system of algebraic equations. This method is expected to converge asymptotically with quadratic order. For preconditioning, the linear solution routines of the single-disciplinary solvers are embedded as block solvers. The disciplines are structured in (sub)groups which allows a flexible association of solution algorithms and HPC resources.

The CFD is usually the most costly simulation component and the biggest scalability challenge within the multidisciplinary problem. This paper focuses on the framework integration of the next-generation CFD code *CODA* [5] – a joint development of DLR, ONERA and Airbus. *CODA*'s software design allows for a granular framework integration that exceeds the modularity level of many other CFD software tools for industrial use. Unlike black-box tools, *CODA* makes exact derivative information, obtained by algorithmic Differentiation (AD), transparent to the framework. Accurate derivatives are considered a key ingredient to overcome efficiency and robustness shortcomings of finite-difference approximations in analysis and optimization [6,7]. The individual algorithms and the framework approach discussed here are well known [8,9], and the aeroelastic test cases use a simple spring as placeholder for an advanced structure mechanics models. However, this paper is believed to be unique in the way a modern CFD software – which is capable of industrial aerodynamic analyses, provides AD-based derivatives, and supports unstructured grids – is integrated into the framework to enable strong implicit solution algorithms on the multidisciplinary level that yield deep convergence of the multidisciplinary problem.

2 Methodology

We consider a general problem from static fluid-structure interaction; however, the method described here is generally applicable to various kinds of multidisciplinary problems. After spatial discretization, we get an autonomous index-1 semi-explicit differential-algebraic system of nonlinear equations

$$M\frac{\mathrm{d}u}{\mathrm{d}t} = r_{\mathrm{CFD}}(u, x) \stackrel{!}{=} 0, \tag{1a}$$

$$q = f_{\mathrm{post}}(u), \tag{1b}$$

$$r_{\mathrm{CSM}}(x, q) \stackrel{!}{=} 0, \tag{1c}$$

where u is the flow state vector (i.e. density, momentum, etc.), x is a grid information vector (e.g. grid coordinates, far-field angle of attack, etc.), and q is an integral quantities vector (e.g. aerodynamic lift, surface forces, etc.). The problem is partitioned such that

- a CFD solver for u given x, with residuals vector r_{CFD} and mass-matrix M, iterating in pseudo-time $t \to \infty$ with linearized implicit-Euler for the steady-state solution, corresponds to (1a),
- a CFD post-processor for q given u, with explicit function f_{post}, corresponds to (1b), and
- a CSM solver for x given q, with residuals vector r_{CSM}, solving with Newton's method, corresponds to (1c).

To solve the partitioned problem (1), the single-disciplinary programs have to be coupled.

2.1 Baseline Method

The commonly used nonlinear block-Gauß-Seidel iteration scheme [10, Eq. 7.4.26, NR 7.4-10] is given in Algorithm 1: from an initial guess, let the CFD solver solve the CFD problem for u, then execute the post-processor for q, then let the CSM solver solve the CSM problem for x, and repeat. This method is implemented easily, requires no derivatives and runs matrix-free. However, like all generalized linear methods, it has severe limitations regarding algorithmic scalability and convergence: by [10, Th. 10.3.5, NR 10.3-3], the method converges asymptotically with linear order at best and diverges if the coupling becomes too dominant. Under- or over-relaxation in the sense of nonlinear block-SOR [10, Th. 10.3.5, NR 10.3-3] cannot overcome this issues. Also, inexact iteration in the sense of [10, Th. 12.2.1] may reduce computational costs, but the convergence issue persists.

Algorithm 1: nonlinear block-Gauß-Seidel

Input: initial guess for u, x

Output: stationary solution u, q, x

while *stopping criteria not met* **do**
- solve $r_{\mathrm{CFD}}(u, x) \overset{!}{=} 0$ for u via the CFD solver;
- calculate $q \leftarrow f_{\mathrm{post}}(u)$ via CFD post-processor;
- solve $r_{\mathrm{CSM}}(x, q) \overset{!}{=} 0$ for x via CSM solver;

end

return u, q, x;

2.2 Newton-Krylov Method

We implemented a globalized preconditioned full Newton-Krylov method: executing Newton's method for the coupled problem (1) with pseudo-transient continuation (PTC), as in [11], for globalization, and solving the linearized system for updates with a Krylov method block-preconditioned by the single-disciplinary solvers. By [11, Th. 2.1, Th. 2.2], for sufficiently small initial CFL number, this method is expected to converge globally and asymptotically with quadratic order, and may also converge for problems with dominant coupling. Note that

Algorithm 2: Jacobi operator

Function $Jacobian([\Delta u^{\mathsf{T}}, \Delta q^{\mathsf{T}}, \Delta x^{\mathsf{T}}]^{\mathsf{T}})$:

  calculate $\Delta r_{\mathrm{CFD}} \leftarrow \left(\frac{-M}{CFL} + \frac{\partial r_{\mathrm{CFD}}}{\partial u}\right)\Delta u + \frac{\partial r_{\mathrm{CFD}}}{\partial x}\Delta x$ via CFD solver;

  calculate $\Delta r_{\mathrm{post}} \leftarrow \Delta q - \frac{\partial f_{\mathrm{post}}}{\partial u}\Delta u$ via CFD post-processor;

  calculate $\Delta r_{\mathrm{CSM}} \leftarrow \frac{\partial r_{\mathrm{CSM}}}{\partial q}\Delta q + \frac{\partial r_{\mathrm{CSM}}}{\partial x}\Delta x$ via CSM solver;

  return $[\Delta r_{\mathrm{CFD}}^{\mathsf{T}}, \Delta r_{\mathrm{post}}^{\mathsf{T}}, \Delta r_{\mathrm{CSM}}^{\mathsf{T}}]^{\mathsf{T}}$;

end

PTC can be combined with additional globalization techniques, such as line-search backtracking by Armijio-Goldstein condition.

The linearized system of the coupled problem (1) by PTC with the CFL number CFL reads

$$\begin{bmatrix} \Delta r_{\mathrm{CFD}} \\ \Delta r_{\mathrm{post}} \\ \Delta r_{\mathrm{CSM}} \end{bmatrix} = \begin{bmatrix} \frac{-M}{CFL} + \frac{\partial r_{\mathrm{CFD}}}{\partial u} & 0 & \frac{\partial r_{\mathrm{CFD}}}{\partial x} \\ -\frac{\partial f_{\mathrm{post}}}{\partial u} & I & 0 \\ 0 & \frac{\partial r_{\mathrm{CSM}}}{\partial q} & \frac{\partial r_{\mathrm{CSM}}}{\partial x} \end{bmatrix} \begin{bmatrix} \Delta u \\ \Delta q \\ \Delta x \end{bmatrix}. \tag{2}$$

Since products of Jacobians with vectors are equivalent to directional derivatives, which can be calculated matrix-free either by finite-differences or algorithmic differentiation, (2) can be implemented via the single-disciplinary solvers as in Algorithm 2. Note that the single-disciplinary programs can all be run in parallel.

As preconditioner to the linearized system (2), the linear solution routines from the single-disciplinary solvers can be used. Since they represent the matrix-free inverse operators of the diagonal block elements of (2), a matrix-free i-step zero-initialized block-Jacobi preconditioner can be defined as in Algorithm 3. Note that the single-disciplinary solvers can use any – possibly again preconditioned – method to solve their linear systems. Again, the single-disciplinary solvers can all be run in parallel. Block-Gauß-Seidel could be used alternatively, if the solvers may not necessarily have to run in parallel, e.g. if the CFD problem is much larger than the CSM problem.

The Newton-Krylov algorithm is given in Algorithm 4. As Krylov method the most sensible choice is the flexible general minimal residual (FGMRES) method, since variable preconditioners may be involved and no further information on the linear system (2) is given. Note that the linearization of (1b) is only needed to perform the chain-rule between (1a) and (1c); hence the update Δq is not applied and q is always computed with f_{post}, introducing no linearization error.

Algorithm 3: zero-initialized block-Jacobi(i) preconditioner

Function $blkJacobi^{(i)}(b)$:
  if $k \neq 0$ **then**
    $[v_{\text{CFD}}^{\mathsf{T}}, v_{\text{post}}^{\mathsf{T}}, v_{\text{CSM}}^{\mathsf{T}}]^{\mathsf{T}} \leftarrow blkJacobi^{(i-1)}(b)$;
    $[b_{\text{CFD}}^{\mathsf{T}}, b_{\text{post}}^{\mathsf{T}}, b_{\text{CSM}}^{\mathsf{T}}]^{\mathsf{T}} \leftarrow b - Jacobian([v_{\text{CFD}}^{\mathsf{T}}, v_{\text{post}}^{\mathsf{T}}, v_{\text{CSM}}^{\mathsf{T}}]^{\mathsf{T}})$;
    solve $\left(\frac{-M}{CFL} + \frac{\partial r_{\text{CFD}}}{\partial u}\right) \Delta v_{\text{CFD}} \overset{!}{=} b_{\text{CFD}}$ via CFD solver;
    set $\Delta v_{\text{post}} \leftarrow b_{\text{post}}$; // no solving with identity as system
    solve $\frac{\partial r_{\text{CSM}}}{\partial x} \Delta v_{\text{CSM}} \overset{!}{=} b_{\text{CSM}}$ via CSM solver;
    update $v_{\text{post}} \leftarrow v_{\text{post}} + \Delta v_{\text{post}}$;
    update $v_{\text{CFD}} \leftarrow v_{\text{CFD}} + \Delta v_{\text{CFD}}$;
    update $v_{\text{CSM}} \leftarrow v_{\text{CSM}} + \Delta v_{\text{CSM}}$;
    return $[v_{\text{CFD}}^{\mathsf{T}}, v_{\text{post}}^{\mathsf{T}}, v_{\text{CSM}}^{\mathsf{T}}]^{\mathsf{T}}$;
  else
    return $[0, 0, 0]^{\mathsf{T}}$;
  end
end

Algorithm 4: Newton-FGMRES-block-Jacobi(i) with PTC

Input: initial guess for u, x
Output: stationary solution u, q, x
while *stopping criteria not met* **do**
  calculate $r_{\text{CFD}} \leftarrow r_{\text{CFD}}(u, x)$ via CFD solver;
  calculate $q \leftarrow f_{\text{post}}(u)$ via CFD post-processor;
  calculate $r_{\text{CSM}} \leftarrow r_{\text{CSM}}(x, q)$ via CSM solver;
  calculate $CFL \leftarrow SER(\|r_{\text{CFD}}\|)$ via switched evolution relaxation;
  solve $Jacobian\left(\begin{bmatrix}\Delta u \\ \Delta q \\ \Delta x\end{bmatrix}\right) \overset{!}{=} \begin{bmatrix}-r_{\text{CFD}} \\ 0 \\ -r_{\text{CSM}}\end{bmatrix}$ for Δu, Δq, Δx via FGMRES with right-preconditioner $blkJacobi^{(i)}$;
  update $u \leftarrow u + \Delta u$;
  /* drop update Δq as q can be computed error free */
  update $x \leftarrow x + \Delta x$;
end
return u, q, x;

3 Framework Implementation

The method from Sect. 2.2 is implemented in a framework approach: a top-level framework performs Algorithm 4, passing and receiving data from the single-disciplinary programs by accessing their routines.

3.1 CODA

As CFD solver and post-processor for (1a) and (1b), *CODA* is used. Via the *FlowSimulator* [see 2,3] HPC infrastructure, *CODA* is given as *Python*-wrapped *C++* SWIG object instance which provides the routines

- `get_flow_dof`, `get_grid_dof` and `get_forces_dof` to return the number of degrees of freedom of the flow u, the grid x and the integral quantities q, respectively,
- `compute_res` to return the residual vector r_{CFD} given u and x,
- `prepare_jacobian` to set the linearization point and the CFL relaxation of the linearized system $A := -M/CFL + \partial r_{\text{CFD}}/\partial u|_u$ given CFL, u and x,
- `compute_d_res` to return the product $A\Delta u$ as directional derivative by algorithmic differentiation for given Δu,
- `solve` to return the iterative solution v to the linear system $Av = b$ given a right-hand side b,
- `compute_forces` to return the results $q = f_{\text{post}}(u)$ given u, and
- `compute_d_forces` to return the product $\partial f_{\text{post}}/\partial u|_u \Delta u$ as directional derivative by algorithmic differentiation for given u and Δu.

These routines take *NumPy* float arrays as parameters, pass their values to the *C++* instance for further processing, and pass the results back to *Python* which returns them again as *NumPy* float arrays. In *CODA*, algorithmic differentiation is implemented using operator overloading in forward mode.

3.2 OpenMDAO

Algorithm 4 is implemented via the open-source high-performance multidisciplinary analysis and optimization *Python* framework *OpenMDAO* [4]. Its main advantages are modular integration of disciplines and objective functions, easy setup of various nested coupling and optimization methods, and extensive off-the-shelve MPI capability for distributed computing.

Analogous to the partition of (1), we implemented two *OpenMDAO* components wrapping the appropriate routines from a given *CODA* and CSM solver instance:

- The *CODA* solver is wrapped by the `CodaSolverComponent` class implementing the abstract `ImplicitComponent` class from *OpenMDAO*. As variables, u is defined as `output` and x as `input`.
- The *CODA* post-processor is wrapped by the `CodaPostprocessorComponent` class implementing the abstract `ExplicitComponent` class from *OpenMDAO*. As variables, q is defined as `output` and u as `input`.
- The CSM solver is wrapped by the `CsmSolverComponent` class implementing the abstract `ImplicitComponent` class from *OpenMDAO*. As variables, x is defined as `output` and q as `input`.

Based on these components, Listing 1 shows an excerpt of a simplified implementation of the top-level script for the Newton-Krylov Algorithm 4: the single-disciplinary programs are initialized and distributed across MPI processes, the

OpenMDAO model is set up and the instances are wrapped by the components, the component variables are connected, and the solver hierarchy is defined. With the top-level linear solver `PETScKrylov`, *OpenMDAO* wraps the MPI-capable FGMRES implementation in *PETSc*. The solver hierarchy can easily be changed: for example, instead of using the block-Jacobi method as preconditioner, the block-Gauß-Seidel method could be used – so the preconditioner could run sequentially (the parallel CFD component followed by the parallel CSM component) while the top-level linear solver runs in parallel.

Listing 1. Simplified *OpenMDAO* problem and solvers definition.

```
import openmdao.api as om
from flowsim import Coda
# single-disciplinary program instances
coda = Coda(n_proc=255) # distributed across 255 MPI processes
csm_solver = CSMSolver(n_proc=1)
# framework setup
prob = om.Problem()
model = prob.model
pg = model.add_subsystem('PG', om.ParallelGroup())  # its subsystems can run in parallel
pg.add_subsystem('CODA_solver', CodaSolverComponent(coda))       # (1a)
pg.add_subsystem('CODA_posprocessor', CodaSolverComponent(coda)) # (1b)
pg.add_subsystem('CSM_solver', CsmSolverComponent(csm_solver))   # (1c)
pg.connect('CODA_solver.flow', 'CODA_posprocessor.flow')
pg.connect('CODA_posprocessor.forces', 'CSM_solver.forces')
pg.connect('CSM_solver.grid', 'CODA_solver.grid')
pg.nonlinear_solver = om.NewtonSolver(...)  # could be changed to om.NonlinearBlockGS()
pg.linear_solver = om.PETScKrylov(ksp_type='fgmres', ...)
pg.linear_solver.precon = om.LinearBlockJac(...)  # could be changed to om.LinearBlockGS()
```

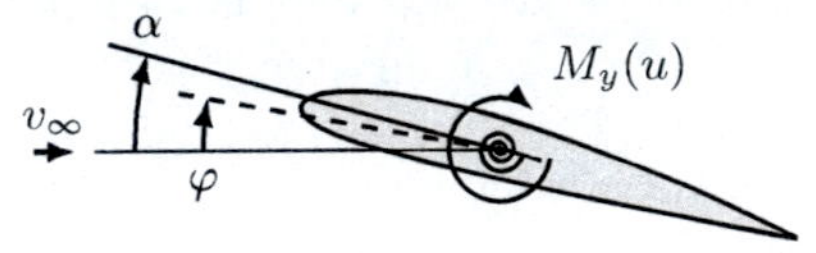

Fig. 1. Wing's root with torsional spring attachment.

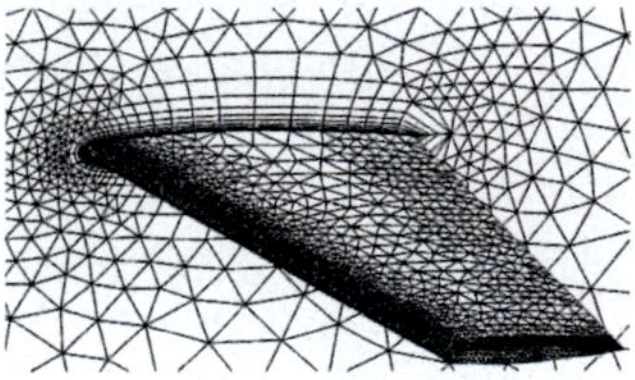

Fig. 2. CFD grid of ONERA M6 Wing.

4 Performance

To test the Newton-Krylov method, we consider a simple problem from static aeroelasticity: The ONERA M6 rigid half-wing (Fig. 2) is attached at its root in its 80%-point to a torsional spring with stiffness k and resting angle φ (Fig. 1). The problem then reads: find an angle of attack α and a flow state u such that

$$\frac{\mathrm{d}u}{\mathrm{d}t} \overset{!}{=} 0 \qquad \text{stationarity,} \tag{3a}$$

$$M_y(u) \overset{!}{=} k \cdot (\alpha - \varphi) \qquad \text{equality of moments.} \tag{3b}$$

The coupling strength is characterized by v_∞/v_{div}, the free-stream velocity over the aeroelastic divergence velocity. For v_∞/v_{div} approaching unity, $\alpha-\varphi$ diverges. In this particular example, physical divergence is identical to numerical divergence, though this is not always the case. Problem (3) is equivalent to (1) with the choices

$$x := \alpha, \qquad q := M_y, \qquad r_{\text{CSM}}(\alpha, M_y) := M_y - k \cdot (\alpha - \varphi),$$

and is solved as such by the implementation.

We run the ONERA M6 half-wing [12] with $1.3{\cdot}10^6$ degrees of freedom distributed across 256 parallel MPI processes (255 for *CODA* and 1 for the scalar CSM solver) in turbulent compressible flow ($Re = 14.6{\cdot}10^6$, $Ma = 0.84$). The negative Spalart-Allmaras turbulence model is used. The implementation of the scalar CSM component is trivial. The top-level FGMRES method (Algorithm 4:6) and the (non)linear solvers of both *CODA* (Algorithm 3:5, Algorithm 1:2) and the CSM solver (Algorithm 3:7, Algorithm 1:4) are all set to reduce the relative residual norm – i.e. against initial norm – to 10^{-14}. Newton's method could be run with variable tolerance in the sense of [13], and FGMRES could be run with variable tolerance for its preconditioner in the sense of [14], to reduce computational cost – investigating their influence on the overall performance is beyond the scope of this study. *CODA* internally uses its own FGMRES linear solver with its own block-Jacobi preconditioner executing 100 iterations based on the LU-decomposition of its disciplinary diagonal block elements. Alternatively, any other combination of solver and preconditioner could be applied. With strict convergence tolerances for the disciplinary linear solvers, the block-Jacobi preconditioner is set to run only one iteration, i.e. $i := 1$. Doing more preconditioner iterations would pay off if it would reduce the amount of inner FGMRES iterations accordingly – this was not observed in our case. We would expect a different picture with more relaxed convergence settings for the disciplinary linear solvers, which was not studied systematically in this work. For the Newton-Krylov method, PTC is set with initial CFL number 1.0, SER exponent 1.0 and switchover to steady state form (i.e. $CFL \leftarrow \infty$) at $CFL = 10^3$. In the baseline method, *CODA* uses the same SER configuration, solving the nonlinear system by iterating the ordinary differential equation (1a) to the stationary solution using linearized implicit Euler method and its internal linear solver stack. As initial guess, α is set to φ and the flow state u is set to free-stream.

Figure 3 shows the nonlinear iteration logs of both the Newton-Krylov method and the baseline method for the case $v_\infty/v_{\text{div}} := 0.61$ and $\varphi := 1.86°$. By [10, Th. 10.3.5, NR 10.3-3] the baseline method converges with linear order and rate $(v_\infty/v_{\text{div}})^2$. As expected, the Newton-Krylov method converges with linear order in the initial phase, but converges asymptotically with quadratic order in the terminal phase where $CFL = \infty$ (i.e. without relaxation) – until round-off errors due to machine precision become dominant.

Figure 4 shows the linear iteration logs from the top-level FGMRES method (Algorithm 4:6) for some of the 18 nonlinear iterations of the aforementioned run with the Newton-Krylov method. As expected, the block-Jacobi preconditioner (Algorithm 3) is an effective preconditioner such that only 3 to 4 FGMRES

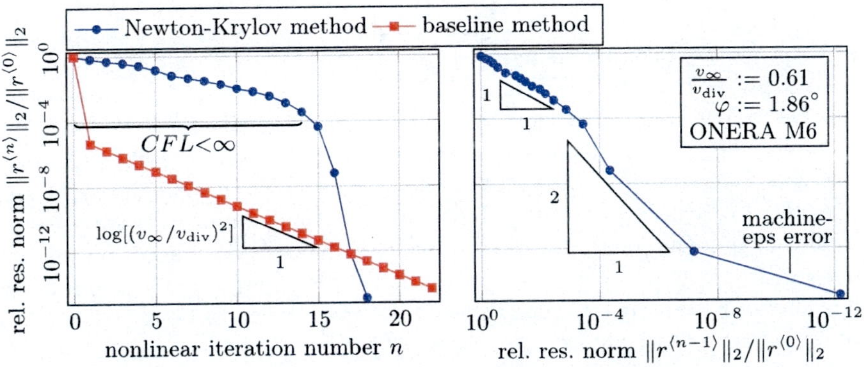

Fig. 3. Relative norm of the nonlinear residual $r := [r_{\mathrm{CFD}}^{\mathsf{T}}\ r_{\mathrm{CSM}}^{\mathsf{T}}]^{\mathsf{T}}$ against nonlinear iteration number n (left) and relative residual norm of the previous iteration (right).

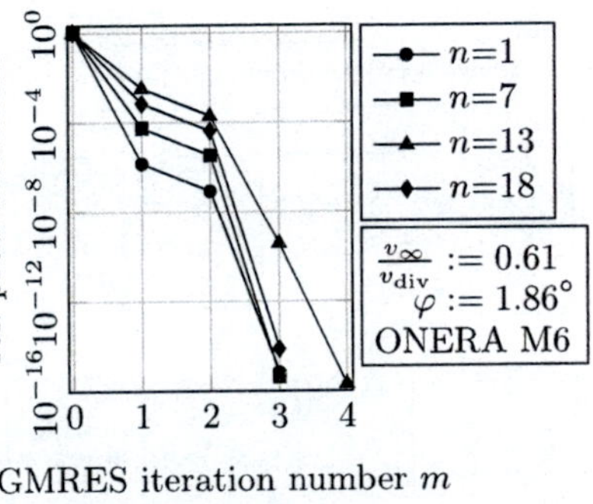

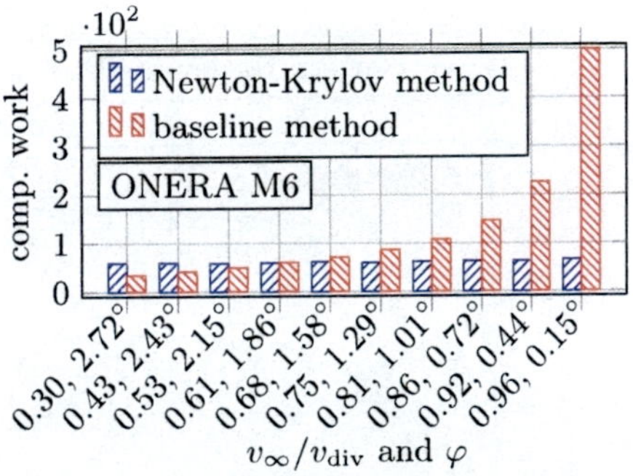

Fig. 4. Relative norm of the preconditioned residual against iteration number m from the top-level FGMRES method (Algorithm 4:6) for nonlinear iterations n of the Newton-Krylov method.

Fig. 5. Computational work – i.e. how often the CFD solver had to solve a linear system – for cases of rising coupling strength $v_\infty/v_{\mathrm{div}}$. The steady-state solution is identical for all cases.

iterations are required to solve the linear systems. Hence, memory usage is small enough for FGMRES to be run without restarting.

Figure 5 shows the computational work of both the Newton-Krylov method and the baseline method for various cases of rising $v_\infty/v_{\mathrm{div}}$. The resting angle φ is chosen such that the steady-state angle of attack is in all cases 3°. For the baseline method, the required work increases drastically as $v_\infty/v_{\mathrm{div}}$ approaches unity. The Newton-Krylov method requires a roughly constant amount of work. As expected, for tightly coupled cases, the Newton-Krylov method requires less work than the baseline method.

The majority of the time, about 90%, is spent by *CODA* in its *C++* instance, especially during executions of Algorithm 3:5. Overhead from *OpenMDAO*, executing the Newton-Krylov method, and from data transfers from and to *Python* is hence marginal compared to the former.

To demonstrate the scalability of the framework approach, we replaced the ONERA M6 wing by the larger NASA DPW5-CRM half-plane [15] with $18{\cdot}10^6$ (L1) and $144{\cdot}10^6$ (L3) degrees of freedom distributed across 4096 parallel MPI processes (4095 for *CODA* and 1 for the scalar CSM solver) in cruise condition ($Re = 5{\cdot}10^6$, $Ma = 0.85$). The same solver configuration is used but with the SER exponent set to 0.5. Figure 6 shows the nonlinear iteration log of the Newton-Krylov method for the case $v_\infty/v_{\text{div}} := 0.18$ and $\varphi := 0.5°$. As expected, the Newton-Krylov method converges with linear order in the initial phase, but converges asymptotically with superlinear order in the terminal phase. While the convergence for this case is sub-quadratic, this shows that the framework is capable of handling computationally large problems distributed among many HPC cluster nodes with parallel MPI processes.

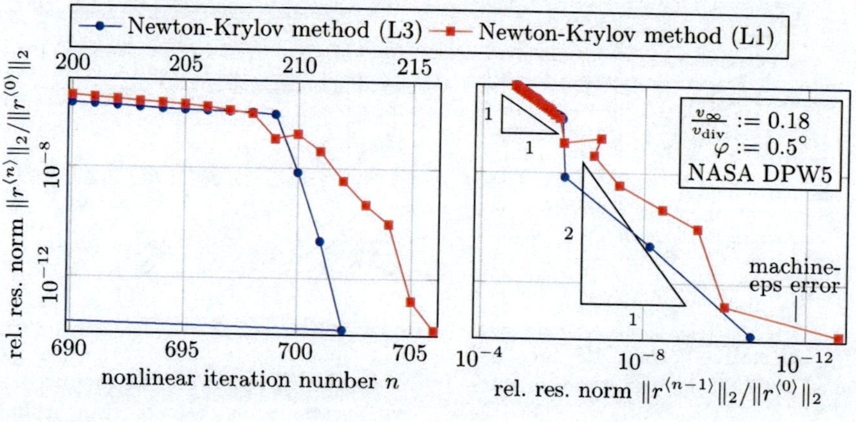

Fig. 6. Relative norm of the nonlinear residual $r := [r_{\text{CFD}}^\top \; r_{\text{CSM}}^\top]^\top$ against nonlinear iteration number n (left) and relative residual norm of the previous iteration (right).

5 Conclusion

We successfully demonstrated the feasibility of solving multidisciplinary problems by coupling *CODA* and other single-disciplinary solvers using *OpenMDAO* and the *FlowSimulator* infrastructure. The Newton-Krylov method is highly scalable, as the single-disciplinary solvers can run in parallel with their domains decomposed and distributed among MPI processes. Thus, the framework integration fully retains the computational efficiency of the single-disciplinary solvers. For a sample problem, we observed global and asymptotically quadratic convergence with the Newton-Krylov method, and we showed the reduction in computational work against the baseline method (nonlinear block-Gauß-Seidel). Moreover, while the latter diverges for tightly coupled problems, the Newton-Krylov method can still converge. We expect that our findings carry over to problems that are heterogeneously composed of more disciplines.

In ongoing work, we will extend the sample problem from Sect. 4 by replacing the scalar CSM equation (3b) with a FEM problem and a general-purpose solver. Moreover, we intend to run a parameter study investigating the influence of the solver settings on the total computational work. In future work, we will include mesh deformation and load transfer between the CFD and CSM grid to solve aeroelasticity problems for flexible wings. The framework will be optimized to reduce memory redundancy and extended to perform time-accurate iteration (e.g. by Runge-Kutta methods) and design optimization.

References

1. Chourdakis, G., et al.: preCICE v2: a sustainable and user-friendly coupling library. Open Res. Europe **2**(51) (2022). https://doi.org/10.12688/openreseurope.14445.2
2. Reimer, L.: The FlowSimulator—a software framework for CFD-related multidisciplinary simulations. In: European NAFEMS Conference Computational Fluid Dynamics (CFD) - Beyond the Solve (2015). elib.dlr.de/100536/
3. Backhaus, T., Gottfried, S., Merle, A., Hwang, J., Stück, A.: Modularization of high-fidelity static aeroelastic MDO enabling a framework-based optimization approach for HPC. In: AIAA Scitech 2021 Forum (2021). https://doi.org/10.2514/6.2021-1236
4. Gray, J.S., Hwang, J.T., Martins, J.R.R.A., Moore, K.T., Naylor, B.A.: OpenMDAO: an open-source framework for multidisciplinary design, analysis, and optimization. Struct. Multidisc. Optim. **59**(4), 1075–1104 (2019). https://doi.org/10.1007/s00158-019-02211-z
5. Leicht, T., et al.: DLR-project digital-X - next generation CFD solver FLUCS. In: DLRK (2016). elib.dlr.de/111205/
6. Brown, P.N., Saad, Y.: Hybrid Krylov methods for nonlinear systems of equations. SIAM J. Sci. Stat. Comput. **11**(3), 450–481 (1990). https://doi.org/10.1137/0911026
7. Abu-Zurayk, M., et al.: Sensitivity-based multifidelity multidisciplinary optimization of a powered aircraft subject to a comprehensive set of loads. In: AIAA Aviation 2020 Forum (2020). https://doi.org/10.2514/6.2020-3168
8. Knoll, D., Keyes, D.: Jacobian-free Newton-Krylov methods: a survey of approaches and applications. J. Comput. Phys. **193**(2), 357–397 (2004). https://doi.org/10.1016/j.jcp.2003.08.010
9. Hwang, J.T., Martins, J.R.: A computational architecture for coupling heterogeneous numerical models and computing coupled derivatives. ACM Trans. Math. Softw. **44**(4), 1–39 (2018). https://doi.org/10.1145/3182393
10. Ortega, J.M., Rheinboldt, W.C.: Iterative Solution of Nonlinear Equations in Several Variables. SIAM (2000). https://doi.org/10.1137/1.9780898719468
11. Coffey, T.S., Kelley, C.T., Keyes, D.E.: Pseudotransient continuation and differential-algebraic equations. SIAM J. Sci. Comput. **25**(2), 553–569 (2003). https://doi.org/10.1137/S106482750241044X
12. Rumsey, C.: 3D ONERA M6 Wing Validation Case. NASA, 10 November 2021. http://turbmodels.larc.nasa.gov/onerawingnumerics_val.html. Accessed 14 Aug 2022

13. Dembo, R.S., Eisenstat, S.C., Steihaug, T.: Inexact Newton methods. SIAM J. Numer. Anal. **19**(2), 400–408 (1982). https://doi.org/10.1137/0719025
14. Van Den Eshof, J., Sleijpen, G.L., Van Gijzen, M.B.: Relaxation strategies for nested Krylov methods. J. Comput. Appl. Math. **177**(2), 347–365 (2005). https://doi.org/10.1016/j.cam.2004.09.024
15. Morrison, J.H.: 5th AIAA CFD Drag Prediction Workshop. NASA, 10 June 2015. aiaa-dpw.larc.nasa.gov/Workshop5/workshop5.html. Accessed 27 Feb 2023

Surrogate Modeling of Hybrid Laminar Wing Aerodynamic Coefficients

Mauricio M. Jentys(✉) and Christian Breitsamter

Chair of Aerodynamics and Fluid Mechanics, Technical University of Munich, Boltzmannstr. 15, 85748 Garching, Germany
mauricio.jentys@tum.de

Abstract. This contribution presents the implementation and results of a Gaussian Process regression based surrogate model for prediction of the aerodynamic drag coefficient C_D of a transonic transport aircraft wing. The wing contains a hybrid laminar flow control system, alongside variable camber integration via an Adaptive Dropped Hinge Flap. The latter enters the envisaged input parameter space of the surrogate through specification of its deflection angle, alongside variations in cruise lift coefficient and altitude. The model builds upon high-fidelity computational fluid dynamics (CFD) results and aims at incorporating them into an overall aircraft design workflow. The predicted drag coefficients agree well when compared to CFD validation data, showing the model being suitable for the present application case.

1 Introduction

The design and optimization process of an aircraft is a highly interdisciplinary and computationally expensive process, mainly stemming from the vast amount of computations required. Therefore, overall aircraft design (OAD) toolboxes typically employ low-fidelity (LowFi) methods, like pre-computed databases or handbook formulae, omitting the prohibitively high computational expenses connected to high-fidelity (HiFi) methods such as computational fluid dynamics (CFD).

To make use of the higher fidelity levels connected to the results of CFD simulations, while still satisfying the requirement for near real-time data queries from OAD, a lot of effort is directed towards the formulation of reduced order or surrogate models based on HiFi data, for instance presented in Ref. [11].

An attractive technique, based on the level of abstraction of complex aerodynamic phenomena into corresponding coefficients, is presented by Gaussian Process (GP) regression models. Application of GPs for regression tasks generalize well with little training data, provide an inherent uncertainty quantification and include an automatic trade-off between the model fitting the data and its associated complexity (see [10]).

Numerous recent publications dealing with surrogate modeling in an aerodynamic context build upon GP regression. In Ref. [3], the authors apply GP

A. Dillmann et al. (Eds.): STAB/DGLR Symposium 2022, NNFM 154, pp. 231–240, 2024.
https://doi.org/10.1007/978-3-031-40482-5_22

regression models for multi-fidelity aerodynamic datafusion, using the thereby constructed surrogate to assess stability and control limits and properties of a blended wing body aircraft. Within Ref. [14], a GP regression model is used to enhance linearized dynamics models, accounting for aerodynamic and propulsive forces at pre-fixed anchor points within the flight envelope, where the enhanced model is applied to an electric quad-rotor air-taxi concept vehicle.

Within this paper, a GP regression model based on HiFi CFD data for parametric prediction of aerodynamic coefficients is presented, the focus lying upon the drag coefficient C_D of the wing of a reference transonic transport aircraft. The framework and reference configuration used within this contribution stem from the LuFo VI-1 project CATeW (*Coupled Aerodynamic Technologies for Aircraft Wings*). The project focuses on the assessment of the potential for aerodynamic drag reduction through synergistic application of a hybrid laminar flow control (HLFC) system together with a variable camber (VC) wing, where investigations are performed on the two aforementioned fidelity levels. The LowFi part is represented by the toolchain MICADO (see [12,13]) of the Institute of Aerospace Systems (RWTH Aachen University), while CFD simulations of the technology coupling at the Chair of Aerodynamics and Fluid Mechanics (Technical University of Munich) form the HiFi complement. As indicated above, the bilateral structure of the project aims at assessing the full synergistic potential on system level using OAD tools, while complex aerodynamic interaction effects require detailed investigations based on CFD. Therefore, a key requirement within the project is formed by implementing an efficient method of introducing HiFi-data into the OAD toolchain via reduced order models or surrogates, the GP surrogate presented in this paper belonging to the latter category. For a more detailed introduction to the project the reader may refer to Ref. [7].

2 Methods

This section provides a brief overview considering the basics of Gaussian Process regression and the CFD-framework applied for generation of training data, with respect to the wing of the reference configuration CATeW-02.

2.1 Gaussian Process Regression

As introduced in Sect. 1, Gaussian Process (GP) regression is a supervised machine learning technique, based upon probabilistic principles. A GP is defined as a distribution over functions, where the corresponding function values $f(\boldsymbol{x})$ are jointly Gaussian distributed. Correspondingly, a GP can be characterized by its mean $\mu(\boldsymbol{x})$ and covariance $k(\boldsymbol{x}, \boldsymbol{x'})$ functions (see [1]):

$$f(\boldsymbol{x}) \sim \mathcal{GP}(\mu(\boldsymbol{x}), k(\boldsymbol{x}, \boldsymbol{x'})). \tag{1}$$

Without providing any observations to the GP, one speaks of the GP prior, which relates the outputs of the function $f(\boldsymbol{x})$ solely to the input parameter vector $\boldsymbol{x}$.

The mean of the prior is typically set to be zero $\mu(\boldsymbol{x}) = 0$, while the covariance function, also referred to as the kernel of a GP, determines the structure of the output functions to be considered.

To illustrate the aforementioned, an example of a GP prior for a one-dimensional case is depicted in the upper panel of Fig. 1, where 20 function realizations (light red lines) are drawn at random from the prior of the GP. The samples provided in Fig. 1 are generated applying a squared-exponential kernel (see [1]):

$$k(\boldsymbol{x}, \boldsymbol{x}') = \sigma_f^2 \exp\left(-\frac{1}{2}\sum_{i=1}^{d} \frac{(x_i - x_i')^2}{\ell_i^2}\right), \tag{2}$$

where the hyper-parameters ℓ_i represent length-scale parameters for each input dimension and σ_f^2 denotes the marginal variance of the GP.

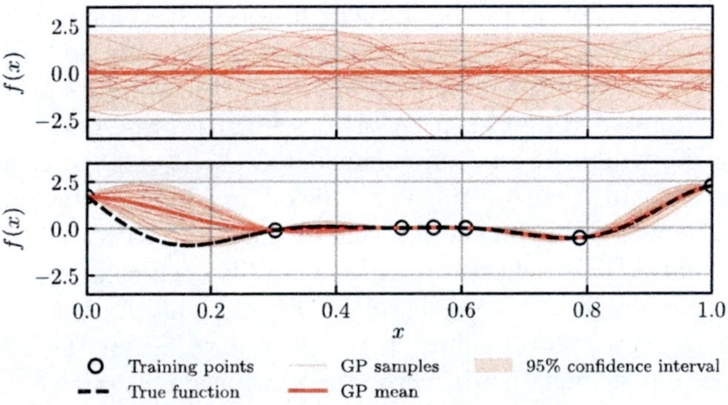

Fig. 1. Upper panel: Mean (red line) and 95% confidence interval (light red shaded area) of the GP prior, alongside 20 random samples (light red line). Lower panel: Mean and 95% confidence interval of the GP posterior, conditioned upon seven training points (black circles) from $f(x) = (3x - 1.5)^2 \sin(12x - 4)$ (black dashed line), alongside 20 random function samples from the posterior.

In a general regression setting, one seeks to determine a function $f : \mathbb{R}^d \rightarrow \mathbb{R}$, fulfilling $f(\boldsymbol{x}_i) = y_i$ at a discrete set of observation or training points $\{(\boldsymbol{x}_i, f(\boldsymbol{x}_i))|i = 1, \ldots, n_{Samples}\}$. Since per definition of a GP, function values $f(\boldsymbol{x})$ are jointly Gaussian distributed, the distribution of training outputs $f(\boldsymbol{x}_i)$ (index i) and prediction outputs $f(\boldsymbol{x}_*)$ (index $*$) can be expressed as (see [10]):

$$\begin{bmatrix} f(\boldsymbol{x}_i) \\ f(\boldsymbol{x}_*) \end{bmatrix} \sim \mathcal{N}\left(\boldsymbol{0}, \begin{bmatrix} k(\boldsymbol{x}, \boldsymbol{x}) & k(\boldsymbol{x}, \boldsymbol{x}_*) \\ k(\boldsymbol{x}_*, \boldsymbol{x}) & k(\boldsymbol{x}_*, \boldsymbol{x}_*) \end{bmatrix}\right). \tag{3}$$

This joint prior distribution, allows for formulation of the conditional distribution, namely through conditioning the unknown prediction values $f(\boldsymbol{x}_*) = \boldsymbol{f}_*$ upon the given training outputs $f(\boldsymbol{x}_i) = \boldsymbol{f}$ and kernel matrices $k(\cdot, \cdot)$ (see [10]):

$$\begin{aligned}(\boldsymbol{f}_*|\boldsymbol{x}_*, \boldsymbol{x}, \boldsymbol{f}) \sim \mathcal{N}\big(&k(\boldsymbol{x}_*, \boldsymbol{x})k(\boldsymbol{x}, \boldsymbol{x})^{-1}\boldsymbol{f}, \\ &k(\boldsymbol{x}_*, \boldsymbol{x}_*) - k(\boldsymbol{x}_*, \boldsymbol{x})k(\boldsymbol{x}, \boldsymbol{x})^{-1}k(\boldsymbol{x}, \boldsymbol{x}_*)\big).\end{aligned} \tag{4}$$

The conditioned distribution is referred to as the posterior of a GP. An example of 20 samples from a GP posterior (light red lines), conditioned on seven observation points (black circles) drawn from the function $f(x) = (3x-1.5)^2 \sin(12x-4)$, alongside the resulting mean function (red line) and true function (black dashed line) is illustrated in the lower panel of Fig. 1. Reflected in the samples is the graphical intuition behind conditioning a noise-free Gaussian distribution upon known function values, namely reducing the possible set of functions contained in the prior to only include those exactly matching the training data points.

2.2 Computational Fluid Dynamics and Reference Configuration

Training data for implementation of the GP regression is generated via CFD simulations applying the DLR Tau code (see [15]). An overview of the wing geometry of the reference configuration CATeW-02 is presented in Ref. [6], alongside a more detailed description of the implementations used for modeling of active boundary layer suction within the suction panel of the wing and VC integration. Variable camber capabilities are integrated into the wing using deflections of an Adaptive Dropped Hinge Flap (ADHF, see [16]), presented in Fig. 2 middle for $\delta_{ADHF} = 4°$, while the suction coefficient $C_q = V_n/U_\infty$ is set to a constant value of $C_q = -12 \cdot 10^{-4}$ within this paper (see Fig. 2 left). The resulting pressure coefficient distribution on the suction side of the wing, with the aforementioned suction level and ADHF deflection angle, for a lift coefficient of $C_L = 0.5$ at an altitude of $H = 35.000$ ft, is presented in Fig. 2 right. As the flow conditions envisage the transonic flow regime, a shock is forming on the suction side of the wing, losing in strength with increasing outboard stations along the span of the wing. Positive deflections of the ADHF generally result in an alignment of the shock front with the streamwise starting point of the spoiler, whereas for an undeflected case the shock front is generally positioned more upstream with respect to the deflected cases. For a more detailed overview, the reader may refer to Ref. [6].

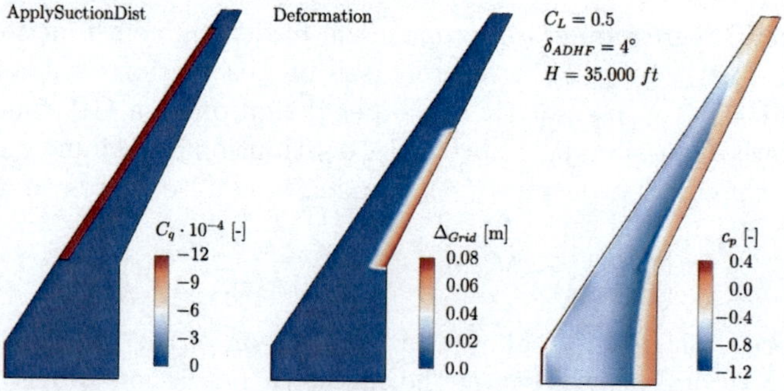

Fig. 2. Left: Suction coefficient distribution, Middle: Grid deformation according to ADHF deflections, Right: Pressure coefficient distribution $p = [0.5, 4°, 35000 \text{ ft}]$.

The simulations are performed applying the $\gamma - Re_\theta$ transition turbulence model, with a helicity Reynolds number based extension to include cross-flow induced transition, as described in Ref. [5].

Computational grids are generated using CentaurSoft Centaur in an unstructured manner, the final grid consisting of approx. $55 \cdot 10^6$ elements. A second order central scheme is applied for discretization of convective and viscous fluxes, time-stepping is performed applying a dual-time scheme, while a 3w multigrid algorithm is chosen for convergence acceleration.

3 Results

Applying the methodical background introduced in Sect. 2, training data generation, alongside the included input parameter space and model conditioning is detailed in Sect. 3.1. Section 3.2 presents selected results of the implemented surrogate.

3.1 Surrogate Model Conditioning

In agreement with requirements from the LowFi-Aerodynamic toolchain, the envisaged input parameter space for surrogate predictions consists of three dimensions, namely the lift coefficient C_L of the wing, the ADHF deflection angle δ_{ADHF} and the flight altitude H. The input parameter space spans the ranges $\mathcal{P} = C_L \times \delta_{ADHF} \times H = [0.4, 0.6] \times [-2°, 4°] \times [33000\text{ ft}, 39000\text{ ft}]$, covering the required parameter combinations derived for cruise flight. As denoted earlier, the suction coefficient is held constant at $C_q = -12 \cdot 10^{-4}$ within this paper, as previous studies indicate maximum extent of laminar flow for this suction level given the applied modeling techniques, see Ref. [6] for details. Furthermore, simulations are performed at a constant cruise Mach number of $Ma_{cr} = 0.83$, which is used together with the altitude parameters to set the Reynolds number within the simulation according to the ISA atmosphere model as published in Ref. [8].

The implementation of the GP regression model is performed in python, making use of the GP regression module implemented in scikit-learn (see [9]). To train the GP regression model, an initial training set of 50 samples is simulated for parameter combinations $\boldsymbol{p}_{Train} \in \mathcal{P}$ determined via a Latin Hypercube sampling (LHS) strategy (see Fig. 3 left panel, black dots).

Training in the context of GP models refers to optimizing the hyper-parameters of the GP kernel. For the present model, the earlier introduced squared-exponential kernel (see Sect. 2.1, Eq. 2) is chosen, leading to the hyper-parameter vector $\boldsymbol{\theta} = [\sigma_f^2, \ell_1, \ell_2, \ell_3]$. Optimization of the hyper-parameters is performed through maximizing the log-marginal-likelihood of the GP, see Ref. [10] for details.

Furthermore, an adaptive sampling strategy utilizing the inherent uncertainty quantification of the GP model is implemented. Based upon the prediction parameters $\boldsymbol{p}_*$, infill points $\boldsymbol{p}_{Adapt}$ are chosen maximizing the estimated mean square error (MSE) (see [4]):

$$\hat{s}^2(\boldsymbol{p}_*) = \sigma^2 \left(1 - \boldsymbol{\psi}^{\mathrm{T}}\boldsymbol{\Psi}^{-1}\boldsymbol{\psi} + \frac{1 - \mathbf{1}^{\mathrm{T}}\boldsymbol{\Psi}^{-1}\boldsymbol{\psi}}{\mathbf{1}^{\mathrm{T}}\boldsymbol{\Psi}^{-1}\mathbf{1}}\right), \tag{5}$$

where σ^2 denotes the variance of the GP, $\boldsymbol{\Psi}$ the correlation matrix and ψ the corresponding correlation vector. In a practical implementation, the third term within the brackets is usually dropped due to its relatively small magnitude, whereas the variance σ^2 is set to one, as it acts in scaling the expected MSE, thus not altering the position of the maximum (see [4]).

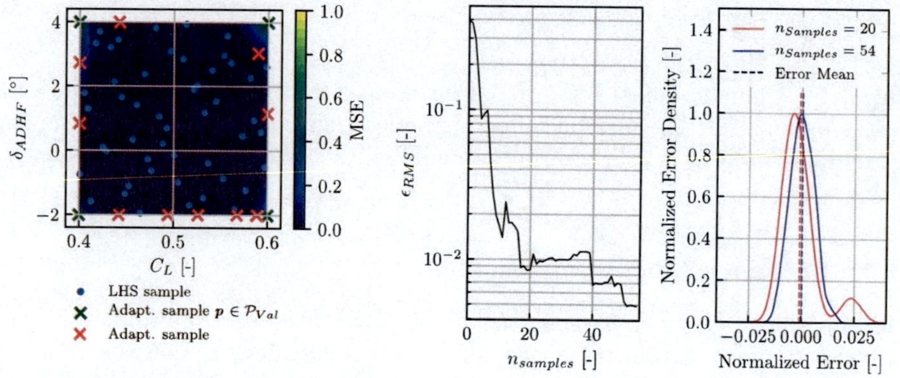

Fig. 3. Left Panel: Sample points used for surrogate model training, projected onto the $C_L - \delta_{ADHF}$ plane, with additional contour plot indication of the (normalized) expected mean square error (MSE) for $n_{samples} = 50$. Mid panel: Development of the root mean square error ϵ_{RMS} with increasing sample size $n_{samples}$, Right panel: Normalized error density distribution for two different sample sizes, alongside the means of the error distributions.

Using the above described technique allows for iterative addition of sampling parameters $\boldsymbol{p}_{Adapt}$ to the training set. Considering the requirement of performing multiple simulations in parallel, the iterative adaptation process computes batches of ten adaptive sample points at a time, via extraction of multiple optima from the MSE distribution (see Fig. 3, left).

To assess the performance of the surrogate, in addition to the training set a full factorial test set consisting of 36 samples is simulated. The parameter combinations for the test set are set to $\mathcal{P}_{Test} = \{0.4, 0.5, 0.6\} \times \{-2°, 0°, 2°, 4°\} \times \{35000\,\text{ft}, 37000\,\text{ft}, 39000\,\text{ft}\}$, thus enclosing the cruise C_L of 0.5 and the extremes of the cruise envelope derived for the reference configuration. These parameter combinations are chosen, as the predictive capabilities of GP based surrogates typically degrade when extrapolating, which, due to the base training set being derived through a LHS strategy, is the case for the extremes of $\mathcal{P}$. This is correspondingly reflected in the adapted sample points (Fig. 3 left, crosses), where the infill strategy initially seeks to place new sample points on the edges of the parameter space. Especially the four extremes $\{-2°, 4°\} \times \{0.4, 0.6\}$ show the highest associated MSE values, nevertheless are omitted from the adapted training set due to them being elements of the test set (see Fig. 3 left, green crosses).

The course of the root mean square error between predictions at test set parameter combinations and the corresponding observations over the number of included samples $n_{samples}$ is shown in the mid panel of Fig. 3. The termination criterion for model training, or addition of adaptively chosen sample points is set to $\epsilon_{RMS} < 5 \cdot 10^{-3}$, which the model reaches after 54 sample points are included into the training. Considering the overall error development of the model, already after 20 samples ϵ_{RMS} falls below values of 10^{-2}. Nevertheless, regarding the distribution of the error as depicted in the right panel of Fig. 3, ultimate statements about the generalization properties of the model with respect to the test data set should include the full base training set of $n_{samples} = 50$. An extracted sample subset from a LHS is itself not optimally space filling and shows strong sensitivity to the locations of the sample points with respect to the test set. The full sample size of $n_{samples} = 54$ results in a zero-centered error density distribution with respect to the full factorial test data set.

3.2 Surrogate Predictions

Prediction results for the drag coefficient of the wing, extracted for an altitude of $H = 35000$ ft under variation of the lift coefficient C_L and ADHF deflection angle δ_{ADHF} are shown in the upper panel of Fig. 4. The above assessed quantitative agreement is qualitatively reflected in the course of the output surface predicted by the model when comparing it to the test data set, marked with black circles in Fig. 4.

Optimizing the hyper-parameters of the squared-exponential kernel results in $\boldsymbol{\theta} = [0.00551^2, 0.459, 0.872, 60.6]$ after data normalization. The values of the hyper-parameters allow for interpretation of to the predictive behaviour of the model. Connected to the marginal variance σ_f^2 is the standard deviation of the function σ_f, determining the amplitude of the response surface. Furthermore, the length-scales ℓ_1 to ℓ_3 determine the width of the kernel function in the respective dimension. Connected to a short length-scale is a rapid change in the corresponding dimension, as opposed to a long length-scale indicating little or slow changes in the corresponding dimension (see [10]). Thus the model correctly displays the physical effects reflected in the training data, with a variation in lift coefficient C_L showing the largest variation in associated drag coefficients C_D (corresponding to the lowest value of ℓ), a reduced impact of the ADHF deflection angle δ_{ADHF}, while the flight altitude shows only small effect upon the drag coefficient, accordingly reflected in the numerical training data.

In line with the qualitative aspects described above, a trained GP allows not only for absolute value prediction but also for quantitative sensitivity assessment by prediction of the gradients of the response surface. The predicted partial derivatives of the wing drag coefficient with respect to the lift coefficient and ADHF deflection angle are depicted in Fig. 4, left and right lower panels, respectively. Considering $\partial C_D / \partial C_L$ the typical quadratic behaviour is reflected within C_D for increasing C_L, nevertheless increasing δ_{ADHF} from $-2°$ to $4°$ shows a reduction in sensitivity of C_D with respect to C_L. This is reflected in the reduced range considering the absolute values of $\partial C_D/\partial C_L$. The partial derivative of the drag coefficient with respect to the ADHF deflection angle $\partial C_D/\delta_{ADHF}$, reflects the possible drag reduction through integration of a VC system. Furthermore, the prediction of the optimal ADHF deflection angle considering $\partial C_D/\delta_{ADHF} = 0$ (see Fig. 4, right lower panel, dashed line), alongside the predicted shift in optimal ADHF deflection angle over the course of C_L is reflected in the predicted gradients. In overall agreement with project partners OAD results, as published in Ref. [2], the optimal value for δ_{ADHF} is found to lie within a range of approx. $[1.4°; 2.7°]$ for the reference mission of the aircraft.

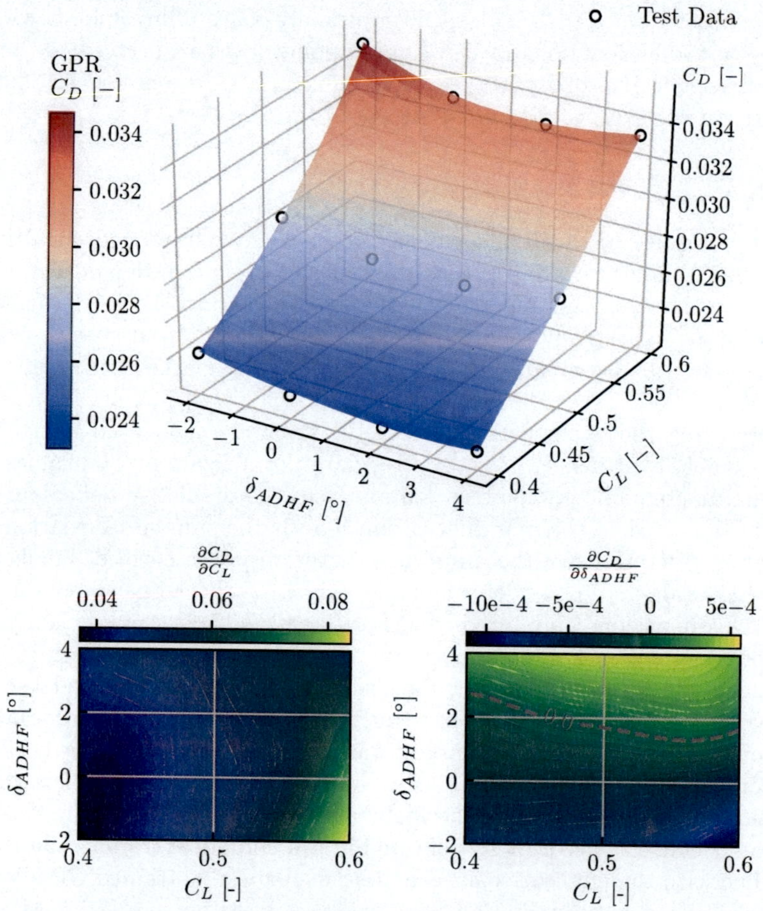

Fig. 4. Upper Panel: Predicted drag coefficient development for $H = 35000$ ft over ADHF deflection δ_{ADHF} [°] and lift coefficient C_L [-], alongside test data points. Lower left Panel: Predicted $\partial C_D/\partial C_L$ gradient. Lower right Panel: Predicted $\partial C_D/\partial \delta_{ADHF}$ gradient, alongside the corresponding C_D minimum (dashed line).

4 Conclusions

Within this paper, the implementation of a Gaussian Process (GP) regression based surrogate model is presented. The model is applied to predict the aerodynamic drag coefficient C_D of the wing of a reference configuration in cruise flight considering multiple operation parameters.

The reference wing is derived from the LuFo VI-1 project CATeW, where a synergistic application of a hybrid laminar flow control (HLFC) system with a variable camber (VC) system, realized through deflections of an Adaptive Dropped Hinge Flap (ADHF), to the wing of a transonic transport aircraft is assessed. The assessment is performed on low-fidelity (LowFi) level, using overall aircraft design tools, and high-fidelity (HiFi) level, applying computational fluid dynamics (CFD). Therefore, the goal of the surrogate is incorporating results with comparable accuracy of HiFi computations into the LowFi workflow.

The input dimensions for the surrogate are set as the lift coefficient C_L of the wing, the ADHF deflection angle δ_{ADHF} and the flight altitude H. Based upon a sample size of 54 training points, the root mean square error of the model based upon a dedicated test set is below $5 \cdot 10^{-3}$, indicating the GP based regression model being suited for integration into a LowFi aerodynamic toolchain.

Furthermore, the model allows for automatic prediction of the associated gradients connected to the multi-dimensional response surface, for which the parameter settings associated to minimal drag coefficients are directly output from the model. In connection to the relatively low amount of training data required and inherent quantitative uncertainty quantification, this presents main advantages of a GP regression based surrogate, as shown in the present contribution.

For future work, especially variations in the suction coefficient C_q are of core interest for the prediction of the drag coefficient. Nevertheless, simulation results used to train the surrogate show only minor synergistic effects between the HLFC and the VC system, due to the predicted extent of laminar flow being limited for the present configuration. Therefore, in parallel future work is also focused on assessment of the transition turbulence model used in the context of the CFD simulations.

Apart from that, GPs allow for inclusion of multiple data sources and efficient multi-fidelity optimization frameworks, which enables for efficient connection of the results of the LowFi-toolchain with CFD results in future applications within the project.

Acknowledgments. Funding of these investigations within the LuFo VI-1 project CATeW (Coupled Aerodynamic Technologies for Aircraft Wings, FKZ: 20E1917B) by the German Federal Ministry for Economic Affairs and Climate Action (BMWK) is gratefully acknowledged. Furthermore, the authors deeply appreciate the German Aerospace Center (DLR) for providing the DLR TAU Code and the Gauss Centre for Supercomputing e.V. (www.gauss-centre.eu) for providing computing time on the GCS Supercomputer SuperMUC-NG at Leibniz Supercomputing Center (www.lrz.de)

References

1. Duvenaud, D.: Automatic model construction with gaussian processes. Ph.D. thesis (2014)
2. Effing, T., Schmitz, V., Schültke, F., Peter, F., Stumpf, E.: Combined application of hybrid laminar flow control and variable camber in preliminary aircraft design. In: International Council of the Aeronautical Sciences (ed.) 33rd International Congress of the Aeronautical Sciences (2022)
3. Feldstein, A., Lazzara, D., Princen, N., Willcox, K.: Multifidelity data fusion: application to blended-wing-body multidisciplinary analysis under uncertainty. AIAA J. **58**(2), 889–906 (2020)
4. Forrester, A.I.J., Sóbester, A., Keane, A.J.: Engineering Design via Surrogate Modelling. Wiley, Hoboken (2008)
5. Grabe, C., Shengyang, N., Krumbein, A.: Transport modeling for the prediction of crossflow transition. AIAA J. **56**(8), 3167–3178 (2018). https://doi.org/10.2514/1.J056200
6. Jentys, M., Breitsamter, C.: Transitional flow modelling of a hybrid laminar flow control and variable camber wing. In: International Council of the Aeronautical Sciences (ed.) 33rd International Congress of the Aeronautical Sciences (2022)
7. Jentys, M., Effing, T., Breitsamter, C., Stumpf, E.: Numerical analyses of a reference wing for combination of hybrid laminar flow control and variable camber. CEAS Aeronaut. J. **13**, 989–1002 (2022). https://doi.org/10.1007/s13272-022-00598-y
8. NASA, National Oceanic and Atmospheric Administration and United States Air Force: U.S. Standard Atmosphere. Technical report NASA-TM-X-74335 (1976)
9. Pedregosa, F., et al.: Scikit-learn: machine learning in Python. J. Mach. Learn. Res. **12**, 2825–2830 (2011)
10. Rasmussen, C.E., Williams, C.K.I.: Gaussian Processes for Machine Learning. The MIT Press, Cambridge (2006)
11. Ripepi, M., et al.: Reduced-order models for aerodynamic applications, loads and MDO. CEAS Aeronaut. J. **9**, 171–193 (2018). https://doi.org/10.1007/s13272-018-0283-6
12. Risse, K., Anton, E., Lammering, T., Franz, K., Hoernschemeyer, R.: An integrated environment for preliminary aircraft design and optimization. In: American Institute of Aeronautics and Astronautics (ed.) 53rd AIAA/ASME/ASCE/AHS/ASC Structures, Structural Dynamics and Materials Conference: SciTech 2012. AIAA 2012, p. 1675 (2012)
13. Schueltke, F., Aigner, B., Effing, T., Strathoff, P., Stumpf, E.: MICADO: overview of recent developments within the conceptual aircraft design and optimization environment. In: Deutsche Gesellschaft für Luft-und Raumfahrt - Lilienthal - Oberth e.V. (ed.) 69. Deutscher Luft-und Raumfahrtkongress (2020)
14. Schuet, S., Malpica, C., Aires, J.: A Gaussian process enhancement to linear parameter varying models (2021)
15. Schwamborn, D., Gardner, A., von Geyer, H., Krumbein, A., Lüdeke, H., Stürmer, A.: Development of the TAU-Code for aerospace applications. In: 50th NAL International Conference on Aerospace Science and Technology (2008)
16. Strüber, H.: The aerodynamic design of the A350 XWB-900 high lift system. In: International Council of the Aeronautical Sciences (ed.) 29th International Congress of the Aeronautical Sciences (2014)

Development of a Coupled Implicit Incompressible Flow Solver in CODA

Deepak Kunhappan(✉) and Johannes Löwe

German Aerospace Center (DLR), Insitute of Aerodynamics and Flow Technology, Bunsenstraße 10, 37073 Göttingen, Germany
{deepak.kunhappan,johannes.loewe}@dlr.de

Abstract. The discretization of the incompressible Navier-Stokes equations leads to a saddle-point problem in which the discrete equations for the pressure and momentum are coupled. Traditionally, this coupling is handled by segregated methods in which separate equations for momentum and pressure are solved with the constraint of obtaining a divergence-free velocity field. In the fully coupled approach, a single block system of equations for the momentum and pressure is formulated and solved simultaneously. Furthermore, the solution of such a coupled system has been shown to improve robustness and convergence properties. In the present work, a coupled approach for the solution of the incompressible Navier-Stokes and RANS equations have been developed in the finite volume framework of the CFD software CODA (CFD for ONERA, DLR and AIRBUS). The implementation is verified and validated with various test cases.

Keywords: computational fluid dynamics · incompressible flows · implicit methods

1 Introduction

Flow around small drones, high lift devices, wind turbine blades and internal flows are some of the commonly faced problems in aerodynamics. These flows are characterized by high Reynolds numbers, low Mach numbers and predicting the characteristics of these flows is often challenging. In the low Mach number or weakly compressible regime, the density variations in the flow can be very small and one can assume the flow to be incompressible. The resulting incompressible Navier-Stokes (NS) equations are relatively simple compared to the compressible counter part. However, the difficulty arises when obtaining the numerical solutions of the incompressible NS or the Reynolds Averaged Navier-Stokes equations (RANS), especially when numerical methods designed for high Mach number flows are used. The main difficulty is due to the lack of a natural equation of the state that relates the constant density to the pressure, as the pressure variable is no longer a thermodynamic quantity, but acts as a lagrangian multiplier. Traditionally a class of pressure-based solvers have been

A. Dillmann et al. (Eds.): STAB/DGLR Symposium 2022, NNFM 154, pp. 241–251, 2024.
https://doi.org/10.1007/978-3-031-40482-5_23

designed to tackle this problem and two solution methodologies exist. In the segregated approach, one sequentially solves each component of the momentum equations neglecting the pressure or using the old pressure field to obtain an intermediate velocity field. Afterwards, a separate pressure correction equation based on the incompressibility constraint is formulated and solved. The resulting pressure field is then used to correct the velocity and pressure fields ensuring the incompressibility constraint. For a given iteration the exact coupling between the momentum components and pressure is not considered because the previous/"lagged" values are used to solve either of these fields and thus resulting in slow convergence rates. A remedy to the issue faced in the segregated solvers is to simultaneously solve the momentum and pressure equation leading to the coupled solution approach. Compared to the segregated approach, we solve for a block system of equations that includes the implicit coupling between pressure and momentum components. Such coupled pressure-based all speed solvers have been introduced in the work of Denner *et al.* [3]. In this paper, we outline the details of the development of a pressure-based coupled-solver within the finite volume framework of the next generation CFD software CODA [6], which is jointly developed by ONERA, DLR and AIRBUS.

2 Governing Equations

The incompressible momentum equation written in conservative form is:

$$\frac{\partial (\rho \boldsymbol{U})}{\partial t} + \nabla \cdot (\rho \boldsymbol{U}\boldsymbol{U}) = -\nabla p \mathbf{I} + \nabla \cdot (2\mu \mathbf{S}) \tag{1a}$$

where $\mathbf{S} = \frac{1}{2}\left[(\nabla \boldsymbol{U}) + (\nabla \boldsymbol{U})^T\right]$ is the rate of deformation tensor, and $\boldsymbol{U}$ is the velocity vector. The continuity equation or the incompressibility constraint is

$$\nabla \cdot (\rho \boldsymbol{U}) = 0 \tag{1b}$$

$\boldsymbol{W} = (\rho u, \rho v, \rho w)^T$ is the vector consisting of the conservative variables, and let $\boldsymbol{R} = f(\boldsymbol{W}, \nabla \boldsymbol{W}, p, \nabla p)$ be the residual vector containing the momentum and the additional pressure variable. In semi-discrete form equations (1a), (1b) are written as:

$$\frac{d\boldsymbol{W}}{dt} + \boldsymbol{R} = 0 \tag{2}$$

2.1 Segregated Approach

Based on the method of lines, one can treat the temporal discretization independent of the spatial discretization as an ODE, and therefore in this section we mainly focus on the residual which is tied to the spatial discretization alone.

Re-writing the residual from equations (1a) and (1b) with a set of discrete and suitably linearized operators yields the following equation:

$$\boldsymbol{R} = \bar{\mathbf{M}}\bar{\boldsymbol{X}} = \begin{bmatrix} \mathbf{A} & \mathbf{G} \\ \mathbf{D} & \mathbf{0} \end{bmatrix} \begin{bmatrix} \boldsymbol{U} \\ p \end{bmatrix} = \begin{bmatrix} \boldsymbol{R}_U \\ 0 \end{bmatrix} \tag{3}$$

where in (3), $\bar{\mathbf{M}}$ is the block matrix with dimensions $(n+m) \times (n+m)$, in which **A is the convection and diffusion operator with dimensions of** $n \times n$. $\mathbf{G}$ is the gradient operator with dimensions $n \times m$. $\mathbf{D}$ is the discrete divergence operator with dimensions $n \times m$. The vector of unknowns $\bar{\boldsymbol{X}}$ with dimensions $(n+m)$ consists of the velocity vector $\boldsymbol{U}$, and the pressure variable p. The vector $\boldsymbol{R}_U$ at the RHS consists of the residuals of the velocity variables, obtained when including the boundary conditions, forcing term(s) and explicit terms from a time integration method. Equation (3) shows the saddle point structure of the incompressible Navier-Stokes equations system, as there is a zero in the diagonal. In the segregated class of approaches, the solution for the momentum and pressure are obtained based on a predictor-corrector method. For example, in the popular SIMPLE algorithm, Eq. (3) is factorized into:

$$\begin{bmatrix} \mathbf{A} & \mathbf{0} \\ \mathbf{D} & -\mathbf{D}\mathbf{A}^{-1}\mathbf{G} \end{bmatrix} \begin{bmatrix} \mathbf{I} & \mathbf{A}^{-1}\mathbf{G} \\ \mathbf{0} & \mathbf{I} \end{bmatrix} \begin{bmatrix} \boldsymbol{U} \\ p \end{bmatrix} = \begin{bmatrix} \boldsymbol{R}_U \\ 0 \end{bmatrix} \tag{4}$$

where Eq. (4) solved in two separate steps, as shown in (5). In the first part, an intermediate solution for the velocity vector ($\boldsymbol{U}^*$) while neglecting the pressure is obtained. In the next step the pressure (p^*) is obtained from the solution of a Poisson equation.

$$\begin{bmatrix} \mathbf{A} & \mathbf{0} \\ \mathbf{D} & -\mathbf{D}\mathbf{A}^{-1}\mathbf{G} \end{bmatrix} \begin{bmatrix} \boldsymbol{U}^* \\ p^* \end{bmatrix} = \begin{bmatrix} \boldsymbol{R}_U \\ \mathbf{0} \end{bmatrix} \tag{5}$$

Once the pressure field is obtained, the velocity fields are then corrected using the following correction equation:

$$\begin{bmatrix} \mathbf{I} & \mathbf{A}^{-1}\mathbf{G} \\ \mathbf{0} & I \end{bmatrix} \begin{bmatrix} \boldsymbol{U}^{n+1} \\ p^{n+1} \end{bmatrix} = \begin{bmatrix} \boldsymbol{U}^* \\ p^* \end{bmatrix} \tag{6}$$

Since the Schur complement $-\mathbf{D}\mathbf{A}^{-1}\mathbf{G}$ is approximated, this procedure needs to be iterated.

2.2 Coupled Approach

The approximation of the Schur complement term is similar to the Rhie-Chow stabilization term that is commonly used in the pressure-based solvers [4,9] to suppress spurious oscillations in the pressure field. In the Rhie-Chow scheme, **the inverse of the convection and diffusion operator** $\mathbf{A}^{-1}$ is simplified by only taking the inverse of the diagonal of $\mathbf{A}$. In the present implementation of the coupled approach, we construct an equation for the pressure using a simplified variant of this Schur complement term. Hence the system of equations in the coupled approach is as follows:

$$\begin{bmatrix} \mathbf{A} & \mathbf{G} \\ \mathbf{D} & -\mathbf{D}\gamma\mathbf{G} \end{bmatrix} \begin{bmatrix} U \\ p \end{bmatrix} = \begin{bmatrix} \boldsymbol{R_u} \\ \boldsymbol{R_p} \end{bmatrix} \tag{7}$$

In Eq. (7), rather than using the inverse of the diagonal of $\mathbf{A}$ which often is not easy to obtain, we use the stabilization term γ. Details on this stabilization term are presented in the next section.

2.3 Implementation in CODA Flow Solver

The incompressible solver is implemented in the collocated cell-centered finite volume discretization framework of CODA. In such a spatial discretization the residual for an element E in the mesh is obtained by integrating the face fluxes. For any component of the velocity/momentum vector, using an upwind difference (in practice a linear upwind scheme is used) for the convection term and a central differencing scheme for the diffusion term, the following expression for the residual is obtained:

$$R_{E,\phi} = \sum_{fnb} \left(\left(\max\left(\dot{m}_f, 0\right) \phi_E + \min\left(\dot{m}_f, 0\right) \phi_{nb}\right) - \mu \left(\frac{\phi_E - \phi_{nb}}{\Delta} \right) + \bar{p} \right) \cdot \boldsymbol{n}_{E,nb} A_{f_{E,nb}} \tag{8}$$

where in (8), ϕ denotes any component of the velocity vector, $\dot{m}_f$ is the mass flux across a particular face. The subscript E denotes any generic element and nb is the face neighbour of this element. $\boldsymbol{n}_{E,nb}$ and $A_{f_{E,nb}}$ are the face normal vector and the face area, respectively. The residual of the pressure equation obtained from the Schur complement from (7) is:

$$R_{E,p} = \sum \left(\bar{U} + \gamma \left(\overline{\nabla p}_{P_E} - \overline{\nabla p}_f \right) \right) \cdot \boldsymbol{n}_{E,nb} A_{f_{E,nb}} \tag{9}$$

where in (9), γ is the stabilization parameter, ∇p_{P_E} is the pressure gradient at the elements obtained by Green-Gauss or least squares approximation, $\overline{\nabla p}_f$ is the averaged pressure gradients at the cell faces. The stabilization parameter γ is derived from the stabilized FEM discretization (SUPG/PSPG), which prevents spurious pressure oscillations when equal order interpolation functions/elements are used for both the pressure and velocity variables [8]. Since the original formulation of the stabilization parameter is based on an element stencil and in order to use it with the present FVM discretization, a face-based formulation of this stabilization parameter is used. The stabilization length scale $h_{E,f}$ for a face f connected to an element E is obtained from:

$$h_{E,f} = \min\left(\sqrt[3]{V_E}, V_E / A_f \right) \tag{10}$$

where in (10) V_E, is the volume of the element E, A_f is the area of face f. For an inner face, the stabilization length scale is calculated for both the owner and neighbour element. The stabilization parameter $\gamma_{E,f}$ is defined as:

$$\gamma_{E,f} = \left[\left(\frac{2||U||_E}{h_{E,f}} \right)^2 + \left(\frac{4\nu_E}{h_{E,f}^2} \right)^2 \right]^{-1/2} \tag{11}$$

where, $||U||_E$ is the norm of the velocity from the element E, and ν_E is the value of the effective viscosity in element E. Similar to the stabilization length scale, this quantity is also defined for the neighbouring element. The stabilization parameter used in the pressure Eq. (9) is the minimum of the two stabilization parameters of the elements E and G which a face f connects:

$$\gamma = \min\left(\gamma_{E,f}, \gamma_{G,f}\right) \tag{12}$$

Using a first-order Euler backward method and a Newton-like linearization for the residual, Eq. (2) reads:

$$\left(\frac{\mathbf{M}_\Omega}{\Delta t} + \frac{d\boldsymbol{R}}{d\boldsymbol{W}} \right) \Delta \boldsymbol{W} = -\boldsymbol{R} \tag{13}$$

where M_Ω is the mass matrix, and $\frac{dR}{dW}$ is the Jacobian of the residual vector. In CODA, the Jacobian in matrix form is computed using Automatic Differentiation (AD), the matrix-free variants can also be obtained by AD or finite difference based approximation. The matrix version of the Jacobian is used for preconditioning, and the matrix-free variant is used for the GMRes linear solver. Steady-state solutions are obtained for large value of Δt. The efficiency of the time-stepping is further improved using a residual-exponent based time-step ramping and as $\Delta t \to \infty$, Newton-like convergence is achieved. The linear solution of the block sparse system arising from (13) is obtained using the linear solver library SPLISS [5] developed by DLR.

2.4 Turbulence Modelling

Several turbulence models have been implemented and coupled with the compressible solver in CODA. The generic and abstract programming model used in CODA permits the re-use of these existing turbulence models with the incompressible solver with slight modifications. The Spalart-Allmaras negative (SA-neg) and the the $k-\omega$ SST turbulence models have been coupled with the incompressible NS equations. The full incompressible RANS equations are then solved in the coupled framework.

2.5 Boundary Conditions

The standard no-slip wall, specified-velocity inflow and zero normal stress outflow boundary conditions have been implemented. A special farfield-like boundary condition was implemented in order to extend the capability of the incompressible solver to perform external aerodynamic simulations. This boundary condition was obtained by combining the inflow and outflow boundary conditions: if the dot product of the surface normal and the cell velocity is < 0 the face is treated as an inflow BC. Otherwise, it acts as an outflow BC.

3 Verification and Validation

This section deals with the verification and validation of the incompressible solver with the SA-neg and $k - \omega$ SST turbulence models. Steady-state test cases were considered, and all the test cases presented were solved using a GMRes method with relaxed lines inversion preconditioning. In the works of Uroić [11] and Darwish [2], it was shown that the use of a linear multigrid preconditioning tremendously improves the convergence of the coupled system. However, only a preliminary implementation of the linear multigrid method is currently available in CODA, and this implementation was not used for all the test cases presented in this paper.

3.1 Flow Around Three Element DLR F15 Airfoil

In order to demonstrate the coupled approach and the farfield-like boundary condition, the 2D flow around the DLR F15 three element high-lift airfoil was simulated. An angle of attack, $\alpha = 6^o$, with $Re = 2 \times 10^6$, and $Ma = 0.15$ was the specified flow condition. A hybrid grid consisting of quadrilateral and triangular elements shown in Fig. 1(a) was used. Convergence near to machine precision was obtained using the $k - \omega$ SST turbulence model as shown in Fig. 1(b).

Figures 1(c) and 1(d) show the streamlines around the DLR F15 three element airfoil using the $k-\omega$ SST and the SA-neg turbulence models. The obtained streamlines from both the turbulence models are qualitatively similar and have captured the main flow physics of the shear layer formation behind the slat trailing edges, the local geometry induced separations behind the slat, wing, and the reattachment of the shear layer. The main difference can be seen at the trailing edge of the flap where the flow separation due to the adverse pressure gradient occurs, in which the $k - \omega$ turbulence model predicts a larger separation compared to the SA-neg model. The results for the pressure coefficient C_p is shown in Fig. 1(e). The predicted C_p is in qualitative agreement with results of the TAU solver [10] and the experimental data. It should be noted that the experimental results are not at exact $\alpha = 6^o$, as these results were corrected based on the SA computations from TAU, and hence the good agreement for the SA models. From the figure it is observed that the present implementation of the solver recovers the modelling behaviour of the well established legacy TAU solver. Figure 1(f) shows the comparison for the skin friction coefficient C_f between the two turbulence models and the SA-model obtained from TAU.

3.2 Flow Around Ahmed Body Geometry

Preliminary 3D computations were performed on the Ahmed body geometry corresponding to the 35° rear slant angle case. The Reynolds number based on

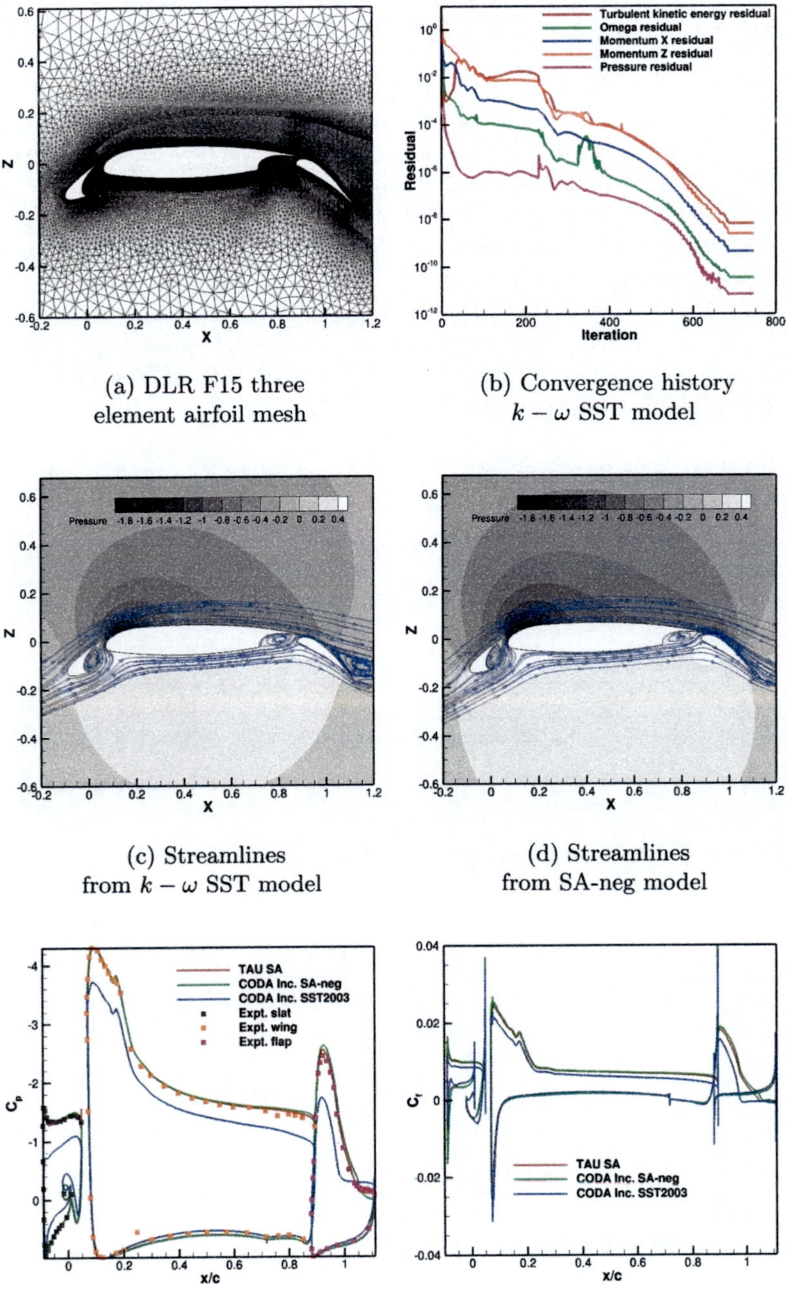

(a) DLR F15 three element airfoil mesh

(b) Convergence history $k - \omega$ SST model

(c) Streamlines from $k - \omega$ SST model

(d) Streamlines from SA-neg model

(e) C_p comparison

(f) C_f comparison

Fig. 1. Results for the flow around DLR F15 three element airfoil.

the body length was $Re = 2.76 \times 10^6$, and the 3D mesh consisted of 2.4×10^6 elements, with $y^+ = 1$ in the wall-normal direction of the bluff body. Turbulence was modeled using the SA-neg model. Residual convergence of 10^{-6} was obtained as shown in Fig. 2(a). Figure 2(b) shows the streamlines at the slant. The streamlines are in good qualitative agreement with the results of Ahmed *et al.* [1]. The solid black lines denote the positions along which the velocity profiles were extracted and compared. Figures 2(c) and 2(d) shows the comparison of the velocity profiles at $x = -103$ mm, and $x = -3$ mm respectively, on the slant and it is a direct comparison of the re-circulation region with respect to the experiments of Lienhart *et al.* [7]. Overall, there is reasonable agreement with the experimental results and the discrepancy stems mainly from the use of the SA-neg turbulence model. From this simulation it is shown that the present implementation of the incompressible solver is capable of performing simulations of complex separated flows over 3D geometries.

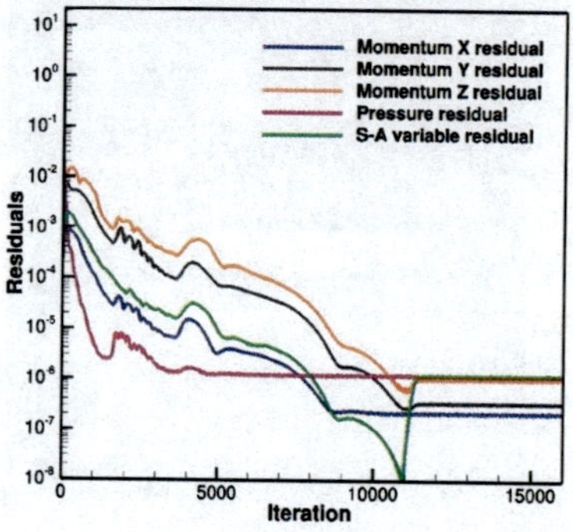

(a) Convergence history

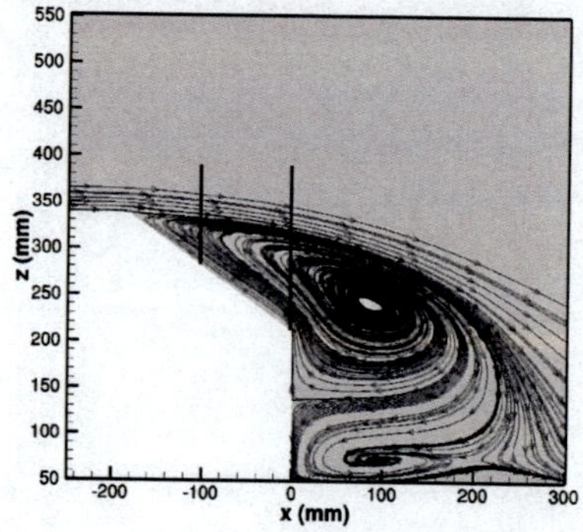

(b) Streamlines and profile locations

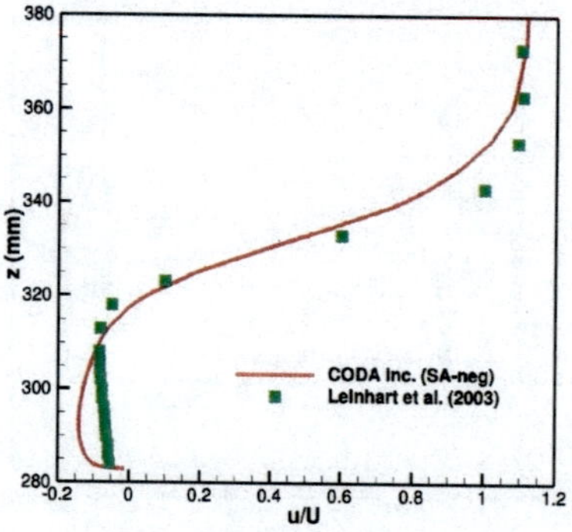

(c) Velocity profile at x = -103mm

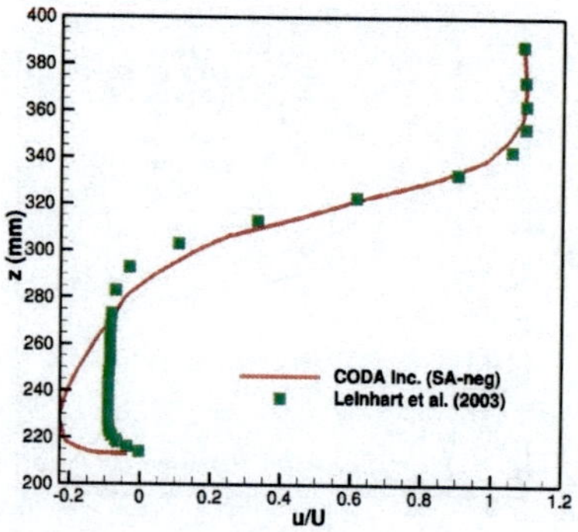

(d) Velocity profile at x = -3mm

Fig. 2. Results for the flow around the Ahmed Body.

However, the outer iteration numbers are still high which is caused by insufficient convergence of the linear solver, due to inefficient preconditioning.

3.3 Towards Linear Multigrid Preconditioning

As mentioned in the beginning of Sect. 3, previous studies [11] have shown that the use of a linear multigrid preconditioning reduced the number of outer iterations and the overall wall clock time. In this section, we present the first experiments of using the linear multigrid (LMG) preconditioning with the incompressible solver. The lid-driven cavity problem at $Re = 3200$ was solved with and without the linear multigrid preconditioning. The grid consists of uniform quadrilateral cells with a resolution of 128×128. The coarse level discretizations were obtained by geometric agglomeration of the elements. Figure 3 shows the comparison of the convergence histories with (dashed lines) and without the LMG preconditioning (solid lines), and the number of linear iterations per iteration. It is seen that on utilizing the LMG preconditioning, the number of outer iterations have been reduced by a factor of 4. When comparing the number of inner/linear iterations per outer iteration (iterations used by the GMRes solver), it is observed that the LMG preconditioning reduces the number of linear iterations by a factor of 20 to reach the specified linear solver tolerance of 1×10^{-4}. For the case without LMG, this tolerance could not be reached and hence the stagnation at 200 linear iterations. The LMG preconditioning is a promising tool to further improve the robustness and the convergence rate of the incompressible solver, and first tests with the SA-neg turbulence model and 3D geometries are ongoing.

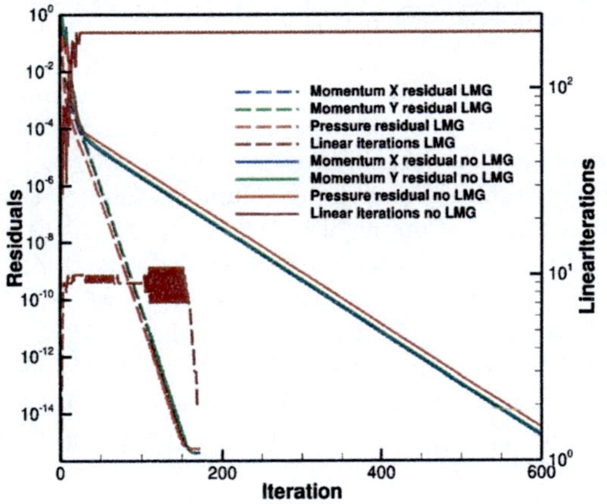

Fig. 3. Convergence history comparison with and without linear multigrid preconditioning.

4 Conclusion and Future Work

The formulation of a fully-coupled implicit pressure-based solver and its implementation in the finite volume framework of the CFD software CODA was presented. The implementation was validated against the 2D flow around the three element DLR F15 airfoil. The results obtained from the simulation were in good agreement with the reference and experimental data. Preliminary 3D flow computations around the Ahmed body geometry were performed, obtaining physically consistent results and stable convergence behaviour. First experiments with the linear multigrid preconditioning on the lid driven cavity case shows large improvements in the linear solver performance. In the future work, we intend to perform in-depth studies on the performance comparison between the compressible and incompressible solvers in CODA for more realistic 3D geometries. Furthermore, it is envisioned to explore and improve the linear multigrid preconditioning, and further develop the incompressible solver for wind energy applications.

Acknowledgements. We would like to thank Dr. Axel Probst from DLR for providing the DLR F15 mesh, reference data and many discussions with respect to the solver development. We also thank Dr. Simone Crippa from AIRBUS for providing the Ahmed body mesh.

References

1. Ahmed, S.R., Ramm, G., Faltin, G.: Some salient features of the time-averaged ground vehicle wake. SAE Trans. **93**, 473–503 (1984)
2. Darwish, M., Sraj, I., Moukalled, F.: A coupled finite volume solver for the solution of incompressible flows on unstructured grids. J. Comput. Phys. **228**(1), 180–201 (2009)
3. Denner, F., Evrard, F., van Wachem, B.G.: Conservative finite-volume framework and pressure-based algorithm for flows of incompressible, ideal-gas and real-gas fluids at all speeds. J. Comput. Phys. **409**, 109348 (2020)
4. Klaij, C.M.: On the stabilization of finite volume methods with co-located variables for incompressible flow. J. Comput. Phys. **297**, 84–89 (2015)
5. Krzikalla, O., Rempke, A., Bleh, A., Wagner, M., Gerhold, T.: Spliss: a sparse linear system solver for transparent integration of emerging HPC technologies into CFD solvers and applications. In: Dillmann, A., Heller, G., Krämer, E., Wagner, C. (eds.) STAB/DGLR Symposium 2020. NNFMMD, vol. 151, pp. 635–645. Springer, Cham (2021). https://doi.org/10.1007/978-3-030-79561-0_60
6. Leicht, T., et al.: DLR project Digital-X-Next generation CFD solver "FLUCS" (2016)
7. Lienhart, H., Becker, S.: Flow and turbulence structure in the wake of a simplified car model. SAE Trans. **112**, 785–796 (2003)
8. Lube, G.: Stabilized galerkin finite element methods for convection dominated and incompressible flow problems. Banach Center Publ. **29**(1), 85–104 (1994)
9. Rhie, C.M., Chow, W.L.: Numerical study of the turbulent flow past an airfoil with trailing edge separation. AIAA J. **21**(11), 1525–1532 (1983)

10. Schwamborn, D., Gerhold, T., Heinrich, R.: The DLR TAU-code: recent applications in research and industry (2006)
11. Uroić, T., Jasak, H.: Block-selective algebraic multigrid for implicitly coupled pressure-velocity system. Comput. Fluids **167**, 100–110 (2018)

An Analytical View on Data-Driven Turbulence Modeling and a Realization via a regularized Newton Method

Stefan Langer(✉)

Member of the Helmholtz Association, Institute of Aerodynamics and Flow Technology, German Aerospace Center, Lilienthalplatz 7, 38108 Braunschweig, Germany
stefan.langer@dlr.de

Abstract. Field Inversion and Machine Learning is an active field of research in **C**omputational **F**luid **D**ynamics (CFD). This approach can be leveraged to obtain a closed-form correction for a given turbulence model to improve the predictions. The fundamental approach is to insert a parameter into the system of RANS equations and determine it in a way such that, for example, a given pressure dissipation is better approximated compared to the one obtained with the original set of equations. The goal of this article is twofold. Numerical arguments are presented that these kinds of problems can be severely ill-posed. Second, instead of introducing a field parameter into a given turbulence model, an approach is presented to directly reconstruct the eddy viscosity field along with an example of an RAE2822 airfoil in transonic conditions. A regularized Gauss-Newton method is used for a realization. Finally, an outlook is presented in which way this approach can be used to possibly modify or improve turbulence models such that not only one, but a larger number of test cases are considered.

1 Introduction

Formally, the RANS equations for a linear turbulence model differ from the Navier-Stokes equations by the additional unknown eddy viscosity μ_t. For closure, a turbulence model given by an additional set of equations is used to determine μ_t, such that

a) a steady-state solution of the RANS equations exists and it is unique,
b) this steady-state solution represents important features compared to a solution of the original Navier-Stokes equations,
c) it is possible to compute numerically an approximation to this steady-state solution.

Commonly used turbulence models for external aerodynamics are for example the one of Spalart and Allmaras [1,2], the k-ω model of Wilcox [3,4] and the SST model proposed by Menter [5,6].

A. Dillmann et al. (Eds.): STAB/DGLR Symposium 2022, NNFM 154, pp. 252–261, 2024.
https://doi.org/10.1007/978-3-031-40482-5_24

To adapt a given turbulence model for one or several flow cases, a **F**ield **I**nversion and **M**achine **L**earning (FIML) approach can be leveraged to obtain a closed-form correction for the given turbulence model to improve the predictions [7]. For example, for the Spalart-Allmaras turbulence model

$$\frac{\mathrm{d}}{\mathrm{d}t}\int_{\Omega}\tilde{\nu}\,\mathrm{d}\mathbf{x}+\int_{\partial\Omega}(\mathbf{f}_{c,turb}\cdot\mathbf{n}-\mathbf{f}_{v,turb}\cdot\mathbf{n})\,\mathrm{d}s=\int_{\Omega}Q\,\mathrm{d}\mathbf{x}. \tag{1a}$$

the main idea, which can be found in the literature, is, to manipulate a field parameter β weighting the production term in the turbulence model

$$Q=\beta Pr-De+Di, \tag{1b}$$

where $\beta\equiv 1$ yields the original unmodified model of Spalart and Allmaras, and Pr, De and Di denote the production, destruction and diffusion source term. Then, one determines β such that the corresponding solution of the RANS equations fits some c_p-distribution "better" with respect to some measure (norm). Such an approach has been applied successfully for several examples [8]. In this article, instead of trying to modify a given turbulence model, we directly try to reconstruct an eddy viscosity field μ_t from given data. To do so, we consider the eddy viscosity field as a parameter in the RANS equations, which needs to be identified, such that the criteria mentioned above are satisfied.

Formally, the problem is formulated as follows: We consider R to be the residual operator corresponding to the set of equations for a boundary-value problem for the RANS equations, μ_t is the field parameter of interest and W a state vector. Then these are linked via the equation

$$R(W;\mu_t)=R_{\mu_t}(W)=0. \tag{2}$$

The solution of this state equation for a given fixed parameter μ_t can be interpreted as the direct problem.

For the inverse problem, we are not interested in solving (2) for a given μ_t, but instead, to determine the field parameter μ_t with respect to observed data related to a given state $W=W(\mu_t)$. This observation operator could be, for example, either the identity (distributed measurement), or a trace operator to boundary values of the solution (boundary measurements). In CFD, these are often the c_p-distribution and/or the skin friction coefficient c_f on the surface of the geometry. Such viewpoint is, for example, discussed in [7,8]. Typically the data, that is for example these surface distributions,

a) come from measurement,
b) or are artificially generated for test purposes.

In both cases, the data is typically noisy with an unknown noise level δ, for example

$$\|c_p^{\delta}-c_p^{\dagger}\|\leq\delta,$$

where the superscript δ indicates the noisy and $\dagger$ the exact data, respectively.

In this article, we show with an example of transonic flow over an RAE2822 airfoil that this inverse problem is severely ill-posed. A regularized Gauss-Newton method is presented and applied to approximate a solution of this problem. Evidence is given suggesting that many, if not all, similar problems share similar mathematical properties. This article concludes with a perspective on how the approach proposed here can be used to modify existing turbulence models so that they may show more suitable results for a large number of test cases.

2 An Inverse View on Data Driven Turbulence Modeling

2.1 Inverse Problem as Operator Equation

To formulate the inverse problem as an operator equation, and to have a sufficient amount of data, we consider a distributed measurement. To avoid an inverse crime, the data is generated from a finer mesh and then projected. That is, a complete field state $W^\dagger$ is given, and the corresponding eddy viscosity $\mu_t^\dagger$ is the field parameter to be reconstructed. Then, the inverse problem is to find a root of the equation

$$F(\mu_t) = W(\mu_t) - W^\dagger. \tag{3}$$

When the given data are a surface distribution, then (3) is replaced by

$$F(\mu_t) = P\left(W(\mu_t)\right) - P\left(W^\dagger\right), \tag{4}$$

and P denotes the trace operator corresponding to the given surface data. For example, if a c_p- or a c_f-distribution are the given data, then (4) reads as

$$F(\mu_t) = P_{c_p}\left(W(\mu_t)\right) - c_p^\dagger \quad \text{or} \quad F(\mu_t) = P_{c_f}\left(W(\mu_t)\right) - c_f^\dagger$$

Even both, as a system of equations, can be considered,

$$F(\mu_t) = \begin{pmatrix} P_{c_p}\left(W(\mu_t)\right) \\ P_{c_f}\left(W(\mu_t)\right) \end{pmatrix} - \begin{pmatrix} c_p^\dagger \\ c_f^\dagger \end{pmatrix}.$$

Note that the state $W = W(\mu_t)$ satisfies (2), so the Eq. (3) involves the solution of the RANS equations given the parameter μ_t, that is

$$F(\mu_t) = R_{\mu_t}^{-1}(0) - W^\dagger. \tag{5}$$

Here the symbol $R_{\mu_t}^{-1}(0)$ means in practice to approximate a solution of the discretized RANS equations. Considering problem (1), equation $F(\beta) = R_\beta^{-1}(0) - W^\dagger$ can be derived in a completely analogous way.

2.2 Ill-Posed Character of the Field Reconstruction

Using the implicit function theorem, we can compute the derivative

$$F'[\mu_t] = \frac{dR_{\mu_t}^{-1}(0)}{d\mu_t} = \frac{dW}{d\mu_t} = -\left(\frac{\partial R(\mu_t, W)}{\partial W}\right)^{-1} \frac{\partial R}{\partial \mu_t}. \tag{6}$$

Both required derivatives $\frac{\partial R(\mu_t,W)}{\partial W}$ and $\frac{\partial R}{\partial \mu_t}$ can be determined explicitly. To understand the character of problem (5) we recall the following classification of ill-posed problems [9–11]. An ill-posed problem is called

a) **mildly ill-posed**, if the singular values σ_j of $F'[\mu_t]$ decay at an polynomial rate, i.e.

$$c_m j^{\alpha} \leq \sigma_j \leq C_m j^{-\alpha}, \qquad \alpha, c_m, C_m > 0$$

b) **severely ill-posed**, if the singular values σ_j of $F'[\mu_t]$ decay at an exponential rate, i.e.

$$c_e e^{-c_s j^{\beta}} \leq \sigma_j \leq C_e e^{-c_s j^{\beta}} \qquad \beta, c_s, c_e, C_e > 0$$

The degree of ill-posedness depends on the differential operator of the underlying differential equation.

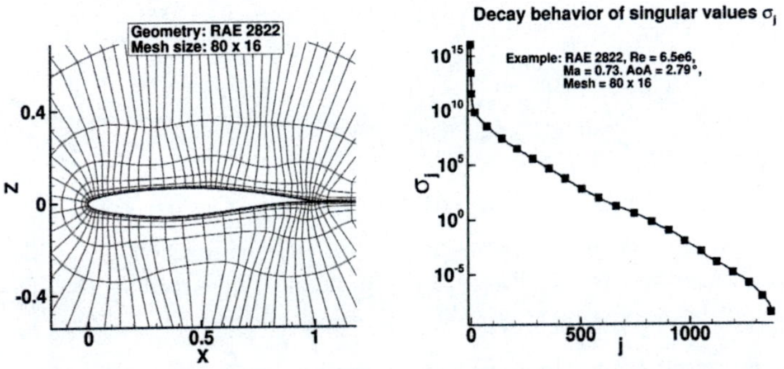

Fig. 1. Used mesh and decay behavior of the singular values of $F'[\mu_t] = \frac{dW}{d\mu_t}$

For the example presented in Sect. 4, Eq. (6) is used to compute the singular values σ_j. Figure 1 (right) shows the decay behavior of $\sigma_j\,(F'[\mu_t])_{j=1,2,...}$ for an iterate in a neighborhood of the solution. The decay at an exponential rate of these singular values is obvious. This observation serves as an indicator that solving (3) is a severely ill-posed problem and that the stability condition for a well-posed problem is violated.

3 An Iterative Regularized Gauss-Newton Method

The derivation and realization of the regularized Gauss-Newton method follows [10,12–14].

To approximately solve the operator Eq. (5) we substitute F by its linear approximation. Given an initial guess $\mu_{t,0}$, in each Newton step the least squares problem

$$\|F'[\mu_{t,n}]h_n + F(\mu_{t,n})\|^2 = \min_{h_n}! \tag{7}$$

needs to be solved and the update is computed by $\mu_{t,n+1} = \mu_{t,n} + h_n$. This approach is well-known as Gauss-Newton method. In the case where (5) is ill-posed, it is important to consider the situation where the given data $W^\dagger$ is replaced by noisy data W^δ satisfying

$$\|W^\dagger - W^\delta\| \le \delta.$$

In this case a straightforward implementation of the Gauss-Newton method usually fails and does not lead to a good reconstruction of the solution after several Newton steps. One reason for the failure of the Gauss-Newton method in this situation is the ill-posedness of the least squares problem (7), which is inherited from the original operator Eq. (5). Thus, to perform one Newton step some sort of regularization has to be employed when solving (7). In this article we consider a regularized Gauss-Newton method, where instead of (7), the regularized least squares problem

$$\|F'[\mu_{t,n}^\delta]h_n + F(\mu_{t,n}^\delta)\|^2 + \gamma_n\|h_n\|^2 = \min_{h_n}! \tag{8}$$

is solved in each Newton step. Here

$$\gamma_n = \gamma_0\gamma^{-n} > 0, \quad 0 \le \gamma \le 1, \quad \gamma_0 > 0 \tag{9}$$

denotes a regularization parameter. This iterative regularization method can be interpreted as a common Newton method where in each step Tikhonov regularization is applied to the linearized equation. The problem (8) is well-posed, in particular there exists a uniquely defined minimizer $h^\dagger$.

The least squares problem (8) can be equivalently reformulated by the linear equation

$$G_n^* G_n h_n = G_n^* g_{\mu_t}^\delta, \tag{10}$$

where

$$G_n := \begin{pmatrix} F'\left[\mu_{t,n}^\delta\right] \\ \sqrt{\gamma_n} I \end{pmatrix}, \quad g_{\mu_{t,n}}^\delta := \begin{pmatrix} -\left(F'[\mu_{t,n}^\delta]\right)^* F(\mu_{t,n}^\delta) \\ 0 \end{pmatrix}$$

with the self-adjoint strictly coercive operator $G_n^* G_n$. A natural choice to solve this problem is the conjugate gradient method, which can be implemented for (10) in the following way:

Conjugate Gradient Algorithm

$h_n^0 = 0;\ d^0 = g_n^\delta;\ r^0 = G_n^* d^0;\ p^1 = r^0;\ k = 0;$
while $\|r^k\| > \varepsilon$
$\quad k = k + 1;$
$\quad q^k = G_n p^k;$
$\quad \alpha_k = \|r^{k-1}\|^2 / \|q^k\|^2;$
$\quad h_n^k = h_n^{k-1} + \alpha_k p^k;$
$\quad d^k = d^{k-1} - \alpha_k q^k;$
$\quad r^k = G_n^* d^k;$
$\quad \beta_k = \|r^k\|^2 / \|r^{k-1}\|^2;$

$$p^{k+1} = r^k + \beta_k p^k.$$

In this article, so far, nothing has been said about the evaluation of the norms in the image space and in the pre-image space, that is $\|F'(\mu^\delta_{t,n})h_n + F(\mu^\delta_{t,n})\|$ and $\|h_n\|^2$. Such distinction is important and it has an effect on both the accuracy of the approximate solution and the success of the algorithm. The choice of these norms has an effect on the implementation of the identity operator I and the adjoint $F'\left[\mu^\delta_{t,n}\right]^*$. Due to a lack of theory we choose for the image space an Euclidean norm. For the pre-image space we choose as a norm

$$\|h\|^2 = \sum_{i=1}^{N_{elem}} \left(\frac{1}{\mathrm{vol}(\Omega_i)^2} + \mathrm{vol}(\Omega_i) \right) h_i^2. \tag{11}$$

This can be interpreted as some kind of discrete realization of an H^1-Sobolev norm, when the gradients are realized using a Green-Gauss formulation. This choice of the norm is not perfect at all, and further research is required. But without such weighting the algorithm almost always failed. Then, Using (6), the linear operator G_n is given by,

$$G_n := \begin{pmatrix} \left(\frac{\partial R(\mu_t, W)}{\partial W}\right)^{-1} \frac{\partial R}{\partial \mu_t} \\ \sqrt{\gamma_n}\tilde{I}, \end{pmatrix}. \tag{12}$$

and $\tilde{I}$ is understood in the sense of (11). As a consequence, inside the CG-method the evaluations of $G_n p^k$ and $G_n^* d^k$ need to be realized, that is

$$G_n p^k = \begin{pmatrix} \left(\frac{\partial R(\mu_t, W)}{\partial W}\right)^{-1} \frac{\partial R}{\partial \mu_t} p^k \\ \sqrt{\gamma_n}\tilde{I} p^k \end{pmatrix}$$

$$G_n^* d^k = \frac{\partial R}{\partial \mu_t}^T \left(\frac{\partial R(\mu_t, W)}{\partial W}\right)^{-T} d_1^k + \sqrt{\gamma_n}\tilde{I} d_2^k,$$

where we have used the notation $d^k = (d_1^k, d_2^k)^T$. Hence, for one step of the CG-method we need to solve the linear systems

$$\frac{\partial R(\mu_t, W)}{\partial W} u = \frac{\partial R}{\partial \mu_t} p^k, \tag{13a}$$

$$\left(\frac{\partial R(\mu_t, W)}{\partial W}\right)^T v = d_1^k. \tag{13b}$$

Soving the linear systems (13) is the main complexity of the algorithm. In this work we have applied a preconditioned GMRes method to solve these linear systems. The preconditioner applied can be found in [15–17]. Figure 2 (left and mid) compares exemplarily but representative for some linear systems the difference between the application of a non-precondtioned and the preconditioned GMRes method. The non-preconditioned GMRes requires about 6000 iterations, the preconditioned only about 200 iterations. In our example this factor of about 30 translates roughly one-to-one into runtime.

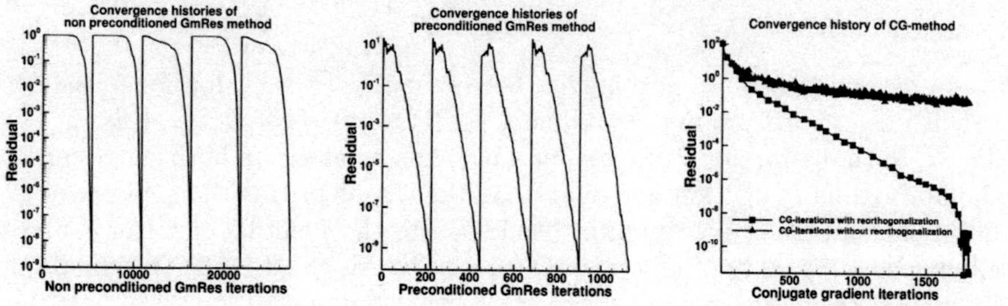

Fig. 2. Effect of preconditioner in GMRes method and of complete reorthogonalization

In summary, the main complexity of the proposed implementation of Algorithm (7) is given by:

a) To determine $F(\mu^{\delta}_{t,n})$ in (8) we need to solve the RANS equations for the given iterate $\mu^{\delta}_{t,n}$.
b) For each step in the CG-method we need to solve the linear Eq. (13a).
c) For each step in the CG-method we need to solve the linear Eq. (13b).

A further important ingredient for a successful implementation of the algorithm is the incorporation of a complete reorthogonalization procedure [18] for the residual vectors r^k in the CG-iteration. It is well-known from other applications that for ill-posed problems there is a loss of orthogonality which can be prevented by reorthogonalization [13]. Figure 2 (right) shows the effect of reorthogonalization. Without this method a successful realization was not possible in our code environment.

4 Numerical Example

As an example we consider turbulent flow over the RAE 2822 airfoil at $M = 0.73$, angle of attack $\alpha = 2.79°$ and $Re = 6.5 \cdot 10^6$. This case features a shock on the upper part of the airfoil. To avoid an inverse crime the right hand side W^{δ} is computed as the projection of the solution from one finer mesh level. On the one hand, these data might be viewed as exact, since they stem from an exact procedure, on the other hand due to the projection these data include noise. At this early stage of these investigations the computations have been done on the rather coarse mesh shown in Fig. 1 (left).

As initial guess we use the eddy viscosity $\mu_{t,0}$ computed by the same geometry and the flow conditions $M = 0.75$, angle of attack $\alpha = 2.81°$ and $Re = 6.2 \cdot 10^6$. The regularization parameter (9) is chosen by $\gamma_0 = 1000$, $\gamma = \frac{1}{2}$.

Figures 3 and 4 give an impression of the change of the c_p- and c_f-distribution during the iteration. After about 15 iterations the c_p-distribution is almost perfectly reconstructed. On the other hand the c_f-distribution is not captured at all and shows a significant deviation, and there are at least two reasons to explain it. Since the c_f-distribution is based upon the gradients, it is important to formulate the inverse problem in adequate Hilbert spaces. To take care of this we have introduced (11), but this choice is only a first guess and surely sub-optimal. Further research is required to find better suited norms. On the other hand, due to the severe ill-posedness and resulting instability of the problem, we still need to include into our process a stopping criterion, for example Morozov's discrepancy principle, to stop at the first index M when

$$\|R^{-1}_{\mu_{t,M}}(0) - W^{\delta}\| \leq \tau\delta, \qquad \tau > 1.$$

A violation of such stopping criterion may lead to oscillations and instabilities in the solution. Furthermore, to improve results a stochastic modeling of the noise can be included, requiring often a deep understanding of the underlying problem and additionally the solution of many inverse problems.

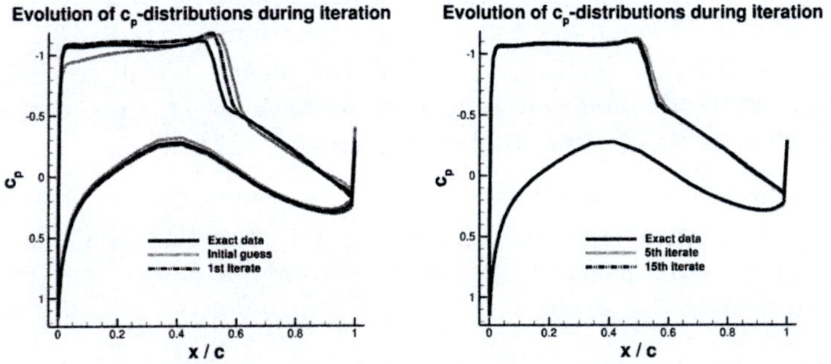

Fig. 3. Evolution of c_p-distribution during the iteration

5 Outlook

From a general point of view, one is interested in learning an eddy viscosity or improving an existing turbulence model for a large number of different test cases. For example, this could be either for the same geometry and a number of different flow conditions or even a change in the geometry can be considered. Then we can generalize the considered idea in this article in the following way. Assume that we want to reconstruct (learn) the field parameters $\mu_1, \ldots, \mu_s$, $s \in \mathbb{N}$ and that T sets of data $g_1, \ldots, g_T$ are given. Similar as (4) we can write these as

$$F(\mu_1, \ldots, \mu_s) = \begin{pmatrix} \left(P_1 \left(R^1_{\mu_1,\ldots,\mu_s}\right)^{-1}(0)\right) \\ \vdots \\ \left(P_T \left(R^T_{\mu_1,\ldots,\mu_s}\right)^{-1}(0)\right) \end{pmatrix} - \begin{pmatrix} g_1 \\ \vdots \\ g_T \end{pmatrix}, \tag{14}$$

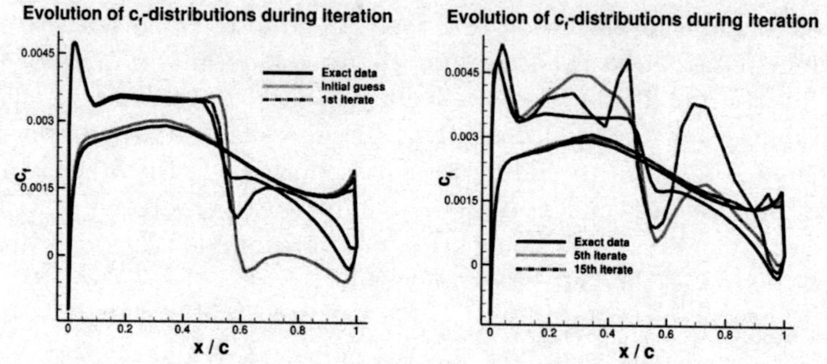

Fig. 4. Evolution of c_f-distribution during the iteration

which represents in general a large-scale, ill-posed nonlinear system of equations. For example, following either the FIML approach or the reconstruction of an eddy viscosity field, then one parameter either β or μ_t needs to be determined such that T equations must be satisfied simultaneously. To consider a natural generalization, instead of a linear turbulence model we could ask for reconstructing the Reynolds stresses. Then we have not only one parameter but six parameters to choose from to find a solution of (14).

Furthermore, if we want to carry over the determined parameters $\mu_1, \ldots, \mu_s$ to other geometries, then we have to get away from considering this as a field parameter for a given mesh. Instead we need to find a suitable ansatz function. That is we consider

$$\mu_j = \mu_j(\omega_{s_1}, \ldots \omega_{s_N}), \qquad j = 1, \ldots, s,$$

and $\omega_{s_1}, \ldots, \omega_{s_N}$ represent the coefficients in the ansatz function. For example, the approach presented here can be adapted for this directly. Then the derivative needs to be adjusted to the ansatz coefficients, that is $\frac{\partial R}{\partial \mu_t}$ is replaced by

$$\frac{\partial R}{\partial(\omega_1, \ldots, \omega_d)} = \frac{\partial R}{\partial \mu_t} \frac{\partial \mu_t}{\partial(\omega_1, \ldots, \omega_d)}, \quad \mu_t = \mu_t(\omega_1, \ldots, \omega_d), \quad d \in \mathbb{N}.$$

Hence, additionally the derivative of the ansatz function with respect to its coefficients is required. Possible ansatz functions which can be considered in future work are for example convolutional neural networks.

References

1. Spalart, P.R., Allmaras, S.R.: A one-equation turbulence model for aerodynamic flows. In: AIAA Computational Fluid Dynamics Conference, number 1992–439 in Conference Proceeding Series. AIAA (1992)

2. Allmaras, S.R., Johnson, F.T., Spalart, P.R.: Modifications and clarifications for the implementation of the Spalart-Allmaras turbulence model. In: International Conference on Computational Fluid Dynamics 7, Hawaii, number ICCFD7-1902 in Conference Proceeding Series (2012)
3. Wilcox, D.C.: Reassessment of the scale-determining equation for advanced turbulence models. AIAA J. **26**(11), 1299–1310 (1988)
4. Wilcox, D.C.: Formulation of the k-ω turbulence model revisited. AIAA J. **46**(11), 2823–2838 (2008)
5. Menter, F.R.: Two-equation eddy-viscosity turbulence models for engineering applications. AIAA J. **32**, 1598–1605 (1994)
6. Menter, F.R.: Review of the shear-stress transport turbulence model experience from an industrial perspective. Int. J. Comput. Fluid Dyn. **23**(4), 305–316 (2009)
7. Jäckel, F.: Sensitivity analysis of discrepancy terms introduced in turbulence models using field inversion. In: Dillmann, A., Heller, G., Krämer, E., Wagner, C. (eds.) 22. STAB/DGLR Symposium 2020, volume 151 of Notes on Numerical Fluid Mechanics and Multidisciplinary Design, pp. 625–634. Springer, Cham (2021). Hardcover ISBN 978-3-030-79560-3 Softcover, ISBN 978-3-030-79563-4 eBook, ISBN 978-3-030-79561-0 Series, ISSN 1612-2909 Series, E-ISSN 1860-0824, https://doi.org/10.1007/978-3-030-79561-0_59
8. Jäckel, F.: A closed-form correction for the Spalart-Allmaras turbulence model for separated flows. AIAA (2022)
9. Engl, H.W., Hanke, M., Neubauer, A.: Regularization of Inverse Problems. Kluwer Academic Publisher, Dordrecht, Boston, London (1996)
10. Langer, S.: Preconditioned Newton methods for ill-posed problems. PhD thesis, University of Göttingen (2007)
11. Langer, S.: Complexity analysis of the iteratively regularized Gauss-Newton method with inner CG-iteration. J. Inverse Ill-posed Probl. **17**, 871–890 (2009)
12. Hohage, T.: On the numerical solution of a three-dimensional inverse medium scattering problem. Inverse Prob. **17**, 1743–1763 (2001)
13. Langer, S.: Investigation of preconditioning techniques for the iteratively regularized Gauss-Newton method for exponentially ill-posed problems. SIAM J. Sci. Comput. **32**(5), 2543–2559 (2010)
14. Hohage, T., Langer, S.: Acceleration techniques for regularized Newton methods applied to electromagnetic inverse medium scattering problems. Inverse Prob. **26**(7), 074011 (2010)
15. Langer, S.: Preconditioned newton methods to approximate solutions of the Reynolds averaged Navier-Stokes equations. Technical report, Institut für Aerodynamik und Strömungstechnik (2018)
16. Langer, S.: Agglomeration multigrid methods with implicit Runge-Kutta smoothers applied to aerodynamic simulations on unstructured grids. J. Comput. Phys. **277**, 72–100 (2014)
17. Langer, S., Schwöppe, A., Kroll, N.: The DLR flow solver TAU - status and recent algorithmic developments. In: Proceedings of the 52nd Aerospace Sciences Meeting, 2014, number 2014–0080 in Conference Proceeding Series. AIAA (2014)
18. Golub, G.H., van Loan, C.F.: Matrix Computations, 2nd edn. The John Hopkins University Press, Baltimore (1983)

Comparison and Unification of Finite-Volume Discretization Strategies for the Unstructured Node-Centered and Cell-Centered Grid Metric in TAU and CODA

Stefan Langer(✉), Axel Schwöppe, and Tobias Leicht

German Aerospace Center,
Member of the Helmholtz Association Institute of Aerodynamics and Flow Technology, Lilienthalplatz 7, 38108 Braunschweig, Germany
{stefan.langer,axel.schwoeppe,tobias.leicht}@dlr.de

Abstract. 2nd order finite-volume discretization schemes generally offer a good compromise between computational effort and accuracy. For a given mesh, the accuracy of an approximate solution often depends significantly on the discretization applied to the underlying grid. Therefore, the discretization has to be adapted to the chosen grid metric, including the types of elements present in the mesh. To keep the number of degrees of freedom acceptable, in regions of the grid, where steep gradients need to be resolved, quadrilateral, hexahedral or prismatic elements with an extreme cell-stretching are required. Hence, one is now forced to develop unstructured discretizations that reliably produce accurate solutions for mixed grids with elements of high anisotropies. In this work we review and demonstrate, in which way these anisotropies can be meaningfully incorporated into different, commonly used discretization schemes and how the schemes can be extended for unstructured grids in order to achieve good quality solutions.

1 Introduction

Finite-volume methods can be distinguished in node-centered (often also called vertex-based) and cell-centered formulations (see e.g. [1,2]). In a node-based formulation, the flow variables W_i can be considered to be stored at the nodes of the mesh. The discretization involves stencils that are defined by the nearest and/or next-to-nearest neighbors of each mesh point. Cell-centered schemes store the flow variables at the centers of the cells or elements of the mesh. The stencils used for discretization then involve the nearest and/or next-to-nearest neighboring cells. Formally, the correlation between node-centered and cell-centered schemes can be demonstrated with the concept of a dual mesh. If a dual mesh point is created at each cell center and dual mesh edges are created by joining neighboring cell centers, the cell-centered scheme can be seen to be topologically

A. Dillmann et al. (Eds.): STAB/DGLR Symposium 2022, NNFM 154, pp. 262–272, 2024.
https://doi.org/10.1007/978-3-031-40482-5_25

(but not necessarily geometrically) equivalent to a node-centered scheme operating on the dual mesh. Figures 1 and 2 show examples for a 2-D quadrilateral and triangular mesh as well as the corresponding dual mesh.

For a purely quadrilateral or hexahedral mesh, the dual mesh contains the same elements, and the number of nodes and edges in the original and dual meshes is identical (neglecting boundaries). When considering other types of elements this is no longer the case. A triangular mesh contains about twice as many triangles as vertices, and a tetrahedral mesh five to six times more elements than vertices, for example. Generally speaking, a cell-centered scheme on unstructured mixed-element meshes is expected to have a larger number of unknowns while generating relatively simple stencils due to a smaller average number of faces per element compared to edges per node. A node-centered scheme is expected to have a smaller number of unknowns with larger stencils.

The question of whether the additional degrees of freedom translate into an increase in accuracy is still not satisfactorily answered, with an interesting but unclear effect of the increased number of fluxes due to larger stencils for node-based schemes. Already the definition of an 'equivalent' grid for comparing cell-based and node-based discretizations at equivalent accuracy levels still seems to be an open question. An interesting study [3] for isotropic meshes found that matching the number of surface grid variables for both grids achieved similar accuracy for aerodynamic quantities in transonic flow cases. Since there are twice as many triangular elements on the surface as vertices in a given grid, this results in node-based grid sizes which on average contain approximately three times more vertices overall than the equivalent meshes for cell-centered discretizations. However, such results might not automatically generalize to other types of flows, particularly for cases where off-body flow-physics are of primary importance.

The use of face-based data-structures for both node-based and cell-centered discretizations makes it possible to construct a solver which can be operated in either mode, with relatively minor modifications mostly concerning boundary conditions, at least for inviscid flows. For the 2-D results for the RAE2822 test case, we apply such an implementation (based on the DLR TAU-Code [4]) to both node-centered and cell-centered grid metric using exactly the same discretization strategy to obtain a consistent comparison of the results. Furthermore, the cell-centered finite-volume discretization of the computational fluid dynamics (CFD) software CODA is used to compare extensions of established cell-vertex discretization schemes applied to an unstructured cell-centered discretization to retain accuracy on unstructured meshes.

2 General Roe-Type Upwind Schemes

Using a finite-volume scheme, the discretization of the Euler equations gives

$$\mathrm{vol}\,(D_i)\,\frac{dW_i}{dt} = -\sum_{j\in\mathcal{N}(i)} a_{ij}\mathcal{H}\left((W_i,\nabla W_i),(W_j,\nabla W_j),n_{ij}\right), \tag{1}$$

where D_i denotes the control volume, $\mathcal{H}_{ij} = \mathcal{H}((W_i, \nabla W_i), (W_j, \nabla W_j), n_{ij})$ is a numerical flux function, a_{ij} is the surface area and n_{ij} is the normal. Here, the neighbors of a node i are denoted by $\mathcal{N}(i)$. For a Roe-type upwind scheme [5], the numerical flux function takes the form

$$\begin{aligned}\mathcal{H}_{ij} = &\frac{1}{2}(f(W_i, \nabla W_i, n_{ij}) + f(W_j, \nabla W_j, n_{ij})) \\ &-\frac{1}{2}\left|\frac{\partial f}{\partial W}((W_i, \nabla W_i), (W_j, \nabla W_j), n_{ij})\right| d((W_i, \nabla W_i), (W_j, \nabla W_j), n_{ij}) \quad (2)\end{aligned}$$

For example, $\mathcal{H}_{ij}^{\mathrm{Roe,1st}} := \frac{1}{2}(f(W_i, n_{ij}) + f(W_j, n_{ij})) - \frac{1}{2}\left|\mathbf{A}_{ij}^{\mathrm{Roe}}\right|(W_j - W_i)$ is a first order Roe scheme where Roe-averaging and the following notation is used: $\mathbf{A}_{ij}^{\mathrm{Roe}} = \frac{\partial f}{\partial W}(W_i, W_j, n_{ij})$.

3 Central Schemes

In the literature, there does not exist one, uniquely defined central scheme

$$\mathcal{H}_{ij} = \frac{1}{2}(f(W_i, \nabla W_i, n_{ij}) + f(W_j, \nabla W_j, n_{ij})) - d_{ij}\left(W_{\mathcal{N}(i,j)}, n_{ij}\right), \quad (3)$$

$$d_{ij}\left(W_{\mathcal{N}(i,j)}, n_{ij}\right) := \frac{1}{2}\left|\mathbf{A}_{ij}^{\mathrm{Roe}}\right|\{\Psi_{ij}(W_j - W_i) - \xi s_{ij}(1 - \Psi_{ij})(L_j - L_i)\}. \quad (4)$$

Typically, these kinds or schemes differ in the way the blending between the second and fourth differences based on the computation of the pressure sensor Ψ_{ij}, Roe-matrix and entropy fix are realized [6]. Furthermore, we optionally consider the gradients ∇W_i and ∇W_j in (3) to be able to reconstruct W linearly left and right at the face integration-point.

3.1 Role of Undivided Laplacians

To start with an investigation of central schemes, we shortly discuss the effect of the undivided Laplacians of a node k, $L_k = \sum_{j\in\mathcal{N}(k)}(W_j - W_k)$. Using the notation of Fig. 1, we obtain for the node (i, j)

$$\begin{aligned}L_{i,j}^{\mathrm{quad}} &= (W_{i+1,j} - 2W_{i,j} + W_{i-1,j}) + (W_{i,j+1} - 2W_{i,j} + W_{i,j-1}) \\ &\approx h_x^2\frac{\partial^2 W}{\partial x_1^2} + h_y^2\frac{\partial^2 W}{\partial x_2^2}. \quad (5)\end{aligned}$$

That is, for a quadrilateral mesh, the undivided Laplacian represents an approximation to the second derivative evaluated at i, j. Considering (1), and assuming that $\mathbf{A}_{ij}^{\mathrm{Roe}}$ is constant, we obtain

$$\sum_{j\in\mathcal{N}(i)} a_{ij}\mathbf{A}_{ij}^{\mathrm{Roe}}(W_j - W_i) \approx \mathbf{A}_{ij}^{\mathrm{Roe}}\left(h_y h_x^2\frac{\partial^2 W}{\partial x_1^2} + h_x h_y^2\frac{\partial^2 W}{\partial x_2^2}\right). \quad (6)$$

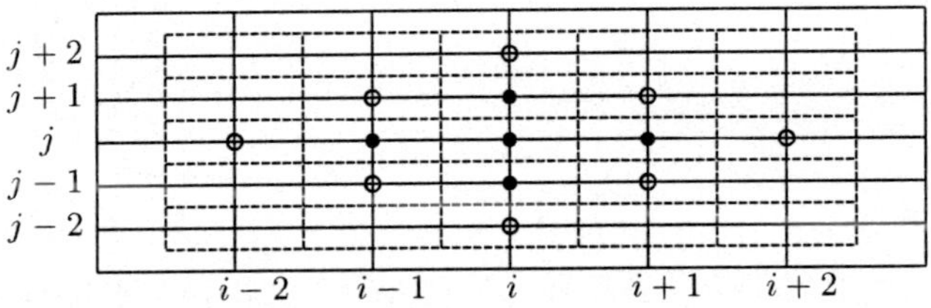

Fig. 1. Example of a quadrilateral mesh and its dual mesh

Division by the volume $\text{vol}(D_i) = h_x h_y$ gives us a term satisfying $O(\max\{h_x, h_y\})$. For the undivided Laplacians of the undivided Laplacians we can do a similar argumentation,

$$\sum_{j \in \mathcal{N}(i)} a_{ij} \mathbf{A}_{ij}^{\text{Roe}} (L_j - L_i) \approx \mathbf{A}_{ij}^{\text{Roe}} \left(h_y h_x^2 \left(h_x^2 \frac{\partial^4 W}{\partial x_1^4} + h_y^2 \frac{\partial^4 W}{\partial x_1^2 \partial x_2^2} \right) \right.$$
$$\left. + h_x h_y^2 \left(h_y^2 \frac{\partial^4 W}{\partial x_2^4} + h_x^2 \frac{\partial^4 W}{\partial x_1^2 \partial x_2^2} \right) \right). \tag{7}$$

Here, division by $\text{vol}(D_i)$ gives a term satisfying $O\left(\max\{h_x^2 h_y, h_x h_y^2\}\right)$. For a node centered scheme on a triangular mesh, we can compute the undivided Laplacian using Fig. 2,

$$\begin{aligned} L_{i,j}^{\text{tri}} &= (W_{i+1,j} - 2W_{i,j} + W_{i-1,j}) + (W_{i,j+1} - 2W_{i,j} + W_{i,j-1}) \\ &\quad + (W_{i-1,j-1} - 2W_{i,j} + W_{i+1,j+1}) \\ &\approx h_x^2 \left(\frac{\partial^2 W}{\partial x_1^2} \right) + h_y^2 \left(\frac{\partial^2 W}{\partial x_2^2} \right) + \left(h_x^2 + h_y^2 \right) \nabla_n \cdot \nabla W. \end{aligned} \tag{8}$$

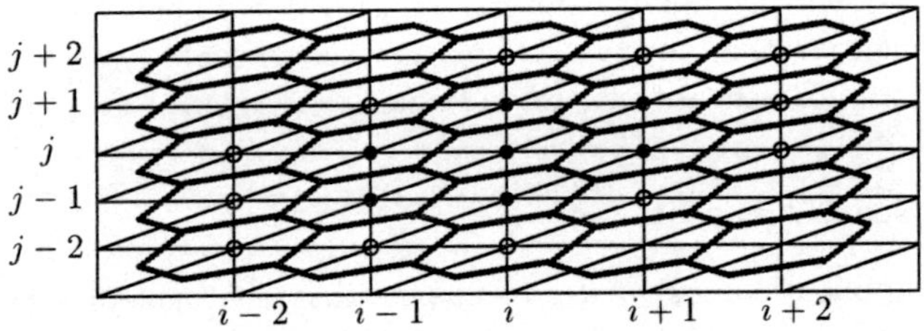

Fig. 2. Example of a triangular mesh and its dual mesh

For the additional terms, we have approximated the gradients on the edge connecting nodes (i, j) and $(i + 1, j + 1)$, and $(i - 1, j - 1)$, respectively, by

$$\nabla W_{(i+1,j+1;i,j)} \approx \frac{1}{\sqrt{h_x^2 + h_y^2}} \left(W_{i+1,j+1} - W_{i,j} \right) (n_x, n_y)^T,$$
$$\nabla W_{(i-1,j-1;i,j)} \approx \frac{1}{\sqrt{h_x^2 + h_y^2}} \left(W_{i-1,j-1} - W_{i,j} \right) (n_x, n_y)^T,$$

and hence we get

$$\nabla_n \cdot \left(\nabla W_{(i+1,j+1;i,j)} - \nabla W_{(i-1,j-1;i,j)}\right) = \frac{1}{h_x^2 + h_y^2} \left(W_{i-1,j-1} - 2W_{i,j} + W_{i+1,j+1}\right).$$

Following (6) using (8) we estimate

$$\sum_{j \in \mathcal{N}(i)} a_{ij} \mathbf{A}_{ij}^{\mathrm{Roe}} \left(W_j - W_i\right) \approx \mathbf{A}_{ij}^{\mathrm{Roe}} \left(h_x^2 h_y \frac{\partial^2 W}{\partial x_1^2} + h_x h_y^2 \frac{\partial^2 W}{\partial x_2^2}\right) + \left(h_x^2 + h_y^2\right) \sqrt{h_x^2 + h_y^2} \nabla_n \cdot \nabla W. \quad (9)$$

Estimating the volume of the polygon by $h_x h_y$, we obtain a term which behaves like $O(\max\{h_x, h_y, h_x^2/h_y, h_y^2/h_x\})$. The undivided Laplacians of the undivided Laplacians for the node-centered metric on the triangular mesh gives

$$\sum_{j \in \mathcal{N}(i)} a_{ij} \mathbf{A}_{ij}^{\mathrm{Roe}} \left(L_j - L_i\right) = O\left(\max\{h_x^4/h_y, h_y^4/h_x\}\right). \quad (10)$$

For an isotropic mesh, when $h_x/h_y \approx 1$, the differences in (6) and (9) are possibly negligible. But for a highly stretched mesh, needed to efficiently resolve a high Reynolds number boundary layer, when $h_x \sim 10^d h_y$ and $d = 3, 4, 5$ can easily be satisfied, there can be detrimental effects on accuracy. However, for a systematic mesh refinement the method is consistent with $O(h_x)$. Obviously, it is desirable that the low order terms in (4) corresponding to (6) and (9) are only active when necessary, which is controlled by a pressure switch $\Psi_{ij} \in [0, 1]$ (see [6]).

Finally, for a cell-centered scheme on a triangular mesh, we observe for example for the triangle $\Delta_{cc} = (i-1, j-1), (i, j-1), (i, j)$ that the stencil has only three neighbors, and hence the stencil operations can represent a lower quality approximate derivative. This heuristic observation raises the expectation for a loss in accuracy without further modifications of the scheme. For completeness, we want to mention that these results carry over to the 3-D case. For example, for a Cartesian hexahedral this can be found in [6].

3.2 Grid Scaling Factor

Due to the applications, we need to ensure that the central schemes considered, retain their accuracy when applied to highly stretched grids required to efficiently resolve a high Reynolds number boundary layer. To ensure that the schemes reveal good damping properties for high frequency modes, a certain degree of smoothing needs to be incorporated. Therefore, in (4) the scaling factor s_{ij} is introduced. In the literature different strategies are proposed to construct an adequate measure. For example, in [7] a factor based on geometric data is used, in [4,8] a factor based on local flow data is describes, [6] gives an overview of different methodologies. All these factors have in common, that their effect is negligible for isotropic cells, whereas for anisotropic cell they mimic an appropriate weighting yielding improved accuracy.

4 Roe Scheme

To construct a 2nd order Roe scheme from the prototype in (2) with good accuracy for mixed-element type meshes, the accurate construction of the cell-gradients ∇W is of high importance. Here we shortly describe the widely used Green-Gauss (GG) and least-squares (LS) approaches along with modifications which yield improved accuracy in particular for the cell-centered grid metric.

4.1 Least-Squares Gradient

The LS cell-gradient is computed by solving a system of linear equations for the gradient values. The system results from the minimization of the functional

$$\sum_{j \in \mathcal{N}(i)} w_{ij}^2 \left(\nabla W_i \cdot (\mathbf{x}_j - \mathbf{x}_i) - (W_j - W_i)\right)^2 \rightarrow \min . \tag{11}$$

Here $\mathbf{x}_i$ is the coordinate vector of point i and w_{ij} is a weighting factor chosen as $w_{ij} = 1/\left|\mathbf{x}_j - \mathbf{x}_i\right|^p$, where p denotes the weighting exponent. $p = 1$ yields the so-called inverse distance weighting [9]. This weighted LS method is known to improve gradient accuracy on certain high aspect ratio grids [9]. While LS cell-gradients exactly represent linear functions for node-based and cell-centered discretizations, Mavriplis [9] noted that this is not a sufficient criterion for accuracy of the overall finite volume scheme. The accuracy depends on the choice of stencils, i.e. the set of points involved in the sum of (11).

A description of various LS stencils can be found in [10]. The nearest neighbor (NN) stencil includes only face-neighbors. The NN stencil does not necessarily provide accuracy and robustness [9,11], mainly due to a non sufficient number or distribution of stencil points. The full augmentation (FA) stencil includes all neighbors that share a vertex with the given volume. The FA stencil normally leads to accurate and robust solutions. Nevertheless, it is recommended to build the stencil dependent on the contribution of points to the condition of the LS system [10]. However, in this paper we use the LS gradients within modified Green-Gauss gradients based on the FA stencil only and control robustness and accuracy through the weighting exponent.

4.2 Green-Gauss Gradient

Following the Green-Gauss theorem, the cell gradient is approximated as a discrete surface integral, a sum of scalar values W_f reconstructed at the face integration-point multiplied by the face area and normal vector

$$\nabla W_i = \frac{1}{\Omega_i} \sum_{j \in \mathcal{N}(i)} W_f a_{ij} \mathbf{n}_{ij}, \qquad W_f = \alpha W_i + (1 - \alpha) W_j . \tag{12}$$

The weight α is computed regarding the spatial distribution of $\mathbf{x}_i$, $\mathbf{x}_j$ and the face integration-point $\mathbf{x}_f$. For a node-centered based gradient computation the interpolation corresponds to the average, i.e. $\alpha = 0.5$. Because of the approximation properties of the node-centered integration scheme, the GG gradient is exact for a linear function only on tetrahedral or triangular meshes, although reasonable accuracy has been demonstrated in computations on mixed grids [9]. For the cell-centered metric, the GG gradient is not generally exact for a linear function. Accuracy is achieved only if the interpolation point on the edge coincides with the face-integration point [9], otherwise a non-negligible integration error is introduced. To reduce this error, we compute W_f by a gradient-based, piecewise-linear reconstruction, for which pre-computed LS gradients based on the FA stencil are used to avoid robustness and accuracy reduction

$$W_f = \frac{1}{2}\left(W_i + (\mathbf{x}_f - \mathbf{x}_i)\,\psi_i \nabla^{LS} W_i + W_j + (\mathbf{x}_f - \mathbf{x}_j)\,\psi_j \nabla^{LS} W_j\right), \tag{13}$$

where ψ denotes a limiter value, that can be computed by any usual gradient limiter [1]. Substituting (13) into (12) yields the *extended Green-Gauss* gradient which, in principle, can be applied to both node-based and cell-centered grid metric and hence can be seen as a unification of the Green-Gauss gradient formulation to reduce metric-dependent integration-errors.

5 Numerical Results

Restricting ourselves to compressible Euler flow, as a measure for accuracy we use throughout the article the total pressure loss given by $1 - p_{\text{tot}}/p_{\text{tot},\infty}$. For all test cases we considered the inflow Mach number $M_\infty = 0.3$ and angle of attack $2.0°$. Figure 5 (left) shows the quadrilateral mesh of size 320×64, and the triangular mesh generated by the quadrilateral one by subdividing the quadrilaterals in triangles. The cell stretching ratio for cells near the boundary of the airfoil is about $\frac{h_x}{h_y} \in [500, 600]$, in the wake of the airfoil there are cells with $\frac{h_x}{h_y} > 100000$.

For a central scheme (3) on quadrilateral meshes, both the node- and cell-centered scheme yield the expressions (5) and (7). This theoretical equivalence can be observed in practice as well in Fig. 3 (left) showing almost identical distributions. Estimates (8) and (9) explain the deviation of the total pressure loss shown in Fig. 3 (mid) for the node-centered scheme on the triangular mesh compared to the quadrilateral mesh. For the cell-centered scheme on the triangular mesh, see Fig. 3 (right), we get a further and stronger deterioration of accuracy. The total pressure loss is more than an order of magnitude larger compared to the node-centered scheme for both the triangular and the quadrilateral mesh. In contrast to the deviation in Fig. 3 (mid), this deterioration cannot be explained by the construction of the unstructured cell-centered Laplacian alone.

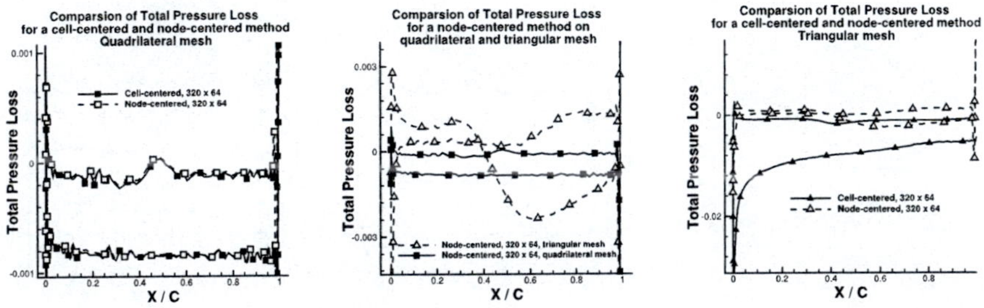

Fig. 3. Comparison of total pressure loss for quadrilateral and triangular mesh

Next, we use the CODA software, that allows to apply extended Green-Gauss gradients in (3) to linearly reconstruct W at a face integration-point instead of averaging W_i and W_j. Figure 4 (left) shows the resulting elimination of the deterioration in total pressure loss for the cell-centered scheme on the triangular mesh. Of course, the deviation due to the construction of the unstructured Laplacian remains.

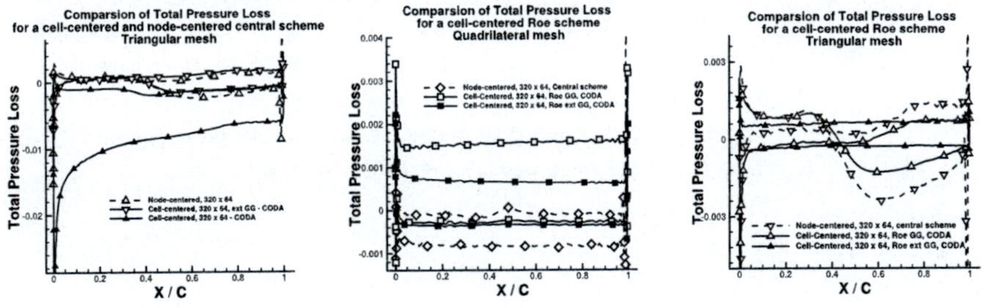

Fig. 4. Comparison of total pressure loss for quadrilateral and triangular mesh (usage of extended Green-Gauss gradients)

As a final aspect concerning central schemes we investigate the effect of inclusion of the grid scaling factor. Figure 5 shows the influence of such a factor for a node-centered scheme on a quadrilateral (mid) and triangular (right) mesh. In both cases, the accuracy can be significantly improved with such a scaling.

In summary, the analysis carried out in Sect. 3 is supported by the observed results. At least parts of the deterioration of the accuracy of the central scheme on triangular meshes can be explained by additional terms approximating cross derivatives. As a consequence, the cell-stretching ration h_x/h_y is the dominant describing factor for accuracy. To improve the accuracy for a given mesh, a methodology needs to be incorporated into the method such that these cross terms do not appear in (8) and (9). Instead of trying to modify and improve these terms, one can also directly consider a Roe scheme, where the accuracy is based on the approximation quality of the gradients.

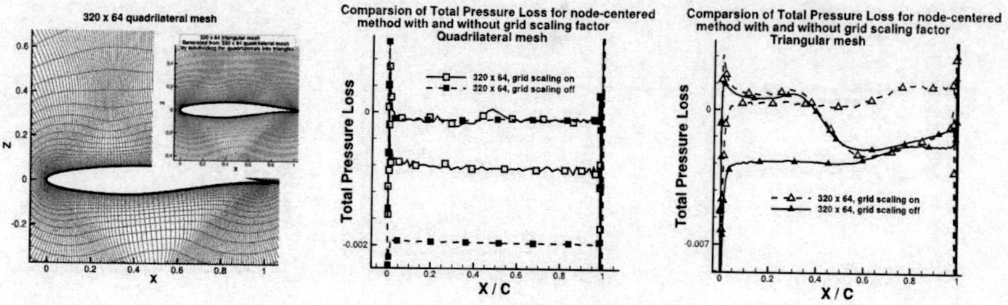

Fig. 5. Comparison of influence of grid scaling factor

A Roe scheme (2) based on standard (12), and extended (13) Green-Gauss gradients yields smooth distributions of total pressure loss on quadrilateral meshes as shown in Fig. 4 (mid). The lower integration error introduced by extended Green-Gauss gradients improves accuracy. Compared to a central scheme we observe an offset, that indicates lower dissipation introduced by the central scheme on quadrilateral meshes. On triangular meshes we get a similar deviation of the total pressure loss for the Roe scheme based on standard Green-Gauss gradients as observed for the node-based central scheme, see Fig. 4 (right). For the Roe scheme based on extended Green-Gauss gradients no deviation in accuracy can be observed on triangular meshes. The total pressure loss distribution remains smooth. Therefore, we can conclude that the approximation quality of the extended Green-Gauss gradients avoids the inaccuracy introduced by estimate (8) to a central scheme on triangular meshes.

6 Conclusion

In this short article we reviewed some Roe-type upwind schemes. In the DLR TAU Code, a Roe-type scheme has been implemented in the form of a central scheme. The dissipative part is realized by a blending of second and fourth differences, and the blending is realized by a pressure switch. Applied to a node-centered and cell-centered metric on a highly-stretched quadrilateral mesh, typically used for high Reynolds number turbulent flows, and subsonic inviscid flow over an RAE2822 airfoil, comparable accuracy is reached, when a measure for accuracy is given by the total pressure loss. Both, theoretically and in practice, evidence is given, that for triangular meshes there is a significant loss in accuracy. This loss in accuracy is even greater for a cell-centered grid metric. One reason for the loss is due to the construction of the undivided Laplacians. To improve the accuracy for triangular meshes, at least an adequate concept for improving the accuracy of the undivided Laplacians on triangular meshes is required. The reason for the severe loss in accuracy of the cell-centered scheme could be related to the computation of the central part of the scheme. An accurate, piecewise linear reconstruction is needed instead of the commonly used average to avoid integration errors when applying a cell-centered grid metric on triangular meshes.

As we have shown, undivided Laplacians are similar to constructing gradients and derivatives. Hence, instead of trying to improve the accuracy of these terms, it is possible to directly improve the accuracy of gradients required for a Roe scheme. For the cell-centered software CODA standard and extended Green-Gauss gradients were compared. It was shown that using extended Green-Gauss gradients improves accuracy compared to the implementations of central schemes in both the node-centered code TAU and the cell-centered CODA software on triangular meshes, even if in CODA also the accuracy of the central scheme could be significantly improved via the linear reconstruction as mentioned above. It also needs to be mentioned that the extended Green-Guass gradients need a stencil that goes beyond the neighbors of the neighbors. This means that the gain in accuracy of these gradients is also based on an increase in the computing effort. If this increased computing effort is allowed in the node-centered scheme, it can be expected that the error can be further reduced there as well.

Of course, in this short article many aspects could not be discussed, such as the implementation of an entropy fix, realization of scaling factors, blending of first and second order terms as well as limitation of gradients. Therefore, the results presented in this article are restricted to inviscid, smooth flows, where many of these just mentioned ingredients possibly play a minor role. However, to design an accurate scheme for high Reynolds number transonic flow, it is of major importance to ensure that the scheme works well for basic flow cases.

Acknowledgment. CODA is the computational fluid dynamics (CFD) software being developed as part of a collaboration between the French Aerospace Lab ONERA, the German Aerospace Center (DLR), Airbus, and their European research partners. CODA is jointly owned by ONERA, DLR and Airbus.

References

1. Blazek, J.: Computational Fluid Dynamics: Principles and Applications. Elsevier Science Ltd., (2001)
2. Mavriplis, J.: Unstructured mesh Discretizations and solvers for computational aerodynamics. AIAA 2007–3955 (2007)
3. Levy, D., Thacker, M.: Comparison of unstructured cell- and node-based schemes for the Euler equations, pp. 606–618. AIAA 99–33409 (1999)
4. Langer, S., Schwöppe, A., Kroll, N.: The DLR flow solver TAU - status and recent algorithmic developments. In: Proceedings of the 52nd Aerospace Sciences Meeting, number 2014–0080 in Conference Proceeding Series. AIAA (2014)
5. Roe, P.L.: Approximate Riemann solvers, parameter vectors, and difference schemes. J. Comput. Phys. **43**, 357–372 (1981)
6. Langer, S., Schwöppe, A.: TAU Hyperflex Report. Technical report (2022). https://elib.dlr.de/188909
7. Moinier, P.: Algorithm Developments for an Unstructured Viscous Flow Solver. PhD thesis, University of Oxford (1999)
8. Langer, S.: Agglomeration multigrid methods with implicit Runge-Kutta smoothers applied to aerodynamic simulations on unstructured grids. J. Comput. Phys. **277**, 72–100 (2014)

9. Mavriplis, D.: Revisiting the least-squares procedure for gradient reconstruction on unstructured meshes. In: 16th AIAA Computational Fluid Dynamics Conference, p. 3986 (2003)
10. Schwöppe, A., Diskin, B.: Accuracy of the cell-centered grid metric in the DLR TAU-Code. In: Notes on Numerical Fluid Mechanics and Multidisciplinary Design, vol. 121, pp. 429–437 (2013)
11. Diskin, B., Thomas, L.T.: Comparison of cell-centered and node-centered unstructured finite-volume discretizations: inviscid fluxes. In: Proceedings of the 48th Aerospace Sciences Meeting, 2010, number 2010–1079 in Conference Proceeding Series. AIAA (2010)

Deformation of CFD Meshes with Anisotropic Cells in a Viscous Boundary Layer Using Line-Implicit Methods

Arne Rempke(✉)

German Aerospace Center (DLR), Institute of Software Methods for Product Virtualization, Zwickauer Straße 46, 01069 Dresden, Germany
arne.rempke@dlr.de

Abstract. Various numerical applications in the context of multidisciplinary high-fidelity aircraft design, e.g. fluid-structure interaction or shape optimization, require the consideration of changes in the geometry of the aircraft shape. Such geometry changes can be realized using mesh deformations, and we demonstrate such a volume mesh deformation method based on elasticity analogy in conjunction with line-implicit solvers. Line-implicit solvers have already been successfully used in CFD simulations with strongly anisotropic cells in a viscous boundary layer, and are now applied to mesh deformation problems of such meshes. In this approach, the solution of a block-tridiagonal subsystem of the Jacobian matrix is computed exactly using the Thomas algorithm. Here, the selection of the tridiagonal part – representing the lines – is of crucial importance for the speed of convergence. A new algorithm for line identification is presented and compared with existing ones. Also, a comparison and combination of these methods with multigrid methods is performed and the implications for industrially relevant test cases are demonstrated.

Keywords: mesh deformation · CFD mesh · anisotropic cells · viscous boundary layer · line-implicit method · Thomas algorithm · line identification · numerical method · linear algebra · multigrid

1 Introduction

Mesh deformation is an important ingredient for multidisciplinary simulation or geometry optimization [6,10]. Since it is crucial for succeeding simulation parts that the resulting mesh stays intact (i.e. no collapsing elements), choosing a robust implicit method makes sense especially for large deformations or difficult meshes. The robust mesh deformation method we examine in this work is based on linear elasticity analogy [11] and will be described in Sect. 2.

We describe the numerical algorithms that are employed in order to speed up solving the linear equation systems occuring in this method in Sect. 3. These

A. Dillmann et al. (Eds.): STAB/DGLR Symposium 2022, NNFM 154, pp. 273–283, 2024.
https://doi.org/10.1007/978-3-031-40482-5_26

comprise line-implicit methods in general [5], but also a new algorithm for line detection (Sect. 3.1), that is particularly well suited for finite-element-based discretizations, and the algebraic multigrid setup (Sect. 3.2), that is important for an efficient handling of large meshes.

In Sect. 4 we will present the results that were obtained from using, combining and comparing these methods on two industry-relevant test cases. Finally, Sect. 5 contains a summary and some closing remarks.

2 Volume Mesh Deformation Using Linear Elasticity Analogy

Mesh deformation using elasticity analogy is a robust form of volume mesh deformation. The method we use is based on the idea that the spacial CFD domain behaves like an artificial body under elastic deformation. We will shortly wrap up the idea in this section and refer to [1,9,11] for a more detailed description.

Following a linear approach, the strain tensor $\boldsymbol{\varepsilon}$ is written in dependency of the deformations $\mathbf{u}$ by

$$\boldsymbol{\varepsilon}(\mathbf{u}) = \frac{1}{2}\left((\nabla\mathbf{u}) + (\nabla\mathbf{u})^{\top}\right). \tag{1}$$

The assumption of a homogenous, isotropic material law gives us the following stress $\boldsymbol{\sigma}$:

$$\boldsymbol{\sigma}(\boldsymbol{\varepsilon}) = \frac{E}{1+\nu}\boldsymbol{\varepsilon} + \frac{\nu E}{(1+\nu)(1-2\nu)}\mathrm{tr}(\boldsymbol{\varepsilon})\mathbf{I}, \tag{2}$$

where E and ν are material parameters of the virtual elastic body, tr is the trace operator and $\mathbf{I}$ is the identity. Then we provide a deformation of the boundary $\partial\Omega$ of our CFD domain Ω using Dirichlet boundary conditions and solve the deformations according to the system of partial differential equations

$$\begin{aligned} \nabla\cdot\boldsymbol{\sigma}(\boldsymbol{\varepsilon}(\mathbf{u})) &= \mathbf{0} \quad \text{on } \Omega, \\ \mathbf{u} &= \mathbf{u_D} \text{ on } \partial\Omega. \end{aligned} \tag{3}$$

2.1 Artificial Jacobian-Based Stiffening

In order to ensure the geometrical properties of the elements like orthogonality, aspect ratio, and, most importantly, non-overlapping positive volume are conserved, especially the cells in the viscous boundary layer should behave stiff under the deformation.

This is achieved by choosing a higher Young's modulus E for those elements to be conserved, while other elements that are not that critical can be modeled using a smaller E. The criticalness is indicated from the volume size of each element, as suggested by [11]: When choosing $E = \mathrm{vol}(e)^{-\chi}$ for each element e of the mesh, the smaller elements obtain a larger stiffness, while larger elements behave more flexible. Such a model is very convenient, since the size of the element is

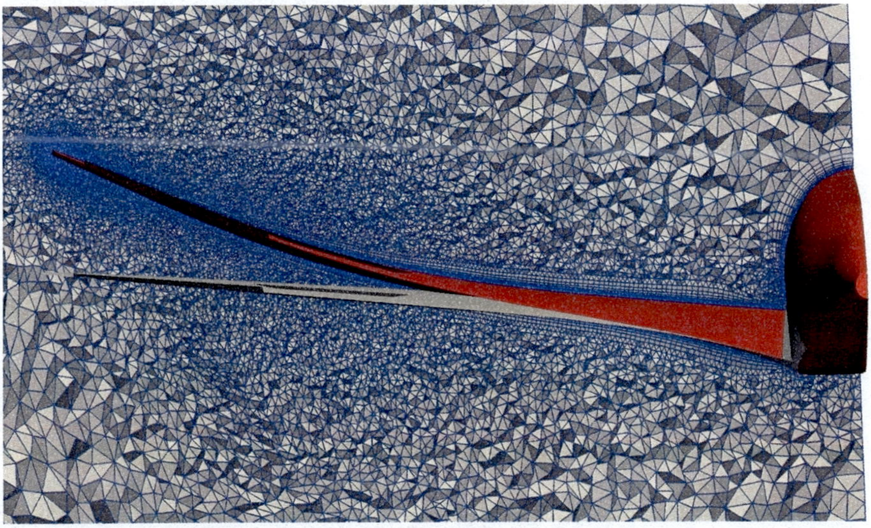

Fig. 1. Example of a CFD volume mesh deformation using linear elasticity analogy for the deflection of the wing of an aircraft: Undeformed (gray) and deformed (red) aircraft surface and unstructured deformed volume mesh (blue) with boundary layer.

directly obtained from the transformation of the mesh element to a reference element in order to evaluate the quadrature formulas of the discretization.

For the computations presented in this paper, we used $\chi = 1$ and $\nu = 0.3$ for the Poisson ratio ν.

Using this approach, mesh deformations like presented in Fig. 1 can be achieved, where the wing of an aircraft is bend quite a large amount and the volume mesh is successfully transformed to match the new surface geometries while preserving positive cells and the properties of the viscous boundary layer. The implementation operates on unstructured meshes with various element types.

3 Line-Implicit Methods

When solving the linear equation system resulting from a discretization of (3), iterative methods usually work quite well for the elliptical problem. However, when the discretization mesh contains very anisotropic cells with aspect ratios of more than $1000 : 1$, these anisotropic problems are difficult to solve with standard methods. This is well known for the finite volume method in computational fluid dynamics, see [2,4,5,8,12].

The idea of line-implicit methods is to combine multiple cells to a line of cells, and to invert the tridiagonal part of the equation system instead of a purely diagonal part. Solving a tridiagonal equation system is an exact method when following the Thomas algorithm with no additional infill or approximation.

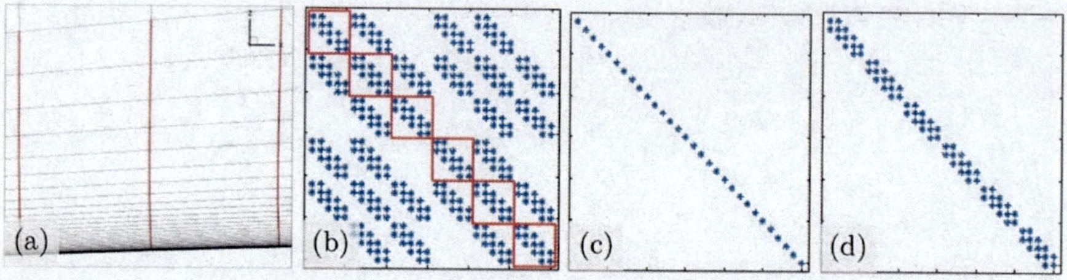

Fig. 2. Possible lines within the viscous boundary layer in the mesh (a) and the sparsity structure (after permutation) of a typical resulting matrix (b). Entries corresponding to the lines are marked by red boxes. Sparsity structure of the approximating matrices that are used as preconditioner in point-implicit (c) or line-implicit (d) methods.

Figure 2 illustrates the idea of the line-implicit method in contrast to the point-implicit method. In fact, we actually combine the three degrees of freedom (displacement in 3D space) for each vertex to a dense block and solve the block-tridiagonal system

$$\mathbf{A}_i \mathbf{x}_{i-1} + \mathbf{B}_i \mathbf{x}_i + \mathbf{C}_i \mathbf{x}_{i+1} = \mathbf{d}_i \quad i = 1, \ldots, n \tag{4}$$

with $\mathbf{A}_1 = \mathbf{C}_n = \mathbf{0}$ using the block Thomas algorithm [13]:

$$\begin{aligned}
\mathbf{B}'_1 &:= \mathbf{B}_1^{-1} \\
\mathbf{d}'_1 &:= \mathbf{B}'_1 d_1 \\
\left.\begin{aligned}
\mathbf{C}'_{i-1} &:= \mathbf{B}'_{i-1}\mathbf{C}_{i-1} \\
\mathbf{B}'_i &:= (\mathbf{B}_i - \mathbf{A}_i \mathbf{C}'_{i-1})^{-1} \\
\mathbf{d}'_i &:= \mathbf{B}'_i(\mathbf{d}_i - \mathbf{A}_i \mathbf{d}'_{i-1})
\end{aligned}\right\} & \; i = 2, \ldots, n \\
\mathbf{x}_n &:= \mathbf{d}'_n \\
\mathbf{x}_{i-1} &:= \mathbf{d}'_{i-1} - \mathbf{C}'_{i-1}\mathbf{x}_i \qquad i = n, \ldots, 2
\end{aligned} \tag{5}$$

3.1 Lines Detection

For identifying good lines in an anisotropic mesh, algorithms like described in [4,5] are well known. They select the strongest anisotropy in the mesh as a seeding point to form a new line and then greedily append new elements to the line in both directions until all possible continuations seem to belong to an isotropic mesh region. This process is then looped until no starting point is available, because all anisotropic elements have been incorporated to a line.

In this work we introduce a different algorithm to detect lines, and we will describe it in this paragraph. It is also based on the criterion to detect anisotropy using the inverse Euclidean distances as weights in the graph to identify stronger couplings that should form a line. However, instead of the greedy approach to grow a single line, it first identifies for each vertex those two neighbors with the highest edge weight as line continuation candidates. Then, a tridiagonality check is done whether those two candidates are also directly connected within the graph, in which case only the highest edge weight is considered for a line segment and the connection to the second highest edge weight is rejected. If now

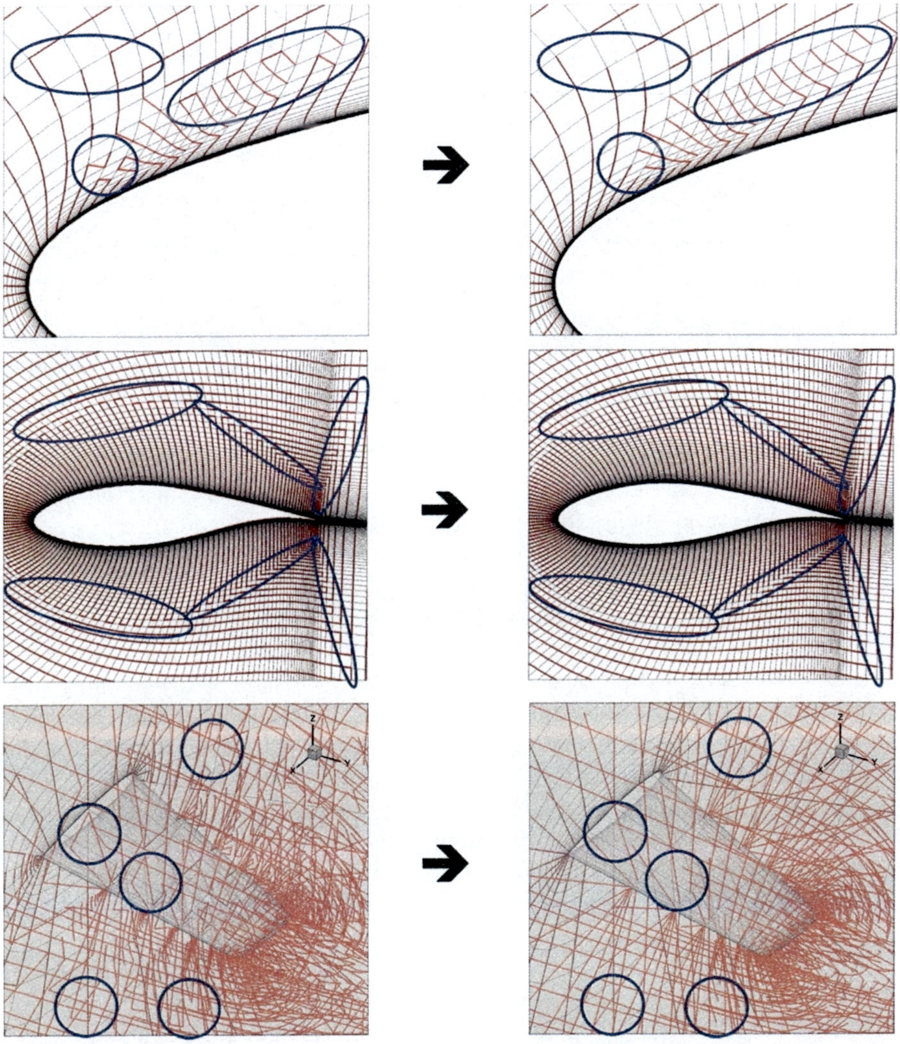

Fig. 3. Comparison of the old (left) and the new (right) lines identification algorithm on three different meshes. Regions of differently selected lines are highlighted. While the greedy selection chooses bends and turns, the new algorithm leads to mostly parallel lines.

two vertices did select each other as candidates and potential line neighbors, and the corresponding edge weight is the highest of all neighbors for at least one of these vertices, a line segment between them is formed. If both vertices are only connected by the second highest edge weight, the line segment is only formed, if and only if the two vertices with the highest edge weights are unconnected. Now, all line segments have been created and can finally be collected and followed to form lines. Special care needs to be taken to avoid cycles of line segments: Although the weights from typical meshes usually do not result in such specialities, the algorithm did not prohibit them until now. Either the line solver needs to be able to solve a cyclic tridiagonal system (e.g. using a transformation based on the Sherman-Morrison formula), or all cyclic lines should be split up to classical lines with two ends. In our observations it was sufficient to just split cyclic lines at the lowest edge weight.

The algorithm introduced in the last paragraph will be referred to as the new one, while we compare it to the greedy one described in [4,5]. Comparing the two algorithms leads to some observations:

1. The new algorithm does not stop at isotropic regions. Although it is possible to add a respective check, it turned out, that in general it is more beneficial for solver convergence rates to continue the lines and reach into isotropic regions of the mesh.
2. The new algorithm does involve local tridiagonality checks of the mesh. A tridiagonality check as described by [7] ensures that the submatrix that contains all relevant entries for a line is actually tridiagonal and therefore can be inverted exactly by the Thomas algorithm. In a non tridiagonal submatrix, there still exist nonzero entries apart from the tridiagonal regardless of permutations. As [7] described, ensuring tridiagonality helps to gain better solver convergence rates. However, instead of full tridiagonality checks for the whole line, we observed that local tridiagonality checks that involve at most three potential line segments are sufficient enough. This reduces the complexity and runtime of the line detection algorithm compared to a full tridiagonality check.
3. The new algorithm can be implemented in parallel, since it does not require to process one line after another, and only needs knowledge of a quite local part of the mesh when selecting the relevant line segments. Just the final step, the collection of constructed line segments to lines, needs to be consolidated between threads.
4. Performance-wise, both algorithms basically scale with the number of vertices in the mesh. However, while the greedy one requires to maintain and update a sorted list of the not-yet-incorporated vertices, the new one gets along without such central data structure, which also could become a bottleneck especially for parallel runs.

The improved line detection algorithm helps to find good lines, especially for FEM-based connectivities. See Fig. 3 for a comparison of the results of the algorithm presented in [4,5] and the new line identification algorithm. Note that [5] uses a minimum aspect ratio of $\alpha = 4$ to stop the continuation of lines to isotropic regions. However, we observed a better convergence rate with long lines also reaching to isotropic regions, which is why we used $\alpha = 1$ for Fig. 3. This makes the old algorithm also produce longer lines, but introduces artifacts like lines snapping off or even turning around that are handled much better with the new algorithm.

3.2 Multigrid

Since both the line-implicit and even more the point-implicit methods operate on a relatively local scope, reducing global errors can take many iterations on large meshes. In order to improve the convergence rates, a widespread method is to apply a multigrid technique, which is also what we did. We employed an isotropic agglomeration based coarsening, where, in hexahedral structured regions of the mesh, approximately 8 vertices of the fine mesh are combined to one vertex of the next coarser level. The restriction operator just adds the different contributions of all agglomerated vertices with equal weight. Therefore, in contrast to semi-coarsening, we get quite a nice downsizing of the matrix with each level. Note, that this is a purely algebraic multigrid technique, since the agglomerates are identified via analysis of the nonzero sparsity pattern, and the coarse operators are created as restriction from the fine level matrix.

As smoother we use the line-implicit method only on the finest level. It was completely sufficient to only use a point-implicit method on the coarser levels. Depending on the mesh size we use up to 8 levels, and a V-cycle without pre-smoothing and 2 iterations of a multi-color Gauss-Seidel scheme is used as post-smoothing. This multigrid method is used as preconditioner for a GMRES Krylov subspace solver as illustrated in Fig. 4.

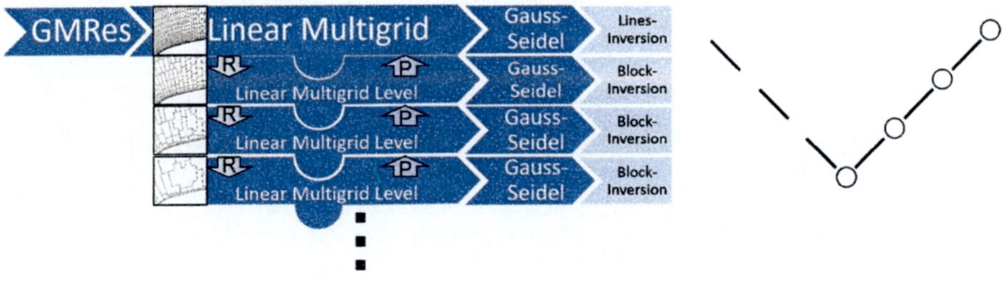

Fig. 4. Schematic solver chain (only four of the eight levels of multigrid are shown) used for the multigrid setup using line-implicit method on the finest level and point-implicit smoothing on coarser levels. V-cycle without pre-smoothing illustrated on the right.

4 Results

This section shows the results from combining the previously described methods. The investigations are realized by extending the implementation FSMeshDeformation [9] and using the Spliss linear algebra library [3]. Note that all computations converge to the same deformed meshes (except for small floating-point errors), which do not contain any collapsed cells. The main focus of this comparison is on compute time and effectiveness of the different methods to solve the large and distributed linear equation systems. All computations are performed on a cluster of two chips AMD EPYC 7601 (32 cores; 2,2 GHz) per node with 64 processes per node.

4.1 NASA Common Research Model

The first testcase is a 3D mesh of the Common Research Model (CRM) from the Drag Prediction Workshop, containing 29M nodes and aspect ratios of 4000:1 in the viscous boundary layer. A comparison of the convergence of the different methods for solving the linear equation system on 512 processes in parallel can be found in Fig. 5. While one can see that the single level point-implicit method gets stuck after a reduction of approximately 12 orders of magnitude, both adding multigrid or line-implicit method help to converge further. Multigrid always gives a speed-up of a factor of at least 2 compared to single grid when reducing to 14 orders of magnitude. The new lines detection algorithm also reduces the runtime compared to the old line detection algorithm by 20–50%.

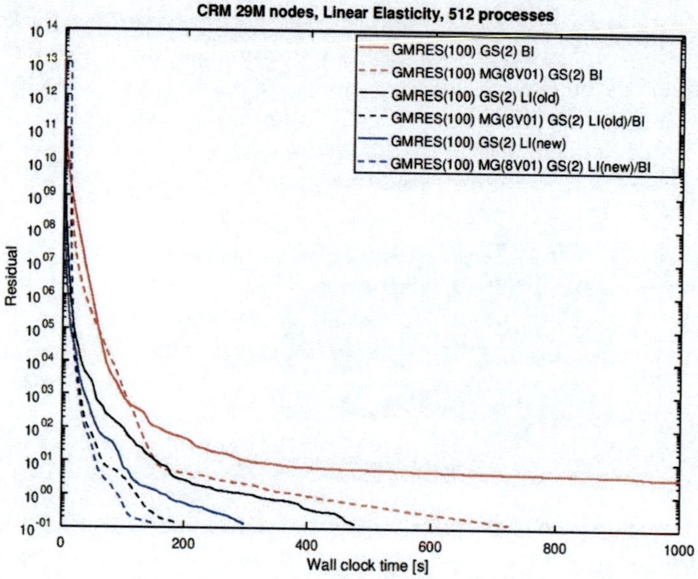

Fig. 5. Convergence time of different methods for the CRM test case.

4.2 ONERA M6 Wing

The second testcase is the ONERA M6 wing from NASA's Turbulence Modelling Resource. It is a 3D mesh with 61M nodes, 52M hexahedron and 17M prism cells, and aspect ratios of 18000:1 in the viscous boundary layer. A comparison of the convergence of the different methods for solving the linear equation system on 2048 processes in parallel can be found in Fig. 6. For the single level computations, both point-implicit and the line-implicit methods using the old line detection behave quite comparable, while the method using the new line detection is significantly faster.

For the multigrid methods, the additional time of setting up the transfer operators and coarse systems is more significant than for the CRM case, so at first we see no progress in the residual reduction at all. But after the setup phase, the multigrid methods surpass the single grid methods easily and show similar characteristics as for the CRM case: The method using line-implicit smoother on the finest grid saves about 70% of the time of point-implicit, and the new lines detection reduces the runtime by another 30% compared to the old lines detection.

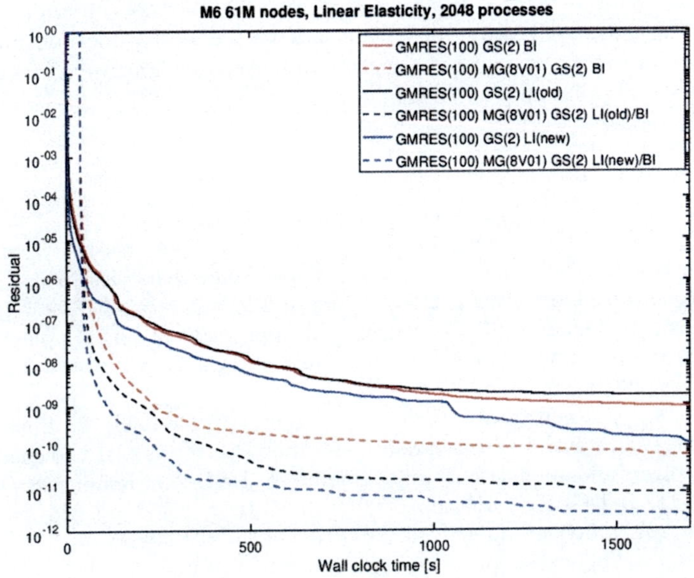

Fig. 6. Convergence time of different methods for M6 wing.

4.3 Comparison

Comparing both test cases, for M6 the gain by multigrid is much larger than that of line-implicit, compared to CRM. This is probably due to the fact that the partitioning did not take lines into account, and our line-implicit implementation is only able to handle lines that are fully available locally. Since the partitioner was not aware of the lines and the M6 test case contains fewer elements per domain than the CRM case, it is more likely that more lines were split by the partitioning, and thus the algorithmic strength is weakened. However, this explanation currently is just an assumption and needs to be confirmed in the future.

5 Conclusion

We have investigated the effectiveness of line-implicit methods for mesh deformation problems. In particular, these methods are very beneficial for CFD meshes with highly anisotropic cells in a viscous boundary layer. We have presented a new line identification algorithm that is particularly useful for node-based discretizations, and have found lines that lead to improved convergence rates when used in line-implicit solvers, reducing solution time by 20–50%. We also compared these methods to multigrid methods and combined them, which improves convergence as well, especially for large grids. The multigrid line-implicit method using the new line detection algorithm is a very effective method and leads to a 90–95% reduction in solution time for industry-relevant meshes compared to the single-level point-implicit method.

References

1. Dwight, R.P.: Robust mesh deformation using the linear elasticity equations. In: Deconinck, H., Dick, E. (eds.) Computational Fluid Dynamics 2006, pp. 401–406. Springer, Heidelberg (2009). https://doi.org/10.1007/978-3-540-92779-2_62
2. Eliasson, P., Weinerfelt, P., Nordström, J.: Application of a line-implicit scheme on stretched unstructured grids. In: Proceedings of the 47th AIAA Aerospace Sciences Meeting (2009)
3. Krzikalla, O., Rempke, A., Bleh, A., Wagner, M., Gerhold, T.: Spliss: a sparse linear system solver for transparent integration of emerging HPC technologies into CFD solvers and applications. In: Dillmann, A., Heller, G., Krämer, E., Wagner, C. (eds.) STAB/DGLR Symposium 2020. NNFMMD, vol. 151, pp. 635–645. Springer, Cham (2021). https://doi.org/10.1007/978-3-030-79561-0_60
4. Langer, S.: Application of a line implicit method to fully coupled system of equations for turbulent flow problems. Int. J. Comput. Fluid Dyn. **27**(3), 131–150 (2013)
5. Mavriplis, D.J.: Multigrid strategies for viscous flow solvers on anisotropic unstructured meshes. J. Comput. Phys. **145**, 141–165 (1998)
6. Ojha, V., Fidkowski, K.J., Cesnik, C.E.S.: Adaptive high-order fluid-structure interaction simulations with reduced mesh-motion errors. AIAA J. **59**(6), 2084–2101 (2021)
7. Okusanya, T.O.: Algebraic Multigrid for Stabilized Finite Element Discretizations of the Navier-Stokes Equations. PhD dissertation, M.I.T., Department of Aeronautics and Astronautics (2002)
8. Pierce, N.A., Giles, M.B., Jameson, A., Martinelli, L.: Accelerating three-dimensional Navier-Stokes calculations. In: 13th AIAA Computational Fluid Dynamics Conference, no. 1997–1953 in Conference Proceeding Series, AIAA (1997)
9. Rempke, A.: Netzdeformation mit Elastizitätsanalogie in multidisziplinärer FlowSimulator-Umgebung. In: Jahresbericht 2016 zum 20. DGLR-Fach-Symposium der STAB (2016)
10. Secco, N.R., et al.: Efficient mesh generation and deformation for aerodynamic shape optimization. AIAA J. **59**(4), 1151–1168 (2021)

11. Stein, K., Tezduyar, T., Benney, R.: Mesh moving techniques for fluid-structure interactions with large displacements. J. Appl. Mech. **70**(1), 58–63 (2003)
12. Swanson, R.C., Turkel, E., Rossow, C.-C.: Convergence Acceleration of Runge-Kutta Schemes for Solving the Navier-Stokes Equations. J. Comput. Phys. **224**, 365–388 (2006)
13. Thomas, L.H.: Elliptic problems in linear difference equations over a network, Watson Scientific Computing Laboratory Report, Columbia University (1949)

Numerical Simulation of Vortex-Dominated Flows Using Advanced Physical Models

Vamshi Togiti(✉)

DLR, Institute of Aerodynamics and Flow Technology, Branch: $C^2A^2S^2E$, 38108 Braunschweig, Germany
vamshi.togiti@dlr.de

Abstract. The capabilities of Reynolds stress model compared to eddy-viscosity RANS turbulence models and scale-resolving simulation methods for the flows characterized by separation from smooth surfaces and subsequent vortex formation are assessed. To this end, the flow over a delta wing and a diamond wing at the incidence angle of 13° and 12°, respectively, are investigated. Predictions obtained for the aforementioned configurations by different approaches are compared to each other and to experiments. In this paper, improvements and detriments observed in Reynolds stress model predictions compared to RANS predictions and to scale-resolving simulations are presented.

Keywords: Vortex flows · Reynolds stress model · Scale-resolving simulation

1 Introduction

Significant enhancements in computational fluid dynamics (CFD) have led to the use of numerical simulation in the design process of both civil and military aircraft, especially for nominal cruise performance where flows are generally characterized by attached flows. However, military vehicles routinely operate well outside of the steady, attached flow regimes often dominated by separated vortex flows. State-of-the-art CFD methods lack the ability to predict onset and progression of separated vortex-dominated flows, especially from smooth surfaces. Investigations carried out in several AVT task groups of the NATO Science and Technology Organization (STO) [1] indicate that the shortcomings for these predictions mainly stem from deficiencies of the models of turbulence. In most numerical investigations for slender wings with low sweep angle and blunt leading edges [2] linear eddy-viscosity turbulence models are employed. These models tend to deliver acceptable predictions at moderate incidence angles. However, at higher incidence angles, force coefficients and pitching moments were observed to be off compared to experiments. This is due to incorrect (earlier or delayed) prediction of incipient separation and differences in the strength of the subsequent development of vortices. The reason for the aforementioned failures of the

A. Dillmann et al. (Eds.): STAB/DGLR Symposium 2022, NNFM 154, pp. 284–293, 2024.
https://doi.org/10.1007/978-3-031-40482-5_27

linear eddy-viscosity models (EVM) can be attributed to the underlying Boussinesq approximation used to model the Reynolds stresses terms. In contrast to this, full differential Reynolds stress models (RSM) provide individual transport equations for the Reynolds stresses and hence they are considered particularly suitable for vortex-dominated flows. Investigations conducted in [3] demonstrate the applicability of RSM for vortex dominated flows.

Vortex separated flows often are mildly or massively separated and highly unsteady. Numerical investigations of such flows involving complex flow phenomena often may not be well predicted by advanced RANS models, including RSM, and may require scale-resolving simulations (SRS) in which a small portion of turbulence is modeled and the rest is resolved. However, such methods require more computational resources compared to pure RANS simulations.

In this work, the objective is to assess the capabilities of advanced turbulence models for the flows involving separation from a smooth body and subsequent vortex formation. To this end, first the capabilities of a Reynolds stress model for vortex separated flows are evaluated. Later, scale-resolving hybrid RANS/LES simulations based on eddy-viscosity-models are employed to the aforementioned flows and the advantage of applying such methods over Reynolds stress models is assessed.

2 Details of the Investigated Configurations

In the current study the flow over a delta wing, and a diamond wing are investigated. The flow over the delta wing is based on the experimental investigations carried out in the Vortex Flow Experiment-2 [4] in which a 65° swept delta wing configuration with different rounded and sharp leading was investigated. In the present study, the delta wing with a medium radius rounded leading edge of $r/c_{mac} = 0.15\%$ case at a freestream Reynolds number based on the mean aerodynamic chord (c_{mac}) of 3×10^6 and Mach number of 0.4 and the incidence angle of 13.3° is investigated.

The other case investigated is the flow over a diamond wing which is based on the experimental study conducted in the NATO/AVT-183 task group [5]. The geometry is a diamond wing with leading and trailing-edge sweeps of 53° and −26.5°, respectively. In the experiments, the geometry is mounted on a tunnel floor wall with a peniche with standoff distance of $0.075c_r$ where c_r is the root chord which is $1.2m$ and tests were conducted at a free stream Mach number of 0.15 and Reynolds number based on the mean aerodynamic chord of 2.7×10^6 and the incidence angle range of 10° to 15° with an increment of 1°. For the present numerical investigation, the incidence angle of $\alpha = 12°$ is chosen as the flow separates midway along the blunt leading edge.

3 Turbulence Modeling

Separated vortex flows investigated in the current study involve separation from a smooth body surface and the formation of vortices. Often the vortex interaction with the attached boundary layer exposed to an adverse-pressure gradient leads to secondary vortices where the anisotropy of normal Reynolds stresses persists. Over the years, extensions to eddy viscosity models have been introduced to improve the predictions of vortex flows and normal Reynolds stress anisotropy. In the present study, the Spalart-Allmaras one-equation [6] turbulence model coupled to a rotation/curvature correction (RC) [7] and the Quadratic Constitutive Relation (QCR) [8] which improve the prediction of the vortex strength and the anisotropy of normal Reynolds stresses is employed and in the discussion it is denoted as SA-RC-QCR. To assess scale-resolving simulation (SRS) methods, the SA-RC-QCR-based delayed detached eddy simulation (DDES) [9] is considered. In DDES, the model switches from RANS to LES mode based on the modified length scale $l_{DDES} = l_{RANS} - f_d \max(0, l_{RANS} - l_{LES})$, where $l_{RANS} = d_w$ is the nearest wall distance and $l_{LES} = C_{DES}\,\Delta$, where $C_{DES} = 0.65$ is the models constant and Δ is the maximum grid spacing in all three directions. Here f_d is the delay function which avoids early switching of model to LES mode inside boundary layer on very fine grids. The delay function f_d is defined as

$$f_d = 1 - tanh([C r_d{}^m]), r_d = \frac{\nu_t + \nu}{\kappa^2 d^2 \sqrt{U_{i,j} U_{i,j}}} \tag{1}$$

where κ is von Karman constant and $U_{i,j}$ velocity gradient tensor. For a SA based DDES, the multiplier $C = 8$ and exponent $m = 3$ in f_d function are used. In the current study, the same values are employed for the SA-RC-QCR based DDES. A recalibration of the C and m may be required for the SA-RC-QCR based DDES. However, in this study such investigation has not been conducted. In the discussion the SA-RC-QCR based DDES predictions are denoted as SA-RC-QCR-DDES.

Concerning Reynolds stress model (RSM), often the application of differential RSMs to flows involving complex flow phenomena leads to numerical stability problems due to numerical stiffness. In the present work, the SSG/LRR-RSM [10] coupled to $\ln(\omega)$ length-scale equation [11] which improves the stability of RSM for complex flows is employed and in the discussion it is denoted as RSM-$\ln(\omega)$.

4 Numerical Methods and Grids

In the current investigation, the unstructured compressible DLR flow solver TAU is used. Details about the DLR TAU code can be found in [12]. The inviscid fluxes of the main-flow equations are calculated by a central scheme with matrix artificial dissipation. For the turbulence equations, the convective fluxes are approximated with a second-order Roe scheme in the RSM investigations and an Average-of-flux scheme in SA-RC-QCR and SA-RC-QCR-based DDES investigations. The viscous fluxes of the main flow and the diffusion fluxes of the turbulence equations are discretized using central differences.

For the DDES computations, low numerical dissipation is required in order to avoid excessive damping of resolved turbulent structures. In the current work, for the SA-RC-QCR-DDES investigations the hybrid low-dissipation/low-dispersion (LD2) scheme is used, details can be found in [13].

For RANS investigations, steady computations are performed using a semi-implicit lower-upper symmetric Gauss-Seidel (LUSGS) method with a 2v multi-grid cycle. However, with the RSM oscillatory force coefficients are observed in steady computations and hence time-accurate computations are carried out with single grid. For SRS and unsteady RSM computations, time-accurate computations are carried out with a 2nd-order dual time-stepping scheme. In RSM computations, a non-dimensional time step size of 0.01, which is about 1/100th CTU (CTU: convective time unit calculated from wing root chord length and reference velocity; c_r/U_{ref}) is used while for SRS investigations, a non-dimensional timestep of 0.0005 is employed. During the inner iterations of the dual-time stepping, the inner iterations are stopped based on Cauchy-convergence criteria for lift and drag–in the present investigations, it needed around 40 single grid iterations per time step. In the present study, hybrid grids consisting of prisms in the near-wall region and tetrahedrons in the remainder of the computational domain are employed. The grids employed here are the finest grids obtained from a grid refinement study conducted for both configurations. The grid used for delta wing investigation consists of 44 million grid nodes while the grid for the diamond wing has about 43 million grid nodes. In the current study, the grid employed in the SRS investigations is used in the RANS investigations. An overview of the grid is displayed in Fig. 1.

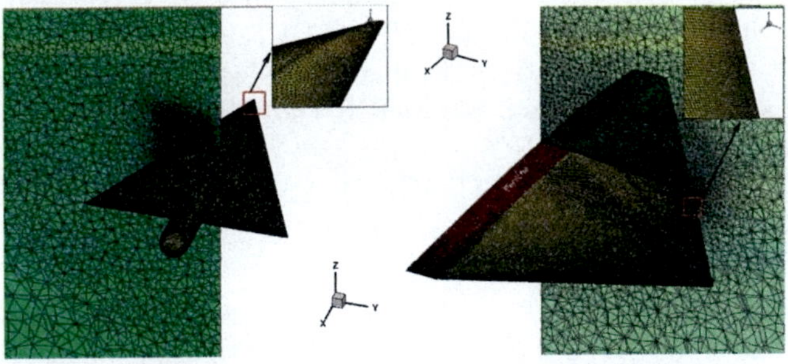

Fig. 1. Overview of grids (resolution coarsened by factor 2 for visualization). Left: delta wing, right: diamond wing.

5 Results

In this section, the predictions obtained for the flow over a delta wing and diamond wing using different turbulence modeling approaches are presented.

5.1 Delta Wing

To evaluate the turbulence model capabilities, investigations are carried out with the RSM, SA-RC-QCR, and SA-RC-QCR-DDES. In Fig. 2 the iso-surface of λ_2

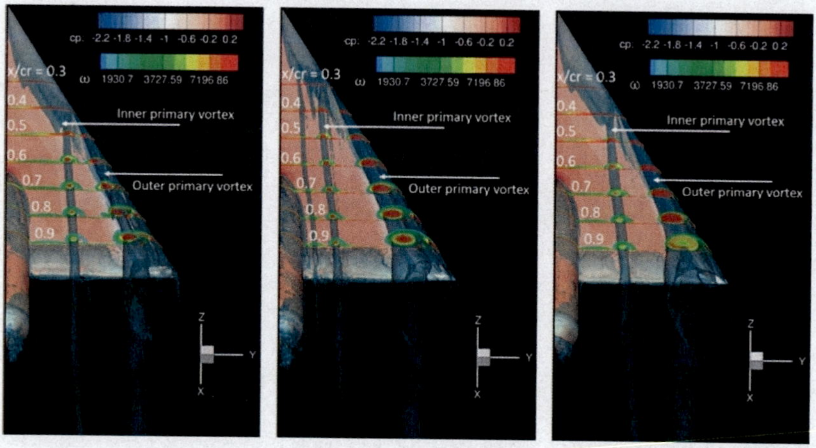

Fig. 2. Iso-surface of $\lambda_2 = -10[-]$ and the contours of vorticity magnitude delivered by RSM(left), SA-RC-QCR-DDES(center), and SA-RC-QCR(right) for the delta wing.

along with the vorticity contours at several streamwise locations are shown to unveil the resolved vortices. As shown in the figure, in the RSM predictions the flow separates from the leading edge at about $x/c_r = 0.4$, and forms inner and outer primary vortices. In the SA-RC-QCR predictions, flow separation from leading edge and formation of inner and outer primary vortices occur at about $x/c_r = 0.35$ while in the SA-RC-QCR-DDES predictions the separation is predicted upstream of the SA-RC-QCR location, at about $x/c_r = 0.3$. The reason for the earlier separation in SA-RC-QCR-DDES compared to the corresponding RANS model predictions is the modeled stress depletion which is caused by earlier switch of the model to LES mode with in boundary layer which lead to lower turbulent stresses and that caused earlier separation. The possible reason is the delay function f_d used in DDES in which the multiplier and exponent from SA-DDES formulation are employed. To extend the RANS region within the boundary layer, the multiplier and exponent in f_d need to be recalibrated for the SA-RC-QCR-DDES. In the experiments the flow separation and formation of vortices are observed at about $x/c_r = 0.48$. Among the applied models, the predicted separation location from RSM is closer to the experiments.

To further assess the performance of the applied turbulence models, sectional c_p distributions delivered by the RSM, SA-RC-QCR-DDES and SA-RC-QCR are compared to the experimental data in Fig. 3. At $x/c_r = 0.2$, all the models predict attached flow and deliver almost identical pressure distribution and agree well with the experiments. At $x/c_r = 0.4$, the flow is still attached in the experiments. This trend is reproduced in the RSM predictions. However, in the case of SA-RC-QCR and SA-RC-QCR-DDES, the flow separation occurs upstream of this location and subsequent vortex formation begins. As a result, low pressure in the vortex core is exerted on the surface which is rendered in the c_p

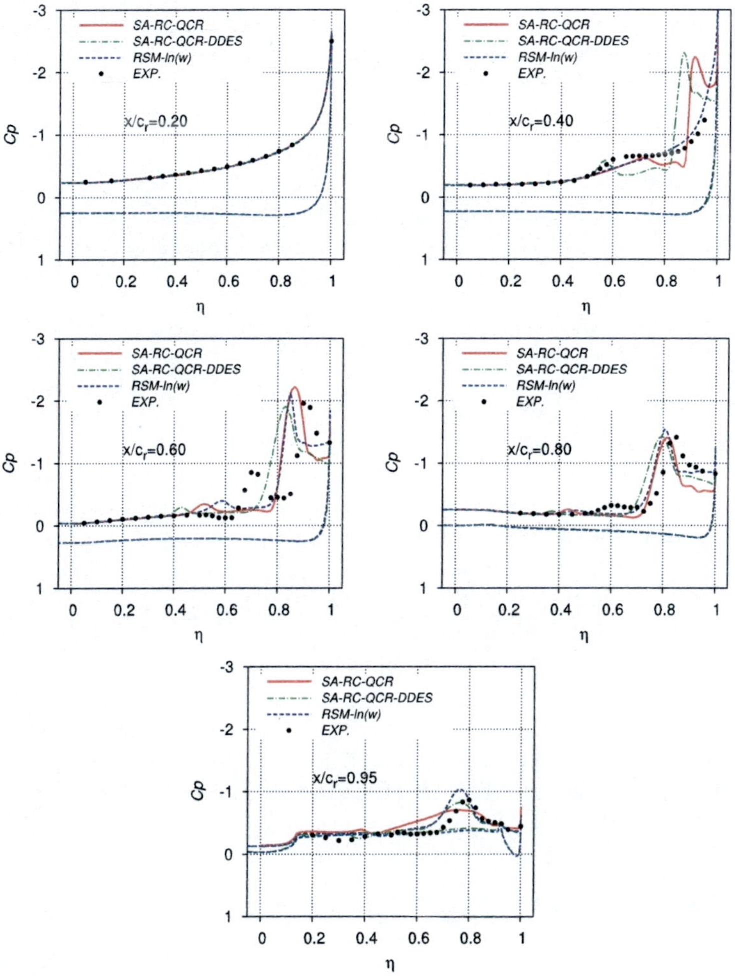

Fig. 3. Sectional c_p on the delta wing.

spike at $\eta = 0.8$ to 0.9. In the SA-RC-QCR-DDES, the is spike is located more inboard compared to the SA-RC-QCR due to earlier separation. At $x/c_r = 0.6$ and 0.8, the local suction peaks in the experimental pressure distribution indicate the location and associated pressure distribution of the inner primary and outer primary vortices. Low pressure associated with the outer primary vortex is predicted well with the SA-RC-QCR, SA-RC-QCR-DDES and RSM, however, the vortex is located too far inboard in the predictions. In the experiments, the inner primary vortex location and the associated pressure is rendered with the local $c_{p,min}$ in the region $\eta = 0.6$ to 0.8. With all the applied models the pressure associated with the primary vortex is underpredicted. Concerning the location, both the SA-RC-QCR and SARC-QCR-DDES predict the vortex too far inboard while RSM delivers the location which is in between experiments and the other predictions and, thus, closer to the experiments. At $x/c_r = 0.95$, in the SA-RC-QCR predictions an almost flat c_p-distribution is delivered due to the weaker outer primary vortex while in the RSM and SA-RC-QCR-DDES predictions the c_p associated with the outer primary vortex is reproduced as in the experiments.

Overall, the RSM shows the tendency to represent the flow better than the other methods employed by delivering separation locations closer to experiments and stronger outer and inner primary vortices.

5.2 Diamond Wing

Flow separation from the leading edge and subsequent vortex formation on the diamond wing is investigated using the RSM, SA-RC-QCR and SA-RC-QCR-DDES approaches. To visualize the predicted vortices on the suction side of the wing, the iso-surface of λ_2 and vorticity contours at several streamwise sections are shown in Fig. 4. As can be seen in the RSM predictions the flow separation and subsequent vortex formation occurs at about $x/c_r = 0.3$ while in the other predictions it occurs in between streamwise stations of $x/c_r = 0.2$ and 0.3. Similar to the delta wing predictions, in the SA-RC-QCR-DDES predictions the flow separation occurs earlier than the SA-RC-QCR predictions. The reason for this trend is again attributed to the modeled stress depletion caused by the delay function f_d. As mentioned earlier, a recalibration of the multiplier and the exponent in f_d for the SA-RC-QCR-DDES may improve the separation prediction. On the inboard section of the wing, similar to the experiments, an inner primary vortex is delivered in all the predictions. From the contours of the

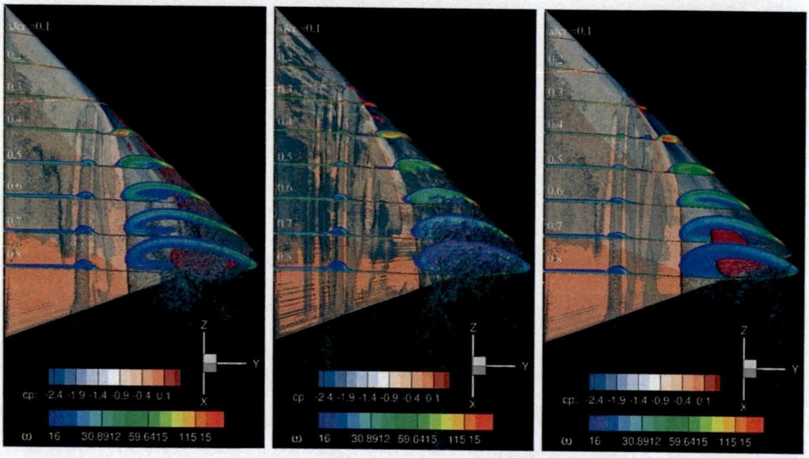

Fig. 4. Iso-surface of $\lambda_2 = -1[-]$ and the contours of vorticity magnitude delivered by RSM(left), SA-RC-QCR-DDES(center), and SA-RC-QCR(right) for the diamond wing.

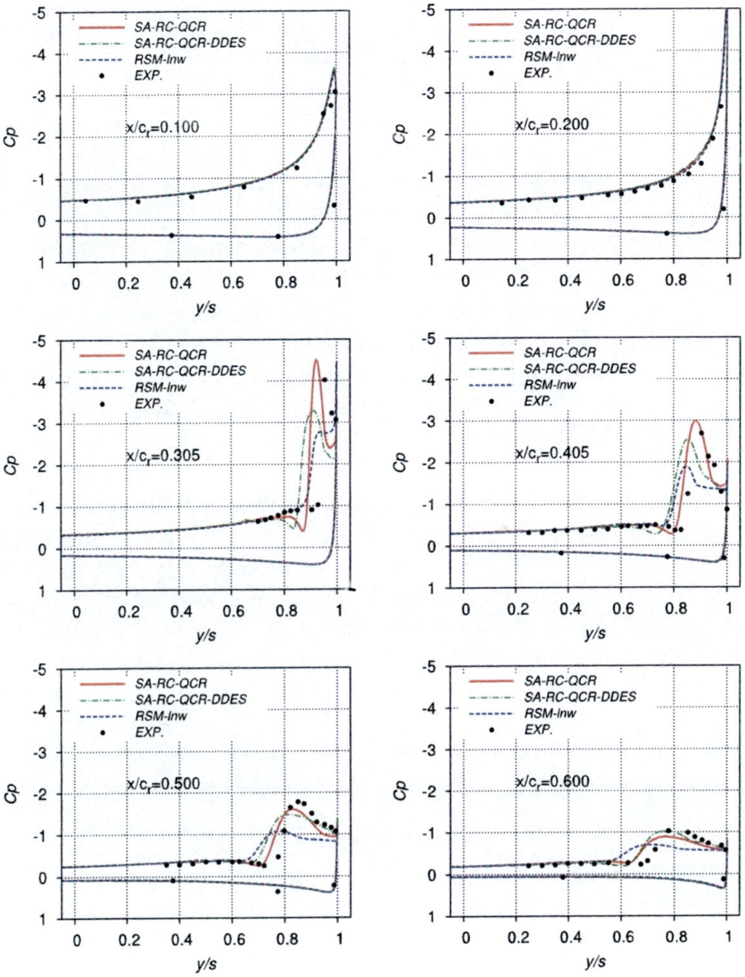

Fig. 5. Sectional pressure on diamond wing.

vorticity magnitude it is apparent that the predicted outer primary vortex in RSM is weaker than the other predictions.

To assess the effect of the vortex strength on the flow filed, the comparison of the sectional pressure distribution is displayed in Fig. 5. At the streamwise sections of $x/c_r = 0.1$ and 0.2 the experiments show the trend of an attached flow. This trend is reproduced in the predictions with all the turbulence models applied. At $x/c_r = 0.305$, in the experiments the inboard movement of the suction peak from the leading edge indicates flow separation and subsequent vortex formation. The location of the suction peak also indicates the vortex axis. As can be seen in Fig. 5(middle left), the RSM predicts attached flow at this location while the SA-RC-QCR and SA-RC-QCR-DDES deliver the experimental trend, however, the predicted vortex axis is more inboard compared to the experiments which is because of the earlier vortex separation than in the experiments. At $x/c_r = 0.405$, in the experiments a lower suction peak compared to the previous location is observed. This is because with the downstream distance the size of the vortex grows and its strength decreases. The location of $c_{p,min}$ indicates that the vortex core axis moves inboard with the downstream distance. At this streamwise station the predictions of the SA-RC-QCR are in good agreement with experimental data with regard to the vortex position and the c_p level associated with it. The SA-RC-QCR-DDES predicts the vortex more inboard and delivers lower c_p compared to the experiments as a weaker vortex is predicted. In RSM the position of the vortex is predicted more inboard and delivers lower c_p compared to the experiments. At farther downstream locations the vortex core axis and the c_p levels are well predicted by the SA-RC-QCR. Both the SA-RC-QCR-DDES and RSM deliver the vortex more inboard and lower c_p compared to the experiments.

Overall, for this test case, the SA-RC-QCR shows good agreement with experiments with respect to the pressure associated with the outer primary vortex and its location. Concerning the RSM, the downstream development of the vortex is underpredicted. With the SA-RC-QCR-DDES, improvements in the predictions compared to the base RANS model are not observed due to early switch of the model to LES mode with in the boundary layer that lead to lower turbulent stresses (modeled stress depletion) compared to RANS which caused earlier flow separation.

6 Summary

The predictive capabilities of the RSM are assessed by comparing the predictions to the predictions delivered by the SA-RC-QCR, SA-RC-QCR-DDES and experimental data for the flow over a delta wing and a diamond wing with round leading edge.

For the delta wing, among the applied models, RSM predicted the flow better by delivering the flow separation location closer to the experiments and stronger outer and inner primary vortices. While for the diamond wing, the SA-RC-QCR predicted the flow better than the other models by delivering the pressure associated with the outer primary vortex and its location in line with the experiments.

Concerning the applied DDES model, for both flow cases, improvements in the predictions compared to the underlying RANS model are not observed due to earlier flow separation compared to the base RANS model. This is due to the early switch of model to LES mode with in the boundary layer, which led to lower turbulent stresses (modeled stress depletion) and that caused earlier flow separation compared to the underlying RANS model predictions. To improve the predictions of SA-RC-QCR-DDES, a recalibration of multiplier and exponent in the delay function f_d may be needed.

Acknowledgements. The research presented in this report is funded by the Air Force office of Scientific Research (AFOSR), Award No. FA9550-18-1-7019. The funding is gratefully acknowledged. The author would like to thank A. Krumbein and A. Probst of DLR Goettingen for managing activities conducted in the AFOSR grant and reviewing the manuscript.

References

1. Frink, N., Tomac, M., Rizzi, A.: Collaborative study of incipient separation on 53° swept diamond wing. Aerosp. Sci. Technol. **57**, 76–89 (2016)
2. Ghoreyshi, M., Ryszka, K., Cummings, R., Lofthouse, A.: Vortical flow prediction of a diamond wing with rounded leading edges. Aerosp. Sci. Technol. **57**, 103–117 (2016)
3. Togiti, V., Ciobaca, V., Eisfeld, B., Knopp, T.: Numerical simulation of steady blowing active flow control using a differential reynolds stress model. In: CEAS/KATnet II, Bremen, Germany (2009)
4. Hummel, D., Redeker, G.: A new vortex flow experiment for computer code validation. In: RTO-AVT Symposium on Vortex Flow and High Angle of Attack, Loen, Norway, 7–11 May 2001 (2001)
5. Hoevelmann, A., Grawunder, M., Buzica, A., Breitsamter, C.: AVT-183 diamond wing flow field characteristics Part 2: experimental analysis of leading-edge vortex formation and progression. Aerosp. Sci. Technol. **57**, 31–42 (2016)
6. Spalart, P.R., Allmaras, S.R.: A one-equation turbulence model for aerodynamic flows. Recherche Aerospatiale (1994)
7. Shur, M.L., Strelets, M.K., Travin, A.K., Spalart, P.R.: Turbulence modeling in rotating and curved channels: assessing the Spalart-Shur correction. AIAA J. **38**(5), 784–792 (2000)
8. Mani, M., Babcock, D.A., Winkler, C.M., Spalart, P.R.: Predictions of a supersonic turbulent flow in a square duct. In: AIAA Paper 2013, p. 860 (2013)
9. Spalart, P., Deck, S., Shur, M., Squires, K., Strelets, M.K., Travin, A.: A new version of detached-eddy simulation, resistant to ambiguous grid densities. Theor. Comput. Fluid Dyn. **0935–4964**, 181–195 (2006)
10. Eisfeld, B., Rumsey, C., Togiti, V.: Verification and validation of a second-moment-closure model. AIAA J. **54**(5), 1524–1541 (2016)
11. Braun, S.: Implementation of a ln(w) based SSG/LRR reynolds stress model into the DLR-TAU code. Institutsbericht, DLR-IB-AS-BS-2019-37, Report of the Institute of Aerodynamics and Flow Technology, Braunschweig (2019)
12. Schwamborn, D., Gerhold, T., Heinrich, R.: The DLR TAU-code: recent applications in research and industry. In: Europen Conference on CFD, ECCOMAS CFD (2006)
13. Probst, A., Lowe, J., Reuß, S., Knopp, T., Kessler, R.: Scale-resolving simulations with a low-dissipation low-dispersion second-order scheme for unstructured flow solvers. AIAA J. **54**(10), 2972–2987 (2016)

Preliminary CFD Validation of Unsteady Onflow Effects on Model Train Drag

Keith A. Weinman(✉) and Klaus Ehrenfried

DLR Göttingen, Institute of Aerodynamics and Flow Technology, Bunsenstr. 10, 37073 Göttingen, Germany
keith.weinman@dlr.de

Abstract. The DLR New Generation Train Cargo concept (NGT-Cargo) aims to increase the rail share of the European freight traffic market. Computational Fluid Dynamics (CFD) is a useful tool in analyzing high-speed operations which are an important part of this concept. The development of aerodynamic forces under unsteady on-flows should be analyzed within the context of industrial standards, for example guidelines are provided for CFD assessments at cross-wind conditions using the RANS (Reynolds-averaged Navier-Stokes) equations. A major challenge arises because the multi-scale nature of these flows are characterized by a large range of energetically significant flow scales. Wind-tunnel measurements, focused on the development of the aerodynamic forces acting on a train, are being currently performed on a 1:25 scaled NGT model in the cross-wind facility *Seitenwindversuchsanlage* Göttingen (SWG) located at the DLR Göttingen. This paper presents preliminary work assessing the ability of CFD to reproduce both the observed on-flow conditions as well as the aerodynamic drag of the vehicle.

1 Introduction

Stringent safety requirements over a wide range of operational conditions are applied to modern high-speed trains. Achieving this requires an understanding of the aerodynamic forces under cross-wind conditions. Measurement of force coefficients for full-scale vehicles is optimal but it is expensive. Normal practice is geared towards the use of small-scale models that can be tested inexpensively in wind-tunnel experiments or by using full-scale in-service vehicles [1]. Vehicle stability is of paramount importance, but accurate assessments of vehicle aerodynamic forces under realistic atmospheric conditions are now becoming relevant. The use of computational methods to assess the aerodynamic loading on trains has been recognized by the transport industry. The German standard EN 14067-6 [3] permits evaluation of aerodynamic forces for cross-wind conditions using CFD simulations for full-scale or reduced model geometries. Guidelines in EN 14067-6 for Reynolds Averaged Navier-Stokes (RANS) methods are stringent: computed integral forces cannot be accepted for certification work if variations against an accepted reference value differs by more than three percent. Computational assessments of the flow about a train, traditionally undertaken on the

A. Dillmann et al. (Eds.): STAB/DGLR Symposium 2022, NNFM 154, pp. 294–304, 2024.
https://doi.org/10.1007/978-3-031-40482-5_28

basis of RANS methods, have not been satisfactory. For example [10] and [11] compared computational estimates of integral forces and moments using well-resolved meshes against the NGT2 experiment of [12]. Computations were performed for Reynolds numbers in the range $R \in [250000, 750000]$ with cross-wind conditions of up to 30°. Computed aerodynamic force and moment coefficients, particularly drag, demonstrated differences against experimental measurements of up to 15%. Improvements in vehicle force moment estimations have been demonstrated with improvements in physical modeling [8,11,13], however drag estimates remain unsatisfactory. A significant challenge is provided by the multi-scale nature of flows characterized by a large range of energetically significant length scales. Geometrical features of a train, such as the underflow region, the train base [4,5], the inter-car gaps (ICG's) and bogie cavities, generate small-scale unsteady flow structures which interact with larger flow scales and can thereby influence the development of the aerodynamic forces. [4] observed that traditional RANS methods will retain their importance for the foreseeable future. This observation provided the motivation for a study on meshing requirements for RANS methods [9]. This study identifies the front nose of the vehicle as a critical refinement area with other regions being of secondary importance. Minor modifications of these recommendations, with the aim of extending them to unsteady flows, are made in the present paper.

Vehicle aerodynamic loading under unsteady conditions were studied experimentally in [14] and [15]. Measurements were collected in-field under real conditions. Wind-tunnel experiments using a flapped system to control model on-flow length and time scales were also undertaken. Two CFD methods were developed and validated against both of these experimental approaches [16]. The first method uses mesh free reconstruction methods to construct on-flow boundary conditions that are representative of measured flow length and time scales. A second method, upon which the work in this paper is based, used sliding mesh techniques to model a set of oscillating flaps. The paper is organized as follows: A brief overview of the train model is presented in Sect. 2. The numerical method is discussed in Sect. 3. A short overview of the mesh is then introduced in Sect. 4. This is following by a discussion on the comparison of the computational and experimental results. Conclusions are presented in Sect. 6 respectively.

2 Details on the Train Model and Experiment

For this paper a model consisting only of the main train aerodynamic surfaces and the wind tunnel is considered. Features, such as the inter-car gaps and bogey cavities, are also included. Figure 1(a) illustrates the computational model. The main components include the model, the wind tunnel (with nozzle, test section, and diffuser), as well as the room within which the wind tunnel is located. The origin of the coordinate system used both for the experiment and the CFD is illustrated at the start of the test section at $\mathbf{x} = (\mathbf{0}, \mathbf{0}, \mathbf{0})$). Figure 1(b) illustrates the flapped system used to control on-flow conditions. The leading edge of the wing is located at $\mathbf{x} = (\mathbf{0.97}, \mathbf{0}, \mathbf{0})$). The flapped system introduces an onflow

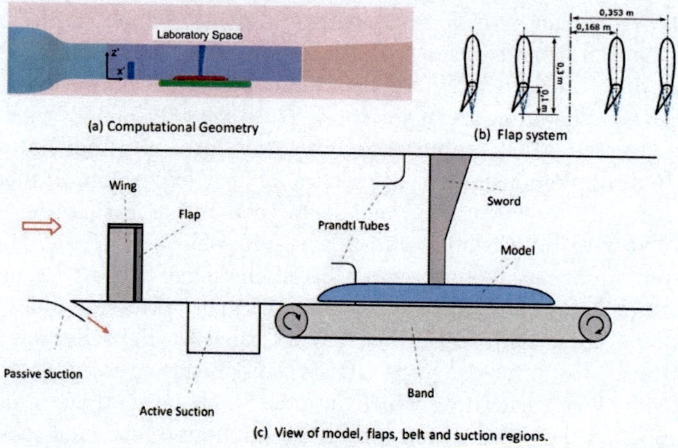

Fig. 1. (a) The computational domain in the x-z symmetry plane of the wind tunnel. The wind tunnel co-ordinate system, shown at the start of the test section, is used for both experiment and CFD. (b) The flap system is aligned with the tunnel symmetry axis. (c) Components of the experimental setup are illustrated.

condition represented by controlled oscillations about the mean train velocity. For the present work, the term "stationary flap" indicates a non-oscillating flap fixed in at the minimum drag (or neutral) position. The flap is described using the parameters of frequency (30 Hz.) and amplitude (3°) of oscillation. Figure 1(c) illustrates the model positioned in the wind tunnel test section over the moving belt. The train nose is positioned at $\mathbf{x} = (\mathbf{0.97}, \mathbf{0}, \mathbf{0})$. The moving belt velocity matches the bulk on-flow velocity which simulates an inertial reference system moving at this velocity. The model is an 1:25 scale NGT-Cargo train with a center wagon and two end wagons. The length of the model is 2.58 m. The model train height (L_h) is 200 mm. The train base has a width (L_b) of 125 mm. The reference area of the model (given by the projected area of the train head in the mean flow direction) is 0.02 m^2. Passive/active suction removes the boundary layer approaching the flaps and model and allows the experiment to better match a moving reference system with no mean oncoming wind. A passive ventilation system, driven by the local pressure differential between the wind tunnel interior and exterior, is located upstream of the flap system. An additional active system, driven by a compressor, is located in front of the moving belt. Suction removes the oncoming boundary layer which is consistent with an inertial reference frame moving at the train velocity. The head of the static probe shown in the figure are 400 mm above the tunnel floor or the belt. The definition of the characteristic length for the Reynolds number typically uses a reference width of 3 m at full scale in Europe. The scaled reference width provides the reference length used to compute the Reynolds number for the wind-tunnel experiment. The Reynolds number is defined as

$$R = \rho L_b U / \mu, \tag{1}$$

where U, ρ, μ, and L_b are the free stream velocity, the fluid density, the fluid dynamic viscosity and the Reynolds length scale respectively. For this work the Reynolds number, which is based on L_b, is 200,000 for the selected test section bulk velocity of 30 m/s (U_{ref}). Both the reference pressure and temperature varies slightly about 101325 Pa. and 298.5 °K respectively. Table 1 shows the set of experimental configurations selected from the experimental program for comparison against the CFD. A complete description of the measurement system used in the experiments is not within the scope of this paper but can be found in [15] and references therein. The wind tunnel turbulence level is approximately 0.15%. Spatial and temporal variations of the test section velocity profile are less than 0.5% with errors in both steady and unsteady force measurements being comparable. The pressure gradient across the head and tail of the model is of the order of three Pascals per meter. A description of the measured test conditions against which the CFD has been compared can be seen in Table 1.

Table 1. Test conditions against which CFD results are compared. Measured drag counts for the model (dcnts) are provided. At present only the RMS drag count for experiment 0047 is available. All experiments (except 0018) use the moving belt.

Experiment	Passive Suction	Active Suction	Flaps/Freq. (Hz.)	dcnts	rms(dcnts)
0018	Off	Off	No/NA	43	-
0020	Off	On	No/NA	47	-
0021	On	Off	No/NA	45	-
0022	On	On	No/NA	48	-
0023	On	Off	Yes/0	44	-
0024	On	On	Yes/0	49	-
0047	On	On	Yes/30	50	3

3 The Numerical Method

The current work uses the Engys OpenFOAM releases 3.5.0/3.5.1. Justification for use of the incompressible form of the Navier-Stokes equations is provided by the Mach number of the wind-tunnel flow (M = 0.09). Turbulence effects are conventionally modeled using an effective viscosity model [18]. For RANS solutions the Semi Implicit Method for Pressure Linked Equations (SIMPLE) [17] algorithm is used. Unsteady calculations use the Pressure Implicit Method with Splitting of Operators (PISO) [22]. The Courant-Friedrichs-Lewy (CFL) number is limited to O(1), leading to time steps of the order 10^{-6} s for unsteady calculations. The discrete operators are second order in both space and time. Steady solutions are terminated when the residuals of the pressure correction equation, momentum equations, and turbulent viscosity equations are of O(1e-10, 1e-8, 1e-7) respectively. This convergence criterion is also used for the unsteady calculations however an additional check is used to assess when the averaged drag vector variation lies within a 99% confidence interval (CI) about an error of

1.0e-6 drag units. Based on the comparative performance of other turbulence models for similar problems [9–11] the baseline turbulence model chosen is the Menter k-ω SST turbulence model [19]. The hybrid RANS-LES calculations use the Delayed Detached Eddy Simulation (DDES) version of this turbulence model [20]. The DDES filter width chosen as 2Δ [23], where Δ is the cube root of the cell volume. Approximately 60–70% of the total turbulent kinetic energy is resolved. The intention is to reduce dissipation effects inside vortex cores and free-shear layers without impacting on computational costs. A blended form of the inviscid flux operator is used to assist in reducing numerical dissipation effects for hybrid RANS-LES. Boundary conditions are specified in [9]. In this paper a drag count (dcnt) is defined as $1000C_d$ where C_d is the drag force normalized by the product of the reference dynamic pressure and the reference area. Wall clock time for these calculations scale at about 5.0e-4 s per grid point per iteration on the DLR CARO cluster.

4 The Computational Mesh

It is considered that the presence of significant vortical and separated flow regions about the train (under-body, inter-car gaps) require additional refinement and the train model was embedded inside a high resolution refinement block. The refinement block extends from $6L_b$ upstream of the train head to $30L_b$ downstream of the train tail respectively. The height of the refinement block is set at $2L_h$. Surface length scales are of O(2) mm, while control volume length scales vary from 2 mm near the vehicle to 15 mm inside the wind tunnel space. Wake regions for the flaps are resolved to a length scale of 3 mm up and including the refinement block. Specification of the mesh near-wall resolution (in terms of y+) follows [9] however y+ was reduced by a factor of two and the mean value of y+ about the model was set to y+ 16. The concept of an optimum y+ implies that near wall numerical solutions are representative of near wall physics. The wall function models used in this work replicate zero and mild pressure gradient boundary layer behavior sufficiently well but are expected to deviate from experiment for unsteady flows with stronger unsteady adverse pressure gradient and streamline curvature effects. Investigation of a universal wall function approach is not within the scope of this investigation however these calculations fall within the envelope of published industrial applications.

5 Assessment of the Computed Flow Fields

Current CFD assessments of the integral forces are given in Table 2. This table is divided into 6 groups with first 5 groups representing the CFD cases computed, while the last presents data derived from the experiments. Data sets (m1, m2, ...) are denoted here as M(m1, m2, ...). By comparing M(0018,0020) it can be seen that active suction with a moving belt increases the measured model drag by 4 dcnts. This observation is consistent with M[0023,0024], which includes the additional influence of a stationary flap system. The active suction increases the

Table 2. Comparison of the drag force estimates returned by the CFD.

Case	CFD Method	Matching Experiment	dcnts	rms(dcnts)
A	RANS	0018	52	-
B	RANS	0020	56	-
C	URANS	0024	69	1
D	DDES	0024	57	1
E	DDES	0047	59	4

model drag by 5 dcnts. The mean drag count differences between M[0018,0021] and M[0020, 0022] suggest that passive suction increases the model drag count by 1.5 dcnts on average. Application of passive and active suction could introduce changes of up to 6.5 dcnts. Comparison of M(0018,0020) and M(0023,0024) suggests that the moving belt increases the drag by 4 to 5 dcnts. M[0021,0023] and M[0022,0024] indicate that the stationary flap system reduces drag by 1 dcnt. This is expected since the flap generated wake fields which partially block the model's onflow (see Figs. 2 and 3). Differences in drag for M[0021,0023] shows an reduction of 1 drag count due to the presence of the stationary flaps, however activation of the moving belt for M[22,24] increases the drag count by one for this configuration. The moving belt and passive/active suction mechanisms remove the oncoming boundary layer so that the modeled drag coefficient will increase - this is seen in the experimental data. Neglecting the passive/active suction mechanisms can introduce an error of about 6 dcnts with an additional error of up to 5 if no moving belt is present.

The CFD results are compared against the EN 14067-6 standard (despite the fact that the standard is accepted for RANS only). The experimental values are considered as reference values for the remainder of this paper since they are the only independent source for comparison against the CFD. The percentage difference of the CFD against a reference value is presented together with the drag count as a data pair (represented by E[dcnts, percentage difference]). Calculation A is a RANS calculation based on the Menter-SST turbulence model. No boundary layer removal mechanisms are modeled. Differences for the pair M[A,0018] are given by E[9,20]. Calculation B uses a RANS method however the moving belt system is now modeled. Differences between [B,0020] are given by E[9,19]. This is comparable to the difference between M[A,0018]. However drag differences between M[A,B] and M[0018,0020] are 4 dcnts respectively. Groups (C) and (D) evaluate the effect of an improved physical models with the moving belt included together with the flaps fixed in the neutral position. Comparison of M[C,0024] returns E(22,47) for the URANS result while the comparison M[D,0024] returns E[8,16] for the DDES method. This result is consistent with other studies which suggest that flow resolving methods should be obligatory for these calculations. In further discussions non-resolved calculations (A,B,C) are discarded. M[E,0047] returns E(9,18) while the differences between M[D,E] and M[0024, 0047] are two dcnts and one dcnt respectively. This illustrates a typical observation in industry: CFD results are usually more accurate in estimating drag deltas across design changes then when comparing a drag assessment at a single design condition. Industrial flows are highly complex. The simplifications required to make cost effective numerical methods for these flows can introduce systematic errors. These errors appear to cancel when differences between solutions (based on identical numerical schemes) are computed. The observed error in neglecting both passive and active suction (6.5 dcnts) suggests that, if these are included in the CFD model, the drag count differences between CFD and experiment could be reduced to less than 4 dcnts with the certification standard used in this paper being achieved.

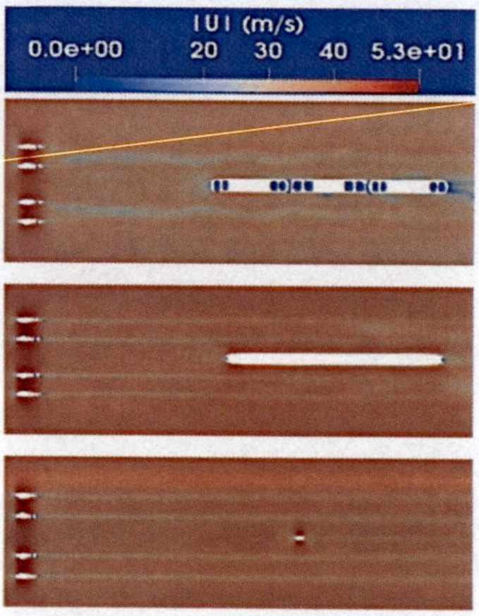

Fig. 2. In the figure the wake development behind both the stationary flap system is shown. Upper figure: $z/L_h = 0.2$; Middle figure: $z/L_h = 0.94$; Upper figure: $z/L_h = 2.4$.

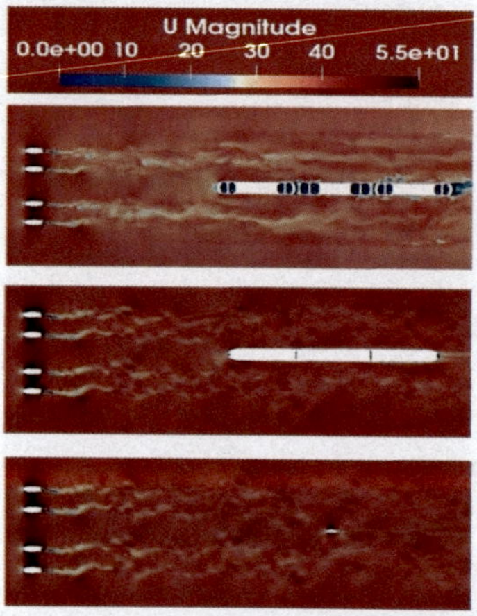

Fig. 3. In the figure the wake development behind both the oscillating flap system is shown. Upper figure: $z/L_h = 0.2$; Middle figure: $z/L_h = 0.94$; Upper figure: $z/L_h = 2.4$.

Figure 2 illustrates the wake development downstream of the stationary flap at three different values of height above the wind-tunnel floor. Close to the wind tunnel floor the twin wakes, seen to the left and right of the symmetry axis, merge. The model is then bounded by two merged wakes to the left and right of the model's symmetry axis. As distance from the floor is increased the wake interactions with the model reduce until the wake travels undisturbed through the test section. The situation changes when the flaps oscillate as can be seen in Fig. 3 where significant interaction between the wakes and the model are seen. This figure shows that the left and right wake pairs have not merged into two pairs. Investigations of the flow field at other time instances suggest that wake merging is time dependent which adds additional disturbances to the onflow condition. Both figures show evidence of interaction between the flow and ICG's and under-body.

Figure 4(a) shows the computed power spectral densities for the computed drag force components (case D). The spectra are averaged over 30 convective time units with a low pass Hanning filter being applied. Similar post-processing is also applied to the experimental data. At the time of writing the measured force time series for cases with the steady flaps requires additional post-processing and direct comparison of measured and computed spectra will be provided in later work. All force components exhibit a peak 3 Hz (St = 0.007) with indications of higher harmonics of up 9 Hz (St = 0.07). F_z shows a peak at 25 Hz (St = 0.12) while F_y contains two modes at about 12 (St = 0.06) 45 Hz (St = 0.23). T25 Hz peak is believed to relate to the moving belt. A timescale derived from the model length and U_{ref} matches 12 Hz (St = 0.06) mode implying that this mode is flow related. however this mode matches the rotational Eigenfrequency of the model-spear combination suggesting coupled flow-structure interaction. T45 Hz (St = 0.22) mode is close to the Strouhal number observed for vortex shedding about bodies of aspect ratio greater than 2, however the CFD overestimates this mode by about 14% which is characteristic for non-resolving methods. Figure 4(b) illustrates the development of the fluctuating aerodynamic force coefficients over 25 convective time units. Since $L_b = 1/8$ m and both the forcing frequency and the bulk velocity share the same magnitude (30 Hz/30 m/s), a total angular rotation of 2π radians corresponds to eight convective time units and it is easy to see that the period of F_Y is slightly shorter than the flap oscillation frequency. As noted earlier the CFD results show that wakes dynamically interact downstream of the flaps, thus adding additional dynamic processes which influence the model's aerodynamic force response. However the aerodynamic forces respond directly to on-flow changes, supporting the observations of [14,15] and [16].

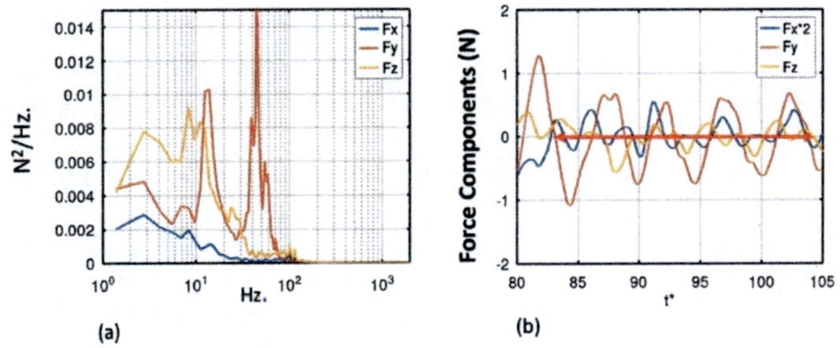

Fig. 4. (a) Power spectral density of the model's aerodynamic force components. (b) A time interval sample of the model aerodynamic force development under an unsteady onflow condition created by the flapped system. Note that $t* = tU/L_h$ is the time normalized by the flow convective time unit based on model height.

6 Conclusions

Investigations are undertaken to validate CFD methods against experiments by studying the influence of a time-dependent inflow on the aerodynamic drag of a 1:25 scaled model train. Both steady and unsteady on-flow conditions are considered. An oscillating flap system is used to control the amplitude and frequency of the on-flow oscillations. Non-resolving DDES methods shows superior performance to URANS at an equivalent computational cost. Drag changes predicted by the CFD due to changes of the computational model (inclusion of flaps and moving belt) closely match measured drag changes. An analysis of the experimental data suggests that the inclusion of the active/passive suction into the CFD model will reduce the difference between CFD and measurements. The fluctuations in computed drag match measured the fluctuation under unsteady onflow conditions. The CFD indicates that wake interactions upstream of the model can occur. These interactions should have the potential to generate additional flow dynamics (which are not yet understood) and will influence the aerodynamic forces frequency response to the upstream flow. The flow examined is highly complex and the CFD appears to contain systematic errors which make it difficult to return accurate drag assessments at arbitrary onflow conditions. Future research will need to identify and remove these systematic errors while retaining an efficient numerical method. The next step is in this work is to include both passive and active suction into the computational model. An important observation arising from this work is that CFD and experiment are complementary in achieving a deeper understanding of these flows.

Acknowledgment. The authors would also like to acknowledge Frau Annika Köhne for her assistance in the preparation of this paper. The resources of the Cluster for Advanced Research in Aerospace (Dresden (CARA), Göttingen (CARO)) are used for the work presented in the paper.

References

1. Baker, C.: The flow around high speed trains. J. Wind Eng. Ind. Aerodyn. **98**(6), 277–298 (2010)
2. Orellano, A.: Aerodynamics of High Speed Trains. Vehicle Aerodynamics lecture, Stockholm KTH (2010)
3. DB Netz AG., Richtlinie 807.04 Bautechnik, Leit-, Signal- und Telekommunikationstechnik: Ausgewählte Maßnahmen und Anforderungen and das Gesamtsystem Fahrweg/Fahrzeug Aerodynamik/Seitenwind, Frankfurt (2010)
4. Sima, M., Gurr, A., Orellano, A.: Validation of CFD for the flow under a train with 1:7 scale wind tunnel measurements. In: BBAA IV International Colloquium on Bluff Bodies Aerodynamics and Applications (9), pp. 1638–1649 (2008)
5. Jönsson, M., Wagner, C., Loose, S.: Particle image velocimetry of the underfloor flow for generic high-speed train models in a water towing tank. Proc. Inst. Mech. Eng. Part F J. Rail Rapid Transit **228**(2), 194–209 (2014)

6. Muld, T.W.: Slipstream and flow structures in the near wake of high-speed trains. Royal Institute of Technology, Department of Aeronautical and Vehicle Engineering, Stockholm, Sweden (2012)
7. Baker, C.: A review of train aerodynamcics: part 1 - fundamentals. Aeronaut. J. **118**(1201), 201–228 (2014)
8. Hemida, H.N.: Large-Eddy Simulation of the Flow around Simplified High-Speed Trains under Side Wind Conditions, Engng, Division of Fluid Dynamics, Dept. of Applied Mechanics, Chalmers University of Technology, Göteborg, Sweden (2006)
9. Weinman, K.A., et al.: Assessment of the mesh refinement influence on the computed flow-fields about a model train in comparison with wind tunnel measurements. J. Wind Eng. Ind. Aerodyn. **179**, 102–117 (2018)
10. Weinman, K.A, Fey, U., Loose, S., Wagner, C., Deiterding, R., Fragner, M.: Comparison between CFD and Wind Tunnel experiment for slender bodies of aspect-ratio O1 in the presence of cross-wind. In: 10th World Conference on Railway Research (2013)
11. Fragner, M.M., Weinman, K.A., Deiterding, R., Fey, U., Wagner, C.: Comparison of industrial and scientific CFD approaches for predicting cross wind stability of the NGT2 model train geometry. Int. J. Railways Technol. **4**(1), 1–18 (2015)
12. Haff, J., Richard, U., Kowalski, T., Loose, S., Wagner, C.: Wind tunnel experiments with a high-speed train model subject to cross-wind conditions. In: Proceedings on the First International Conference on Railway Technology: Research, Development and Maintenance (24), J. Pombo, Civil-Comp Press, Stirlingshire, Scotland (2012)
13. Fragner, M.M., Deiterding, R.: Investigating cross-wind stability of high speed trains with large-scale parallel CFD. Int. J. Comput. Fluid Dyn. **30**(6), 402–407 (2016)
14. Wilhelmi, H., Jessing, C., Bell, J., Heine, D., Wagner, C., Wiedemann, J.: Aerodynamic characterisation of a compact car overtaking a heavy vehicle. In: Dillmann, A., Heller, G., Krämer, E., Wagner, C., Tropea, C., Jakirlić, S. (eds.) DGLR 2018. NNFMMD, vol. 142, pp. 794–804. Springer, Cham (2020). https://doi.org/10.1007/978-3-030-25253-3_75
15. Wilhelmi, H., et al.: Simulation of transient on-road conditions in a closed test section wind tunnel using a wing system with active flaps. SAE Int. J. Passeng. Cars Mech. Syst. (0688), 1–13 (2020). ISSN 1946-3995
16. Weinman, K.A., Wilhelmi, H., Bell, J.R., Heine, D., Wagner, C.: On the simulation of a heavy vehicle wake in OpenFOAM with real-world data. In: Dillmann, A., Heller, G., Krämer, E., Wagner, C. (eds.) STAB/DGLR Symposium 2020. NNFMMD, vol. 151, pp. 504–513. Springer, Cham (2021). https://doi.org/10.1007/978-3-030-79561-0_48
17. Patankar, S.V., Spalding, D.B.: A calculation procedure for heat, mass and momentum transfer in three-dimensional parabolic flows. Int. J. Heat Mass Transf. **15**(10), 1787–1806 (1971)
18. Wilcox, D.C.: Turbulence Modeling for CFD, 3rd edn. DCW Industries, Canada, CA, USA (2006)
19. Menter, F.R., Kuntz, M., Langtry, R.: Ten years of industrial experience with the SST turbulence model. In: Proceedings of the 4th International Symposium on Turbulence, Heat and Mass Transfer, Begell House Inc., West Redding, pp. 625–632 (2003)
20. Strelets, M.: Detached eddy simulation of massively separated flows. In: 39th AIAA Aerospace Sciences Meeting and Exhibit, Reno, NV (2001)
21. Bardina, J.E., Huang, P.G., Coakley, T.J.: Turbulence modeling validation, testing, and development. NASA Technical Memorandum, vol. 110446 (1997)

22. Issa, R.I.: Solution of the implicitly discretized fluid flow equations by operator-splitting. J. Comput. Phys. **62**, 40–65 (1985)
23. Weinman, K., et al.: A study of grid convergence issues for the simulation of the massively separated flow around a stalled airfoil using DES and related methods. In: Wesseling, P., Oñate, E., Périaux, J. (eds.) Proceedings of the European Conference on Computational Fluid Dynamics ECCOMAS CFD 2006, Egmond aan Zee, The Netherlands (2006)

HyperCODA – Towards High-Performing Time-Resolving Flow Simulations

Johannes Wendler[1(✉)], Immo Huismann[1], Ronny Tschüter[1], and Stefan Fechter[2]

[1] German Aerospace Center (DLR), Institute of Software Methods for Product Virtualization, Dresden, Germany
{Johannes.Wendler,Immo.Huismann,Ronny.Tschueter}@DLR.de

[2] German Aerospace Center (DLR), Institute of Aerodynamics and Flow Technology, Göttingen, Germany
Stefan.Fechter@DLR.de

Abstract. The present work focuses on the performance analysis of the DG-SEM implementation of the CFD solver CODA. The turbulent Taylor-Green vortex is employed as a simple testcase for scaling behavior, while for a more detailed node-level performance analysis more granular kernel benchmarks are used. Bottlenecks in the implementation are highlighted and possible solutions proposed.

Keywords: DG-SEM · CFD · performance analysis · optimization

1 Introduction

While steady-state RANS simulations are the state of the art for flow simulation in industry, they often mispredict the resulting flow patterns [1,2]. Scale-resolving and time-resolving simulations offer a way out of the predicament at the expense of resources. Not only does the flow have to be resolved on a per-timestep basis, but low diffusivity is also required. This is where high-order methods can shine.

From the available high-order methods, the nodal discontinuous Galerkin spectral-element method (DG-SEM) offers speed and scalability while allowing for shock capturing and coupling with finite-volume methods [3,4]. The speed results from two factors: On the one hand, tensor-product operators can be well optimized such that the code stays compute-bound, not memory bound [5]. On the other hand, well-crafted DG-SEM methods can scale up to one element per core [6]. In order for these methods to jump the gap from academia to industry, CODA implements a DG-SEM method. CODA is the computational fluid dynamics (CFD) software being developed as part of a collaboration between the French Aerospace Lab ONERA, the German Aerospace Center (DLR), Airbus, and their European research partners. CODA is jointly owned by ONERA, DLR and Airbus.

A. Dillmann et al. (Eds.): STAB/DGLR Symposium 2022, NNFM 154, pp. 305–314, 2024.
https://doi.org/10.1007/978-3-031-40482-5_29

While the DG-SEM method allows for reaching peak performance, it is often not reached in practice. Meticulous performance engineering is essential here [8]. For this purpose we analyze the current DG-SEM implementation of CODA and HyperCODA [9] with the turbulent TAYLOR-GREEN vortex as a simple testcase for scaling behavior and perform more granular benchmarks to evaluate the node level performance of the most relevant kernels using Score-P [12] and LIKWID [13]. This analysis reveals some inefficiencies and we propose possible optimization strategies.

2 Test Case

The focus of this paper is the performance evaluation of CODA to enable efficient simulations of complex use cases with the DG-SEM method. As a simple test case we chose the turbulent TAYLOR-GREEN vortex. It is a benchmark from the 1st International Workshop on High-Order CFD methods, test problem 3.5 [14]. The time-resolved simulation of the TAYLOR-GREEN vortex transitions to fully turbulent at the chosen REYNOLDS number of $Re = 1600$ [10,11].

The TAYLOR-GREEN vortex test case is simulated with the flow solver CODA, using the DG-SEM discretization implemented using the methods for shock capturing and multispecies extension by Renac [17] and Marmignon et. al. [18]. As discretization, we chose DG-SEM of order $n = 4$, meaning each element contains four points per direction. The computation is performed on a cubical domain of 64^3 elements until $20T$ was reached. $T = 1/u_0$ represents the characteristic time scale of the flow, where u_0 is the amplitude of the periodic initial velocity. For time stepping a fourth-order, time-explicit RUNGE-KUTTA scheme with constant time step size during the whole simulation is employed. As typical with DG-SEM, an "implicit LES" is used, i.e. the inherent dissipation of the discretization serves as turbulence model [6]. Furthermore, the Roe scheme in conjunction with the second scheme of Bassi and Rebay (BR2) [15,16] were employed.

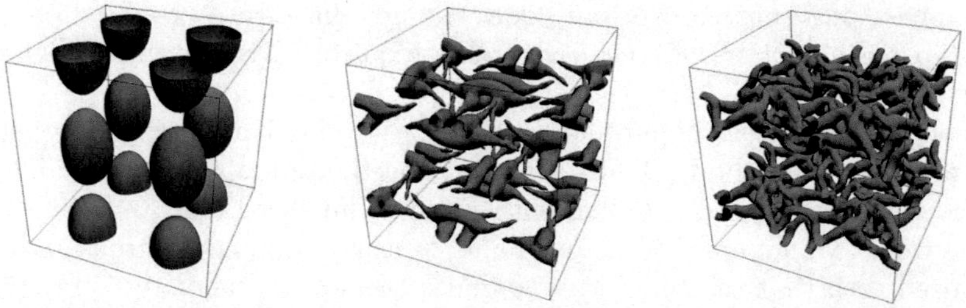

Fig. 1. Vortex structures in the TAYLOR-GREEN vortex at different times, visualized via isoplanes of negative relative pressure of $p - p_0 = -0.002$. Left: $t = 0 \cdot T$, middle: $t = 7 \cdot T$, right: $t = 11 \cdot T$.

Figure 1 depicts the resulting vortex structures, their decay and generation of smaller structures. Here, isoplanes of negative pressure visualize the vortex structures. Initially, the periodic, analytic pressure distribution can be observed, until $t = 5T$, where vortices form and join. At $t = 10T$, the flow becomes fully turbulent with the large-scale vortices decaying into ever-smaller ones.

3 Performance Analysis

To investigate the performance of the DG-SEM implementation in CODA we performed benchmarks on the DLR cluster CARO. The machine has two sockets per node, each with an AMD EPYC 7702, running at a fixed clock speed of 1.8 GHz for all test runs. All used binaries were compiled with GCC 10.3.0, using

```
-O3 -march=native --param inline-min-speedup=1 --param
    inline-unit-growth=200 -DNDEBUG
```

as additional compile flags and with disabled asserts in the build system for CODA version `2022.07.0-137`.

3.1 Scaling Behavior

As a baseline, we performed a weak scaling analysis with 10 timesteps of the turbulent TAYLOR-GREEN vortex test case from 1 to 32 nodes. The number of elements per node is kept constant and the size is roughly 2.5 times the size of the L3 caches to stay in memory.

Figure 2 depicts the runtime per element and the parallel efficiency for the DG-SEM orders $n = 4$, $n = 8$ and $n = 16$. Each node having the same amount of data should, in the best case, keep the runtime constant. The linear increase of total elements w.r.t. the number of nodes then leads to a linear decrease in runtime per element. This exact behavior is represented in the plot of runtime over the number of nodes. Also the parallel efficiency stays at the ideal 100% mark. From these two metrics of this benchmark, we conclude that the parallelization and communication parts of the implementation are no bottleneck to consider for optimization.

3.2 Identification of Critical Kernels

For the identification of critical kernels, we use a version of CODA instrumented by Score-P [12]. The turbulent TAYLOR-GREEN vortex benchmark was rerun with a DG-SEM of order $n = 4$ on a cubic domain of size 64^3. Profiling data (aggregated runtime of functions) and a trace (time resolved data about communication and computation patterns) were extracted. The runtime overhead compared to the original simulation was below 2%, meaning the data depicts an accurate representation of the simulation.

Figure 3 depicts a time-resolved stack of the computation, where the lowermost bar is the currently running function. It can be seen, that the most time is spent in the mesh loopers (light blue), which are all called by the residual calculation (turquoise), which is in turn called four times by the RUNGE-KUTTA scheme.

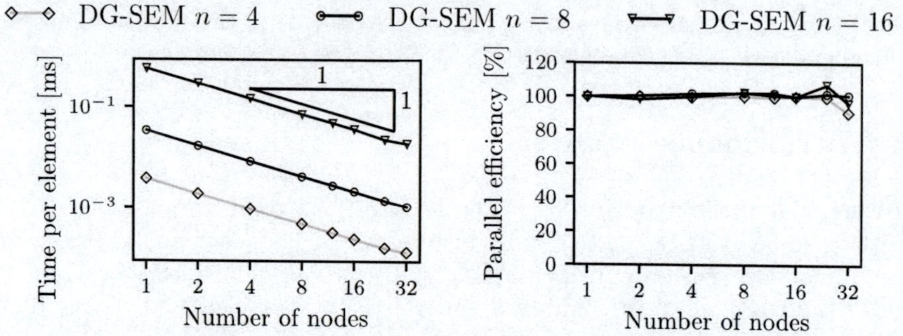

Fig. 2. Left: Runtime per element and right: parallel efficiency for different numbers of nodes for the TAYLOR-GREEN vortex, with roughly the same amount of data per node.

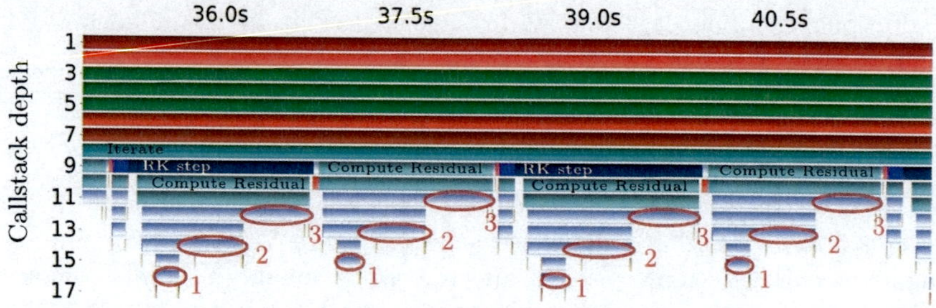

Fig. 3. Trace of one timestep, showing the callstack. The red ellipses highlight the critical kernels (1: Lifting Operator, 2: Convection and Diffusion face flux, 3: Chandrashekar [7] element contribution), which are called by the residual calculation and loop over the mesh. (Color figure online)

Table 1 summarizes the aggregated runtime data of the most relevant kernels. Three of the mesh loopers stand out in particular: the convection and diffusion face flux functor, the element contribution functor and the lifting operator. These three plus a few smaller functor loopers make up the residual evaluation, which accounts for more than 87% of the total runtime.

3.3 Analysis of Critical Kernels

With the relevant parts of the code identified, their behavior for different DG-SEM orders and how these kernels scale with it is the focus of this next section. The next step then is the analysis of node level performance of the residual calculation to grasp how well the implementation utilizes the capabilities of the hardware.

To have an estimation of how the kernels should behave, we took a look at the loops and operations performed in the residual calculation. It gets triggered four times during one time step, once for each RUNGE-KUTTA stage. Inside, one loop over the faces and two loops over the elements of the mesh. All the functors

Table 1. The most relevant function calls with their respective percentage of total runtime on a cubic domain with a width of 64 elements with DG-SEM of order $n = 4$.

Kernel	runtime	percentage
Residual evaluation	246.6	87.7
· Convection and Diffusion face flux	81.7	29.1
· Chandrashekar element contribution [7]	74.1	26.4
· Lifting Operator	65.6	23.3
· Zero Setter Functor (gradient)	18.1	6.4
· Inverse mass matrix	7.1	2.5
Explicit Runge Kutta	18.6	6.6
Other	16.0	5.7
Total	281.2	100

of these loops use the BR2 scheme with a lifting operator when accessing the state of the simulation. The lifting operator is triggered when the state gradient is accessed for the first time by any operator. The lifting operator itself sets the gradient inside the element to zero (order n^3), then computes a face integral contribution to the gradients for each point on the element faces, which is of order n^2. Lastly it applies the inverse mass matrix on the gradients for the element, again order n^3.

The face loop of the residual calculation, which computes the convection and diffusion flux, is called first and triggers the lifting operator of order n^3. For the flux computation on the faces (order n^2), the gradient is needed for each point, which is of order n per point, resulting in order n^3 for the faces.

Of the two element loops, the first called functor calculates the volume term contributions to the residual. First the entropy conservative and kinetic energy preserving numerical convection flux is calculated using the scheme of Chandrashekar [7], which is of order n per point, leading to order n^4 for the numerical flux. Then the convection and diffusion flux is evaluated for each point inside the element (order n^3), which triggers the gradient computation for each point, exactly as in the face loop, resulting in order n^4. Lastly the source terms get evaluated in this element functor, but as there are none in the TAYLOR-GREEN vortex test case, we leave this one out.

The last step for the residual calculation is the application of the inverse mass matrix on the residual on all points inside an element, simply of order n^3.

Figure 4 shows the scaling behavior w.r.t DG-SEM order of the critical kernels. Due to the fact that the lifting operator is triggered by the convection and diffusion functor, they are measured together here. As can be seen in the runtime per element plot, the convection and diffusion face flux plus lifting operator scales nearly with order n^3 and the element contribution scales almost perfectly with order n^4, just as expected. Also, the whole residual calculation gets dominated by the Chandrashekar contribution for higher orders, which is apparent in both the time per element and throughput diagram.

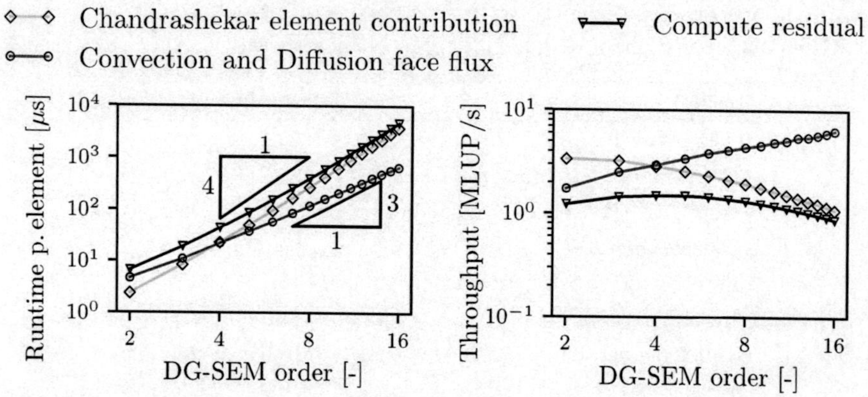

Fig. 4. Left: Runtime per element and right: throughput as computed degrees of freedom per second in Mega Lattice Updates per Second (MLUP/s) for the different operators.

3.4 Roofline Model

As the parts of the residual evaluation scale as expected, we analyze how well the hardware is actually utilized. The roofline model is a well established analysis tool for node level performance. It uses the concept of arithmetic intensity

$$I = \frac{\text{flops performed}}{\text{bytes loaded or stored}}, \tag{1}$$

which is a measure of the code balance between computation and memory transfer.

Table 2 shows the ideal number of flops performed and bytes transferred through the memory interface of the CPU for the three most relevant kernels. To obtain these values we made a few assumptions: All constant values and local buffer data structures reside in the caches, special functions (i.e. log, sqrt) are counted as a single flop and index calculations are ignored. This theoretical code balance of the implementation proves unattainable in practice, but provides an upper limit, where it could be, when no unnecessary data is loaded or stored.

Table 2. Flops performed and bytes loaded/stored by each kernel.

Scaling order	Flops			Bytes	
	n^2	n^3	n^4	n^2	n^3
Convection and Diffusion face flux	853	12	-	640	-
Lifting Operator	24	5	-	88	80
Chandrashekar Element contribution	-	459	414	-	320

The roofline itself is then defined as

$$P_{\text{Roofline}} = \min(\beta * I, P_{\text{Peak}}), \tag{2}$$

where β is the memory bandwidth of the CPU and P_{Peak} the peak performance of the CPU. The peak performance is given by

$$P_{\text{Peak}} = \text{clock} \cdot \text{n}_{\text{AVX}} \cdot \text{n}_{\text{FMA}} \cdot \text{n}_{\text{ports}} \cdot \text{n}_{\text{cores}}. \tag{3}$$

The AMD EPYC 7702 CPU is running at a clock speed of 1.8 GHz, has AVX registers that fit 4 double precision float values, has two execution ports that can perform FMA instructions (multiply and addition in one cycle) and CARO has 128 cores per node. This results in a peak performance of 3.69 TFlop/s per node and 0.92 TFlop/s without AVX vectorization. The memory bandwidth was determined with a sum stream benchmark to be 360 GB/s.

To measure the achieved flops and memory throughput, we use LIKWID [13], a lightweight tool for accessing hardware counters on CPUs during runtime of an application. The AMD Zen2 architecture provides counters for flops and for cachelines transferred through the memory interface, which together with the runtime provides measured data for arithmetic intensity and computation performance.

Figure 5 depicts the roofline analysis, once with the measured and once with the computed arithmetic intensity (Table 2). On the left, in the classic roofline plots, it can be seen that the measured intensity is much lower than the computed one. On the right are the percentages of the roofline performance that the kernels reach. Assuming our computed arithmetic intensity, the element contribution functor, which is the more relevant one for higher orders, achieves less than 10% for orders $n > 4$, with order $n = 16$ reaching 9.0%. Even for the measured intensity of the current implementation it is only around 25% of the hardware limit.

The most conspicuous thing to note is the unexpected behavior of the element contribution kernel: Its arithmetic intensity should be higher, which suggests unnecessary data transfer happening. Also, the low performance hints at a low utilization of vectorized instructions, especially towards higher orders, where it should be compute bound.

Lastly, to get a different angle of the element contribution kernel, its generated assembly code was analyzed. Almost no vectorized instructions were present and many functions were not inlined by the compiler, even with the compile flags allowing more aggressive inlining. Instead of AVX instructions, the assembly code shows many mov and push instructions, i.e. moving and handling of data. In general it seems the compiler is not able to optimize much from the very complex DG-SEM implementation in CODA, at least for the configuration chosen here.

4 Optimization Potential

The above performance analysis leads to a few possible optimizations that can be considered down the line.

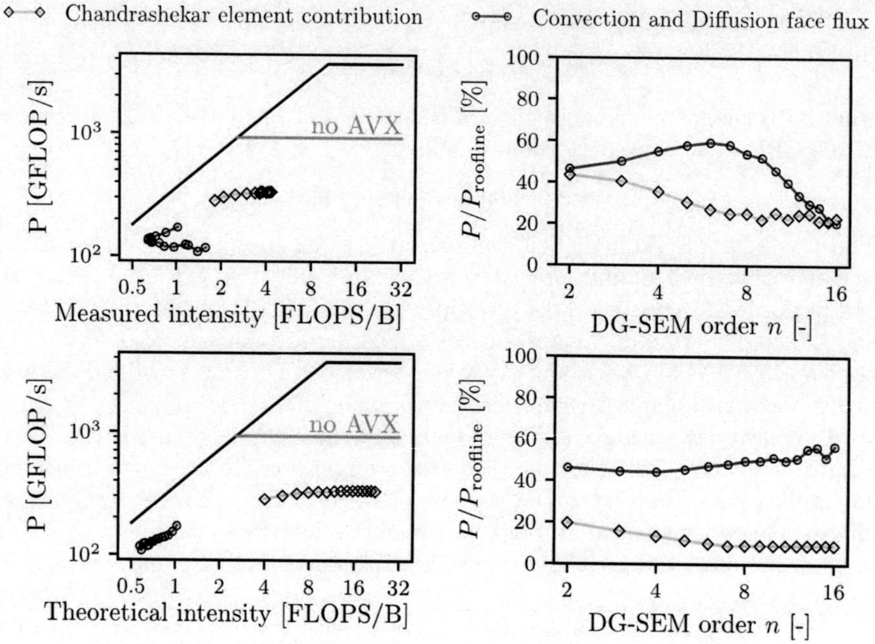

Fig. 5. Top left: Measured Roofline for both critical kernels, top right: percentage of roofline performance reached over the DG orders, bottom left: Roofline of the critical kernels with their respective performance at the theoretically optimal intensity (Table 2) and bottom right: percentage of roofline performance reached for optimal intensity.

First and foremost, the low utilization of vectorized instructions and function inlining is the most pressing concern. As long as the compiler is not capable of properly interpreting and optimizing the code, most optimizations will prove fruitless. The first focus here should be the Chandrashekar computation of the numerical flux, with its n^4 complexity. It is purely dependent on the element itself and does only require local restructuring of the data access patterns to enable the compiler to vectorize the code.

Secondly, the gradient gets computed multiple times, once for each functor requiring it. Also the convection and diffusion face flux kernel would be of order n^2 instead of n^3 without the gradient computation inside. With some reordering and the gradient being saved after being calculated when needed for the first time this can be implemented more efficiently.

Lastly, the lifting operator looping once over the mesh, setting the gradient to zero is unnecessary when the implementation of the actual computation is changed accordingly.

A long term consideration would be to introduce a different memory layout, with better cache locality and clearer access patterns for the compiler to vectorize with more ease. Figure 6 depicts the current memory layout and a reordered variant. Placing the states of four points of four different elements, all with the same index and the same physical variable, next to each other can bring the compiler to calculate 4 elements in parallel during one loop, each instruction working on 4 points at the same time.

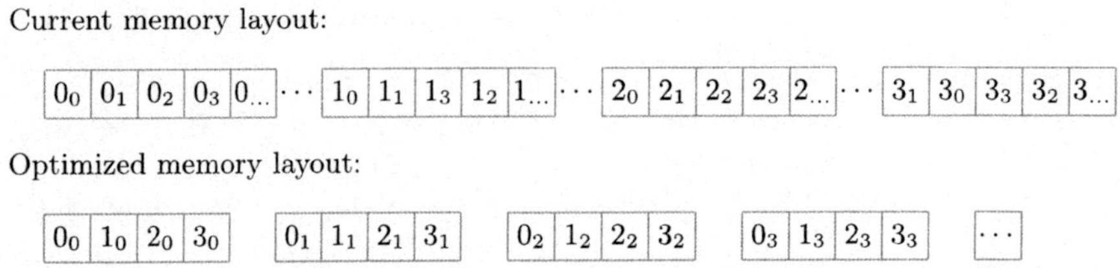

Fig. 6. Top: current memory layout with data of all integration points of an element locally grouped in memory and bottom: optimized memory layout with integration points of the same index of four different elements grouped together for better vectorization.

A second step can be to evaluate how the number of abstraction layers in the heavily templated code could be reduced. This can lead to more inlining, possibly also more vectorization and other optimization methods the compiler can utilize.

5 Conclusion

The performed benchmarks of CODA using the turbulent TAYLOR-GREEN vortex and a more granular kernel benchmark show, that the current DG-SEM implementation scales well, but the node level performance leaves much to be improved upon. Through the performance analysis using the roofline model we found some inefficiencies in the gradient computation and the lifting operator that can be removed. The assembly code additionally revealed optimization potential in more inlining and more vectorization. The next step consists of assessing the possible optimization strategies and choose the desired approach.

Acknowledgements. This work is supported by Fabio Naddei (ONERA) and his work on DG-SEM methods in CODA.

References

1. Grabe, C., Probst, A., Horchler, T., Shishkin, A., Werner, M.: Potential, perspectives and HPC-requirements of scale-resolving simulation techniques. HPCN, DLR-AS (2022)
2. Slotnick, J.P., et al.: CFD vision 2030 study: a path to revolutionary computational aerosciences. Technical report NASA/CR-2014-218178 (2014)

3. Krais, N., et al.: FLEXI: a high order discontinuous Galerkin framework for hyperbolic-parabolic conservation laws. Comput. Math. Appl. **81**, 186–219 (2021)
4. Marmignon, C., Naddei, F., Renac, F.: Energy relaxation approximation for compressible multicomponent flows in thermal nonequilibrium. Numer. Math. **151**, 151–184 (2022)
5. Cantwell, C.D., Sherwin, S.J., Kirby, R.M., Kelly, P.H.J.: From h to p efficiently: strategy selection for operator evaluation on hexahedral and tetrahedral elements. Comput. Fluids **43**(1), 23–28 (2011)
6. Beck, A.D., Gassner, G.J., Bolemann, T., Frank, H., Hindenlang, F., Munz, C.-D.: Underresolved turbulence simulations with stabilized high order discontinuous Galerkin methods. In: Fröhlich, J., Kuerten, H., Geurts, B.J., Armenio, V. (eds.) Direct and Large-Eddy Simulation IX. ES, vol. 20, pp. 103–108. Springer, Cham (2015). https://doi.org/10.1007/978-3-319-14448-1_14
7. Chandrashekar, P.: Kinetic energy preserving and entropy stable finite volume schemes for compressible Euler and Navier-Stokes equations. Commun. Comput. Phys. **14**(5), 1252–1286 (2013)
8. Huismann, I., Stiller, J., Fröhlich, J.: Efficient high-order spectral element discretizations for building block operators of CFD. Comput. Fluids **197**, 104386 (2020)
9. Huismann, I., Fechter, S., Leicht, T.: HyperCODA – extension of flow solver CODA towards hypersonic flows. In: Dillmann, A., Heller, G., Krämer, E., Wagner, C. (eds.) STAB/DGLR Symposium 2020. NNFMMD, vol. 151, pp. 99–109. Springer, Cham (2021). https://doi.org/10.1007/978-3-030-79561-0_10
10. Gassner, G.J., Beck, A.D.: On the accuracy of high-order discretizations for underresolved turbulence simulations. Theoret. Comput. Fluid Dyn. **27**(3), 221–237 (2013)
11. Fehn, N., Wall, W.A., Kronbichler, M.: Efficiency of high-performance discontinuous Galerkin spectral element methods for under-resolved turbulent incompressible flows. Int. J. Numer. Meth. Fluids **88**(1), 32–54 (2018)
12. Knüpfer, A., et al.: Score-P: a joint performance measurement run-time infrastructure for Periscope, Scalasca, TAU, and Vampir. In: Brunst, H., Müller, M., Nagel, W., Resch, M. (eds.) Tools for High Performance Computing 2011, pp. 79–91. Springer, Heidelberg (2012). https://doi.org/10.1007/978-3-642-31476-6_7
13. Treibig, J., Hager, G., Wellein, G.: LIKWID: lightweight performance tools. In: Bischof, C., Hegering, H.G., Nagel, W., Wittum, G. (eds.) Competence in High Performance Computing 2010, pp. 165–175. Springer, Heidelberg (2011). https://doi.org/10.1007/978-3-642-24025-6_14
14. 1st International Workshop on High-Order CFD Methods. https://cfd.ku.edu/hiocfd.html
15. Bassi, F., Rebay, S., Mariotti, G., Pedinotti, S., Savini, M.: A high-order accurate discontinuous finite element method for inviscid and viscous turbomachinery flows. In: Proceedings of the 2nd European Conference on Turbomachinery Fluid Dynamics and Thermodynamics, pp. 99–109 (1997)
16. Bassi, F., Grivellini, A., Rebay, S., Savini, M.: Discontinuous Galerkin solution of the Reynolds-averaged Navier-Stokes and k-ω turbulence model equations. Comput. Fluids **34**(4–5), 507–540 (2005)
17. Renac, F.: Entropy stable, robust and high-order DGSEM for the compressible multicomponent Euler equations. J. Comput. Phys. **445**, 110584 (2021)
18. Marmignon, C., Naddei, F., Renac, F.: Energy relaxation approximation for compressible multicomponent flows in thermal nonequilibrium. Numer. Math. **151**(1), 151–184 (2022)

Reproducing Vortex Topology Mode Alternation of a Generic Combat Aircraft Model by Means of Hybrid RANS-LES Simulations

Jonas Zastrow[1(✉)] and Axel Probst[2]

[1] Institute of Aeroelasticity, DLR Göttingen, Göttingen, Germany
jonas.zastrow@dlr.de
[2] Institute of Aerodynamics and Flow Technology, DLR Göttingen, Göttingen, Germany
axel.probst@dlr.de

Abstract. The unsteady flow phenomena of a delta wing configuration with multiple leading edges are investigated by means of high-resolution numerical simulation and comparisons with experimental data. The goal is to establish a validated numerical baseline for further research on potential mode alternations due to topological instabilities of the complex vortical flow, which have been observed on the DLR-F23 combat aircraft wind tunnel model during previous experiments. In order to draw conclusions from the underlying flow physics, it needs to be shown first, that coherent and plausible flow field data can be calculated by numerical simulations. Hence, a challenging test case at an angle of attack of $\alpha = 21°$ and a transonic Mach number of 0.85 is chosen. Comparisons between experimental and numerical data show, that the hybrid-RANS-LES approach reproduces the flow to a satisfying degree.

Keywords: multi delta wing · vortex burst/breakdown · continuous wavelet · IDDES

1 Introduction

The design of aerodynamic configurations demands the assessment of their performance and handling. Especially, the borders of the flight envelope or maneuvers are usually the limiting factor of structural and also of aerodynamic design. The reason is, that the flow topologies show severe levels of unsteadiness. As a matter of fact, investigations on unsteady flows come at a great computational cost and analysis efforts. At the same time, the research investments into new configurations need to remain reasonable. Therefore, a thorough assessment of the desired outcome determines the means of numerical simulations. The unsteadiness of an aerodynamic system may be described globally, but consists of interacting local phenomena that oscillate in their own modes, or

A. Dillmann et al. (Eds.): STAB/DGLR Symposium 2022, NNFM 154, pp. 315–325, 2024.
https://doi.org/10.1007/978-3-031-40482-5_30

reduced frequencies respectively [1]. The entity of all phenomena modes, yields the global topology mode characteristics. In detail, small-scale high-frequency structures may influence and feedback into large-scale low-frequency structures. Hence, the numerical set-up needs to compromise between cost on the one side and resolving or neglecting parts of the flow physics on the other side. Besides many other technical disciplines, the evaluation of combat aircraft aerodynamics offers challenging flow physics, in order to test and expand numerical capabilities. Especially, strongly separated transonic unsteady vortical flows are of interest for this work. A broad knowledge base on combat aircraft aerodynamics has been developed over the last decades. Even though, many phenomena are understood and can be resolved numerically, there are still many challenges at hand. One of these is the understanding of the vortex topology mode alternation, which is very prominent in some flow cases of the DLR-F23 combat aircraft wind tunnel model [2]. Only a numerical simulation, that is capable to reproduce this phenomenon, will enable an investigation and eventually an explanation. This work presents a numerical approach, which achieves that at reasonable cost. To be specific, the set-up employs the DLR-TAU flow solver in a domain that includes approximated parts of the wind tunnel and employs a hybrid RANS-LES model.

1.1 Flow Topology Mode Alternation

Usually, unsteady flows feature several distinct flow topologies, which comprise their own reduced frequencies, or modes respectively. Seen as an entity, this results in the overall flow topology modes. In case parts of the flow topology experience instabilities, these parts alter their modes. As a consequence, the complete flow topology modes alternate. In order to separate this phenomenon from pure unsteadiness, two requirements may be full-filled. First of all, the alternation needs to be expressed by a temporal change of the topological structure. Secondly, this process should occur randomly. Within the scope of this work, the objective is to numerically reproduce a vortical flow topology mode alternation, that was observed in an experiment. In particular, the phenomenon can be seen in results of time-resolved pressure sensitive paint (iPSP) [3] and in piezo-electric balance data. A description of the phenomenon follows in Sect. 3.1.

2 Numerical Method

The simulations presented in this paper use the unstructured compressible flow solver DLR-TAU [4] which employs the finite-volume approach on mixed-element (hybrid) grids. It is of 2nd order in space and time, and applies implicit dual-time stepping for time-accurate simulations. Besides different RANS turbulence models, TAU offers scale-resolving capabilities using hybrid RANS-LES and appropriate spatial discretization schemes.

2.1 Hybrid RANS/LES Model

In order to overcome potential deficiencies of pure RANS modeling in the complex vortical flow, a hybrid model with local LES behavior called Improved

Delayed Detached-Eddy Simulation (IDDES) [5] is applied. The IDDES uses local blending functions to switch automatically between local RANS, LES and wall-modeled LES (WMLES) modes, which is controlled by the hybrid length scale:

$$l_{hyb} = \tilde{f}_d\,(1 + f_e)\,l_{\mathrm{RANS}} + \left(1 - \tilde{f}_d\right) l_{\mathrm{LES}} \tag{1}$$

Here, l_{RANS} is the integral turbulence length scale defined by the RANS model, whereas l_{LES} is the LES length scale. The function $\tilde{f}_d = \max\{(1 - f_{dt}), f_B\}$ provides the main blending switch between the different modeling modes, where f_{dt} and f_B depend on local grid and flow properties [5]. The elevating function f_e acts near the wall-normal RANS-LES interface to avoid a log-layer mismatch in boundary layers. In the outer parts of the flow, l_{hyb} reduces to $l_{\mathrm{LES}} = C_{\mathrm{DES}}\Delta_{max}$, which approximates the behavior of a Smagorinsky-type sub-grid model for LES.

2.2 Hybrid Low-Dissipation Low-Dispersion Scheme

The accuracy of hybrid RANS-LES strongly relies on low numerical errors in the LES regions on a given grid. Thus, besides choosing a sufficiently small physical time step for the temporal discretization, the spatial scheme should be designed to provide low numerical dissipation and dispersion errors, as well. To this end, we apply a hybrid low-dissipation low-dispersion scheme (HLD2) [6], which combines a skew-symmetric kinetic-energy-preserving flux scheme with small amounts of matrix-valued 4th-order artificial dissipation and a gradient-based extrapolation scheme for the central face values. While the former provides low-dissipation properties, the extrapolation is adjusted to minimize the dispersion error even on distorted, unstructured grids. To retain stability in regions with e.g. grid imperfections, an additional hybrid blending is applied, which activates the LD2 scheme predominantly in the well-resolved vortical LES regions, whereas coarser outer-flow regions are treated with a more robust standard scheme. Moreover, flow discontinuities are detected by a classic pressure-based sensor, so that the scheme is locally turned into a first-order upwind scheme near shocks.

3 Test Case

The hybrid-RANS-LES simulation is employed to resolve the unsteady flow physics of the DLR-F23 semi-span wind tunnel model (Fig. 1a). This model is part of the DLR project Diabolo, which investigates on combat aircraft related challenges in research and engineering. The first wind tunnel experiment with the DLR-F23 was conducted in the Transonic Wind Tunnel Göttingen and covered an angle of incidence range of 0° to 30° and a Mach-number range of 0.5 to 1.2. A detailed description of the wind tunnel model and this experiment is available in [2].

3.1 General Flow Topology

As discussed before, the phenomenon of interest is vortex topology mode alternation and it occurs at moderate to high angles of incidence. The chosen test case for numerical reproduction is at $\alpha = 21°$ and Mach-number 0.85. An outline of the mean flow topology is presented in Fig. 1b, in which the vortical structures are expressed by the λ_2-criterion [7]. Additionally, $\lambda_{2,mean}$ is colored by the mean Mach-number (averaged for 200 ms).

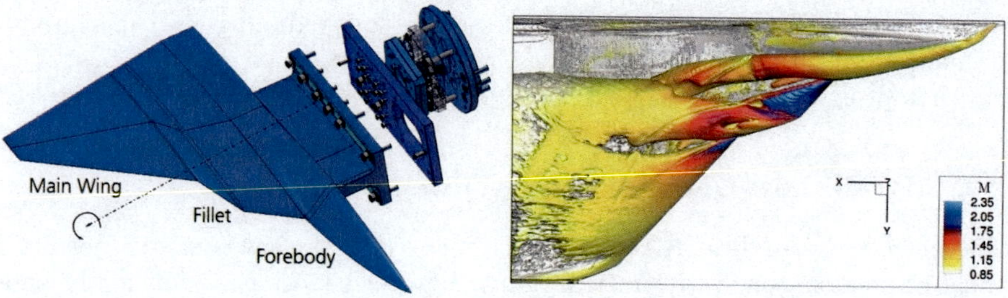

(a) Model with 6c-piezo-electric balance (b) $M = 0.85$, $\alpha = 21°$, $\lambda_{2,mean}$ in M

Fig. 1. DLR-F23 test case

With regards to the test case, the vortex system consists of three primary and three secondary vortices. In detail, the forebody vortex intertwines with the fillet vortex and both experience breakdown at a similar position. Additionally, the main wing vortex experiences heavy breakdown, and as will be discussed later on, is the driving factor for the topology mode alternation. With regards to secondary separations, only the forebody and the fillet secondary vortices are visible in the current visualization. Apparently, both structures find a abrupt reduction in $\lambda_{2,mean}$ and vanish. This can be explained, when the shock system of this transonic flow is considered. An implicit localization of the first shock is available through the steep gradient in Mach-number. Upstream of the intertwining, this shock passes perpendicular trough both, the forebody vortex core and the fillet vortex core. Interestingly, the secondary forebody vortex diminishes near the same position, which suggests a termination by the first shock. Similarly, the secondary fillet vortex seems to be terminated as well, however, the second shock oscillates heavily over a large length of the planforms chord. Thus, it is not seen so clearly by a gradient in the mean Mach-number. Nevertheless, the standard deviation of surface pressure $\sigma(p)$ in the numerical data (Fig. 7b) and the standard deviation of the surface pressure coefficient $\sigma(cp)$ in experimental data (Fig. 7a), show the large area of unsteadiness within the second shock's oscillation. Furthermore, the surface data suggest, that not only the second shock's unsteadiness, but also that the breakdown positions of all three primary vortices and even of the secondary fillet vortex oscillate as well. A closer look at the breakdown locations of the main wing vortex reveal, that those are situated either directly at the leading edge, or just downstream of this position. In case of the latter, the main wing vortex is expected to experience an oscillating breakdown, that should imprint its reduced frequency into the topology modes. In contrast to this, the area of severe unsteadiness at the leading edge suggests a complete burst of the main wing vortex. Thereby, the first requirement for a topology mode alternation is full-filled. The second requirement demands, that this process occurs randomly. As will be discussed in Sect. 4 and presented in Fig. 8, the alternation between breakdown and burst does occur arbitrarily.

3.2 Numerical Set-Up

There are some detrimental considerations, which a set-up of a numerical simulation demands. Therefore, the choices for spatial discretization, temporal discretization, boundary conditions and convergence criteria are presented. Since this work is focused on a hybrid-RANS-LES approach, the turbulence modeling and solver specific aspects are discussed in Sect. 2.1 and 2.2, respectively. Furthermore, general considerations for the set-up of detached eddy simulations are taken into account [8]. The applied numerical boundary conditions should match the experiment to begin with. However, it is not possible to match the Mach-number directly. The reason is, that wind tunnels usually utilize a control law, that is based on empty test sections. This applies to the Transonic Wind Tunnel Göttingen as well. Thus, the actual Mach-number during the experiment may deviate from the desired target Mach-number in dependence of geometrical blockage by the model, through aerodynamic blockage by vortices, wakes and shocks, and further effects. A proper procedure to approach this, is to set-up a numerical wind tunnel, that utilizes the same control laws. In order to reduce the cost for the design of such a set-up and the corresponding computations, a non-controlled Euler-tunnel is used, to test the hybrid-RANS-LES approach first. This means, that the wind tunnel model is mounted on a viscous-wall, whereas the other three surrounding walls are Euler-walls. The quadratic test section is $1m$ by $1m$ (see Fig. 2a). As stated before, the Euler-tunnel is not controlled, so inflow and outflow are set to the assumed Mach-number of 0.85. With the DLR-F23's mean aerodynamic chord (MAC) of 411 mm, this corresponds to an assumed Reynolds-number of 4.2×10^6. The generation of the numerical mesh is conducted with ANSA (by Beta CAE Systems) and results in a hybrid mesh, that consists of surface quadrilaterals, fine-resolution hexahedra in the near-wall (boundary layer) regions and larger hexahedra in the outer flow domain, as well as unstructured elements to bridge the former and the latter. In total, the mesh features 35×10^6 points, or 41×10^6 elements. A block around the model, that consists of 2 mm elements (0.5% of MAC), ensures adequate resolution of small scale flow topologies. It encloses the major part of the unsteady vortex flow topology above the DLR-F23. Additionally, the leading edge with a radius of 0.5 mm is discretized with 25 elements in circumferential direction (0.0628 mm) and with 0.5 mm increments in span-wise direction. In contrast to these small elements, the rest of the Euler-tunnel comprises 25 mm elements. The transition from the rather fine 2 mm block to these coarser 25 mm elements is clearly seen through $\lambda_{2,mean}$. An impression of the grid design is seen in Fig. 2b. At last, the prism layer consists of 33 layers, which start at a height of $1.5\,\mu\text{m}$ ($y^+_{max} < 1$) and have a growth rate of 1.2.

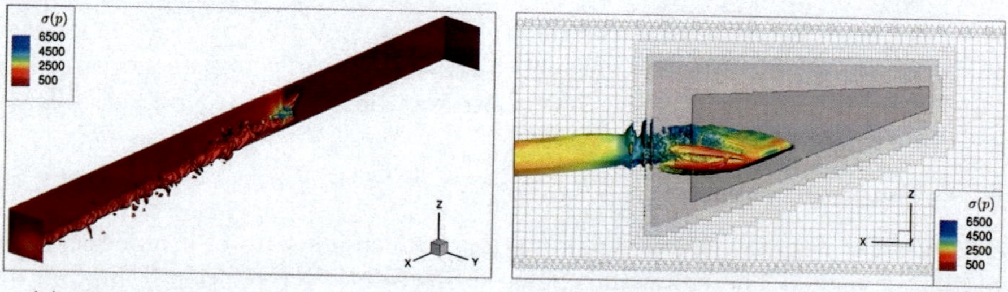

(a) **Numerical domain; $\lambda_{2,inst}$ in $\sigma(p)$** **(b)** **Mesh at $y = 0.25$; $\lambda_{2,mean}$ in $\sigma(p)$**

Fig. 2. Numerical set-up

After the definition of the spatial discretization Δl, the time-step size Δt can be derived. In order to resolve the unsteady phenomena explicitly, a convective Courant-Friedrich-Lewys (CFL) number smaller than unity is needed:

$$CFL < \frac{\Delta l}{u_{max} \times \Delta t} \tag{2}$$

It has been shown, that $u_{max} \approx 3 \times u_0$ in delta wing vortices [9]. This maximum velocity is found at the leading edges and in the vortex cores, which is seen in Fig. 1b as well. Additionally, the unsteadiness in these areas can lead to occasional and large velocity overshoots, which are even greater than $\approx 3 \times u_0$ (comparison in Fig. 3a and Fig. 3b). As a consequence, this maximum instantaneous flow velocity should find consideration, for the definition of the time step size. Therefore, it was decided to choose a time step that would account for $5 \times u_0$ and that the convective CFL-number should refer to the 2 mm elements in the field above the suction side. After all, this yields a time step size of $1.25 \times 10^{-6} s$. Also, this means that unsteady effects at the leading edges are not resolved explicitly with a convective CFL-number smaller than unity (see Fig. 4a), because the leading edge elements are more than an order of magnitude smaller than in the 2 mm field (see Fig. 4b). Finally, the set-up applies relative Cauchy convergence control for the (inner) pseudo-time iterations of the implicit dual-time scheme: If in 10 consecutive inner iterations, both lift and drag stay within a (normalized) interval that is smaller than 10^{-7}, convergence is reached.

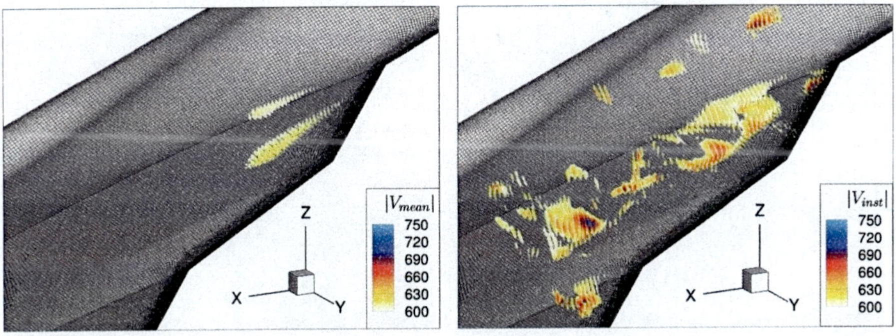

(a) Slices of mean velocity magnitude (b) Slices of instant velocity magnitude

Fig. 3. Local velocities to consider for time step size

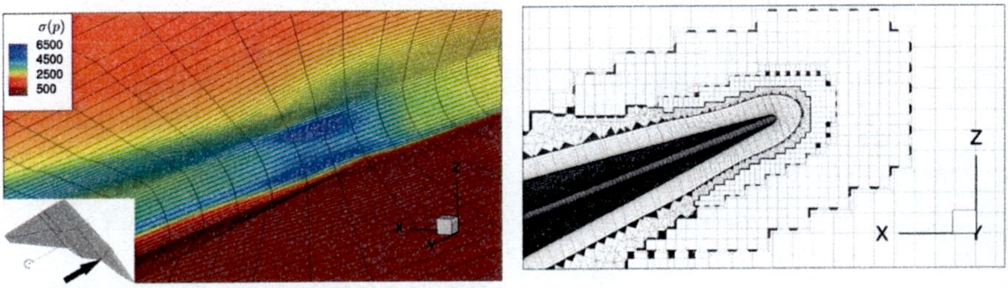

(a) Unsteady leading edge kink in $\sigma(p)$ (b) Sizes of field and leading edge elements

Fig. 4. Details on grid design

4 Results

As expected, the hybrid RANS-LES simulation shows LES regions for most parts of the numerical domain, except near viscous walls. Even there, the LES region protrudes into the lower parts of the boundary layer, which is even more pronounced at the leading edges. An impression of the RANS-LES blending is shown in Fig. 5a, in which $1 - \tilde{f}_d \approx 1$ means LES-region and $1 - \tilde{f}_d \approx 0$ means RANS-region. Furthermore, an instantaneous flow field is presented in Fig. 5b and utilizes $\lambda_{2,inst}$ to reveal vortical structures of various sizes and differing orientations. By coloring $\lambda_{2,inst}$ in $\sigma(p)$, the severe unsteadiness of the flow becomes apparent. A coherent time series of this numerical data will enable the analysis of the underlying phenomena. However, the first step is to provide strong indicators, that the numerical simulation compares well to the experimental data.

In a first step, numerical data and experimental data are compared with regards to the mean pressure coefficient of the DLR-F23's suction side (see Fig. 6). Overall, both data-sets show a very similar cp_{mean} distribution, even though there are differences. Those are found in the area of vortex breakdown, in the strength of the second shock's oscillation, and at the leading edge of the

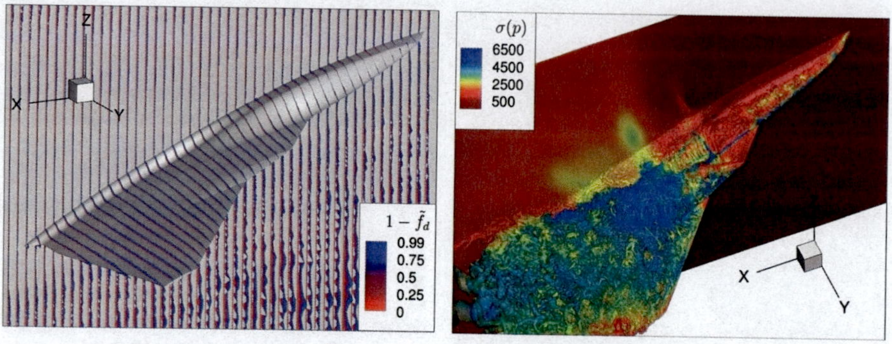

(a) Blending function $\tilde{f}_d$ (b) Unsteady vortical flow; $\lambda_{2,inst}$ in $\sigma(p)$

Fig. 5. Hybrid RANS-LES results

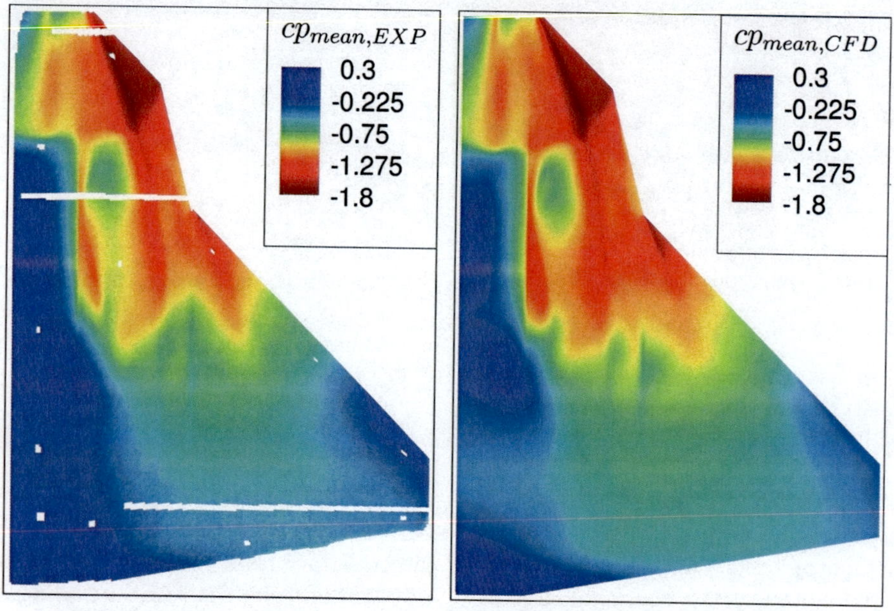

(a) Experimental results for cp_{mean} (b) Numerical results for cp_{mean}

Fig. 6. Mean values of pressure coefficient

outer main wing. However, the focus of this work is on unsteady flow topologies. Hence, a comparison of the standard deviation of the surface pressures is presented. Whereas the numerical results are presented in $\sigma(p)$, it is different for iPSP, which is presented in $\sigma(cp)$. For a direct comparison, either a modification of the TAU-code's statistics utility, or the evaluation routine of iPSP would need to be modified, but these two options remain tasks for future work. Still, it is seen in Fig. 7, that both data-sets compare well in their distribution

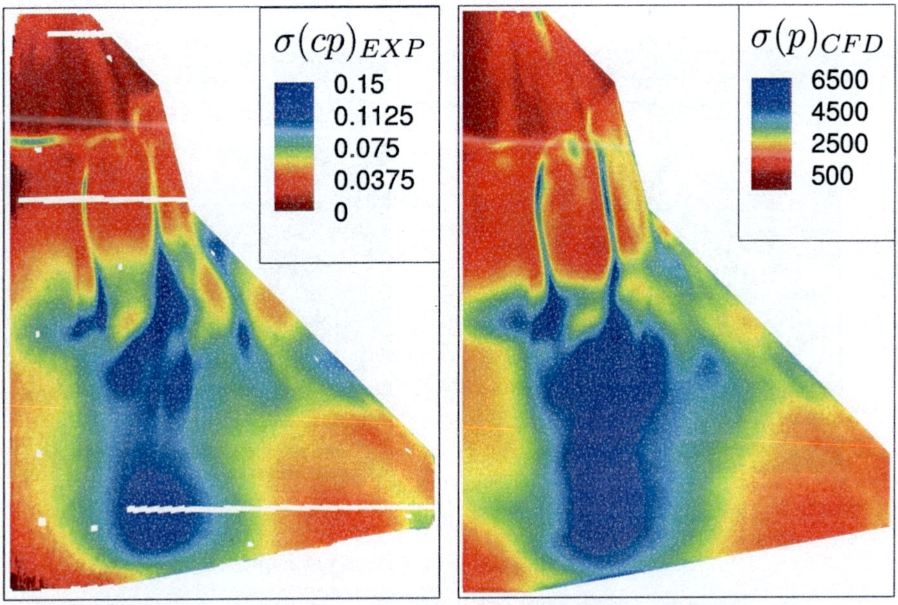

(a) Experimental results for $\sigma(cp)$ (b) Numerical results for $\sigma(p)$

Fig. 7. Standard deviations of pressure (coefficient)

of unsteadiness. As can be expected, the positions of larger differences match the three areas, that were discussed for cp_{mean}. Additionally, further differences are apparent directly at the leading edge of the fillet and further inboard a the position of the fillet's secondary vortex. These areas of unsteadiness occur in the numerical simulation, but not in the experiment. Thus, it can be assumed, that the experiment could lack the temporal resolution to resolve these phenomena. In detail, iPSP is measured at 3.2 kHz, but the numerical sampling rate is 800 kHz. In contrast to this, there is another major difference in the numerical simulation. It is located at the kink of the main wing, were stronger unsteadiness is shown in the experimental data. As discussed before, the chosen time step size is still to large, to resolve the unsteady leading edges explicitly. After all, the driving factor for the topology mode alternation is the main wing's vortex instability. Apparently both, CFD and experiment, show a similar phenomenon near the main wing's leading edge. This suggests that indeed, the alternation between complete burst and oscillating breakdown is resolved numerically, even though much less pronounced. At this point a comparison of the pitching momentum coefficient c_{my} between experiment and numerical simulation is needed. Since a change of the frequency spectra over time is expected, a continuous wavelet transform is applied to c_{my} (see Fig. 8). At first glance, experiment and CFD seem to show few commonalities. Nevertheless, it should be considered, that the experimental c_{my} shows a larger amplitude and thereby larger intensities in the wavelet transform. With regards to the mode alternation, it expresses itself in

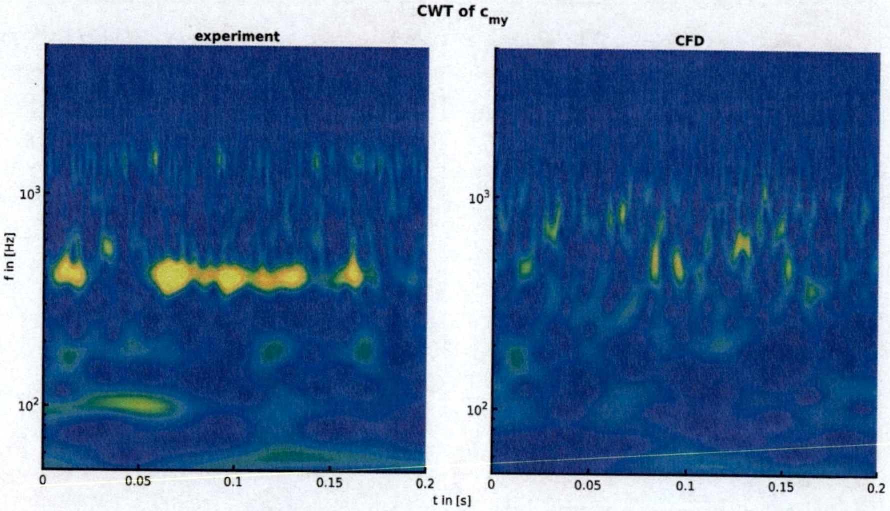

Fig. 8. Wavelet transform of c_{my} for experiment and CFD

the randomly changing intensities around the ≈ 400 Hz band. A high intensity there refers to the topology with breakdown oscillation, whereas low intensities refer to a complete burst of the main wing vortex. As it seems, there are even more reduced frequencies, that change in an arbitrary fashion over time and may be connected to further alternating parts of the flow topology. All in all, the presented numerical set-up and results prove to replicate the underlying phenomenon of interest, but as can be expected still deviate from the experimental data in some aspects. Nonetheless, a thorough analysis of the numerical results should yield new insights into the vortex topology mode alternation of the DLR-F23.

5 Conclusion and Outlook

The DLR-F23 combat aircraft wind tunnel model shows severe unsteadiness in transonic vortical flow. Inspired by the observed unsteady phenomena, the objective of this work is to reproduce these by means of numerical simulations. A hybrid-RANS-LES approach in an Euler-tunnel is chosen and proves to reproduce the flow sufficiently. As a consequence, the numerical data can provide a baseline now, in order to investigate on the vortex flow topology mode alternation. In the next steps, the vortical flow will be analyzed by invariant analysis [10], with regards to vorticity flux [11] and by mode decomposition. Also, the Euler-tunnel will be accompanied by an enhanced controlled numerical wind tunnel, that comprises the perforated test section of the Transonic Wind Tunnel Göttingen. As a consequence, this set-up will lead to better representations of the wind tunnel experiment in CFD, but also enable the simulation of high transonic and supersonic flows. Last but not least, this new set-up could be exploited to derive a control law for the Euler-tunnel, which could lead to numerical simulations, that are cost effective and precise at the same time.

Acknowledgments. The authors thank our colleague Mr. Stefan Melber-Wilkending for tailoring an Ansa-Grid to our needs. Furthermore, the authors gratefully acknowledge that this work could be conducted within the DLR project *Diabolo*, which is funded by the German Ministry of Defense and the BAAINBw.

References

1. Breitsamter, C.: Turbulente Strömungsstrukturen an Flugzeugkonfigurationen mit Vorderkantenwirbeln. PhD - TU München, Utz Verlag (1997)
2. Zastrow, J., Oberdieck, F., Henne, U., Klein, C.: Numerical and experimental investigations on the DLR-F23 combat aircraft wind tunnel model. In: ICAS Congress (2022)
3. Liu, T., Sullivan, J.P., Asai, K., Klein, C., Egami, Y.: Pressure and Temperature Sensitive Paints, 2nd edn. Experimental Fluid Mechanics (2021)
4. Schwamborn, D., Gardner, A., von Geyr, H., Krumbein, A., Lüdeke, H., Stürmer, A.: Development of the tau-code for aerospace applications. ICAS Congress (2008)
5. Shur, M.L., Spalart, P.R., Strelets, M.K., Travin, A.K.: A hybrid RANS-LES approach with delayed-des and wall-modelled les capabilities. Int. J. Heat Fluid Flow **29**(6), 1638–1649 (2008)
6. Probst, A., Löwe, J., Reuß, S., Knopp, T., Kessler, R.: Simulations with a low-dissipation low-dispersion second-order scheme for unstructured flow solvers. AIAA J. **54**(10), 2961–2971 (2016)
7. Jeong, J., Hussain, F.: On the identification of a vortex. J. Fluid Mech. **285**, 69–94 (1995)
8. Spalart, P.R.: Young-person's guide to detached-eddy simulation grids. Technical report, NASA (2001)
9. Payne, F.M., Ng, T.T., Nelson, R.C., Schiff, L.B.: Visualization and wake surveys of vortical flow over a delta wing. AIAA **26**(2), 137–143 (1988)
10. Rütten, M., Werner, M., Zastrow, J.: Invariant analysis of vortical delta wing flow using the extended optimal triple tensor decomposition. In: Stab Symposium (2022)
11. Rütten, M., Zastrow, J.: Vortex interaction, vorticity flux and torsion over a generic double-delta wing configuration. In: AIAA AVIATION (2022)

Aeroacoustic Noise

Relevant Turbulent Time Scale for Broadband Sound and the Prediction of the Grid Cut-Off Frequency for LES Based Aeroacoustic Simulations

Eman Bagheri, Florian Wachter(✉), and Stefan Becker

Institute of Fluid Mechanics LSTM, Friedrich-Alexander Universität Erlangen-Nürnberg, 91058 Erlangen, Germany
{eman.m.bagheri,flo.wachter,stefan.becker}@fau.de
https://www.lstm.tf.fau.de/

Abstract. In this work, we establish a relation between the generation of broadband sound for the inertial subrange in turbulent flows and the turbulence cascade mechanism. A dimensional analysis of the perturbed convective wave equation (PCWE) source term is used to obtain the time scale of the aeroacoustic source term and its relation to the turbulent length scales. Based on these findings, a new method to estimate the grid cut-off frequency (f_{GCO}) is proposed. The frequency f_{GCO} determines the maximum frequency up to which the turbulent structures are fully resolved by the grid of the flow simulation. For hybrid aeroacoustic simulations, this maximum resolvable frequency of the flow simulation also determines the maximum frequency content of the aeroacoustic sources and therefore the highest frequency that can be resolved in the subsequent acoustic simulation of broadband sound. The ability of the new method to estimate the f_{GCO} is tested against two common methods frequently used in literature. The test case of a forward-facing step is chosen which is a common aeroacoustic benchmark case and known to generate broadband sound. The new method shows promising results when the estimated grid cut-off frequency is compared against the estimations from the two reference methods.

Keywords: broadband sound · turbulent time scale · grid cut-off frequency

1 Relevant Turbulent Time Scale for Broadband Sound

1.1 Energy Cascade and Kolmogorov's Hypothesis

The idea of the energy cascade of turbulence was introduced by Richardson in 1922 and describes the transfer of energy from large scales of turbulent motion to smaller scales. The first basic concept is that turbulence can be considered to be composed of eddies of different sizes. The term eddy is not clearly defined

A. Dillmann et al. (Eds.): STAB/DGLR Symposium 2022, NNFM 154, pp. 329–338, 2024.
https://doi.org/10.1007/978-3-031-40482-5_31

but is commonly seen as a turbulent motion, localized in a region of size l which is at least moderately coherent over this region [1]. The second basic concept of the energy cascade is that large eddies are unstable, break up and transfer their energy to smaller eddies. This break-up process and energy transfer repeat for every level of smaller eddies until the eddy size and local Reynolds number become so small that molecular viscosity is effective in dissipating the kinetic energy of the smallest eddies.

For sufficiently high Reynolds numbers and within the universal equilibrium range, the transfer of energy to successively smaller scales is equal to the turbulent dissipation rate ϵ [1]. When the size of the eddy l is within the inertial subrange, an associated timescale τ_l can be defined. The time scale τ_l is referred to as the eddy turnover time, and it describes the time it takes for an eddy of length-scale l to break up into eddies of the next smaller level (say lengthscale $l/2$). From this definition and using dimensional analysis, an associated velocity scale u_l can be introduced

$$u_l \sim \frac{l}{\tau_l} \tag{1}$$

The dissipation rate of the turbulent kinetic energy ϵ can be obtained and related to u_l and τ_l by

$$\epsilon \sim \frac{u_l^2}{\tau_l} \sim \frac{u_l^3}{l} , \tag{2}$$

and by combining Eq. 1 and 2 a relation between the time and length scales can be acquired

$$\tau_l \sim \epsilon^{-\frac{1}{3}} l^{\frac{2}{3}} . \tag{3}$$

The pressure fluctuations associated with a turbulent motion of size l can therefore be defined as

$$p_l \sim \rho u_l^2 \sim \rho \epsilon^{\frac{2}{3}} l^{\frac{2}{3}} . \tag{4}$$

Thereafter, the derivative of pressure p_l for an eddy in the Lagrangian frame can be approximated. Considering the break-up of an eddy with lengthscale l into smaller eddies of lengthscale $l/2$ over the period τ_l, it yields

$$\frac{dp_l}{dt} \sim \frac{\rho \epsilon^{\frac{2}{3}} l^{\frac{2}{3}} - \rho \epsilon^{\frac{2}{3}} (\frac{l}{2})^{\frac{2}{3}}}{t + \tau_l - t} \sim \frac{\rho \epsilon^{\frac{2}{3}} l^{\frac{2}{3}}}{\tau_l} . \tag{5}$$

The turbulent kinetic energy is distributed among the eddies of different sizes. The turbulent motions of lengthscale l correspond to the wavenumber $\kappa = 2\pi/l$. The total turbulent kinetic energy k is obtained by integrating the energy spectrum over the whole wavenumber space

$$k = \int_0^\infty E(\kappa) d\kappa . \tag{6}$$

From Kolmogorov's second similarity hypothesis it follows that the energy spectrum function $E(\kappa)$ can be described in the inertial subrange as

$$E(\kappa) = C\epsilon^{2/3}\kappa^{-5/3} , \tag{7}$$

where C is a universal constant with $C \approx 1.5$.

1.2 Dimensional Analysis of the PCWE Acoustic Source Term

In this section, a dimensional analysis of the source term of the *perturbed convective wave equation* (PCWE) is carried out to relate the turbulent energy cascade to the primary mechanisms of broadband sound generation for the inertial subrange. The PCWE fully describes acoustic sources generated by incompressible flow structures and their wave propagation through flowing media. By introducing the acoustic scalar potential Φ_a, the speed of sound c_0, and the density ρ_0 of the acoustic medium, the perturbed convective wave equation is defined by

$$\frac{1}{c_0^2}\frac{D^2\Phi_a}{Dt^2} - \frac{\partial^2\Phi_a}{\partial x_i^2} = -\frac{1}{\rho_0 c_0^2}\frac{D\hat{p}}{Dt}\,, \quad \frac{D}{Dt} = \frac{\partial}{\partial t} + \overline{U_i}\frac{\partial}{\partial x_i}\,. \tag{8}$$

Considering that the turbulent velocity fluctuations u_i are much smaller than the mean flow velocity $\overline{U_i}$, the derivative of the incompressible pressure fluctuation $D\hat{p}/Dt$ can be approximated as the time derivative of the pressure fluctuation $d\hat{p}/dt$ in the Lagrangian frame, so that the PCWE source term S_{pcwe} can be approximated as

$$S_{pcwe} = -\frac{1}{\rho_0 c_0^2}\frac{D\hat{p}}{Dt} \approx -\frac{1}{\rho_0 c_0^2}\frac{d\hat{p}}{dt}\,. \tag{9}$$

A similar relation was derived by [2] for the noise source term of fine-scale turbulent structures using the gas kinetic theory analogy and considering fine-scale turbulence as small blobs of fluid moving randomly. Using Eq. 3 and Eq. 5, the dimensional analysis of the acoustic source term S_{pcwe} associated with the turbulent motion of lengthscale l results in

$$S^l_{pcwe} \sim \frac{1}{\rho_0 c_0^2}\frac{dp_l}{dt} \sim \frac{\rho}{\rho_0 c_0^2}\frac{\epsilon^{\frac{2}{3}} l^{\frac{2}{3}}}{\tau_l} \sim \frac{\rho}{\rho_0 c_0^2}\frac{l^2}{\tau_l^3} \sim \frac{\rho\epsilon}{\rho_0 c_0^2} \tag{10}$$

and reveals the dependency of the acoustic source term on the dissipation rate ϵ and the eddy-turnover time τ_l. This dependency, which was established from the perspective of dimensional analysis, implies that the primary mechanism of broadband sound generation in the inertial subrange is related to the turbulence cascade mechanisms.

2 Estimation of the Grid Cut-Off Frequency for LES-Based Aeroacoustic Simulations Regarding Broadband Sound

2.1 Problem Description

The *hybrid* aeroacoustic approach utilizes a separate treatment of the flow and the acoustic fields. In the first step, a scale-resolving flow simulation computes the near-field flow quantities from which acoustic source terms are obtained. Subsequently, the acoustic source term is used to compute the acoustic wave propagation in a separate simulation.

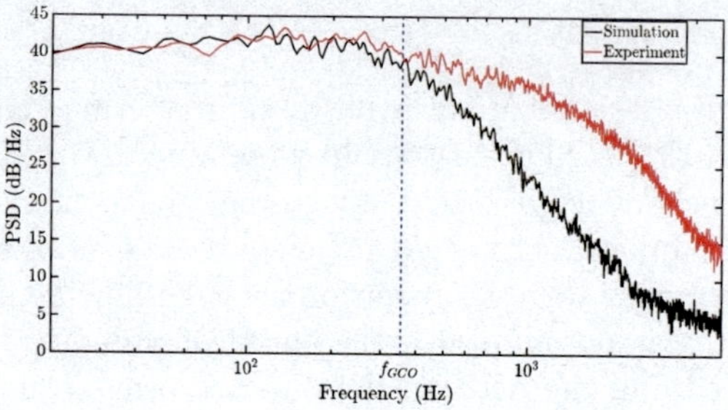

Fig. 1. Comparison of broadband sound spectra from simulation data and experiment. (Color figure online)

Large-eddy Simulation (LES) is widely used to compute the transient, turbulent flow fields. However, LES only resolves the turbulent eddies up to a certain lengthscale, often implicitly determined by the numerical grid of the flow simulation. The effect of turbulence that is not resolved by the grid is usually incorporated by a subgrid-scale model. Therefore, the maximum resolved wavenumber of the turbulent structures is set by the grid resolution. Besides the time-step increment in combination with the *Nyquist* theorem, the maximum resolved wavenumber in the flow simulation determines the maximum frequency content of the acoustic source terms, hence the maximum resolvable frequency in a subsequent acoustic simulation. Figure 1 exemplifies the broadband sound spectrum obtained from a hybrid aeroacoustic simulation in comparison to measurement data. Despite a sufficiently fine time-step and a well-refined acoustic grid, there exists a cut-off frequency beyond which the sound spectrum from the simulation trails off from the measurement data. This cut-off frequency is attributed to the resolution of the flow simulation grid. Therefore, such a behaviour indicates a lack of spatial resolution in the flow computation which leads to an underprediction of the sound pressure level [3]. Since in most acoustic applications, the maximum frequency of interest is known prior to the simulation; it is necessary to choose the grid resolution of the flow simulation such that the grid cut-off frequency (e.g. f_{GCO} in Fig. 1) matches the highest frequency of interest.

2.2 Two Frequently Used Approaches in Literature

There are two common methods available in the literature to determine the grid cut-off frequency f_{GCO} which is used to ensure that the acoustic source region is resolved up to the frequency of interest. Ideally, such methods should predict the f_{GCO} based on RANS (Reynolds-Averaged-Navier-Stokes) simulation data prior to the expensive LES and acoustic simulations. For both methods, the grid cut-off wavenumber $\kappa_{GCO} = 2\pi/2\Delta = \pi/\Delta$ is obtained from the *Nyquist* theorem in which Δ is the grid cell length. According to the *Nyquist* theorem, at least two grid cells per lengthscale are required to resolve a given wavenumber [4]. The first method (M1) is based on Taylor's frozen turbulence hypothesis. Thus, it relates the maximum resolved wavenumber and frequency to the convection velocity U_c and estimates the f_{GCO} via Eq. 11.

$$f_{GCO} = \frac{U_c \, \kappa_{GCO}}{2\pi} = \frac{U_c}{2\Delta} \tag{11}$$

The convection velocity U_c is defined as the velocity by which the turbulent eddy structures are transported downstream. Considering the mechanism of broadband sound generation explained in the previous section using Eq. 10, the frozen turbulence hypothesis is unsuitable for estimating the timescale in the inertial subrange. Furthermore, Eq. 11 assumes a wave-like propagation with a constant velocity for turbulent fluctuations, thus establishes a linear relationship between frequency and wavenumber. This is incompatible with the physics of turbulence and contrary to findings of the dimensional analysis presented in the previous section in Eq. 3. As a result, method M1 typically overestimates the cut-off frequency.

The second method (M2) was introduced by Mendonça [5] (see also [3]). This method attempts to address the problem of the frozen turbulence hypothesis in M1 by incorporating the turbulent kinetic energy into the equation. The grid cut-off frequency based on M2 is given in Eq. 12.

$$f_{GCO} = \frac{\sqrt{\frac{2}{3}k}}{2\Delta} \tag{12}$$

Although turbulence is taken into account by method M2, it still suffers from the assumption that the wavenumber and frequency are linearly related. This is simply because a wave-like propagation is assumed as in M1, but the propagation speed is changed to be the root mean square of the velocity fluctuations. Furthermore, the spectral distribution of the turbulent kinetic energy across different lengthscales is ignored. As velocity fluctuations are orders of magnitude smaller than the convection velocity, this method predicts much smaller cut-off frequencies compared to M1, often underestimating the f_{GCO} as it is shown in the upcoming Sect. 2.4.

2.3 A New Method to Estimate the Grid Cut-Off Frequency

In this section, we propose a new grid cut-off estimation method (M3) using the findings from dimensional analysis and considering the spectral distribution of turbulent quantities. The dependency of the acoustic source term on the dissipation rate and the eddy-turnover time in Eq. 10 implies that the primary mechanism of broadband sound generation in the inertial subrange is the turbulence cascade. Therefore, the frozen-turbulence hypothesis used in method M1 is invalid for the aeroacoustic sound generation of the inertial scales. The relation from Eq. 10 also explains why acoustic sources are dominant on solid walls where the turbulence dissipation rate is at its maximum. Therefore, the local turbulence intensity should not be considered as an indicator of the acoustic source intensity. Instead, the dimensional analysis shows that the source intensity for the inertial scales is attributed to the local turbulence dissipation rate. This is partly the reason why Eq. 12 leads to large errors in proximity to the wall where production and dissipation are not locally in balance. Moreover, Eq. 10 reveals that the eddy turnover time is indeed the relevant timescale for the acoustic source term generated by the inertial scales. The relation between the time and length scales in Eq. 3 shows that the assumption of a linear relationship between time and length scales (frequency and wavenumber) in M1 and M2 is inaccurate. As l is located in the inertial subrange, Eq. 3 can be rewritten as a function of wavenumber $\kappa = 2\pi/l$ using Kolmogorov's model for the energy spectrum of a fully turbulent flow (Eq. 7) to obtain the eddy turnover time τ_κ as a function of wavenumber:

$$\tau_\kappa = [\kappa^3 E(\kappa)]^{-\frac{1}{2}} = C^{-\frac{1}{2}} \epsilon^{-\frac{1}{3}} \kappa^{-\frac{2}{3}} . \tag{13}$$

The maximum wavenumber resolved by the grid in an LES with implicit filtering is $\kappa_{GCO} = 2\pi/2\Delta$, in which Δ is the local grid spacing. Thus, the grid cut-off frequency $f_{GCO} = {\tau_{GCO}}^{-1}$ follows from Eq. 13:

$$f_{GCO} = C^{\frac{1}{2}} [\epsilon \, \kappa_{GCO}^2]^{\frac{1}{3}} = C^{\frac{1}{2}} [\epsilon \, (\pi/\Delta)^2]^{\frac{1}{3}} . \tag{14}$$

Our proposed cut-off estimation method M3 in Eq. 14 expresses the dependency of the cut-off frequency on the turbulence dissipation rate. This is consistent with our analysis in Eq. 10 in which ϵ proves to play the key role in broadband sound generation by the inertial eddies. Furthermore, the spectral distribution of the turbulent kinetic energy and timescale of the turbulence cascade is built into M3. Consequently, this criterion provides an accurate estimation for the grid cut-off frequency, as will be discussed in the next sections.

2.4 Approach to Compare the Methods

To verify the capability of the proposed method for an accurate estimation of f_{GCO}, we performed wall-resolved large-eddy simulations on three numerical grids with different resolutions. The open-source code OpenFOAM was used for the simulations. A forward-facing step (FFS) was chosen as the testcase, which is a common aeroacoustic benchmark case and is known to generate broadband sound. As the present work considers transient, low Mach number flows ($\mathrm{Ma}^2 \ll 1$), we chose a finite volume-based LES with an incompressible, isothermal formulation. To model unresolved scales, the WALE (Wall-Adapting Local Eddy-Viscosity) subgrid-scale model was used. Additionally, RANS simulations were carried out with the k-ω-SST turbulence model with no wall-functions applied.

In Fig. 2, a lateral view of the simulation domain is shown. The step has a height h of 12 mm and faces an incoming turbulent boundary layer with a thickness $\delta \approx 11$ mm. The Reynolds number based on h is about 8000. A no-slip boundary condition is applied on the boundaries indicated with w in Fig. 2. Slip conditions are used at the boundaries labeled with s and periodic conditions are utilized at the lateral boundaries (perpendicular to the image plane). At the outlet o, a Neumann condition is applied. The method to develop a boundary layer with thickness δ was adopted from [6]. A recycling plane was used (indicated in red in the left part of Fig. 2) to map the velocity field back to the inlet boundary m. At the other inlet boundary u, a constant freestream velocity is set. 200 probes are distributed in the flow field domain for further spectral analyses. For the sake of brevity, we focus on the results extracted from probes P1 and P2 shown in Fig. 2 as they represent the spectral behaviour observed in other probes. Probe P1 is positioned in the region where an attached fully developed turbulent boundary layer is present and P2 is located in the wake region of the step.

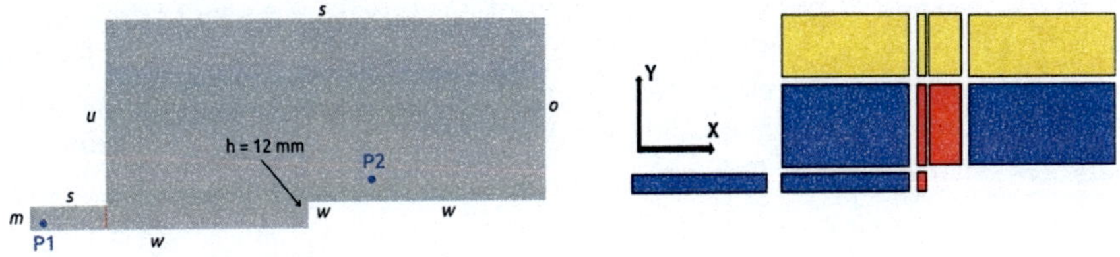

Fig. 2. Left: Simulation domain in side view with probe positions (P1, P2) and boundary conditions, e.g. (m) inlet velocity mapped from the recycling plane shown with a red dotted line, (u) fixed freestream inlet velocity, (o) outlet with Neumann condition, (s) walls with slip and (w) no-slip condition. Right: Block divisions used for the block-structured grid. The colours of the blocks indicate different refinement strategies. (Color figure online)

A fully orthogonal, block-structured grid was used for the simulations. The different block types are sketched in Fig. 2 (right). The yellow blocks indicate the turbulent-free regions where a relatively coarse grid resolution is required. The grid is uniform in the streamwise direction except in red blocks where a streamwise refinement was applied near the step. The *coarse* grid has a dimensionless cell width of $\Delta x^+ = 150$ in streamwise, $\Delta y^+ = 0.8 - 8$ in wall-normal, and $\Delta z^+ = 20$ in spanwise directions. The *medium* grid has a dimensionless cell width of $\Delta x^+ = 80$ and for the *fine* grid $\Delta x^+ = 30$. The spanwise and wall-normal resolutions are identical for all three grids. Note that for the comparison of the methods, the LES filter width Δ is set to be the maximum edge length of each cell. This means that the filter width is equal to the streamwise cell width except for a few cells near the step.

The local mean velocity $\overline{U}$ and the turbulent kinetic energy k computed from the LES of the *fine* grid are used as the input data for methods M1 and M2 respectively. The turbulence dissipation rate ϵ needed for method M3 is also obtained from the *fine* grid LES and is later compared to the result from the k-ω-SST RANS simulation. The dissipation rate was determined from the RANS simulation by the relation $\epsilon = \beta^* \omega k$, with $\beta^* = 0.09$ [7].

2.5 Results

Figure 3 (left) shows the resulting energy spectra at point probe P1 from LES with three different grids. The spectrum obtained from the *coarse* grid clearly trails off from that of the *medium* and *fine* grids. For the *coarse* grid, method M3 ("cutoff_Spectral") estimates the grid cut-off frequency to be at 362 Hz quite accurately. Method M1 ("cutoff_Taylor") results in an overestimated cut-off frequency 973 Hz, conversely, method M2 ("cutoff_Mendonça") underestimates it to be 74 Hz. A similar tendency can be found when the spectra from the *medium* and *fine* grids are compared. Again, the predicted $f_{GCO} = 539$ Hz from method M3 fits best to the frequency where the spectrum of the *medium* grid starts to dwindle away. Once more, M1 overestimates and M2 underestimates the cut-off frequency.

In Fig. 3 (right), the M3 criterion using LES and RANS results are compared for probe P1. The criterion using the turbulence dissipation rate obtained from k-ω-SST somewhat overestimates the grid cut-off frequency in comparison to the accurate prediction based on the LES data. This error is mainly due to the eddy-viscosity hypothesis as well as other simplifications made in the RANS model.

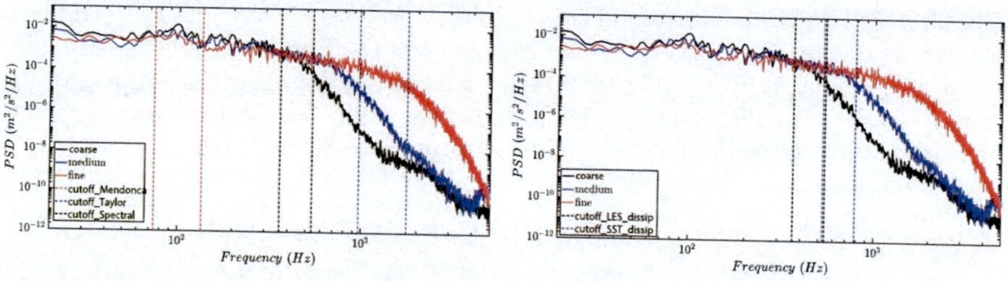

Fig. 3. Energy spectra from LES at probe position P1. Estimated f_{GCO} from the different methods are indicated based on LES data (left). Estimated f_{GCO} based on M3 from RANS data in comparison to f_{GCO} from LES data (right). (Color figure online)

Figure 4 (left) presents the energy spectra obtained from probe P2 located after the step in which a similar trend regarding the performance of the different methods is observed.

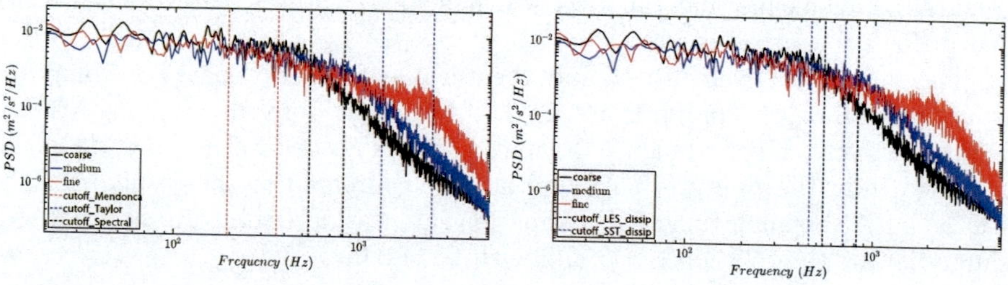

Fig. 4. Energy spectra from LES at probe position P2. Estimated f_{GCO} from the different methods are indicated based on LES data (left). Estimated f_{GCO} based on M3 from RANS data in comparison to f_{GCO} from LES data (right). (Color figure online)

Figure 4 (right) is a similar plot as in Fig. 3 (right) but for the probe P2, where the method M3 based on the RANS data slightly underestimates the cut-off frequency in comparison to the LES based approach. However, even with incorporating less accurate RANS modelled data into the M3 criterion, it still outperforms the other two methods fed with accurate LES-resolved data (compare both plots of Fig. 3 against each other, and both plots in Fig. 4).

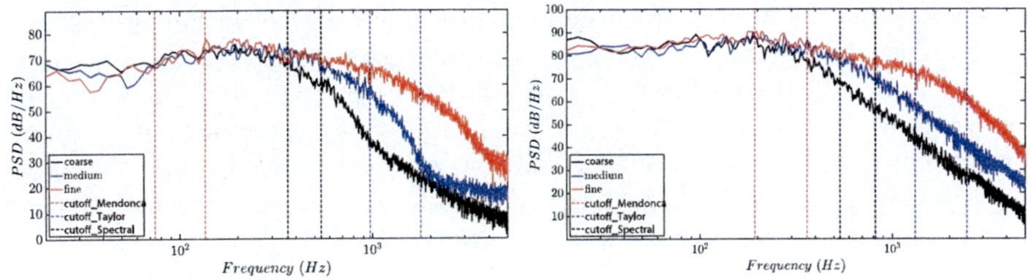

Fig. 5. Pressure spectra from LES at probe position P1 (left) and P2 (right). Estimated f_{GCO} from the different methods are indicated based on LES data. (Color figure online)

In Fig. 5, the obtained pressure spectra from position P1 and position P2 are shown together with the f_{GCO} predictions based on LES data. The pressure spectra at P1 (Fig. 5, left) show that the method M3 accurately estimates the f_{GCO}. Obviously, method M1 overestimates and method M2 underestimates the f_{GCO}. The pressure spectra at position P2 (Fig. 5, right) do not show such a clear superiority of method M3 over M2. While the spectra reveal a f_{GCO} of 600 Hz for the medium mesh, method M3 predicts $f_{GCO} \approx 717$ Hz and M2 $f_{GCO} \approx 363$ Hz. For the coarse mesh, the spectra show a f_{GCO} of 300 Hz. Method M3 estimates $f_{GCO} \approx 538$ Hz and M2 predicts $f_{GCO} \approx 194$ Hz. Regarding the pressure spectra at P2, both methods, M2 and M3, predict the f_{GCO} with an offset of about the same magnitude, with M2 underestimating and M3 overestimating the f_{GCO}. It is noteworthy that in comparison to M2, method M3 predicts the f_{GCO} for the medium mesh more accurately where the LES grid resolves finer turbulent structures and the assumption of the LES filter lying in the inertial subrange is more substantiated. Method M1 shows the worst performance in predicting the f_{GCO} in Fig. 5 (right).

2.6 Summary and Final Remarks

Using the dimensional analysis of the PCWE source term, an aeroacoustically relevant timescale for the inertial subrange was obtained and its relation to turbulence lengthscales was derived. Subsequently, the grid cut-off estimation method M3 was proposed. As expected from the dimensional analysis, the new method performs best for the FFS test case and the prediction of f_{GCO} always lies between that of the M1 and M2. However, method M3 requires reasonably accurate data for the local turbulence dissipation rate to provide valid estimations. As the accuracy of RANS models, especially in regard to ϵ varies from case to case, choosing an appropriate turbulence model is essential for accurate cut-off frequency estimations. As the energy spectra reveal, our proposed method based on RANS data still predicts the f_{GCO} more accurately than the other methods M1 and M2 do with LES data.

As a final remark, the authors want to emphasize that the method to estimate the mesh cut-off frequency is only an additional mesh resolution criterion to ensure that the highest frequency of interest is resolved within the relevant acoustic source region (see Sect. 2.1). All other essential criteria required for a proper large-eddy simulation should be additionally met. To give some examples: the LES filter width should lie within the inertial subrange as suggested, among others, by [1] and [4]. As a rule of thumb, Pope [1] recommends to resolve about 80% of the turbulent kinetic energy. For wall-bounded flows, near-wall mesh resolution requirements are given and discussed in [3].

References

1. Pope, S.: Turbulent Flows. Cambridge University Press, Cambridge (2000)
2. Tam, A.: Jet mixing noise from fine-scale turbulence. In: 4th AIAA/CEAS Aeroacoustics Conference (1998)
3. Wagner, C., Hüttl, T., Sagaut, P.: Large-Eddy Simulation for Acoustics, vol. 20. Cambridge University Press, Cambridge (2007)
4. Davidson, L.: Fluid mechanics, turbulent flow and turbulence modeling. Div. of Fluid Dynamics, Dep. of Applied Mechanics, Chalmers University of Technology, Göoteborg, Sweden (2019)
5. Mendonça, F., Read, A., Caro, S., Debatin, K., Caruelle, B.: Aeroacoustic simulation of double diaphragm orifices in an aircraft climate control system. In: 11th AIAA/CEAS Aeroacoustics Conference (2005)
6. Scheit, C.: Hybrid aeroacoustic methods for broadband noise calculation. PhD thesis, Friedrich- Alexander-Universität Erlangen-Nürnberg (FAU) (2016)
7. Wilcox, D.C.: Turbulence Modeling for CFD, 2nd edn. DCW Industries Inc. (2004). ISBN 1-928729-10-X

Numerical Study of Individuals Exposure to Road-Noise in Urban Environments

Amin Nabikhani(✉), Tobias Simon Müller, and Arne Henning

German Aerospace Center (DLR), Institute of Aerodynamics and Flow Technology, Bunsenstraße 10, 37073 Göttingen, Germany
amin.nabikhani@dlr.de

Abstract. In this paper a method is presented, which allows to simulate the traffic noise exposure of people living in an urban area. This is in contrast to classical noise maps, where the noise exposure of locations is calculated. A temporally resolved noise mapping is used in combination with an agent model, which simulates the location of persons in an urban area. Noise exposure is then assigned to agents based on their specific locations throughout the day, and averaged over time. The question is answered what difference in noise exposure the method shows in comparison to classical noise-maps. The method was successfully applied for a part of Berlin. It is shown that the average noise to which the agents are exposed during the day is increased due to their motion. This is not always due to the fact that people spend more time in places with high noise exposure. Rather, the usual consideration of the noise exposure in a log scale is responsible for this. This always results in a dominance of the noisier locations in the daily mean of the considered agents. It is to be discussed whether this shift towards higher noise exposure corresponds to the actual subjective perception of persons.

1 Introduction

The adverse impact of environmental noise on individuals are of interest to many medical and behavioral studies addressing insomnia, cognitive performance and even cardiovascular diseases. Civil noise caused particularly by transportation means are among the most significant sources. According to an estimation of the World Health Organization (WHO), at least one million healthy disability-adjusted life-years (DALYs) are lost annually due to environmental noise [1]. An accurate method for simulating the exposure of individuals to road-noise during a typical working day in urban environments can help to investigate the physiological impact of noise pollution on the population. It can also be used as an efficient means for planing urban environments with optimal balance between traffic flow efficiency and noise pollution. Studies in the past have shown that if the individual mobility of people is ignored, there may be errors in the calculation of the actual exposure to environmental effects like noise. The effect associated with ignoring these effects are called the neighborhood effect averaging problem (NEAP) [2]. Research in the past has investigated these effects by

A. Dillmann et al. (Eds.): STAB/DGLR Symposium 2022, NNFM 154, pp. 339–348, 2024.
https://doi.org/10.1007/978-3-031-40482-5_32

means of tracking individuals using personal noise dosimeters [3,4] or combining mobility data sets with pollution maps [5,6]. The present work combines three state-of-the-art simulation tools in order to provide an accurate account for the noise level perceived by a portion of the inhabitants of Berlin during a typical working day. Our study explores the impact of mobility on the exposure to environmental noise explicitly. The first tool in the simulation chain is the Travel Activity Pattern Simulation (TAPAS) which determines the traffic demand on the basis of a synthetic population of individuals [7]. This synthetic population is divided into different socio-economic and socio-demographic groups and daily activity plans are constructed for each individual. It is worth mentioning that the activity plans are not constructed randomly, but they are based on 35.000 diaries of 12.600 people who participated in a survey of the Federal Statistical Office of Germany in 2001/2002. After the traffic demand has been determined by TAPAS in form of origin, destination and arrival time for each individual, the second stage is the microscopic simulation of the traffic flow using SUMO (Simulation of Urban Mobility) [8]. As input to SUMO the true-to-scale road network and the corresponding schedules of the traffic-light systems are provided for the considered region in the center of Berlin. On the basis of this data SUMO simulates the traffic flow in the considered region during a full day. The output of SUMO is composed of the trajectories of all vehicles inside the considered region with a temporal resolution of one second. The last step in the simulation chain is the generation of a time-resolved noise map, which will be described in the following section in more detail. Finally, the resulting time-resolved noise map is combined with the results of TAPAS in order to generate a personal noise diary for every individual in the simulated population.

2 Generation of Time-Resolved Noise Maps

In order to evaluate the noise exposure of each individual in the simulated population, a time-resolved noise map is generated which covers a time frame of 24h for a complete district of Berlin city. The generation of this time-resolved noise map is based on the CNOSSOS-EU guideline [9] and is divided into a three-step process which is described in the following.

2.1 Geometry Processing

In a first step the relevant geometries are parsed, which includes the road network and the buildings inside the district. The corresponding data are provided respectively in form of a SUMO network [8] by the department of "Design & Bewertung von Mobilitätslösungen" of the DLR in Braunschweig and in CityGML format by the "Senate Department for Urban Development and Housing". The buildings are represented in LOD1, i.e., they are described by a 2D polygonal outline which is extruded to the buildings height. After the road network has been parsed, static emitter points with a mutual distance of 2 m are placed along the individual lanes [see Fig. 1(a)]. Similarly, static receiver points are placed

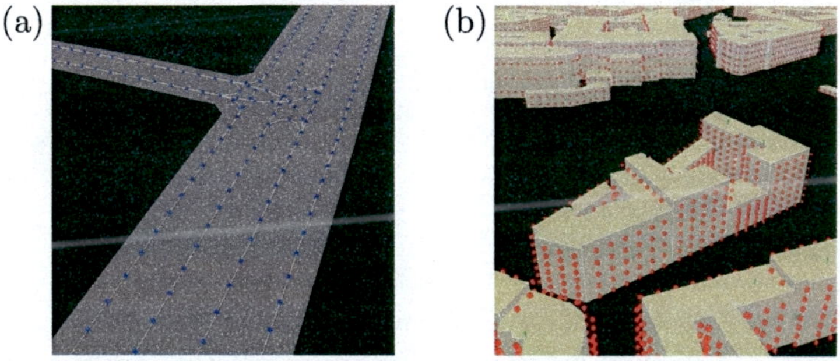

Fig. 1. **(a)** Static emitter points (blue) placed along the individual lanes of the road network. **(b)** Static receiver points (red) placed in front of building façades. (Color figure online)

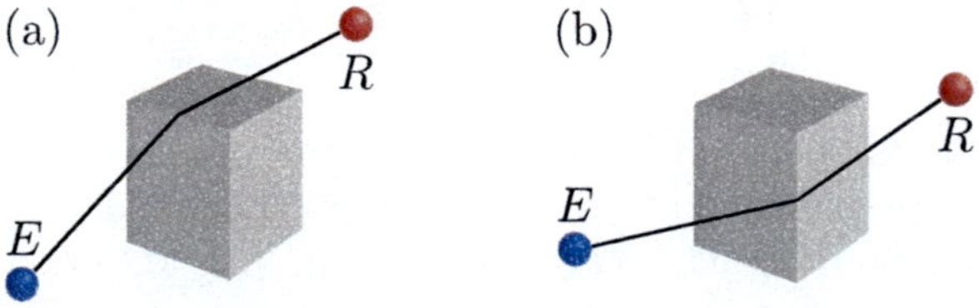

Fig. 2. The two kinds of diffraction patterns considered during the construction of the rays between an emitter E and a receiver R: **(a)** diffraction on the horizontal edges of obstacles and **(b)** "lateral" diffraction on the vertical edges of obstacles.

4 m above the ground on a regular grid covering the complete district. Static receiver points are also placed 2 m in front of building façades [see Fig. 1(b)]. As described in more detail later, the static emitter and receiver points are used as an approximation for the location of individual vehicles (emitters) and agents (receivers) respectively. Using static emitters and receivers allows us to precompute the rays connecting individual pairs of emitters and receivers in the ray-tracing phase (see Sect. 2.2), which can then be used to generate noise maps for arbitrary many traffic scenarios on the given geometry.

2.2 Ray Tracing

In the second step we search for rays which connect individual pairs of emitters and receivers. To this aim four different types of rays are considered, which are defined in the CNOSSOS-EU directive [9]. The most important difference between these ray types is the diffraction pattern considered. Type 1 and 2 rays allow for diffractions on the horizontal edges of buildings [see Fig. 2(a)], while Type 3 and 4 rays allow for diffractions on vertical edges [see Fig. 2(b)]. The difference between Type 1 and 2 rays and Type 3 and 4 rays respectively is that the former (Type 1 and 3) allow for diffractions only while the latter (Type 2 and 4) may include reflections on the building façades. Due to the different

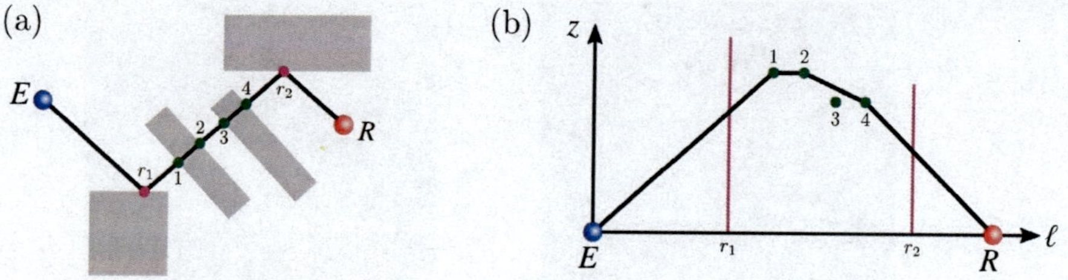

Fig. 3. **(a)** Top-down view of a Type-2 ray between an emitter E and a receiver R. The ray contains two reflections, r_1 and r_2, and intersects building edges at the points 1–4. **(b)** Construction of the ray profile in the vertical view obtained by "unfolding" the ray at the reflections. The two reflections r_1 and r_2 are indicated by vertical lines having the height of the building on which the respective reflection takes place. Furthermore the points 1–4 mark the locations of the roof edges where intersections have been observed in the top-down view of the ray.

nature of the diffractions involved, the procedure of constructing Type 1 and 2 rays differs to some extend from the one for Type 3 and 4 rays. Type 1 and 2 rays are constructed by using the image-source method in order to obtain a top-down view of all possible rays connecting an emitter to a given receiver by a series of reflections. These rays are permitted to intersect building edges in the top-down view, since the corresponding intersection points are candidates for horizontal diffractions at the roof edges. An example for such a top-down view of a single ray connecting an emitter E to a receiver R is shown in Fig. 3(a). In order to obtain the profile of the ray one constructs a vertical plane along each segment of the ray and one "unfolds" these planes at the reflection points like a Chinese screen. In the thus constructed "unfolded" vertical scheme reflections are marked by vertical lines having the height of the building the reflection takes place, and the locations of the roof-edge intersections encountered in the top-down view of the ray are marked by single points [see Fig. 3(b)]. The profile of the ray is then obtained by constructing the convex hull of the emitter, receiver and the roof-edge intersection points. Finally, the ray is validated by checking whether the ray intersects all vertical lines which mark the reflections. A simple example for this procedure is provided in Fig. 3. In order to construct Type 3 and 4 rays one first constructs the propagation plane, which is a tilted horizontal plane[1] containing receiver and emitter. The intersection of the propagation plane with the building outlines yields a 2D geometry in which the actual rays are constructed by an adaptation of the image-source methods which is able to incorporate diffractions at corners of the 2D building outlines. In both methods used for the ray construction the number of permitted reflections and lateral diffractions must be limited. The present results are obtained by allowing for a single reflection and up to four lateral diffractions. Furthermore we have not considered rays whose path length exceeds 350 m. The final step in the ray-tracing

[1] By this we mean that one of the two vectors defining the plane is perpendicular to the z-direction.

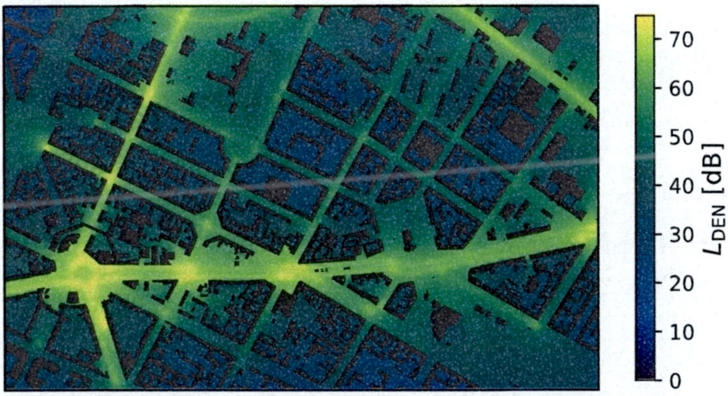

Fig. 4. Day-evening-night sound-pressure level L_{DEN} for a portion of Berlin Kreuzberg obtained from a time-resolved noise map which covers a full representative day.

algorithm is to compute the attenuation A of noise during its propagation along the individual rays. The attenuation rules used in the present implementation are described in detail in Sec. VI.4 of the CNOSSOS-EU directive [9].

2.3 Noise Calculation

The final step in the process is to determine the actual time-resolved sound-pressure level at the individual receiver points. To this aim we first parse the time-resolved traffic data, which are generated by means of a SUMO simulation [8] and are provided to us by the department of "Design & Bewertung von Mobilitätslösungen" of the DLR in Braunschweig. This data contains the type (passenger car, delivery van, etc.) of each vehicle present in the scene at a given time step, along with its location and velocity. Using these data one computes the sound-power emission $L(t, f_k)$ for the individual vehicles at each time step t and for all center frequencies f_k of the eight octave bands ranging from 125 Hz to 4 kHz, as described in detail in Sect. III.2 of the CNOSSOS-EU directive [9]. This sound power is then distributed across the two emitters which encompass the vehicles position on the lane it is driving on. In order to account for the position of the vehicle relative to these two emitters the sound-pressure level $L_i(t, f_k) = w_i L(t, f_k)$ ascribed to emitter $i = 1, 2$ is weighted by $w_i = 1 - s_i/(s_1 + s_2)$, where s_i is the distance between emitter i and the vehicles momentary position. Subsequently, we use the precomputed rays and propagate the noise from the emitters to the receivers. Let R_j be one of the receivers which is connected by a given ray to the emitter E_i. Assume that emitter E_i emits a total sound power $L_i(t, f_k)$ at time t, then the contribution

$$\Delta p_j(t + \ell/c) = \sum_{k=1}^{8} 10^{A_k/20}\, p_i(t, f_k) \tag{1}$$

is added to the sound-pressure p_j perceived at time $t + \ell/c$ at the receiver location R_j. Here c is the speed of sound and A_k is the attenuation for a sound wave with frequency f_k propagating along the ray which has a total path length ℓ. Summing all contributions $\Delta p_j(t)$ given in Eq. (1) for a given time t yields the time-resolved sound-pressure level $L_j(t) = 20\,\mathrm{lg}[p_j(t)/p_0]$ at the receiver site R_j. As an example, Fig. 4 shows the day-evening-night sound-pressure level L_{DEN} computed from the time-resolved noise-pressure level for a portion of Berlin Kreuzberg. The data used to generate this figure cover a full day with traffic conditions that are representative for an average day within this region.

3 Implementation

The demand simulation program TAPAS simulates the daily routines of the individual agents, from which the input data for the traffic simulation SUMO are generated. These included, whereabouts, travel times and choice of transport mode. It is therefore possible to calculate the noise exposure of the respective agents depending on their whereabouts. The starting points and destinations of the agents' trips are available as coordinates. These are assigned to individual buildings for which façade levels are available for coupling to the city's noise map. Before noise diaries are created, a specific façade immission location is assigned for each individual agent, at each of its locations. This is done by random selection. Each location is assigned a building, which in turn is assigned a number of facade immission locations. From these, an immission point is selected by drawing a random number. This assignment remains valid throughout the day. For example, for the time from midnight until the first time one leaves one's home, the same immission point is important as for the time from the last time one enters one's home until the end of the day. The random selection is due to a lack of information. Since there can be several apartments in a building, it is not known from the available data in which part of the building an agent lives. By randomly selecting the immission locations, the influence of this missing information should be kept small. In order to be able to make statements about the whole day, only agents can be considered which have their residence in this area and do not leave the area within the simulated day. For each of the scenarios discussed here, the daily schedules of 3,168,181 agents were created. Of these agents, 88,099 reside within the area. 15,320 of these agents do not leave the area. The agents that do not leave the area visit a total of 18,906 different locations. 18,192 of these locations are indoors. Of the eligible agents, 2,782 visit locations that are not in buildings and are therefore excluded. This leaves 12,538 included agents.

Two energy-equivalent average levels from two different level time courses are determined for each agent. One average level is determined at the facade immission point of the dwelling. The second one results from the level time course considering the different residence locations. Since the locations are given with a temporal resolution of one minute, the level time histories are also formed from levels averaged over one minute. A temporal weighting is omitted for the time being. If the energetic average over time is formed for each agent, a pair of values of energy-equivalent averaging levels results in each case. In the following, the average level at the facade of the agents' apartment is denoted by L_{home} and the average level resulting from the noise diary by L_{diary}.

4 Result

If the probability distribution in Fig. 5(a) or the probability density function of the average levels in Fig. 5(b) is considered, two distributions with different characteristics can be seen. The distribution of the levels at the apartment facades L_{home} is a bimodal distribution, where the lower maximum at 35–40 dB(A) with about 2250 agents is more pronounced than the upper maximum at 55–60 dB(A) with about 2000 agents. This bimodality can be explained by the location of the facade immission points. In the investigated area, the facade immission locations can be divided mainly into two categories. On the one hand there are facades, which are located directly at a street and on the other hand there are facades, which are located at the side of a building turned away from the street, for example in a backyard. Due to the attenuating effect of the obstacles, the road traffic noise level is self-explanatory significantly lower on the sides of the buildings facing away from the road. In the probability distribution of the average levels L_{diary} from the noise diaries this clear expression of two maxima is not to be seen. The upper maximum of the distribution here is also at 55–60 dB(A) but with 2800 agents it is much more pronounced than in the distribution of L_{home}. In the somewhat higher resolution probability density, a local maximum can be seen slightly above 35 dB(A), i.e. at the point where the probability density of L_{home} is maximum. However, this is not more pronounced than other local maxima between 45 and 50 dB(A), for example. A plateau is seen between 35 and 45 dB(A). It can be assumed that this is also a bimodal distribution, in which the lower maximum is much weaker than the higher one. Thus, by taking into account the location of the agents, there is a shift from lower averaging levels to higher averaging levels. This can also be seen in the mean values of the distributions. L_{home} has a mean value of 46.3 dB(A) and is thus 3.8 dB(A) lower than the mean value of L_{diary} with 50.1 dB(A). The medians of the two distributions differ slightly more than the mean, with a difference of 5.1 dB(A). The median of L_{home} is lower with 46.4 dB(A) than that of L_{diary} with 51.5 dB(A). Figure 6 (left) shows the pairs of values of the average levels L_{diary} over L_{home} from each agent in a scatter plot. Additionally, a red line is shown where the levels L_{home} and L_{diary} would be identical. It can be clearly seen that the scatter of the values depends on the value L_{home}. At lower levels, the largest distances to the red straight line above it are larger than the largest distances at higher levels. It is also clearly visible that the largest distances below the red straight line are smaller than the largest distances above the red straight line.

In Fig. 6 (right), the differences between L_{diary} and L_{home} over L_{home} are shown in a scatter plot for better illustration. Here, too, a red straight line is drawn at a difference of 0. In this diagram it can be seen that the maximum

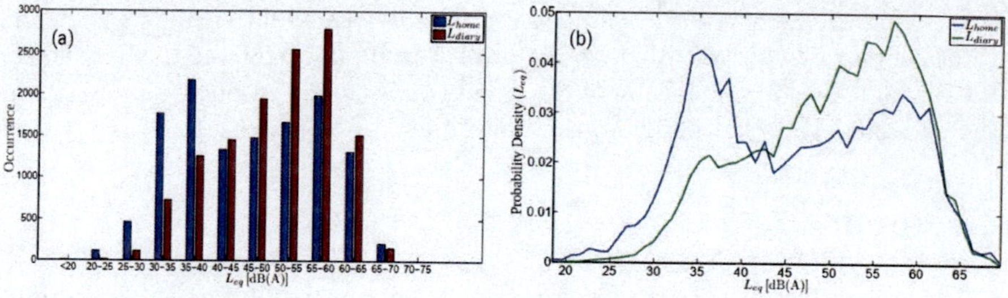

Fig. 5. (a) Occurrence distribution and (b) probability density function for the average sound level perceived by stationary (L_{home}) and vagrant agents (L_{diary}) in a span of a day.

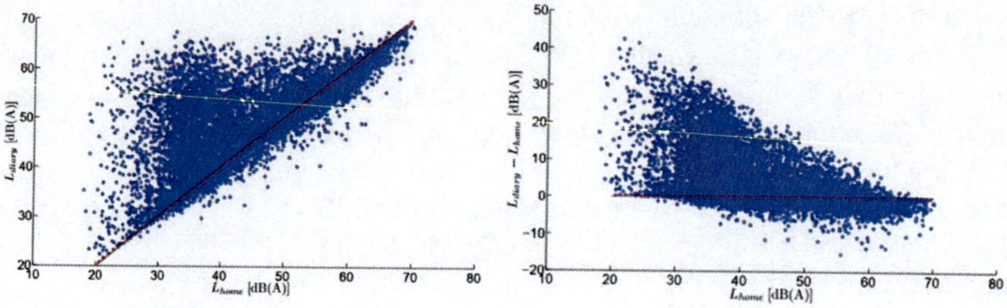

Fig. 6. (left) Scatter plot of the average level L_{diary} over the average level L_{home}. (right) Scatter plot of the differences of the averaging levels L_{diary} and L_{home} over the averaging level L_{home}. The values of the individual agents are represented by blue circles. The red line corresponds to a level difference of 0. (Color figure online)

magnification, i.e. values above the red straight line, falls with L_{home}. The maximum reduction, on the other hand, increases with L_{home}, but is much smaller in magnitude. The largest deviation into the positive range is about 42 dB(A), with the largest deviation into the negative range being about -15 dB(A). The decrease of the deviations into the positive range with increasing L_{home} or the decrease of the deviations into the negative range with decreasing L_{home} can be easily explained. An agent has to move from one environment to a louder environment to increase the mean level. However, if the agent is already in a noisy environment, it is more likely to move to a quieter environment. Analogously, if agents are in a quiet environment. They would have to move to an even quieter environment to achieve a lower averaging level. Again, the opposite is more likely. Agents who are already in a quiet environment move more often to a louder environment than to an even quieter one. In this case, there are 7392 agents whose L_{diary} value is higher than the L_{home}. This corresponds to 58.96%. For two agents L_{home} and L_{diary} are identical, for the remaining 5144 agents, i.e. 41%, L_{diary} is lower than L_{home}. Since the maximum of the frequency distribution of L_{home} is at a relatively low level, it is plausible that more agents move to a noisier environment than to a quieter environment when considering the daily schedule. The reason that the mean value of the L_{home} values is lower than the mean value of the L_{diary} values is on the one hand that more agents move into noisier environments, but on the other hand also that deviations into the positive difference range tend to be larger than deviations into the negative range. This fact can be well illustrated as in the following. If we imagine a quiet environment with an average level of L_1 and a noisy environment with an average level of L_2, we can calculate an average level L_{eq}, which would result if an agent were to stay in the noisy environment for a duration of t_2 and stay in the quiet environment for the rest of the observation time T, i.e. $t_1 = T - t_2$:

$$L_{eq}(t_2) = 10 \log_{10} \left[\frac{1}{T} \left((T - t_2) \cdot 10^{\frac{L_1}{10}} + t_2 \cdot 10^{\frac{L_2}{10}} \right) \right] \tag{2}$$

Figure 7 shows such a curve over one day, i.e. $T = 24h$ with $L_1 = 35$ dB(A) and $L_2 = 60$ dB(A). Due to the logarithmic scale of the level, the increase of the average level with increasing duration of stay is greater for short durations of stay t_2 than for longer durations of stay.

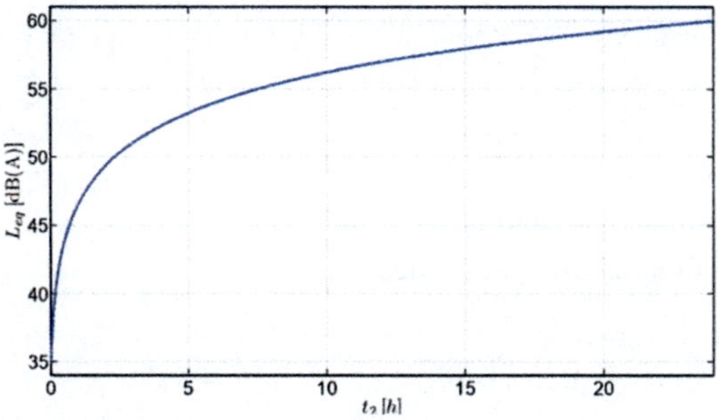

Fig. 7. Resulting average level over the time spent in a noisier area while staying in a quieter area for the rest of the day.

5 Conclusion

We conclude that the present combination of three state-of-the-art simulation tools allows us to account accurately for the noise level perceived by citizens in urban environments. We were able to reproduce the general trends in the noise-level distribution for stationary and vagrant agents and we quantified the impact on the exposure to road noise caused by traveling throughout the city or living in a road-side apartment. Further developments in this field could help to optimize the development of road networks and building blocks in urban environments which allow for a reduced noise pollution in critical areas and thus give rise to a higher overall quality of life.

References

1. Fritschi, L., Brown, A.L., Kim, R., Schwela, D., Kephalopoulos, S.: Burden of Disease from Environmental Noise. World Health Organization, Bonn (2011)
2. Kwan, M.P.: Int. J. Environ. Res. Public Health **15**(9) (2018). https://doi.org/10.3390/ijerph15091841. https://www.scopus.com/inward/record.uri?eid=2-s2.0-85052512806&doi=10.3390%2fijerph15091841&partnerID=40&md5=da530a4ed0152dc52b6d911a91db9d4c. Cited by: 64; All Open Access, Gold Open Access, Green Open Access
3. Kang, T.S., Lee, L.K., Kang, S.C., Yoon, C.S., Park, D.U., Kim, R.H.: J. Acoust. Soc. Am. **134**(1), 822 (2013). https://doi.org/10.1121/1.4807810
4. Ma, J., Li, C., Kwan, M.P., Kou, L., Chai, Y.: Environ. Int. **139**, 105737 (2020). https://doi.org/10.1016/j.envint.2020.105737. https://www.sciencedirect.com/science/article/pii/S0160412019318343
5. Lu, M., Schmitz, O., de Hoogh, K., Hoek, G., Li, Q., Karssenberg, D.: Environ. Model. Softw. **158**, 105555 (2022). https://doi.org/10.1016/j.envsoft.2022.105555. https://www.sciencedirect.com/science/article/pii/S1364815222002559
6. Park, Y.M., Kwan, M.P.: Health Place **43**, 85 (2017). https://doi.org/10.1016/j.healthplace.2016.10.002. https://www.sciencedirect.com/science/article/pii/S1353829216304415
7. Heinrichs, M.: Next Generation Forum 2011, ed. by P. Verkehr (Deutsches Zentrum für Luft und Raumfahrt e.V., 2011), pp. 74–74 (2011)
8. Lopez, P.A., et al.: The 21st IEEE International Conference on Intelligent Transportation Systems. IEEE (2018)
9. Kephalopoulos, S., Paviotti, M., Anfosso-Lédée, F.: Common Noise Assessment Methods in Europe (CNOSSOS-EU). Publications Office of the European Union (2012). https://doi.org/10.2788/32029

Aeroacoustics of High Fidelity URANS Simulations of Distributed Electric Propellers

Robin Wickersheim(✉), Manuel Keßler, and Ewald Krämer

Institute of Aerodynamics and Gas Dynamics, University of Stuttgart, Pfaffenwaldring 21, 70569 Stuttgart, Germany
robin.wickersheim@iag.uni-stuttgart.de
https://www.iag.uni-stuttgart.de/en/

Abstract. A distributed propeller configuration was investigated by using CFD flow field data and evaluating the acoustics in the far field with the Ffowcs Williams-Hawkings approach. The noise sources were localized by implementing two permeable surfaces within the CFD simulation as input for the acoustic formulation. It can be shown that solely obtaining the propeller noise is not sufficient. Since three propellers are present, the propeller-wing interaction noise has a great impact on the overall noise emission. To reduce the noise, the propellers were vertically shifted, which led to a reduction in propeller-wing interaction noise and an overall noise reduction of 2.4 dB in sound power level. Finally, an extensive parameter study was carried out, showed that propeller distance/phase and direction of rotation, had only minor influence on the overall noise emission.

Keywords: Computational Aeroacoustics · Distributed Propulsion

1 Introduction

To reduce the pollutant emission of aircraft propulsion and fulfill the requirements of the Flightpath 2050 [9], the aviation industry explores increasingly towards electric or hybrid electric propulsion. In this field, the configurations of distributed electric propulsion (DEP) have shown several aerodynamic advantages, e.g. an increase of lift due to the propeller stream [3,5,8]. The large number of degrees of freedom in order to layout the propeller positions lead to the need of an aeroacoustic study. Therefore, this paper uses high fidelity CFD results for an acoustic evaluation based on the Ffowcs Williams-Hawkings (FW-H) approach. This will help to understand the trade off between aerodynamic benefits and noise reduction. As the study takes place in the context of the LuFo VI-1 project VELAN, the setup consists of the wing geometry of University of Stuttgart's eGenius-Mod equipped with three propellers.

A. Dillmann et al. (Eds.): STAB/DGLR Symposium 2022, NNFM 154, pp. 349–358, 2024.
https://doi.org/10.1007/978-3-031-40482-5_33

2 Methodology

In this study the finite volume solver FLOWer was used to provide the URANS CFD flow field around the DEP configurations, further coupled to the acoustic FW-H solver ACCO [11]. The structured solver FLOWer was originally developed at the German Aerospace Center (DLR) [16] and significantly extended at the Institute of Aerodynamics and Gas Dynamics [12]. For temporal and spatial discretization a second order scheme is used. To reduce dissipation a sixth order Weighted Essentially Non-Oscillatory (WENO) scheme [4] is applied in the background cells instead of the second order Jameson-Schmidt-Turkel (JST) scheme and the Menter SST model [14] closes the RANS equations. As the focus concentrates on the parameter study of the propeller positions and the coupled acoustic evaluation, an infinite elongated wing behind the propellers is considered. Since the boundaries, like the symmetry boundary condition in spanwise direction, can cause undesired acoustic reflections [15], the cells are stretched along the spanwise direction to maximize the dissipation of any pressure waves before reaching the boundary. Additionally, the blocks containing the stretched grid cells are excluded from the WENO reconstruction to further increase the dissipation and prevent reflections. The full setup as well as the structured mesh can be seen in Fig. 1a. For the flow solution 22 revolutions are simulated, with the time step for the last six revolutions being set from 1° to 0.5°. The last two revolutions are then used as input for the acoustic evaluation.

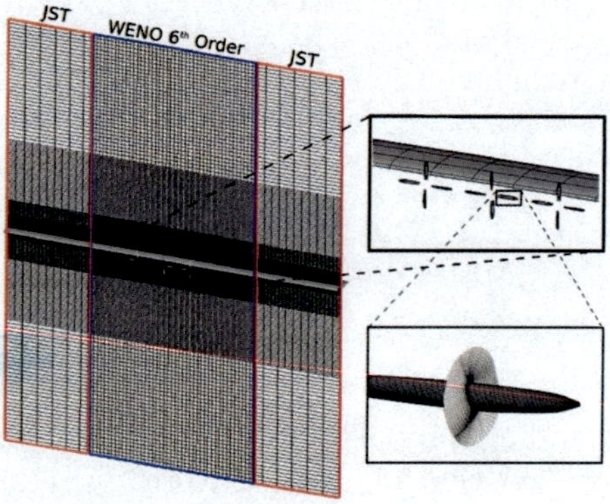

(a) CFD setup including background mesh and surface mesh in detail view

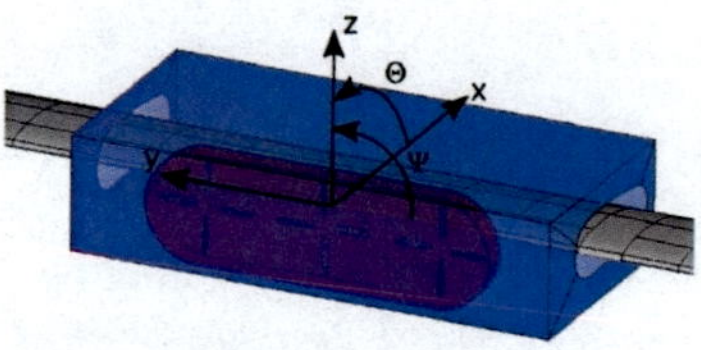

(b) Acoustic integration surfaces: FW-H 1 (blue) and FW-H 2 (red)

Fig. 1. Numerical setup: (a) CFD domain (b) Permeable surfaces

The flow field data of the URANS simulation is used as input for the Ffowcs Williams-Hawkings solver ACCO [11]. Therefore, FLOWer outputs flow data on a permeable surface enclosing the noise sources, called in the following study FW-H surface. Within this study, two FW-H surfaces, the first enclosing the propellers as well as the wing, and the second enclosing solely the propellers, are included, see Fig. 1b.

3 Aeroacoustics of a DEP Configuration

3.1 Geometry and Computational Setup

The baseline configuration consists of three counter rotating propellers (CR) mounted in front of a wing with a vertical distance referred to the wing leading edge (LE) of $z/R = 0\%$, a distance to the wing LE in x-direction relative to the wing chord of $x/c = -40\%$ and a propeller distance of $A/D = 25\%$. The propeller geometry is adapted from an APC 11×9 propeller scaled to a diameter of approximately $D = 0.255\,\text{m}$. Furthermore, the wing geometry is based on University of Stuttgart's eGenius-Mod. In further steps the distance between the propellers is increased to $A/D = 35\%$ and the rotation direction of the inner propeller is changed, which leads to a co-rotation of all three propellers. The rotation speed of all propellers is set to $\Omega = 8000$ RPM and the inflow velocity amounts to $u_\infty = 18\,\text{m}\,\text{s}^{-1}$. The ambient parameters result in a tip Mach number of $\text{Ma}_{\text{Tip}} = 0.3126$ and an advance ratio of $J = 0.53$.

3.2 Noise Generation of Distributed Propellers

At first, the fundamental noise sources of each structure separately are investigated. For this purpose, two different FW-H surfaces are used, the first enclosing the entire structures (FW-H 1) and the second only encasing the propellers (FW-H 2), see also Fig. 1b. Based on this approach, several noise sources can be extracted and the contribution of each structure to the overall noise can be quantified.

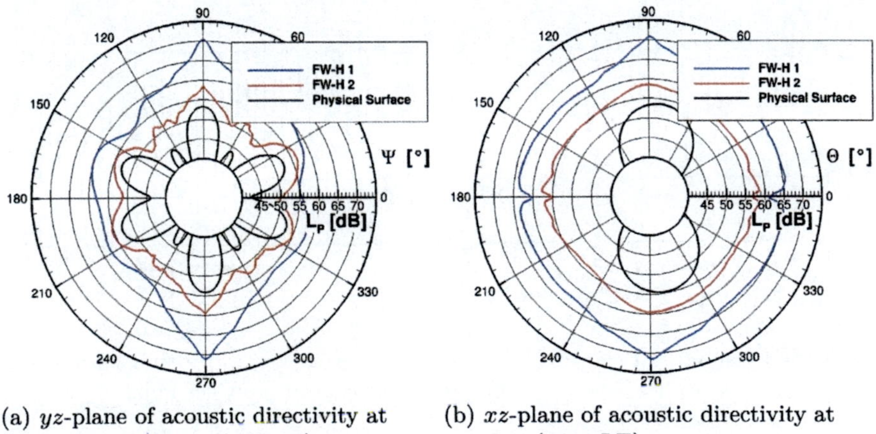

(a) yz-plane of acoustic directivity at $x = -0.13\,\text{m}$ (propeller plane)

(b) xz-plane of acoustic directivity at $y = 0\,\text{m}$ (wing LE)

Fig. 2. L_P-Directivity (observer radius $r_o = 10\,\text{m}$) of configuration 1 for the two FW-H surfaces and the physical propeller surfaces as input for the FW-H formulation.

The application of the FW-H surface determines the noise sources which are included and accordingly a difference between the sound pressure levels (SPL) is expected. Figure 2 shows the polar directivity of the SPL (L_P) for the two mentioned FW-H surfaces as well as for the physical blade surfaces. The Ψ-Directivity plot is shifted from the inertial coordinate system, referred to the leading edge of the wing, to the propeller plane. Whereas FW-H 1 contains all noise sources, as propeller noise and all propeller interaction sources, FW-H 2 takes only the propeller noise and the propeller-propeller interaction noise into consideration. Thus, the deviation between both FW-H surfaces has to be attributed to the propeller-wing interaction noise. By using only the physical surface of the three propellers as input for the acoustic evaluation, solely the monopole and dipole sources generated directly on the surface are determined. When regarding the deviations between FW-H 1 and FW-H 2 on one hand and the physical surface on the other hand, it is apparent that interaction phenomena are present and of significant magnitude in this setup. Furthermore, part of the difference between FW-H 2 and the physical surface can be explained by pressure waves reflected from the wing within the CFD simulation, entering the FW-H 2 domain.

Further insight can be achieved by a decomposition into the frequency domain, using the Fast Fourier Transformation (FFT) of the sound pressure time signal, evaluated in a distance $z = -1$ m below the propeller axis with a global angle of $\Theta = 270°$ (Position 1), see Fig. 3.

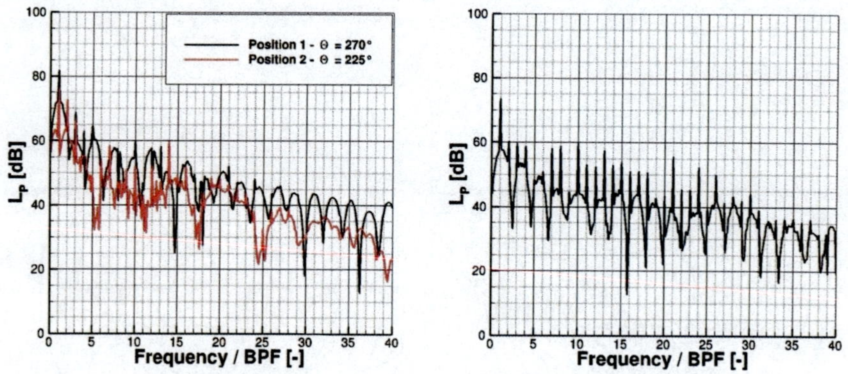

(a) FW-H 1 (b) FW-H 2

Fig. 3. FFT analysis for the two FW-H surfaces normalized to the first harmonic (1. BPF) at an observer position $z = -1$ m below the central propeller hub.

Using both FW-H surfaces for this investigation concludes that the first Blade Passing Frequency (BPF) is dominant. Nevertheless, the first BPF seems to be of a greater magnitude by using FW-H 1 in contrast to FW-H 2. Accordingly, in addition to the propeller noise another noise source has to be present, also emitting at the frequency of the first BPF. Considering the frequency spectrum in

Fig. 3a, it can be seen that the peaks blur by using the FW-H 1 instead of a sharp peak as in Fig. 3b. This phenomenon vanishes by observing a point at $\Theta = 225°$, so the observer is located more in front of the propeller with a distance of 1 m (Position 2). The effect can be referred to the included turbulent flow within the permeable surface and thus the difference in the effective acoustic path from the noise source to the observer, whereby turbulent fluctuations convect the waves differently. This phenomenon is noticeable if the observer position is parallel to the flow direction and is known as hay stacking due to the form of the blurred peak [7].

3.3 Propeller-Wing Interaction Noise

It was shown above that the geometry of the acoustic integration surface is crucial for the noise determination. Since the FW-H 2 surface only encloses the propellers, consequently only the propeller noise and propeller-propeller interaction noise will be considered. The aberration between the FW-H surfaces is therefore referred to the propeller-wing interaction caused by the propeller tip vortices convecting around the wing, which was mentioned above as additional noise source [1,2,6].

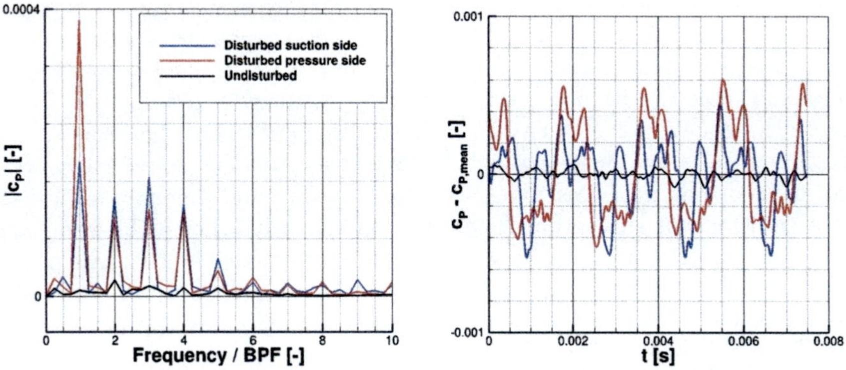

(a) FFT analysis (b) Time signal

Fig. 4. Influence of the propeller wake on c_P in time and frequency - Disturbed behind the propeller and undisturbed on suction side of the wing.

For this purpose, the surface pressure c_P was investigated by plotting the time signal of a point on the suction and pressure side on the wing directly behind a propeller compared to the undisturbed area nearby the symmetry plane. It can be seen in Fig. 4 that the surface pressure fluctuates, especially at the pressure side, with the first BPF as the tip vortex is responsible for the fluctuations.

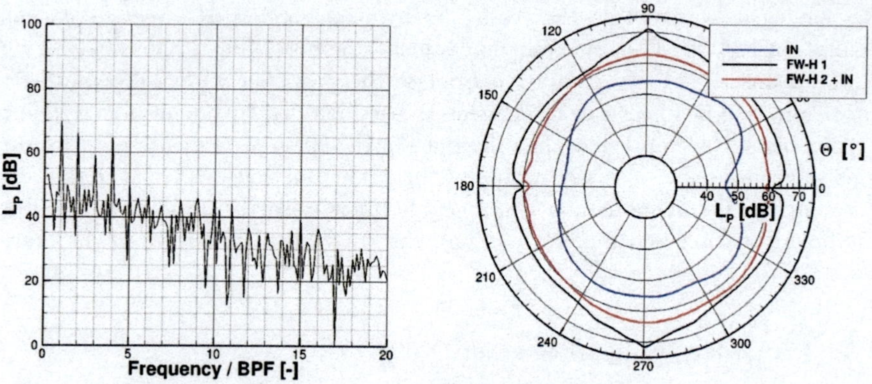

(a) FFT analysis

(b) L_P-Directivity (r_o = 10 m) of additional interaction noise

Fig. 5. Additional noise based on the propeller-wing interaction – FFT analysis $z = -1$ m below the propeller axis and directivity of the interaction noise (IN).

These pressure fluctuations cause the additional noise source due to the propeller - wing interaction and are investigated further within the acoustic evaluation. Therefore, solely the wing surface was specified as integration surface for the acoustic evaluation. By performing an FFT analysis on the acoustic pressure, as expected the noise emitted with the frequency of the first BPF from the wing is dominant, see Fig. 5a. Furthermore, the directivity was investigated since the directivity of FW-H 1 and FW-H 2 differs. As the SPL of the propeller-wing interaction noise has the same magnitude as the propeller noise, the contribution to the overall noise has to be taken into account and therefore the FW-H 2 is not sufficient for this setup. It can be seen that with the combination of the noise from FW-H 2 and the wing interaction noise using the coherent method, most of the noise sources were recognized since only small deviations are present between FW-H 1 and FW-H 2 + Interaction noise (IN), see Fig. 5b.

3.4 Near-Field Acoustics and Spurious Noise

Using the FW-H approach with a permeable surface can cause undesired acoustic signals by convecting vortices through the surface, known as spurious noise [13]. Since three propeller wakes are present in this study and the closing downstream surface of the FW-H 1 and FW-H 2 permeable surfaces cuts the tip vortices, spurious noise is shortly investigated for this setup. Therefore, the downstream surface of FW-H 1 as well as for FW-H 2 was neglected, where spurious noise might occur [10,13,17]. The directivity of the gained results is given in Fig. 6a. It can be seen, that the neglect of the downstream part for each FW-H surface is not evident for the differences between the two FW-H surfaces. Nevertheless, some discrepancies by using the incomplete surfaces are present at the direction of flow ($\Theta = 0°$ and $\Theta = 180°$). However, using the incomplete surfaces can cause more trouble than it prevents, since significant information about the sound field will also be neglected. Finally, it can be seen that spurious noise is present, but in this application is a minor problem in contrast to waive the downstream information.

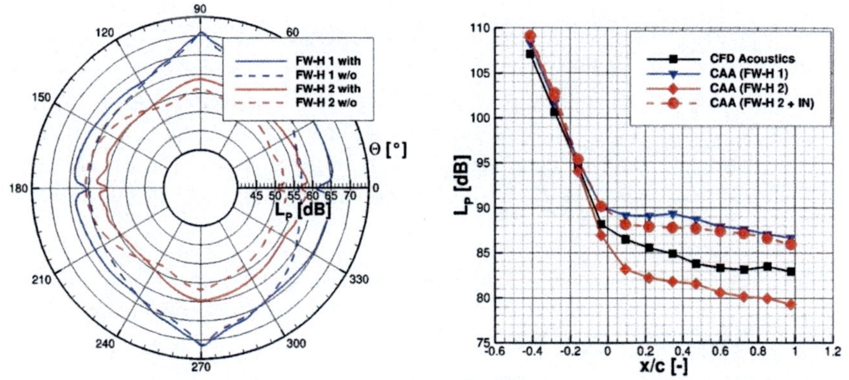

(a) Assessment of spurious noise by neglecting the downstream surface of the FW-H surfaces (xz-plane of acoustic directivity at $y = 0\,\mathrm{m}$ and $r_o = 10\,\mathrm{m}$).

(b) SPL comparison of CFD and CAA acoustics on discrete points ($z = -0.17\,\mathrm{m}$ below the propeller axis)

Fig. 6. Influence of neglecting the downstream surface and near-field acoustics comparing the direct CFD acoustics to the CAA approach.

The fluid domain consists of a fine resolution nearby the part structures, i.e. the propeller blades and the wing. Therefore, the near-field acoustics is directly given by the pressure fluctuations in the CFD solution. A comparison of the direct acoustics evaluated within the fluid domain and the FW-H approach can give further insights for the development of spurious noise. Figure 6b shows the discrepancies among the approaches by plotting the SPL over horizontal positions parallel to the stream direction. Whereas directly below the propeller axis all approaches show the same magnitude, the deviation increases further below the wing. However, it seems reasonable that FW-H 2 shows a lower SPL value below the wing than the CFD acoustics and the FW-H 1. By combining FW-H 2 and the interaction noise resulting from the propeller-wing interaction the curve differs only slightly form FW-H 1. The shown deviation to the CFD acoustics might result from the dissipation due to finite grid resolution. Looking at the frequency domain, a deviation of approximately 2 dB can be found in the first BPF, which underlines the given statement. Considering the near-field investigation, where dissipation results in 2 dB noise reduction, and spurious noise, the difference between FW-H 1 and CFD acoustics can be explained.

4 Variation of the Propeller Positions, Direction of Rotation and Propeller Phase

As shown above, propeller-wing interaction noise is of significant relevance for this setup and therefore an optimization strategy is presented below. To cover a wider range of possible propeller positions, the vertical distance related to the wing LE is varied. This will influence aerodynamics and acoustics. In Table 1 the Sound Power Level (SWL) L_{W} for three positions is given. As expected the noise emission decreases due to lower propeller-wing interaction noise, which will be discussed below, and the same holds for the interaction noise ($L_{\mathrm{W,IN}}$). When considering that the focus of DEP aims towards a lift increase, the lift coefficient was additionally investigated in the mentioned variation of the propeller position.

Table 1. Sound Power Level (O = Overall and IN = Interaction Noise) of several configurations by varying the vertical distance to the wing LE and comparison of thrust and lift (CR = Counter Rotating).

	CR – $z/R = 0\%$	CR – $z/R = 100\%$	CR – $z/R = -100\%$
Thrust [N]	8.81/8.82/8.81	8.6/8.61/8.6	8.74/8.74/8.74
C_{L} [-]	0.2765	0.2659	0.229
$L_{\mathrm{W,O}}$ [dB]	87.576	85.278	85.209
$L_{\mathrm{W,IN}}$ [dB]	75.992	73.064	73.708

Both vertical positions above and below the wing result in a noise reduction. Almost 2.4 dB in SWL can be achieved in overall noise, whereas the thrust decreases slightly for the vertical position above and below. However, while the deviations in thrust are small, the lift coefficient decreases significantly by mounting the propeller below the wing, see Table 1. To identify the reason for the noise reduction, the SWL was investigated for the propeller-wing interaction by evaluating the wing surface as input for the FW-H formulation, as in Sect. 3.3. As expected the noise decreases for the propeller positions above and below the wing, by about 3 dB (SWL).

In addition to the vertical propeller positions, the direction of rotation, propeller distance and relative propeller phase is varied, see Table 2. Compared to the vertical distance, only slight deviations are present. However, increasing the propeller distance to $A/D = 35\%$ decreases the SWL by around 0.5 dB, with nearly the same generated thrust. Setting the propeller phase to $\Delta\psi = 30°$ between the three propellers causes as well a noise reduction of 0.5 dB, whereas co-rotating propellers (CoR) had no influence on the noise. For all propeller settings in Table 2 the propeller thrust was almost constant and the effect on the lift coefficient was not significant.

Table 2. Sound Power Level of different configurations by varying the direction of rotation, propeller distance and phase angle (CR = Counter Rotating, CoR = Co-Rotating) and A/D in parentheses.

	CR – (25%)	CoR – (25%)	CR – (35%)	CR – (25%) Phase
Thrust [N]	8.81/8.82/8.81	8.81/8.81/8.81	8.81/8.82/8.81	8.81/8.82/8.81
C_L [-]	0.2765	0.2771	0.2767	0.2766
$L_{W,O}$ [dB]	87.576	87.568	87.0396	87.075

5 Conclusions

The present study gives insight into the aeroacoustic phenomena of distributed propulsion configurations as investigated in the LuFo VI-1 project VELAN. The results show that the interaction phenomena between propellers and wing can not be neglected and consequently the acoustic integration surface has to be adjusted. Furthermore, propeller-wing interaction noise seemed to be of a high proportion of the overall noise. By separating the acoustic phenomena with several integration surfaces, the noise sources could be determined, which was used to optimize the configuration. Finally, with changing the vertical propeller position, propeller-wing interaction noise could be reduced, affecting also the lift and thrust. The vertical shift of the propellers showed significant noise reduction. However, a decrease in lift and thrust has to be taken into account, where the lift loss was lower for the $z/R = 100\%$ position. Increasing the propeller distance from $A/D = 25\%$ to $A/D = 35\%$ and adapting the propeller phase had also a positive effect by reducing slightly the noise with almost no loss in thrust and lift. In contrast, the direction of rotation had relatively no influence on lift and noise emission.

Acknowledgment. The authors gratefully acknowledge the Federal Ministry for Economic Affairs and Climate Action (BMWK) for funding this work in the framework of the research project VELAN (FKZ: 20E1919).

Gefördert durch: Bundesministerium für Wirtschaft und Klimaschutz aufgrund eines Beschlusses des Deutschen Bundestages

References

1. Akkermans, R.A.D., Pott-Pollenske, M., Buchholz, H., Delfs, J.W., Almoneit, D.: Installation effects of a propeller mounted on a high-lift wing with a coanda flap. Part I: aeroacoustic experiments. In: 20th AIAA/CEAS Aeroacoustics Conference, Atlanta, GA (2014)
2. Akiwate, D.C., Parry, A.B., Joseph, P., Paruchuri, C.: Analytical investigation of propeller-wing interaction noise. In: 28th AIAA/CEAS Aeroacoustics Conference, Southampton, UK (2022)
3. Bongen, D., et al.: Simulation of a Distributed Propulsion System in a Wind Tunnel. AIAA Aviation, Chicago (2022)

4. Borges, R., Carmona, M., Costa, B., Don, W.S.: An improved weighted essentially non-oscillatory scheme for hyperbolic conservation laws. J. Comput. Phys. **227**(6), 3191–3211 (2008)
5. Deere, K., Viken, S., Carter, M., Viken, J., Wiese, M., Farr, N.: Computational analysis of powered lift augmentation for the LEAP tech distributed electric propulsion wing (2017)
6. Dierke, J., Akkermanns, R.A.D., Delfs, J.W., Ewert, R.: Installation effects of a propeller mounted on a high-lift wing with a coanda flap. Part II: numerical investigation and experimental validation. In: 20th AIAA/CEAS Aeroacoustics Conference, Atlanta, GA (2014)
7. Ewert, R., Kornow, O., Delfs, J.W., Yin, J.: A CAA based approach to tone haystacking. In: AIAA/CEAS Aeroacoustics Conference (2009)
8. Firnhaber-Beckers, M., Schollenberger, M., Lutz, T., Bongen, D., Radespiel, R., Florenciano, J.L., Funes-Sebastian, D.E.: CFD Investigation of High-Lift Propeller Positions for a Distributed Propulsion System. AIAA Aviation, Chicago (2022)
9. Flightpath 2050, Report of the High Level Group on Aviation Research. Technical Report, EC Directorate for Research and Innovation Directorate General for Mobility and Transport, European Union (2011)
10. Ikeda, T., Enomoto, S., Yamamoto, K., Kazuhisa, A.: Quadrupol corrections for the permeable-surface Ffowcs Williams-Hawkings equation. AIAA J. **55**(7), 2307–2320 (2017)
11. Keßler, M., Wagner, S.: Source time dominant aeroacoustics. Comput. Fluids **33**(5–6), 791–800 (2004)
12. Kowarsch, U., Oehrle, C., Hollands, M., Keßler, M., Krämer E.: Computation of helicopter pheonomena using a higher order method. In:Nagel W.E., Kröner D.H., Resch M.M., High Performance Computing in Science and Engineering 2013, pp. 423-438 (2013)
13. Lopes, L.V., Boyd, D.D.Jr., Nark, D.M., Wiedemann K.E.: Identification of spurious signals from permeable Ffowcs Williams and Hawkings surfaces. In: AHS 73rd Annual Forum, Fort Worth, Texas, USA (2017)
14. Menter, F.R.: Two-equation eddy-viscosity turbulence models for engineering applications. AIAA J. **32**, 1598–1605 (1994)
15. Peric, R.: Minimizing undesired wave reflection at the domain boundaries in flow simulations with forcing zones. PhD thesis, Technical University of Hamburg (2019)
16. Raddatz, J., Fassbender, J.K.: Block structured Navier-Stokes solver FLOWer. In: Kroll, N., Fassbender, J.K. (eds.) MEGAFLOW-Numerical Flow Simulation for Aircraft Design, Notes on Numerical Fluid Mechanics and Multidisciplinary Design (NNFM), vol. 89, pp. 27–44. Springer, Berlin, Heidelberg (2005). https://doi.org/10.1007/3-540-32382-1_2
17. Ricciardi, T.R., Wolf, W.R., Spalart, P.: On the application of incomplete FWH surfaces for aeroacoustic predictions. AIAA J. **60**(3) (2022)

Flow Control

A New Approach of Using Porous Bleed Boundary Conditions - Application of Local Porosity

Julian Giehler(✉), Pierre Grenson, and Reynald Bur

DAAA, ONERA, Université Paris Saclay, 92190 Meudon, France
{julian.giehler,pierre.grenson,reynald.bur}@onera.fr

Abstract. Porous bleed systems are widely used to mitigate the shock-induced boundary layer separation, e.g., in supersonic air intakes. However, the complex geometry makes simulations expensive and motivates the application of suitable models. Existing models are based on applying a continuous blowing/suction boundary condition (continuous porosity) along the porous plate, leading to an overestimation of the boundary layer thinning. The new method using distributed suction by a local porosity is applied on a Mach 1.6 supersonic flow and improves the prediction of the effect on the boundary layer significantly. In contrast to the existing methods, the wall shear stress is not overestimated, and the boundary layer profiles downstream of the bleed region are better reproduced. Moreover, all expansion fans and barrier shocks caused by the bleed holes are captured.

Keywords: Porous bleed · Suction · Flow control · Boundary condition · Supersonic flow

1 Introduction

Porous bleed systems are a commonly used and proven technology to mitigate shock-induced boundary layer separations in supersonic air intakes. The general principle is the removal of low-momentum flow close to the wall by placing a porous plate below the shock foot to suck air through tiny holes in a lower-pressured cavity. However, estimating the so-called bleed mass flow rate and forecasting the impact on the boundary layer are challenging since simulations of the complete system are expensive in terms of computational costs. Typical systems consist of hundreds of holes with a diameter approximately the size of the (compressible) boundary layer displacement thickness. This results in a large number of cells required to model the geometry and makes even Reynolds-Averaged Navier-Stokes (RANS) simulations too cost- and time-consuming for a pre-design of an air intake.

As early as the first numerical simulations on porous bleed systems were performed, bleed models were proposed as the computational resources were insufficient to simulate the flow inside the holes [1–3]. These models are utilized as a sucking/blowing boundary condition generally dependent on the cavity plenum

A. Dillmann et al. (Eds.): STAB/DGLR Symposium 2022, NNFM 154, pp. 361–371, 2024.
https://doi.org/10.1007/978-3-031-40482-5_34

pressure and the external wall pressure. In recent years, the modeling approaches improved due to more extensive computational resources and (numerical) studies investigating the flow behavior through the porous plate [4–6]. Nevertheless, existing models still lack reproducing the flow field obtained in reality. Geometrical parameters or inflow conditions (boundary layer properties) are often not considered in the computation of the bleed mass flow rate.

Moreover, the effect on the boundary layer needs to be taken into account, which can result in a bad design of the bleed system. Consequently, a new modeling approach is needed to mimic the complex flow physics correctly. In the current study, we propose a new approach to implement bleed boundary conditions using a more realistic local distribution of suction.

This paper is divided into six sections. Section 2 introduces the investigated physical problem. Sect. 3 gives a brief overview of the drawbacks of using a continuous porosity for all cells. The new methodology of the local porosity is described in Sect. 4 before showing the results of the new approach in Sect. 5. Our conclusions and outlooks to future work are drawn in the final section.

2 Investigated Physical Problem

The investigated problem is a turbulent supersonic flow on a flat plate with a porous bleed region. In this scenario, the porous bleed aims to thin the boundary layer, similar to the application on a supersonic air intake upstream of a shock. Figure 1 illustrates the investigated domain. A turbulent boundary layer profile with a Mach number of $M = 1.6$, generated in a previous flat plate simulation, is applied as an inflow condition. The porous plate has a length of $L_p = 40$ mm, and is located with an offset of 40 mm to the inlet. In the span, the domain size corresponds to the distance between two lines of holes. On the center planes of the holes, symmetry is assumed.

The boundary layer properties (boundary layer thickness δ_{99}, displacement thickness δ_1, momentum thickness δ_2, and shape factor $H = \delta_1/\delta_2$) are extracted 5 mm upstream of the first hole and shown in Table 1. Moreover, the total pressure p_t and the total Temperature T_t are given.

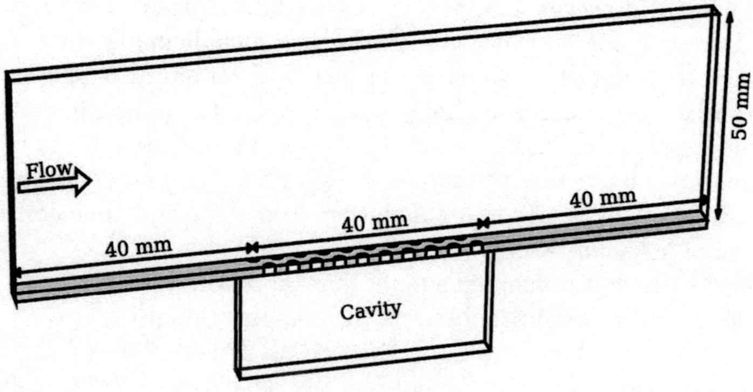

Fig. 1. Schematic of investigated numerical domain

Table 1. Flow properties 5 mm upstream of the first bleed hole

M	p_t	T_t	Re_{δ_2}	δ_{99}	δ_1	δ_2	H
1.6	93 000 Pa	300 K	5450	4.20 mm	0.57 mm	0.42 mm	1.36

In most applications, the bleed mass flow rate is determined by fixing a plenum exit area that is choked. As a result, the static pressure inside the cavity plenum changes until a mass balance between the mass flow rate entering the cavity through the porous plate and the mass exiting through the nozzle sets in. For the simulations, a throat ratio TR is defined, which corresponds to the ratio

$$TR = \frac{A_{pl,ex}}{A_{bl}} \tag{1}$$

of the plenum exit area $A_{pl,ex}$ and the bleed area A_{bl}, which is the sum of all hole areas.

3 Drawbacks of Continuous Porosity Boundary Conditions

In our previous study [7], several bleed models were implemented within the compressible multi-block structured Safran-ONERA finite-volume solver **elsA**.

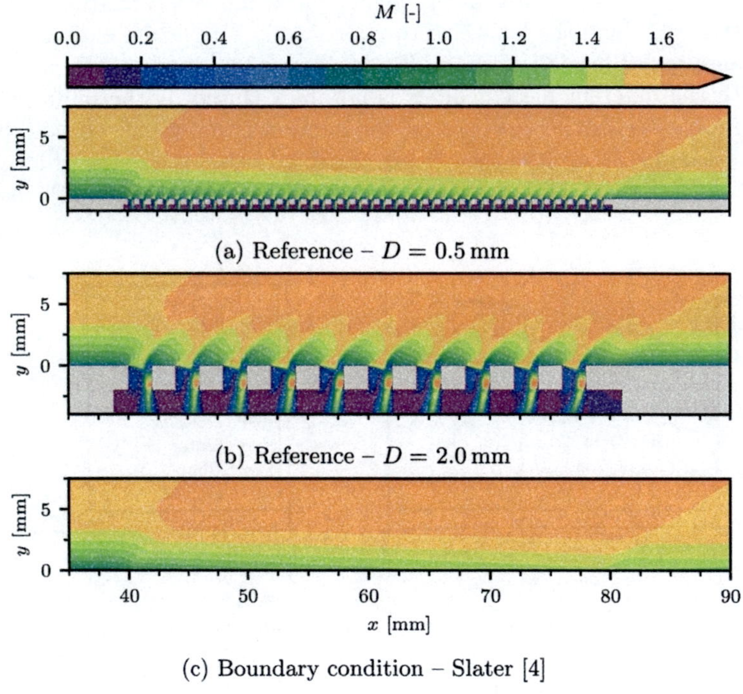

Fig. 2. Mach contour fields for two reference simulations and applied boundary condition

The turbulence is modeled using the Spalart-Allmaras turbulence model with constitutive relation [8]. In Fig. 2, the implemented boundary condition using the model of Choe et al. [4] (Fig. 2c) is compared to two reference simulations with hole diameters $D = 0.5$ mm (Fig. 2a) and $D = 2.0$ mm (Fig. 2b) for a throat ratio $TR = 0.7$. In the reference simulations, the hole contour is taken into account, and the holes are included in the mesh (see Giehler et al. [9]). In contrast, the boundary conditions are applied on a solid wall using source terms in the first layer of cells. All simulations are performed under steady-state conditions.

At first glance, all simulations show the same flow topology: a thinning of the boundary layer provoking an expansion fan at the beginning and a trailing shock wave at the end of the bleed region. While only one expansion fan and trailing shock are apparent in the simulation with the boundary condition, multiple shock waves and expansions caused by the holes are visible in the reference simulations. The larger the holes, the more pronounced these phenomena. A closer look at the flow downstream of the plate reveals a higher momentum in the vicinity of the wall compared to the reference simulations. Also, the diameter strongly influences the shape of the boundary layer, respectively the momentum close to the wall, as shown in our previous paper [9]. However, the model of Slater [4] does not consider any geometrical parameter other than the porosity level, which is the wall area fraction covered by the bleed holes. Thus, the flow field is assumed to be equal regardless the hole diameter or hole pattern.

The differences between boundary conditions and reference simulations are shown in more detail in Fig. 3 for several bleed models [4–6,10–12]. The wall shear stress τ_w along the length of the porous plate L_p is detailed in Fig. 3a. A significant overestimation of the wall shear stress by using the boundary condition with the different bleed models is notable, caused by the uniform suction along the plate. Contrary to reality, the suction is not locally distributed but uniformly applied, and its mass flux is corrected using the porosity level of the plate. The higher the estimated mass flux of the models (see Table 2), the higher

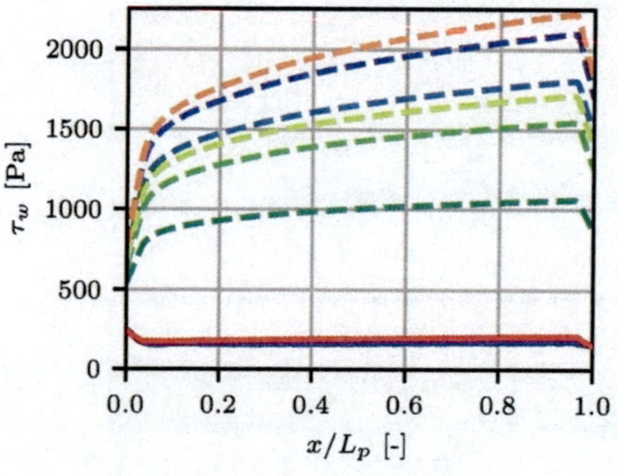

(a) Wall shear stress

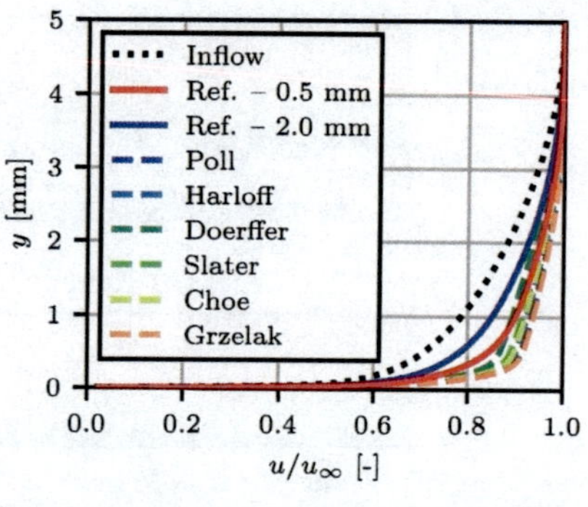

(b) Boundary layer 10 mm downstream

Fig. 3. Comparison of continuous porosity boundary conditions with reference simulations

Table 2. Boundary layer shape factor H extracted 10 mm downstream of the bleed region, bleed mass flux q_{bl}, and plenum pressure p_{pl}

	Reference		Boundary condition					
	0.5 mm	2.0 mm	Poll	Harloff	Doerffer	Slater	Choe	Grzelak
H [-]	1.27	1.30	1.25	1.24	1.23	1.24	1.24	1.25
q_{bl} [kg/m^{2}s]	19.8	19.3	24.4	21.2	13.0	18.3	20.1	25.8
p_{pl} [Pa]	12 110	12 060	14 880	12 890	7930	11 120	12 260	15 680

the obtained wall shear stress in the simulations. Moreover, the trailing shock is stronger for higher bleed mass fluxes resulting in a more significant drop of the wall shear stress at the end of the plate.

A look at the boundary layer profiles in Fig. 3b reveals a substantial deviation from the reference simulations for the boundary conditions. Independently of the used model, the boundary layer profiles downstream of the bleed region are significantly fuller. As indicated in [7], the flow momentum close to the wall is overestimated. Moreover, we see a strong variation caused by the hole diameter, which is not captured by the continuous porosity boundary conditions.

Table 2 evidences the overperformance of the continuous porosity boundary condition. The shape factor is underestimated for all boundary conditions compared to the reference simulations, which indicates a more effective control than in reality. This effect is independent of the bleed mass flux. No matter if the bleed mass flux is over- or underestimated, applying the continuous porosity boundary condition leads to fuller profiles downstream of the bleed region. The prediction of the plenum pressure is in line with the prediction of the bleed mass flux. An overestimation of the mass flux leads to overestimating the plenum pressure.

The underestimation of the shape factor is a severe problem for pre-designing a supersonic air intake, where the porous bleed is applied to mitigate the shock-induced flow separation. The estimated overperformance of the system may lead to an undersized system, which could be unable to achieve the desired control effect.

4 Methodology for the Local Porosity

The principle of the local porosity is the discrete modeling of the hole contour with a pre-defined number of cells as schematically illustrated in Fig. 4. All cells with the cell center inside the hole contour are categorized as hole cells which means that suction is applied. All other patch cells are viscous walls. This procedure aims to enable the discrete modeling of the bleed holes on a simple, structured mesh.

For the current study, one hole is modeled by 10 cells streamwise and 6 cells in the spanwise direction. The higher streamwise resolution aims to accurately capture the flow phenomena as the expansion fan at the hole opening and the barrier shock at the rear edge of the hole. In the spanwise direction, a lower impact of the mesh resolution is assumed. Moreover, a low number of cells is

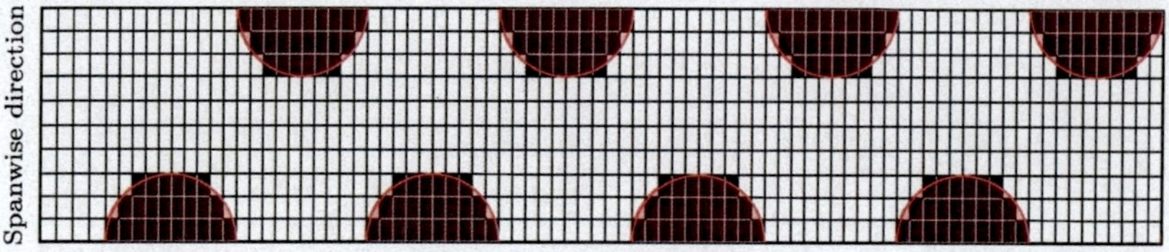

Fig. 4. Schematic of local porosity: black cells correspond to suction cells, white cells to solid wall; red circle illustrates theoretical hole position (Color figure online)

preferred to increase the savings in computational costs compared to the reference simulations. The total mesh size depends on the hole diameter and increases for smaller holes (similar to the three-dimensional reference simulations). Table 3 shows the number of cells for the different configurations. The savings using the local porosity are approximately one order of magnitude compared to the reference simulations. However, the new method requires approximately one order of magnitude more cells than the simulations with continuous porosity, which are only two-dimensional.

The structured mesh is automatically generated by the in-house mesh-generator **Cassiopee**. Similarly to our previous work [7], the following source terms are added to the compressible Navier-Stokes equations to apply suction in the first row of cells. For the computation of the source terms, the cell size is taken into account by using the cell patch area A_c and the cell volume V_c.

$$\text{Continuity:} \qquad \frac{q_{bl} A_c}{V_c}$$

$$\text{Momentum:} \qquad \frac{q_{bl}^2 A_c}{\rho V_c}$$

$$\text{Energy:} \qquad \frac{q_{bl} c_p T_t A_c}{V_c}$$

The bleed mass flux q_{bl} is computed in an external python script executed by the flow solver every 10th iteration. The bleed model applied for the local porosity boundary condition is the model of Choe et al. [5] since it has the best agreement with the reference simulations regarding the bleed mass flux and the plenum pressure. The required physical quantities for the model (static wall pressure and temperature) are extracted at the patch interface using the implemented extractors in **elsA**. The plenum pressure is found by fixing the (choked) plenum exit area and iteratively adjusting the plenum pressure until the mass balance is achieved.

Table 3. Comparison of the mesh sizes for the different methods

Hole diameter	Reference	Local porosity	Continuous porosity
0.5 mm	5.1 M.	0.8 M.	<0.1 M.
2.0 mm	2.4 M.	0.3 M.	

5 Results and Discussion

Both reference cases from Sect. 3 are simulated using the local porosity boundary condition. Figure 5 shows the flow fields for the simulations, including the holes and the local porosity. The Mach number contours for a hole diameter of $D = 0.5$ mm are compared in Fig. 5a and 5b. The application of the local porosity shows a good fit with the reference simulation. The impact of the local suction is apparent in the vicinity of the wall, where the expansion fans and barrier shocks generated by the holes are well reproduced. Compared to the continuous porosity (see Fig. 2c), the near-wall flow momentum is lower at the end of the bleed region, indicating a less full boundary layer profile.

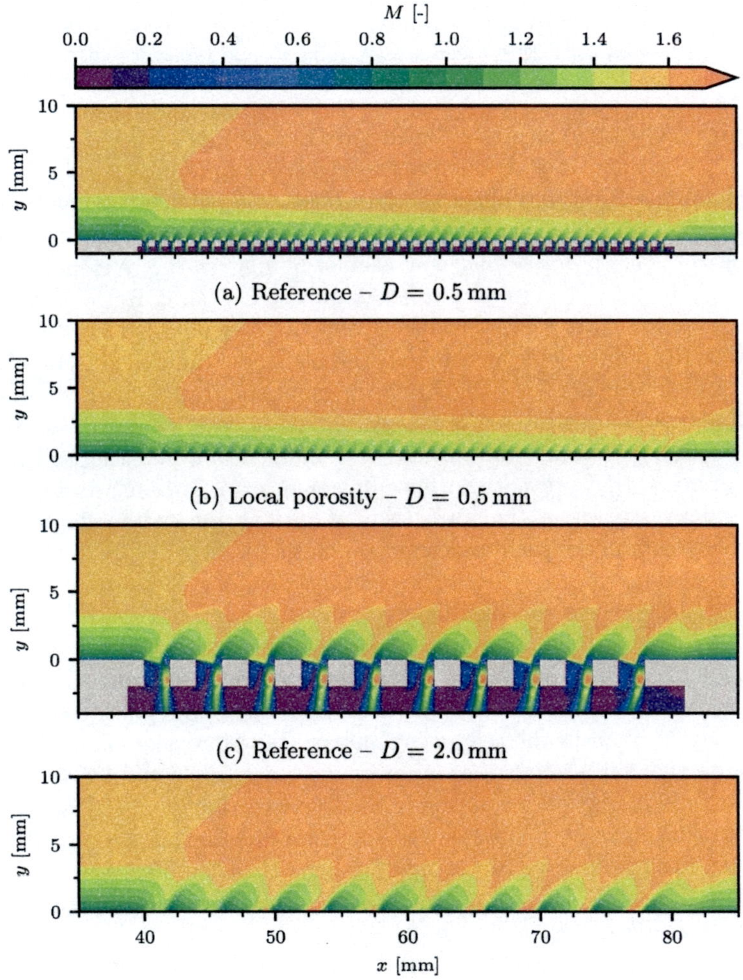

(d) Local porosity – $D = 2.0$ mm

Fig. 5. Mach number contour fields for reference and local porosity

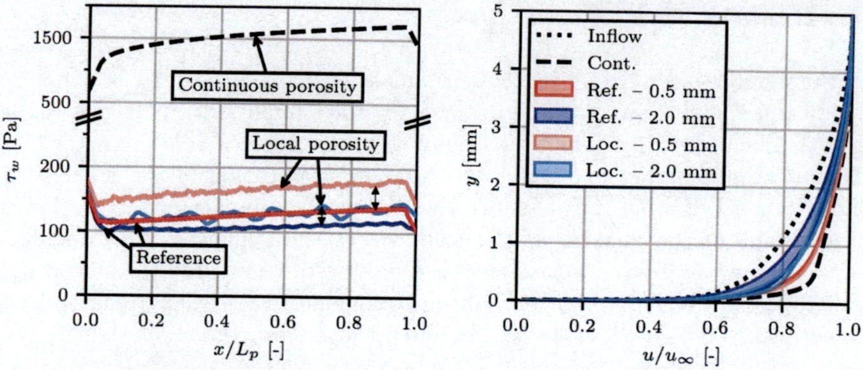

(a) Wall shear stress

(b) Boundary layer 10 mm downstream

Fig. 6. Comparison of the local porosity boundary condition with the reference simulations

More noticeable is the impact of the local porosity for a diameter of $D = 2$ mm in Fig. 5d. For larger hole diameters, the expansion fans and shock penetrate the whole boundary layer and affect the flow field even outside the boundary layer. While the continuous porosity boundary condition cannot mirror this effect, applying the local porosity makes it visible. Similar to the reference simulation (Fig. 5c), only a slight boundary layer thinning is notable. Both expansion fans and barrier shocks are well reproduced, and the flow field fits very well.

However, a closer look at the wall reveals slight differences. For the application of the local porosity, higher Mach numbers are found at some holes in the vicinity of the wall. This is caused by an interaction of the source term, which generates a strong wall-normal flow momentum, and the wall boundary condition below that prevents outflow and, thus, a wall-normal flow component. For future investigations, the application of outflow/inflow boundary conditions is planned to eliminate such undesired (numerical) effects.

Further comparison of the local porosity boundary condition and the reference simulations are shown in Fig. 6. In contrast to the continuous porosity boundary condition, the prediction of the wall shear stress is close to the reference simulations by applying the local porosity (Fig. 6a). However, its amount is slightly overestimated. Already at the beginning of the plate, the wall shear stress is higher than in the reference simulations. Along the plate, the positive gradient of the wall shear stress is approximately the same. A reason for this behavior could be the low resolution in the spanwise direction. Further tests with higher resolutions in both stream- and spanwise directions are required to investigate the mesh influence. However, a higher resolution corresponds to higher computational costs. Thus, a compromise between accuracy and practicability needs to be made.

A view on Fig. 6b reveals a better fit of the boundary layer downstream of the bleed region than for the continuous porosity. The profile is only slightly fuller than in the reference simulation for the smaller hole diameter of $D =$

Table 4. Boundary layer shape factor H extracted 10 mm downstream of the bleed region, bleed mass flux q_{bl}, and plenum pressure p_{pl}; relative errors are given in percentage of the reference simulations

	Reference		Local porosity		Continuous porosity
	0.5 mm	2.0 mm	0.5 mm	2.0 mm	
H [-]	1.27	1.30	1.26	1.28	1.24
q_{bl} [kg/m^{2}s]	19.8	19.3	20.6 (+4.4%)	21.1 (+9.3%)	20.1 (+1.5%/+4.2%)
p_{pl} [Pa]	12 110	12 060	12 300 (+1.6%)	12 120 (+0.5%)	12 260 (+1.2%/+1.7%)

0.5 mm. Also, for $D = 2$ mm, the boundary layer profile computed using the local porosity is closer to the reference simulation but still fuller. Moreover, the variation of the boundary layer in the spanwise direction can be compared. Since the holes are locally distributed, three-dimensional effects are induced. These effects result in a profile variation along the span and can be modeled by the local porosity. Therefore, the range of all profiles is visualized. Compared to the reference simulations, the variation of the boundary layer profiles in the spanwise direction is lower for the local porosity. Again, the low spanwise resolution can be the reason and needs to be investigated.

The difference between the boundary layer profiles is also found in the shape factor, shown for the local porosity and the reference simulations in Table 4. For both hole diameters, the shape factor found downstream of the bleed region is slightly smaller than in the reference simulations. However, the predicted profiles are significantly closer to the reference than the continuous porosity.

Moreover, the predicted mass flux of the model of Choe et al. [5] is slightly higher by using the local porosity. Interestingly, the predicted mass flux is higher for large hole diameters, in contrast to the reference simulations and previous findings [9]. The larger the hole diameter, the stronger the barrier shock and, thus, the higher the static wall pressure downstream of the hole. Consequently, the model predicts a higher mass flux since the pressure difference from the external wall to the plenum is larger. The prediction of the static plenum pressure is only slightly different from the continuous porosity boundary condition and shows a good agreement.

6 Conclusion and Future Work

In this paper, a new approach to applying bleed boundary conditions has been implemented in the flow solver **elsA**, tested, and compared to reference simulations. The suction through the holes is locally applied by discretely modeling the hole contour and utilizing suction only on these pre-defined cells. In contrast to standard methods using continuous porosity on the patch, all expansion fans and barrier shocks induced by the holes are mirrored.

Using the local porosity boundary condition enables a more accurate prediction of the boundary layer profiles downstream of the bleed region. While the profiles are significantly fuller for the continuous porosity boundary condition, the present method only slightly overestimates the boundary layer thinning. Moreover, the wall shear stress along the porous plate is in the range of the reference simulations and not strongly overestimated like for the continuous boundary condition.

Future work is required to investigate the influence of the resolution of the holes. Therefore, a compromise between the saving of cells and accuracy needs to be found. Furthermore, the replacement of the source term and the viscous wall by local inflow and outflow boundary conditions is planned to eliminate numerical issues in the vicinity of the wall. Besides, the application of the new approach will be tested to control of a shock-boundary layer interaction to examine the accuracy improvement in this complex flow pattern.

Acknowledgments. This project has received funding from the European Union's Horizon 2020 research and innovation programme under grant agreement No EC grant 860909.

References

1. Abrahamson, K., Brower, D.: An empirical boundary condition for numerical simulation of porous plate bleed flows. In: 26th Aerospace Sciences Meeting Reston, Virigina: American Institute of Aeronautics and Astronautics (1988)
2. Chyu, W.J., Howe, G.W., Shih, T.I.-P.: Bleed-boundary conditions for numerically simulated mixed-compression supersonic inlet flow. J. Propuls. Power **8**(4), 862–868 (1992)
3. Paynter, G.C., Treiber, D.A., Kneeling, W.D.: Modeling supersonic inlet boundary-layer bleed roughness. J. Propuls. Power **9**(4), 622–627 (1993)
4. Slater, J.W.: Improvements in modeling 90-degree bleed holes for supersonic inlets. J. Propuls. Power **28**(4), 773–781 (2012)
5. Choe, Y., Kim, C., Kim, K.: Effects of optimized bleed system on supersonic inlet performance and buzz. J. Propuls. Power **36**(2), 211–222 (2020)
6. Grzelak, J., Doerffer, P.P., Lewandowski, T.: The efficiency of transpiration flow through perforated plate. Aerosp. Sci. Technol. **110**, 106494 (2021)
7. Giehler, J., Grenson, P., Bur, R.: Porous bleed boundary conditions for shock-induced boundary layer separation control. In: 56th 3AF International Association Aéronautique et Astronautique de France, Conference on Applied Aerodynamics Paris (2022)
8. Spalart, P.: Strategies for turbulence modelling and simulations. Int. J. Heat Fluid Flow **21**(3), 252–263 (2000)
9. Giehler, J., Grenson, P., Bur, R.: Parameter Influence on the Porous Bleed Performance with & without Shock-Boundary Layer Interaction. In: AIAA Aviation: Forum, p. 2022. American Institute of Aeronautics and Astronautics, Reston, Virginia (2022)
10. Poll, D.I.A., Danks, M., Humphreys, B.E.: Aerodynamic Performance of Laser Drilled Sheets. University of Manchester. Aeronautical Engineering Group, Technical report (1992)

11. Harloff, G.J., Smith, G.E.: Supersonic-inlet boundary-layer bleed flow. AIAA J. **34**(4), 778–785 (1996)
12. Doerffer, P.P., Bohning, R.: Modelling of perforated plate aerodynamics performance. Aerosp. Sci. Technol. **4**(8), 525–534 (2000)

Simulation of Suction Chambers with Variable Pressure Distribution for HLFC Applications

Heinrich Lüdeke(✉)

DLR, Institute of Aerodynamics and Flow Technology, Braunschweig, Lilienthalplatz 7, 38108 Braunschweig, Germany
heinrich.luedeke@dlr.de

Abstract. For a simplified but effective chamber concept of boundary layer suction for hybrid laminar flow control, different velocity distributions inside suction chambers under consideration of pressure gradients are simulated. The flow inside these chambers is calculated on 2D and 3D grids to predict suction velocity as well as the wall pressure distribution at different suction pressures. The classical concept of a microporous metal with a substructure of stringers and single chambers for adapted inner pressure will be investigated, as well as modern chamberless applications with micro etched metal foils at varying porosity for each surface part, stabilized by metal fabric below. By the latter approach, extensively long suction regions result in non-negligible pressure gradients inside the porous sheet and the pressure distribution on the outside wall, which is also taken into account. These effects require an additional pressure gradient boundary condition, which was added and successfully evaluated by comparison with experimental data, allowing the future prediction of suction distribution for an optimal transition suppression.

1 Introduction

Modern commercial aircraft design criteria are mainly oriented by economical and ecological aspects and particularly the necessity of fuel consumption reduction, for example by reducing the total aerodynamic drag. For a common wing-profile under subsonic conditions, the friction-drag is a dominant part and its reduction will result in significant savings. Since the friction-drag of a laminar Boundary Layer (BL) is generally lower compared to a turbulent one, an effective approach for drag-reduction is the use of laminar wings or artificial laminarization. Different HLFC (Hybrid Laminar Flow Control) working principles have been demonstrated by means of wind-tunnel [1,2] and flight tests [3] on a vertical tail-plane (VTP). For a laminar BL on the VTP, the flow needs stabilization against amplified instabilities, which is generally achieved by an appropriate pressure distribution along the surface or suction of the BL to reduce amplifications of instabilities. Both approaches delay transition in comparison with the original BL.

For the chamber concept of boundary layer suction for HLFC, velocity distributions inside suction chambers under consideration of pressure gradients will

A. Dillmann et al. (Eds.): STAB/DGLR Symposium 2022, NNFM 154, pp. 372–381, 2024.
https://doi.org/10.1007/978-3-031-40482-5_35

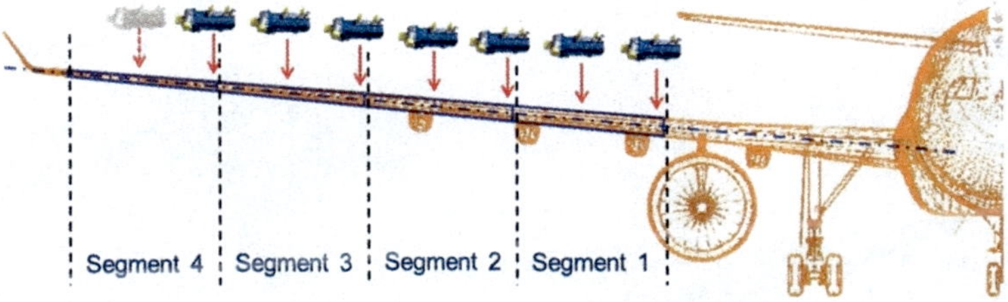

Fig. 1. HLFC segments and compressor distribution on a wing for the XRF1 configuration from [4].

be simulated by highly accurate steady CFD studies. The flow inside these suction chambers is calculated by the structured DLR FLOWer code in 2D and 3D geometries to predict suction velocity as well as pressure distribution inside the chambers. Furthermore, variations of the suction pressure are carried out. FLOWer is a finite volume CFD solver for inviscid and viscous flow. A central numerical scheme, including turbulence modeling for the 3D geometries, is used for this study of steady flow fields. The ALTTA concept of an external laser drilled porous metal with a substructure of stringers and single chambers at different inner pressure [5,6] will be investigated at first. This concept was derived within the European ALTTA project, including flight experiments. Furthermore, modern chamberless applications with TSSD (Tailored Skin Single Duct) surfaces and micro etched metal foils [2] are taken into account. For the latter ones, only interconnected suction ribs without individually designed orifices share a single constant pressure plenum. This new approach results in significant pressure losses inside the chamber and also at the suction wall. Consequently, a pressure gradient was added along the wall by an extended boundary condition (BC) in the CFD code, which was validated in comparison with experiments. This BC will allow an improved design of future suction systems.

In the scope of the European Clean Sky 2 project HLFCWin, the benefits and integration aspects of HLFC on a long-range wing have been investigated. Figure 1 gives an overview of HLFC segments on a wing, including compressor distribution. In order to investigate these pressure losses and the potential impact on HLFC performance, 3D steady CFD simulations have been carried out [7].

1.1 Global Setup

The approach of a porous wall with stringer sub-structure and a second inner sheet provides a double-wall shell with a number of integrated suction chambers [7] will be called ALTTA concept, as described by Rohardt and Schrauf [5,6] and used by the VTP wind tunnel model. Inside these individual suction chambers circular orifices allow the choice of the suction pressure to extract enough material for a stabilized boundary layer on the wing-surface with significant transition delay (Fig. 2). For this study the inner chamber geometry is simplified as a rectangular space without curvature which is nearly infinitely long in spanwise direction. Recent concepts apply micro etched metallic foils of about 50 micron

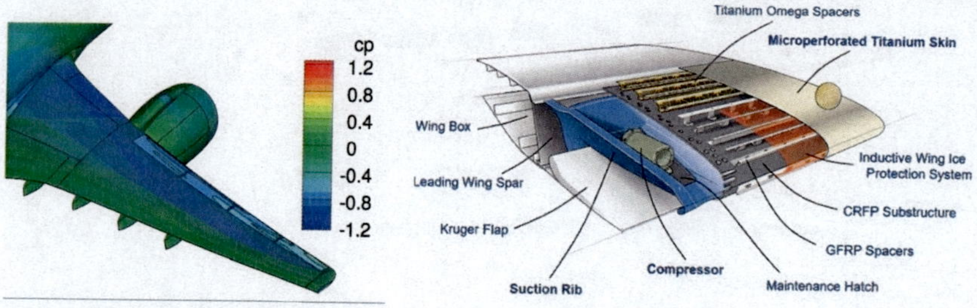

Fig. 2. XRF1 configuration with suction panels (left) and internal suction rib overview for the TSSD wing concept (right)

thickness, welded on fine metal fabric, also used to adjust local pressure loss and suction velocities. Consequently, TSSD concepts allow a simplified chamberless design. The following investigations are intended as an overview and validation of different CFD results with 2D and 3D simplifications of various levels and the implemented pressure gradient BC [8].

2 Configurations

2.1 Geometric Data

The separated chambers for the ALTTA concept are generally of rectangular shape with a wall-normal dimension of 10 mm, an average chordwise diameter of 100 mm and a spanwise length of 2 m. Circular orifices with diameters of 3 mm are located in a row on the duct-wall along these 2 m at a distance between orifices of 0.25 m [7]. In the following, simplifications will be described for adequate 2D calculations between respective orifices, the flow is expected perfectly periodic in spanwise direction [7].

In addition 3D simulations from different TSSD single duct configurations, including settling chambers or boxes, are investigated and compared with experimental data [4]. As a 3D validation test case, pressure losses in a wooden suction model of 2.75 m without external flow have been determined. This simplified

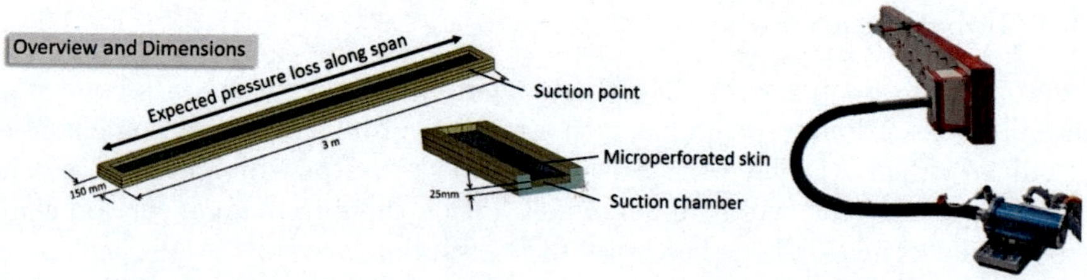

Fig. 3. instrumented simplified experimental setup (SSD) of a wooden chamber with suction surface.

setup is shown in Fig. 3 and comprises a chamber covered by the suction sheet. The included compressor is connected to one end via a settling box, representing the suction rib below, which will be included in the CFD modeling. Experimental and numerical results of this study are presented along with an assessment of the potential impact on the HLFC wing. Internal chamber coordinates for 2D and 3D will be called streamwise along the porous wall or wall-normal respectively.

2.2 2D Idealizations

For 2D chamber simulations of the ALTTA configuration, including orifices, the averaged suction velocity is determined by the local pressure loss through the porous sheet. For this purpose, the surface pressure at the BL wall is taken from simulations of the VTP under wind tunnel conditions.

Instead of circular orifices on the chamber bottom, suction slots were generated for the 2D meshes. To keep the mass flow through the 2D representation realistic, the slot diameter is fitted in a way, that spanwise periodic orifice-surfaces are identical to the periodic suction surface of the slots. Consequently, their lateral dimension is much smaller than the orifice diameter (Fig. 4) [7]. Two symmetry planes are defined between adjacent slots and also in the middle, resulting in a rectangular domain between porous and duct wall. The duct on the inner side of the orifices is replaced by a coarse grid at given exit-pressure, provided by the fraction of outer and duct-pressure from the ALTTA configuration. So the flow-field is expected as an infinitely spanwisely repeated structure, which is justified by the large spanwise chamber extend of 2m. The grid resolution for 2D approaches vary between 64×128 cells for an inviscid flow and 104×144 cells for viscous chamber flow. Grids are strongly refined near the slots in both directions. Consequently, very quick design studies are possible for such geometries.

Externally added pressure gradients on 2D grids are carried out by special periodic boundary treatment This periodic condition requires a doubled mesh around the 2D orifice-slot, where the 2D geometry is taken from VTP.

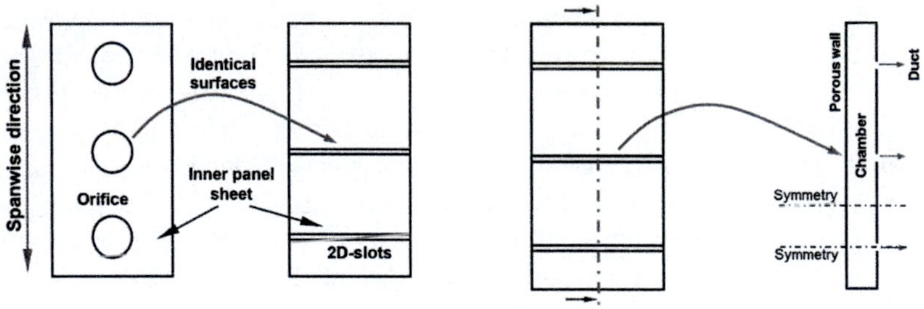

Fig. 4. ALTTA approach: Idealized circular taps in a row by slots of identical surface (left) and 2D periodic panel with slots and symmetry planes (right). internal chamber coordinates are called along the porous wall.

2.3 3D HLFCWin Chambers

In addition, 3D simulations are carried out in comparison with experimental data. The experiments examine effects of an extensively long chamber with porous metal sheet, measuring the inner pressure distribution without using orifices but a direct connection to the compressor by a cubic settling box [4]. Pressure taps inside this simplified model allow comparison with numerical results. Due to a stronger suction and an increased channel Reynolds number, the chamber flow will become turbulent shortly after inlet. For this reason a linear $k-\omega$ turbulence model was chosen, since steady recirculation regions appear inside the settling box, which was also added to the 3D mesh. For simplicity, non-linear terms for the prediction of stress anisotropies were not considered in the $k-\omega$ model. Furthermore, the fully quadratic correlation between suction pressure and velocity for the porous material was taken into account, due to the higher suction velocities.

Finally, external pressure gradients from the XRF1 wing configuration without orifices, but 3m long chambers are added. A grid resolution of $208 \times 384x32$ cells is chosen for all 3D simulations.

3 Flow-Conditions and Numerical Setup

3.1 Geometry and Flow Conditions

For the experimental setup of the porous wall, a micro-perforated titanium skin is mounted on the sealed wooden channel, connected to the compressor by a cubic settling box. This setup is called the Small Scale Demonstrator (SSD) from HLFCWin. Chamber dimensions of 2850 mm × 25 mm are chosen at two suction area widths of 150 mm or 50 mm. This reduced cross section results in medium to very high suction velocities in comparison with former ALTTA configurations. The full chamber is shown in Fig. 3, while 1/3 cross section is tested by adding non symmetrical inserts to the SSD. The reduced cross section was necessary to increase the mass flow at a limited compressor power. Results of both cross sections will be used for validations of the steady 3D chamber simulations. Reduction of the chamber width was simply carried out by rescaling the chamber grid in this direction. Along the channel span and opposite to the suction wall, five pressure sensors are mounted for comparison with CFD output. The mass flow for the experimental setup, as measured at the compressor ranges between 19 g/s at 30000 RPM and 62 g/s at 130000 RPM. Both configurations are simulated by 3D meshes. Difference pressures $\Delta p = p - p_{ref}$ are always referenced by the external reference pressure p_{ref} of 1013 hPa. This is also chosen for the plenum pressure at the outlet, called p_{ex}, which is denoted as a percentage of p_{ref}. A wall and reference temperature of 288.15K and a Prandtl number of 0.72 was chosen for all simulations. The reference Mach number of 0.294 is just used to keep the calculation globally compressible, since the chamber flow is just determined by the suction wall properties, the suction slot geometry and the plenum pressure.

3.2 Numerical Setup

All simulations were carried out on structured 2D or 3D grids of the suction chamber and the respective plenum or settling box below. For the suction BC, a wall-normal velocity component was added to the slip or no-slip conditions, depending on the difference pressure through the porous wall. To calculate this suction velocity, a quadratic approach with two suction coefficients between pressure difference and velocity was chosen. For the ALTTA chambers, a constant reference pressure was prescribed outside the wall (see last section). For the wing configuration this external pressure is varied by a prescribed tangential pressure gradient in the suction wall boundary condition. It has to be noted, that the quadratic part of the suction velocity is omitted for the ALTTA chambers, due to the very small gradients for these test cases. The numerical scheme and solution algorithms used in this work are described in more detail in [7].

4 Results

4.1 2D Chamber Simulations

Inviscid and viscous 2D results of the chamber flow, as a good representation of the ALTTA configuration, are demonstrated in this section. A refined grid for the no-slip suction BC is generated to resolve the BL and the flow through the suction slot for Navier-Stokes calculations, while coarser near wall grids are generated for Euler simulations.

For a nominal unit Reynolds-number of $7 \cdot 10^6$ 1/m, the viscous simulations have shown similar small velocities inside the chamber as well as for inviscid cases (see [7]). In this chamber, the flow between inner walls gets a very small representative unit Re-number of about 1400 1/m, in contrast to single duct configurations of the SSD model. For this reason laminar internal flow is expected. Chamber- and duct walls are defined by non-slip BC. Numerically, convergence

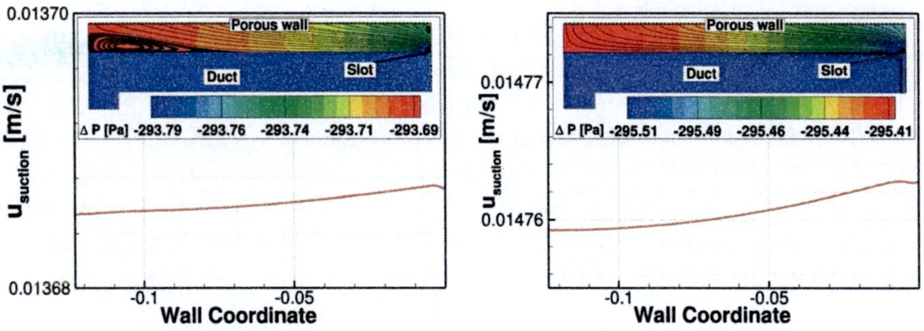

Fig. 5. 2D inviscid and viscous velocity distribution along porous panel wall. Contours of difference pressure $p - p_{ref}$ inside the periodic chamber with orifices of the ALTTA configuration at $p_{ref} = 1013$hPa

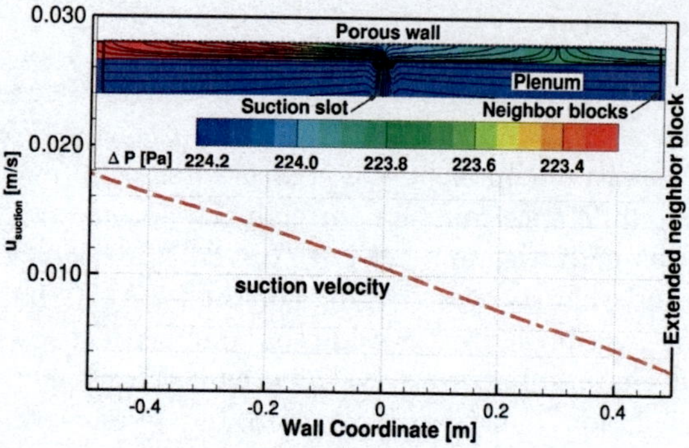

Fig. 6. 2D velocity distribution with external wall pressure gradient for ALTTA chamber. Contours from difference wall pressure $p - p_{ref}$

over more than 8 orders of magnitude is necessary until the final steady flowfield is established. Due to the strong gradients near the 2D suction slot, the extensively small pressure and velocity variation at the porous wall and consequently due to the reduced numerical time-steps for the small near-wall cells of the refined grid, this is an expected result.

In Fig. 5, streamlines in the suction chamber, pressure contours and suction velocity along the wall are depicted. Since the mean values of chamber pressure and suction velocity show strong variations, the $u_{suction}$ range is chosen identical but shifted for both cases, as well as the contour plots for Δp with a range of 0.1 Pa.

For inviscid simulations in the left diagram of Fig. 5, the velocity vector of the flow, entering through the porous wall is deflected by the sign of the pressure gradient, and tangential velocity-components are visible on the wall. In contrast, these tangential velocity-components vanish for the no-slip condition at the viscous wall (Fig. 5, right diagram) and change the streamline topology in the chamber by this way. Nevertheless, suction velocity variations are nearly negligible for both cases.

For validation of the external pressure-gradient condition, another 2D test case was mirrored around the suction slot and a streamwise gradient was added. The resulting suction velocity is shown in Fig. 6, its magnitude is following the constant gradient by showing a nearly linear behavior, except slight deviations near the connection slot between chamber and constant pressure duct. It has to be noted, that this approximately linear curve seems to be superimposed by a periodic perturbation, which is just a result of the slot influence on the chamber flow.

4.2 Suction Chamber 3D Calculations in Comparison with Generic Experiments

The pressure loss in the small scale demonstrator (SSD) was investigated for different compressor power in conjunction with steady CFD simulations at a

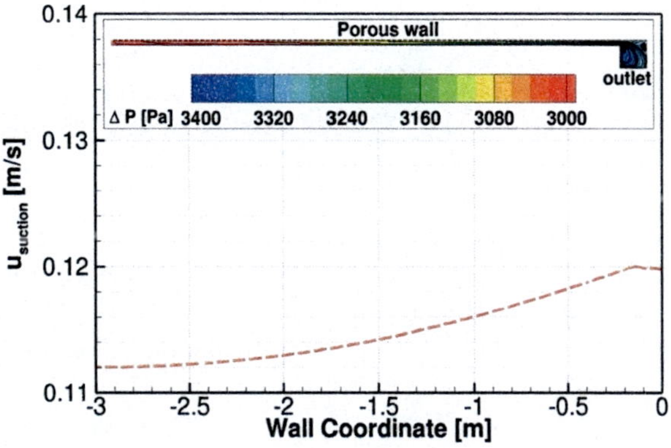

Fig. 7. Suction velocity along chamber wall from CFD, including settling box [4]

maximum distance to the compressor. For simplified 3D CFD meshes, including porous wall and settling box, the suction velocity shown in Fig. 7. Since no external flow is apparent for the SSD, no external pressure gradient was expected. In the following diagrams, experimental tolerances are added from sensor accuracy in Fig. 8. Only at a high compressor power (130000 RPM), larger differences between CFD and sensor data are visible, due to the fact, that the sensors are located opposite to the wall, where turbulent recirculations appear. In contrast, all CFD results are taken from the porous wall without such influences. For these diagrams, difference pressure on the y-axis is given in hPa.

4.3 Suction Chamber Flow with External Pressure Gradient from XRF1 Configuration

For the XRF1 configuration with suction panels, shown in Fig. 2, external pressure gradients are extracted from CFD of the full configuration [4]. As a result,

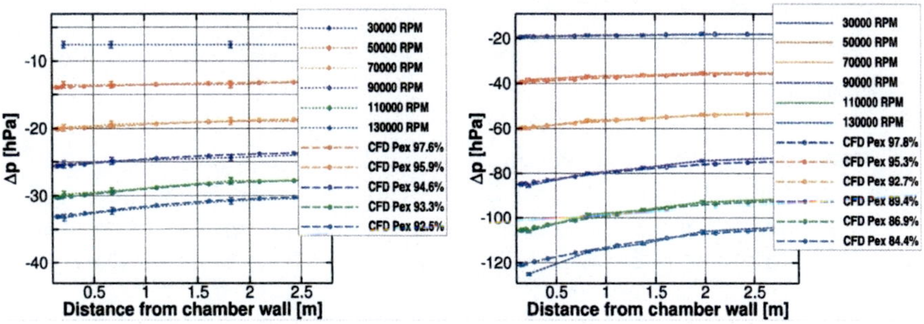

Fig. 8. Pressure loss along 3D chambers from HLFCWin experimental data [4] in comparison with CFD. Left: symmetrical setup, right: 1/3 chamber with Δp along the y-axis scaled by an appropriate factor of 3

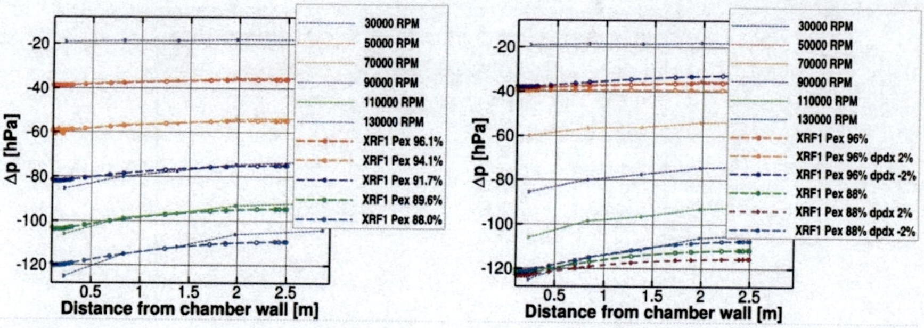

Fig. 9. Pressure loss in 3D chambers from XRF1 configuration [4]. Left: no pressure gradient and datasets from experimental setup, right: 2 % pressure gradient per meter added for two suction pressures

these gradients of about 2 % external pressure per meter are added along the suction panels BC. Inside the panel walls, the pressure distribution is mainly influenced by the linear external gradient, which can be added to the constant pressure for suction velocity prediction. We must keep in mind the shorter XRF1 chamber with 50 mm connections to suction rib (Fig. 2), so the following comparisons with the SSD configuration results are expected as a qualitative overview and differences are not unexpected (Fig. 9). The maximum pressure loss along 2.5 m symmetrical chamber is even smaller than the SSD outcome at 130000 RPM.

For the compressor speed of 50000 and 130000 RPM two pressure gradients of +2 % and −2 % are added. The right diagram in Fig. 9 shows deviations from the wall inner pressure, where the external gradient of about 2 kPa is intentionally omitted for the post processing to fit all data into the diagram, keeping pressure distributions comparable. As mentioned, the external pressure gradient can simply be added to the original chamber results for a good approximation of suction velocities for stability calculations. This statement only holds for moderate gradients, as calculated for the XRF1 configuration, otherwise strong non-linearities will appear near the suction rib.

5 Conclusion

For simplified suction chamber concepts for hybrid laminar flow control by boundary-layer suction, the 2D and 3D flowfield inside different chamber geometries is simulated. Steady CFD of the internal flow are carried out by the structured DLR FLOWer code to predict velocity as well as pressure distribution behind the porous sheet at varying suction pressure.

The suction through a micro-porous surface is approximated by using an analytic relation between difference-pressure and suction velocity at the wall. The results show very small variations of the relevant flow data like suction velocity and chamber pressure, (ALTTA chambers) by simulating viscous laminar flow

with non-slip suction walls. Variations of the difference between external and duct pressure have shown a shift in the mean data but only little changes in the global behavior [7]. Spanwise pressure loss throughout long chambers of TSSD configurations needs to be accounted for HLFC wing design using suction ribs instead of separated plenum ducts. The internal pressure loss was successfully quantified by using a dedicated demonstrator, simulated and validated by 3D CFD in very good agreement with the experiments. By this way, a detrimental impact on HLFC wing design will be possible and needs to be assessed w.r.t. suction rates, chamber geometry and surface pressure distribution in future studies.

Acknowledgements. The author acknowledges Alexander Bismark for chamber experiment pressure data. Major parts of this study were carried out under the scope of the German Federal research program for aeronautics LuFo-V InFlyTec and the Clean Sky 2 HLFCWin Joint Undertaking under the European Union's Horizon 2020 programme.

References

1. Joslin, R.D.: Overview of laminar flow control. NASA/TP-1998-208705 (1998)
2. Horn, M., Frhr. H., Geyr, V., Weddig, T., Bruns, J., Schradick, M.: Auslegung, Fertigung und Funktionsdemonstration einer HLFC Leading Edge. DLRK 2016, Braunschweig, Germany (2016)
3. Schrauf, G., Geyr, H.V.: Simplified hybrid laminar flow control for the A320 fin aerodynamic and system design - first results. In: AIAA Science and Technology Forum and Exhibition - SCITECH, Orlando, Florida (2020)
4. Kilian, T., Bismark, A., Lüdeke, H., Kamp, B.V.D., Schröder, A.: The influence of high spanwise chamber extent on HLFC performance. In: DLRK 2021, Bremen Germany (2021)
5. Rohardt, C.-H., Seitz, A., et al.: Simplified-HLFC/ entwurf eines seitenleitwerks mit hybrid- laminarhaltung für den airbus A320. In: Deutscher Luft- und Raumfahrt Kongress, 27–29 Sept 2011, Bremen (2011)
6. Schrauf, G., Horstmann, K.H.: Simplified hybrid laminar flow control. In: ECCOMAS 2004, Jyväskylä, Finland, 24–28 July 2004. CD-Proceedings (2004)
7. Lüdeke, H.: Interior flow simulation of suction chambers for HLFC profiles. In: Deutscher Luft- und Raumfahrtkongress 2014, 16–18. September 2014, Augsburg, Germany (2014)
8. Kilian, T., Horn, M.: Verification of a chamberless HLFC design with an outer skin of variable porosity. In: DLRK 2019, Darmstadt, Germany (2019)

Some Specific Challenges of HLFC Design on a Long-Range Wing

Geza Schrauf(✉) and Thomas Kilian

DLR, 38108 Braunschweig, Germany
contact@schrauf.de, thomas.kilian@dlr.de

Abstract. We describe the aerodynamic design of an HLFC system for the wing of a large transport aircraft and compare it to the HLFC system installed in the VTP of the DLR A320 ATRA aircraft. For the VTP we used the ALTTA-design with a laser-drilled microperforated titanium sheet with only one microperforation and small suction chambers beneath it. For the wing we designed a system with only one large plenum and controlled the suction with the help of variable microperforation. Some specific challenges of an HLFC design on a wing in contrast to a VTP application are highlighted here.

Keywords: Hybrid Laminar Flow Control (HLFC) · suction system design · variable microperforation · A320 VTP · HLFC wing

1 Introduction to the HLFC-WIN Project

The European project HLFC-WIN is developing a hybrid-laminar-flow-control (HLFC) system for the wing of a long-range aircraft as part of Clean Sky 2. The configuration considered for the project is shown in Fig. 1. The wing is based on the turbulent XRF1 research configuration provided by Airbus.

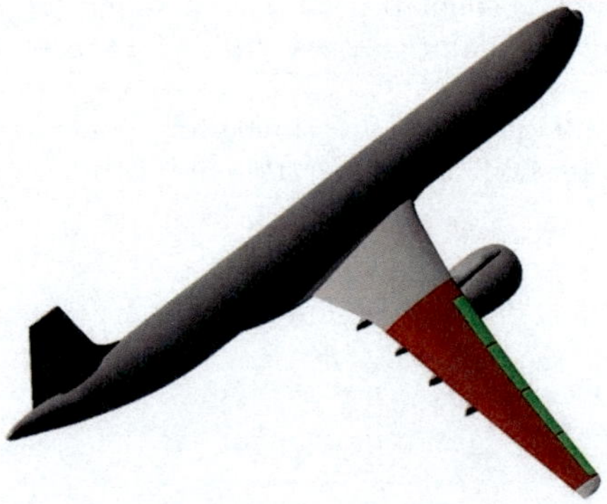

Fig. 1. Configuration for HLFC-WIN project. (Color figure online)

A. Dillmann et al. (Eds.): STAB/DGLR Symposium 2022, NNFM 154, pp. 382–391, 2024.
https://doi.org/10.1007/978-3-031-40482-5_36

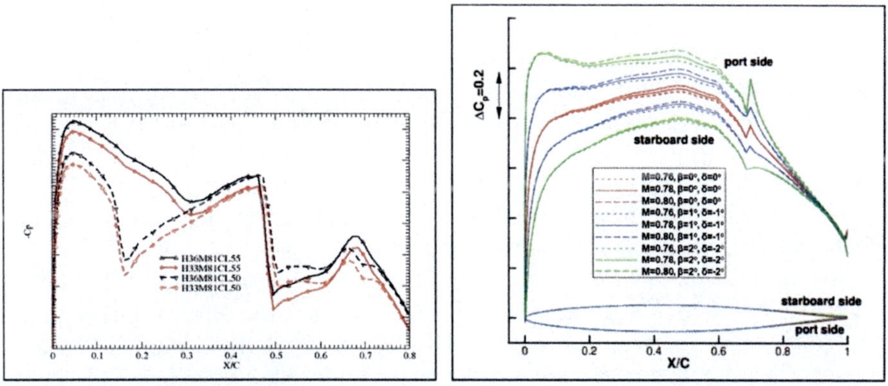

(a) HLFC wing. (b) A320 VTP.

Fig. 2. Comparison of two pressure distributions.

Fig. 3. The A320 aircraft with the HLFC system installed in the VTP.

For the purpose of this project, the wing planform was not altered, only the airfoil sections outboard of the engine (cf. red area in Fig. 1) were changed to favor HLFC [1].

The wing pressure distribution has a suction peak followed by a pressure recovery as shown in Fig. 2a. This differs from the pressure distribution considered for earlier HLFC applications, as, for example, in that of the A320 VTP presented in Fig. 2b. The design of the HLFC system for the VTP and some flight test results are described in [2,3]. The installed VTP system is shown in Fig. 3.

The pressure distribution of the HLFC wing also differs from that of the HLFC wing designed in the nineties within the ELFIN II project. This system was built and successfully tested in the ONERA S1MA wind tunnel [4].

For the HLFC-WIN project, we used the design idea of a pressure recovery over the suction panel (ending at $x/c = 0.29$) to allow for a region of decreasing pressure, i.e. with an accelerated boundary layer, behind the panel and before the shock. This enables a delay of the Tollmien-Schlichting (TS) transition without lowering the pressure level before the shock. The shock strength, and with that, the wave drag, is therefore not increased. The disadvantage of this pressure distribution is the need for stronger suction, not only to suppress, as usual, cross-flow (CF) transition, but also to avoid TS-transition and, possibly, laminar separation in the pressure recovery region. The size of this region varies strongly throughout the flight envelope. For some cases it even extends beyond the front spar location which usually limits the suction area.

We have 27 design cases for the HLFC wing shown in Table 2. The cases in which the pressure distributions differ most are shown in Fig. 2a. Other important differences between the current HLFC wing and the A320 VTP are the higher Mach number and the lower leading-edge sweep. A comparison of these two design parameters is given in Table 1:

Table 1. Mach numbers and sweep angles for wing and VTP.

	Wing	VTP
Mach number	0.83 ± 0.02	0.78 ± 0.02
Leading edge sweep	33°	40°

For the wing we consider a higher Mach number in combination with a smaller sweep angle, and a larger radius of curvature. Thus we need to apply stronger suction at the leading edge to keep the attachment line laminar.

2 HLFC Design Requirements

Within HLFC design we control the laminar-turbulent transition with the help of boundary layer suction. At the same time, we have to make sure that no early transition is provoked by local reverse flow or by other detrimental effects such as "over-suction," i.e. equivalent roughness[1].

The general requirements for any HLFC design are listed here:

R1 Keep the attachment line laminar, i.e. satisfy the K-criterion [5,6].
R2 Apply enough suction where boundary layer is close to laminar separation.
R3 Apply enough suction to suppress CF- and delay TS-transition.
R4 Avoid outflow.
R5 Allow for local blockage caused by supporting structural elements, such as stringers.

[1] The flow into a suction hole forms an ending stream tube [7, Figure 1], which acts like a 3D roughness element. The height h of the roughness corresponds to the height r of the stream tube. Therefore, we can apply the criterion for a 3D roughness element shown in [8].

R6 Avoid transition by equivalent roughness: $Re_{rr} < 400 - 450$.
R7 Avoid choking in the suction holes: $M_{hole} < 0.3$.
R8 Minimize mass flow and suction power.
R9 Aim for a robust design, i.e. allow for a large variation of the microperforation and other manufacturing deviations.

The requirements R1 to R5 are are strict, meaning that if one of them is violated, laminarity is lost. Requirements R7 to R9 are good design guidelines. The values given here can be relaxed.

The design of the HLFC system for the A320 VTP is described in [2]. Here we repeat only some aspects explaining the design differences between the HLFC wing and the VTP.

3 Some Aspects of the Design of the A320 HLFC VTP

For the VTP, the computation of the suction velocities for the aerodynamic design was based on the pressure-loss characteristic of the microperforated panel

$$\Delta p = A \frac{\mu_s}{\mu_0} w_s + B \frac{\rho_s}{\rho_0} w_s^2, \tag{1}$$

and not on any geometric quantities of the suction holes, such as hole diameter or hole pitch. To be able to proceed with the design simultaneous with the development of the laser-drilling process, we defined a range of allowable pressure-loss characteristics shown in Fig. 4,

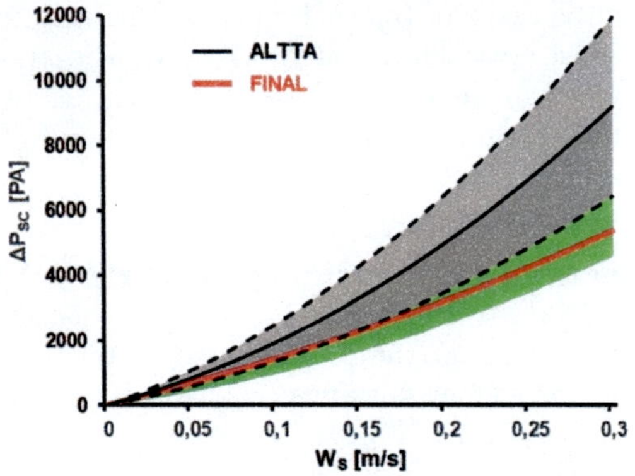

Fig. 4. Range of allowable pressure loss characteristics for the VTP design.

which is a reproduction of [2, Figure 3]. The manufacturer has to make sure that the pressure-loss characteristic of the large microperforated titanium sheet is, everywhere on the sheet, within this range. This allows for some inaccuracies occurring when producing large sheets.

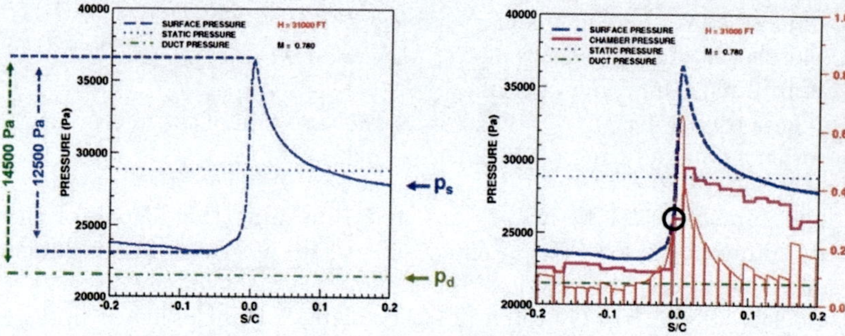

(a) Outer pressure.

(b) Chamber pressures and suction velocities.

Fig. 5. Dimensional pressures and suction velocities for 31000 ft, Mach = 0.78, side slip angle $\beta = 2^o$, rudder deflection angle $\delta = -2^o$.

In Fig. 5a we present the dimensional outer pressure for the case with the largest variation of the outer pressure. The difference between the stagnation pressure and the pressure of the local minimum is 12500 Pa. It is a good design practice to work with a duct or plenum pressure p_d which is 2000 – 3000 Pa below the minimum, so that the difference between outer and plenum pressure is, in the worst case, 14500 Pa. Using one microperforation, we control the suction velocities with the help of the chamber pressures, which are set by suitable throttle holes (cf. [2, Sect. III.C]. The chamber pressures are shown in Fig. 5b with the pink line, and the suction velocities are inserted in red with a second scale. Please note the small no-suction zones between the chambers. They are a result of the local blockage of the suction caused by the stringers. The black circle in the figure is the critical location for this case: here, the chamber pressure is only slightly lower than the outer pressure. An increase of the chamber size or of the duct pressure p_d would result in a reverse flow out of the chamber and trigger laminar-turbulent transition.

4 Some Aspects of the Design of the HLFC Wing

In order to limit the length of the leading edge parts, the wing outboard of the engine was divided into four segments. In this paper, we concentrate on the middle section of segment 3 at 74% span or section S74 shown in Fig. 6. The design points for the HLFC wing are listed in Table 2:

Table 2. Design points for the HLFC wing.

Flight level	330, 360, 390
Mach number	0.81, 0.83, 0.85
Lift coefficient	0.45, 0.50, 0.55

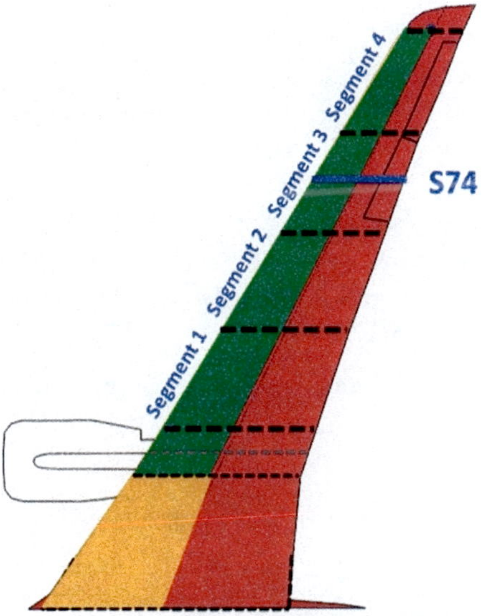

Fig. 6. Top view on wing, segment 3 with section S74.

The dimensional outer pressure for the case 33000 ft, Mach = 0.81, C_L = 0.50 is shown in Fig. 7a. In this case, the difference between the stagnation pressure and the pressure in the local minimum is 20100 Pa. This high pressure difference is due to the relatively larger lift requirements of a wing compared to a VTP application and directly impacts on the system requirements. As before, we have to consider a plenum or duct pressure P_d which is 2000 Pa lower than this minimum.

Furthermore, we notice the strong pressure recovery marked with the red circle in Fig. 7a. Here, the boundary layer exhibits not only strong TS-amplification, but even worse, laminar separation. Fortunately, the separation can be avoided by applying sufficiently strong suction as shown in Fig. 7b. Thus, we have to apply much stronger suction to prevent separation than would be necessary to only dampen TS-amplification. In a future project, we should reconsider the balance between the pressure recovery and the shock strength as one critical design feature of this specific airfoil design for HLFC.

The envisaged structure of the HLFC wing is shown in Fig. 8. The difference between a wing and a VTP is that the leading edge of a wing is not empty, but houses both high-lift and de-icing system. As usual for a laminar wing, a Krueger system is installed to keep the upper side of the wing undisturbed, thus allowing laminar flow.

Due to the space needed for the Krueger system, the space for the HLFC system is very limited, making a system with small suction chambers discharging the sucked air via throttle holes into a large plenum or duct unfeasible. The solution is to design fewer and larger suction chambers allowing for a sim-

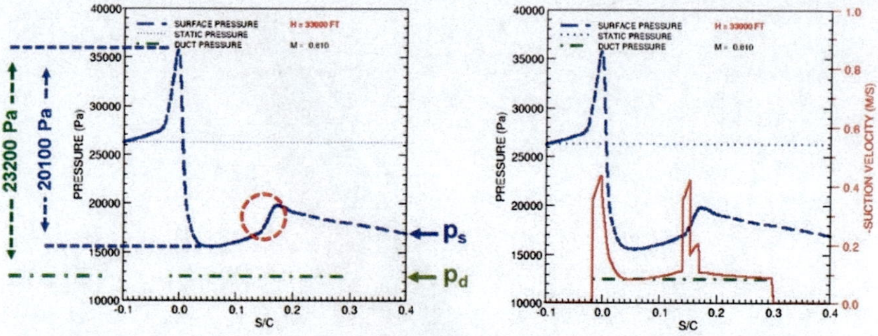

(a) Outer pressure and plenum pressure. (b) ... with suction velocity.

Fig. 7. Pressures and suction velocity for 33000 ft, Mach = 0.81, $C_L = 0.50$.

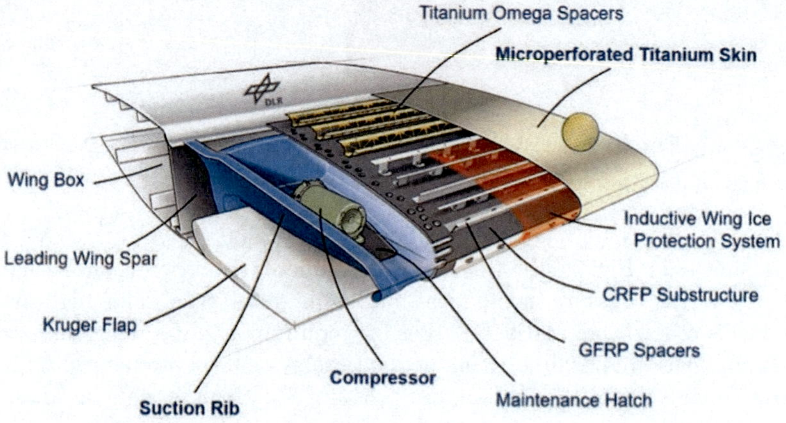

Fig. 8. Proposed structure and system solution for the wing.

pler structure to support the microperforated sheet. Our solution is based on a microperforated skin with porosity variation in chordwise direction. This would make an HLFC system with only one large suction chamber or plenum possible. The functionality of this one-chamber approach has been validated by DLR in a large scale wind tunnel test [9].

If a laser-drilling process is used to produce the microperforated suction sheet, the easiest method for varying the porosity is to vary the hole pitch. This allows the manufacturer to keep the settings of the laser constant, so that all suction holes have approximately the same geometry. Based on samples, which were tested with the laminar flow meter [10], twenty possible microperforations were defined together with the manufacturer. Considering these manufacturable microperforations, we propose a design[2] with five regions and the four different

[2] Another design based on ten regions and nine different microperforations can be found in [1].

Table 3. Manufacturable microperforations used for the design. Pitch in μm.

Pitch	[μm]	500	800	900	1100
A	[Pa m/s]	9990	25574	32367	48351
B	[Pa m/s^2]	25606	167814	268805	599845

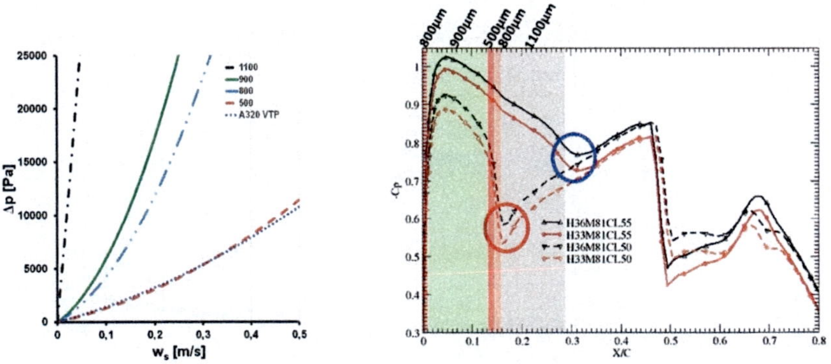

(a) Pressure-loss characteristics. (b) C_p-distributions and microperforations.

Fig. 9. Five regions with four different microperforations.

microperforations presented in Table 3. Their pressure-loss characteristics are presented in Fig. 9a, which also shows the pressure-loss charcteristic used for the A320 HLFC VTP. We see that the A320 characteristics is comparable to the most open microperforation with pitch 500 μm.

The regions in which the different microperforations are used, are indicated with different colors in Fig. 9b. We notice a short region with pitch 800 μm near the stagnation point. For the three-dimensional wing, this short region translates into a long and narrow strip with pitch 800 μm along the leading edge. Furthermore, there are two additional short regions with pitch 500 μm and 800 μm where the boundary layer is close to separation, indicated by the red circle. Here we need some extra suction as previously mentioned.

In Fig. 10 we present the Reynolds number Re_{δ_1} and the form factor H_{12} for 33000 ft, Mach $= 0.81$, $C_L = 0.50$, i.e. in the case of Fig. 7 obtained with the boundary layer code COCO [11]. We observe the rapid increase in H_{12} at $X/C = 0.14$, which is limited by strong suction. Without applying suction that is strong enough, the laminar boundary layer would separate at this location.

In Fig. 9b we also show the C_p-distribution with the longest pressure-recovery zone, which, in this case, extends beyond the end of the suction panel marked by the blue circle.

In Fig. 11 we present some N-factor results obtained with the stability code LILO [12]. We see that the Tollmien-Schlichting N-factors are very small. If no problem with separation arises, we could reduce the suction without causing early TS-transition.

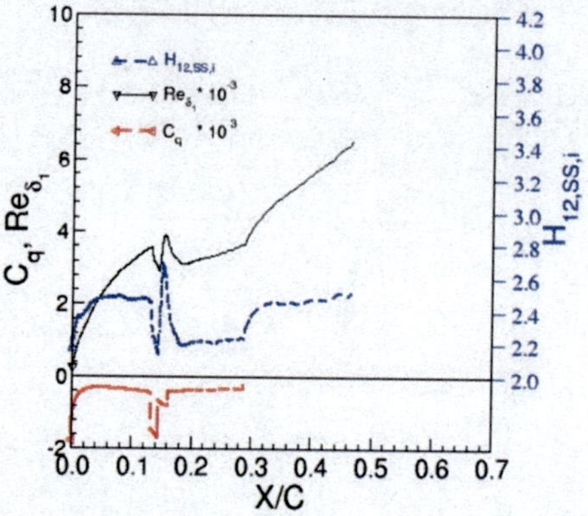

Fig. 10. Re_{δ_1} and H_{12} for the case of Fig. 7.

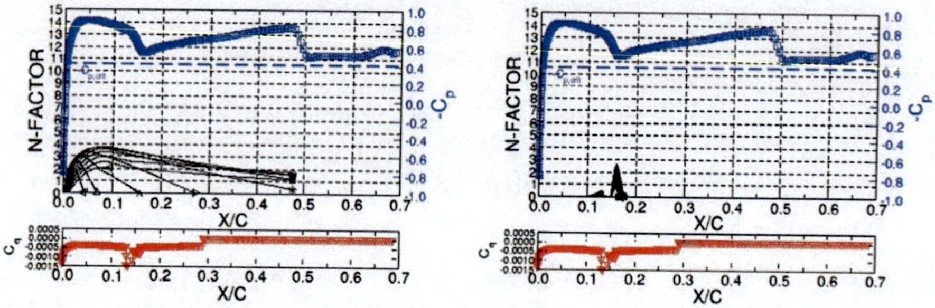

(a) Crossflow *N*-factors. (b) Tollmien-Schlichting *N*-factors.

Fig. 11. *N*-factor results for the case of Fig. 7.

5 Conclusions

In this paper, we compared the A320 HLFC VTP design with the design of an HLFC system for the wing of a long-range transport aircraft in the context of the HLFC-WIN project. In the first design, we used one microperforation and adjusted the suction strength with the help of suction chambers. In the second design, we applied the concept of variable microperforation and showed that this concept is feasible. This concept definitely deserves more attention. Future work should address the following points:

1. Reconsideration of the balance between pressure recovery and the shock strength.
2. Replacement of each pressure-loss characteristic by a range of allowable ones to allow for inaccuracies in the laser-drilling process.
3. Inclusion of no-suction zones caused by the supporting structure in the design calculations.

Acknowledgments and disclaimer. The authors thank Martin Radestock for his support and Nancy Schrauf for editing.

The A320 flight test activities received funding from the European Community's Seventh Framework Programme FP7/2007-2013, under grant agreement no 604013, AFLoNext project.

The HLFC-WIN project received funding from the Clean Sky 2 Joint Undertaking (JU) under grant agreement CS2-LPA-GAM-2020-2023-01. The JU receives support from the European Union's Horizon 2020 research and innovation programme and Clean Sky 2 JU members not in the Union.

The results, opinions, conclusions, etc. presented in this work are those of the authors only and do not necessarily represent the position of the JU; the JU is not responsible for any use made of the information contained herein.

References

1. Streit, T., Kruse, M., Kilian, T., von Geyr, J., Petropoulos, I.: Aerodynamic design and analysis of HLFC wings within the European project HLFC-WIN. In: 33rd Congress of the International Council of the Aeronautical Sciences (ICAS), Stockholm, Sweden, 4-9 September 2022 (2022)
2. Schrauf, G., von Geyr, H.: Hybrid laminar flow control on A320 fin: retrofit design and sample results. AIAA J. Aircr. **58**(6), 1272–1280 (2021). https://doi.org/10.2514/1.C036179
3. Schrauf, G., von Geyr, H.: Simplified hybrid laminar flow control for the A320 fin - part 2: evaluation with the e^N-method. In: AIAA SciTech Forum, 11-15 & 19-21 January 2021, Virtual Event. AIAA Paper AIAA-2021-1305 (2021). https://arc.aiaa.org/doi/10.2514/6.2021-1305
4. Schrauf, G.: Large-scale laminar-flow tests evaluated with linear stability theory. AIAA J. Aircr. **41**(2), 224–230 (2004). https://doi.org/10.2514/1.C036179
5. Arnal, D., Juilien, J.C., Reneaux, J., Gasparian, G.: Effect of wall suction on leading-edge contamination. Aerosp. Sci. Technol. **8**, 505–571 (1997)
6. Methel, J., Méry, F., Forte, M., Vermeersch, O.: Laminar flow control along the attachment line of a swept wing. In: ECCOMAS 2022, Oslo, Norway (2022)
7. MacManus, D.G., Eaton, J.A.: Flow physics of discrete boundary layer suction - measurements and predictions. J. Fluid Mech. **417**, 47–75 (2000)
8. Schrauf, G.: On allowable surface tolerances for laminar flow. DLR-IB-AS-BS-2022-70 (2022). https://elib.dlr.de/187060/
9. Kilian, T., Horn, M.: Verification of a chamberless HLFC design with an outer skin of variable porosity. CEAS Aeronaut. J. **12**(4), 835–845 (2021). https://doi.org/10.1007/s13272-021-00528-4
10. Seitz, A.: Messkörper, Durchflussmesssystem und Computerprogramm dafür," Offenlegungsschrift DE 2015 113 999 A1 2016.03.03, Anmeldetag 24.08.2015, Offenlegungstag 03.03.2016, Deutsches Patent- und Markenamt
11. COCO - A Program to Compute Velocity and Temperature Profiles for Local and Nonlocal Stability Analysis of Compressible, Conical Boundary Layers with Suction. ZARM Technik report (1998)
12. Schrauf, G.: LILO 2.1 User's Guide and Tutorial, Bremen, Germany, GSSC Technical report, No. 6, originally issued 2004, modified for Version 2.1 (2006)

Parametric Study of Vortex Generator Jets by Direct Numerical Simulation

Björn Selent[1(✉)], Ulrich Rist[1], and Oliver T. Schmidt[2]

[1] Institute of Aerodynamics and Gas Dynamics, University of Stuttgart, 70569 Stuttgart, Germany
bjoern.selent@iag.uni-stuttgart.de

[2] Department of Mechanical and Aerospace Engineering, University of California San Diego, La Jolla, CA 92093, USA

Abstract. Direct numerical simulations are used to gain insight into the effect of pitch and skew angle variations of vortex generator jets. The skew is investigated for a range from 30° to 150° for a pitch of 30° and 45°. The baseline consists of a laminar wall shear flow at a low Reynolds number of 165 based on the displacement thickness and freestream velocity. The resulting compound flows have been classified by integral values such as the friction coefficient and the momentum gain coefficient. The dependency on the pitch is small except at a small region close to the jet exit. Transition to turbulence occurs for skew angles above 90°. The jet trajectories are extracted from the data and a polynomial model that characterizes the trajectories as a function of skew and pitch is proposed.

Keywords: DNS · Vortex Generator Jets · pitch · skew angle · turbulent jets · friction coefficient · momentum gain

1 Introduction

Vortex generator jets (VGJ) are pneumatic devices consisting of a pipe or square duct that exits onto an aerodynamic surface. They insert a transverse jet into the surface's boundary layer flow which interacts with the latter to form a vortex system on the surface. The resulting compound flow is complex and can contain a wide range of flow features, such as near-wall counter-rotating vortex pairs, areas of separated flow, laminar streaks, large vortices reaching out of the boundary layer, and areas of almost fully turbulent flow. These flow features can achieve various desirable effects and make VGJ a promising technology for a wide range of industrial applications. For example in combustion chambers the mixing of fuel and oxidant can be enhanced, the trailing edge noise of the wind turbine blades can be reduced or flow detachment on airfoils and high-lift devices can be prevented. These benefits can furthermore be achieved without any additional drag for off-design-point operation. The complexity of the flow is a direct result of the number of design parameters that are involved, i.e. (in case

A. Dillmann et al. (Eds.): STAB/DGLR Symposium 2022, NNFM 154, pp. 392–401, 2024.
https://doi.org/10.1007/978-3-031-40482-5_37

of round jet actuators with constant blowing) the Reynolds number of the crossflow Re_{δ^*}, the jet-to-crossflow velocity ratio R, the boundary layer thickness δ, the pipe diameter D, the spacing of actuator arrays l_z as well as the pitch α and skew angle β of the jet. The number of influencing parameters makes the flow unfeasible to similarity analysis or likewise simplified solution approximations. Nonetheless, in order to control a flow of interest knowledge of the influence of the specific parameters of vortex generators is desirable. Accordingly, significant research effort was undertaken to both understand the evolution of the actuator flow and identify the best parameters for the chosen application.

An early parametric study for flow separation control by jet vortex generators was done by Selby et al. [8]. They already identified the ability to prevent separation to be a function of the jet speed and orientation and jet location with strong intermingling dependencies of the parameters with each other. Godard et al. [3] continued the work in a detailed study. They measured the effects of co- as well as counterrotating round jet actuators on a decelerating boundary layer over a wide variety of parameters including the velocity ratio V_{Jet}/V_e, skew angle, hole diameter and the number of jets and provided ranges of values for optimal operation in separation control. Ortmanns et al. [7] did stereoscopic particle image velocimetry experiments to gain deeper insight into the evolution of coherent structures and their interaction with a turbulent boundary layer. They concluded that the vortex core must be close to the surface in order to move high momentum close to the wall. Common to the above studies is that they investigate the effects of the jets for the specific purpose of separation control. A more general approach was taken by Keffer et al. [5] who provided a scaling law for the jet trajectory based on the constant density momentum flux ratio R^2 for vertical jets in crossflow ($\alpha = 90°$). Muppidi et al. [6] extended this work and performed high-fidelity numerical simulations of vertical jets. They proposed an exponential scaling including the vertical height of the trajectory. Similar efforts to describe the vortex flow of jets in a crossflow were introduced by Jones [4]. He developed an analytical solution to compute the vortex path from potential flow theory although for vane vortex generators only. More recent work from v. Stillfried et al. [10] provided a statistical modeling approach for the effect of jet vortex generators. This allowed replacing the actual geometry in CFD calculations with additional forcing terms.

Goals. This contribution aims to demonstrate the feasibility to model the jet trajectory of an exemplary jet vortex actuator flow configuration under arbitrary pitch and skew angle. To do so a set of high-fidelity numerical simulations is performed to provide training data for the model. The variations are chosen to produce distinguished flows which exhibit prototypical features such as laminar and turbulent jets, vortex pairs of different sizes and locations, and differently affected near-wall wakes. Global parameters are extracted to characterize the overall effect of the actuators. Finally, the derived model validity is shown for data from both training and additional testing simulations.

2 Numerical Method

2.1 Flow Parameters and Computational Domain

The numerical set-up of the steady jet vortex actuator flows is illustrated in Fig. 1. An array of pipes of diameter D and length L is flush mounted on a flat plate at position x_0 with spanwise spacing l_z. l_z is chosen large enough to allow an unperturbed evolution of the jet flow with the dimensions of the computational domain. The pipe flow is laminar. Both the pitch α and the skew angle β can be varied arbitrarily. The crossflow along the plate is assumed laminar with zero pressure gradient. The relevant dimensions and flow parameters are listed in Table 1. The values for Re, Re_{δ^*} and R match the set-up of Bagheri et al. [1] and are suitably low for direct numerical simulations. Nonetheless, the results for these low values can be regarded generally enough to allow for the transfer to larger Reynolds numbers (cf. [2]).

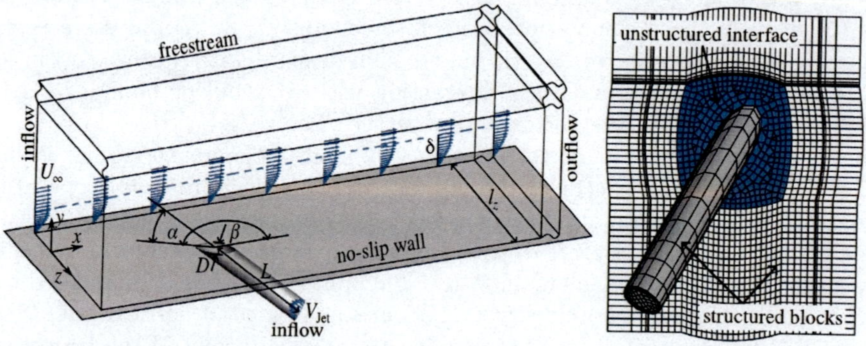

Fig. 1. Definition of VGJ parameters, computational domain, and sketch of computational grid

Altogether 11 simulations have been done covering a range of $\alpha \in [30°, 90°]$ and $\beta \in [30°, 150°]$. The combinations are listed in Table 1. For $\alpha = 90°$, β is invariant though. Furthermore, a baseline simulation without jet was done to obtain an initial solution for the simulations containing the pipe and to get reference values for the wall friction coefficient and undisturbed velocity field.

2.2 Direct Numerical Simulation

The open-source CFD software OpenFOAM has been used to perform the numerical simulations. The governing equations are the incompressible continuity and momentum equations

$$\begin{aligned} \nabla \cdot \vec{u} &= 0 \\ \frac{\partial}{\partial t}\vec{u} + (\vec{u} \cdot \nabla)\,\vec{u} &= -\nabla p + \frac{1}{Re}\Delta \vec{u}. \end{aligned} \tag{1}$$

Table 1. Parameters for numerical simulations and matrix of simulated cases

$Re_{\delta^*,x_0} = U_\infty \delta^*/\nu$	$Re = V_{\text{Jet}} D/\nu$	$R = V_{\text{Jet}}/U_\infty$	L	l_x	l_y	l_z
165	1485	3	5	25	12	10
α	β					
	0°	30°	60°	90°	120°	150°
30°		×	×	×	×	×
45°		×	×	×	×	×
90°	×					

Here $\vec{u}^T = (u, v, w)$ are the velocities in streamwise, wall-normal, and spanwise directions, (x, y, z). p denotes the static pressure and t the time. The respective variables are normalized using the reference velocity U_∞ and the characteristic length $D = \delta_{x_0}$. The equations are solved using the Pressure-Implicit with Splitting of Operators (PISO) algorithm. Both pressure and velocity equations are solved by a conjugate gradient method. The spatial schemes and time integration are second-order accurate.

The grid consists mostly of structured blocks except for the pipe-box-interface region where an unstructured mesh is used (Fig. 1). The box mesh is identical for all simulations with dimensions $l_x \times l_y \times l_z$ and the inflow positioned at $x = -8$. The pipe mesh with dimension $L \times D$ is built once for each α and rotated by the regarded β. A grid convergence study was conducted to ensure mesh-independent solutions. The final mesh contained approximately 42M cells.

The flow field above the wall was initialized with the steady-state solution of the baseline case. The velocity inside the pipe was set to zero initially. Dirichlet boundary conditions were used at the box and pipe inflow which prescribed the respective laminar boundary layer and parabolic flow. von Neumann boundary conditions at the freestream and an advective boundary condition at the outflow prevented reflections into the domain. The bottom wall and pipe wall were set to no-slip boundary conditions and the box was set to be periodic in the spanwise direction.

3 Results

3.1 Actuator Flow

In order to get a broad understanding of the specific features of the different configurations, the mean velocity field and the instantaneous vortex structures were investigated. The mean values were sampled after the initial transient left the domain at $T = 60$ for $\Delta T = 120$ time units.

Figure 2 depicts two prototypical flow states encountered. In case of a skew angle into the streamline ($\beta < 90°$) the vortex system is stable containing two distinct longitudinal vortices. The outer vortex (OV) persevers for a longer distance downstream while moving away from the wall and spiraling around the

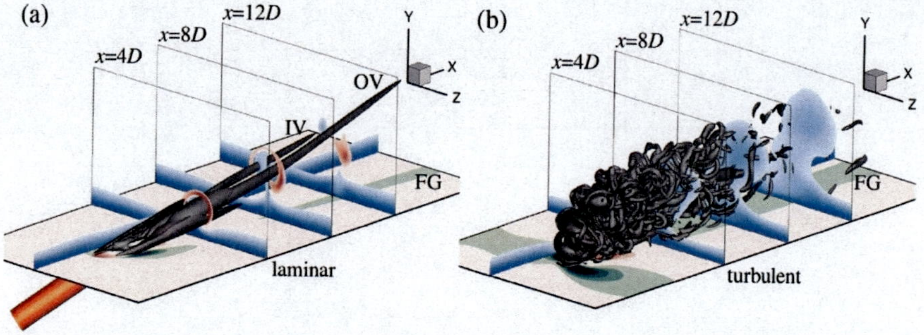

Fig. 2. Flow features for $\alpha = 30°$. (a) $\beta = 30°$, (b) $\beta = 150°$. Mean values for τ_W (bottom wall), mean downstream velocity ($x - y$ and $y - z$ planes), and instantaneous vortices (visualized by the Q-criterion).

inner vortex (IV). The change in wall friction is small and confined to a thin stripe and which be regarded as a footprint of the jet. For a skew angle of ($\beta \geq 90°$) no stable vortex system is upheld but the strong shear layer instability in the top side of the jet leads to a fast transition to turbulence. The mean downstream velocity shows a large region of influence behind and above the jet. Close examination does still show two regions of vorticity stacked above one another. A wide area of increased friction (FG) at the wall indicates that the vortices have a larger effect on the near-wall boundary layer compared to the former case.

In view of any application of the vortex actuator, a categorization is desirable. In this work, it was opted to look into parameters that are associated with separation control, i.e. the wall friction coefficient

$$c_f(x) = \frac{2}{\rho_\infty U_\infty^2} \int_{l_z/2}^{l_z/2} \tau_W \, \mathrm{d}z, \quad \text{where} \quad \tau_W(x, z) = \nu_\infty \partial U/\partial y, \tag{2}$$

and the gain in wall friction compared to the baseline value, c_{f_0},

$$\Delta c_f(x) = c_f - c_{f_0}. \tag{3}$$

The one-dimensional friction coefficient can be reduced to a scalar value by either integrating over x or finding the maximum value along x. The outcome is shown in Fig. 3. To avoid outliers due to the local effects in the vicinity of the jet exit, the integration and maximum search starts at $x = 2$. Both the maximum value and the integral value show a strong dependency on the chosen angle. A symmetry to some degree can be seen for a varying β at $\alpha = 30°$ with clear peaks at $\beta = 90°$. As this does not scale with α though, no straightforward relation $c_f = f(\alpha, \beta)$ is revealed. The friction coefficient does not take into account any effects of the vortex system outside of the wall gradient. To overcome this shortcoming and include the fully induced secondary flow field, the momentum

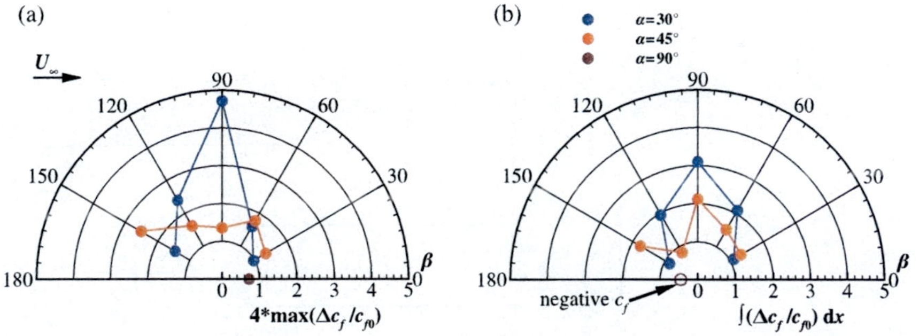

Fig. 3. Maximum and integral values of c_f

gain coefficient (cf. [9]) can be computed with

$$\Delta u(x, y, z) = U - U_0 \tag{4}$$

$$\Delta I(x, y)/l_z = \int_{-l_z/2}^{l_z/2} \Delta u |\Delta u| \, \mathrm{d}z \tag{5}$$

where U is the mean value of u. The momentum gain coefficient allows identifying regions in two dimensions where the flow is potentially energized. With the goal in mind to establish close relations between the induced flow and the geometric parameters, here the two-dimensional field is reduced analogously to c_f. The maximum value of ΔI in wall-normal direction y is shown in Fig. 4. The maximum was determined for the full domain height as well as inside the boundary layer only.

The further the jet is skewed into the main flow direction the more regions outside the boundary layer are affected. This is a direct result of the vortex system prevailing in the flow with its upward motion (cf. Fig. 2(a)). Overall, the largest gain in momentum is obtained for these flows, too. In the vicinity of the jet exit $x < 5D$, the momentum gain entering through the jet exit is upheld. With the jet skewed ever more in the opposite flow direction the jet instabilities diffuse the jet's momentum into the lateral directions. The net momentum gain decreases with increasing skew angle. For $\beta > 90°$, the maximum momentum gain occurs inside the boundary layer. Compared to the friction coefficient, distinct trends in the evolution of the momentum gain with the jet's pitch and skew can be seen which are encouraging in terms that a dependency on the jet's main trajectory exists.

3.2 Model Computation

Based on the preceding results it was attempted to build a model for the mean jet with the pitch and skew as sole parameters. This would allow us to predict the trajectory for arbitrary pitch α and skew β. The model was to be derived from the given set of simulated data. The focus was put on the angles albeit knowing

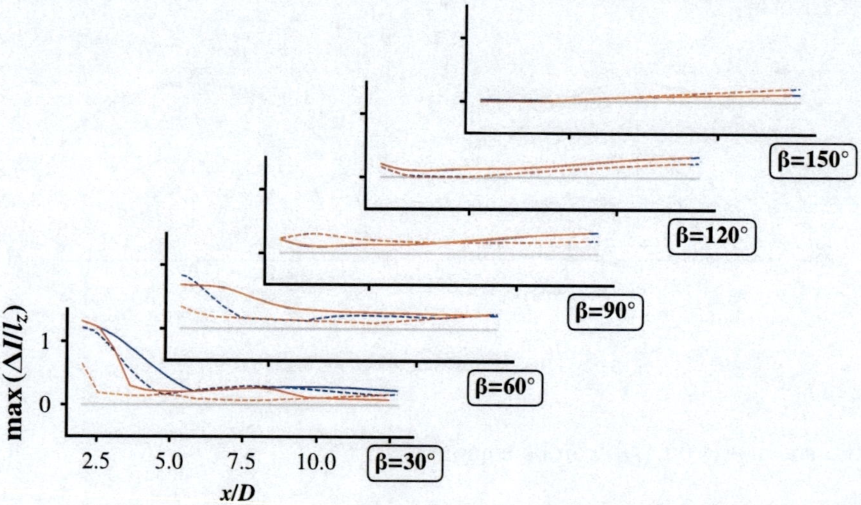

Fig. 4. Maximum values of the momentum gain coefficient. (solid lines) $\alpha = 30°$, (dashed lines) $\alpha = 45°$, (blue) all domain. (orange) inside boundary layer (Color figure online)

that this is a simplification and more advanced models should incorporate further parameters such as the velocity ratio and/or the boundary layer thickness, too. Hence it is an alternative to models of the jet trajectory in the classical jet-in-crossflow configuration as discussed in [6]. The jet's spatial coordinates are the variables that are to be described in closed form.

$$p_x(s) = \sum_{i=0}^{5} c_{x,i} s^i, \quad p_y(s) = \sum_{i=0}^{5} c_{y,i} s^i, \quad p_z(s) = \sum_{i=0}^{5} c_{z,i} s^i \tag{6}$$

with the body coordinate $s(x, y, z)$ along the streamline originating from the center of the orifice. The coordinates x, y and z are plotted against this streamline coordinate in Fig. 5. It can already be seen that x is linear versus s except for the region close to the exit. When the jet enters against the main flow, it is deflected by the momentum of the crossflow. This deflection follows a parabolic form. y exhibits exponential or hyperbolic behavior. For $\beta \leq 90°$, an inflection point is identified further downstream which coincides with the helical motion that twists the inner and outer vortex about each other. z exhibits similar behavior to y except that the inflection is less pronounced. It is worth noting that somewhat unintuitively the trajectory for $\alpha = 120°$ shows the widest spread rather than $\alpha = 90°$. As the trajectories of the simulated flows showed several inflection points, a 5th-order polynomial base was chosen. The polynomial coefficients c_i are computed by applying a general interpolation ansatz,

$$\mathsf{S}\vec{c} = \vec{f}, \tag{7}$$

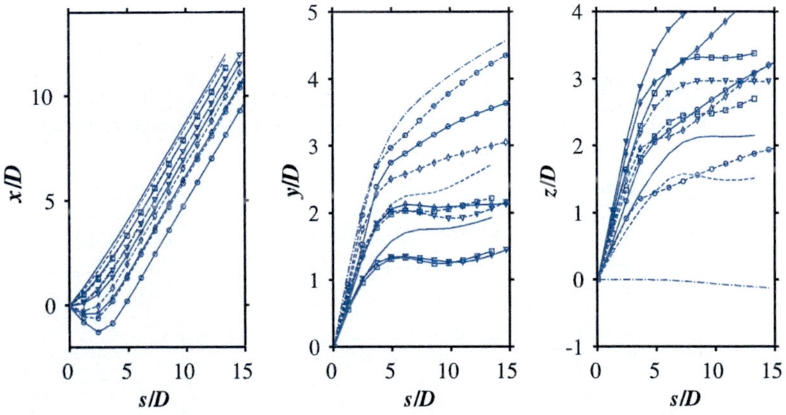

Fig. 5. Jet trajectory coordinates vs. streamline coordinate for $\alpha = 30°$ (solid line), $\alpha = 45°$ (dashed line) and $\alpha = 90°$ (dashed-dotted line) ((no symbol) $\beta = 30°$, (□) $\beta = 60°$, (▽) $\beta = 90°$, (◇) $\beta = 120°$, (○) $\beta = 150°$).

where S is the Vandermonde-matrix

$$\mathsf{S} = \begin{bmatrix} 1 & 0 & 0 & 0 & 0 & 0 \\ | & | & | & | & | & | \\ 1 & s & s^2 & s^3 & s^4 & s^5 \\ | & | & | & | & | & | \\ 0 & 0 & 2 & 6s_\infty & 12s_\infty^2 & 20s_\infty^3 \end{bmatrix} \tag{8}$$

and $\vec{f}$ are the coordinates at the respective s-position. The general ansatz is constrained by introducing boundary conditions at the beginning and end of the streamline, i.e. $\vec{p}(0) = 0$ and $\partial^2\vec{p}(s_\infty = 100)/\partial s^2 = 0$. Equation 7 is finally solved for the coefficients in the least square sense

$$\vec{c} \approx \mathsf{S}^\dagger \vec{f} = \mathsf{V}\boldsymbol{\Sigma}^{-1}\mathsf{U}^T \vec{f} \tag{9}$$

where $\mathsf{S}^\dagger$ is the pseudoinverse. A singular value decomposition of S yields the rotation matrix U, reflection matrix V^T and the singular values $\boldsymbol{\Sigma}$. This procedure is done for the three spatial directions separately. To close the model the analogous interpolation ansatz is used to determine the dependency of the coefficients c_i on the angles. Here it proved to be sufficient to use second order polynomials in 2D

$$\mathsf{D}\vec{d} = \vec{c} \tag{10}$$

$$\mathsf{D} = \begin{bmatrix} | & | & | & | & | & | \\ 1 & \alpha & \alpha^2 & \beta & \alpha\beta & \beta^2 \\ | & | & | & | & | & | \end{bmatrix} \tag{11}$$

The final set of coefficients are listed in Tables 2–4. Using these values (α, β measured in radians) the model was tested by both recovering the trajectories of the simulated flows and predicting the trajectory of a new (α, β)-combination. The x-coordinate can be modelled with confidence as can be seen in Fig. 6. The y- and z-reconstructions are not as good. The limit case of $\alpha = 90°$ depicts strong deviations. Further optimization of the coefficient computation to account for physical boundary conditions, e.g. no z-coordinate for a jet aligned with the downstream direction, is necessary.

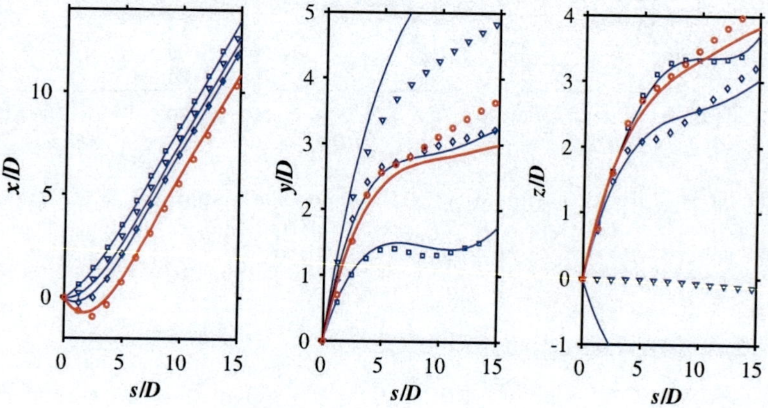

Fig. 6. Jet trajectory approximation: (solid line): model equation (6), (symbols): simulation data. (▽) $\alpha = 90°, \beta = 0°$, (◊) $\alpha = 45°, \beta = 120°$, (□) $\alpha = 30°, \beta = 30°$ (○) $\alpha = 30°, \beta = 135°$. (blue) data used in training set, (orange) new data. (Color figure online)

Table 2. Coefficient for vortex generator trajectory model: x-coordinate

	const.	α	α^2	β	$\alpha\beta$	β^2
$c_{x,1}$	1.721	−8.765E−1	−1.708E−1	−1.245	9.06E−1	−5.648E−2
$c_{x,2}$	−2.14E−1	3.074E−1	−3.588E−2	2.365E−1	−1.563E−1	3.412E−3
$c_{x,3}$	2.086E−02	−3.175E−2	6.506E−3	−1.832E−2	1.117E−2	1.653E−4
$c_{x,4}$	−6.739E−04	1.063E−03	−2.609E−04	4.881E−04	−2.941E−04	−4.999E−06
$c_{x,5}$	1.351E−06	−2.532E−06	5.841E−07	6.117E−07	8.173E−07	−6.311E−07

Table 3. Coefficient for vortex generator trajectory model: y-coordinate

	const.	α	α^2	β	$\alpha\beta$	β^2
$c_{y,1}$	−4.543E−01	2.084	−6.532E−01	1.976E−01	−1.679E−01	2.255E−02
$c_{y,2}$	1.385E−01	−3.723E−01	1.448E−01	−8.787E−02	1.465E−02	2.014E−02
$c_{y,3}$	−1.370E−02	2.796E−02	−1.137E−02	8.190E−03	−6.475E−04	−2.211E−03
$c_{y,4}$	3.906E−04	−6.431E−04	2.719E−04	−2.148E−04	−1.161E−05	7.063E-05
$c_{y,5}$	3.366E−06	−4.720E−06	9.738E−07	−2.587E−06	1.779E−06	−1.926E−08

Table 4. Coefficient for vortex generator trajectory model: z-coordinate

	const.	α	α^2	β	$\alpha\beta$	β^2
$c_{z,1}$	3.255E−02	2.114E−02	−2.593E−01	1.156	−1.408E−01	−2.972E−01
$c_{z,2}$	1.638E−01	−2.002E−01	9.029E−02	−1.796E−01	4.286E−02	3.499E−02
$c_{z,3}$	−1.939E−02	2.193E−02	−7.638E−03	1.208E−02	−4.323E−03	−1.579E−03
$c_{z,4}$	5.375E−04	−6.361E−04	2.033E−04	−1.867E−04	9.918E−05	5.842E−06
$c_{z,5}$	5.319E−06	−5.751E−07	−1.043E−06	−6.463E−06	2.223E−06	9.460E−07

4 Conclusion

High fidelity numerical simulations of active vortex generator jet flows within laminar boundary layers at low Reynolds numbers and modest velocity ratios have been performed. The simulations focused on the variation of the involved geometric angles while keeping further parameters at prototypical values of separation control applications. The flows have been categorized in a stable and unstable jet regime with the skew $\beta = 90°$ being the distinguishing parameter. The friction coefficient and the momentum gain were measured. In contrast to c_f ΔI showed clear trends especially with the skew of the jet. Value of $60° < \beta < 120°$ lead to an energizing of the near wall flow.

In order to establish a model to predict the jet mean flow for varying angles, the simulation data was used as base for polynomial interpolation. Using this model the mean trajectory of a test set was computed and compared to simulation data. The results albeit leaving room for optimization proved the concept to be generally feasible.

References

1. Bagheri, S., Schlatter, P., Schmid, P.J., Henningson, D.: Global stability of a jet in crossflow. J. Fluid Mech. **624**, 33–44 (2009)
2. Camussi, R., Gui, G., Stella, A.: Experimental study of a jet in a crossflow at very low Reynolds number. J. Fluid Mech. **454**, 113–144 (2002)
3. Godard, G., Stanislas, M.: Control of a decelerating boundary layer. Part 3: optimization of round jets vortex generators. Aerosp. Sci. Tech. **10**(6), 455–464 (2006)
4. Jones, J.P.: The Calculation of the Paths of Vortices from a System of Vortex Generators, and a Comparison with Experiments. A.R.C., C.P. 361 (1955)
5. Keffer, J.F., Baines, W.D.: The round turbulent jet in a cross wind. J. Fluid Mech. **15**, 481–496 (1963)
6. Muppidi, S., Mahesh, K.: Study of trajectories of jets in crossflow using direct numerical simulations. J. Fluid Mech. **530**, 81–100 (2005)
7. Ortmanns, J., Bitter, M., Kahler, C.J.: Dynamic vortex structures for flow-control applications. Exp. Fluids **44**, 397–408 (2008)
8. Selby, G.V., Lin, J.C., Howard, F.G.: Control of low-speed turbulent separated flow using jet vortex generators. Exp. Fluids **12**, 394–400 (1992)
9. Scholz, P., Radespiel, R., Kähler, C.J.: Performance Optimization of Jet Actuator Arrays for Active Flow Control. CEAS/KATnet (2005)
10. von Stillfried, F., Wallin, S., Johansson, A.V.: Statistical vortex-generator-jet model for turbulent flow separation control. AIAA J. **51**(5), 1119–1129 (2013)

Direct Numerical Simulation of Counter-Rotating Cylindrical Roughness Pairs for Laminar Flow Control

Yongxiang Wu(✉), Christoph Wenzel, and Ulrich Rist

Institute of Aerodynamics and Gas Dynamics,
Pfaffenwaldring 21, 70569 Stuttgart, Germany
yongxiang.wu@iag.uni-stuttgart.de

Abstract. Counter-rotating cylindrical roughness pairs are used to control the Tollmien-Schlichting instability-induced laminar-turbulent transition. Direct numerical simulations are performed to study this active flow-control method's potential for delaying an H-type transition, which is indicated by a linear stability analysis by the current authors [15]. A complete stabilization of the TS-instability-induced transition in the investigated integration domain is obtained. A Fourier analysis of the streamwise evolution of instabilities reveals that both fundamental and subharmonic components of the controlled TS waves are attenuated. The present work thus proposes and investigates an active technique using wall-normal finite-length cylinders, a widely used structure, to control boundary layer transition.

Keywords: Laminar-turbulent transition · boundary-layer control · rotating cylindrical roughness elements

1 Introduction

Delay the laminar-turbulent transition by means of laminar flow control (LFC) has long been a focused area since it reduces the skin-friction coefficient C_f of a boundary layer by one order of magnitude from a turbulent state to a laminar state for relevant Reynolds numbers. Therefore, a successful LFC technology offers a significant benefit in terms of reducing operational costs and decreasing pollutant emissions for engineering parts and moving vehicles.

However, boundary-layer stability is endangered by numerous sources, one type of which is the presence of two-dimensional (2D) or three-dimensional (3D) surface roughness elements. Whereas the effect of 2D roughness elements, such as gaps and steps, is well-known to amplify the growth rate of the 2D Tollmien-Schlichting (TS) waves [8], the influence of 3D roughness elements is much more complicated. The induced flow topology is characterised by a horseshoe vortex wrapping around the 3D roughness element and two elongated counter-rotating streamwise vortex legs [5]. Depending on the level of external disturbances, the

A. Dillmann et al. (Eds.): STAB/DGLR Symposium 2022, NNFM 154, pp. 402–411, 2024.
https://doi.org/10.1007/978-3-031-40482-5_38

alternating high-low-high velocity streaks can be subjected to different instability mechanisms. If the amplitude of external disturbances is low, two types of inviscid secondary mode due to the inflectional velocity profile can be supported by the low-speed streak, i.e. sinuous (anti-symmetric) and varicose (symmetric) instabilities [1]. Another instability mechanism occurs at a moderate or large amplitude of external disturbances, which is characterized by an algebraic growth of the perturbations followed by either an exponential decay or breakdown to turbulence [11]. This transient growth of boundary layer instability can be explained theoretically by the non-normality of the governing differential stability operator [14], that the algebraic growth is a result of supposition of decaying linear or nonlinear instability modes.

In contrast to the effects discussed above, surface roughness elements may also stabilize the boundary layer. Boiko *et al.* [2] observed that finite amplitude unsteady boundary-layer streaks can stabilize the TS waves. A linear stability analysis revealed the underlying mechanism, that the interaction between the TS waves and the streaks results in a negative spanwise perturbation kinetic energy production [3]. It together with the viscous dissipation overweighs the positive wall-normal production term, which leads to an overall stabilisation effect. This theoretical prediction was verified experimentally by Fransson *et al.* [4] with streaks generated by spanwise narrowly spaced circular cylinders. The present authors have previously studied the linear stability property of boundary-layer flows governed by rotating cylindrical roughness elements and discovered that the produced velocity streaks may also stabilize TS waves. However, the mechanism is recognized as a decrease in wall-normal production [15–17]. Experimental testing in a laminar water channel supports the theoretical prediction, even when only one pair of rotating cylinders is employed [9]. It should be mentioned that the stabilization effect only applies to low rotation rates. At medium to high rotation rates, the induced boundary layer is subject to various instabilities, especially a deceleration mechanism that occurs on the cylinder's decelerating side [18], which can directly trigger a transition.

In this work, the potential of counter-rotating cylinder pairs to control the TS-instability-induced transition is investigated by a direct numerical simulation. A blowing and suction boundary condition is used to stimulate TS-instability-triggered transition. The configuration of Fransson *et al.* [4], which places the blowing and suction strip behind the rotating cylindrical roughness elements, is used to demonstrate the viability of this method. At the current stage, the study is more to demonstrate the proof-of-concept. In the future, more complicated flow scenarios like boundary-layer with free-stream turbulence or transition induced by cross-flow can be used to study the feasibility of the method. For a summary of our previous work on boundary layer destabilization behind roughness elements, the reader is referred to the dissertation by the first author [19].

2 Numerical Methods

Depending on the rotational sense of the cylinders, there are two types of setups for the roughness array, namely co-rotating and counter-rotating cylinder pairs.

In Wu *et al.* [15], it is discovered in a linear stability theory frame that the velocity streaks produced by counter-rotating pairs, which speed up the flow in between the cylinder pair, are capable of effectively stabilising TS instabilities. In this work, this effect is investigated with a direct numerical simulation, whose numerical set-up is illustrated in Fig. 1(a). The rotation strength is measured by the ratio of the tangential velocity at the top of the cylindrical roughness element to the incoming local velocity, i.e.

$$\Omega_u = \frac{\Omega D}{2u(k)}, \tag{1}$$

where Ω is the angular velocity of the cylinder around its axis, D and k the diameter and height of the cylinder, respectively. Then, $u(k)$ is the undisturbed Blasius streamwise velocity component at the height of the cylinder.

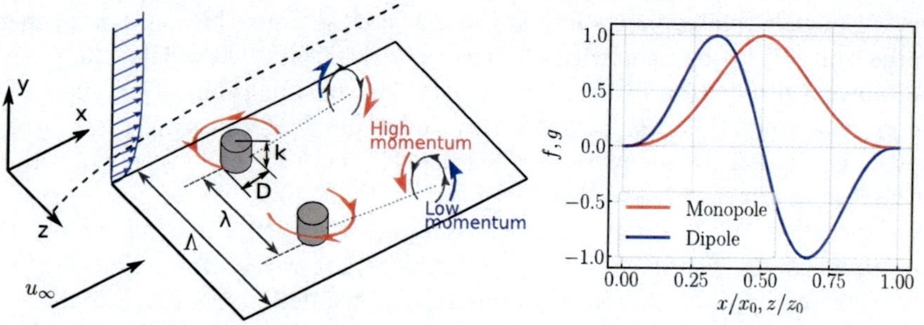

(a) (b)

Fig. 1. (a) Numerical set-up of counter-rotating roughness pair mounted on a flat-plate. High- and low-momentum flows induced by streamwise vortices are colored in red and blue, respectively. (b) Spatial distribution of the monopole and dipole function of the blowing and suction boundary condition. (Color figure online)

The H-type transition scenario over a flat-plate with and without rotating cylindrical roughness arrays is studied. Based on the distance from the leading edge x, the Reynolds number at the inlet of the integration domain is $Re_x = u_\infty x/\nu = 0.8 \times 10^5$. Here, the freestream velocity is $u_\infty = 15\,(\mathrm{m/s})$ and the kinematic viscosity is $\nu = 1.5 \times 10^{-5}$ $(\mathrm{m^2/s})$. The integration domain has a dimension of $0.8 \leq Re_x/10^5 \leq 8$ in the streamwise direction and $0 \leq Re_y/10^5 \leq 4$ in the wall-normal direction while $-1.5 \leq Re_z/10^4 \leq 1.5$ in the periodic spanwise direction. At the inlet, a Blasius velocity profile is prescribed and an advective boundary condition is set at the outlet patch to avoid any possible reflection. For the top boundary, a zero-pressure condition is used. For cylinders, the rotational wall boundary condition calculated as $\boldsymbol{u}_w = \boldsymbol{\Omega} \times \boldsymbol{r}$ is applied, being $\boldsymbol{r}$ the vector from the rotational axis of the cylinder to the wall surfaces. Five pairs of counter-rotating cylinders with $k = 0.00075\,(\mathrm{m})$ and $\eta = D/k = 0.5$ which are spaced by $\lambda/k = 2$ (between cylinders within a pair) and $\Lambda/k = 8$ (between pairs) are placed at $Re_x = 1.2 \times 10^5$. At the cylinder station, the boundary layer displacement thickness is $\delta^*/k = 0.779$, and the roughness-height-based Reynolds numbers are $Re_k = u_\infty k/\nu = 750$ and $Re_{kk} = u(k)k/\nu = 540.9$.

To excite the H-type transition, a blowing & suction strip located at $1.42 \leq Re_x/10^5 \leq 1.58$ according to the following equation

$$v = fA_1cos(\omega_1 t) + gA_{1/2}cos(\kappa z + \omega_{1/2}t) + gA_{1/2}cos(-\kappa z + \omega_{1/2}t), \tag{2a}$$

$$f = \begin{cases} \frac{81}{16}(2\frac{x}{x_0})^3[3 \cdot (2\frac{x}{x_0})^2 - 7(2\frac{x}{x_0}) + 4], & \frac{x}{x_0} \in [0; 0.5] \\ -\frac{81}{16}(2 - 2\frac{x}{x_0})^3[3(2 - 2\frac{x}{x_0})^2 - 7(2 - 2\frac{x}{x_0}) + 4], & \frac{x}{x_0} \in [0.5; 1] \end{cases} \tag{2b}$$

$$g = \begin{cases} -3(2z)^4 + 4(2z)^3, & \frac{z}{z_0} \in [0; 0.5] \\ -3(2 - 2z)^4 + 4(2 - 2z)^3, & \frac{z}{z_0} \in [0.5; 1] \end{cases} \tag{2c}$$

is used to generate the TS and oblique waves, the frequencies of which are ω_1 and $\omega_{1/2}$, respectively. Here, x_0 and z_0 are the streamwise and spanwise lengths of the blowing & suction strip, respectively. As shown in Fig. 1(b), f (dipole function) and g (monopole function) are the spatial distribution of the fundamental and subharmonic disturbances in the blowing and suction strip. $A_1/u_\infty = 0.001$ and $A_{1/2}/u_\infty = 0.0001$ are the amplitudes for fundamental and subharmonic waves, respectively. The non-dimensional fundamental and subharmonic frequencies are $F_1 = \omega_1\nu/u_\infty^2 = 1.24 \times 10^{-4}$ and $F_{1/2} = \omega_{1/2}\nu/u_\infty^2 = 0.62 \times 10^{-4}$, respectively. The spanwise wavelength of the oblique wave $\beta_{1/2} = 2\pi/\kappa$ is 1/2 of the spanwise dimension of the computational domain z_0, where κ is the fundamental spanwise wavenumber of the oblique wave. These parameters are selected for comparison with the experiments of Kachanov and Levchenko [7] and direct numerical simulation of Sayadi *et al.* [12]. The frequencies are plotted with the neutral stability curve of the corresponding boundary layer in Fig. 2(a). The locations of the cylinder pair array and that of the TS generator are also displayed. In order to promote flow randomization and asymmetry in the post-transitional turbulent region, the blowing and suction wall-normal velocities are also combined with a temporal broadband disturbance with an amplitude two orders of magnitude smaller than the oblique wave.

Following Sayadi *et al.* [12], the grid is evenly spaced in the streamwise and spanwise direction with $\Delta x/x_0 = 0.002$ and $\Delta z/x_0 = 0.001$, respectively. In the wall-normal direction, the grid is clustered toward the wall. At $Re_x = 6 \times 10^5$, the grid spacings in wall units correspond to $\Delta x^+ = 8$ and $\Delta z^+ = 4$. The minimum grid spacing in the wall-normal direction is $\Delta y^+ = 0.3$, which results in a total of 150 million grid cells. Figure 2(b) shows the validation of the spatial development of the Fourier components of primary and subharmonic frequency of the H-type transition. Here, the Fourier components are denoted as (n, p), representing a disturbance with frequency $n\omega_1$ and spanwise wavenumber $p\kappa$. For comparison, the Fourier analysis is performed at a constant wall-normal plane of $y/\delta = 0.26$. Good agreement between experimental measurement of [7], DNS of [12] and the present simulation is obtained for both the fundamental and the subharmonic modes.

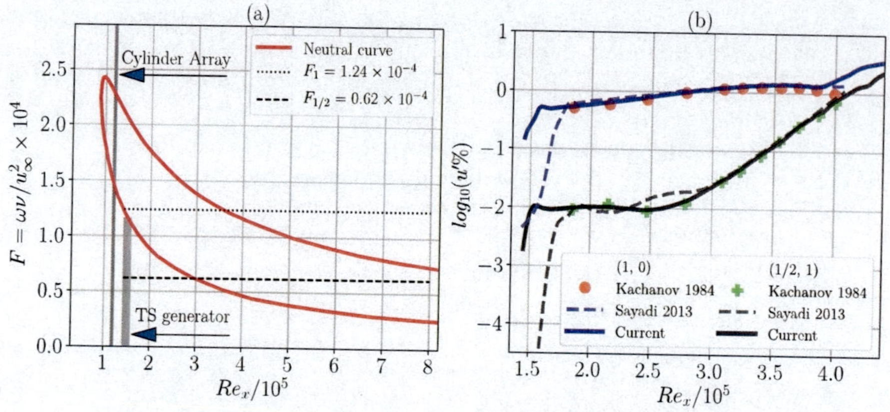

Fig. 2. (a) Neutral stability curves of the Blasius boundary layer. The vertical grey line marks the location of the cylinder arrays and the location of the TS generator. (b) Comparison of the primary (1,0) and subharmonic (1/2, 1) modes of H-type transition with [7] and [12]. (Color figure online)

3 Results

Linear stability analysis in Wu *et al.* [15] indicates that the stabilization effect is evident at low rotation rates. Following case C1 in table 1 of [15], where cylinders have an aspect ratio $\eta = 0.5$, three rotation speeds are investigated in current work, i.e. $\Omega_u = 0, 0.1, 0.2$. Further increasing the rotation rate might lead to premature transition of the boundary layer [18], which is outside the scope of this work's discussion.

The effect of the counter-rotating cylindrical roughness pair arrays on the boundary layer can be visualized by the changes in meanflow. Figure 3 depicts the streamwise time-averaged mean velocity $\bar{u}$ at two representative streamwise locations, i.e. $Re_x/10^5 = 1.4$ for the near-wake and $Re_x/10^5 = 3$ for the far-wake, for three rotating cases. For the non-rotating case $\Omega_u = 0$, the typical wavy spanwise modulations (or velocity streaks) are almost unnoticeable for both near-wake and far-wake. It should be emphasized that effective generation of velocity streaks is necessary for the laminar-turbulent transition delay employing cylindrical roughness elements since the method depends on the negative perturbation kinetic energy (PKE) production brought in by the spanwise velocity modulation [3,4]. The non-rotating case presented in the current study is evidently unable to generate such necessary velocity streaks that are able to stabilize the TS instability. Instead, it serves as a reference case for the rotating cases. As shown in Fig. 3(c,d) for $\Omega_u = 0.1$ and Fig. 3(e,f) for $\Omega_u = 0.2$, the rotating cylindrical roughness pairs are very effective in creating velocity streaks in both near-wake and far-wake. Moreover, the induced high-speed streak is more evident in the current spanwise configuration of the positive counter-rotating setup whereas the low-speed streak is almost suppressed. As a consequence, the benefits are two-sided. On the one hand, the high-speed streaks promote the

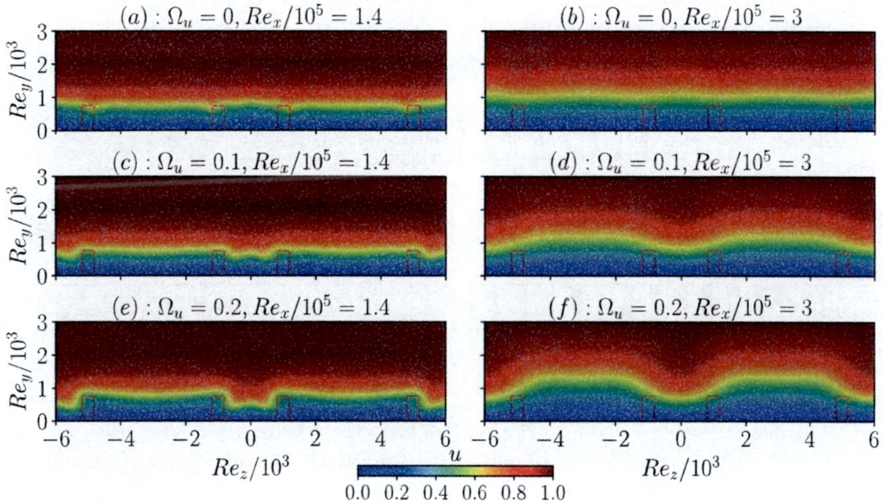

Fig. 3. Presentation of meanflow at near-wake ($Re_x/10^5 = 1.4$) in the left column and far-wake ($Re_x/10^5 = 3$) in the right column for non-rotating case $\Omega_u = 0$ (a,b), rotating cases $\Omega_u = 0.1$ (c,d) and $\Omega_u = 0.2$ (e,f). The spanwise location of the cylinders are marked by the dashed rectangles.

spanwise modulation of the velocity profile, which can result in the formation of a negative PKE production. On the other hand, the typically accompanying low-speed streak for cylindrical roughness elements is substantially inhibited, allowing for the best possible avoidance of the secondary inflectional instability.

The overall effects of laminar-turbulent transition delay can be presented by the instantaneous vortical structures as shown by the Q-criterion [6] in Fig. 4, and the streamwise evolution of the temporal and spanwise averaged skin-friction coefficients c_f versus Re_x in Fig. 5. The H-type transition on a flat-plate is illustrated in Fig. 4(a). The characteristic staggered Λ vortices start to form at $Re_x = 4 \times 10^5$. Prior to this location, the 2D TS wave in the form of a spanwise vortex gradually evolves into 3D Λ vortices. Such streamwise evolution is also represented in Fig. 5, where the skin friction deviates from the laminar correlation at the same location. The instantaneous vortical structure shown in Fig. 4(b) as well as the streamwise evolution of the skin-friction coefficient for the non-rotating case $\Omega_u = 0$ is similar. This is not surprising for the static cylindrical roughness elements setup given that the chosen cylinders are thin ($\eta = 0.5$), since their capability to produce downstream velocity streak is constrained to the very near wake and therefore has little influence on the evolution of TS-instability.

The instantaneous vortices' evolution for the rotating cases in Fig. 4(c)(d) are all modulated by the streamwise vortices produced by the counter-rotating cylindrical pair arrays. The transition delay effect is evident in both Fig. 4 and Fig. 5, resulting in a full stabilization or relaminarization in the investigated

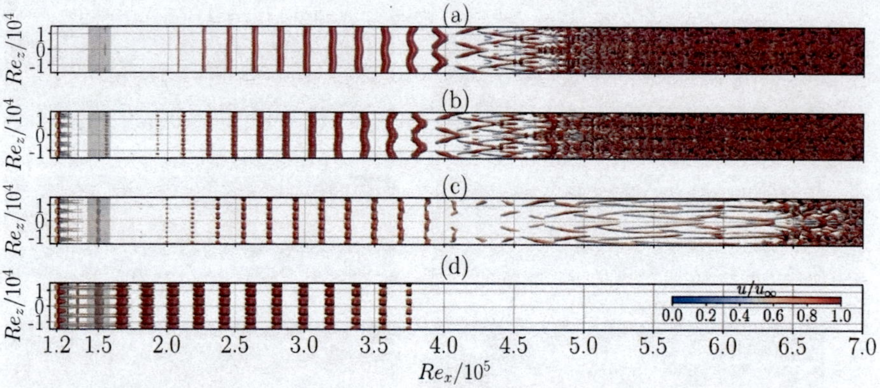

Fig. 4. Iso-surfaces of the Q-criterion with $Q = 8 \times 10^4$ for (a): Flat-plate reference, (b): $\Omega_u = 0$, (c): $\Omega_u = 0.1$ and $Q = 8 \times 10^3$ for (d): $\Omega_u = 0.2$. Color represents the streamwise velocity u/u_∞. The shaded region at around $Re_x = 1.5 \times 10^5$ marks the location of the TS generator.

domain for the case $\Omega_u = 0.2$, which confirms the prediction of Biglobal LST in Wu *et al.* [15].

A Fourier analysis of the spatial evolution of instabilities and perturbations for cases $\Omega_u = 0$ and $\Omega_u = 0.1$ are shown in Fig. 6. The modal amplitudes with rotating cylindrical roughness pairs are compared with that of the flat-plate. H-type transition is normally triggered by subharmonic resonance, which is active at amplitude in the order of 0.1% of the freestream velocity. The fundamental modes (1,0) and subharmonic modes (1/2,1) are highlighted for clarification. Higher harmonics of the flat-plate case are also plotted. For the flat-plate case, the initial amplitude of the subharmonic mode (1/2,1) is one order of magnitude lower than the fundamental mode (1,0). The subharmonic mode begins to rise rapidly at $Re_x/10^5 \approx 2.5$ and reaches its maximum value at $Re_x/10^5 \approx 4.2$. Further downstream, the growth rate of almost all modes turns slightly smaller for a short distance until reaching nonlinear saturation at $Re_x/10^5 \approx 5.2$.

For the case with non-rotating cylindrical pairs, i.e. $\Omega_u = 0$, the fundamental mode (1,0) almost overlaps with the reference flat-plate case until $Re_x/10^5 \approx 3.8$, see Fig. 6(a). The subharmonic mode sets in slightly upstream as compared to the flat-plate case and saturates at a similar amplitude. This observation complies with the overlap of the C_f curves in Fig. 5. The effect of rotating case $\Omega_u = 0.1$ is very evident on both fundamental and subharmonic modes in Fig. 6(b), where the growth rate of both modes are reduced and the nonlinear saturation location is delayed to $Re_x/10^5 \approx 7$. As discussed in linear stability analysis [15], this stabilization effect can be attributed to the negative spanwise PKE induced by the velocity streaks. Once the rotational speed is increased to $\Omega_u = 0.2$, the flow is completely stabilized and a relaminarization in the investigated integration domain is achieved, see Fig. 4(d).

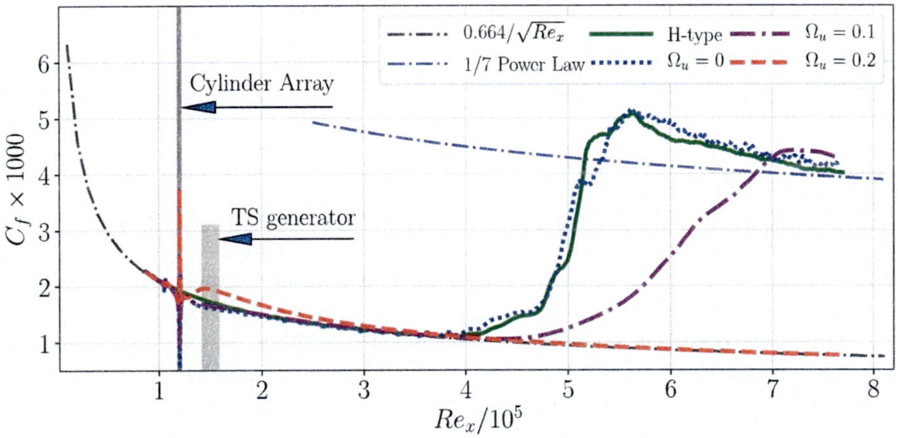

Fig. 5. Local skin friction versus Re_x on a flat-plate. 1/7 power law with experimental calibration [13]. The locations of the rotating cylinder array and the TS generator are indicated by arrow.

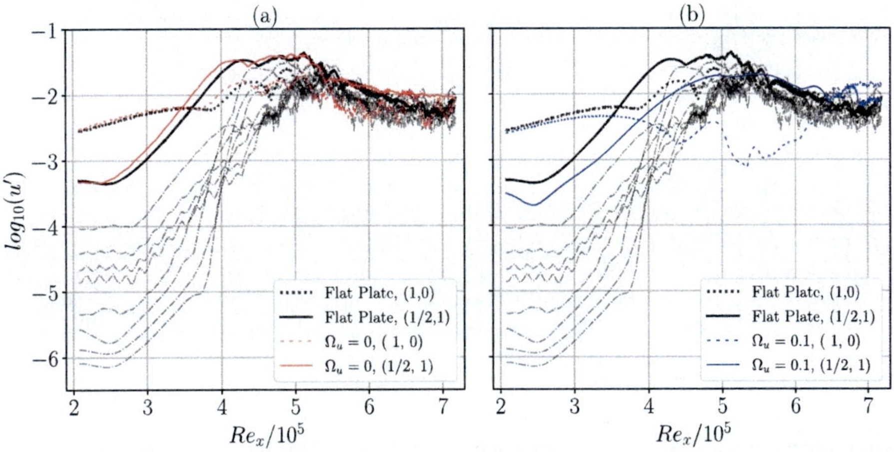

Fig. 6. Streamwise evolution of modal amplitudes for case (a) $\Omega_u = 0$ and case (b) $\Omega_u = 0.1$ from Fourier analysis, with comparison to flat-plate. The unannotated curves are higher harmonics of the TS wave in the flat-plate case.

4 Concluding Remarks

Counter-rotating wall-normal cylinder pairs embedded in a boundary layer over a flat-plate are investigated by direct numerical simulations. This active control method is effective in inducing high-speed streak, which is beneficial for stabilizing TS-instabilities. Meanwhile, the byproduct of cylindrical roughness elements, i.e. the low-speed streaks which are likely to support secondary inflectional instabilities, is suppressed to the greatest extent. With a proper rotating rate ($\Omega_u = 0.2$ for the current setup), it is found that the H-type laminar-turbulent transition over a flat-plate can be delayed, in the absence of strongly destabilising nonlinear interaction between the boundary layer instability waves and the induced streaks. The current study provides proof of the theoretical prediction in Wu *et al.* [15]. It is to be pointed out that transition delay techniques using surface roughness elements are exclusively directed to the exploitation of the passive method, as investigated by the Fransson group [4,10]. The current work provides the possibility to achieve transition delay using an active control technique, particularly, with a rotating thin cylinder. However, the complex receptivity process of incoming waves by the roughness array is untouched in the current setup since the TS waves are generated downstream of the cylinder-pair arrays. Furthermore, if any of the current setup parameters are changed, for instance, the streamwise location of the cylinder-pair arrays x, the spanwise spacing of cylinder-pair (λ, Λ) or the frequencies of the H-type transition (ω_1, $\omega_{1/2}$), the effectiveness of the present control method should be investigated.

Acknowledgements. The authors gratefully acknowledge the financial support from the Deutsche Forschungsgemeinschaft under project numbers RI 680/39-1 and RI 680/48-1. Computational resources provided by the federal high-performance computing center Stuttgart (HLRS) is kindly acknowledged.

References

1. Andersson, P., Brandt, L., Bottaro, A., Henningson, D.S.: On the breakdown of boundary layer streaks. J. Fluid Mech. **428**, 29–60 (2001)
2. Boiko, A.V., Westin, K.J.A., Klingmann, B.G.B., Kozlov, V.V., Alfredsson, P.H.: Experiments in a boundary layer subjected to free stream turbulence. Part 2. The role of TS-waves in the transition process. J. Fluid Mech. **281**, 219–245 (1994)
3. Cossu, C., Brandt, L.: On Tollmien-Schlichting-like waves in streaky boundary layers. Eur. J. Mech. B Fluids **23**, 815–833 (2004)
4. Fransson, J.H., Talamelli, A., Brandt, L., Cossu, C.: Delaying transition to turbulence by a passive mechanism. Phys. Rev. Lett. **96**, 064501 (2006)
5. Gregory, N., Walker, W.S.: The effect on transition of isolated surface excrescences in the boundary layer. Aero. Res. Counc. R&M **2779**, 1–10 (1956)
6. Hunt, J.C., Wray, A.A., Moin, P.: Eddies, streams, and convergence zones in turbulent flows. Studying turbulence using numerical simulation databases, 2. In: Proceedings of the 1988 Summer Program (1988)
7. Kachanov, Y.S., Levchenko, V.Y.: The resonant interaction of disturbances at laminar-turbulent transition in a boundary layer. J. Fluid Mech. **138**, 209–247 (1984)
8. Klebanoff, P.S., Tidstrom, K.D.: Mechanism by which a two-dimensional roughness element induces boundary-layer transition. Phys. Fluids **15**(7), 1173–1188 (1972)
9. Römer, T., Schulz, K., Wu, Y., Christoph, W., Rist, U.: Delay of laminar-turbulent transition by counter-rotating cylindrical roughness elements in a laminar flat-plate boundary layer. Exp. Fluids **64**, 42 (2023)
10. Siconolfi, L., Camarri, S., Fransson, J.: H: stability analysis of boundary layers controlled by miniature vortex generators. J. Fluid Mech. **784**, 596–618 (2015)

11. Reshotko, E.: Transient growth: a factor in bypass transition. Phys. Fluids **13**(5), 1067–1075 (2001)
12. Sayadi, T., Hamman, C.W., Moin, P.: Direct numerical simulation of complete H-type and K-type transitions with implications for the dynamics of turbulent boundary layers. J. Fluid Mech. **724**, 480–509 (2013)
13. Schlichting, H., Gersten, H.: Boundary-Layer Theory. Springer, Heidelberg (2003). https://doi.org/10.1007/978-3-662-52919-5
14. Trefethen, L.N., Trefethen, A.E., Reddy, S.C., Driscoll, T.A.: Hydrodynamic stability without eigenvalues. Science **261**, 578–584 (1993)
15. Wu, Y., Axtmann, G., Rist, U.: Linear stability analysis of a boundary layer with rotating wall-normal cylindrical roughness elements. J. Fluid Mech. **915**, A132 (2021)
16. Wu, Y., Rist, U.: Boundary layer stability with embedded rotating cylindrical roughness element. In: Dillmann, A., Heller, G., Krämer, E., Wagner, C., Tropea, C., Jakirlić, S. (eds.) DGLR 2018. NNFMMD, vol. 142, pp. 274–283. Springer, Cham (2020). https://doi.org/10.1007/978-3-030-25253-3_27
17. Wu, Y., Rist, U.: Stability analysis of boundary layer flow with rotating cylindrical surface roughness elements. In: Sherwin, S., Schmid, P., Wu, X. (eds.) IUTAM Laminar-Turbulent Transition. IB, vol. 38, pp. 815–824. Springer, Cham (2022). https://doi.org/10.1007/978-3-030-67902-6_71
18. Wu, Y., Römer, T., Axtmann, G., Rist, U.: Transition mechanisms in a boundary layer controlled by rotating wall-normal cylindrical roughness elements. J. Fluid Mech. **945**, A20 (2022)
19. Wu, Y.: Numerical investigation of boundary layer instability and transition induced by rotating wall-normal cylindrical roughness elements. Verlag Dr. Hut (2022)

Turbulence and Transition

Turbulent Wedge Modeling in Local Correlation-Based Transition Models

Michael Fehrs[1(✉)] and Sebastian Helm[2]

[1] DLR, Institute of Aeroelasticity, Bunsenstr. 10, 37073 Göttingen, Germany
michael.fehrs@dlr.de

[2] DLR, Institute of Aerodynamics and Flow Technology, Bunsenstr. 10, 37073 Göttingen, Germany

Abstract. This article presents a novel approach to include turbulent wedges in local correlation-based γ transition models. The turbulent wedge is modeled by increasing the intermittency at the wedge apex. The wedge develops downstream without further interference in the transition model behavior. The method is demonstrated for an experimental high Reynolds number test case on the NASA CRM-NLF configuration. Wedges are successfully created, but the wedge angles are too large compared to the experimental data. Grid spacing, initial disturbance size, and scaling of the diffusion term only have a minor effect on the wedge angles.

Keywords: turbulent wedges · transition tripping · local correlation-based transition models · intermittency transport models

1 Introduction

The development and assessment of future eco-efficient natural laminar flow aircraft require comprehensive numerical transition models that cover a wide variety of flows (steady and unsteady) and offer an insight into transition induced flow phenomena. Turbulent contamination in the form of turbulent wedges impedes model validation based on high Reynolds number wind tunnel tests and can lead to flight scenarios due to partial loss of laminarity that need to be covered in the aircraft assessment.

A first description of a spanwise turbulent contamination of a laminar boundary layer resulting in turbulent wedges is given by Charters [1]. Schubauer & Klebanoff [2] observed turbulent wedges with a fully turbulent core growing at a half-angle of about 6.4°. The turbulent core is surrounded by an intermittent flow region increasing the wedge half-angle to about 10.6°. Different authors report similar half-angles for zero-pressure gradient flows [2]. The half-angles for turbulent spots [3] and turbulent wedges [4] increase for adverse pressure gradients and vice-versa for favorable pressure gradients.

The exact mechanisms of lateral turbulent wedge spreading are subject of numerical and experimental investigations [5,6]. These studies show that the

A. Dillmann et al. (Eds.): STAB/DGLR Symposium 2022, NNFM 154, pp. 415–424, 2024.
https://doi.org/10.1007/978-3-031-40482-5_39

growth of the turbulent wedge is likely caused by the creation of high- and low-speed streaks on the lateral edges of the wedge in a self-regenerating mechanism. This adds to the early observation that the lateral growth of turbulent wedges is not driven by turbulent mixing/entrainment alone but by some additional destabilizing mechanism [7].

Local correlation-based transition models offer a fast and simple way to include transitional flows within common RANS architectures. To further increase the predictive capabilities, a functionality for intermittency transport models to model turbulent wedges is presented within this article. The specific growth mechanisms of turbulent wedges are beyond the modeling depth of an eddy viscosity model, but it might be possible to model the lateral spreading by the transport behavior of the γ equation itself.

2 Methods

2.1 DLR TAU-Code, Transition Modeling and Numerical Settings

The DLR TAU-Code [8] is used in this investigation. The novel transition tripping to model turbulent wedges is implemented for the γ and γ-CAS transition models [9,10]. Both models are used in combination with the SST k-ω turbulence model [11]. For all computations, a central scheme with artificial matrix dissipation is used for the convective flux discretization of the mean flow equations. The convective fluxes of the turbulence equations are discretized with a second-order Roe upwind scheme. A local time step is used in combination with an implicit Backward-Euler scheme to accelerate convergence employing an LU-SGS scheme for the steady computations.

2.2 Numerical Transition Tripping

At high Reynolds numbers, γ transition models are very prone to undergo transition at the slightest increase in intermittency. This high Reynolds number sensitivity can be used to model turbulent wedges by tripping the boundary layer at the wedge apex without any further prescription of the downstream wedge properties.

For the γ transition models in the DLR TAU-Code, a limiter exists to ensure that the intermittency γ stays bounded within certain values:

$$\gamma = \min\left(\max\left(\gamma, 0\right), 1\right). \tag{1}$$

This limiter can be used to set the intermittency at any location within the laminar boundary layer to $\gamma = 1$ representing a disturbance or tripping. The disturbance location $\mathrm{P}(x, y, z)$ and a disturbance radius δ_r representing e.g. the apex of a turbulent wedge are defined. The intermittency for each point $\mathrm{Q}(x, y, z)$ within the disturbance radius is set to $\gamma = 1$ based on the distance $\overline{\mathrm{PQ}} = |\vec{r}|$:

$$\gamma_w(\mathrm{Q}) = \begin{cases} 1, & \text{if } |\vec{r}| \leq \delta_r, \\ 0, & \text{if } |\vec{r}| > \delta_r, \end{cases} \tag{2}$$

$$\gamma = \min\left(\max\left(\gamma,\, \gamma_w\right), 1\right). \tag{3}$$

In the case of a transition tripping along a polyline, it is analyzed if a given point is located within the disturbance radius around any line segment of the polyline. Points close enough to the tripping line are set turbulent with $\gamma_w = 1$. The points required for the transition tripping are identified in the first solver iteration and then kept for the remaining computation.

The lateral growth of the turbulent wedge is theoretically, among other things, controlled by the size of the diffusion term of the intermittency transport equation. The original diffusion term is expanded by an additional coefficient σ_w:

$$Diff_\gamma = \frac{\partial}{\partial x_j}\left[\sigma_w\left(\mu + \frac{\mu_t}{\sigma_f}\right)\frac{\partial \gamma}{\partial x_j}\right]. \tag{4}$$

The γ and γ−CAS transition model are recovered for $\sigma_f = \sigma_w = 1$. As the wedge needs to spread into a laminar boundary layer that has vanishing values of μ_t, the scaling of the diffusion term by σ_w is supposed to have a larger effect on the model behavior. It should be noted that local correlation-based transition models are calibrated precisely and any change in model coefficients can have a severe impact on the overall predictive capabilities.

2.3 Aerodynamic Test Case

Helm et al. [12] present an investigation into the effects of static deformation and turbulent wedges on the aerodynamics of the CRM-NLF configuration [13]. Test case p2523 is used in the present study to examine the turbulent wedge functionality for both γ transition models. The exact flow conditions as requested for the 1st AIAA Transition Modeling and Prediction Workshop listed in Table 1 are used to set up the computation [14].

Table 1. Angle of attack, total temperature, Reynolds number, Mach number, and turbulence level for test case p2523 [14]

α/deg	T_t/K	Re_{MAC}	M	Tu/%
1.44848	277.10805	14 971 972	0.856489	0.24

Helm et al. [12] define the serrated transition front by a polyline through the TAU transition module [15]. The transition location is identified based on temperature sensitive paint images. For the intermittency transport model computations, only the apex of each turbulent wedge is defined on the upper surface. On the lower surface, a polyline tripping at 5% of the local chord is applied to create a fully turbulent boundary layer downstream of the tripping.

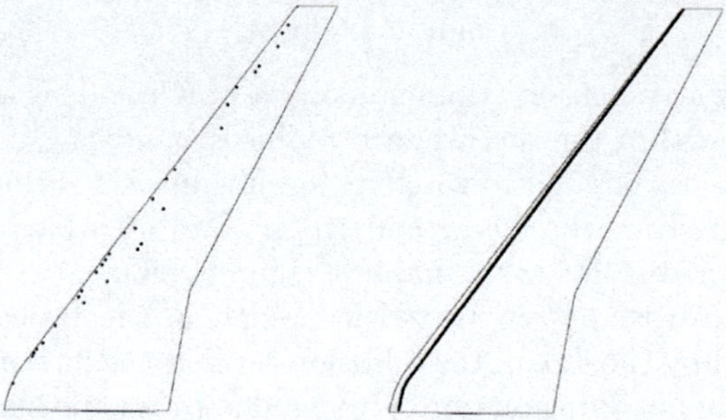

Fig. 1. Turbulent wedge apex locations on upper (left) and polyline tripping on lower (right) wing surface

Figure 1 shows the apex locations on the upper surface and the polyline definition on the lower surface. For this basic computational set-up, a disturbance radius of $\delta_r = 0.005\,\mathrm{m}$ is used for the wedge apices and the polyline tripping. At the wing tip, the polyline definition in combination with the given disturbance radius also affects grid points on and above the upper surface of the wing. This results in a spurious wedge at the wing tip on the upper surface (see Fig. 2).

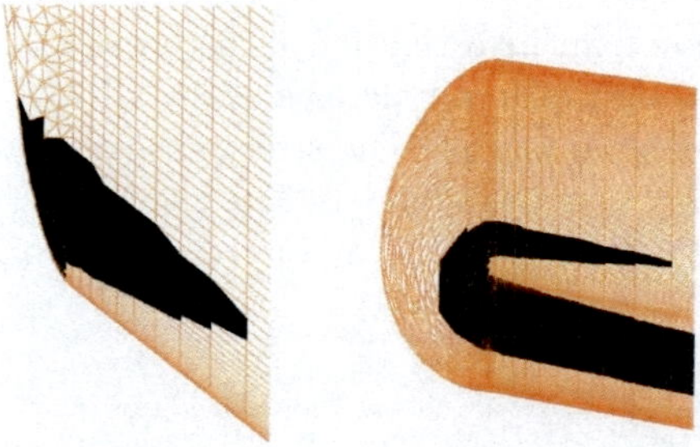

Fig. 2. Top and front view of surface grid with polyline definition intersecting upper surface causing a spurious wedge at the wing tip

2.4 Computational Grids

A grid family for the CRM-NLF configuration is created with a gradually decreasing grid spacing in spanwise direction. A structured surface grid is used on the wing with 300 grid points in chordwise direction with a clustering at the leading and trailing edge. The spanwise number of grid points is 450, 550, 650, and 750 for the respective member of the grid family. Table 2 gives the spacing $\Delta\eta_{LE}$ along the leading edge in terms of the half-span $b/2 = 1.5745715\,\mathrm{m}$. The spacing at the wing root, wing tip, and the spanwise growth rates are the same for all grids.

Table 2. Spanwise grid spacing on leading edge

ID	spanwise no. grid points	$\Delta\eta_{LE}\,/(b/2)$	total no. grid points
1	450	0.25 %	30 706 436
2	550	0.20 %	36 327 531
3	650	0.17 %	41 960 529
4	750	0.15 %	47 712 759

The trailing edge height is meshed with 10 evenly distributed grid points. The fuselage surface is meshed with an unstructured triangular grid. The hexahedral/prismatic boundary layer grid on the wing/fuselage consists of nominally 90 layers. The first cell height results in $y^+_{\max} \approx 0.5$ in all computations presented. A tetrahedral grid is used for the remaining flow field.

3 Results

3.1 Experimental Test Case P2523

The experimental test case p2523 is computed with the γ and γ-CAS transition model with and without turbulent wedges on grid 2. The diffusion term is not modified for these computations. The results are compared with the reference solution by Helm et al. [12] computed with the TAU transition module for free transition (e^N method) and prescribed transition locations (i.e. with turbulent wedges).

Figure 3 shows the skin friction coefficient distribution for the TAU transition module computations. Figures 4 and 5 show the results for the γ and γ-CAS transition model, respectively. The transition location is given by the main increase in skin friction coefficient. The γ-CAS transition model gives a more downstream transition location, which is in better agreement with the reference data in Fig. 3.

The turbulent wedges are assessed based on the resulting skin friction coefficient distribution. The turbulent wedge half-angles of the reference case range from $\Lambda/2 \approx 10°$ to $12.5°$. These values correspond well with other experimental data noting that the turbulent wedges occur only in mildly favorable or adverse pressure gradient regions. The turbulent wedges predicted by both correlation-based models show larger half-angles than the reference case of $\Lambda/2 \approx 17.5°$ to $22.5°$. In addition, the wedges are oriented in a steeper angle to the leading edge.

The γ and γ-CAS transition model are mostly built on the same framework (i.e. scaling/definition of production and destruction term, model coefficients), which explains the similarity in the predicted wedge shape and orientation.

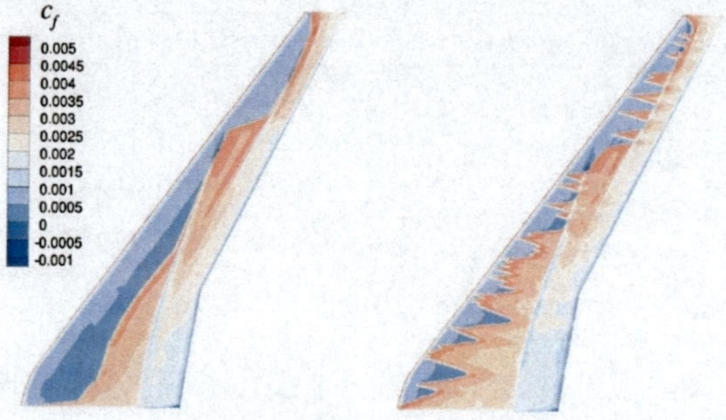

Fig. 3. Reference solution computed with the TAU transition module: free transition (left) and prescribed transition (right) on the upper surface of the wing

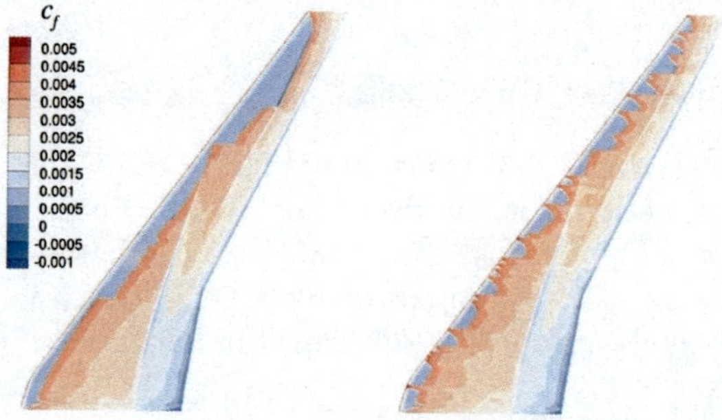

Fig. 4. Solution computed with the γ transition model: free transition without (left) and with (right) turbulent wedges on the upper surface of the wing

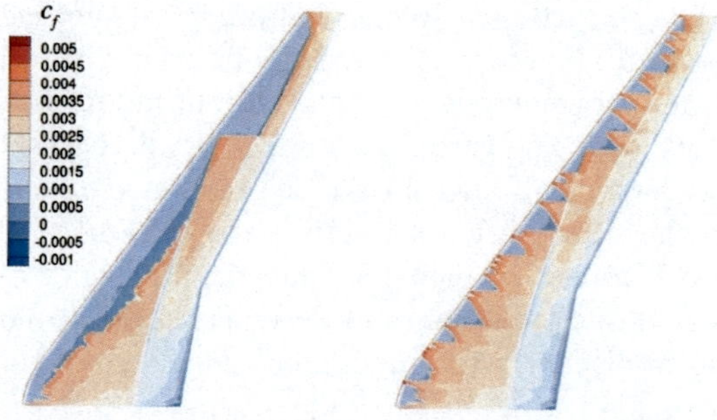

Fig. 5. Solution computed with the γ-CAS transition model: free transition without (left) and with (right) turbulent wedges on the upper surface of the wing

3.2 Spanwise Grid Resolution

Figure 6 shows the skin friction coefficient distribution given by the γ-CAS model for all four grid levels (see Table 2). There is no significant effect of the spanwise grid resolution on the turbulent wedge half-angles. On the inner wing, there is a slight upstream shift of the overall transition location on the finer grids.

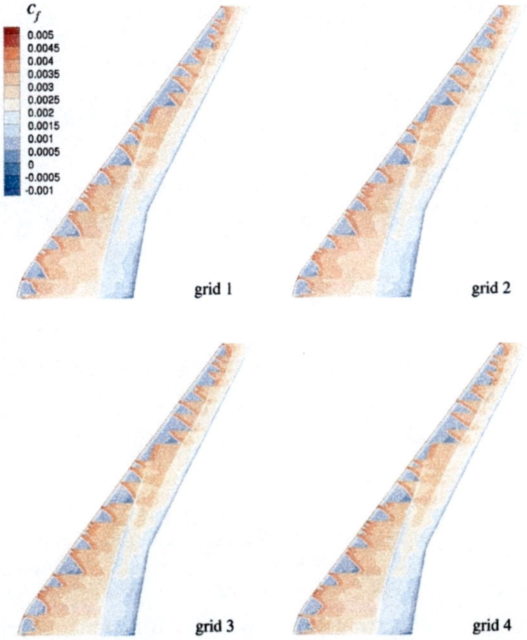

Fig. 6. Results for the γ-CAS transition model for different spanwise grid spacings on the upper surface of the wing

3.3 Disturbance Size

The disturbance radius is changed to investigate the effect on the turbulence wedges. Again, only the γ-CAS model is used on the second grid level for this set of computations. The results are presented in Fig. 7. For a disturbance size with radius $\delta_r > 1.5$ mm, there is always a grid point located within the disturbance for the given spanwise grid resolution. For smaller disturbance radii, a turbulence wedge might not be initialized as it is the case for $\delta_r = 1$ mm.

The overall effect of a reduced disturbance size is small, but especially the apex of the turbulent wedge is represented better as the disturbance size is reduced. For all disturbance radii in Fig. 7, the spurious wedge at the wing tip does not occur as the disturbance radius connected to the polyline on the lower wing surface does not intersect the upper surface anymore. Thus, the disturbance radius should be defined to cover at least one grid point on a given grid size. This approach might not be suited at lower Reynolds numbers as the model will be less sensitive to a local increase in intermittency.

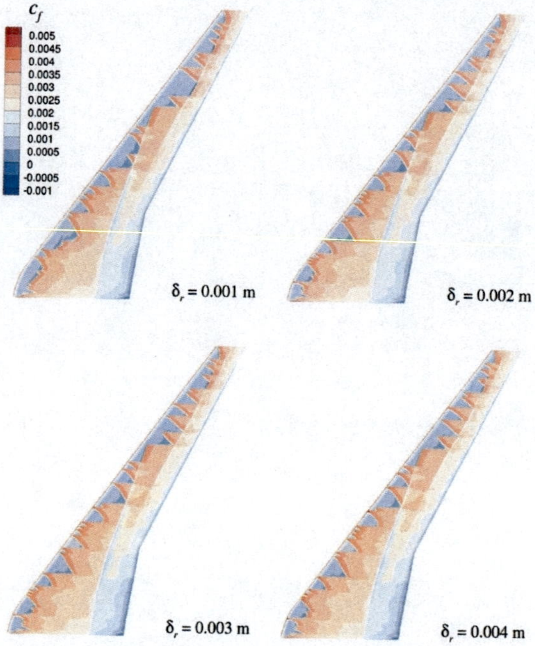

Fig. 7. Results for the γ-CAS transition model for different disturbance radii on the upper surface of the wing

3.4 Diffusion Term

A decrease in diffusion is sought as both γ transition models overpredict the turbulent wedge half-angles. The diffusion term given by Eq. 4 is altered by increasing σ_f for $\sigma_w = 1$ and decreasing σ_w for $\sigma_f = 1$. Computations are performed for the γ-CAS model on grids 2 and 4 with a disturbance radius of $\delta_r = 2\,\text{mm}$. Values of σ_f as high as 1000 have no significant effect on the lateral intermittency distribution. The effect of σ_w is investigated for values as low as $\sigma_w = 0.01$. The effect on a turbulent wedge in the outer wing section is shown in Fig. 8 for $\sigma_w = 1$ and $\sigma_w = 0.01$ on the finest grid level with $\sigma_f = 1$. No significant effect on the lateral growth of the turbulent wedge is observed, but there is a downstream effect on the skin friction coefficient distribution.

The intermittency transport behavior is most likely dominated by advective transport in the turbulent wedge region. A further decrease in diffusion is therefore not suited to decrease the wedge angle. Future investigations need to show

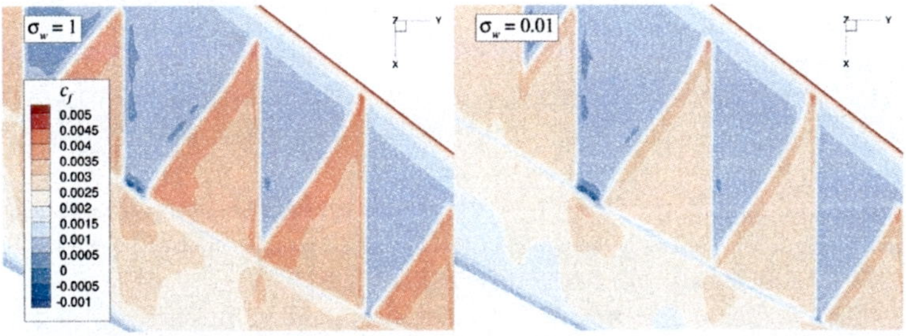

Fig. 8. Influence of σ_w on turbulent wedges

how advected intermittency, production/destruction of intermittency, intermittency distribution, and pressure distortions caused by the initial wedge apex affect the downstream wedge angle.

4 Conclusion and Outlook

This article presents an approach to model turbulent wedges in the framework of local correlation-based γ transition models. Turbulent wedges occur e.g. in laminar high Reynolds number wind tunnel tests. In order to use these experiments to validate CFD models, turbulent wedges need to be modeled for the thorough assessment of a transition model. Furthermore, turbulent contamination in free flight can reduce the overall benefit of laminar flow technology or can have an unfavorable effect on local flow physics (e.g. spanwise lift distribution, shock or separation location) and aeroelastic stability. A numerical turbulent wedge approach can therefore provide a more robust laminar flow technology assessment within and at the borders of the flight envelope.

A local increase in intermittency at the wedge apex is used to create a turbulent wedge downstream without any further interference in the transition model behavior. In a first study based on the CRM-NLF configuration, it is found that turbulent wedges can easily be created, but the half-angles of the modeled wedges are larger than these observed in experiments. Spanwise grid spacing, disturbance size, and intermittency diffusion coefficients only have a minor effect on the lateral spreading. Based on further investigations, an improvement of the lateral turbulence spreading will be developed. In addition, the turbulent wedge approach will be tested in unsteady computations and a TAU and/or FlowSimulator user interface needs to be developed that allows a robust definition of turbulent wedges.

References

1. Charters, A.C., Jr.: Transition between Laminar and Turbulent Flow by Transverse Contamination. NACA-TN-891, NACA, Washington, D.C. (1943)

2. Schubauer, G.B., Klebanoff, P.S.: Contributions on the Mechanics of Boundary-Layer Transition. NACA-TR-1289, NACA, Washington, D.C. (1956)
3. Gostelow, J.P., Melwani, N., Walker, G.J.: Effects of streamwise pressure gradient on turbulent spot development. J. Turbomach. **118**(4), 737–743 (1996). https://doi.org/10.1115/1.2840929
4. Zhong, S., Chong, T.P., Hodson, H.P.: A comparison of spreading angles of turbulent wedges in velocity and thermal boundary layers. J. Fluids Eng. **125**(2), 267–274 (2003). https://doi.org/10.1115/1.1539871
5. Goldstein, D., Chu, J., Brown, G.: Lateral spreading mechanism of a turbulent spot and a turbulent wedge. Flow Turbul. Combust. **98**(1), 21–35 (2016). https://doi.org/10.1007/s10494-016-9748-z
6. Berger, A.R.: Fundamental Mechanism of Turbulent Wedge Spreading. PhD thesis, Texas A&M University, College Station (2020)
7. Gad-El-Hak, M., Blackwelderf, R.F., Riley, J.J.: On the growth of turbulent regions in laminar boundary layers. J. Fluid Mech. **110**, 73–95 (1981)
8. Schwamborn, D., Gerhold, T., Heinrich, R.: The DLR TAU-code: recent applications in research and industry. In: Proceedings European Conference on Computational Fluid Dynamics ECCOMAS, Delft (2006)
9. Fehrs, M.: Boundary Layer Transition in External Aerodynamics and Dynamic Aeroelastic Stability. PhD thesis, TU Braunschweig, ISSN 1434-8454, ISRN DLRFB-2018-11, also NFL-FB 2017-27, Braunschweig (2018)
10. François, D.G., Krumbein, A., Krimmelbein, N., Grabe, C.: Simplified stability-based transition transport modeling for unstructured computational fluid dynamics. In: AIAA SCITECH 2022 Forum, AIAA, San Diego (2022). https://doi.org/10.2514/6.2022-1543
11. Menter, F.R., Kuntz, M., Langtry, R.: Ten years of industrial experience with the SST turbulence model. In: Hanjalić, K., Nagano, Y., Tummers, M. (eds.) Turbulence, Heat and Mass Transfer, vol. 4, pp. , 625–632. Begell House, Inc. (2003)
12. Helm, S., Fehrs, M., Kaiser, C., Krimmelbein, N., Krumbein, A.: Numerical simulation of the common research model with natural laminar flow. J. Aircraft **60**(2), 449–460 (2023). https://doi.org/10.2514/1.C036889
13. Lynde, M.N., Campbell, R.L.: Computational design and analysis of a transonic natural laminar flow wing for a wind tunnel model. In: 35th AIAA Applied Aerodynamics Conference, AIAA, Denver (2017). https://doi.org/10.2514/6.2017-3058
14. 1st AIAA Transition Modeling and Prediction Workshop. Description of Test Cases. https://transitionmodeling.larc.nasa.gov/wp-content/uploads/sites/109/2020/02/TransitionMPW_CaseDescriptions.pdf. Accessed 20 July 2022
15. Krumbein, A., Krimmelbein, N., Schrauf, G.: Automatic transition prediction in hybrid flow solver, part 1: methodology and sensitivities. J. Aircraft **46**(4), 1176–1190 (2009)

First Comparison of CFD Simulation and Wind Tunnel Test of the Forward-Swept Natural Laminar Flow Model NLF-ECOWING-FSW

Sebastian Helm(✉), Kristopher Davies, and Michael Fehrs

DLR, Institute of Aeroelasticity, 37073 Göttingen, Germany
sebastian.helm@dlr.de
https://www.dlr.de/ae/

Abstract. The forward-swept NLF-ECOWING-FSW model is designed to facilitate extensive natural laminar flow (NLF) at a high Mach number of 0.78. A wing-fuselage half-model with a half-span of 1.25 m was tested in the European Transonic Windtunnel (ETW). A first comparison of the experimental data with CFD simulations is presented in this paper. The objective is to understand the steady aerodynamics of the NLF-ECOWING-FSW configuration. This work aims at building the foundation for future unsteady investigations of this model.

Keywords: CFD validation · transonic natural laminar flow · transition modeling · wind tunnel test

1 Introduction

The NLF-ECOWING-FSW model is a forward-swept NLF wing model designed as an advanced derivative of the TuLam configuration [1]. It represents a typical single-aisle transport aircraft, adapted for natural laminar flow. This design shows a promising reduction of fuel consumption compared to a fully turbulent reference configuration of the A320 class.

A major objective of the first wind tunnel test of the NLF-ECOWING-FSW configuration performed in the ETW is to analyze the performance of the model close to the design condition as well as in off-design. Furthermore, the test provides data to validate the numerical capabilities to predict laminar-turbulent transition at high Mach- and Reynolds numbers. In addition, suitable operating conditions for future unsteady testing are to be derived from the analysis of the steady test.

The objective of this paper is to present a first comparison of CFD simulation and wind tunnel data of the NLF-ECOWING-FSW model. This work contributes to the understanding of laminar wings with the perspective of aeroelastic analysis.

A. Dillmann et al. (Eds.): STAB/DGLR Symposium 2022, NNFM 154, pp. 425–434, 2024.
https://doi.org/10.1007/978-3-031-40482-5_40

2 Simulation Set-Up

The DLR TAU code [2] is employed for steady-state RANS simulations. A second-order central scheme is used and matrix artificial dissipation is applied. The implicit lower-upper symmetric Gauss-Seidel scheme is employed for pseudo-time integration. The CFL number is set to a value of 10.0 and a 3-level V-cycle multigrid algorithm is used.

The negative Spalart-Allmaras (SA-neg) one-equation turbulence model is applied for fully turbulent computations and for the turbulent part of the transitional flow. The quadratic constitutive relation (QCR) [3] and a correction for rotation/curvature effects [4] are used in order to improve the turbulence modeling with regard to the wing-fuselage juncture flow.

Laminar-turbulent transition is modeled by the Transition Prediction Module [5] of the DLR TAU code. The turbulence model is deactivated in a laminar region around the wing surface, defined by a laminar height orthogonal to the wall and a streamwise transition location. On the wing upper side, the streamwise transition location is predicted by the transition module. The lower side of the wing as well as the fuselage is considered fully turbulent in the simulations, whereas in the experiment both are tripped close to the leading edge and nose, respectively. The boundary layer tripping at the pressure side of the wing takes the practical limitation of laminar flow due to steps and gaps at a Krüger high-lift device into account.

The linear stability code LILO [6] which uses the e^N method are applied for transition prediction. Transition prediction is applied at 36 line-of-flight cuts (y=const.) on the upper side of the wing. The boundary-layer profiles required by LILO are determined by the compressible conical laminar boundary-layer code COCO [7]. The default integration strategy (prescribed-frequency and prescribed-propagation-direction for Tollmien-Schlichting (TS) waves, prescribed-frequency and prescribed-wavelength for stationary crossflow waves) is used. The transition location is determined where the incompressible N-factors exceed the critical value $N_{crit} = 9.0$ for TS transition according to Perraud et al. [8] and $N_{crit} = 8.0$ for crossflow transition according to Hue et al. [9].

A family of unstructured CFD grids of the wing-fuselage half model is generated using the commercial software CENTAUR. The computational grids yield good agreement with respect to the predicted pressure distribution. The coarsest grid consists of approximately $12.7 \cdot 10^6$ nodes and is used as the default grid for this study. A symmetry boundary condition is used at the fuselage center plane ($y = 0$). The CFD domain is bounded by a semi-sphere with a radius of 30 times the semispan ($b/2 = 1.25\,\text{m}$). The main part of the wing surface is discretized by a structured grid with 250 nodes in spanwise direction and 250 nodes in chordwise direction on the upper side as well as on the lower side. The boundary layer of the wing is resolved by 61 structured layers. A maximum y^+ value of approximately 0.8 is obtained throughout this investigation.

The aerodynamic coefficients determined from the CFD simulations include the global forces of the wing and the fuselage.

3 Wind Tunnel Test

The NLF-ECOWING-FSW model is manufactured from maraging steel and mounted to the half model balance on the top wall of the wind tunnel. The test is performed at the ETW to facilitate high Mach numbers and Reynolds numbers. Furthermore, the ETW provides flow quality suitable for laminar flow experiments [8–10]. A free-stream turbulence intensity of $Tu = 0.1\%$ is applied to represent the turbulence level of the ETW in the numerical model.

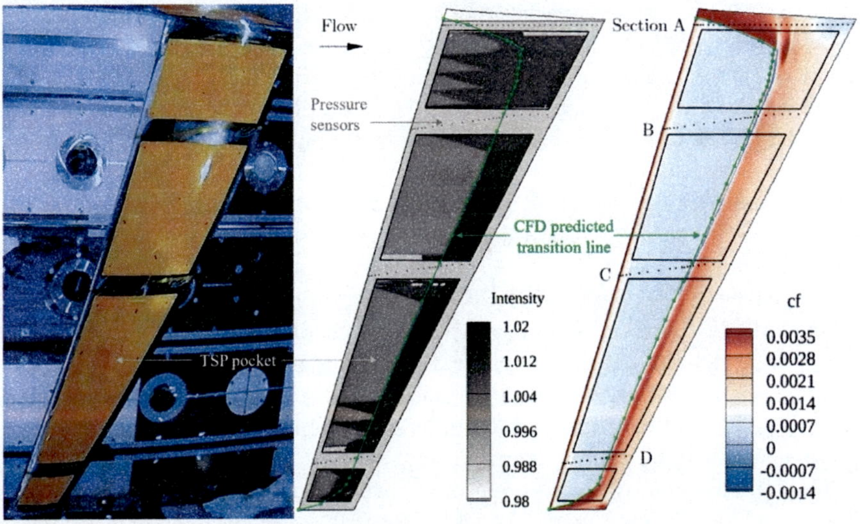

Fig. 1. NLF-ECOWING-FSW model in the ETW (left), TSP image (middle), CFD skin friction distribution (right), TSP point p186 at $\alpha = 1.06°$

The wind tunnel model is equipped with one section (A) of steady pressure sensors at the inner wing in order to capture the wing-fuselage juncture flow (Fig. 1). Section A is oriented in line-of-flight and contains sensors at the upper and lower side of the wing. In contrast, sections B to D contain unsteady pressure sensors, which are limited to the upper side of the wing. These are staggered in order to facilitate laminar flow at each sensor.

Laminar-turbulent transition is detected by means of the temperature-sensitive-paint (TSP) technique [11]. The TSP coating is applied to the upper surface of the wing arranged in four pockets divided by the rows of pressure sensors (Fig. 1). The TSP images shown in this work are acquired based on the temperature-step method facilitating a high contrast with distinguished laminar (lighter) and turbulent (darker) regions.

The wind tunnel data contains a correction for wind tunnel effects in order to obtain the best possible comparison to free-flight CFD simulations. A peniche is used to reduce the influence of the boundary layer at the tunnel wall.

4 Results

4.1 Overview

The present work is restricted to a Mach number of $M = 0.78$ and a Reynolds number of $Re_{MAC} = 16 \cdot 10^6$ at a total temperature of $T_t = 123$ K representing the central conditions of the test campaign. Extensive laminar flow is predicted by the CFD simulations near the design condition (as shown by the low skin friction coefficients in Fig. 1). The corresponding TSP image confirms the presence of large regions of laminar flow in the wind tunnel experiment. Laminar-turbulent transition occurs close to the shock for large portions of the wing span. It is apparent that laminarity is reduced due to the presence of turbulent wedges which might originate from surface irregularities or contamination remaining even after dedicated polishing. This premature transition impedes the detection of TS and CF transition mainly at the most in-board TSP pocket.

A comparison of the lift curves for the wing-fuselage model is shown in Fig. 2. As expected, the experimental lift curve with free transition lies between the CFD results for free transition and the fully turbulent simulations. It is likely that this can be partly attributed to the loss of laminarity due to turbulent wedges, as shown for the CRM-NLF configuration at similar flow conditions by Helm et al. [12]. Furthermore, a steeper gradient of the experimental lift curve can be observed. This can be explained by static aeroelastic deformation of the forward swept wing. The wing deformation was measured in the wind tunnel but is not included in the present numerical simulations. A separate study by Davies et al. includes model deformation [13].

The NLF-ECOWING-FSW exhibits a rather conventional NLF pressure distribution. This means a nearly constant favorable pressure gradient up to the re-compression shock at about 60% of the chord (Fig. 3). A single shock is present for the most part of the wing span and the pressure distribution exhibits relatively small spanwise variation. The inner wing section A deviates from this pattern due to the three-dimensional wing-fuselage juncture flow.

The numerical and experimental pressure distributions show a good agreement up to angles-of-attack of $\alpha \approx 2.5°$. At $\alpha = 1.06°$ the shock is predicted slightly downstream compared to the experiment (Fig. 3). Again, this can be explained partially by turbulent wedges, which lead to a local increase of the displacement thickness and a decrease in effective camber.

4.2 Side-of-Body Separation

Significant differences between experiment and simulation are apparent at angles-of-attack $\alpha > 2.5°$. In this range, the numerical computations exhibit a side-of-body (SoB) separation that leads to a significant loss of lift. The re-compression shock is predicted significantly further upstream at the inner wing (Sections A and B) compared to the experiment (Fig. 4). A large separation bubble arises where this shock impinges on the wing-fuselage juncture (Fig. 5). The separation is obtained with different CFD grids and different turbulence models [13].

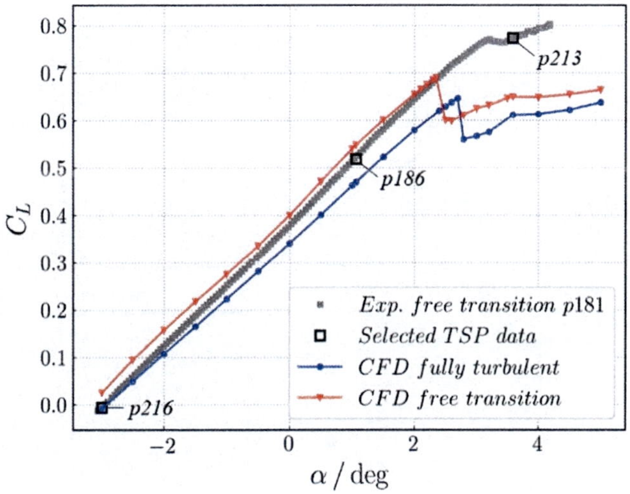

Fig. 2. Lift curve at Mach number $M = 0.78$, Reynolds number $Re_{MAC} = 16 \cdot 10^6$

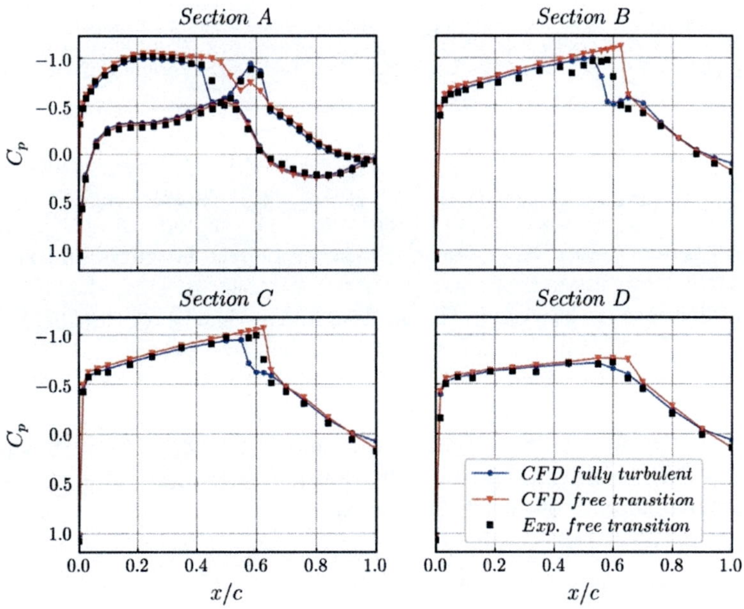

Fig. 3. Pressure distribution at measurement sections, TSP point p186 at $\alpha = 1.06°$

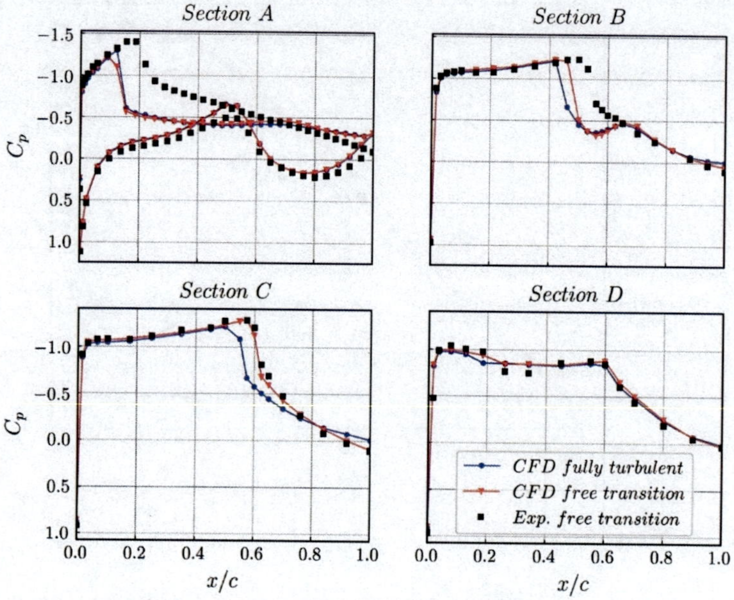

Fig. 4. Pressure distribution at measurement sections, TSP point p213 at $\alpha = 3.59°$

In contrast, it is evident from the experimental data (TSP images, force and pressure data) that the extent of SoB separation predicted by the CFD simulations does not occur in the wind tunnel throughout the test campaign. The wind tunnel data reveals a nearly-linear rise of the lift curve up to $\alpha \approx 3.0°$ before saturation sets in more gently compared to the CFD results.

A similar case is observed in the off-design investigation of the CRM wing at $\alpha = 4.0°$ by Crippa [14]. The SoB separation is reported to have an indirect effect on the shock over the complete wing up to the tip. For the NLF-ECOWING-FSW, this problem is relieved due to the forward sweep leading to less influence of the fuselage on the outer wing. This can be seen at Section C (Fig. 5), where the numerical and experimental pressure distributions show good agreement despite significantly different flow at the inner wing.

4.3 Laminar-Turbulent Transition

The TSP data shows that laminar-turbulent transition occurs in the vicinity of the re-compression (shock) for the most part of the envelope. This confirms the robust laminar wing design [1] in terms of laminar flow extent. According to the CFD simulations, only positive angles-of-attack lead to Tollmien-Schlichting (TS) transition upstream of the shock, whereas crossflow (CF) transition is found at negative angles-of-attack.

At $\alpha \approx 1.0°$ TS transition at the outer wing starts to move upstream of the shock with increasing angles-of-attack. This loss of laminarity at the outer wing

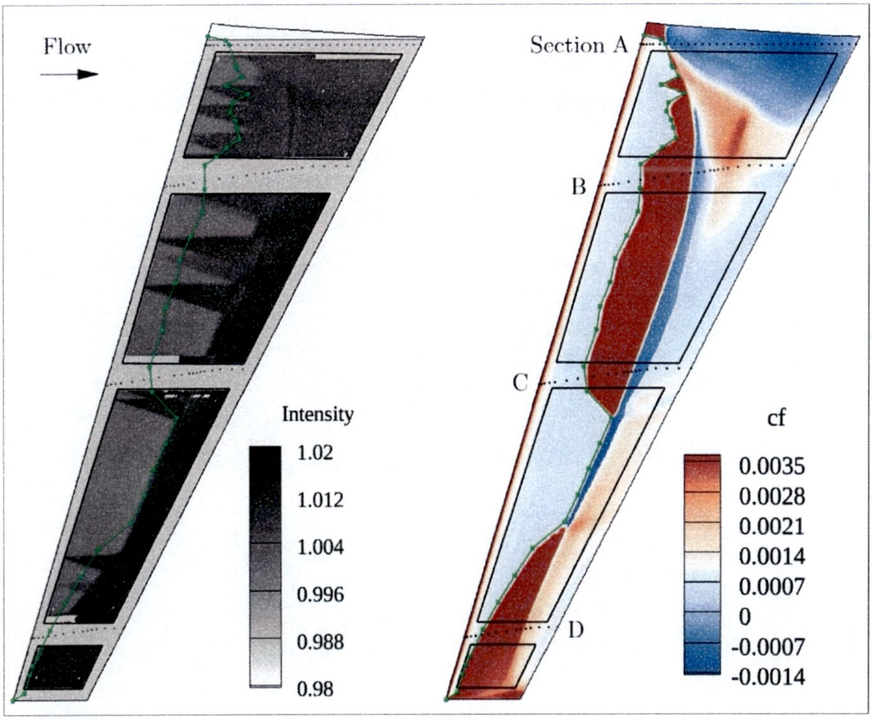

Fig. 5. TSP image (left), CFD skin friction distribution (right), TSP point p213 at $\alpha = 3.59°$

(apparent at $\alpha = 3.59°$ in Fig. 5) is attributed to the weakening of favorable pressure gradient. At high angles-of-attack an adverse pressure gradient is present in the numerical and experimental pressure distribution (up to 30% of the chord at Section D in Fig. 4). At these conditions the TSP image indicates transition further downstream than predicted by the CFD simulation. The present numerical setup over-predicts this loss of laminarity at the outer wing for angles-of-attack of $\alpha > 2.0°$, while the pressure distributions coincide well. Furthermore, TS transition is predicted significantly more upstream at the inner wing for angles-of-attack $\alpha > 2.5°$ (Fig. 5). This might indicate that the flow quality at the ETW is more favorable to avoid TS transition than expected or that the stabilizing effect of compressibility is relevant for these flow conditions.

In contrast, close to the design condition the outboard TSP pocket indicates laminar-turbulent transition slightly upstream of the predicted transition line (Fig. 1). This behavior is, however, restricted to a small region at the wing tip and angles-of-attack $\alpha \approx 1.0°$.

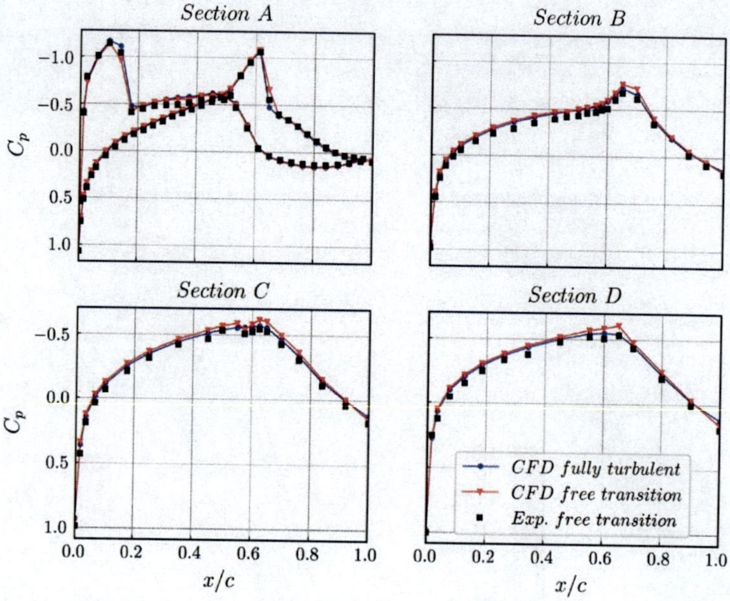

Fig. 6. Pressure distribution at measurement sections, TSP point p216 at $\alpha = -3.0°$

At negative angles-of-attack the strong favorable pressure gradient (Fig. 6) leads to crossflow transition being dominant and leading to reduced laminar flow particularly at the inner wing. With decreasing angle-of-attack CF transition is expanding upstream and towards the outer wing. The sawtooth pattern typical for crossflow transition is well resolved by the TSP image (Fig. 7, left). CF transition is well predicted by the CFD simulation in general. However, the most inner TSP pocket shows further upstream transition. This might be related to the bypass transition appearing in form of turbulent wedges particularly at this TSP pocket throughout the test campaign. At the outer wing, transition is predicted about 5 to 10% of the chord upstream of the shock (Fig. 7 outboard of Section C). Although the critical N-factors are not reached (neither TS nor CF) the automatic transition prediction specifies transition where the boundary-layer code terminates. This limitation of the current setup could be relieved by the use of boundary-layer data from the RANS solution.

It is remarkable that extensive laminar flow remains at the limits of the given angle-of-attack envelope considered within the experiment. Therefore, no distinct drag bucket, typical for NLF airfoils, can be observed.

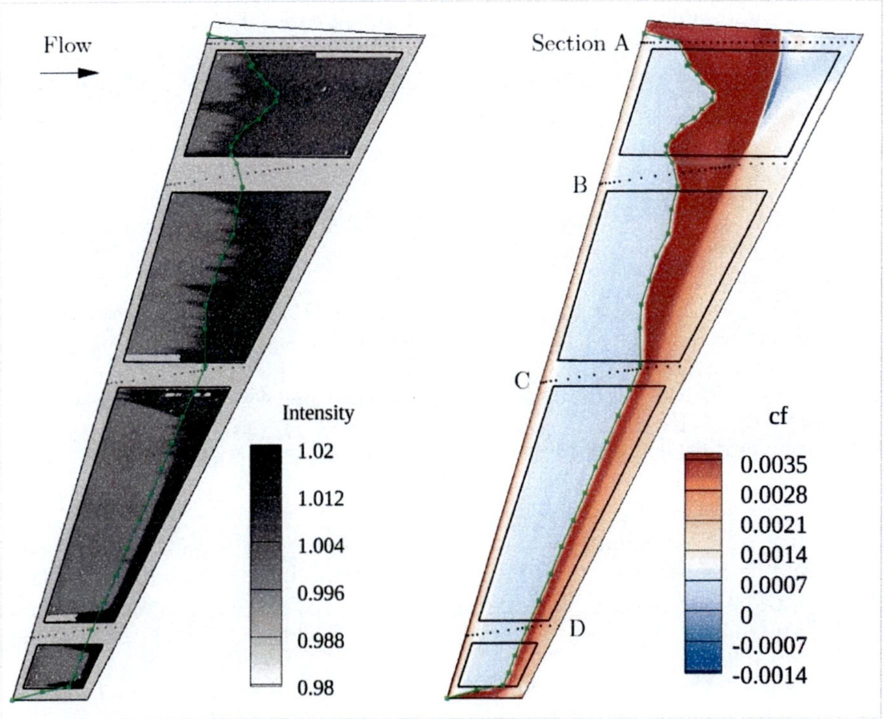

Fig. 7. TSP image (left), CFD skin friction distribution (right), TSP point p216 at $\alpha = -3.0°$

5 Conclusion

For the first time, numerical simulations of the NLF-ECOWING-FSW model are compared with wind tunnel data. In general, the steady RANS simulations by means of the DLR TAU code capture the transition phenomena derived from the experiment well. However, at high angles-of-attack the CFD simulation deviates from the experiment due to the appearance of a large region of separated flow at the wing-fuselage juncture that is not present in the wind tunnel test.

The performance of the NLF-ECOWING-FSW configuration at design conditions is confirmed in terms of laminar flow extent. Furthermore, a remarkable off-design performance is observed in the experiments, in terms of laminar flow extent and in terms of the absence of any large flow separation. A dedicated analysis of the drag is, however, beyond this work.

The comparison of numerical and experimental data confirms the high flow quality in the ETW. The results at high angles-of-attack for which Tollmien-Schlichting transition is dominant might indicate that the flow conditions at the ETW are even more favorable for laminar flow testing than expected from previous work or that compressibility is relevant at these flow conditions.

Acknowledgement. The authors would like to thank the team of the ETW, the team of Airbus, the team of the DLR Department of Aeroelastic Experiments and the team of the DLR Institute of Aerodynamics and Flow Technology for the provision of data.

References

1. Seitz, A., Hübner, A., Risse, K.: The DLR TuLam project: design of a short and medium range transport aircraft with forward swept NLF wing. CEAS Aeron. J. **11**(2), 449–459 (2019)
2. Schwamborn, D., Gerhold, T., Heinrich, R.: The DLR TAU-code: recent applications in research and industry. In: European Conference on Computational Fluid Dynamics, p. 25 (2006)
3. Togiti, V., Eisfeld, B., Brodersen, O.: Turbulence model study for the flow around the NASA common research model. J. Aircraft **51**(4), 1331–1343 (2014)
4. Shur, M.L., Strelets, M.K., Travin, A.K., Spalart, P.R.: Turbulence modeling in rotating and curved channels: assessing the spalart-shur correction. AIAA J. **38**(5), 784–792 (2000)
5. Krumbein, A., Krimmelbein, N., Schrauf, G.: Automatic transition prediction in hybrid flow solver, part 1: methodology and sensitivities. J. Aircraft **46**(4), 1176–1190 (2009)
6. Schrauf, G.: LILO 2.1 Users Guide and Tutorial. Technical report. GSSC Technical Report 6 (2006)
7. Schrauf, G.: COCO - A Program to Compute Velocity and Temperature Profiles for Local and Nonlocal Stability Analysis of Compressible. Conical Boundary Layers with Suction, ZARM Technik report (1998)
8. Perraud, J.,et al.: High reynolds number transition experiments in the ETW test facility with the pathfinder model. In: Proceedings of the V European Conference on Computational Fluid Dynamics, Lisbon, Portugal (2010)
9. Hue, D., Vermeersch, O., Duchemin, J., Colin, O., Tran, D.: Wind-tunnel and CFD investigations focused on transition and performance predictions of laminar wings. AIAA J. **56**, 132–145 (2017)
10. You, J.H., Wright, M.C.N., Quix, H.: Recent activities on flow quality assessment at the European transonic windtunnel. In: 33rd ICAS Congress in Stockholm, Sweden (2022)
11. Klein, C., Weiss, A., Henne, U.: Application of temperature sensitive paint for time resolved investigations of laminar-to-turbulent transition on oscillating airfoils. In: AIAA SCITECH 2022 Forum, San Diego, CA (Virtual), p. 1717 (2022)
12. Helm, S., Fehrs, M., Krimmelbein, N., Krumbein, A.: Transition prediction and analysis of the CRM-NLF wing with the DLR TAU code. In: Dillmann, A., Heller, G., Krämer, E., Wagner, C. (eds.) STAB/DGLR Symposium 2020. NNFMMD, vol. 151, pp. 185–194. Springer, Cham (2021). https://doi.org/10.1007/978-3-030-79561-0_18
13. Davies, K.-M., Fehrs, M., Helm, S.: Static aeroelasticity of a wind tunnel model with a forward-swept laminar wing. In: International Forum on Aeroelasticity and Structural Dynamics, Madrid, Spain (2022)
14. Crippa, S.: Application of chimera with hexahedral blocks in solar meshes. In: Third Symposium "Simulation of Wing and Nacelle Stall", Braunschweig (2012)

Stochastic Modeling and Large-Eddy Simulation of Heated Concentric Coaxial Pipes

Marten Klein[1,2](✉), Pei-Yun Tsai[1], and Heiko Schmidt[1,2]

[1] Lehrstuhl Numerische Strömungs- und Gasdynamik, Brandenburgische Technische Universität Cottbus-Senftenberg, Siemens-Halske-Ring 15A, 03046 Cottbus, Germany
{marten.klein,pei-yun.tsai,heiko.schmidt}@b-tu.de
[2] Scientific Computing Lab, Energie-Innovationszentrum (EIZ), Brandenburgische Technische Universität Cottbus-Senftenberg, 03046 Cottbus, Germany
https://www.b-tu.de/fg-stroemungsmodellierung,
https://www.b-tu.de/en/energie-innovationszentrum

Abstract. Turbulent concentric coaxial pipe flows are numerically investigated as canonical problem addressing spanwise curvature effects on heat and momentum transfer that are encountered in various engineering applications. It is demonstrated that the wall-adapting local eddy-viscosity (WALE) model within a large-eddy simulation (LES) framework, without model parameter recalibration, has limited predictive capabilities as signalized by poor representation of wall curvature effects and notable grid dependence. The identified lack in the modeling of radial transport processes is therefore addressed here by utilizing a stochastic one-dimensional turbulence (ODT) model. A standalone ODT formulation for cylindrical geometry is used in order to assess to which extent the predictability can be expected to improve by utilizing an advanced wall-modeling strategy.

Keywords: Heat and mass transfer · Stochastic turbulence modeling · Spanwise curvature effects · Pipe flow

MSC codes: 76F25 · 76F40 · 80A20 · 82C31 · 82C70

1 Introduction

Robust predictions of turbulent heat, mass, and momentum transfer crucially depend on the accurate numerical modeling of wall-normal transport processes. This is a standing challenge for various engineering applications encompassing internal flows (like heat exchangers [1], chemical reactors [2], or electrostatic precipitators [3]). Wide-gap coaxial configurations are frequently encountered, putting a burden on numerical analysis due to nonuniversal scaling properties of radial boundary layers that govern the transfer processes. Complications arise from variations of molecular-diffusive to curvature length scales, which should

A. Dillmann et al. (Eds.): STAB/DGLR Symposium 2022, NNFM 154, pp. 435–444, 2024.
https://doi.org/10.1007/978-3-031-40482-5_41

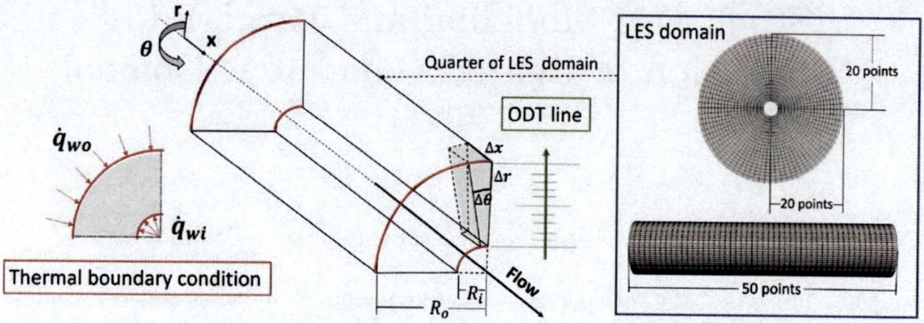

Fig. 1. Schematic of the heated concentric coaxial pipe flow configuration investigated. The computational domains utilized for LES (*blue*) and standalone stochastic ODT simulations (*green*) are highlighted by colored boxes. The flow is heated by the prescribed constant wall heating rates per unit area and wall radius $\dot{q}_{\mathrm{w,i}}$ and $\dot{q}_{\mathrm{w,o}}$ at the inner and outer cylinder wall, respectively (*red*). (Color figure online)

ideally be treated with direct numerical simulation (DNS) (e.g. [4]), among Schmidt and Prandtl number effects (e.g. [5]). The high costs resulting from the resolution requirements severely limit the applicability of DNS. Alternative strategies for modeling and simulation are therefore required.

A promising approach in this regard is map-based stochastic modeling for the near-wall region (e.g. [6]). In this approach, the so-called one-dimensional turbulence (ODT) model [7] is applied as wall model (WM) within a large-eddy simulation (LES) to resolve all relevant scales of the flow but only along the wall-normal physical coordinate. The applicability of ODT for the simulation of pipe flows has been demonstrated previously [8,9]. Here, we extend the previous studies by applying ODT as a standalone tool to pressure-driven concentric coaxial pipe flow of notionally infinite length radially bounded by a heated inner (convex) and heated outer (concave) cylinder. In addition, we compare the model's predictive capabilities with that of an open-source LES solver. Figure 1 shows a sketch of the set-up. Numerical efficiency is addressed by utilizing a dynamically-adaptive cylindrical ODT formulation [8] in which the ODT domain represents an infinitesimal wedge that spans the radial gap.

The rest of this paper is organized as follows. In Sect. 2, an overview of the LES and ODT model formulations and their application to coaxial pipe flow is given. In Sect. 3, the predictive capabilities of ODT and LES are addressed by comparison with available reference DNS results. Last, in Sect. 4, the key results and conclusions of this study are presented.

2 Method

2.1 Overview of the Large-Eddy Simulation Set-Up

The key idea of large-eddy simulation (LES) is to resolve nonuniversal large and meso-scale flow features while modeling the dissipative ensemble effect of small-scale turbulent eddies based on physical relations with the resolved scales [10]. The universality assumptions underlying such subgrid-scale (SGS) closure models are *not* generally justified, for instance, when the flow over a curved instead of the planar wall is considered. In this study, we demonstrate the lack of predictive capabilities for a wall-modeled LES of concentric coaxial pipe flow utilizing the WALE model [11] previously calibrated for plane channel flow. The filtered continuity, momentum, and scalar transport equation (per unit mass) for an incompressible flow are given by

$$\frac{\partial \overline{u}_i}{\partial x_i} = 0 \,, \tag{1}$$

$$\frac{\partial \overline{u}_i}{\partial t} + \overline{u}_j \frac{\partial \overline{u}_i}{\partial x_j} = -\frac{1}{\rho}\frac{\partial \overline{p}}{\partial x_i} + \frac{\partial}{\partial x_j}\left(\nu \frac{\partial \overline{u}_i}{\partial x_j}\right) + \frac{\partial \tau_{ij}}{\partial x_j} + \overline{f}_i \,, \tag{2}$$

$$\frac{\partial \overline{\Theta}}{\partial t} + \overline{u}_j \frac{\partial \overline{\Theta}}{\partial x_j} = \frac{\partial}{\partial x_j}\left(\Gamma \frac{\partial \overline{\Theta}}{\partial x_j}\right) + \frac{\partial q_j}{\partial x_j} + \overline{s}_\Theta \,, \tag{3}$$

where t is time, x_i $(i = 1, 2, 3)$ the Cartesian coordinates of the configuration space, $\overline{u}_i$ the Cartesian components of the filtered velocity vector, $\overline{p}$ the filtered pressure, $\overline{f}_i$ the prescribed filtered momentum sources, like a prescribed streamwise mean pressure-gradient force, $\overline{f}_i = -\rho^{-1}\,(\mathrm{d}P/\mathrm{d}x)\,\delta_{i1}$, where δ_{ij} $(j = 1, 2, 3)$ is the Kronecker delta, $\overline{\Theta}$ the filtered temperature, $\overline{s}_\Theta$ the prescribed filtered heat sources, ρ, ν, and Γ the constant mass density, kinematic viscosity, and thermal diffusivity of the fluid, respectively, $\tau_{ij} = \overline{u_i u_j} - \overline{u}_i\overline{u}_j$ the unclosed SGS stress tensor, and $q_i = \overline{u_i\Theta} - \overline{u}_i\overline{\Theta}$ the unclosed SGS kinematic heat flux (temperature flux). Einstein's summation convention is implied for indices that appear twice. By applying the Boussinesq hypothesis, the SGS stress tensor and SGS heat flux are parameterized as

$$\tau_{ij} = 2\nu_\mathrm{t}\overline{S}_{ij} \,, \quad q_i = \Gamma_\mathrm{t}\frac{\partial \overline{\Theta}}{\partial x_i} \,, \quad \Gamma_\mathrm{t} = \frac{\nu_\mathrm{t}}{Pr_\mathrm{t}} \,, \tag{4a,b,c}$$

where $\overline{S}_{ij} = (\partial\overline{u}_i/\partial x_j + \partial\overline{u}_j/\partial x_i)/2$ denotes the resolved-scale rate-of-strain tensor, ν_t the turbulent eddy viscosity, Γ_t the turbulent thermal diffusivity, and $Pr_\mathrm{t} = \nu_\mathrm{t}/\Gamma_\mathrm{t} \lesssim O(1)$ the turbulent Prandtl number, which needs to be determined empirically. In the WALE model, ν_t is formulated by taking into account the strain rate and rotation effects, and Pr_t is an empirical model constant [11].

For the present study, we use the open-source CFD software *OpenFOAM* [12] (version 8) performing LES by adjusting the LES channel flow tutorial case utilizing the PIMPLE solver. The LES domain is shown in Fig. 1 and made of four blocks, each representing a quarter of the pipe. The periodic section length L was chosen as ten times the outer radius R_o, that is $L = 10R_\mathrm{o}$. This implies $L > 10H$, where $H = R_\mathrm{o} - R_\mathrm{i}$ is the gap width, in terms of which the exact value of L/H depends on the selected radius ratio $\eta = R_\mathrm{i}/R_\mathrm{o}$.

2.2 Overview of the One-Dimensional Turbulence Model

Kerstein's [7] one-dimensional turbulence (ODT) model aims to resolve all relevant turbulent flow scales. This is made feasible for high Reynolds number flows by modeling turbulent advection and associated pressure fluctuations [13] by a stochastically sampled sequence of mapping events that punctuate the deterministic advancement due to molecular diffusion, prescribed forcing terms, initial and boundary conditions. The model aims to resolve molecular-diffusive transport processes and distinguishes them from turbulent-advective ones along its one-dimensional domain ('ODT line' in Fig. 1). Following [5,8], the dimensionally reduced stochastic ODT equations for constant-density temporal pipe flows are given by

$$\frac{\partial \tilde{u}_i}{\partial t} + \sum_{t_\mathrm{e}} \mathcal{E}_i\left(\tilde{u}_x, \tilde{u}_r, \tilde{u}_\theta\right) \tilde{\delta}\left(t - t_\mathrm{e}\right) = \frac{1}{r}\frac{\partial}{\partial r}\left(r\nu \frac{\partial \tilde{u}_i}{\partial r}\right) - \frac{1}{\rho}\frac{\mathrm{d}P}{\mathrm{d}x}\delta_{1i} \,, \tag{5}$$

$$\frac{\partial \tilde{\Theta}}{\partial t} + \sum_{t_\mathrm{e}} \mathcal{E}_\Theta\left(\tilde{u}_x, \tilde{u}_r, \tilde{u}_\theta\right) \tilde{\delta}\left(t - t_\mathrm{e}\right) = \frac{1}{r}\frac{\partial}{\partial r}\left(r\Gamma \frac{\partial \tilde{\Theta}}{\partial r}\right) + \beta \tilde{u}_x \,, \tag{6}$$

where t_e denotes the stochastically sampled eddy occurrences, $\tilde{\delta}(t-t_\mathrm{e})$ the Dirac distribution function, representing a discrete 'delta kick' of infinite rate that becomes a finite 'Heaviside jump' contribution upon time integration across t_e, r the radial coordinate, $\tilde{u}_i(r,t)$ the model-resolved velocity vector with cylindrical components $\tilde{u}_x, \tilde{u}_r, \tilde{u}_\theta$ in axial, radial, and azimuthal direction, respectively, $\tilde{f}_i(t) = -\rho^{-1}\,(\mathrm{d}P/\mathrm{d}x)\,\delta_{1i}$ the prescribed spatially uniform mean pressure-gradient force in axial direction, and $\tilde{s}_\Theta(r,t) = \beta\tilde{u}_x(r,t)$ the momentary heat sources for the model-resolved perturbation temperature $\tilde{\Theta}(r,t)$. Following [5] for the modeling of weakly heated channel flow, but extending the approach to annular geometry [14], we define $\tilde{\Theta}(r,t) = \beta x + T_0(r) - T(x,r,t)$, where $T(x,r,t)$ is an axial-radial slice of the actual temperature field at a given time, $\beta = \mathrm{d}T_\mathrm{m}/\mathrm{d}x$ the constant axial gradient of the bulk mean temperature, and $T_0(r)$ a background diffusive solution. In the fully developed, statistically stationary state, $T_0(r)$ absorbs the physical wall temperature increase in downstream (x) direction so that homogeneous Dirichlet boundary conditions are prescribed for the perturbation field $\tilde{\Theta}(r,t)$. In the available reference DNS [4], the wall heat fluxes are the same ($\dot{q}_\mathrm{w,i} = \dot{q}_\mathrm{w,o}$), whereas the wall heating rates are balanced ($\dot{q}_\mathrm{w,i}R_\mathrm{i} = \dot{q}_\mathrm{w,o}R_\mathrm{o}$) in the present ODT formulation for practical reasons. Since boundary conditions are absorbed in $T_0(r)$, the perturbation temperature $\tilde{\Theta}$ remains as the relevant thermal flow variable.

3 Results and Discussion

This section presents the results of the momentum and heat transfer. We utilize the ODT model calibration based on heat transfer in concentric coaxial pipe flow [14] and also keep the adjustable model parameters fixed as for the LES.

For the assessment of boundary layer similarity, the friction velocity and friction temperature are defined as

$$u_\tau = \sqrt{\nu \left|\frac{\mathrm{d}u}{\mathrm{d}r}\right|_\mathrm{wall}} \,, \quad \Theta_\tau = \frac{\Gamma}{u_\tau}\left|\frac{\mathrm{d}\Theta}{\mathrm{d}r}\right|_\mathrm{wall} \,, \tag{7a,b}$$

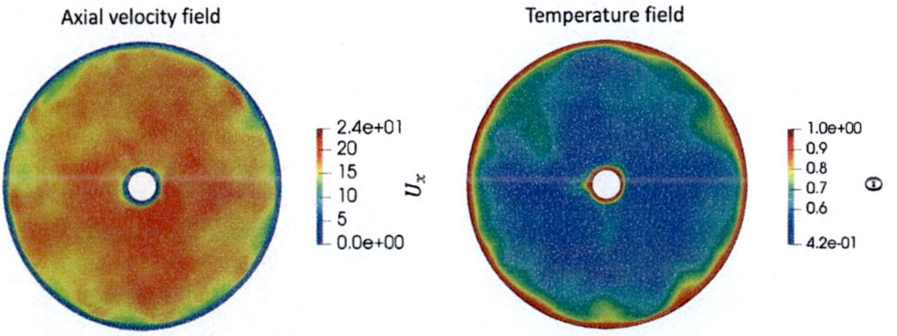

Fig. 2. Contours of the (*left*) momentary axial velocity and (*right*) momentary perturbation temperature in a radial-azimuthal section of the LES results for the periodic concentric coaxial pipe with $\eta = 0.1$ at $Pr = 0.71$. The equivalent circular pipe friction Reynolds number $Re_* = 2U_* R_\mathrm{o}/\nu = 2780$ was kept constant by a prescribed mean pressure gradient $\mathrm{d}P/\mathrm{d}x$, which is related to the equivalent circular pipe friction velocity $U_* = \sqrt{-R_\mathrm{o}/(2\rho)\,(\mathrm{d}P/\mathrm{d}x)}$.

where u_τ denotes the friction velocity and Θ_τ the friction temperature, $u(r) = \langle \tilde{u}_x(r,t)\rangle_t$ and $\Theta(r) = \langle \tilde{\Theta}(r,t)\rangle_t$ denote temporal averaged variables in the case of ODT, but temporal, axial, and azimuthal averaged variables $u(r) = \langle \overline{u}_x(x,r,\theta,t)\rangle_{t,x,\theta}$ and $\Theta(r) = \langle \overline{\Theta}(x,r,\theta,t)\rangle_{t,x,\theta}$ in the case of LES after transformation from prognostic Cartesian to diagnostic cylindrical coordinates.

The scaled (superscript '+') variables are given by the normalized mean axial velocity $u^+ = u/u_\tau$ and the normalized mean perturbation temperature $\Theta^+ = \Theta/\Theta_\tau$ that are evaluated over the inner and outer wall for the boundary layer coordinate $r^+ = |r - R_{\mathrm{i/o}}|\, u_\tau/\nu$. The similarity parameters of the flow are given by the friction Reynolds number $Re_\tau = (H/2)u_\tau/\nu$ and Prandtl number $Pr = \nu/\Gamma = 0.71$ (air).

3.1 Flow Visualization and Computational Resource Requirements

Figure 2 shows a radial-azimuthal section of the momentary LES solution from the statistically stationary state midway downstream at $x/L = 0.5$ for a representative case. The flow is visualized by contours of the filtered velocity component $\overline{u}_x$ and the filtered temperature $\overline{\Theta}$. The results demonstrate that the region over the strongly curved inner cylinder exhibits small-scale turbulence, making the case numerically more challenging than circular pipe flow.

We note that the LES for $\eta = 0.1$ with a resolution of $N_x \times N_r \times N_\theta = 100 \times 40 \times 160$ grid cells in axial, radial, and azimuthal direction required 10.2 h user time for the simulation of 15 advective time units running in parallel on 10 CPU cores on a local cluster equipped with Intel® Xeon® E5-2630 (2.40 GHz) processors. The corresponding standalone adaptive ODT simulation, which is discussed below, required only 0.253 h user time on a single core. This demonstrates the cost reduction achieved by reduced-order modeling with ODT while maintaining full-scale resolution. This makes it feasible to simulate extended

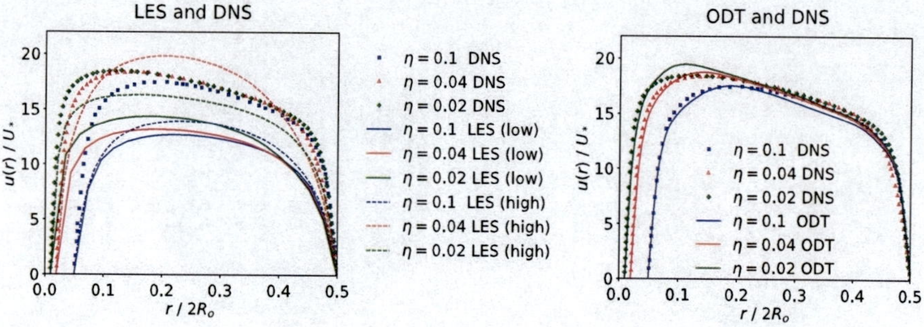

Fig. 3. Mean velocity profiles $u(r)$ normalized by the friction velocity scale U_* as function of the radial coordinate r. (*left*) LES predictions and (*right*) ODT predictions in comparison to reference DNS [15] results for various η for the equivalent circular pipe friction Reynolds number $Re_* = 600$ (defined in the caption of Fig. 2). The label 'low' indicates the coarse and 'high' a finer LES mesh.

time series, higher Reynolds numbers, and broader ranges of Prandtl numbers and radius ratios by utilizing ODT as wall model.

3.2 First-Order Velocity Statistics

Figure 3 shows mean velocity profiles of concentric coaxial pipe flow predicted by a high resolution ($N_x \times N_r \times N_\theta = 100 \times 40 \times 160$) and low resolution ($50 \times 20 \times 80$) LES, as well as mesh-adaptive ODT simulations. The LES (*left*) is not capable of capturing the velocity profile. The wall gradient is notably underestimated, and the mean velocity may reach larger or smaller values than the DNS [15] depending on the radius ratio and mesh size. In fact, the present LES exhibit a strong grid dependence. These results suggest that radial transport processes over walls with spanwise curvature are poorly represented. Even if geometry-adapted wall parameterizations are used to improve the results, the present simulations demonstrate that LES can have limited predictive capabilities if the wall model is not able to capture relevant flow physics, here spanwise curvature.

By contrast, ODT (*right*), even as a standalone model, does not exhibit such limitations. The mean profile of the low-Reynolds number reference DNS [15] is very reasonably captured, particularly in the vicinity of the wall. Modeling errors manifest themselves primarily in the bulk, especially for the lowest radius ratio, $\eta = 0.02$, investigated. We attribute this behavior to unresolved three-dimensional flow structures around a slender inner cylinder in addition to unresolved finite Reynolds number effects, as the model prefers a fully developed turbulent state even at small Reynolds numbers. Based on these results, we only consider ODT for further analysis of the thermal boundary layer discussed next.

3.3 First-Order Temperature Statistics

Figure 4 (*left*) shows radial profiles of the mean perturbation temperature for various radius ratios. It is observed that the ODT results reasonably reproduce the reference DNS [4]. The wall gradients are accurately captured due to the calibration for scalar transfer [14], but modeling errors manifest themselves by larger maximum values. Nevertheless, it is remarkable that finite radius ratio effects, like the radial inner–outer asymmetry of the mean profiles, are reasonably captured by ODT for previously calibrated but now fixed model parameters. The thermal boundary layer is thicker on the outer than on the inner wall as flow features reach down to smaller scales there as a consequence of strong wall curvature. As η increases, the boundary layer thickness over the inner wall increases nonlinearly in the same way as the boundary layer thickness over the outer wall decreases.

Figure 4 (*right*) shows law-of-the-wall plots of the thermal boundary layer (*top right*) over the inner cylinder and (*bottom right*) over the outer cylinder, respectively, for various η. These profiles exhibit the linear conductive sublayer, $\Theta^+(r^+) = Pr\, r^+$, for $r^+ \lesssim 5$ and a turbulent logarithmic region for $r^+ \gtrsim 30$. The latter is parameterized as $\Theta^+(r^+) = \kappa_\Theta^{-1} \ln r^+ + B_\Theta$, where $\kappa_\Theta = 0.35$ is the thermal von Kármán constant and $B_\Theta = 1.34$ the thermal additive constant that describe the weakly turbulent DNS [4] data. For the law-of-the-wall plots, reference DNS results are omitted in order to aid visual clarity. While present ODT data collapses well over the outer cylinder demonstrating inner layer similarity, wall-curvature effects prevent such a collapse for the inner cylinder. This is not a modeling error but a correct representation of flow physics. The origin of this effect is the nonlinear dependence of the boundary layer thicknesses on η as noted above, which is a consequence of the generation of increasingly smaller scales in the flow as the inner cylinder curvature radius decreases. To zeroth order, assuming approximate validity of the analogy between the transport of heat and axial momentum towards the wall, the model results for the thermal boundary layer are consistent with the mixing length arguments put forward in [15] for the momentum boundary layer. This demonstrates that the fidelity and predictive capabilities of the ODT model are a result of the physics-based simultaneous resolution of heat and momentum transport on all relevant scales of the flow, which is consistent with plane channels [5].

3.4 Second-Order Temperature Statistics

Figure 5 (*left*) shows the normalized root-mean-square (rms) fluctuation temperature $\Theta^+_{\rm rms}$ for various radius ratios comparing ODT with reference DNS [4]. ODT significantly underestimates the near-wall turbulent fluctuation peak at $r^+ \approx 15$ but captures the trend for η with respect to the fluctuation magnitude and the asymmetry between the boundary layer over the inner and outer cylinder. The latter is consistent with the mean profiles discussed above, whereas the former is a known modeling artifact [5,8,9,16]. The deficit in fluctuation magnitude does not necessarily imply incorrect radial fluxes. On the contrary, capturing the radial asymmetry requires a physical representation of radial fluxes within the dimensionally reduced model. This is made possible by the map-based turbulence modeling within the eddy events that take the role of advecting radial velocity fluctuations formally denoted by u_r' below.

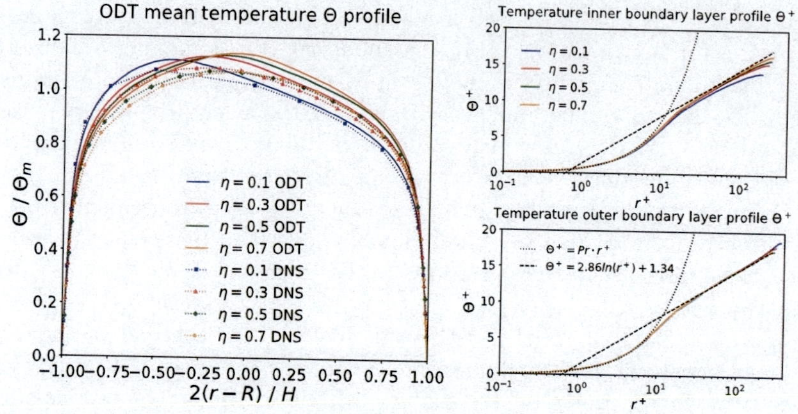

Fig. 4. (*left*) Mean perturbation temperature profiles $\Theta(r)$ for various radius ratios η comparing ODT and reference DNS [4] normalized by the gap width H, mean radius $R = (R_\mathrm{o} + R_\mathrm{i})/2$, and bulk mean temperature $\Theta_\mathrm{m} = 2\int_{R_\mathrm{i}}^{R_\mathrm{o}} u\,\Theta\,r\,\mathrm{d}r/(U_\mathrm{b}\,(R_\mathrm{o}^2 - R_\mathrm{i}^2))$, where $U_\mathrm{b} = 2\int_{R_\mathrm{i}}^{R_\mathrm{o}} u\,r\,\mathrm{d}r/(R_\mathrm{o}^2 - R_\mathrm{i}^2)$ is fixed by a variable momentum source. The Prandtl number is $Pr = 0.71$ and the bulk Reynolds number $Re_\mathrm{b} = 2U_\mathrm{b}H/\nu =$ 17,700. (*right*) Law-of-the-wall plots for ODT. (*top right*) Inner cylinder, and (*bottom right*) outer cylinder.

Figure 5 (*right*) shows radial profiles of the normalized turbulent heat flux, $\langle\Theta' u_r'\rangle^+$, for various radius ratios. Finite radius ratio effects dominate the spatial structure of $\langle\Theta' u_r'\rangle^+$ since wall-curvature effects govern the radial asymmetry of the flow profiles. These geometrical features are reasonably captured by ODT. This is revealed, in particular, by the zero-crossing r_0, where $\langle\Theta' u_r'\rangle^+ = 0$. The radial location r_0 is insensitive to the Reynolds number, which suggests asymptotic turbulence properties in low Reynolds number ODT simulations. Since r_0 corresponds with the location where $\Theta(r)$ reaches its maximum, a consistent representation of spanwise wall-curvature effects and wall-normal radial transport is therefore established. The agreement between ODT and DNS is not perfect for the low Reynolds number investigated. ODT is built for highly turbulent flows, but no coaxial pipe DNS is available for this regime. Slightly weaker turbulent heat fluxes observed in the bulk in comparison to reference DNS [4] are, therefore, attributed to finite Reynols number effects. Weak turbulence only exhibits a small range of scales so that map-based advection within ODT is not carrying the entire radial heat flux due to missing large-scale structural information as signalized by missing buffer layer fluctuations. However, the effect is weak, and the modeling error tends to vanishes for the fluxes for high asymptotic Reynolds numbers [5]. Based on the current model validation, ODT can be applied more confidently to heat exchangers, chemical reactors, and gas cleaning devices that involve concentric pipe flows.

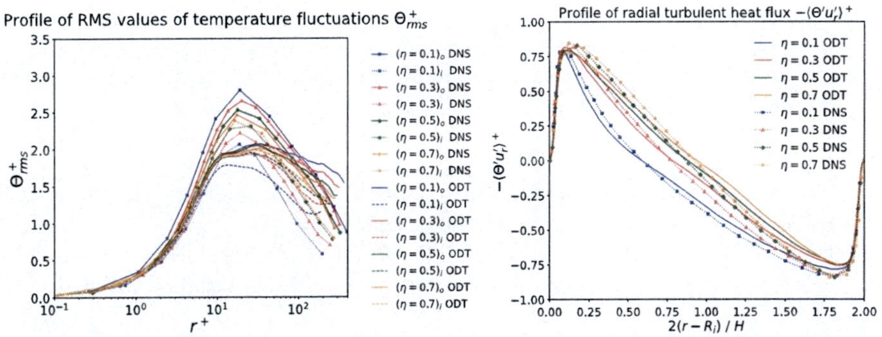

Fig. 5. (*left*) Boundary layer profiles of the normalized root-mean-square fluctuation temperature $\Theta^+_{\rm rms}$ for various radius ratios showing ODT and reference DNS [4] for fixed $Re = 17{,}700$ and $Pr = 0.71$. (*right*) Corresponding bulk profiles of the normalized turbulent radial heat flux $\langle \Theta' u_r' \rangle^+$ across the radial gap.

4 Concluding Remarks

A stochastic modeling strategy based on a one-dimensional turbulence (ODT) model has been utilized for an accurate representation of radial heat and momentum transfer processes in turbulent concentric coaxial pipe flows. It was shown that the standalone ODT formulation can reasonably predict the radius ratio dependence of the mean flow, the boundary layer structure, and the turbulent fluxes, down to moderate Reynolds number. Modeling errors manifest themselves **primarily by a lack of fluctuation variance at finite wall distance in the buffer** layer and missing turbulent fluxes at very low Reynolds numbers. It was demonstrated that ODT may be used to overcome present limitations in the predictive capabilities of wall-modeled large-eddy simulation (WM-LES) of turbulent flows with spanwise curvature.

Acknowledgements. This research is supported by the German Federal Government, the Federal Ministry of Education and Research and the State of Brandenburg within the framework of the joint project EIZ: Energy Innovation Center (project numbers 85056897 and 03SF0693A) with funds from the Structural Development Act (Strukturstärkungsgesetz) for coal-mining regions. M.K. acknowledges support by the BTU Graduate Research School (Conference Travel Grant). All authors thank Yousef Baroudi for initial LES tests within an EUNICE European University student exchange internship.

References

1. Fukuda, T., Tsukahara, T.: Heat transfer of transitional regime with helical turbulence in annular flow. Int. J. Heat Fluid Flow **82**, 108555 (2020). https://doi.org/10.1016/j.ijheatfluidflow.2020.108555

2. Balestrin, E., Valle, J.A.B., da Silva, A., et al.: Sensitivity of the turbulent schmidt number and the turbulence models to simulate catalytic and photocatalytic processes with surface reaction limited by mass transfer. Chem. Eng. Res. Des. **170**, 90–106 (2021). https://doi.org/10.1016/j.cherd.2021.03.035
3. Medina Méndez, J.A., Bacher, C., Riebel, U., Schmidt, H.: Electrohydrodynamically-enhanced drag in a vertical pipe-flow with a concentric electrode: a one-dimensional turbulence study. Eur. J. Mech. - B/Fluids **95**, 240–251 (2022). https://doi.org/10.1016/j.euromechflu.2022.05.008
4. Bagheri, E., Wang, B.C.: Direct numerical simulation of turbulent heat transfer in concentric annular pipe flows. Phys. Fluids **33**(5), 055131 (2021). https://doi.org/10.1063/5.0047531
5. Klein, M., Schmidt, H., Lignell, D.O.: Stochastic modeling of surface scalar-flux fluctuations in turbulent channel flow using one-dimensional turbulence. Int. J. Heat Fluid Flow **93**, 108889 (2022). https://doi.org/10.1016/j.ijheatfluidflow.2021.108889
6. Freire, L.S., Chamecki, M.: Large-eddy simulation of smooth and rough channel flows using a one-dimensional stochastic wall model. Comput. Fluids **230**, 105135 (2021). https://doi.org/10.1016/j.compfluid.2021.105135
7. Kerstein, A.R.: One-dimensional turbulence: model formulation and application to homogeneous turbulence, shear flows, and buoyant stratified flows. J. Fluid Mech. **392**, 277–334 (1999). https://doi.org/10.1017/S0022112099005376
8. Lignell, D.O., et al.: One-dimensional turbulence modeling for cylindrical and spherical flows: model formulation and application. Theor. Comput. Fluid Dyn. **32**(4), 495–520 (2018). https://doi.org/10.1007/s00162-018-0465-1
9. Medina, J.A.M., Klein, M., Schmidt, H.: One-dimensional turbulence investigation of variable density effects due to heat transfer in a low Mach number internal air flow. Int. J. Heat Fluid Flow **80**, 108481 (2019). https://doi.org/10.1016/j.ijheatfluidflow.2019.108481
10. Smagorinsky, J.: General circulation experiments with the primitive equations: I. the basic experiment. Monthly Weather Rev. **91**(3), 99–164 (1963). https://doi.org/10.1175/1520-0493(1963)091⟨0099:GCEWTP⟩2.3.CO;2
11. Nicoud, F., Ducros, F.: Subgrid-scale stress modelling based on the square of the velocity gradient tensor. Flow Turbul. Comb. **62**(3), 183–200 (1999). https://doi.org/10.1023/A:1009995426001
12. Weller, H.G., Tabor, G., Jasak, H., Fureby, C.: A tensorial approach to computational continuum mechanics using object-oriented techniques. Comput. Phys. **12**(6), 620–631 (1998). https://doi.org/10.1063/1.168744
13. Kerstein, A.R., Ashurst, W.T., Wunsch, S., Nilsen, V.: One-dimensional turbulence: vector formulation and application to free shear flows. J. Fluid Mech. **447**, 85–109 (2001). https://doi.org/10.1017/S0022112001005778
14. Tsai, P.Y., Schmidt, H., Klein, M.: Modeling simultaneous momentum and passive scalar transfer in turbulent annular poiseuille flow. Proc. Appl. Math. Mech. **22**, e202200272 (2023). https://doi.org/10.1002/pamm.202200272
15. Boersma, B.J., Breugem, W.P.: Numerical simulation of turbulent flow in concentric annuli. Flow Turbul. Comb. **86**(1), 113–127 (2011). https://doi.org/10.1007/s10494-010-9295-y
16. Lignell, D.O., Kerstein, A.R., Sun, G., Monson, E.I.: Mesh adaption for efficient multiscale implementation of one-dimensional turbulence. Theor. Comput. Fluid Dyn. **27**(3), 273–295 (2013). https://doi.org/10.1007/s00162-012-0267-9

Large Eddy Simulations Towards Industrial Relevant Flows on Graphics Processing Units

Christian Lübon(✉)

SimIoTec GmbH, Telawiallee 2, 88400 Biberach, Germany
christian.luebon@simiotec.com

Abstract. In this publication, an open source numerical flow solver for Large Eddy Simulations (LES) optimized for graphics processing units (GPUs) is analyzed and presented. Its applicability is demonstrated with a series of test cases calculated with different LES models and the results are analysed. In addition, the performance and effectiveness of this approach is examined, the parallelization of the solver is explained and finally the overall system is discussed.

1 Introduction

Since a couple of years the feasibility of LES for industrial relevant flows has been discussed [1]. In order to overcome the limiting factor for LES, the required computing time, especially for wall bounded flows, Wall-Modeled Large Eddy Simulation (WMLES) has been continuously improved and powerful wall functions are available (WFLES).

The computational power increases, especially due to parallel architectures like multi-core processors (CPUs) and GPUs. Consumers are asking for more and more powerful GPUs, which are used in the entertainment industry, in the automotive sector, for virtual reality or artificial intelligence (AI). These GPUs are very powerful and highly parallelized, what makes them very interesting for numerical computing such as computational fluid dynamics (CFD). A lot of CFD solvers have been tackled, in most cases existing and established solvers. Often these solvers are not designed for GPUs and they work in a hybrid mode with calculations on the CPU, like discretization of the conservation laws, and simultaneously with calculations on the GPU, especially solving linear equation systems. Therefore, the performance gain is often not particularly large.

The CFD solver OpenCruncher [2] is being investigated here that calculates exclusively, with an impressive speedup, on the GPU and, using innovative LES models, is also able to calculate wall bounded high Reynolds number flows.

2 Solver

OpenCruncher is a free open-source CFD solver running on CPUs or GPUs. The parallel infrastructure and high flop counts of current GPUs make it possible to perform LES calculations on workstations.

A. Dillmann et al. (Eds.): STAB/DGLR Symposium 2022, NNFM 154, pp. 445–454, 2024.
https://doi.org/10.1007/978-3-031-40482-5_42

The discretization is density based, second order accurate and uses explicit time integration techniques. It solves the spatially-filtered compressible Navier-Stokes equations using the finite-volume method. Various LES models, like the Smagorinsky or WALE model are available to model the physical effects of unresolved turbulence. For high-Reynolds number wall bounded flows wall-models and wall-functions are implemented. Synthetic turbulence at inlets is also available.

To get the maximum performance out of the GPUs, especially from the cheap gaming hardware, the solver can operate in single or double precision. Due to the explicit nature of the solver, without a system of linear equations to be solved, the results in either single or double precision are identical. Therefore the validation results shown in the following chapters are all calculated in single precision. The performance chapter compares between single and double precision.

3 Large Eddy Simulation

More and more LES simulations have been carried out in the last decades and a lot of CFD codes can perform LES simulations in the meantime. The basic idea of LES is that large turbulent structures are problem dependent and can be resolved by the numerical method, whereas small structures are problem independent and can be modeled [3]. LES subgrid scale models are often quite simple, but due to the fact that relavant scales and structures have to be resolved LES is a technology that requires a lot of computational power.

For wall bounded flows the turbulence length scale and the size of the turbulent eddies reduces close to the wall. To resolve these structures a high resolution in space and in time is required, the wall resolved LES (WRLES). Especially for higher Reynolds numbers this becomes very expensive in terms of computational time. When making use of wall functions (WFLES) or wall models (WMLES) this scaling behavior and the computational time can be reduced. But one has to keep in mind that we introduce additional modeling errors.

Fig. 1. Triangular cylinder, vorticity isosurfaces

Beyond simple test cases such as the decaying homogeneous isotropic turbulence test [4], a bluff body triangular cylinder test case, a flameholder (Fig. 1), is ideal to test the numerical scheme and the LES subgrid model. The bluff body creates a huge disturbance which then triggers the turbulence in the wake, the influence of the boundary layer can be neglected for this test case.

The Reynolds number based on the cylinder height is $4.7 \cdot 10^4$. This Volvo Aero corporation triangular bluff body has been extensively researched, both in terms of experiments and simulation. Figure 2 shows a comparison with experimental data [5] and the results of a commercial CFD solver [6].

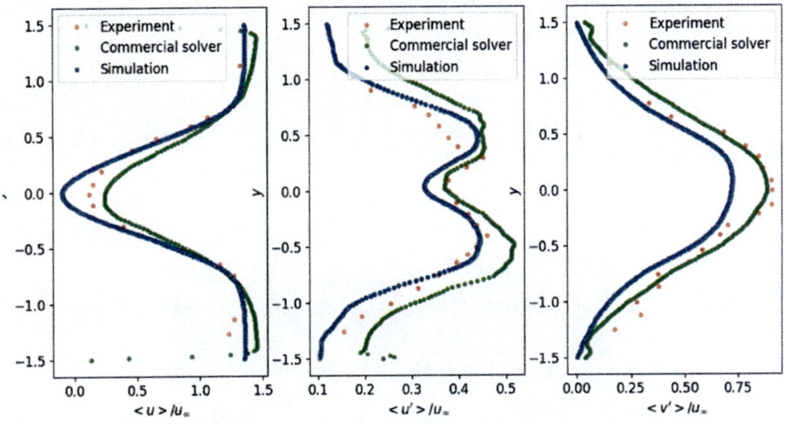

Fig. 2. Comparison of the velocity and velocity fluctuation in the wake (x/l = 1.53)

There is a good agreement between the current results, the experimental data and as well the commercial solver. OpenCruncher is running, with meshes up to 20 million cells, on a local workstation using only two NVIDIA GTX graphic cards. The Mach number is equal 0.05, the flow is almost incompressible, but due to the fact that OpenCruncher is using an explicit time integration technology we do not run into the typical linearization or precondition issues for density based solvers.

3.1 Wall-Modeled LES

WMLES is an extension to the classical LES which couples the simple algebraic Prandtl-van Driest Reynolds Averaged Navier Stokes (RANS) model, used close to the wall in the boundary layer, and the Smagorinsky SGS model, used in the rest of the computational domain. This reduces the Reynolds number dependent grid resolution requirements of classical wall-resolved LES. The turbulent viscosity is defined as:

$$\mu_t = \rho l_H^2 f_d \left|S\right| \tag{1}$$

where $|S|$ is the magnitude of the strain rate tensor, f_d is a wall-damping function [7] that assures the correct behavior of μ_t at the wall and l_H is the hybrid length scale

$$l_H = min\left(l_{RANS}, l_{LES} f_a\right) \tag{2}$$

More details, especially about the length scales, l_{RANS} and l_{LES}, and the damping function can be found in the literature [8].

To test the wall-modeled LES capabilities a turbulent channel flow is performed [8]. The computational domain is rectangular, with two no-slip walls oriented normal to the y-axis, the remaining boundaries are connected using periodic boundary conditions in order to approximate an infinite channel, see Fig. 3. The Reynolds number based on the friction velocity for the three cases is $Re_\theta = 400$, 2400 and 18000. To maintain the friction Reynolds numbers Re_θ, a body force is applied which is proportional to the pressure gradient. This gradient is taken into account via a source term in the momentum equations to drive the flow through the channel. It is important to notice that the bulk velocity u_b is not specified and is therefore part of the solution. The same high aspect ratio mesh spacing in x- and z-direction is always used. In y-direction, the wall normal direction, the grid is becoming more and more coarse. Due to the high aspect ratio cells next to the wall the turbulent viscosity will be modeled with the WMLES approach. For high y^+ values a wall function approach is used [9].

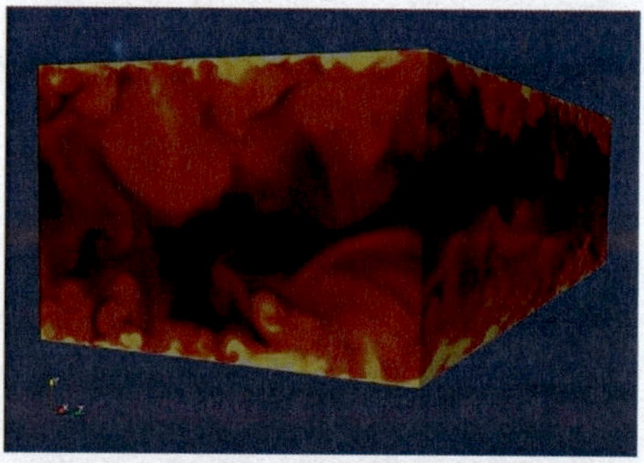

Fig. 3. Turbulent channel flow, velocity contours, $Re_\theta = 18000$

A comparison against the analytical law of the wall and against Prandtl's assumption for the turbulent viscosity ($\mu_t = \kappa y^+$) is done. The agreement is quite good, Fig. 4 (left), also for three different grids with different y^+ values.

3.2 Wall-Function LES

The wall function approach combines the WALE model with a wall function. The WALE model [10] is based on the velocity gradient tensor, which takes into account the shear stress tensor as well as the rotation tensor:

$$\mu_t = \rho l_H^2 \frac{\left(S_{ij}^d S_{ij}^d\right)^{3/2}}{\left(\overline{S}_{ij}\overline{S}_{ij}\right)^{5/2} + \left(S_{ij}^d S_{ij}^d\right)^{5/4}}, \tag{3}$$

l_H is the hybrid length scale. Further details can be found in the literature [10]. For the wall-functions we use a friction velocity approach, which can be calculated iteratively [9] from the analytical wall law [11].

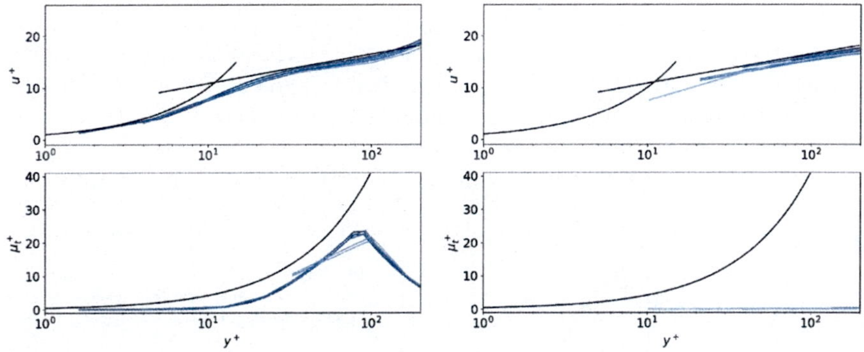

Fig. 4. Law of the wall (top); Prandtl's assumption (bottom), WMLES (left); WFLES (right), $Re_\theta = 2400$, different y^+ values (blue). (Color figure online)

It is important to understand that the near wall resolution of wall function LES (WFLES) requires an isotropic wall resolution of $\Delta x \approx \Delta y \approx \Delta z$. The WALE model does not add eddy viscosity to near wall cells to model the boundary layer. The turbulent structures must be resolved by the grid, but the size of the turbulent eddies scales linearly with the wall distance. This approach works well, but is very expensive for high Reynolds number cases combined with low y^+ values.

When running the turbulent channel case from above with a WFLES and different meshes with different y^+ values there is also a good agreement with the law of the wall, Fig. 4 (right, top). No viscosity has been added near the wall, Fig. 4 (right, bottom).

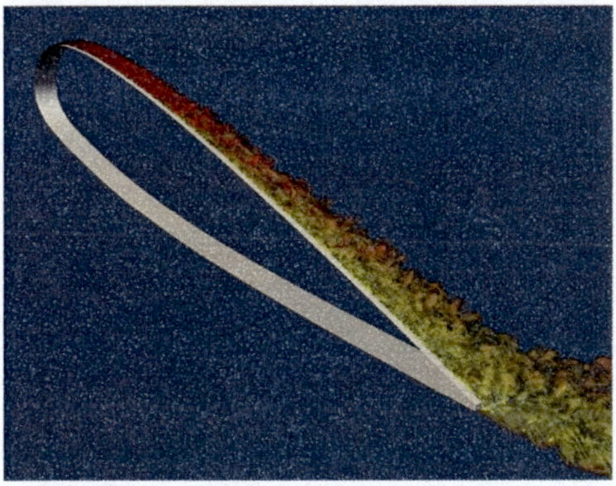

Fig. 5. Aerospatiale A-aerofoil, vorticity isosurfaces

The wall function approach is also beneficial for complex cases with transition and separation like the flow around an Aerospatiale A-aerofoil [12], Fig. 5.

This flow, a high lift configuration, at an angle of attack of $\alpha = 13.3°$, a chord Reynolds number of $2.1 \cdot 10^6$ and a freestream Mach number of 0.15 has been subject of extensive study. It is a challenging case for LES calculations due to the high Reynolds number and the different flow regimes. The strong adverse pressure gradient causes a laminar separation bubble and natural transition, the boundary layer grows and next to the blunt trailing edge there is separation.

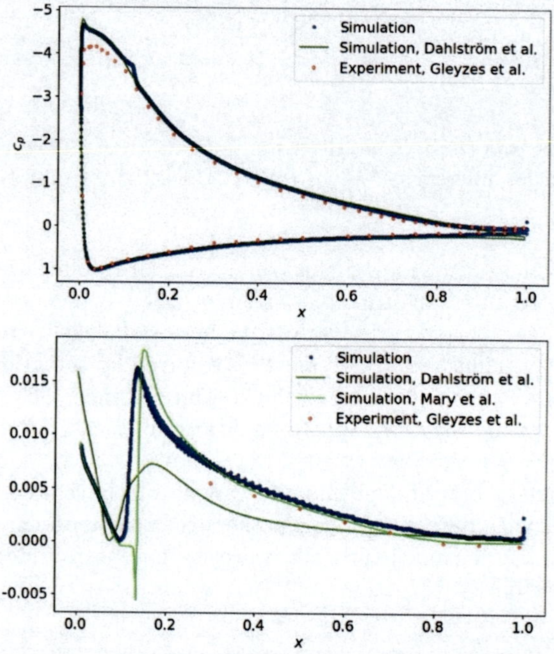

Fig. 6. Pressure and skin friction coefficient

The surface data, pressure coefficient and skin friction coefficient, shows an agreement with the simulation [12] and experiment [13], Fig. 6. It was not possible to match the experiment exactly in the trailing edge separation zone and

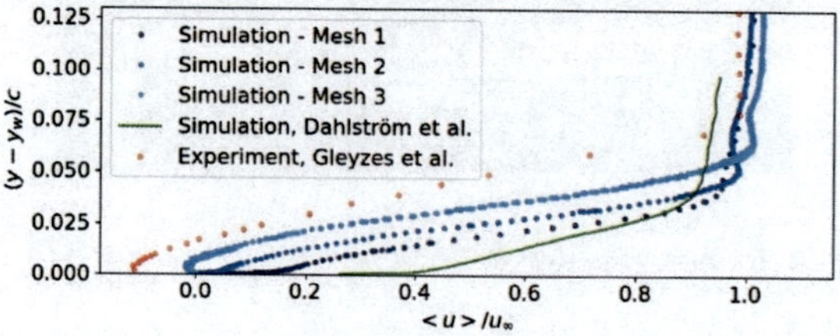

Fig. 7. Velocity

backflow area. Figure 7 shows the results of a mesh study with meshes from 5 million cells (mesh 1) to 63 million cells (mesh 3). There is a clear trend towards even finer grids, no grid convergence has been reached yet.

3.3 Synthetic Turbulence Generation

It is useful to categorize turbulent industrial flows into different types: globally unstable flows, locally unstable flows and finally stable flows. A typical example for globally unstable flows is the triangular cylinder case from above. A strong disturbance exists which triggers turbulence behind the bluff body. It is not so easy to define locally unstable flows because turbulent flows are by definition unstable. But when thinking about shear layers or cavity flows, there are local disturbances that trigger turbulent fluctuations. The last type of flows, stable flows, for example channels or boundary layer flows, are more difficult to calculate with scale resolving turbulence models, like LES. There are no disturbances or instabilities causing turbulent fluctuations. These instabilities or turbulent flow content must be added artificially. Different techniques, like recycling of turbulence or synthetic turbulence generation exist.

Due to the fact that the focus is on industrial relevant flows, a synthetic turbulence generator [14] based on a superposition of Fourier modes [15,16] is used.

To verify the synthetic turbulence generator and the implementation different test cases were run, the zero pressure gradient turbulent flat plate, shown here is a good example, Fig. 8.

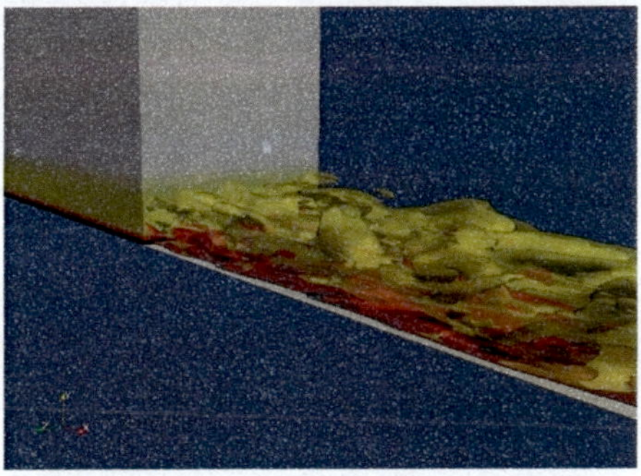

Fig. 8. Vorticity, $Re_\theta = 1000$

As a 'steady state' inlet condition a developed turbulent boundary layer from a precursor RANS simulation is used, here the synthetic turbulence generator is activated, Fig. 8. In spanwise direction there are periodic boundary conditions. The Mach number of the flow is Ma $= 0.1$, the Reynolds number based on the friction velocity is $Re_\theta = 1000$ and $Re_\theta = 10000$ for the second configuration. There is a good agreement with the turbulent wall law and the skin friction coefficient recovers quickly, Fig. 9.

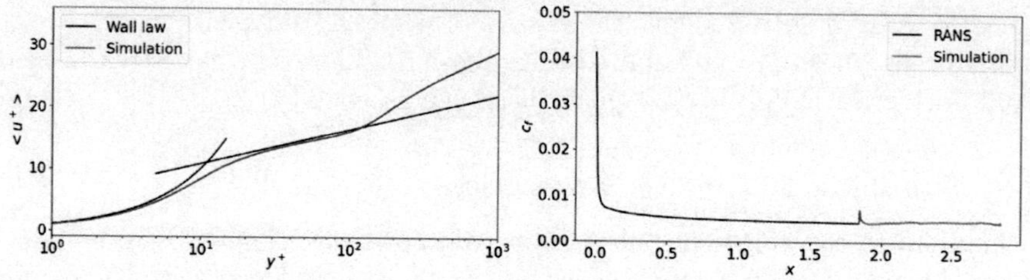

Fig. 9. Law of the wall, WMLES, $Re_\theta = 10000$ (left); skin friction coefficient, WFLES, $Re_\theta = 1000$ (right)

4 Performance

The computer gaming market has grown rapidly in the last couple of years and powerful GPUs have been developed. These GPUs can perform numerical operations quickly and in parallel. With their vectorization, they are somewhat comparable to the earlier supercomputers such as the NEC SX8, of course on a much smaller scale. An entry-level GPU, like the NVIDIA GeForce RTX 2080 Super, already has 3072 arithmetic logic units (ALUs) and a impressive peak performance of 11.2 TFlops. Unfortunately, experience has shown that the performance bottleneck is mostly memory bandwidth and memory access time.

The solver OpenCruncher is density based, explicit and uses a global time step for the time integration scheme. The global time step is determined by the smallest cell, or the cell with the worst quality, and can be correspondingly small for these kind of explicit solvers. On the other hand for LES calculations high quality meshes are required and an explicit solver can be blazingly fast on GPUs. The numerical integration in space and time runs completely on the GPU, that means no device (GPU) and host (CPU) interaction and memory copies are necessary. This has a very positive effect on the performance.

The performance test case is the flow past a sphere at a Mach number of $Ma = 0.1$, Table 1. All available CPU and GPU cores are used. For the sake of completeness, the speed is also compared with another open-source CFD solver, OpenFoam (rhoSimpleFoam).

Table 1. Sphere flow, hardware performance comparison

Solver	Precision	Hardware	Runtime [s]	Speedup
OpenFoam	double	AMD EPYC 7443P	$1.4 \cdot 10^{-7}$	0.46
OpenCruncher	double	AMD EPYC 7443P	$6.5 \cdot 10^{-8}$	1.00
OpenCruncher	double	NVIDIA GeForce RTX 2080 Super	$2.3 \cdot 10^{-8}$	2.82
OpenCruncher	double	NVIDIA A100	$3.3 \cdot 10^{-9}$	19.70
OpenCruncher	single	NVIDIA GeForce RTX 2080 Super	$5.2 \cdot 10^{-9}$	12.50
OpenCruncher	single	NVIDIA A100	$2.3 \cdot 10^{-9}$	28.26

The OpenCruncher CPU implementation is more then twice as fast as OpenFoam, but even more interesting is that an impressive 28× acceleration is possible when running on GPU hardware. As already mentioned, the solution or numerical results are identical in single and double precision. Therefore it is possible to use cheap GPUs, developed for the gaming market (NVIDIA GeForce RTX 2080 Super), as they are only efficient in single precision.

High-performance systems require a lot of electrical energy, what also determines the total costs in the product live cycle. The power consumption of the A100 GPU is 400 W compared to the EPYC CPU with 200 W. With the same performance, however, this means that the A 100 card only needs 7% of the electrical energy of the CPU.

Strictly speaking, a GPU is a highly parallel working vector machine, but unfortunately the amount of memory is limited, especially for the gaming GPUs. Therefore there is a lot of interest in doing multi-GPU calculations. Similar to classic high performance (HPC) CPU systems, a distinction can be made between shared and distributed memory systems and parallelization. OpenCruncher is successfully parallelized for shared memory systems, distributed memory parallelization is almost done, but this is an ongoing activity.

Table 2. Speedup, periodic hill flow, NVIDIA RTX A4000 GPUs

GPUs	Strong scaling	**Efficiency**
1	1.00	100.0 %
2	1.82	91.2 %
4	3.70	92.6 %

To partition the computational domain between the GPUs, the proven approach of domain decomposition based on multilevel graph partitioning is used [17]. For shared memory systems the interaction between the GPUs is done via direct memory access. The RTX A4000 GPUs don't have a high-speed connection like NVIDIA NVLINK, so they communicate via the slower PCI Express bus. GPUs usually have multiple copy engines, data can be exchanged with several other GPUs at the same time, what leads to an acceptable speedup, Table 2.

5 Conclusion

Various LES models have been demonstrated and successfully validated, which are important for industrial relevant high Reynolds number flows. With these models, WFLES or WMLES, in combination with synthetic turbulence a wide range of applications and problems can be covered.

Furthermore, it was shown that an explicit density-based solver on the GPU can achieve a remarkable speedup compared to established methods and solvers. This makes it possible to compensate disadvantages, such as the strong time step limit, and this is an important step towards industrially relevant problems that can be solved with high accurate Large Eddy Simulations.

References

1. Slotnick J., et al.: CFD vision 2030 study: a path to revolutionary computational aerosciences. Langley Research Center, NASA CR-2014-218178 (2014)
2. OpenCruncher. https://www.opencruncher.com
3. Sagaut, P., Deck, S., Terracol, M.: Multiscale and Multiresolution Approaches in Turbulence. Imperial College Press, London (2006)
4. Comte-Bellot, G., Corrsin, S.: Simple Eulerian time correlation of full- and narrow-band velocity signals in grid-generated isotropic turbulence. J. Fluid Mech. **48**, 273–337 (1971)
5. Sjunnesson, A., Nelsson, C., Max, E.: LDA measurements of velocities and turbulence in a bluff body stabilized flame. Laser Anemometry **3**, 83–93 (1991)
6. Menter, F.R.: Best practice: scale resolving simulations in ANSYS CFD. Ansys Germany GmbH (2015)
7. Piomelli, U., Moin, P., Ferzinger, J.H.: Model consistency in large-eddy simulation of turbulent channel flows. Phys. Fluids **31**, 1884–1894 (1988)
8. Shur, M.L., Spalart, P.R., Strelets, M.K., Travin, A.K.: A hybrid RANS-LES approach with delayed-DES and wall-modelled LES capabilities. Int. J. Heat Fluid Flow **29**, 1638–1649 (2008)
9. Kalitzin, G., Medic, G., Iaccarino, G., Durbin, P.: Near-wall behavior of RANS turbulence models and implications for wall functions. J. Comput. Phys. **204**, 265–291 (2005)
10. Nicoud, F., Ducros, F.: Subgrid-scale stress modelling based on the square of the velocity gradient tensor. Flow Turbul. Comb. **62**, 183–200 (1999)
11. Spalding, D.B.: A single formula for the "Law of the Wall". J. Appl. Mech. **28**, 455–458 (1961)
12. Dahlström, S., Davidson, L.: Large eddy simulation of the flow around an Aerospatiale A-aerofoil. In: Proceedings of ECCOMAS 2000, Barcelona (2000)
13. Gleyzes, C., Capbern, P.: Experimental study of two AIRBUS/ONERA airfoils in near stall conditions. Aeros. Sci. Technol. **7**, 439–449 (2003)
14. Bechara, W., Bailly, C., Lafon, P., Candel, S.: Stochastic approach to noise modeling for free turbulent flows. AIAA J. **32**, 455–463 (1994)
15. Adamian, D., Travin, A.: Assessment of an approach to generating inflow synthetic turbulence for large eddy simulations of complex turbulent flows. Progr. Flight Phys. **5**, 43–54 (2013)
16. Shur, M.L., Spalart, P.R., Strelets, M.K., Travin, A.K.: Synthetic turbulence generators for RANS-LES interfaces in zonal simulations of aerodynamic and aeroacoustic problems. Flow Turbul. Comb. **93**(1), 63–92 (2014). https://doi.org/10.1007/s10494-014-9534-8
17. Karypis, G., Kumar, V.: A fast and high quality multilevel scheme for partitioning irregular graphs. SIAM J. Sci. Comput. **20**, 359–392 (1998)

Contribution of Vortices to Laminar-Turbulent Transition in a Boundary Layer Disturbed by a Roughness Element

Tristan M. Römer[(✉)], Christoph Wenzel, and Ulrich Rist

Institute of Aerodynamics and Gas Dynamics, University of Stuttgart, Pfaffenwaldring 21, 70569 Stuttgart, Germany
roemer@iag.uni-stuttgart.de

Abstract. This paper presents analyses of the flow topology around and behind a roughness element in a laminar flat plate boundary layer. Hydrogen-bubble and dye visualization performed in a laminar water channel give insights into primary structures, such as the horseshoe vortex, hairpin vortex and recirculation zone. Moreover, two secondary structures are presented: the Λ_1 vortex, known from classical laminar-turbulent transition, and a novel vortex, here named Λ_2 vortex. The Λ_2 vortex is elongated and orientated like the Λ_1 vortex parallel to the wall. However, it lies near to the wall and is believed to be initiated by the Λ_1 vortex. Hot-film measurements reveal that the fluctuation amplitude of the Λ_1 vortex increases up to the highest value downstream of the roughness element. This indicates that the Λ_1 vortex contributes most to laminar-turbulent breakdown.

Keywords: boundary layer · roughness element · secondary structures

1 Introduction

The classical laminar-turbulent transition in Blasius boundary layers is characterized by the exponential growth of a primary instability, which is the Tollmien-Schlichting (TS) wave for small initial disturbances. If the TS wave is grown to a finite amplitude, there follows a three-dimensional (3D) secondary instability characterized by a resonance type. Common resonance types are the fundamental resonance (K-type, [4]) and the subharmonic (H-type, [3]) resonance, which differ in the alignment of Λ vortices in streamwise and spanwise direction. The secondary instability finally leads to a turbulent boundary layer. Placing a three-dimensional (3D) roughness element in a laminar boundary layer triggers new disturbance modes that distinguish the roughness-induced breakdown to turbulence from the classical TS breakdown. This can be demonstrated at the location of the vortex formation downstream of a cylindrical roughness element, where two major vortices can be found: a horseshoe and hairpin type [1]. The vortices are illustrated in Fig. 1 by dye streakline visualization, where U_e is the

A. Dillmann et al. (Eds.): STAB/DGLR Symposium 2022, NNFM 154, pp. 455–464, 2024.
https://doi.org/10.1007/978-3-031-40482-5_43

free-stream velocity. The horseshoe vortex wraps around the cylinder and forms low-speed and high-speed streaks in the roughness element wake. If the roughness Reynolds number $Re_{kk} = u_k k/\nu$, where u_k is the velocity at the cylinder height k, exceeds a certain value, hairpin vortices periodically separate from the top of the cylinder. Although these primary structures, which are followed by complex secondary structures, have been intensively studied (e.g. [1,6]), not all structures are known. Moreover, their contribution to laminar-turbulent transition is still under discussion. In this publication, we contribute with visualizations to better characterize the vortex topology. Moreover, the question of how these structures contribute to laminar-turbulent transition is addressed.

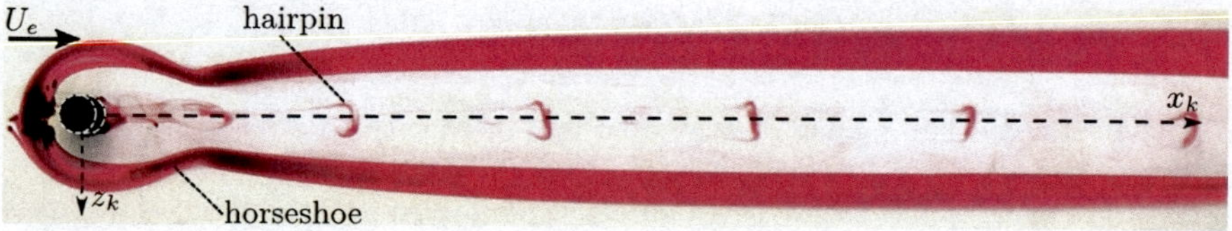

Fig. 1. Roughness element with horseshoe and hairpin vortices.

2 Experimental Setup

To identify primary and secondary structures, visualizations are performed in the Laminar Water Channel (Laminarwasserkanal, LWK) at the Institute of Aerodynamics and Gas Dynamics (IAG) of the University of Stuttgart. In contrast to a laminar wind tunnel, the LWK allows deeper insights into e.g. roughness element studies with different visualization techniques. The LWK is a closed-loop water facility that provides a reproducible measurement environment for flat plate laminar boundary layer studies. A steady two-dimensional (2D) laminar boundary layer forms on the flat plate, where the leading edge has an elliptical nose. A detailed overview of the LWK can be found in [8]. Measurement are carried out using a Dantec 55R15 hot-film probe. The position of the probe can be set by a traverse system. A Dantec Streamware bridge is connected with the hot-film probe and works according to the constant temperature anemometry principle. The output voltage of the bridge is recorded with a 16-bit National Instruments USB-6216 A/D converter and converted to velocity u through King's law. A digital bandpass filter is applied between 0.1 Hz and 10 Hz on u to filter out measurement noise and low-frequency disturbances in the LWK. Every position is measured 60 s with a sampling rate of 100 Hz.

The experimental setup is given in Fig. 2. A cylindrical roughness element with height $k = 10$ mm equal to width d (aspect ratio $\eta = d/k = 1$) is placed 570 mm downstream of the leading edge. The origin of the (x_k, y_k, z_k) - coordinate system (non-dimensionalized by k) is at the center of the roughness element, where x_k gives the distance to the cylinder in streamwise direction and y_k gives the wall-normal distance. The roughness Reynolds number is $Re_{kk} = 716$, large enough to trigger primary and secondary structures [6].

Two visualization techniques are used: hydrogen bubble and dye. Figure 2 shows the hydrogen bubble setup with the top view (left) and the side view (right). In the top view setup, the hydrogen bubbles are illuminated by two light emitting diodes (LEDs) or by a laser sheet in the x_k/z_k plane. In the side view setup, the bubbles are illuminating by a laser sheet in the y_k/z_k plane. Since the camera cannot be placed in the water, a mirror is used in the side view setup and is placed downstream of the laser sheet. The mirror is placed not too close to the visualization plane, so that the vortices are not affected, and not too far away, so that the camera can image the vortices with a high resolution. To produce a continuous sheet of hydrogen bubbles, the current flow from the DC power supply stimulates an almost steady electrolytic process at the generating wire.

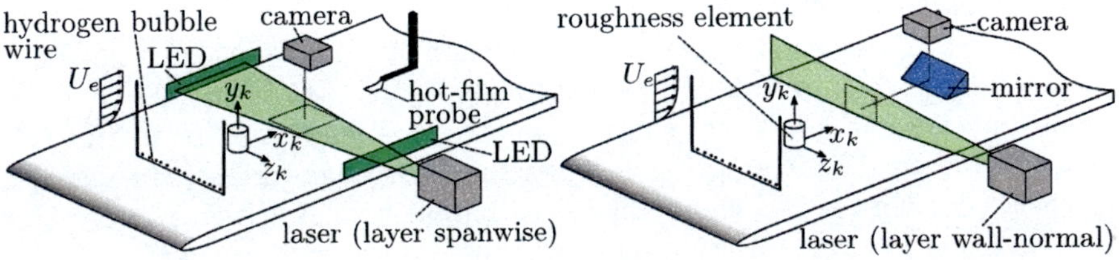

Fig. 2. Experimental setup of hydrogen bubble visualization and hot-film measurement. Left: top view setup. Right: side view setup. Either the LED or the laser is used.

3 Results

3.1 Primary Structures

In the following, primary structures are presented, which are essential structures in the roughness-disturbed boundary layer, from which secondary structures originate. The primary structures are defined here as structures, which originate directly at or around the roughness element. These are: horseshoe vortex, recirculation zone and hairpin vortex. By placing the hydrogen bubble wire at $y_k = 0.4$ upstream of the roughness element, the horseshoe vortex becomes visible in the hydrogen bubble visualization. The vortex is shown in Fig. 3 where the camera exposure time is set much

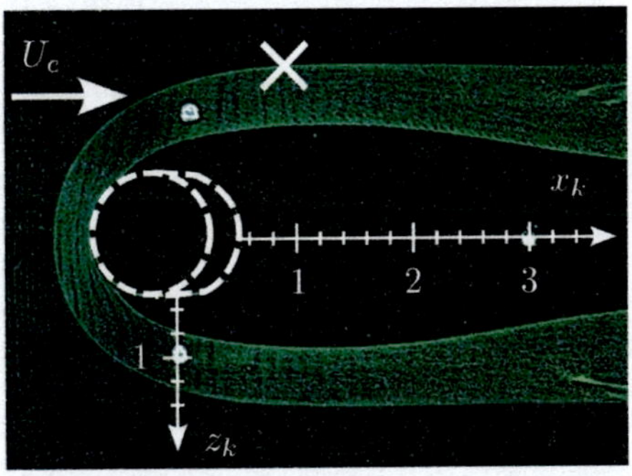

Fig. 3. Top view visualization showing the horseshoe vortex. The hydrogen bubble wire is placed at $y_k = 0.4$ and the bubbles are illuminated by an LED. The cross (×) indicates the outer edge of the horseshoe vortex.

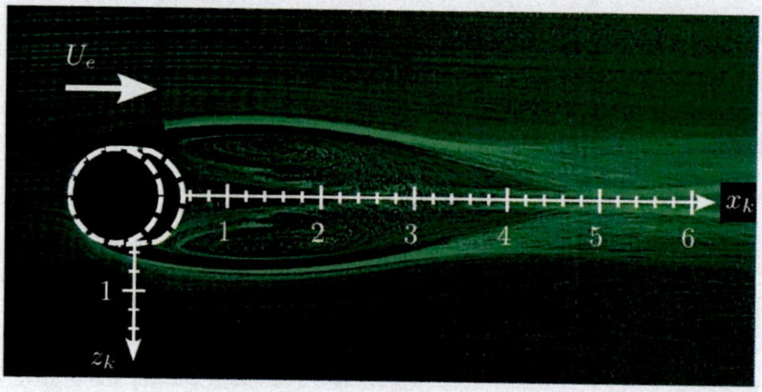

Fig. 4. Side view visualization showing the recirculation zone. The hydrogen bubble wire is placed at $y_k = 0.8$ and the bubbles are illuminated by an LED.

higher than the shedding time of the hairpin vortex. The horseshoe vortex wraps around the cylinder and dissolves downstream of the roughness element. In Fig. 1, the horseshoe vortex pushes the dye on the flat plate away from the cylinder in radial direction, whereas the hydrogen bubbles in Fig. 3 are directed into the horseshoe vortex. Therefore, Fig. 1 shows the outer edge (cross × in Fig. 3) of the horseshoe vortex. Placing the hydrogen bubble wire even higher at $y_k = 0.8$, the horseshoe vortex is not visible anymore. Instead, the recirculation zone downstream of the cylinder becomes visible, as shown in Fig. 4. It can be concluded, that the upper fluid of the incoming boundary layer is mostly directed into the recirculation zone, while the lower part of the incoming flow is directed into the horseshoe vortex. It can be also seen from Fig. 4 that the recirculation zone extends up to $x_k = 4.4$. This zone depends on Re_{kk} and is important to know when performing hot-film measurements with a single-sensor normal probe downstream of the roughness element. In the recirculation zone, the streamwise velocity component u is in its magnitude not everywhere the dominant velocity component, resulting in an incorrect velocity caused by the measurement principle.

Figure 5 illustrates a dye visualization, where dye is injected into the flow through injection holes in the cylinder. With this method, the hairpin vortex can be visualized isolated from other structures. While moving downstream, the hairpin head moves away from the wall and the hairpin leg is pushed to the wall. Acarlar and Smith [1] found that the legs of the hairpin are rotating counterwise so that fluid between the legs is pushed from the wall to the boundary layer edge due to mutual induction effects. The induced velocity pushes the forward portion of the legs and the head of the trailing hairpin away from the wall. The head undergoes a higher velocity than the legs due to the lower mean velocity near the wall. This results in a streamwise stretching of the hairpin, which is visible in Fig. 5.

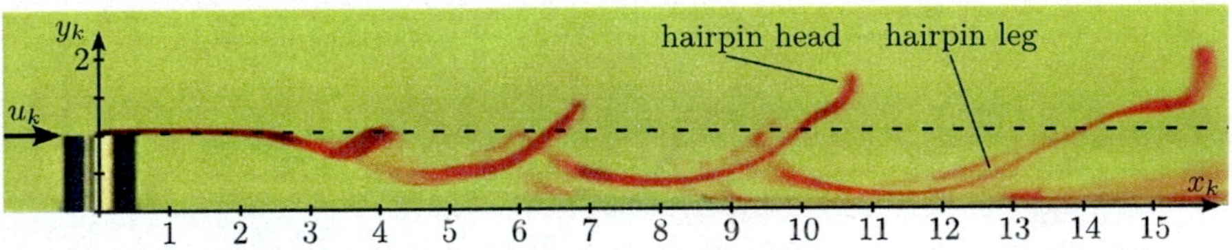

Fig. 5. Side view of dye visualization showing hairpin vortices. Dye is injected into the flow through a hole in the cylinder.

3.2 Secondary Structures

The secondary structures can be considered as the next stage of the laminar-turbulent transition process. Thus, it is necessary to characterize these structures topologically in order to explain the laminar breakdown. Figure 6 illustrates a top view hydrogen bubble visualization snapshot at $y_k = 1$. In contrast to Figs. 3 and 4, the exposure time is set in Fig. 6 and in the following visualization figures to a minimum so that the visualizations can be regarded as snapshots. This allows to get insights into unsteady structures, such as the hairpin vortex and secondary structures. Dye visualization is not appropriate for this investigation, because the diffusion coefficient of the dye is much lower than the diffusion of the vorticities of the secondary structures [1].

At $(x_k, z_k) = (8.5, 0)$ in Fig. 6, the hairpin legs are visible as two circles because the laser plane intersects the legs, which are oriented in streamwise and wall-normal direction (see Fig. 5 dashed line). The intersection of hairpin legs can be also identified at $x_k = 12.5$, $x_k = 17$ and $x_k = 21.5$, resulting in a hairpin shedding wavelength of $\lambda_{x,b} = 45$ mm. To ensure that the visualization captures unsteady structures correctly, a hot-film measurement at $(x_k, y_k, z_k) = (28, 1, 0)$ is performed. The Power Spectral Density analysis of the velocity signal reveals a shedding frequency of $f_h = 1.367$ Hz, resulting with $\lambda_x = \overline{u(x_k, y_k, z_k)}/f$ in a shedding wavelength of $\lambda_{x,h} = 46$ mm, being $\overline{u}$ the mean velocity. The wavelengths determined by visualization $\lambda_{x,b} = 45$ mm and hot-film measurement $\lambda_{x,h} = 46$ mm match very good. The non-dimensional angular frequency $\omega = 2\pi f k/U_e$ is $\omega = 0.82$ and matches with the literature [5]. It can be confirmed that the shedding wavelength is captured by hydrogen bubble visualization and no visual distortion is present.

Between $x_k = 11 - 13.5$, the hairpin is surrounded by a secondary structure that is elongated in the streamwise and slightly in the spanwise direction. Here, this structure is named Λ_1 vortex, because it matches in elongation and vortex rotation with the known Λ vortex from classical boundary layer transition, where the secondary instability is characterized by different aligned Λ vortices [3,4]. Comparing the spanwise position of the horseshoe vortex in Fig. 3 with the spanwise position of the Λ_1 vortex, one might assume that the horseshoe vortex dissolves and forms into the Λ_1 vortex.

Looking at $x_k = 18–22$ in Fig. 6, another secondary structure is visible, here named Λ_2 vortex. The Λ_2 vortex is located between the Λ_1 vortex and the hairpin vortex and is aligned similarly to the Λ_1 vortex. To the authors' knowledge, this structure has not been explicitly described in the literature before. It can be assumed that the structure was found, for instance, in the experimental work of Acarlar and Smith [1] or in the numerical study of Bucci et al. [2], but is not identified as such. This can be attributed to the complexity of this structure and the possible Reynolds number dependence. In contrast to the Λ_1 and the hairpin vortex, the Λ_2 vortex is not visible upstream of $x_k \approx 15$ in the visualization from Fig. 6. Note that this observation is found at $y_k = 1$, as Fig. 6 is a visualization at the wall-normal position $y_k = 1$. In the next section, it is shown by hot-film measurements that at least one of the two vortices is already present at $x_k = 10$ near the wall. Both secondary structures have the same shedding wavelength of $\lambda_{x,h} = 46\,\text{mm}$ due to their connection with the hairpin vortex. This indicates that the secondary structures are only present when the hairpin is present in the boundary layer. Since the hairpin vortex is only present above a certain Reynolds number Re_{kk}, the secondary structures are also only to be expected above this Reynolds number. Ongoing research is investigating the Reynolds number dependency and is not discussed here.

Figure 6 bottom right gives a sketch of the topology of the secondary structures and hairpin vortex. The vortex rotation of the Λ_1 and Λ_2 vortex is determined from hydrogen bubble videos and the hairpin rotation is taken from Acarlar and Smith [1]. However, the rotation of the secondary structures should be validated by e.g. Particle Image Velocimetry measurements. The previous assumption of the vortex direction explains the rotation of the Λ_2 vortex. The Λ_1 vortex and the hairpin induce a rotation of the Λ_2 vortex in the correspondingly opposite direction due to their same direction of rotation. Due to the close connection of the Λ_1 and Λ_2 vortex, both vortices are called Λ vortex and differ by their index 1 and 2. Downstream of $x_k = 32$ all visible structures decay, implying that $(Re_{kk}, \eta) = (716, 1)$ is an unstable boundary layer which agrees with the literature [5,7].

To understand the secondary structures Λ_1 vortex and Λ_2 vortex better, cross-cut view hydrogen bubble visualizations are performed. The visualization setup is sketched in Fig. 2 right. Figure 7 shows hydrogen bubble snapshots at $x_k = 18$ (compare with Fig. 6) for three different time snapshots $t_1 < t_2 < t_3$. At t_1, the hairpin head can be identified. In the next snapshot t_2, the hairpin head has passed and the hairpin legs are visible at $y_k = 1$, which are surrounded by the Λ_1 vortex. At t_3, the Λ_2 vortex is visible. The Λ_2 vortex is located between the Λ_1 vortex and the hairpin, as demonstrated in Fig. 6. The spanwise cross section of both secondary structures is indicated by the red and blue areas. It is observed that the Λ_2 vortex is closer to the wall than the Λ_1 vortex.

3.3 Impact on Laminar-Turbulent Transition

The individual impact of the structures on laminar-turbulent transition is analyzed by hot-film measurements taken in the y_k/z_k plane at the same streamwise position $x_k = 18$ as in Fig. 7. The plane is traversed in evenly spaced intervals of $\Delta y_k = 0.1$ and $\Delta z_k = 0.1$ to gain a measurement grid with $N_y \times N_z = 18 \times 33$

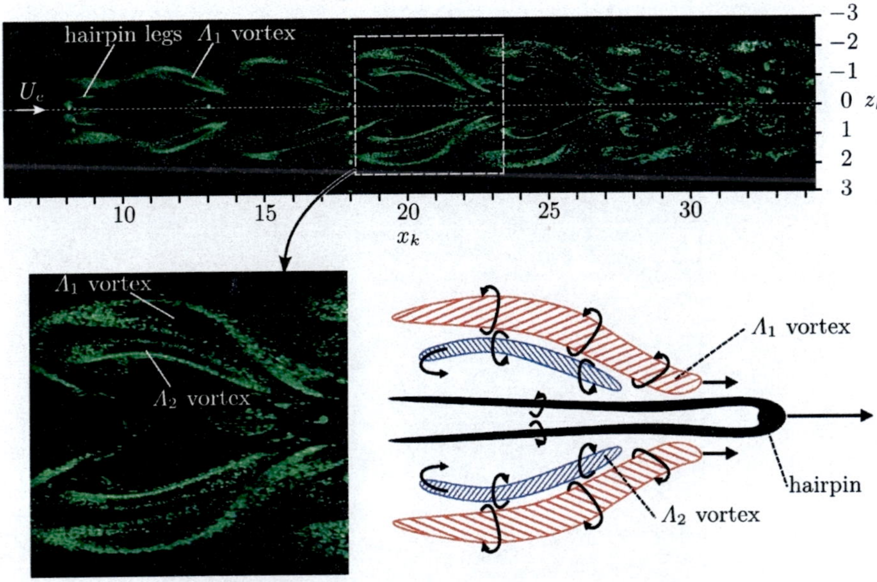

Fig. 6. Top view snapshot at $y_k = 1$, detail view and vortex topology sketch. Hairpin rotation taken from Acarlar and Smith [1].

points. Figure 8 shows the fluctuation amplitude u'_{rms}/U_e, where u'_{rms} is the root mean square (rms) of the velocity fluctuation. Note that the measurements are only performed in spanwise direction beginning in the center of the plate ($z_k \geq 0$), since it is shown in Figs. 6 and 7 that the structures are symmetrical to the x_k/y_k plane at this streamwise position. The red and blue area is extracted from the hydrogen bubble visualization (Fig. 7) and shows the cross section of the Λ_1 and Λ_2 vortex. The Λ_1 vortex can be clearly attributed to the most dominant fluctuation amplitude region, while the hairpin and the Λ_2 vortex have about 50% fluctuation amplitude of the Λ_1 vortex. The high fluctuation amplitude may initiates the Λ_2 vortex, as already pointed out with Fig. 6. Moreover, the high fluctuation amplitude indicates that the Λ_1 vortex plays a key role in roughness-induced transition by contributing most to laminar-turbulent breakdown. This implies that the roughness-induced transition at the final stage is comparable to secondary instability of classical laminar-turbulent transition.

To analyze the fluctuation amplitude in spanwise and streamwise direction, hot-film measurements are taken in the x_k/z_k plane in evenly spaced intervals of $\Delta z_k = 0.1$ at roughness height $y_k = 1$ and near the wall at $y_k = 0.75$. Figure 9 shows fluctuation amplitude profiles in spanwise direction at different x_k positions. The gray area ($0 \leq z_k \leq 0.5$) marks where the hairpin is approximately present and the white area ($z_k > 0.5$) marks where the Λ_1 and Λ_2 vortex is approximately present (compare Fig. 6). The measurements are performed in spanwise direction beginning in the center of the plate.

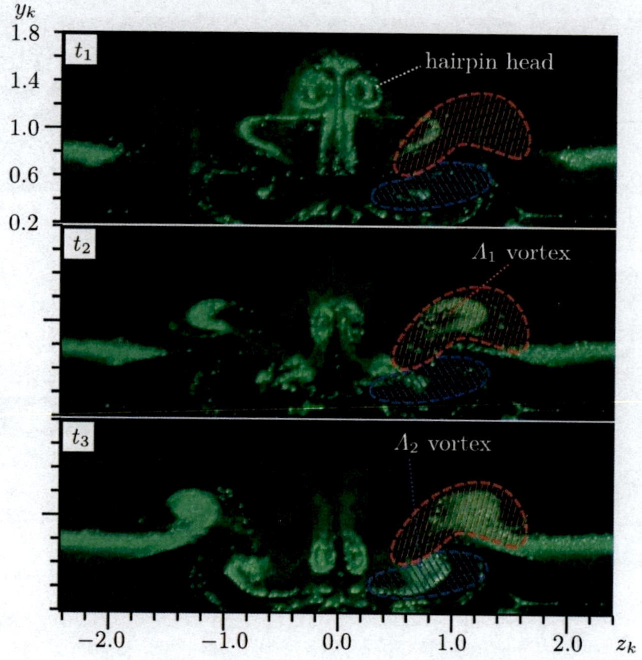

Fig. 7. Front view snapshots showing three time snapshots at $x_k = 18$.

Downstream of the roughness element at $x_k = 2$ is the recirculation zone, as shown in Fig. 4. At this position Fig. 9 shows zero fluctuation amplitude in both planes $y_k = 1$ and $y_k = 0.75$, implying that the recirculation zone must lie below $y_k = 0.75$. At $x_k = 10$ the fluctuation amplitude shows in the gray region a peak in both planes. The peak can be linked to the development region of the hairpin vortex, as pointed out by Puckert [6]. In the lower plane $y_k = 0.75$ two low peaks can be identified in the Λ_1 and Λ_2 vortex area at $z_k = 0.7$ and $z_k = 1.5$. It is not possible to assign each peak to a specific structure because both structures are elongated in spanwise direction and are relatively close to each other. However, position $x_k = 10$ in plane $y_k = 0.75$ can be defined as the position where at least one of the two structures evolves. From Fig. 6 it is suggested that the peak at $z_k = 0.7$ can be attributed to the Λ_2 vortex and the peak at $z_k = 1.5$ can be attributed to the Λ_1 vortex. The fluctuation amplitude profile in plane $y_k = 1$ is almost zero, indicating that the secondary structures originate from near the wall.

At position $x_k = 18$ the hairpin is reduced in its fluctuation amplitude and the Λ_1 and Λ_2 vortex are the dominating structures in both planes. Note that the y_k/z_k plane at this position is shown in Figs. 7 and 8. The maximum fluctuation amplitude of $u_{rms}/U_e = 0.16$ is reached at $(x_k, y_k, z_k) = (26, 1, 1.6)$. Downstream of $x_k > 26$, both planes $y_k = 0.75$ and $y_k = 1$ show similar fluctuation amplitude profiles. This reveals that the secondary structures decay and

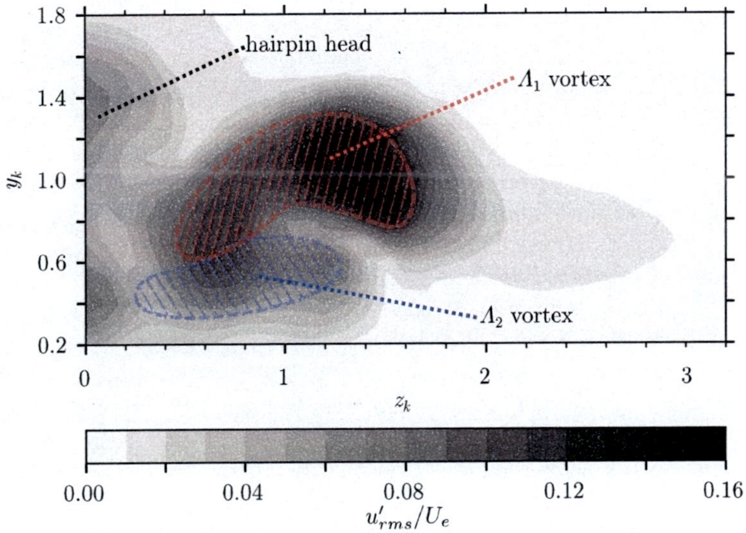

Fig. 8. Fluctuation amplitude at $x_k = 18$.

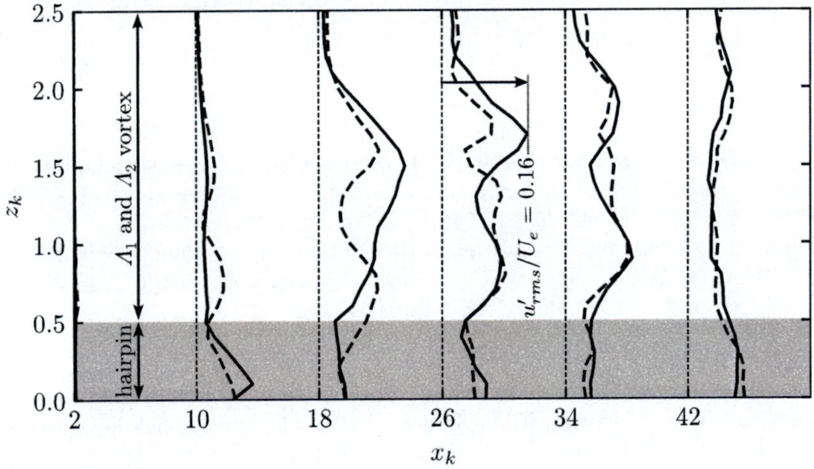

Fig. 9. Fluctuation amplitude profiles at $y_k = 1$ (——) and $y_k = 0.75$ (- - -)

a turbulent boundary layer develops, as can be also seen in Fig. 6. Interestingly, the hairpin vortex at $x_k = 42$ has a high fluctuation amplitude again, indicating that it is not completely decayed.

4 Conclusion

An investigation of primary and secondary structures triggered by a cylindrical roughness element placed in a laminar flat plate boundary layer is performed to understand the complex vortex topology and its influence on laminar-turbulent transition. Hydrogen-bubble and dye visualizations show new insights into the primary structures: horseshoe vortex, hairpin vortex and recirculation zone. The visualizations in the LWK can help to extend the understanding of the roughness vortex topology gained from wind tunnel experiments and numerical studies. It is observed that the lower layer of the incoming flow in the boundary layer is mainly directed into the horseshoe vortex, while the upper layer is directed into the recirculation zone. The Λ_1 vortex is a secondary structure, elongated and orientated in streamwise and spanwise direction. A novel secondary vortex, called Λ_2 vortex, is presented. It is elongated and orientated similar to the Λ_1 vortex and located between the Λ_1 vortex and the hairpin close to the wall. Both secondary structures must be connected to the hairpin as they all share the same shedding wavelength. Hot-film measurements show that the fluctuation amplitude of the Λ_1 vortex is twice as high as the fluctuation amplitude of the Λ_2 and hairpin vortex. This indicates that the Λ_1 vortex may give birth to the Λ_2 vortex and contributes most to laminar-turbulent breakdown. It is of interest how the secondary structures depend on the Reynolds number. In further studies, the dependence of secondary structures on the Reynolds number Re_{kk} is of interest.

Acknowledgments. The funding of the Deutsche Forschungsgemeinschaft (DFG) under grant number RI 680/39-1 is gratefully acknowledged.

References

1. Acarlar, M.S., Smith, C.R.: A study of hairpin vortices in a laminar boundary layer. Part 1. Hairpin vortices generated by a hemisphere protuberance. J. Fluid Mech. **175**, 1–41 (1987). https://doi.org/10.1017/S0022112087000272
2. Bucci, M.A., Cherubini, S., Loiseau, J.-Ch., Robinet, J.-Ch.: Influence of freestream turbulence on the flow over a wall roughness. Phys. Rev. Fluids. **6**(6), 063903 (2021). https://doi.org/10.1103/PhysRevFluids.6.063903
3. Herbert, T.: Secondary instability of plane channel flow to subharmonic three-dimensional disturbances. Phys. Fluids **26**, 871 (1983). https://doi.org/10.1063/1.864226
4. Klebanoff, P.S., Tidstrom, K.D., Sargent, L.M.: The three-dimensional nature of boundary-layer instability. J. Fluid Mech. **12**(1), 1–34 (1962). https://doi.org/10.1017/S0022112062000014
5. Puckert, D.K., Rist, U.: Global instability in a laminar boundary layer perturbed by an isolated roughness element. Exp. Fluids **59**(3), 1–6 (2018). https://doi.org/10.1007/s00348-018-2510-7
6. Puckert, D.K.: Experimental investigation of global instability and critical Reynolds number in roughness-induced laminar-to-turbulent transition. Ph.D. thesis, Universität Stuttgart (2019)
7. Römer, T. M.; Rist, U.: Stability analysis in boundary layer behind a roughness with free-stream turbulence. In: International Symposium on Turbulence and Shear Flow Phenomena 12, Osaka (2022)
8. Wiegand, T.: Experimentelle Untersuchungen zum laminar-turbulenten Transitionsprozess eines Wellenzugs in einer Plattengrenzschicht. Ph.D. thesis, Universität Stuttgart (1996)

Curvature Correction for Turbulent Diffusion Inside a Differential Reynolds-Stress Model

Philip Ströer(✉) and Tobias Knopp

C2A2S2E - Center for Computer Applications in AeroSpace Science and Engineering, German Aerospace Center (DLR), Institute of Aerodynamics and Flow Technology, Bunsenstraße 10, 37073 Göttingen, Germany
philip.stroeer@dlr.de

Abstract. A correction of the turbulent diffusion terms inside a Reynolds-stress model for mean-streamline curvature is presented. The correction is based on work by Zeman and is reformulated in this work to be applicable in general CFD codes. For this purpose, a new and grid-point local gradient Richardson number is derived which is used within the correction term in order to characterize mean-streamline curvature. These quantities are verified for a channel with U-turn and delta wing vortex. The final correction term was finally applied to these test cases: The overall effect of the correction term on the local Reynolds stresses depends the relative effect of diffusion on the turbulent kinetic energy balance. In this sense, no effect on the solution of the mean flow field was observed for the channel test case, whereas lower values of turbulent kinetic energy appear for the delta wing vortex.

Keywords: Differential Reynolds-Stress Model · SSG/LRR-ω · Streamline Curvature · Turbulent Diffusion

1 Introduction

In many flow configurations, mean-streamline curvature can have a significant effect on the turbulent boundary layer, a wake or a vortex core. Flow features such as curved surfaces, curved wakes, or vortices are inevitable in aerodynamic flows. Therefore, reliable simulation capabilities must be provided for these types of flows. In general, mean-streamline curvature can have either a stabilizing or destabilizing effect on a turbulent flow, depending on whether the curvature is characterized as convex or concave. In this respect, the sense of rotation in combination with the direction of shear determines the character of the mean-streamline curvature (cf. Ref. [1]).

It is well known, that eddy-viscosity turbulence models are not able to capture mean-streamline curvature effects. Therefore, a number of curvature corrections exist. These corrections use additional terms that allow for a local adjustment of the eddy-viscosity level to account for amplification or damping. A

A. Dillmann et al. (Eds.): STAB/DGLR Symposium 2022, NNFM 154, pp. 465–474, 2024.
https://doi.org/10.1007/978-3-031-40482-5_44

correction for the Spalart-Allmaras (SA) turbulence model has been presented by Spalart and Shur et al., Refs. [2] and [3]. They derived a quantity that characterizes the strain-stress misalignment angle, which evolves at the same rate as the angular frequency does. This quantity allows a characterization of the mean-streamline curvature and is eventually used within the SA model to scale the production source.

For Reynolds-Stress Turbulence models (RSM) such as the SSG/LRR-ω model [4] and other differential RSM, it is known that the production term captures curvature effects in a natural way. This can be seen by considering the production tensor of the Reynolds stresses in cylindrical coordinates (cf. textbook of Durbin and Petterson -Reif [1]).

However, in cases where the production is weak, turbulent transport may control the balance and the local level of Reynolds stresses [5]. Therefore, according to Zeman [5], a closer look at the third moment terms is needed.

For this purpose, the Reynolds-stress equations are considered:

$$\frac{\partial(\rho R_{ij})}{\partial t} + \frac{\partial(\rho u_k R_{ij})}{\partial x_k} = \rho P_{ij} + \rho \Pi_{ij} - \rho \varepsilon_{ij} + \rho D_{ij}, \tag{1}$$

where R_{ij} is the Reynolds-stress tensor, P_{ij} is the production, Π_{ij} is the pressure-strain correlation and ε_{ij} is the disspation-rate tensor. The turbulent diffusion term D_{ij} terms reads:

$$D_{ij} = \frac{\partial}{\partial x_k}\left(\nu \frac{\partial R_{ij}}{\partial x_k}\right) + \frac{\partial}{\partial x_k}(-\overline{u_i' u_j' u_k'}) + \frac{\partial}{\partial x_k}\left(-\overline{p'(u_i'\delta_{jk} + u_j'\delta_{ik})}\right). \tag{2}$$

The first term of Eq. 2 (viscous part) can be computed analytically, the last term, which is called "pressure-diffusion term" is usually assumed to be small compared to the turbulent transport term. The middle, third-moment term in Eq. 2 characterizes the turbulent transport. This term is usually approximated using a gradient-diffusion hypothesis (GDH), such as the simple gradient-diffusion hypothesis (SGDH, $\partial/\partial x_k\,(\nu_t\,\partial R_{ij}/\partial x_k)$) or the generalized gradient-diffusion hypothesis (GGDH). According to Ref. [5], these approaches are not sufficient to characterize third moments in vortical flows. That is, the mean-streamline curvature effects on the third moments have to be considered directly. For this purpose, Zeman derived a corrected effective viscosity ν_{Tr}, which can be used inside the SGDH (or inside the GGDH) the following way: $\partial/\partial x_k\,(\nu_{Tr}\,\partial R_{ij}/\partial x_k)$. Taking everything into account, the turbulent diffusion term can be written as:

$$D_{ij} \approx \frac{\partial}{\partial x_k}\left(\nu\,\frac{\partial R_{ij}}{\partial x_k}\right) + \frac{\partial}{\partial x_k}\left(\nu_{Tr}\,\frac{\partial R_{ij}}{\partial x_k}\right). \tag{3}$$

Compared to the regular SGDH term, ν_t is replaced by ν_{Tr} in Eq. 3. Within the ω Eq. the correction term is used in the same way. How to calculate the corrected eddy viscosity ν_{Tr} will be shown in the next chapter. It should be noted, that this correction only accounts for mean streamline-curvature effects. System rotation is not taken into account in the current work.

The paper is organized as follows: First, modeling aspects are discussed and the final correction term is derived. Afterwards, these new terms are verified using analytical reference solutions. The fourth chapter presents the application of the correction for a channel with U-turn test case and a delta wing vortex. Finally, conclusions are drawn and an outlook is given.

2 Modeling Aspects

The corrected eddy viscosity introduced in Eq. 3, denoted by ν_{Tr}, was derived by Zeman considering the linearized Equations for Lagrangian fluctuating radial and tangential velocity components (in cylindrical coordinates). In this way, an expression is derived, that includes the eddy-viscosity used in the GDH and a correction for the radial transport:

$$\nu_{Tr} = \frac{c_1 \frac{k}{\varepsilon} \overline{v'^2}}{1 + \left(c_2 \frac{k}{\epsilon}\right)^2 \left(\frac{\partial (Ur)^2}{\partial r}\right) / (2r^3)}, \tag{4}$$

where k is the turbulent kinetic energy, ε is the dissipation rate and $\overline{v'^2}$ is discussed in the following. $U := u_\phi$ is the tangential velocity component and r is the local radius of the curved streamline in polar coordinates.

Equation 4 cannot be evaluated in terms of a general Computational Fluid Dynamics (CFD) code, because it uses quantities that are not Galilean invariant and general. That is, they cannot be evaluated using only quantities that are available at each grid point. Therefore, in the following, a formulation is derived that is grid-point local, general and Galilean invariant.

First, the nominator of Eq. 4 is considered. The nominator represents an eddy viscosity consisting of a length scale and a velocity scale (cf. Ref. [6] for example). The term $\overline{v'^2}$ describes a quantity that is oriented normal to the local mean-streamline direction. However, the computation of $\overline{v'^2}$ based on the mean-streamline is not Galilean invariant. Therefore, instead of the original eddy viscosity term $c_1 \frac{k}{\varepsilon} \overline{v'^2}$ the SGDH formulation of the SSG/LRR-ω RSM of Eisfeld et al. [4] is used:

$$\nu_t = \frac{D^{(SD)}}{C_\mu} \frac{k^2}{\varepsilon}. \tag{5}$$

Replacing the eddy-viscosity definition used in the nominator of Eq. 4 (i.e. $c_1 k/\varepsilon\, \overline{v'^2}$) gives a correction factor ξ that can be used in conjunction with an arbitrary definition of the eddy viscosity ν_t (here Eq. 5):

$$\nu_{Tr} = \frac{\nu_t}{1 + \left(c_2 \frac{k}{\epsilon}\right)^2 \left(\frac{\partial (Ur)^2}{\partial r}\right) / (2r^3)} = \nu_t \frac{1}{1+\Phi} = \nu_t \left(1 + \frac{-\Phi}{1+\Phi}\right) = \nu_t (1+\xi) \tag{6}$$

That is, the term ξ characterizes the effect of the mean-streamline curvature on the effective eddy viscosity used inside the SGDH. Thus, the effective viscosity is increased or decreased depending on Φ, i.e. $\xi = f(\Phi)$.

In a second step, the function Φ has to be evaluated in a grid-point local and Galilean-invariant manner. For this purpose, Φ is rewritten in Eq. 7 using the gradient Richardson number of Bradshaw [7]:

$$\Phi = \left(c_2 \frac{k}{\epsilon}\right)^2 \frac{\partial (Ur)^2}{\partial r} / (2r^3) = c_2 \left(\frac{k}{\epsilon}\right)^2 \frac{1}{2} R_i \left(\frac{\partial U}{\partial r}\right)^2, \tag{7}$$

where the gradient Richardson number reads [7]:

$$R_i := \frac{2 \frac{U}{r^2} \frac{\partial (Ur)}{\partial r}}{\left(\frac{\partial U}{\partial r}\right)^2}. \tag{8}$$

Here, the nominator of Eq. 8 represents an equivalent of the square of the Brunt-Väisälä frequency (buoyancy frequency) that is related to inertial forces in the denominator.

Since the squared radial gradient of the tangential velocity in Φ is equal to the denominator of the Richardson number, the remaining task in deriving a reformulated correction term is to find a grid-point local replacement for R_i (Eq. 8). For this purpose, a new grid-point local representation of Eq. 8 has been derived in Ströer and Knopp [8]:

$$R_i := \frac{-n_\Omega \|\Omega_{ij}\| (\|S_{ij}\| - n_\Omega \|\Omega_{ij}\|)}{(\frac{1}{2}\|S_{ij}\| + \frac{1}{2} n_\Omega \|\Omega_{ij}\|)^2}. \tag{9}$$

Since only absolute values are used in Eq. 9, the directional information is lost. That is, it is not possible to distinguish between amplification and damping (for wall-bounded flows: concave and convex curvature). Therefore, a condition used in a switching function n_Ω (that is either $+1$ or -1) was derived by Ströer and Knopp [8]:

$$n_\Omega := \tanh\left(1000\left(1 - \left[\frac{\tilde{r}}{|\tilde{r}|} \frac{\|S_{ij}\|}{\|\Omega_{ij}\|}\right]\right)\right). \tag{10}$$

This condition ensures that the sign of the nominator of Eq. 9 is always equal to that of the analytical expression of Eq. 8:

$$\mathrm{sgn}\,(2U/r\ \Omega_z) = \mathrm{sgn}\,(-\|\Omega_{ij}\|(\|S_{ij}\| - \|\Omega_{ij}\|)) \tag{11}$$

The term $\tilde{r}$ used in Eq. 10 was derived in the rotation correction of the SA turbulence model by Spalart, Shur et al. (cf. Section II. of Ref. [3]). Their main idea is to consider the angle between the principal axis of the strain rate tensor and the axis of the inertial reference frame. In this way, they are able to distinguish between amplification and damping in a general and Galilean-invariant manner.

It should be noted that for $\mathrm{sgn}\,(2U/r\ \Omega_z) = +1$ and $\|S_{ij}\|/\|\Omega_i\|_2 < 1$ condition Eq. 11 is ambiguous, because in this case $-\|\Omega_i\|_2$ and $\|\Omega_i\|_2$, both satisfy condition Eq. 11 and lead to the correct sign of the nominator of R_i. That is, although the Richardson number may be quantitatively wrong the qualitative behavior (amplification or damping) is captured in any case.

In Ströer and Knopp [8] a detailed derivation of the Eqs. 9, 10 and 11 is given.

3 Verification

3.1 Channel with U-Turn

The reformulated quantities derived in Sect. 2 are verified in the following using the channel with U-turn test case of Manson et al. [10] ($Re = U_{in}D/\nu = 10^6$, D: channel width, $Ma = 0.1$). This test case combines convex curvature at the inner wall and concave curvature at the outer wall for a wall-bounded flow. Moreover, an analytical reference can be given because the geometry approximately represents the curvature of the mean streamline (geometric transformation from Cartesian to polar coordinates, Fig. 1, *left*). In this way, the radius, the polar angle and radial/angular derivatives can be computed directly. These terms are then used to compute an analytical reference solution for quantities such as the Bradshaw Richardson number Eq. 8. The origin of the polar system (i.e. $r = 0$) is defined as the center of the circular U-turn and is shown in Figs. 1 (*left*) and 4 (*left*).

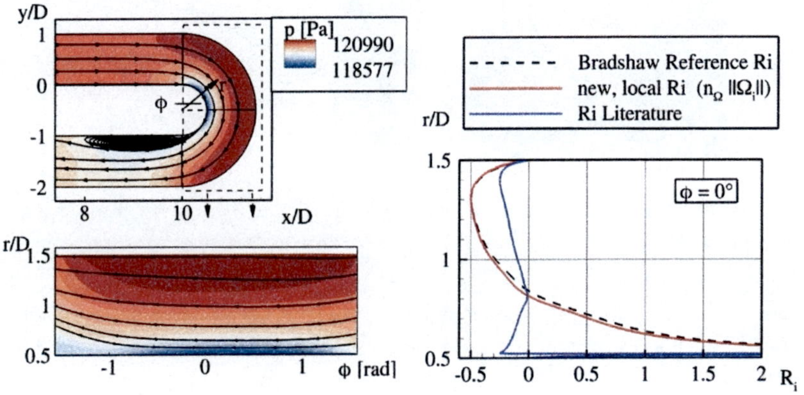

Fig. 1. Coordinate transformation for the channel with U-turn (*left*) and verification of the gradient Richardson number at $\phi = 0° = 0\,\text{rad}$ (*right*)

It should be noted, however, that the flow separates downstream of the U-turn (cf. Fig. 1, *left*). Therefore, the assumption that the geometry approximately represents the curvature of the streamline does not hold in or near this area. Consequently, this polar angle range will not be used for verification purposes in the following.

Figure 1 shows a comparison of the analytical Bradshaw Richardson number Eq. 8 and the new local Rirchadson number Eq. 9. In addition, a grid-point local Richardson number from the literature is given [11]. However, this Richardson number uses a different denominator and therefore differs from the reference and the new formulation. Furthermore, only absolute values are used in Ref. [11]. In this way, the directional information is lost and it is not possible to distinguish between amplification and damping. To tackle this issue, the new Richardson number uses condition Eq. 11, which is used in the function n_Ω

Eq. 10. It can be seen that the new formulation achieves a very good agreement with the analytical reference. A more detailed derivation and verification of the new Richardson number is given in Ströer and Knopp [8].

3.2 Delta Wing

Since the turbulent diffusion correction of Zeman [5] essentially aims at targeting vortical flows, a vortex downstream of a delta wing ($Re = 1.98 \times 10^6$, $Ma = 0.155$) is considered in the following. For this purpose, a slice at a constant streamwise position ($x/c_\Delta \approx 2$, Fig. 2, *right*) is transformed into cylindrical coordinates. In this way, the Richardson number can be verified at a constant angle (dashed-dotted line in Fig. 2, *left*) similar to Sect. 3.1.

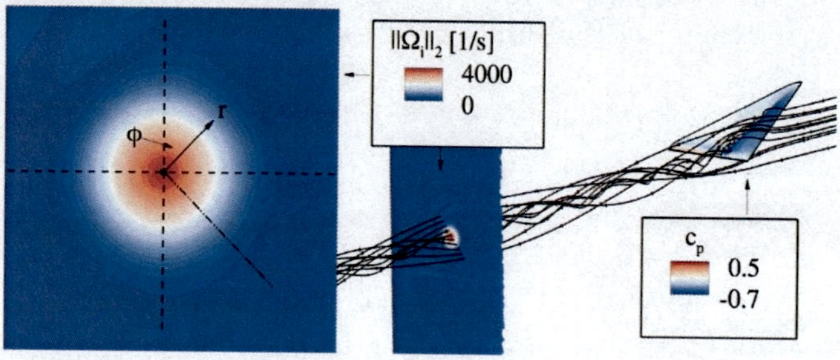

Fig. 2. Geometrical description of a vortex past a delta wing at a constant stream-wise position $x/c_\Delta \approx 2$

The evaluation of the analytical Richardson number is based on a 2D projection of the vortex in the (defined) vortex plane. This means that it does not take into account any streamwise gradients and depends on where this 2D vortex plane is located. A justification of this procedure can be given by analyzing the components of the strain rate tensor (Fig. 3, *middle*). Apart from the radial

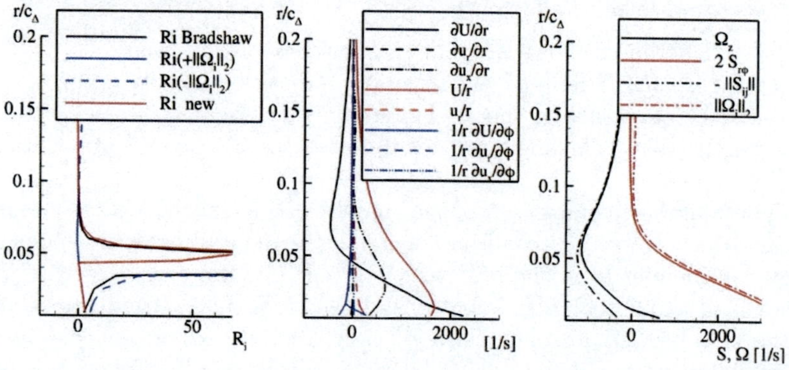

Fig. 3. Verification of the gradient Richardson number (*left*) and effect of different components of the strain rate tensor (*middle, right*) for the delta wing test case

gradient of the streamwise velocity the components U/r (angular frequency) and $\partial U/\partial r$ clearly dominate (components not shown here are negligibly small). Therefore, the strain rate and vorticity magnitude are in good agreement with the $S_{r\phi}$ and Ω_z components, respectively (Fig. 3, *right*). Consequently, the final approximation of the reference Richardson number using the new definition is very good in radial sections (Fig. 3, *left*). For small radii, the approximations become slightly worse due to the effect of the radial gradient of the streamwise velocity.

For r/c_Δ approx. < 0.04 it can be observed that the sign of n_Ω is ambiguous (discussed in Sect. 2). In this radial section $\mathrm{sgn}\left(2\frac{U}{r}\Omega_z\right) = +1$ and $\|S_{ij}\|/\|\Omega_i\|_2 < 1$. Therefore, $+\|\Omega_i\|_2 < 1$ and $-\|\Omega_i\|_2 < 1$ both satisfy condition Eq. 11 and the switching function n_Ω is not activated. Although the Richardson number in this section is quantitatively wrong, the qualitative behavior (amplification or damping) is preserved since the sign of the nominator is correct in any case.

4 Application of the Correction

In the previous section basic quantities were derived and verified so that the final correction ξ (Eq. 6) can now be applied to general cases. For this study the constants of the correction [5] are used without modification. Since the effect of the correction depends on the relative weight of diffusion compared to production and dissipation, the balances of the k equation are analyzed.

All simulations are performed with the compressible, finite-volume, Navier-Stokes solver DLR-TAU [9]. The convective terms are discretized using a central scheme with artificial dissipation. The time integration is done with an implicit scheme in conjunction with a LUSGS linear solver. The convective terms of the additional turbulence equations are discretized using a Roe-like numerical scheme. The new correction term was applied within the SSG/LRR-ω RSM of Eisfeld et al. [4].

4.1 Channel with U-Turn

A contour of the correction factor ξ is depicted in Fig. 4 (*left*) for the channel with U-turn test case. It can be seen that the effective viscosity is reduced near the convex inner wall. Near the concave wall $\xi > 0$. That is, a qualitatively reasonable contour is obtained that represents damping at the convex wall and amplification at the concave wall. Figure 4 (*middle*) shows the results of the final correction factor ξ and the k balance for a constant polar angle ($\phi = 40°$). It can be seen that in regions where ξ has its largest values, the influence of the diffusion on the balance is weak or there is no production at all. Therefore, the overall effect of the correction on the final Reynolds stresses, the length scale and thus the overall solution is negligible (cf. k and ω in Fig. 4, *right*). That is, although there are values of $\xi \neq 0$ that are significant, the overall influence of the corrected diffusion term is determined by the effect of diffusion on the balance.

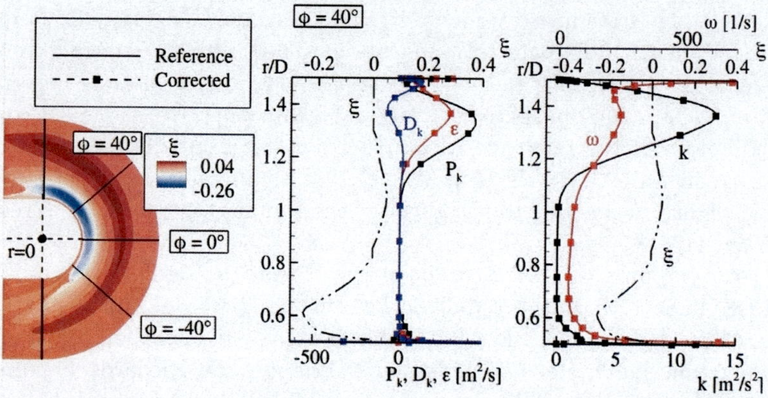

Fig. 4. Final diffusion correction ξ, k balance (*middle*, *right*) and turbulent quantities k and ω at $\phi = 40°$ for the channel with U-turn test case (Note: The lines for the reference and the corrected results are almost identical).

4.2 Delta Wing

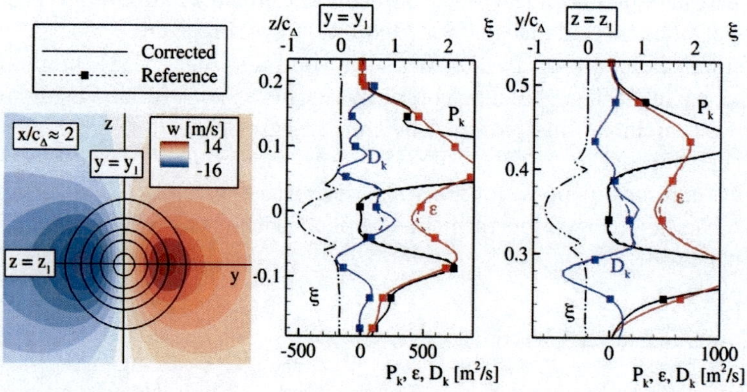

Fig. 5. Final diffusion correction ξ and k balance for two slices (*middle, right*) at $x/c_\Delta \approx 2$ for the delta wing test case

Identical to Sect. 3.2 the same plane at a constant streamwise position $x/c_\Delta \approx 2$ is considered in the following (cf. Fig. 2, *right* and Fig. 5, *left*). Figure 5 shows the final correction factor ξ and the k balance for two slices at constant y and constant z can be seen. In regions where ξ has its largest values the effect of D_k compared to production and dissipation is considerable. Therefore, an overall effect of the correction on the local levels of Reynolds stresses can be expected.

The effect on the Reynolds stresses, the length scale and finally the velocity field can be seen in Fig. 6. The dissipation ω is not affected. The turbulent kinetic

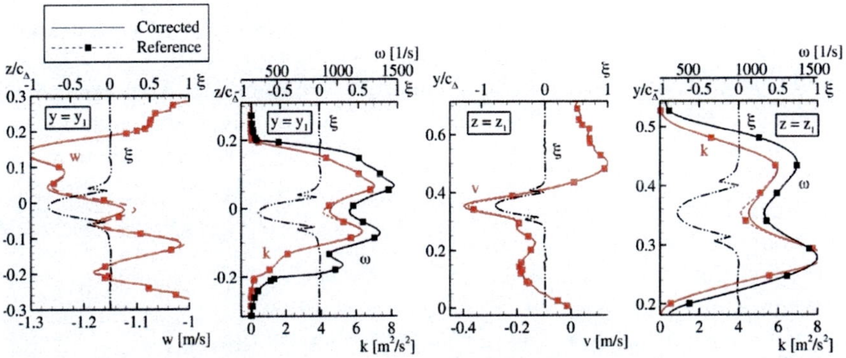

Fig. 6. Effect of the diffusion correction ξ on the turbulence quantities k and ω and the radial velocity components at $x/c_\Delta \approx 2$ for the delta wing test case

energy, which represents the normal Reynolds stresses, is reduced inside the vortex core. This leads to an increase in the radial velocity components. Whether this affects the vortex dynamics needs to be investigated in future projects. There is no effect on the circumferential components. This seems reasonable since the circumferential momentum is clearly dominated by the bulk vortex motion.

5 Conclusions

It is well known that mean-streamline curvature and rotation affect the stability properties of turbulent flows. Research by Zeman [5] indicates that a gradient diffusion model may be too inaccurate to represent third moments in case of curved streamlines. For this purpose, a correction of the effective viscosity was derived in Ref. [5]. In order to apply this correction to general CFD codes, a reformulation is presented in the current research. In this sense, a new, grid-point local and Galilean-invariant gradient Richardson number and eventually a final correction term have been derived. This new Richardson number has been verified for a channel with U-turn test case and a vortex downstream of a delta wing showing a very good agreement of the reformulated quantities in comparison with the analytical reference.

The correction of the GDH was finally applied to these two test cases using the SSG/LRR-ω RSM turbulence model. It can be shown that the correction term is activated qualitatively correct. That is, the amplification and damping are reproduced in a physically meaningful way. Since the correction only affects the SGDH term, the balance of the turbulent kinetic energy equation was analyzed. Only in regions where mean-streamline curvature is detected and the influence of diffusion relative to production and dissipation is large, an overall effect on the level of Reynolds stresses can be expected. Although, mean-streamline curvature is detected for the channel with U-turn test case, the effect of turbulent diffusion is too weak. Taking everything into account, no effect on the local level of Reynolds stresses was detected. For the delta wing test case the correction leads to lower levels of k in the vortex center. Altogether, this leads to a small change in the radial velocity component. How and if this affects the vortex dynamics needs to be investigated in future projects. The results for the channel and the delta wing test cases are consistent with the reasoning of Zeman [5] and are a confirmation of his hypothesis that in case of weak turbulent production, "the turbulent transport may control local levels of the Reynolds stresses" and the model may benefit from a correction.

For vortical flows it is known that some physical phenomena (e.g. vortex bursting) cannot be captured or can only be captured to a very limited extent, and for some phenomena it is not clear yet if they can be treated within the RANS framework (e.g. vortex wandering). Some of these limitations are related to too much turbulent content inside the vortex core. In this sense, it would be important to study the effect of turbulent diffusion on the vortex dynamics. In addition, it is not clear yet, if the level of correction is sufficient. That is, a recalibration of the correction term may be necessary.

Acknowledgment. The funding of the project "Complex Wake Flows in Adverse Pressure Gradient" by DFG (Grants No. RA 595/26-2 and KN 888/3-2) is gratefully acknowledged. The authors gratefully acknowledge fruitful discussions with Dr. Johannes Löwe.

References

1. Durbin, P.A., Pettersson Reif, B.A.: Statistical Theory and Modeling for Turbulent Flows, vol. 2. Wiley, Chichester (2011)
2. Spalart, P.R., Shur, M.: Aerosp. Sci. Technol. **1**(5), 297–302 (1997)
3. Shur, M.L., et al.: AIAA J. **38**(5), 784–792 (2000)
4. Eisfeld, B., Rumsey, C., Togiti, V.: AIAA J. **54**(5), 1524–1541 (2016)
5. Zeman, O.: Phys. Fluids **7**(1), 135–143 (1995)
6. Durbin, P.A.: AIAA J. **33**(4), 659–664 (1995)
7. Bradshaw, P.: J. Fluid Mech. **36**(177) (1969)
8. Ströer, P., Knopp, T.: General formulation of the gradient Richardson number for RANS modelling. In: AIAA SciTech 2023 Forum, National Harbor, MD, USA (2023)
9. Schwamborn, D., et al.: The DLR TAU-code: recent applications in research and industry. In: Wesseling, P., Oñate, E., Périaux, J. (eds.) European Conference on Computational Fluid Dynamics, ECCOMAS CFD 2006, ECCOMAS CFD 2006 Proceedings. Delft University of Technology, Egmond aan Zee (2006)
10. Monson, D.J., et al.: Comparison of experiment with calculations using curvature-corrected zero and two equation turbulence models for a two-dimensional U-duct. In: AIAA Paper 90-1484 (1990)
11. Hellsten, A.: Some improvements in Menter's $k-\omega$ SST turbulence model. AIAA Paper 98-32817 (1998)

A Surrogate-Based e^N Transition Prediction Method for Three-Dimensional Compressible Boundary Layers

Alexander Theiss(✉) and Stefan Hein

German Aerospace Center (DLR), Institute of Aerodynamics and Flow Technology, Bunsenstr. 10, 37073 Göttingen, Germany
{alexander.theiss,stefan.hein}@dlr.de

Abstract. Local stability theory (LST) combined with an e^N method is a well-suited approach for predicting transition. However, simpler or surrogate LST models are often used in an automated transition prediction framework for robustness and efficiency reasons. There are, however, few surrogate-based stability methods for the instabilities encountered in three-dimensional compressible flows. To address this problem, this paper proposes a radial-basis-function-based surrogate model capable of reproducing the instability characteristics of two-dimensional Tollmien-Schlichting waves and stationary cross-flow instabilities. The basic flows necessary to set up the model stem from compressible local Falkner-Skan-Cook similarity solutions. The suitability of the proposed method is demonstrated by N-factor computations for three ATTAS flight test points at transonic conditions.

Keywords: boundary-layer transition · surrogate model · compressible Falkner-Skan-Cooke · radial basis function

1 Introduction

The state of the boundary layer (laminar, transitional, or turbulent) significantly affects the skin-friction drag of objects moving through the air. Therefore, determining the laminar-to-turbulent transition position is essential to improve modern aircraft design and thus reduce fuel consumption. Existing transition prediction methods vary from correlation-based techniques used in Reynolds-averaged Navier-Stokes (RANS) simulations [1] to physics-based approaches. A commonly used physics-based transition prediction technique is the e^N method [2,3] employing linear local [4] or nonlocal [5] stability theory and semi-empirical transition criteria. Transition is assumed when the integrated amplification rate of a boundary-layer instability reaches a threshold value, the critical N-factor. Although the local linear stability theory (LST)-based e^N method has been widely used for transition prediction by expert users, it is still quite complicated to handle in an automated transition prediction framework, e.g., within

A. Dillmann et al. (Eds.): STAB/DGLR Symposium 2022, NNFM 154, pp. 475–485, 2024.
https://doi.org/10.1007/978-3-031-40482-5_45

a RANS simulation. Therefore, efforts have been made for simplified e^N methods that circumvent the growth rate determination by LST calculations. These methods must be robust with no human-in-the-loop iteration to yield reasonable results.

Drela's method [6], for instance, directly models the envelope of the most amplified modes' N-factor. ONERA's [7] parabola method, on the other hand, uses two half parabolas to approximate the growth rate curve of the considered mode. Another class of simplified e^N methods rests on surrogate-based instability analysis in which no assumptions are made about the shape of the envelope or amplification curves; instead, the direct stability properties of the boundary layer are modeled. Surrogate-based instability analyses typically involve a neural network [8,9] or radial basis function (RBF) interpolation within a precomputed sample set [10,11] to rapidly predict amplification rates. However, most surrogate-based e^N methods have been proposed for two-dimensional compressible or three-dimensional incompressible flows.

This work proposes a radial basis function (RBF)-based surrogate model for three-dimensional compressible boundary layers. The model aims to reproduce the instability characteristics of two-dimensional (2D) Tollmien-Schlichting instabilities (TSI) and stationary cross-flow instabilities (CFI) typically encountered in boundary layers on transonic swept-wing aircraft.

2 Methodology

This section outlines the transition prediction method utilizing RBF-based surrogate models. The approach is similar to the concepts published by Nie et al. [11] for two-dimensional flows, while the present paper focuses on three-dimensional boundary layers. The concepts fundamental to the method, such as the local compressible Falkner-Skan-Cooke equations [12,14], the local stability analysis, the radial basis function interpolation, and the surrogate models to predict amplification rates of 2D TSI and stationary CFI, are briefly presented separately.

2.1 Surrogate-Based e^N Transition Prediction Method

The proposed e^N transition prediction method consists of three steps. In the first step, the instability characteristics of TSI and CFI are precomputed with local stability theory for various three-dimensional (3D) compressible boundary layers. The 3D boundary layers are obtained from compressible Falkner-Skan-Cooke local similarity solutions [12,14] for Mach numbers typically encountered in boundary layers on transport aircraft. The instability results are then used to precalculate RBF-based surrogate models for TSI and CFI. The time required to build and evaluate the models is highly dependent on the number of sample points. To minimize the computational effort, a sample partitioning method is employed where the full data set is divided into several subsamples. Thus, the RBF surrogate models are computed for each of these subsample sets, i.e., ignoring all sample points not belonging to the current subsample.

The second step of the surrogate-based e^N transition prediction method involves the evaluation of the surrogate model within the corresponding subsample set and the prediction of amplification rates.

In the last step, the local growth rates are integrated into N-factors, and transition can be predicted at the position where the critical N-factor is reached.

2.2 Compressible Local Falkner-Skan-Cooke Equations

In this paper, the compressible local Falkner-Skan-Cooke (FSC) equations [12, 14] are solved to provide the laminar basic flows for the local stability analysis. The local FSC equations can be formulated as

$$\left.\begin{aligned} (Cf'')' + ff'' + \beta_H \left(\frac{\rho_e}{\rho} - f'^2\right) = 0, \\ (Cg')' + fg' = 0, \\ fq' + \left(\frac{C}{Pr}q'\right)' + \left(1 - \frac{1}{Pr}\right)\frac{U_e^2}{H_e}\left[C\left(f'f'' + \frac{V_e^2}{U_e^2}gg'\right)\right]' = 0, \end{aligned}\right\} \tag{1}$$

where C is the Chapman-Rubesin factor ($C = \frac{\rho\mu}{\rho_e\mu_e}$ with μ being the dynamic viscosity), β_H is the Hartree pressure gradient parameter, ρ is the density, U and V are velocity components in the streamline-aligned coordinate system, Pr is the Prandtl number, H stands for the total enthalpy, prime denotes a wall-normal derivative, and the subscript e represents a boundary-layer edge value. The implementation of the isentropic relationship and the assumption of an ideal gas lead to the following relations

$$\left.\begin{aligned} \frac{\rho_e}{\rho} = \left(1 + \frac{\gamma - 1}{2}Ma_e^2\right)q - \frac{\gamma - 1}{2}Ma_e^2\cos^2\chi_e\left(f'^2 + g^2\tan^2\chi_e\right), \\ \frac{U_e^2}{H_e} = \frac{(\gamma - 1)\,Ma_e^2\cos^2\chi_e}{1 + \left(\frac{1}{2}(\gamma - 1)Ma_e^2\right)}, \end{aligned}\right\} \tag{2}$$

where Ma is the Mach number, γ is the specific heat ratio, and χ_e represents the local sweep angle at the boundary-layer edge.

In this work, the local FSC are solved with MATLAB's bvp4c routine employing no-slip adiabatic boundary conditions

$$\left.\begin{aligned} \text{At the wall}(\eta = 0) : f = f' = g = q' = 0 \\ \text{Farfield}(\eta \to \infty) : f' = g = q = 1. \end{aligned}\right\} \tag{3}$$

In the case of zero pressure gradient and sweep angle ($\beta_H = 0$, $\chi_e = 0°$), the compressible local FSC equations reduce to the well-known similarity solution for compressible flat-plate boundary layers (here referred to as the compressible Blasius solution). A comparison of the current implementation of Eq. 1 with the velocity and temperature profiles obtained with an in-house compressible Blasius solver is shown in Fig. 1 for various Mach numbers. The in-house compressible Blasius code has been intensively verified and is frequently used to generate basic flows for instability analyses (e.g., Ref. [13]). The implementation of the compressible FSC equations is further verified with Qiao et al.'s [14] solution for a three-dimensional boundary-layer profile obtained for $Ma_e = 6$, $\beta_H = 0.182$, and $\chi_e = 55°$ (depicted in Fig. 2).

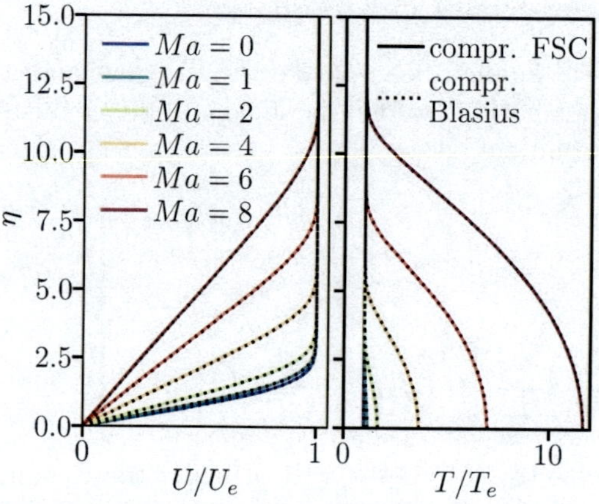

Fig. 1. Comparison of velocity and temperature profiles from the solution of the compressible FSC equations and an in-house compressible Blasius code with $\beta_H = 0$ and $\chi_e = 0°$.

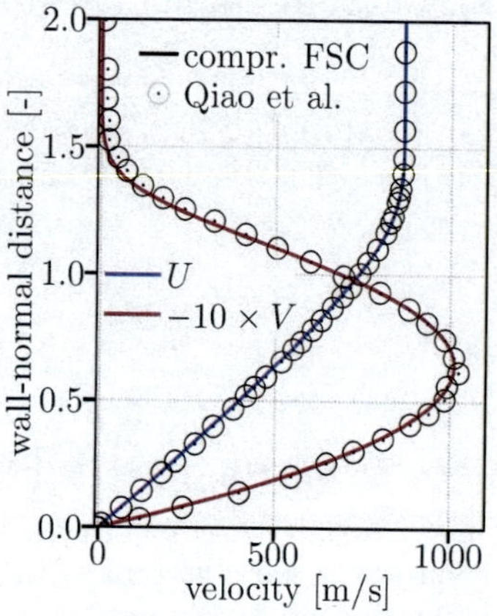

Fig. 2. Velocity profiles of present and Qiao et al.'s [14] compr. FSC solution (digitized); $Ma_e = 6$, $\beta_H = 0.182$, and $\chi_e = 55°$.

2.3 Local Linear Stability Theory and e^N Method

To predict the instability properties of the three-dimensional boundary layers generated by the compressible FSC code, the nonlocal linear stability code NOLOT [13] is used. However, this work utilizes only the classical quasi-parallel local linear approach. In the linear stability theory, a flow quantity $\mathbf{q}$ is divided into a steady basic flow $\bar{\mathbf{q}}$ and a disturbance flow $\tilde{\mathbf{q}}$. A modal disturbance approach is made to model the perturbed flow, which expresses the disturbance as a product of amplitude function $\hat{\mathbf{q}}$ and phase function Θ:

$$\left.\begin{aligned} \tilde{\mathbf{q}}(x_1, x_2, x_3, t) &= \hat{\mathbf{q}}(x_3)\Theta(x_1, x_2, t) + c.c. \\ \Theta(x_1, x_2, t) &= \exp\left(i\left[\alpha x_1 + \beta x_2 - \omega t\right]\right) \end{aligned}\right\} \tag{4}$$

with (x_1, x_2, x_3) being the chordwise, spanwise and wall-normal directions in the infinite swept-wing coordinate system, t the time, *c.c.* denotes the complex conjugate, and (α, β) are the wavenumber components in x_1 and x_2 direction, respectively. In this work, the spatial framework is used; therefore, ω is real and related to the circular frequency of an eigenmode $\tilde{\mathbf{q}}$, while the real part of α, α_r, is related to the physical wavenumber of the disturbance, whereas its imaginary part α_i represents the growth/damping rate in the x_1 direction. The spatial growth rate in the local instability framework is defined as $\sigma = -\alpha_i$ and the waveangle is determined by $\Psi = \arctan(\beta/\alpha_r)$.

The transition location can be determined by the semi-empirical e^N method [2,3]. In the e^N method, the spatial amplification rate of each amplified mode is integrated into a logarithmic amplification factor (N-factor)

$$N(x_1, \omega, \beta) = \int_{x_{1,0}}^{x_1} \sigma(x_1, \omega, \beta) \mathrm{d}x_1, \tag{5}$$

where $x_{1,0}$ denotes the position where first disturbances start to grow at a given frequency and transverse wavenumber. Transition is assumed at the location where the most amplified modes' N-factor reaches the prescribed threshold value N_{crit}, which, e.g., can be determined experimentally.

2.4 Radial Basis Function Interpolation

The surrogate model proposed in this paper rests on radial basis functions (RBF) [15] to predict the instability characteristics of disturbances in three-dimensional boundary layers (i.e., σ, and α_r). The RBF method is one of the simplest techniques for multidimensional data interpolation. Nevertheless, it features excellent isotropic approximation properties and is universally applicable regardless of the dimension d. For a given scattered data set with the sample points $\mathbf{p} \in \mathbb{R}^d$ and known function values $F(\mathbf{p})$, an RBF interpolant for an unknown query point $\mathbf{p_c}$ can be derived of the form

$$s(\mathbf{p_c}) = \sum_{i=1}^{n} w_i \phi\left(||\mathbf{p_c} - \mathbf{p}_i||_2\right), \tag{6}$$

where $s(\mathbf{p_c})$ is the interpolated function value at the query point, $||\cdot||_2$ is the Euclidean norm, w_i are the interpolation weights, and ϕ is the radial basis function. The basis function used in this work is a thin-plate spline function defined as $\phi(r) = r^2 \log r$.

The interpolation coefficients can be precomputed with the original data set by solving

$$\mathbf{F}(\mathbf{p}_j) = \phi\left(||\mathbf{p}_i - \mathbf{p}_j||_2\right) \mathbf{w}_j. \tag{7}$$

Hence, the interpolation weights solely depend on the Euclidean distance (i.e., r) between the i-th and j-th sample points. To avoid ill-conditioning of the basis function matrix, the sample points have been normalized with their maximum absolute value during the precomputation phase.

2.5 Sample Sets and Surrogate Models for TSI and CFI

This paper presents two separate databases, one for two-dimensional Tollmien-Schlichting waves (i.e., $\Psi = 0°$) and one for stationary cross-flow instabilities.

Thus, the databases can be used to determine the amplification rates and corresponding N-factors for two characteristic disturbance types typically encountered in the boundary layer of a swept-wing transport aircraft. Moreover, the capability to estimate separate N-factors for CFI and TSI is compatible with the two-N-factor transition prediction method, e.g., used by DLR [4].

In both databases, the instability results predicted with local stability theory for different compressible FSC similarity solutions are stored in terms of precomputed RBF interpolation weights ($\mathbf{w}$) for the relevant flow parameters ($\mathbf{p}$). However, the databases are presently limited to adiabatic wall temperatures, and the boundary-layer edge temperature is $T_e = 295.15\,\mathrm{K}$.

TSI Surrogate Model. To build a surrogate model for TSI characteristics, precalculated samples are required. The 2D TSI surrogate-based model input parameters include the boundary-layer edge Mach number Ma_e, the incompressible shape factor H_{12}, the displacement thickness Reynolds number Re_{δ_1}, and the nondimensional circular frequency ω. The output parameters are the spatial growth rate and the chordwise wavenumber $(\sigma, \alpha_r) = f\,(Ma_e, H_{12}, Re_{\delta_1}, \omega)$. The samples are generated by performing local stability analyses on a series of compressible two-dimensional local FSC solutions (i.e., $\chi_e = 0°$).

The considered range and the number of sampling levels of each input parameter are listed in Table 1. Overall, the TSI surrogate model consists of 2249775 sample points.

The precalculation, as well as the evaluation of the surrogate model, includes distance determinations to all sample points. Thus, the computational effort to solve Eq. 6 and Eq. 7 increases with sample size. To reduce the time required for surrogate model precalculation and evaluation, the samples are divided into different partitions according to their value. Specifically, the TSI database is split into 500 partitions, and the distribution of sample points among the different partitions is also given in Table 1.

Table 1. Input parameters and distribution of sample and partition levels of the TSI and CFI databases.

	TSI Database				CFI Database			
	Min	Max	Nr. Samples	Nr. Partitions	Min	Max	Nr. Samples	Nr. Partitions
Ma_e	0	1.4	15	5	0	1.4	15	6
χ_e	0°	0°	1	1	2°	88°	37	6
H_{12}	2.22	3.5	33	4	not used			
β_H	not used				−0.15	1.5	27	7
Re_{δ_1}	200	13000	45	5	not used			
Re_{CF}	0	0	1	1	11	15000	34	5
ω	0	0.5	101	5	0	0	1	1
$\beta\delta_1$	0	0	1	1	0	2.95	25	5

CFI Surrogate Model. For stationary CFI, the surrogate model can be represented as $(\sigma, \alpha_r) = f(Ma_e, \chi_e, V_{max}, Re_{CF}, \beta)$, where V_{max} is the maximum cross-flow velocity, and Re_{CF} is the cross-flow Reynolds number based on maximum cross-flow velocity and the boundary-layer height where the cross-flow velocity reduced to 10% of the maximum.

Although V_{max} is given as input to the surrogate model, instability results obtained for compressible FSC profiles at different β_H values were used to precompute the interpolation weights. The advantage of using V_{max} over β_H as an input is that V_{max} is a local flow quantity and, therefore, easier to extract than a pressure gradient at the boundary-layer edge. Moreover, if the basic flow consists of solely one boundary-layer profile, no pressure gradient can be determined. Since the surrogate model is based on compressible FSC equations, V_{max} is uniquely determined for given Ma_e, χ_e, and β_H.

The RBF-based surrogate model for CFI consists of 12737250 sample points divided into 5250 partitions. The considered range, the number of sampling levels of each input parameter, and the distribution of the partitions are provided in Table 1.

An advantage of the presented RBF surrogate methodology for CFI is that the N-factor integration with a fixed β value is compatible with wave propagation kinematics (at least for infinite swept wings).

To validate the proposed method, Fig. 3 shows the growth rates of stationary CFI predicted by the RBF surrogate model and LST for a 3D compressible local similarity boundary-layer profile not included in the sample set to build the surrogate model.

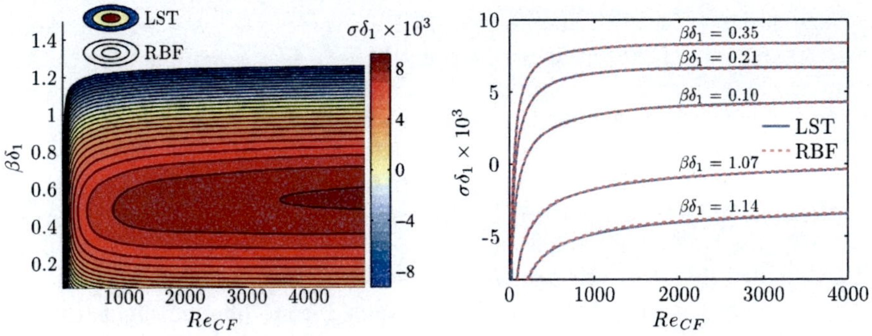

Fig. 3. Comparison of LST and RBF CFI growth rates for a 3D compressible local similarity solution with $Ma_e = 0.55$, $\chi_e = 33°$, and $V_{max}/U_e = -0.0355$.

3 Transition Prediction for ATTAS Flight Experiments

This section evaluates the proposed surrogate-based stability analysis and transition prediction methods for TSI and CFI for different ATTAS flight experiment test points. The ATTAS flight experiment [17] was conducted in 1989 with a glove mounted on the starboard wing of the ATTAS aircraft specially designed to develop and calibrate N-factor-based transition prediction tools. An overview of the flight test points studied in this paper with the respective flight conditions is given in Table 2. The selected test points in this work were chosen to cover different transition scenarios (i.e., KK20357: CFI dominated transition, KK20350: possible interaction of CFI and TSI, and KK19344: TSI dominated transition).

Table 2. Overview of ATTAS flight test points and freestream conditions.

Test point	effective sweep angle	Ma_∞	T_∞ [K]	Re
KK20357	23.1°	0.70	253.42	21.1×10^6
KK20350	20.1°	0.58	253.75	23.1×10^6
KK19344	21.9°	0.35	249.05	12.2×10^6

The laminar basic flows for the considered test points stem from boundary-layer computations with the COCO [16] code based on the measured pressure coefficients. The input parameters for both surrogate models (e.g., Ma_e, χ_e, H_{12}, or V_{max}) are directly extracted/computed from the COCO boundary-layer profiles.

Figure 4 shows the CFI and TSI N-factor curves obtained from RBF surrogate models and full LST computations, including the experimentally determined transition locations for all considered flight test points (x_{tr}). Overall, a good agreement of the N-factor results can be identified, especially for the cases with strongly amplified instability modes. Note that the RBF surrogate model is solely designed for 2D TSI with $\Psi = 0°$. However, modes with constant transverse wavenumbers are integrated for the LST N-factor computation so that Ψ varies along the x/c direction. Furthermore, the transverse wavenumbers are chosen so that at least $\Psi = 0°$ is valid at one point in the amplified region. Nevertheless, the agreement between LST and RBF TSI N-factors is very good over a wide range of the chord and overall deviations are acceptable. At the transition location, the difference in critical N-factor values is $\Delta N_{crit} \approx 0.27$ for KK19344 and $\Delta N_{crit} \approx 0.43$ for KK20350, which is considered as an acceptable deviation for surrogate-based transition prediction methods [10] and is well below the typical scatter of transition N-factor values obtained from different flight test data points. On the other hand, ΔN_{crit} for the cases with strongly amplified CFI is slightly higher, specifically $\Delta N_{crit} \approx 0.65$ for KK20357 and $\Delta N_{crit} \approx 0.58$ for KK20350. The larger difference is essentially due to the fact that in the attachment line region, the RBF amplification rates are underestimated, leading to lower maximum N-factors. This, in turn, is certainly a consequence of the basic flow's nonlocality, which cannot be addressed with the local similarity solutions used to build the surrogate model.

The time required with the RBF surrogate model to obtain the N-factor curves displayed in Fig. 4 is about 5% and 10% of the equivalent LST computing time for CFI and TSI, respectively. The time discrepancy between the two

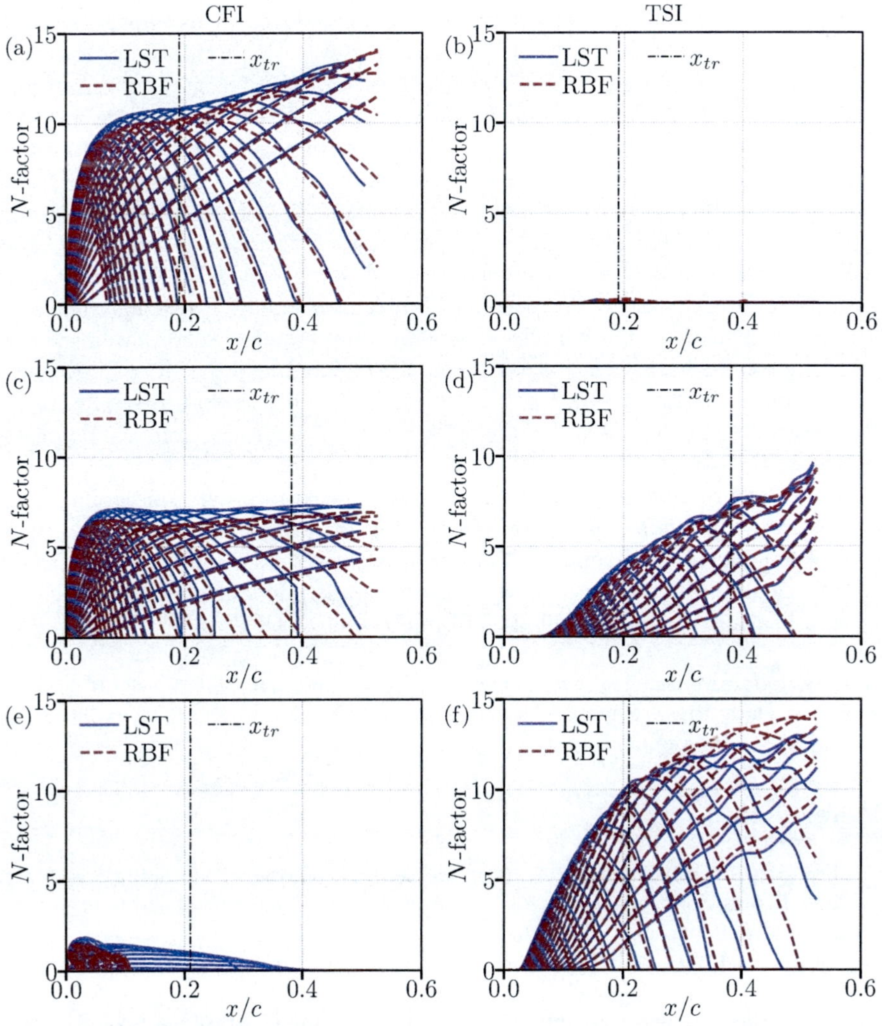

Fig. 4. Comparison of calculated (a, c, e) CFI N-factors and (b, d, f) TSI N-factors by RBF surrogate models and LST for three ATTAS flight test points; (a, b) KK20357, (c, d) KK20350, (e, f) KK19344.

databases is related to the different number of samples per partition. Therefore, the methodology can still be optimized in terms of prediction speed.

4 Conclusion

An efficient and robust surrogate method for transition prediction in three-dimensional compressible boundary layers is presented in this paper. The technique rests on radial basis function (RBF) interpolations within a database of precomputed instability results for two-dimensional Tollmien-Schlichting waves (TSI) and stationary cross-flow instabilities (CFI) where the basic flow originates from compressible local Falkner-Skan-Cooke (FSC) similarity solutions. Given characteristic boundary-layer values, the RBF surrogate model provides amplification rates, which integrated together with an e^N method, can be used for transition prediction.

The suitability of the presented surrogate model is demonstrated by comparing RBF CFI and TSI N-factors to results obtained from local stability analyses (LST) for different ATTAS transonic flight test points. The RBF N-factors at the transition onset locations deviate only slightly from the LST solution. The deviations are attributed to the underlying local compressible FSC profiles needed to build the surrogate model (both TSI and CFI) and a slightly different N-factor integration strategy (TSI only). Future work should therefore precompute the RBF surrogate model based on instability results obtained for nonlocal boundary layers.

Overall, the presented methodology is well suited to provide instability characteristics for boundary layers on swept-wing transonic aircraft at a fractional cost of a complete LST analysis. Moreover, the presented RBF surrogate models can easily be extended to cover more relevant physical parameters such as surface heating/cooling or oblique TSI and traveling CFI.

Acknowledgements. The work has been performed within the DLR project: Adaptive Data-driven Physical Modeling towards Border of Envelope Applications (ADaMant). The support is gratefully acknowledged.

References

1. François, D.G., Krumbein, A., Krimmelbein, N., Grabe, C.: Simplified stability-based transition transport modeling for unstructured computational fluid dynamics. AIAA Paper 2022-1543 (2022)
2. Smith, A.M.O., Gamberoni, N.: Transition, pressure gradient and stability theory, Technical report ES 26388, Douglas Aircraft Co. (1956)
3. van Ingen, J.: The e^N method for transition prediction. Historical review of work at TU Delft. AIAA Paper 2008-3830 (2008)
4. Krimmelbein, N., Krumbein, A.: Automatic transition prediction for three-dimensional configurations with focus on industrial application. J. Aircr. **48**(6), 1878–1887 (2011)
5. Halila, G.L.O., Fidkowski, K.J., Martins, J.R.R.A.: Toward automatic PSE-based transition to turbulence prediction for aerodynamic flows. AIAA J. **59**(2), 462–473 (2021)
6. Drela, M.: Implicit implementation of the full e^N transition criterion. AIAA Paper 2003-4066 (2003)
7. Guillaume, B., Deniau, H., Vermeersch, O., Casalis, G.: Database approach for laminar-turbulent transition prediction: Navier-Stokes compatible reformulation. AIAA J. **55**(10), 3648–3660 (2017)
8. Crouch, J.D., Crouch, I.W.M., Ng, L.L.: Transition prediction for three-dimensional boundary layers in computational fluid dynamics applications. AIAA J. **40**(8), 1536–1541 (2002)

9. Zafar, M.I., et al.: Convolutional neural network for transition modeling based on linear stability theory. Phys. Rev. Fluids **5**(11), 113903 (2020)
10. Pinna, F., Zanus, L., Demange, S., Olazabal-Loumé, M.: Reduced model for transition prediction in hypersonic flows. AIAA Paper 2018-3697 (2018)
11. Nie, H., Song, W., Han, Z., Chen, J., Tu, G.: A surrogate-based e^N method for compressible boundary-layer transition prediction. J. Aircr. **59**(1), 89–102 (2022)
12. Liu, Z.: Compressible Falkner-Skan-Cooke boundary layer on a flat plate. Phys. Fluids **33**, 126109 (2021)
13. Hein, S., Bertolotti, F.P., Simen, M., Hanifi, A., Henningson, D.: Linear nonlocal instability analysis - the linear NOLOT code. DLR-IB 223-94 A56 (1995)
14. Qiao, L., Xu, J., Bai, J., Zhang, Y.: Fully local transition closure model for hypersonic boundary layers considering crossflow effects. AIAA J. **59**(5), 1692–1705 (2021)
15. Buhmann, M.D.: Radial basis functions. Acta Numer. **9**, 1–38 (2000)
16. Schrauf, G.: COCO: a program to compute velocity and temperature profiles for local and nonlocal stability analysis of compressible, conical boundary layers with suction. ZARM, Center of Applied Space Technology and Microgravity (1998)
17. Horstmann, K.H., Redeker, G., Quast, A., Dreßler, U., Bieler, H.: Flight tests with a natural laminar flow glove on a transport aircraft, AIAA Paper 90-3044-CP (1990)

Overview of the Wind Tunnel Tests with Suction in the DNW-NWB

Richard von Soldenhoff(✉)

German Aerospace Center (DLR), Institute of Aerodynamics and Flow Technology, Lilienthalplatz 7, 38108 Braunschweig, Germany
Richard.vonSoldenhoff@dlr.de

Abstract. In the low speed wind tunnel Braunschweig (DNW-NWB), three measurement campaigns with sucked boundary layer for laminarisation have taken place. Their aim was to improve effectiveness and efficiency of HLFC systems on the path to industrial applications. The present article gives an overview over the modular wind tunnel model, the conducted measurement series and the obtained results. Multiple suction surfaces were used. Conventional approaches with laser drilled cylindrical perforations are compared with recently developed suction surfaces, using a fine etched metal foil and a supporting mesh structure. Also two layouts of suction slots in contrast to the cylindrical perforations are considered. All investigated surfaces turned out to be suitable for transition delay. Differences of the effectiveness between the suction panels could be found depending on the suction rate. Moreover, numerical investigations are presented, allowing a deeper insight into aerodynamical processes of unstable boundary layer modes. The focus was on the deviation of experimentally measured transition lines and numerical predictions by an e^N method with regard to effectiveness of suction and predictability.

Keywords: Hybrid Laminar Flow Control · TSSD · Slotted Suction Surfaces

1 Introduction

Despite the fact that the theoretical background of boundary layer suction for laminarization is well known since the 1940s [12], the technology is far from being mature for wider industrial application. Since the reduction of fuel consumption and emissions is becoming more and more important, hybrid laminar flow control (HLFC) is a promising method to achieve this goal. Major challenges are among others the manufacturing of effective suction surfaces, ensuring a suitable suction distribution along the chord and operational issues like clogging of pores.

At DLR, wind tunnel experiments are carried out in the low speed wind tunnel Braunschweig (DNW-NWB) to investigate aeroynamic characteristics of various suction surfaces and to widen the knowledge about the aerodynamic behaviour of a boundary layer stabilized by suction. A laser drilled surface as well

A. Dillmann et al. (Eds.): STAB/DGLR Symposium 2022, NNFM 154, pp. 486–495, 2024.
https://doi.org/10.1007/978-3-031-40482-5_46

as fine etched and slotted surfaces are taken into account. Future suction systems can be improved to achieve a better overall efficiency based on the outcomes of the presented studies. The wind tunnel model is a modular flat plate which allows rapid exchange of suction panels and instrumentation. Different suction panels can be easily compared at exactly the same conditions with respect to their technological efficiency. The tests are well suited to compare experimental and numerical results. In this way, linear stability theory (LST) methods can be improved for better transition prediction with the use of boundary layer suction.

The initial tests have taken place in 2019. Their aim was to investigate the overall aerodynamic effects of the boundary layer with suction and the development of Tollmien-Schlichting waves (TS waves), which lead to transition in the two-dimensional case. Two laser drilled suction panels were used (see Sect. 2). A detailed evaluation of the campaign can be found in [8].

In 2021, suction efficiency tests with a newly developed suction surface were conducted, mainly as a risk mitigation test [15]. The TSSD (Tailored Skin Single Duct) surfaces have been designed at DLR to simplify the overall suction system by eliminating the need for seperate suction chambers. Any variable suction along the chord is realised by a variable pressure loss through the surface [6]. The TSSD surfaces will be described in detail in Sect. 2.2. The wind tunnel campaign showed that the TSSD surfaces have a better effectiveness than the previously investigated laser drilled surfaces at lower suction rates.

In the first two campaigns, it was found that transition in cases with suction occurs earlier than predicted by LST methods. The destabilizing effects are an interaction between the Tollmien-Schlichting waves and the suction chamber flow as well as three-dimensional disturbances caused by every discrete suction orifice. The results led to the development of slotted TSSD surfaces, successfully tested in the campaign in 2022, showing even better performance than microperforated TSSD surfaces for certain slot layouts. Future tests will investigate the observed effects in detail. In addition, optimized new suction surfaces will be considered.

2 Experimental Setup

In the following part, the wind tunnel test section, the model and the measuring techniques are described. The TSSD suction surface concept, which was investigated in the wind tunnel tests, is briefly presented.

2.1 DNW-NWB Wind Tunnel

The DNW-NWB is a Göttingen type wind tunnel, operated either with a closed or an open test section. The closed test section, used for the presented tests, has a width of 3.25 m, a height of 2.8 m and a length of 8 m [3]. Inflow velocities up to 90 m/s are possible, resulting in a maximum unit Reynolds number of $Re = 6.2 \times 10^6$/m. Due to the low turbulence level of this facility, it is well suited for transition measurements. The critical N-factor (see [7]) for TS instabilities is close to the one for free flight experiments [11].

2.2 Modular Flat Plate Model and Suction Panels

The wind tunnel model, as shown in Fig. 1, is a flat plate with an aerodynamically shaped nose and an adjustable trailing edge flap. The chord length is 5 m and the span 2.8 m. At the maximum chosen speed of 60 m/s, the chord Reynolds number amounts to $Re = 20.5 \times 10^6$. In the 500 mm wide central section, surface elements can be flexibly exchanged (see Fig. 2). This way, different suction surfaces can be inserted at different chord positions. Furthermore, this modularity enables inserting a hot wire instrumented panel at various positions, which can also be seen in Fig. 1.

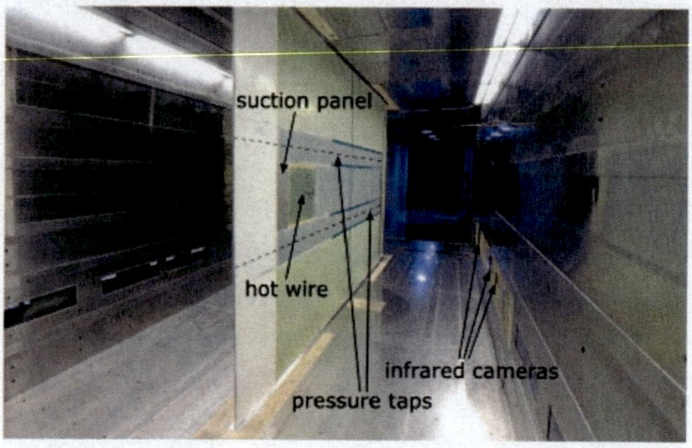

Fig. 1. HLFC flat plate model in the DNW-NWB, from [8]

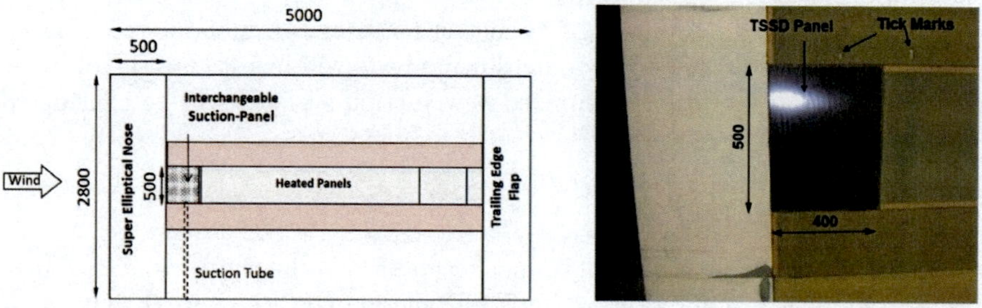

Fig. 2. Sketch of the flat plate with dimensions (left), inserted suction panel (right), from [15]

The inner model structure consists of an aluminium frame, on which GRP (glass reinforced plastic) surface elements and suction panels are mounted. All instrumentation apart from infrared cameras is located inside the model.

Standard suction panels dimensions are 400 mm × 500 mm. Their inner structure can be seen in Fig. 3(a). Below the porous surface, internal suction chambers are separated by brass profiles. The pressure is equal in each suction chamber. This leads to a homogeneous suction distribution.

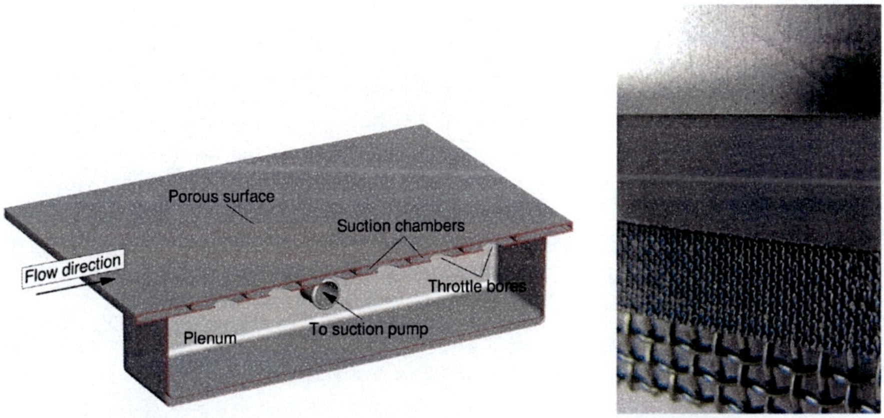

(a) Cross section of a suction panel

(b) Layers of a TSSD surface, according to [6]

Fig. 3. Suction surface concepts for DNW-NWB tests

As suction surfaces, two different types are used. For the first campaign, only a 0.6 mm thick laser drilled sheet metal surface was used. In the subsequent experiments, also various TSSD surfaces including microperforated and slotted surfaces were investigated. The TSSD concept is developed at DLR in the scope of the EU project NACOR. It consists of a 50 µm thick fine etched foil, which is stiffened by multiple wire mesh layers (see Fig. 3(b)). The finest wire mesh with pore sizes of ~10 µm leads to the desired pressure loss to ensure the suction distribution along the chord through a variation of the mesh size [6]. All panels have a porosity of 0.785%. The cylindrically perforated panels have pores with a diameter of 50 µm, while the slotted panels have 60 µm wide slots with 2 mm spanwise extent.

2.3 Measurement Procedures and Post-processing

For the detection of transition lines, infrared measurements are used. Three infrared cameras are located outside the wind tunnel wall (seen in Fig. 1). Turbulent boundary layers show stronger heat transfer to the wall compared to laminar boundary layers. These characteristics result in different wall temperatures in laminar and turbulent regions. The contrast of infrared images is increased by using wall heating. In the central section of the model, as seen in Fig. 2, heated GRP panels are inserted. An example infrared image will be shown in Sect. 4.

In the boundary layer, hot wire anemometry measurements are conducted. For measuring at different wall distances, a traverse is used, which is mounted on an additional GRP panel (see Fig. 1). The hot wire is positioned over the suction panel, directly behind the suction region as well as further downstream.

The suction rate is determined by using a Laminar Flow Meter (LFM). This device measures the pressure loss in a laminar flow tube in a highly accurate way

[5]. For the calculation of the suction coefficient $c_Q = v_W/U_\infty$ compressibility in the LFM tube is taken into account instead of using the linear Hagen-Poiseulle law, including additional tube-end corrections.

To be able to recalculate instability growth by LST methods, the pressure distribution is detected along the chord in two cross sections. The boundary layer profiles along the surface of the model are recalculated by the boundary layer code *COCO* based on a pressure distribution extracted from numerical simulations. These BL profiles are inputs for the stability calculations with the compressible LST solver *LILO* [14], determining the amplification rates. The transition position is predicted by integrating local amplification rates along the chord. An assumption made in [7] and others is that transition occurs at a given ratio of initial TS wave amplitude and the amplitude at transition location. The amplitude factor e^N at which transition occurs, is empirically determined for each test facility. In Fig. 4, an exemplary resulting N-factor diagram is shown for a case at $U_\infty = 60\,\mathrm{m/s}$ without suction. The transition position, which is determined from infrared measurements, is drawn into the diagram. A mode with a frequency of 1100 Hz leads to transition.

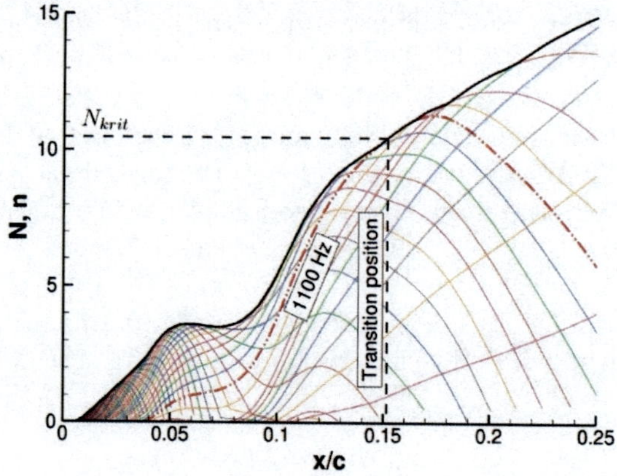

Fig. 4. Determination of the critical N-factor with *LILO*

3 Measurement Series

The measurement series of all wind tunnel campaigns are basically similar to each other. In a first step, the angle of attack and the flap deflection angle are adjusted to match the numerically calculated pressure distribution. For reference cases, a closed surface panel is inserted to measure the natural transition line without a suction panel. Critical N-factors N_{crit} are determined for these cases to compare with those with suction. The considered inflow velocities are $20\,\mathrm{m/s}$, $30\,\mathrm{m/s} \ldots 60\,\mathrm{m/s}$.

For all tested suction panels, a series of suction variations is conducted for each velocity. The suction coefficients range up to $c_Q = 16 \times 10^{-4}$, allowing a comparison between individual suction panels.

Additionally, tapes between 60 µm and 80 µm are placed spanwisely at different streamwise positions on the model nose and in the suction region. A tape can be seen as a combination of a forward facing and a backward facing step. The measurements allow the determination of an additional ΔN resulting from the steps. These ΔN are afterwards compared to the numerical investigations in [9].

Hot wire data is only acquired for single cases, because a transition measurement with the infrared cameras is not always possible with the hot wire installed. Cases with and without suction are considered. For the investigation of slotted TSSD surfaces, a series of hot wire measurements was conducted including the intermittent transitional region of the boundary layer by increasing suction to shift the transition position beyond the hot wire position.

4 Results

In this section, the major outcomes of all three wind tunnel campaigns are presented. Results from the initial tests (2019) have been published in [8], findings from the TSSD risk mitigation tests (2021) are mainly taken from [13].

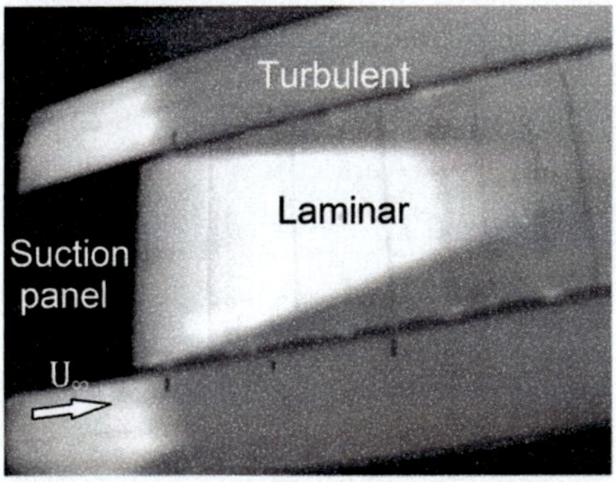

Fig. 5. Infrared image; $U_\infty = 50\,\text{m/s}$, $c_Q = 2.0 \times 10^{-4}$

Figure 5 shows the transition line of a case with suction through a slotted TSSD panel. It is clearly demonstrated that the transition is significantly delayed compared to the natural transition line, which is visible above and below the suction panel. Furthermore, an increase of suction moves transition further downstream, until a maximum delay is reached at strong suction rates above $c_Q = 2 \times 10^{-4}$. This behaviour holds for all investigated suction surfaces and matches the expectations from transition predictions with LST methods. The transition Reynolds numbers Re_{trans} are shown in Fig. 6 for two different suction surfaces.

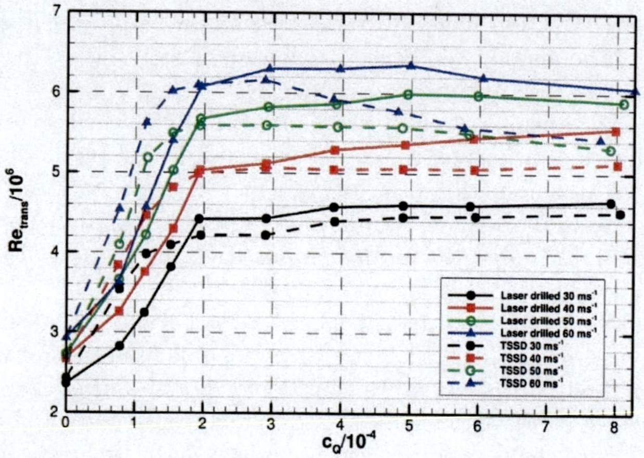

Fig. 6. Transition Reynolds number for different suction surfaces and different velocities, for laser-drilled surface and TSSD panel, according to [13]

For suction rates up to $c_Q = 2 \times 10^{-4}$, Re_{trans} increases nearly linearly with increasing suction, especially for TSSD panels. Transition Reynolds numbers are larger for the TSSD panel, consequently, the TSSD panel is more effective at lower suction rates than the laser drilled surface. The increase of Re_{trans} levels off at higher suction rates and reaches a maximum. For the TSSD surface, Re_{trans} even decreases with increasing suction; it is less effective at high suction rates.

The measured transition positions are compared with LST calculations with the boundary layer solver *COCO* and the stability solver *LILO*, see Sect. 2.3. Critical N-factors are determined for different velocities and suction rates. The results are shown in Fig. 7. For the laser drilled surface, N_{crit} drops significantly as soon as suction is applied. For higher suction rates, the value reaches a level at about $\Delta N_{crit} = 2$ below the critical N-factor without suction. For TSSD surfaces, the critical N-factor slightly increases for low suction rates up to $c_Q = 1.2 \times 10^{-4}$, then drops to even lower values than for the laser drilled surface. At $c_Q = 2 \times 10^{-4}$, the N_{crit} values of both surfaces are nearly equal.

Also noticeable is the fact that the N-factors for 60 m/s are lower than for 40 m/s. A possible reason for this behaviour is a stronger impact of disturbances on the thinner boundary layer, leading to premature transition.

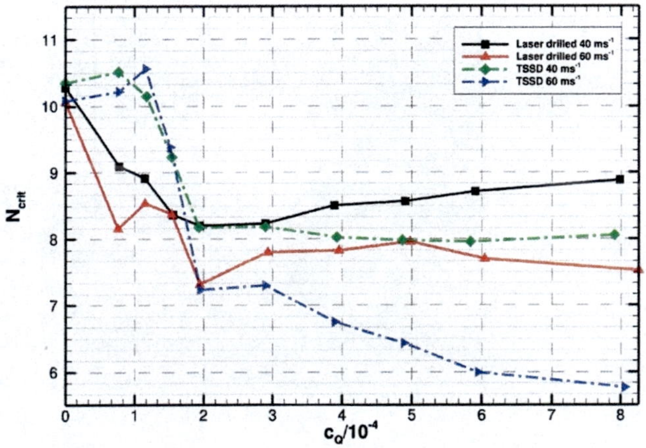

Fig. 7. Critical N-factors with applied suction, according to [13]

Step influences from tapes placed in the nose region are investigated. It shows that even a step height of 80 µm leads to a forward transition shift. The critical N-factor is significantly reduced by values around 1 [13]. Steps near the indifference point have a much higher impact on the transition shift than steps further downstream [8]. If the suction is strong enough to dampen Tollmien-Schlichting waves completely, the effect of the step is nearly cancelled out. In this case, the transition occurs very close to the transition position without the influence of a step.

In the wind tunnel campaign from 2022, slotted TSSD suction panels were investigated. The two different slot layouts show a significantly different effectiveness. While one leads to lower transition Reynolds numbers than the microperforated TSSD surface, the other one moves the transition even further downstream, especially for higher suction rates. The phenomena will be further investigated numerically as well as in a subsequent wind tunnel campaign.

5 Numerical Studies

This section gives a brief overview over numerical investigations related to the wind tunnel tests. More detailed descriptions can be found in [2,10] and [9].

The influence of an interaction between boundary layer and chamber flow on TS wave growth is investigated using temporally resolved 2D simulations. As flow solver, the DLR code *FLOWer* [1,4] in its 4th order version with implicit Padé filtering is used. The computational domain contains the boundary layer, a porous wall with discrete slots and a suction chamber. TS waves are introduced by an initial perturbation and develop temporally in the prescribed boundary layer. This is shown in Fig. 8. The disturbance from the boundary layer extends into the chamber. It can be observed, that the growth rate strongly depends on the slot dimensions. Wider slots and thin surfaces increase TS wave growth.

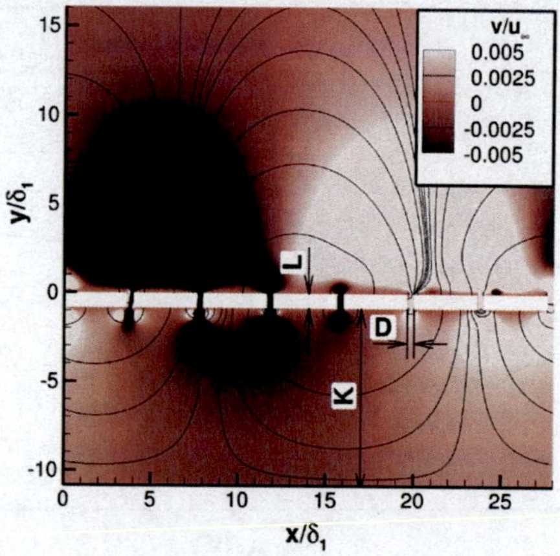

Fig. 8. Wall-normal velocity and isobars of a TS wave and the chamber flow disturbance

These findings can be used for the design of new suction surfaces, where a weak amplification of instabilities is favourable.

Furthermore, 3D RANS simulations of the flow characteristics in the vicinity of suction orifices are conducted. Their aim is to quantify the pressure loss over the surface under consideration of an incoming boundary layer as well as three-dimensional disturbances forming behind the orifices. Derived from this, the effect of longitudinal vortices on TS wave development will be investigated in three-dimensional direct numerical simulations.

6 Conclusions and Outlook

The presented wind tunnel tests allow deeper insight into the characteristics of different suction surfaces. It was shown, that the recently at DLR developed TSSD concept has a good effectiveness, especially at lower suction rates. At high suction rates, the transition position is found further upstream compared to the transition position calculated with LST methods assuming the critical N-factor for a closed surface. The destabilizing effects have been investigated with numerical simulations. An unfavourable interaction between boundary layer and suction chamber was found, which can be mitigated by a suited design of the suction surface. Based on the results, LST methods can be modified to take these effects into account, so that transition predictions become more reliable.

Further wind tunnel tests with the modular flat plate are planned. Besides testing improved TSSD suction surfaces, additively manufactured suction panels will be investigated. Additionally, aerodynamic effects such as suction in the non-linear region of instabilities will be considered. For this, an array of hot film sensors will be installed which allows temporal and spatial resolution of TS wave packets.

Acknowledgements. Many people have contributed to the wind tunnel tests and their evaluation: Heinrich Lüdeke, Arne Seitz, Thomas Kilian, Albert Küpper, Thomas Löser, Jannik Bruns, Jens Friedrich, Martin Bauer, Christian Breitenstein, Simon Schlüter, Florian Siebert and many others.

I would like to acknowledge the funding by the Deutsche Forschungsgemeinschaft (DFG, German Research Foundation) under Germany's Excellence Strategy – EXC 2163/1 - Sustainable and Energy Efficient Aviation – Project-ID 390881007.

References

1. Aumann, P., et al.: FLOWer: installation and user manual. DLR Braunschweig (2008)
2. Bauer, M., Lüdeke, H.: Simulation von Absaugbohrungen für Laminartechnologien unter Berücksichtigung der Anström-Grenzschicht. Deutscher Luft- und Raumfahrtkongress (2017)
3. DNW: Niedergeschwindigkeits-Windkanal Braunschweig. DNW Homepage. https://www.dnw.aero/wind-tunnels/nwb/
4. Enk, S.: Zellzentriertes Padéverfahren für DNS und LES. Institut für Aerodynamik und Strömungstechnik, DLR Braunschweig, Diss. (2015)
5. von Geyr, H.: VER^{2}SUS – Schlussbericht. Technical report, DLR Braunschweig (2015)
6. Horn, M.: Cost-effective HLFC design concept for transport aircraft. Deutscher Luft- und Raumfahrtkongress (2019). https://doi.org/10.25967/490059
7. van Ingen, J. L.: A suggested semi-empirical method for the calculation of the boundary layer transition region. In: Report VTH-74, Vliegtuigbouwkunde, Technische Hogeschool Delft (1956)
8. Lüdeke, H., Breitenstein, C.: Experimental investigation of hybrid laminar flow control by a modular flat plate model in the DNW-NWB. CEAS Aeronaut. J. (2021). https://doi.org/10.1007/s13272-021-00564-0
9. Lüdeke, H., von Soldenhoff, R.: Direct numerical simulation of TS-waves over suction panel steps from manufacturing tolerances. CEAS Aeronaut. J. (2021). https://doi.org/10.1007/s13272-021-00496-9
10. Lüdeke, H., von Soldenhoff, R.: Entwicklung und Erprobung neuartiger Absaugoberflächen zur Laminarhaltung der Grenzschicht. Deutscher Luft- und Raumfahrtkongress (2022)
11. Rohardt, C.-H., Küpper, A.: Determination of the N_{TS}-factor for the DNW-NWB. In: 19th DGLR-Fach-Symposium STAB, 4.-5 (2014)
12. Schlichting, H., Gersten, K.: Grenzschichttheorie, 10th edn. Springer, Berlin (2006). https://doi.org/10.1007/3-540-32985-4
13. Schlüter, S.: Auswertung von TSSD Absaug-Experimenten an einem modularen Plattenmodell im DNW-NWB. DLR Braunschweig, Student research paper (2022)
14. Schrauf, G.: LILO 2.1 users guide and tutorial. GSSC Technical Report 6 (2006)
15. Seitz, A., et al.: Wind tunnel verification of laminar boundary layer control TSSD concept. In: AIAA Aviation Forum (2022). https://doi.org/10.2514/6.2022-3552

Flow in a Concentric Annulus with Inner- or Outer-Cylinder Rotation: A Conventional and Sensitized RANS Modelling Study

Xiaoyu Wang, Jeanette Hussong, and Suad Jakirlic(✉)

Institute of Fluid Mechanics and Aerodynamics (SLA), Technical University of Darmstadt, Alarich-Weiss-Straße 10, 64287 Darmstadt, Germany
jakirlic@sla.tu-darmstadt.de

Abstract. A conventional differential, near-wall Reynolds stress model (RSM) and its eddy-resolving version, sensitized appropriately to account for the turbulence unsteadiness (termed as Improved Instability- Sensitized RSM model - IIS-RSM), are applied within the unsteady Reynolds-Averaged Navier Stokes (RANS) computational framework to simulate the annular flow formed by two concentrically arranged cylinders. Two characteristic situations, in which the rotation of the inner cylinder by a constant angular velocity and that of the outer cylinder were considered individually, were studied in a range of rotational intensities. The corresponding Reynolds numbers based on the hydraulic diameter and axial bulk velocity of amount to 8900 and 12500 for the inner and outer cylinder rotations, respectively. The scale-supplying equation governing the inverse turbulent time scale relies on the 'homogeneous dissipation' rate ($\omega_h = \varepsilon^h/k$). The eddy-resolving capability of the IIS-RSM is enabled by a selective enhancement of the turbulence production by introducing an additional production term in the length-scale determining transport equation, in accordance with the Scale-Adaptive Simulation (SAS) strategy. The results for the cases with inner cylinder rotation obtained by both model versions show substantial mutual agreement and agree well with the reference LES data. The flow driven by the outer cylinder rotation at higher rotational rates is correctly predicted by the IIS-RSM model, in contrast to the conventional RANS approach resulting in premature relaminarization.

Keywords: Turbulent rotating annular flow · Reynolds-stress turbulence model · Eddy-resolving Unsteady RANS

1 Introduction

Flow in a concentric annulus influenced by wall rotation is of importance in various engineering applications, such as heat exchangers, chemically reactive mixers, journal bearings and turbomachinery. Modeling such applications requires

A. Dillmann et al. (Eds.): STAB/DGLR Symposium 2022, NNFM 154, pp. 496–505, 2024.
https://doi.org/10.1007/978-3-031-40482-5_47

physical insight into turbulent rotating flows in concentrically arranged annuli over a range of rotational intensities. The qualitative topology of the flow within the gap between two cylinders resembles the form found in the case of the so-called Taylor-Couette flow (see e.g., Grossmann et al. [3] and Berghout et al. [1]). Depending on the operational conditions - among others, the rotation number $N = U_{\theta,wall}/U_b$ (ratio of the velocity of the rotating wall to the axial bulk velocity) and the radial gap width ($2\delta = R_o - R_i$) in terms of the curvature parameter $R^* = R_i/R_o$, the ratio of the inner (R_i) to the outer (R_o) cylinder radius - the velocity field can be subject of patterns of different complexity. They are characterized by alternately varying velocity gradients, both in terms of intensity and sign. As is well known, the underlying axial pressure gradient has a destabilizing effect on the near wall turbulence structure in a concentric annular flow with the inner rotating wall. In contrast, annular flow is prone to be stable, with a tendency to laminarize as rotational velocities increase in the case of outer cylinder rotation.

In the present study, turbulent flow in a concentric annulus was simulated using a differential near-wall Reynolds stress model employed within a conventional RANS (Reynolds-Averaged Navier-Stokes) framework and its unsteady eddy-resolving version to elucidate the effects of the individual rotation of the inner wall as well as of the outer wall on the near-wall flow structure. A schematic representation of the flow configuration is displayed in Fig. 1. The curvature parameter $R^* = 0.5$ was adopted for all cases considered. Three cases of annular flow with inner cylinder rotation with gradually increased rotational rates ($N = 0.2145, 0.429$ and 0.858) and Reynolds number $Re_{D_h} = U_b D_h/\nu = 8900$, based on the axial bulk velocity (U_b) and the hydraulic diameter ($D_h = 2(R_o - R_i) = 4\delta$), were simulated. While for the case with outer cylinder rotation, the wall was rotated at $N = 0.0, 0.5, 1.0, 2.0$ and 2.8, keeping the Reynolds number constant at the value of $Re_{D_h} = 12500$. Reference Large-Eddy Simulations (LES) carried out under the condition of axially homogeneous flow have been performed by Chung and Sun [2] for the inner cylinder rotation and Hadziabdic et al. [4] for the outer cylinder rotation.

2 Computational Method

2.1 Turbulence Model Formulation

The continuity and momentum equations governing the incompressible flow in the concentric pipe annulus within the unsteady conventional/sensitized RANS framework read:

$$\frac{\partial U_i}{\partial x_i} = 0 \tag{1}$$

$$\frac{\partial U_i}{\partial t} + \frac{\partial (U_i U_j)}{\partial x_j} = -\frac{1}{\rho}\frac{\partial p}{\partial x_i} + \frac{\partial}{\partial x_j}\left(\nu \frac{\partial U_i}{\partial x_j} - \overline{u_i' u_j'}\right) \tag{2}$$

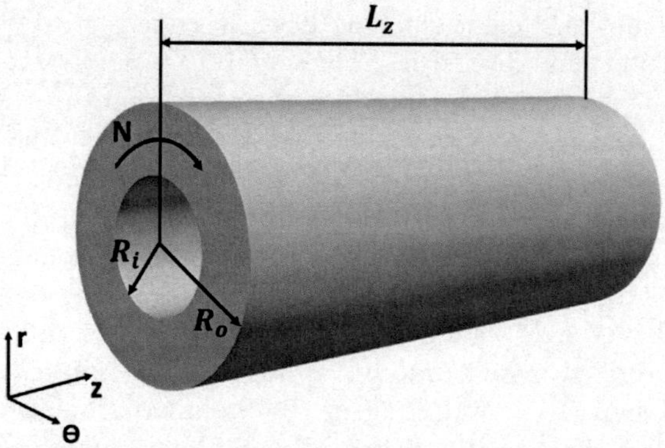

Fig. 1. Schematic representation of the flow configuration considered and the coordinate system adopted; $U_{\theta,wall} = \Omega R_i$ (inner cylinder rotation) and $U_{\theta,wall} = \Omega R_o$ (outer cylinder rotation), with Ω representing the angular velocity of the rotating wall

The Reynolds-stress tensor $\overline{u_i'u_j'}$ represents the solution of a differential near-wall second-moment closure model; furthermore, the corresponding scale-supplying equation for the inverse turbulent time scale $\omega_h = \varepsilon^h/k$ [7] is derived directly from the equation of the homogeneous dissipation rate ε^h ($\varepsilon^h = \varepsilon - 0.5\nu\partial^2 k/(\partial x_l \partial x_l)$, with $k = 0.5\overline{u_i'u_i'}$ and ε representing the kinetic energy of turbulence and its total viscous dissipation rate [5]). In the case of the computations with the conventional RSM, which models the entire turbulence spectrum and leads to a steady-state solution, all dependent variables (U_i, $\overline{u_i'u_j'}$ and ω^h) are actually time-averaged quantities. This RSM formulation, hereafter referred to as 'conventional RSM', has been further extended to adequately capture fluctuating turbulence. The eddy-resolving capability of the model is achieved by a selective enhancement of the turbulence production by introducing an additional production term in the length-scale determining transport equation, denoted by $P^*_{\Delta U}$ in Eqs. (3–5) below. Its functional dependency on the second derivative of the underlying velocity field is motivated by the Scale-Adaptive Simulation (SAS) strategy by Menter and Egorov [8]. The resulting model is denoted as improved instability-sensitive RSM (IIS-RSM). Accordingly, all flow quantities are time-dependent; the conventional RSM equations (Eqs. 3–4, with $P^*_{\Delta U} = 0$) transform into SAS-related equations describing the dynamics of the unresolved sub-scale motion only. The correspondingly large part of the turbulence spectrum is resolved and originates from the non-linear convective term in the equation of motion (Eq. 2).

The model equations governing the dynamics of the Reynolds stress tensor $\overline{u_i'u_j'}$ and the specific dissipation rate ω_h read:

$$\frac{\mathrm{D}\overline{u_i'u_j'}}{\mathrm{D}t} = \frac{\partial}{\partial x_k}\left[\left(\frac{1}{2}\nu + \frac{\nu_t}{\sigma_{\overline{u_i'u_j'}}}\right)\frac{\partial \overline{u_i'u_j'}}{\partial x_k}\right] + P_{ij} - \varepsilon^h_{ij} + \Phi_{ij} \tag{3}$$

$$\frac{\mathrm{D}\omega_h}{\mathrm{D}t} = \frac{\partial}{\partial x_k}\left[\left(\frac{1}{2}\nu + \frac{\nu_t}{\sigma_\omega}\right)\frac{\partial \omega_h}{\partial x_k}\right] - C_{\omega,1}\frac{\omega_h}{k}\overline{u_i' u_j'}\frac{\partial U_i}{\partial x_j} - C_{\omega,1}\omega_h^2$$
$$+ \frac{2}{k}\left(C_{cr,1}\frac{1}{2}\nu + C_{cr,2}\frac{\nu_t}{\sigma_\omega}\right)\frac{\partial \omega_h}{\partial x_k}\frac{\partial k}{\partial x_k} + 2\nu\frac{\nu_t}{k}\frac{\partial^2 U_i}{\partial x_j \partial x_l}\frac{\partial^2 U_i}{\partial x_j \partial x_l}$$
$$+ C_r \max[P^*_{\Delta U}, 0] \tag{4}$$

Here, $P_{ij} = -\overline{u_i u_l}\partial U_j/\partial x_l - \overline{u_j u_l}\partial U_i/\partial x_l$ and $\varepsilon^h_{ij} = \varepsilon_{ij} - 0.5\nu\partial^2\overline{u_i u_j}/(\partial x_l \partial x_l)$ represent the production and viscous dissipation rates of the corresponding Reynolds stress components, respectively; Φ_{ij} is the pressure redistribution term. Interested readers are referred to [7] for the full model specifications.

Instead of the SAS-model-related expression (formulated in conjunction with the $k-\omega$ SST model [8]) of the additional production term $P^*_{\Delta U}$ in the scale-supplying equation for ω^h, which is formulated as a 3-D generalization of the classical boundary layer definition of von Karman length ($L_{vK} \propto \nabla U_i/\nabla^2 U_i$), the nearly original form proposed by Rotta [9], which results from the equation for the integral length scale that depends only on the second derivative of the velocity field ($\nabla^2 U_i$), is currently used as the triggering parameter for generating the resolved motion ($C_r = 0.003$). Such a model term formulation also allows the use of even coarser grid resolutions.

$$P^*_{\Delta U} = 29.1\sqrt{\frac{\partial^2 U_i}{\partial x_j \partial x_k}\frac{\partial^2 U_i}{\partial x_l \partial x_l}}\sqrt{k} - 120k\max\left[\frac{1}{k^2}\frac{\partial k}{\partial x_j}\frac{\partial k}{\partial x_j}, \frac{1}{\omega_h^2}\frac{\partial \omega_h}{\partial x_j}\frac{\partial \omega_h}{\partial x_j}\right] \tag{5}$$

The IIS-RSM model formulation does not include any explicit grid-dependent parameter, such as grid-spacing, representing certainly an advantage over most hybrid LES/RANS models, especially when applying unstructured grids with arbitrary grid-cell topology.

2.2 Numerical Method

The model equations of both the conventional RSM and its eddy-resolving version IIS-RSM are implemented in the OpenFOAM® code (Version 19.12), with which all computations have been performed. The OpenFOAM® code represents an open-source computational fluid dynamics toolbox that utilizes a cell-centered finite volume method on an unstructured numerical grid. Grid generation was performed by using the OpenFOAM mesh utility 'blockMesh'. Appropriate cell grading towards the wall was provided to ensure that the wall-closest computational node is well within the viscous sub-layer, with the dimensionless wall distance y^+ taking values less than one on both walls. The discretization of the convective term is performed by applying a scheme that combines the 2nd order central-differencing (whose fraction corresponds to 90% and more) and 1st order upwind scheme (with the fraction corresponding to 10% and less); the diffusive term was discretized by the second-order central interpolation. For the temporal discretization, the second-order accurate backward scheme was used for the

IIS-RSM-related simulations; the temporal resolution corresponds to the maximum CFL number of 0.7. The steady computations were performed using the conventional RANS model.

For the conventional Steady-RANS model, a wedge-shaped solution domain was used for all computed flow configurations. It was meshed with only one cell in both circumferential and axial directions, with periodic boundary conditions applied in both directions, and with $N_r = 64$ cells in normal-to-wall radial direction. The corresponding length of the annulus was $L_z = \delta = 1/2(R_2 - R_1)$.

For the time-accurate simulations using the IIS-RSM the computational domain was expanded to $L_z = 20\delta$ aiming at correct capturing of the respective vortical structure elongation influenced by the rotational effects. For the cases with inner-cylinder rotation at $N = 0.2145$ and 0.429, the computational domain was adopted to be one-quarter of the full cross-section of the concentric annulus and is meshed with $(N_z, N_r, N_\theta) = (160, 64, 90)$ grid cells. However, for the highest internal rotation rate considered at $N = 0.858$ as well as for all cases with outer-cylinder rotation, the simulations were performed considering the full cross-section of the annulus, aiming at adequately accounting for the increasing complexity of the structural properties of the flow, with the grid resolution corresponding to $(N_z, N_r, N_\theta) = (160, 64, 360)$ cells.

3 Results and Discussion

3.1 Concentric Annulus with Inner-Cylinder Rotation

Figure 2(a) displays the streamwise velocity profiles and Fig. 2(b) the semi-log plots of the 'effective' velocity profile over the range of investigated rotational intensities, with the latter velocity accounting for both streamwise and circumferential velocity components; the modulus of the velocity vector $|U| = \sqrt{U_z^2 + (U_{\theta,wall} - U_\theta)^2}$ is normalized by the corresponding effective friction velocity at the inner rotating wall - $U_f = \sqrt{U_{f,z}^2 + U_{f,\theta}^2}$, $U_{f,z} = \sqrt{\nu(\partial U_z/\partial r)_{wall}}$, $U_{f,\theta} = \sqrt{\nu(\partial U_\theta/\partial r)_{wall}}$. The linear scaling (Fig. 2(a)) reveals a somewhat steeper velocity profile at the rotating wall, indicating a higher velocity gradient compared to that at the upper non-rotating wall. Apart from a slight deviation at the highest rotation number $N = 0.858$ pertinent to the RANS-RSM, the velocity profiles obtained by both model computations show a very good agreement with the reference LES by [2].

Comparison of the velocity profiles in the logarithmic domain in Fig. 2(b) shows a slight discrepancy between the RANS-RSM related profiles and the LES data in the cross-sectional region above $y^+ \approx 10$–20 for two higher rotation rates. The IIS-RSM profiles closely follow those of the LES reference. In general, it can be seen that as the rotation rate increases, the deviation of the velocity profiles in the buffer and logarithmic regions becomes more pronounced compared to the law of the wall adjacent to the rotating inner wall. This can be attributed to the effects of the increase in magnitude of the inner wall rotation, which results in an enhancement of the flow acceleration in the circumferential direction. In contrast,

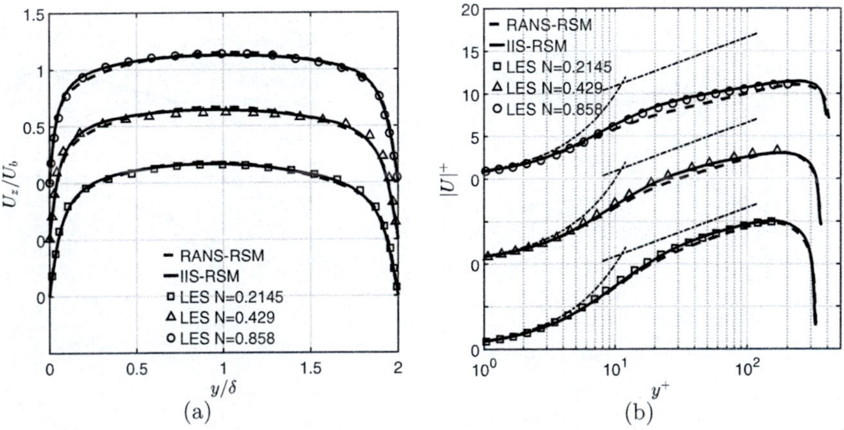

Fig. 2. Annular flow with inner-cylinder rotation: dependence of the axial (a) and 'effective' (b) velocity profile on the rotational intensity, shown in (a) linear and (b) semi-logarithmic (at the rotating wall) scaling

the velocity profiles after normalization by the frictional velocity at the outer wall (not shown here), where the effect of rotation is weak, show a behavior fully consistent with the logarithmic law ($U = \ln Ey^+/\kappa$) for all rotational intensities considered.

Further comparison of the results obtained with the RANS-RSM and IIS-RSM models is depicted in Fig. 3(a), showing the root-mean-square (r.m.s.) profiles of all three Reynolds stress components normalized by the bulk velocity for $N = 0.858$ along with the LES data. The asymmetry in the profiles affected by rotation is correctly reproduced by both turbulence models in very good agreement with the LES results, apart from the turbulence intensity in the radial direction $u'_{r_{rms}}$, calculated with the conventional RANS model exhibiting a significant overprediction near both walls; the application of the eddy-resolving IIS-RSM improved the prediction accordingly. Some slight deviations occur only in the near-wall region in the streamwise $u'_{z_{rms}}$ and azimuthal $u'_{\theta_{rms}}$ directions, where the component maximums are shifted toward the inner wall due to the enhanced destabilizing effect.

The Reynolds shear stress profile ($\overline{u'_r u'_z}$) development in terms of the rotational intensity increase, normalized by the square of the bulk velocity is shown in Fig. 3(b). While the conventional RANS results show acceptable qualitative agreement with the reference LES data, with a certain overprediction in the region near both walls, the IIS-RSM provides a significant improvement with respect to the systematic turbulence level enhancement near the inner cylinder walls; correspondingly, the shear stress component near the upper non-rotating wall experiences an almost constant peak value independent of the rotation rate. The improvement in comparison with the RANS-RSM is especially evident for $N = 0.858$, where the peak value related to the conventional RANS model exhibits large overprediction.

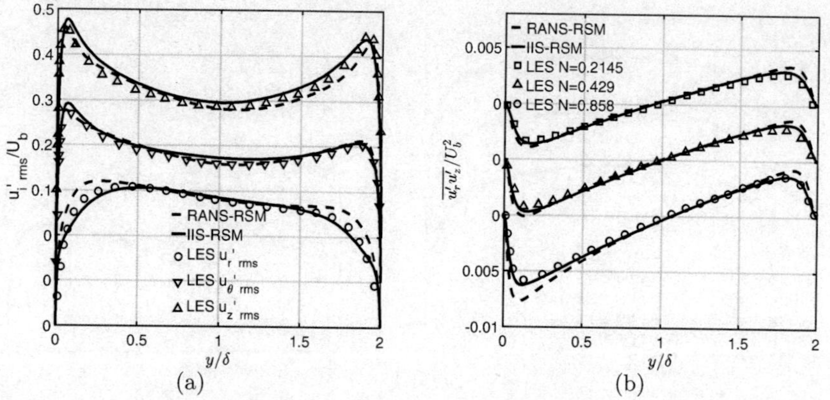

Fig. 3. Annular flow with inner-cylinder rotation: (a) normal Reynolds stress intensities at $N = 0.858$ and (b) Reynolds shear stress $(\overline{u_r' u_z'})$ components over the range of rotational intensities

3.2 Concentric Annulus with Outer-Cylinder Rotation

Figure 4 illustrates the profile development of the streamwise velocity component in the linear scaling and the mean effective velocity in the semi-logarithmic representation, with respect to the rotating outer wall (see discussion related to Fig. 2 for respective definitions), calculated with conventional RANS and IIS-RSM models in comparison with LES data from [4] for different rotation numbers.

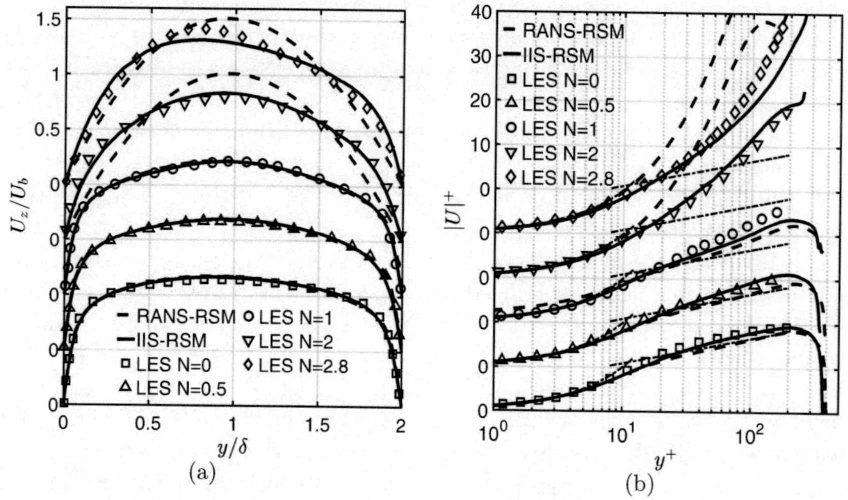

Fig. 4. Annular flow with outer-cylinder rotation: dependence of the velocity profile on the rotational intensity, shown in (a) linear and (b) semi-logarithmic (at the rotating wall) scaling

While the axial velocity distribution at the lower rotation rates $N = 0.5$ and 1 shows a symmetric profile with no significant deviation from the non-rotating case, the higher rotation intensity causes a significant turbulence suppression (Fig. 5) and the velocity profile tends to be quasi-laminar at the highest rotation rate $N = 2$ and 2.8, with the profile shape indicating intensified flow in the core region and weakening of the velocity gradient at the wall, Fig. 4(a). This laminarization tendency implies strong turbulence damping within the flow domain (Fig. 5) and thus poses a major challenge for the RANS method to capture the associated turbulence instability [6]. From Fig. 4(a), it is evident that despite the satisfactory agreement of the conventional RANS results with the LES data at the lower rotation rates $N = 0.5$ and 1, the axial velocity profile tends to become prematurely laminar at higher rotation rates $N = 2$ and 2.8. In contrast, this premature relaminarization was avoided by the IIS-RSM, where the velocity profiles at lower and higher rotation rates show reasonable qualitative and quantitative agreement with the LES data.

The tendency of the velocity profile to transform to a laminar-like form can also be seen in Fig. 4(b), which shows the logarithmic plot of the effective velocity normalized by the friction velocity evaluated at the outer rotating wall. While both turbulence models provide reliable results for the lower rotation rates $N = 0.5$ and 1, the difference in the model's predictive capability in capturing the velocity profiles that gradually approach the laminar limit at higher rotational intensities is obvious. Application of the conventional RANS model results in a fully laminar flow regime at $N = 2$ and 2.8, which is consistent with the previous discussion regarding premature relaminarization. In contrast, the velocity profiles predicted with the IIS-RSM returned a turbulent flow regime with velocity profiles closely resembling the reference LES with respect to both magnitude and profile shape, exhibiting a large overshot compared to the log-law.

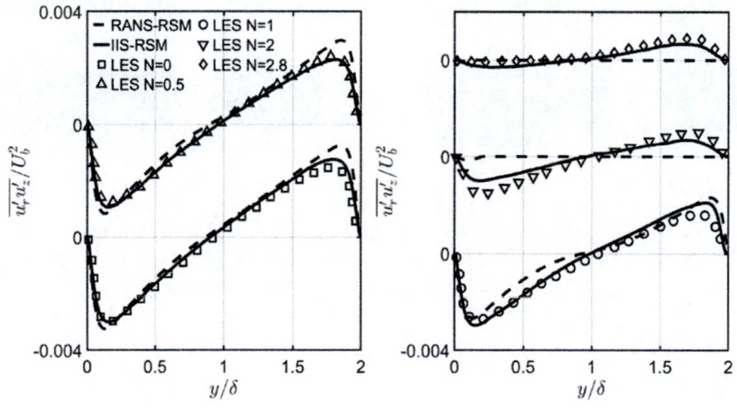

Fig. 5. Annular flow with outer-cylinder rotation: dependence of the Reynolds shear stress profile on the rotational intensity

3.3 Flow Structure, Inner-Cylinder Rotation vs. Outer-Cylinder Rotation

Further insight into the effects of rotation on the turbulent annulus flow is provided by visualizing the iso-contours of the instantaneous axial velocity field at the inner and outer walls at $y^+ \approx 5$ for cases with inner-cylinder rotation at $N = 0.858$ and outer-cylinder rotation at $N = 1$ in Fig. 6. The colors denote the normalized velocity values from 0.1 (blue) to 0.6 (red) to identify the intensity of the generated streaky structures. The qualitatively same pattern is observed in both rotating cases in the region adjacent to the walls, where the underlying vortex structures are suppressed in radial as well as stretched in axial and azimuthal directions due to the rotational effect. There is also an obvious difference between the two cases: in the annulus flow affected by the inner-cylinder rotation, the strength and inclination of the streaks are much more intense than in the flow with outer-cylinder rotation, despite a somewhat lower rotation rate (0.858 vs. 1), indicating the destabilizing effect of internal rotation.

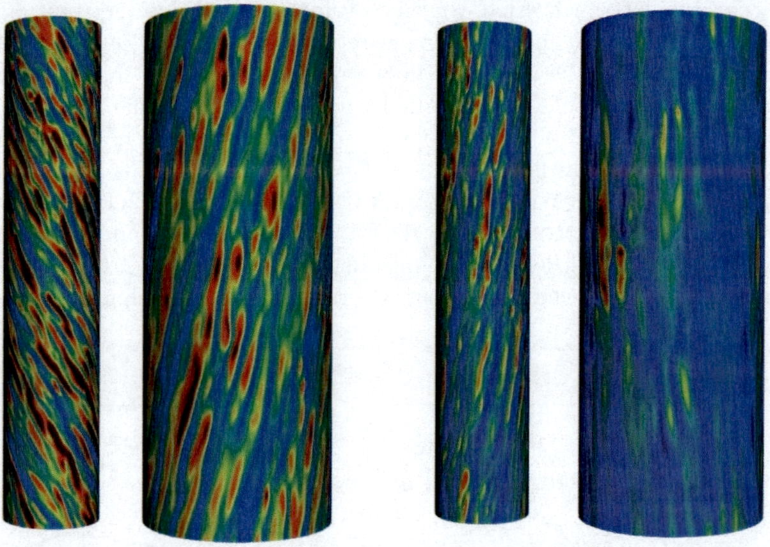

(a) (b)

Fig. 6. Iso-contours of the instantaneous axial velocity field in the annulus at the inner wall and outer wall at $y^+ \approx 5$ obtained by the present eddy-resolving IIS-RSM: (a) inner-cylinder rotation at $N = 0.858$ and (b) outer-cylinder rotation at $N = 1$

4 Conclusion

The predictive performance of the present conventional RANS model and its eddy-resolving version, appropriately modified to account for turbulence instability, was investigated for turbulent flow in a concentric annulus with an inner and outer wall rotation. The conventional RANS is capable of reproducing qualitatively and mostly quantitatively correct results in the case of the two configurations studied at lower wall rotation rates. For the case with rotation of the outer cylinder at higher rotation rates, the conventional RANS model incorrectly predicts the mean velocity distribution and thus the Reynolds shear stress component due to premature relaminarization. In contrast, the eddy-resolving IIS-RSM reproduces the mean flow and second-order statistical properties in good agreement with the LES reference data.

Acknowledgement. The authors gratefully acknowledge the German Federal Ministry of Education and Research (BMBF) and the federal state of Hessen for supporting this project as part of the NHR4CES (National High Performance Computing Center for Computational Engineering Sciences) funding at TU Darmstadt and RWTH Aachen. The authors furthermore acknowledge the computing time granted on the Lichtenberg High Performance Computer of the Technical University of Darmstadt.

References

1. Berghout, P., et al.: Direct numerical simulations of spiral Taylor-Couette flow. J. Fluid Mech. **887**, A18 (2020). https://doi.org/10.1017/jfm.2020.33
2. Chung, S.Y., Sung, H.J.: Large-eddy simulation of turbulent flow in a concentric annulus with rotation of an inner cylinder. Int. J. Heat Fluid Flow **26**(2), 191–203 (2005). https://doi.org/10.1016/j.ijheatfluidflow.2004.08.006
3. Grossmann, S., Lohse, D., Sun, C.: High-Reynolds number Taylor-Couette turbulence. Annu. Rev. Fluid Mech. **48**, 53–80 (2016). https://doi.org/10.1146/annurev-fluid-122414-034353
4. Hadžiabdić, M., Hanjalić, K., Mullyadzhanov, R.: LES of turbulent flow in a concentric annulus with rotating outer wall. Int. J. Heat Fluid Flow **43**, 74–84 (2013). https://doi.org/10.1016/j.ijheatfluidflow.2013.05.008
5. Jakirlic, S., Hanjalic, K.: A new approach to modelling near-wall turbulence energy and stress dissipation. J. Fluid Mech. **459**, 139–166 (2002). https://doi.org/10.1017/S0022112002007905
6. Jakirlic, S., Hanjalic, K., Tropea, C.: Modeling rotating and swirling turbulent flows: a perpetual challenge. AIAA J. **40**(10), 1984–1996 (2002). https://doi.org/10.2514/2.1560
7. Jakirlić, S., Maduta, R.: Extending the bounds of 'steady' RANS closures: toward an instability-sensitive Reynolds stress model. Int. J. Heat Fluid Flow **51**, 175–194 (2015). https://doi.org/10.1016/j.ijheatfluidflow.2014.09.003
8. Menter, F., Egorov, Y.: The scale-adaptive simulation method for unsteady turbulent flow predictions. Part 1: theory and model description. Flow Turbul. Combust. **85**, 113–138 (2010). https://doi.org/10.1007/s10494-010-9264-5
9. Rotta, J.C.: Turbulente Strömungen. Teubner Verlag Stuttgart (1972). ISBN 978-3-322-91207-7. https://doi.org/10.1007/978-3-322-91206-0

Technical Flows

DNS of Aerosol Particle Spreading Emitted by Coughing and Breathing in a Simplified Room

Ege Batmaz[1,2(✉)], Philipp Bahavar[1,2], Daniel Schmeling[1], and Claus Wagner[1,2]

[1] German Aerospace Center (DLR), Institute of Aerodynamics and Flow Technology, Bunsenstraße 10, 37073 Göttingen, Germany
{ege.batmaz,philipp.bahavar,daniel.schmeling,claus.wagner}@dlr.de
[2] Technical University of Ilmenau, Institute of Thermo- and Fluid Dynamics, Helmholtzring 1, 98996 Ilmenau, Germany

Abstract. Besides coughing and sneezing, breathing is the most frequent particle emission event of aerosol droplets carrying the SARS-COV-2 virus or viruses of other airborne diseases. Direct Numerical Simulations (DNS) of 'jet-like' emissions of particle clouds through the mouth caused by coughing and breathing are performed in a cuboidal simplified room to study the spreading of respiratory droplets with different momentum and size. Contrary to coughing, we found that no droplet follows a ballistic trajectory after a breathing event since all the droplets are trapped in the humid puff of air. The detailed analysis and the comparison of the predictions obtained for respiratory droplets emitted by single breathing and coughing events are further discussed. Despite the major difference between the maximum exhalation speeds reached during coughing and breathing, the horizontal propagation distance differs by less than 30%. Additionally, a comparison of the results of the present DNS neglecting aerosol evaporation and considering buoyancy forces with the results of an earlier DNS study from the literature taking evaporation into account but neglecting buoyancy, revealed that buoyancy damps the horizontal propagation of the humid puff and enhances the upward motion.

Keywords: COVID-19 · DNS · aerosol dynamics · coughing/breathing flow · turbulent jet/puff · buoyancy forces

1 Introduction

Research on respiratory aerosol dispersion has gained dramatic importance during the COVID-19 pandemic. From studies conducted before and since the start of the pandemic, it is well-known that virus-carrying aerosol droplets emitted by an infected person by coughing, sneezing, breathing and talking are efficiently dispersed in the air.

There are numerous numerical studies sought to understand the virus spreading through aerosols and the effectiveness of design measures for virus control

A. Dillmann et al. (Eds.): STAB/DGLR Symposium 2022, NNFM 154, pp. 509–518, 2024.
https://doi.org/10.1007/978-3-031-40482-5_48

and mitigation. Rosti et al. (2020) investigated the path and evaporation of respiratory droplets caused by a single cough based on Direct Numerical Simulations (DNS) for generic environments with different relative humidities. They found that the relative humidity of the surrounding air significantly affects aerosol spreading since any droplet evaporates to its dry nuclei or remains liquid under dry or wet environmental conditions. In Rosti et al. (2021) the same group of researchers emphasized that highly resolved DNS are required to precisely track the droplets emitted by a cough due to the existence of turbulent fluctuations. However, buoyancy effects were neglected in these simulations. In contrast, Ng et al. (2021) conducted DNS taking buoyancy forces into account to investigate the droplet dynamics in terms of coughing in different environments with different humidity. They found that droplets in the respiratory vapour puff in a cold and humid environment initially grow before they shrink due to evaporation. Besides, there are many DNS studies focusing on coughing events since they were thought to be the cause of most infections at the beginning of the pandemic. Breathing events, in contrast, did not receive as much attention.

Thus, the present study includes both — a coughing and a breathing event. The main objective is to achieve a better understanding of the physical mechanisms involved in respiratory aerosol transport including the buoyancy effect on the hot moist puff in a generic setup. There are some experimental studies considering coughing and breathing events which provided the information needed to set up the present DNS (Gupta et al. 2009, 2010; Duguid 1946; Pan et al. 2022). Additionally, the size of the computational domain, the grid resolution and the boundary conditions are prescribed in accordance with the studies by Rosti et al. (2020, 2021) which also serve as references.

In the following, the governing equations and simulation details are described. Afterwards, the results are presented, analysed and discussed. The paper ends with a conclusion.

2 Computational Details

2.1 Governing Equations and Discretization

The flow is governed by the incompressible Navier–Stokes equations as well as by transport equations for the temperature and the vapor concentration which are coupled to the former using the Boussinesq approximation:

$$\frac{\partial \mathbf{u}}{\partial t} + (\mathbf{u} \cdot \nabla)\mathbf{u} = -\frac{1}{\rho}\nabla p + \nu \nabla^2 \mathbf{u} + \mathbf{g}\rho_k, \tag{1}$$

$$\nabla \mathbf{u} = 0, \tag{2}$$

$$\rho_k = \rho_{kT} + \rho_{kC} = -\beta(T - T_{ref}) - \gamma(C_{vapor} - C_{ref}). \tag{3}$$

In Eq. (1), ρ is the fluid density, ν is the air kinematic viscosity and ρ_k is the approximated gravitational term to include the buoyancy forces in the momentum equation as detailed by Gray and Giorgini (1976). Further, in Eq. (3), β and

γ are the thermal and vapor molar fraction expansion coefficients, respectively. In addition, the temperature T and the vapor concentration C_{vapor} are determined by solving the transport equation for the energy and the vapor pressure concentration p_{vapor}, where $C_{vapor} = p_{vapor}/p_{atm}$. The equations are

$$\frac{\partial T}{\partial t} = \nabla \cdot (\kappa \nabla T) - \nabla \cdot (\mathbf{u}T) \tag{4}$$

and

$$\frac{\partial p_{vapor}}{\partial t} = \nabla \cdot (D_{vapor} \nabla p_{vapor}) - \nabla \cdot (\mathbf{u} p_{vapor}), \tag{5}$$

where κ and D_{vapor} are the thermal and water vapor mass diffusivity coefficients. The saturation pressure of the water p_{sat} is calculated by means of the August–Roche–Magnus formula (Alduchov and Eskridge 1997) to obtain the relative humidity $RH = p_{vapor}/p_{sat}$ for the gas phase.

The velocity field of the gas phase is directly computed in the DNS using a second-order accurate finite-volume method provided by the OpenFOAM library. Instead of using the pressure coupling approach of OpenFOAM, the projection method by Chorin (1968) implemented by Kath et al. (2016) is applied in combination with a Poisson equation solver. Kath et al. (2016) underlined the need for this more sophisticated pressure-velocity coupling method in their DNS of turbulent channel flows. Additionally, the explicit and second-order accurate Euler-leapfrog time integration scheme is used. The maximum steps in the present DNS of the flows induced by coughing and breathing are restricted in accordance with the von Neumann stability criterion detailed in Shishkina and Wagner (2004) to time steps of the order of 10^{-5} s.

Further, exhaled particles — which were assumed to be discrete spherical points in a continuous gas phase — are tracked using the Lagrangian approach. Thus, trajectories of each droplet are obtained by solving Newton's second law of motion of the particles

$$m_p \frac{\mathrm{d}\mathbf{u}_p}{\mathrm{d}t} = \sum \mathbf{F_i} = \mathbf{F_D} + \mathbf{F_G}, \tag{6}$$

where m_p and $\mathbf{u}_p$ are the particle mass and velocity, respectively. Since there is no evaporation, $\frac{dm_p}{dt} = 0$. In Eq. (6), the drag force $\mathbf{F_D}$ and the gravitational force $\mathbf{F_G}$ represent all the relevant forces acting on the particles. $\mathbf{F_G} = m_p \mathbf{g}(1 - \frac{\rho}{\rho_p})$ where ρ_p is the particle density. Since all the particles are spherical, $\mathbf{F_D} = C_D \frac{\pi D_p^2}{8} \rho (\mathbf{u} - \mathbf{u}_p) |\mathbf{u} - \mathbf{u}_p|$ and C_D is the drag coefficient. Evaporation of the particles is omitted to keep the initial size distribution of the particles unchanged throughout the simulation.

2.2 Simulation Details

The computational domain illustrated in Fig. 1 consists of two segments. The first segment is a cuboidal simplified room without windows or doors with the following dimensions: 4 m long, 2.5 m high and 1.25 m wide. It is discretized with a uniform grid consisting of cells with a length of 3.5 mm, which was used for all DNS. Both the domain and the grid were chosen based on the DNS study by Rosti et al. (2021). The initial field of the DNS in this part of the computational domain reflects a flow at rest, i.e., vanishing velocity values everywhere. Further, no-slip boundary conditions are enforced at the front walls and the floor, whereas a slip-wall boundary condition is prescribed at the ceiling of the room for reasons of simplicity. Besides, Dirichlet boundary conditions for both, the temperature and the vapor pressure are used at all walls such that an ambient room temperature of $T_a = 25$ °C and a relative humidity of $RH_a = 60\%$ is realized. Further, a Neumann boundary condition for the velocity is prescribed at the right wall sketched in Fig. 1. Finally, periodic boundary conditions for the velocity components, the temperature and the vapor pressure are set at the side walls (in z-direction).

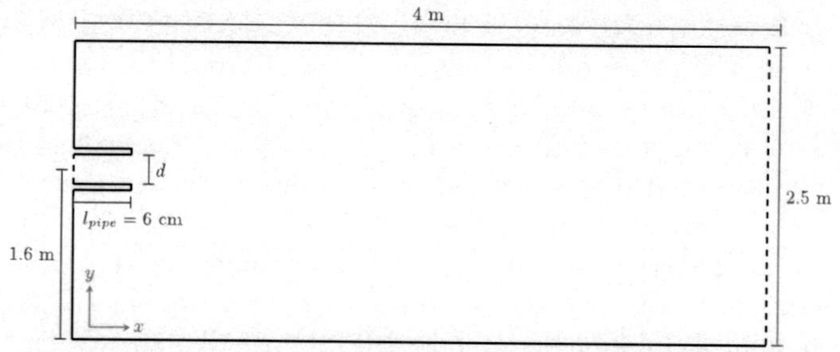

Fig. 1. Computational domain in the x-y-plane with a width of 1.25 m in z-direction. The solid and dashed lines indicate the walls as well as the inlet (left) and outlet (right) boundaries. The pipe diameter d equals 2.4 cm and 1.24 cm for the coughing and breathing cases, respectively. The size of the pipe in the sketch is not to scale.

The second segment is the circular smooth pipe representing the mouth opening. The center line of the pipe segment is located 1.6 m above the floor on the left side of the domain as shown in Fig. 1. The pipe segment is 6 cm long in agreement with the study of Rosti et al. (2021) and has a diameter of $d = d_{cough} = 2.4$ cm and $d = d_{breath} = 1.24$ cm in the coughing and breathing cases, respectively. These pipe diameters are specified following the experimental studies by Gupta et al. (2009, 2010). The grid is refined in the pipe and in the streamwise direction of the pipe where the spacing $\Delta x \approx 1.2$ mm in every direction.

Dekker (1961) reported that the jet-like flows leaving the mouth are probably turbulent for both breathing and coughing. Resolving a turbulent pipe flow at a bulk Reynolds number of 5000 in a DNS requires at least 150 grid points over the diameter if the grid is refined close to the wall where the turbulence intensities and the wall shear stress reach their maximum as reported in Wagner and Friedrich (1998, 2000). However, in the present DNS, a laminar bulk flow with a maximum bulk Reynolds number of 20000 and approximately 20 superimposed eddy-like circular disturbances over the pipe radius trigger the transition from a laminar pipe flow to a turbulent jet after the latter has entered the room. To properly resolve this disturbed pipe flow, at least five cells are specified for each eddy-like disturbance generated with the Divergence Free Synthetic Eddy Method (DFSEM) developed by Poletto et al. (2013). With this method, random positions and intensities are assigned to synthetically generated eddies and their sizes have been determined to reflect typical turbulent length scales. The resulting velocity fluctuations are added to the mean velocity which changes over time in accordance with the experimental studies of Gupta et al. (2009, 2010) as illustrated in Fig. 2. Based on these time series, the maximum Reynolds numbers (with regard to bulk velocity and the pipe diameter) are 20000 and 1800 for coughing and breathing, respectively. Further, the turbulence intensity of the prescribed time-dependent pipe flow is equal to those of Reynolds stresses reflecting the isotropic turbulence with 10% turbulence intensity.

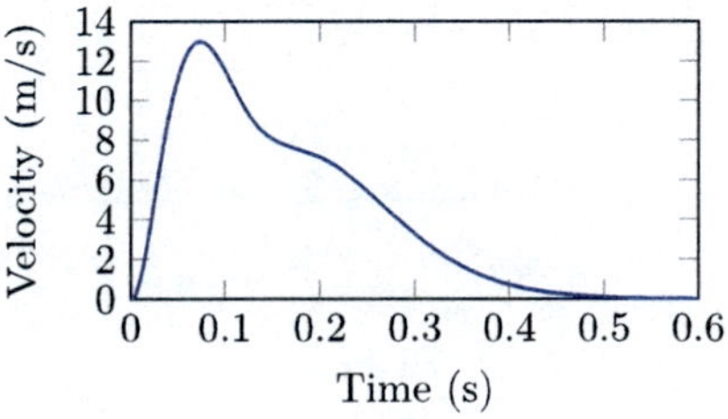

(a) Coughing

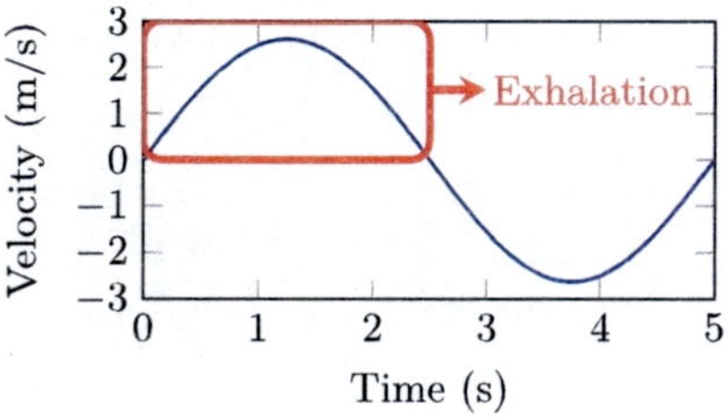

(b) Breathing

Fig. 2. Time-dependent bulk velocities prescribed for the (a) coughing (Gupta et al. 2009) and (b) breathing case (Gupta et al. 2010). The breathing event reflects a single cycle of exhalation and inhalation. However, only the exhalation phase is considered during the simulation.

The DNS are started with 2500 particles randomly positioned within a circle at the outlet plane of the pipe with a diameter of 2 cm for the coughing case and a diameter of 1 cm for the breathing case. These particles are initiated at rest to correctly follow the assumptions made by Rosti et al. (2021). However, it is known that in some other studies, e.g., Ng et al. (2021), particles with a velocity matching the local inlet velocity were injected. It must be further noted that the diameters of the circular regions where the particles were seeded are slightly smaller than the pipe diameters to avoid particles being injected in the viscous boundary layer. Additionally, the initial droplet size distributions shown in Fig. 3 are prescribed for coughing and breathing in accordance with the experimental studies of Duguid (1946) and Pan et al. (2022). The maximum particle Reynolds number of the large particles is $\mathcal{O}(100)$ for coughing and $\mathcal{O}(10)$ for breathing. Further, one-way coupling of the particles to the fluid is realized which means that the fluid affects the particles but not vice versa. The latter is a reasonable assumption for the considered low-volume fractions. Moreover, particle collision and reflection at the walls are neglected but they can stick to the walls and escape the domain through the outlet. All simulation parameters characterizing the flow, the respiration and the particles are summarized in Table 1.

Table 1. List of simulation parameters.

Properties	Symbols	Values
Ambient (reference) temperature	T_a or T_{ref}	25 °C
Ambient (reference) vapor concentration	C_a or C_{ref}	0.0187
Ambient relative humidity	RH_a	60%
Air density	ρ	$1.18\,\mathrm{kg\,m^{-3}}$
Air kinematic viscosity	ν	$1.8 \times 10^{-5}\ \mathrm{m^2\ s^{-1}}$
Thermal expansion coefficient	β	$3.38 \times 10^{-3}\ \mathrm{K^{-1}}$
Thermal diffusivity	κ	$2.19 \times 10^{-5}\ \mathrm{m^2\ s^{-1}}$
Vapor molar fraction expansion coefficient	γ	0.385
Mass diffusivity of vapor	D_{vapor}	$2.5 \times 10^{-5}\ \mathrm{m^2\ s^{-1}}$
Respiratory temperature	T_r	30 °C
Respiratory relative humidity	RH_r	100%
Droplet density	ρ_p	$1007\,\mathrm{kg\ m^{-3}}$
Diameter of mouth for coughing	d_{cough}	2.4 cm
Diameter of mouth for breathing	d_{breath}	1.24 cm

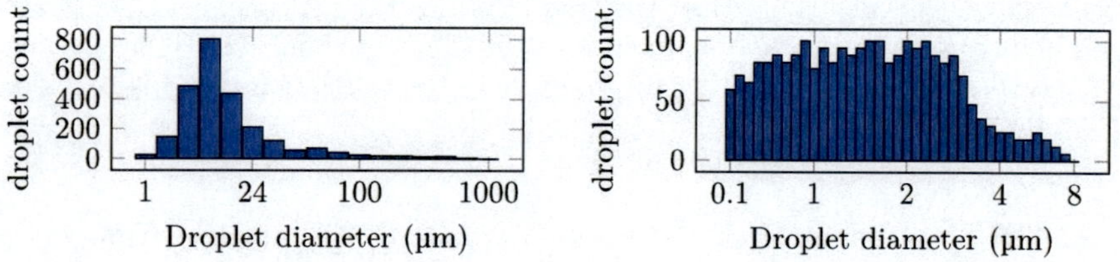

(a) Coughing (b) Breathing

Fig. 3. Particle size distributions in a single exhalation cycle in the coughing (Duguid 1946) and breathing (Pan et al. 2022) cases.

3 DNS of Coughing and Breathing

The instantaneous three-dimensional velocity fields and the particle spread obtained from the DNS at three different times are presented in Fig. 4. Within 60 s, the particle clouds propagate up to 2.7 m and 1.95 m in horizontal direction triggered by coughing and breathing, respectively. Considering that the maximum bulk velocity is nearly 5 times higher in the coughing case than in the breathing case (see Fig. 2), the difference in the penetration length of less than 30% is rather small at first glance. However, taking into account the higher turbulence intensity of the jet developing in the coughing case, which induces a free shear layer with more intense vortices, may explain the higher deceleration of the coughing puff in horizontal direction compared to the breathing case. Additionally, considering the turbulence intensities in terms of the velocity fluctuations — which are up to 15 times stronger in the coughing case — we conclude that turbulence is much more effective for coughing than for breathing.

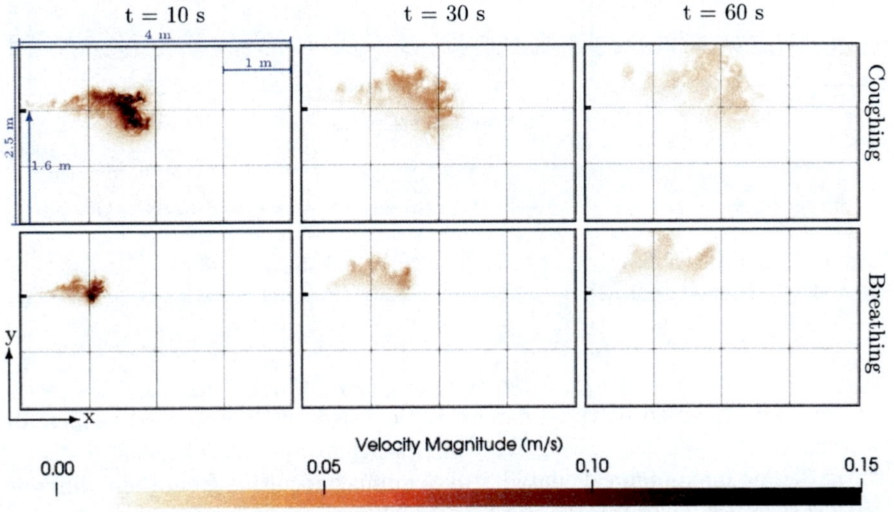

Fig. 4. Velocity field predicted for coughing (top) and for breathing (bottom) after t = 10 s, 30 s and 60 s. The section corresponds to z=0, the cutting through the center of the jet.

Figure 4 further reveals the upward movement of the puffs induced by coughing and breathing. Since the temperature and the relative humidity, $T_a = 25°C$ and $RH_a = 60\%$, of the ambient air are both smaller than $T_r = 30°C$ and $RH_r = 100\%$ of the cloud, buoyancy tends to lift up respiratory clouds in both cases. As a consequence, after 60 s the exhaled clouds have lost most of their potential humidity and their temperature has almost decreased to the ambient temperature.

The particle distributions obtained for coughing and breathing for the previously discussed three time periods are presented in Fig. 5. When comparing the particle spread for both cases, it is observed that the particle densities are much lower in the coughing case since the particles are less packed compared to in the breathing case. Despite the major difference between the maximum exhalation speeds reached after coughing and breathing, the horizontal propagation distances differ by less than 30%. Further, for the breathing case, all particles

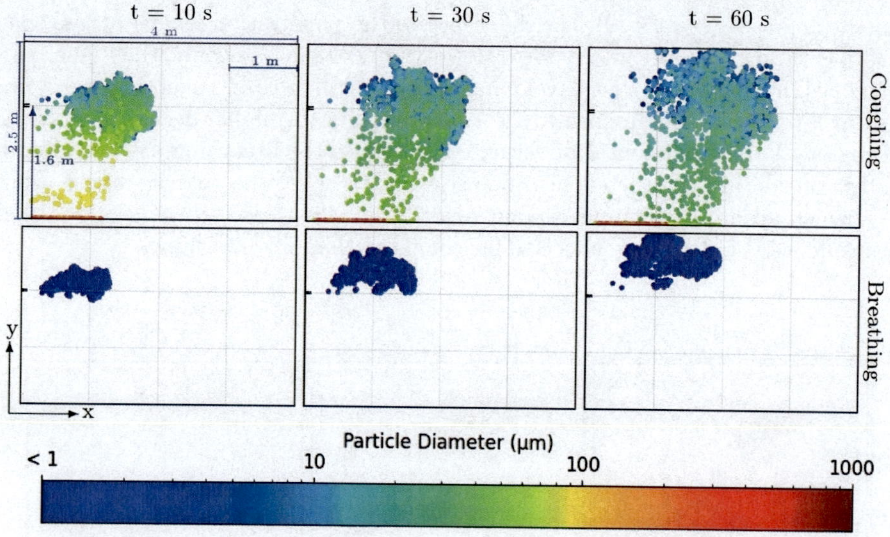

Fig. 5. Particle distributions of the coughing (top) and the breathing simulations (bottom) at t = 10 s, 30 s and 60 s

remain airborne in contrast to the coughing case, where some particles settle. This can be explained by the different particle sizes of the two cases, which are two orders of magnitude larger for coughing than for breathing as illustrated in Fig. 3. The maximum calculated Stokes number resulting from these different particle sizes is smaller than 0.1 for breathing and larger than 2000 for coughing, respectively. As a consequence, the majority of the breathing particles presented in Fig. 5 remain in the cloud and behave like tracers. Thus, these rather small particles are mainly affected by the drag forces with negligibly small gravitational forces, even 60 s after the exhalation process has started. However, in the coughing case, the gravitational forces acting on the particles are larger since the Stokes number exceeds 1. Particles larger than 500 μm follow ballistic trajectories and settle on the floor already after 2 s. After 60 s any particle larger than 39 μm has settled on the floor in the coughing case. The latter applies to 13.5% of all particles. We can further argue that the mid-sized particles emitted by coughing were distributed more in the vertical direction than those emitted by breathing.

Finally, the results obtained for the coughing case are compared to those of Rosti et al. (2020), who considered droplet evaporation but neglected buoyancy. Table 2 shows the cumulative proportion of travelling distances of the airborne particles and the particle proportions of settled particles after 60 s in both studies. The cumulative proportion of traveling distance indicates the percentage of the particles traveled upon that vertical distance from the initial mouth position. The number of particles traveling distances up to 1 m, 2 m and 4 m is smaller if evaporation is neglected and buoyancy is considered. Not a single particle travels up to 4 m in 60 s in the present study. This means that the respiratory cloud propagates less in streamwise direction compared to the study by Rosti et al. (2020) who considered evaporation and neglected buoyancy. A possible explanation for this difference could be the more pronounced viscous and turbulent effects caused by the additional buoyant forces. Table 2 further reflects a two times higher percentage of settled particles for the DNS neglecting evaporation. The reason might be that the larger droplets do not shrink due to the lack of evaporation which is why more particles settle on the floor. This comparison shows that evaporation has unignorable effects on particle sedimentation and propagation, especially for larger particles.

Table 2. Traveling distances of the cumulative proportion of the airborne particles and the settled particle proportions observed in the coughing event after 60 s in comparison with the study of Rosti et al. (2020).

	Present study	Rosti et al. (2020)
Settled (%)	13.5	6
1 m (%)	83	85
2 m (%)	34	45
4 m (%)	0	8

4 Conclusion

The spreading of aerosol droplets in a simplified room is investigated for a breathing and a coughing case by conducting DNS with the Euler-Lagrangian method. Despite the large differences between the maximum coughing and breathing velocities, the horizontal propagation distances are more similar than expected. A possible reason for this is the turbulence being much more effective in the coughing case. This enhances the transversal mixing due to higher Re which slows down the horizontal propagation. In the breathing case, however, all the particles follow the respiratory cloud, i.e. they mainly behave like tracers. Contrary to breathing, particles are transported in various ways after coughing. They either follow ballistic trajectories, are convected in the humid cloud or detach from the latter. The detachment of the particles results in a more pronounced vertical and horizontal spreading than for the breathing. Additionally, the results for coughing obtained in the present DNS study are compared with those of an earlier DNS study performed by Rosti et al. (2020) who included evaporation and neglected buoyancy forces. It is found that buoyancy damps the cloud propagation in horizontal direction and amplifies the particle transport in vertical direction. The comparison with Rosti et al. (2020) further reveals that evaporation reduces the overall particle size of the particle distribution and inevitably the sedimentation proportion. Both buoyant forces and particle evaporation have major effects on the particle transport and will be considered in future projects to be able to analyze the most important physical mechanisms involved in the spreading of aerosol droplets.

Acknowledgement. The work was supported by the Initiative and Networking Fund of the Helmholtz Association of German Research Centres (HGF) under the CORAERO project (KA1-Co-06). The authors would like to thank Ms. Annika Köhne for proofreading the manuscript.

References

Alduchov, O.A., Eskridge, R.E.: Improved Magnus' form approximation of saturation vapor pressure (1997)

Chorin, A.J.: Numerical solution of the Navier-Stokes equations. Math. Comput. **22**(104), 745–762 (1968)

Dekker, E.: Transition between laminar and turbulent flow in human trachea. J. Appl. Physiol. **16**(6), 1060–1064 (1961)

Duguid, J.P.: The size and the duration of air-carriage of respiratory droplets and droplet-nuclei. Epidemiol. Infect. **44**(6), 471–479 (1946)

Gray, D.D., Giorgini, A.: The validity of the Boussinesq approximation for liquids and gases. Int. J. Heat Mass Transf. **19**(5), 545–551 (1976)

Gupta, J.K., Lin, C.-H., Chen, Q.: Flow dynamics and characterization of a cough. Indoor Air **19**(6), 517–525 (2009)

Gupta, J.K., Lin, C.-H., Chen, Q.: Characterizing exhaled airflow from breathing and talking. Indoor Air **20**(1), 31–39 (2010)

Kath, C., Wagner, C.: Highly resolved simulations of turbulent mixed convection in a vertical plane channel. In: Dillmann, A., Heller, G., Krämer, E., Wagner, C., Breitsamter, C. (eds.) New Results in Numerical and Experimental Fluid Mechanics X. NNFMMD, vol. 132, pp. 515–524. Springer, Cham (2016). https://doi.org/10.1007/978-3-319-27279-5_45

Ng, C., Chong, K., Yang, R., Li, M., Verzicco, R., Lohse, D.: Growth of respiratory droplets in cold and humid air. Phys. Rev. Fluids **6**(5) (2021)

Pan, S., Xu, C., Francis Yu, C.W., Liu, L.: Characterization and size distribution of initial droplet concentration discharged from human breathing and speaking. Indoor Built Environ. (2022)

Poletto, R., Craft, T., Revell, A.: A new divergence free synthetic Eddy Method for the reproduction of inlet flow conditions for Les. Flow Turbul. Combust. **91**(3), 519–539 (2013)

Rosti, M.E., Olivieri, S., Cavaiola, M., Seminara, A., Mazzino, A.: Fluid dynamics of Covid-19 airborne infection suggests urgent data for a scientific design of social distancing. Sci. Rep. **10**(1) (2020)

Rosti, M.E., Cavaiola, M., Olivieri, S., Seminara, A., Mazzino, A.: Turbulence role in the fate of virus-containing droplets in violent expiratory events. Phys. Rev. Res. **3**(1) (2021)

Shishkina, O., Wagner, C.: Stability conditions for the Leapfrog-Euler scheme with central spatial discretization of any order. Appl. Numer. Anal. Comput. Math. **1**(1), 315–326 (2004)

Wagner, C., Friedrich, R.: On the turbulence structure in solid and permeable pipes. Int. J. Heat Fluid Flow **19**(5), 459–469 (1998)

Wagner, C., Friedrich, R.: DNS of turbulent flow along passively permeable walls. Int. J. Heat Fluid Flow **21**(5), 489–498 (2000)

Flow Reconstruction of Urban Wind Fields for Wind-Based Path Planning

Carola Ebert(✉) and Julien Weiss

Fachgebiet Aerodynamik, Technische Universität Berlin, Berlin, Germany
carola.ebert@tu-berlin.de

Abstract. Three-dimensional Reynolds-Averaged Navier-Stokes simulations are performed to calculate the wind field of a full-size urban district of 1 km^2 around the campus of Technical University of Berlin with the inlet wind direction as a parameter. Two-dimensional snapshots of the simulation data set are used for model reduction by proper orthogonal decomposition (POD) to reduce the complexity of the system. The POD modes are then used to estimate a high resolution wind field from sparse velocity sensor measurements by using the Gappy POD as a data reconstruction method. The sensor measurements are taken from a simulation test case that is not included in the snapshot basis. The sensor placement problem that needs to be solved to find effective sensor locations is investigated using two different methods. The performance of the data reconstruction is then assessed by calculating the least-squares reconstruction error. An application of the estimated urban wind field is demonstrated by finding a wind-based minimum-energy path from a start to a final location for an operation of unmanned aerial vehicles. The wind-based path is compared to the shortest path for a reduced order model taking only 5 sensor measurements. An overall mean energy reduction of 5.5% was calculated by the full-order model and of 3.9% by the reduced order model. This suggests that trajectory planning may be performed on inexpensive reduced-order models of the urban wind field.

Keywords: CFD simulation · data reconstruction · path planning

1 Introduction

The wind field in a city is characterized by areas that boast strong flow separations, local vortices and formations of large shear layers caused by the flow around buildings and obstacles. These disturbances change with the wind direction and impact the capabilities and flight qualities of unmanned aerial vehicles (UAVs), which in turn causes a high potential risk for a safe and trustworthy flight operation in urban environments [1]. Moreover, areas of high flow acceleration due to narrow buildings can increase the power consumption of UAVs and lead to a reduction of efficiency and sustainability [2]. The integration of the prevailing wind conditions into the flight path planning system of autonomous

A. Dillmann et al. (Eds.): STAB/DGLR Symposium 2022, NNFM 154, pp. 519–528, 2024.
https://doi.org/10.1007/978-3-031-40482-5_49

operating UAVs can provide an important and necessary aspect for a safer and energy efficient flight operation, especially during windy conditions in urban areas [3].

Computational fluid dynamics (CFD) has emerged in recent years as a promising solution to simulate the wind field of an urban environment in several application fields [4–6]. However, high resolution full-size CFD simulations of urban wind fields require very high computational costs that make real-time online predictions not feasible. To circumvent these computational limitations, data reconstruction methods for fast flow field predictions have been investigated in a wide range of applications [7–11].

The Gappy proper orthogonal decomposition (Gappy POD) is an established reconstruction method to generate reduced order models (ROM) based on POD modes [7–9,12]. The POD reduces the complexity of a physical system by deriving a basis from a set of instantaneous flow states, also known as snapshots, that capture the dominant flow features. By using a limited number of sensor measurements in linear combination with the POD basis, the Gappy POD performs a high dimensional flow reconstruction from an incomplete flow field. The data reconstruction performance is strongly dependent on the number of sensors and the choice of their spatial location. Various sensor placement methods have been investigated for several applications [8,13–16].

In this work, a wind field estimation using Gappy POD by constructing a ROM is investigated, together with the associated sensor placement problem. The overall procedure of the ROM is divided in an offline and an online stage. During the offline stage, three-dimensional RANS simulations are performed to calculate the full-order model (FOM) wind fields of a district in Berlin. Thereby, a data set of stationary wind fields, called "snapshots", is created by setting different wind directions as inlet condition. On the one hand the simulations are used to derive a POD basis with the wind direction as a parameter, and on the other hand a couple of simulations are used as test data to validate the flow reconstruction. The fast wind field reconstruction by Gappy POD will be performed during the online stage. The input of the online stage is the POD basis and the velocity data taken from a simulation test case. By varying the number of POD modes as well as the number and the location of the sensors, the accuracy of the wind field reconstruction is influenced. Two different sensor placement methods that find appropriate sensor locations are investigated and compared considering the least-squares reconstruction error. Finally, the estimated wind fields are used for a wind-based path planning of an exemplary UAV flight operation and the energy savings calculated by taking the wind into account are computed.

2 CFD Simulations

The three-dimensional Reynolds-Averaged Navier-Stokes (RANS) simulations were performed using the open-source software OpenFoam®. The numerical domain shows a district around the campus of the Technical University (TU) Berlin with an area of $1\,\text{km}^2$ in full-size. The grid was generated by using the utility snappyHexMesh for unstructured meshing. Due to a variation of the inlet wind direction a circular domain was chosen according to Kastner and Dogan [6] and Hagbo and Giljarhus [17]. The circular domain is shown in Fig. 1(left). As illustrated, the inlet and outlet of the domain are assembled by defining the specific patches, respectively. For a wind direction of $\varphi = 0°$ the patches 11/12/1/2/3 form the inlet, whereas the patches 4/5/6/7/8/9/10 form the outlet of the domain. The advantage of the circular grid is given by a simple redefining of the inlet and outlet rather than a complete re-meshing of the overall grid for a changing wind direction. The inlet wind direction was varied in range of $\varphi = 0 - 360°$ resulting in 36 simulations. The grid generation meets the requirements of the best practice guidelines for urban flow simulations [18] that specify a distance of at least 5H from the inlet to the interesting area and a distance of at least 15H from the interested area to the outlet, where H is the maximum height of the buildings. Within the rectangular area of the domain, a grid refinement was applied to shape the structure of the buildings precisely. Thus, the smallest cell near the buildings measures 0.6 m × 0.6 m with a height of 0.35 m. The inlet and boundary conditions are listed in Fig. 1(right) and described in detail in Ebert et al. [19]. The Reynolds-averaged Navier-Stokes (RANS) equations are solved using the incompressible steady-state solver simpleFoam for turbulent flows. The k-ϵ turbulence model was applied for modeling turbulence effects. The buildings and the ground region are treated as no-slip walls and use wall functions to compute near-wall properties. A velocity profile is defined at the inlet with a reference velocity of 5 m/s at an altitude of $z_{ref} = 10$ m, assuming strong wind conditions. The present work is limited to one inlet wind velocity, but future studies will discuss the influence of different wind speeds, as it is expected that a UAV flight path can be optimized at lower/higher wind speeds. The urban atmospheric boundary layer was estimated assuming dense low-rise buildings in the vicinity of the interesting area. In order to achieve a mesh independent

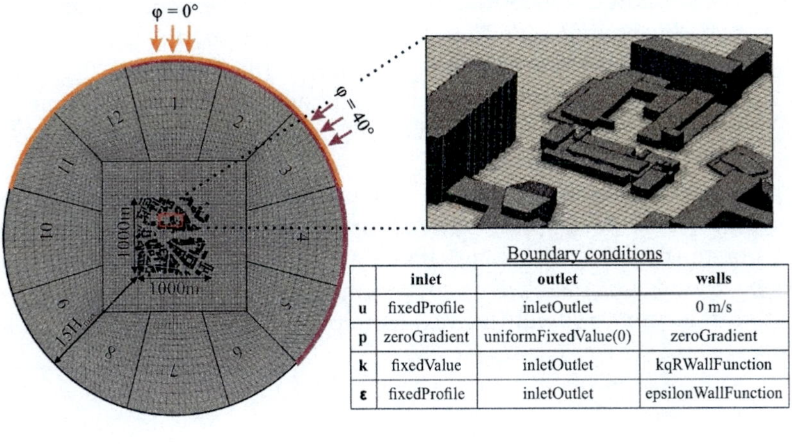

	inlet	outlet	walls
u	fixedProfile	inletOutlet	0 m/s
p	zeroGradient	uniformFixedValue(0)	zeroGradient
k	fixedValue	inletOutlet	kqRWallFunction
ε	fixedProfile	inletOutlet	epsilonWallFunction

Fig. 1. Computational domain of the CFD simulations with two illustrative wind directions. Left: Top view. Right: Isometric view with boundary conditions.

flow simulation, a grid independence study was performed using four different sized meshes. The result of the grid study is shown in Fig. 2(right) where the velocity magnitude along a randomly chosen validation line can be seen. Different validation lines were investigated and showed the same trend. To guarantee a high resolution of the buildings while having an acceptable simulation time, the mesh with 11.5 million cells was chosen for the wind field simulations. In Fig. 2(left) the wind field of the interesting area in the xy-plane at an altitude of $z = 2$ m is shown for the mesh with 11.5 million cells with the wind coming from west ($\varphi = 270°$). It can be seen that the local wind velocity increases along large streets and that flow separations occur behind buildings.

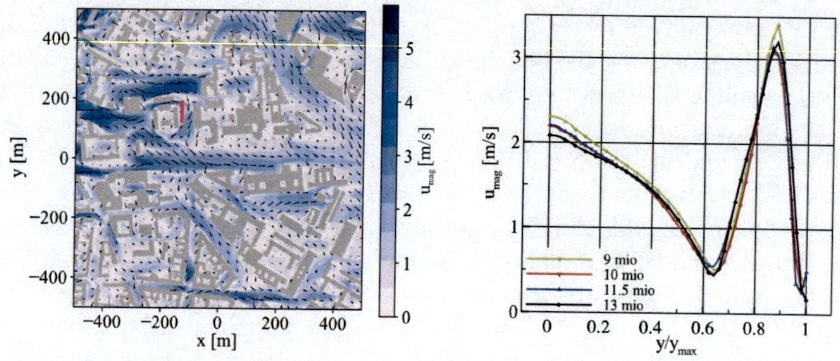

Fig. 2. Grid-independence study. Left: Wind field at an altitude of 2 m (11.5 mio. cell grid). Validation line in red. Right: Velocity along the validation line.

3 POD and Sensor Placement Problem

Proper Orthogonal Decomposition. The POD is performed on the data base of two-dimensional snapshots from the wind field. By varying the inlet wind direction, each of the 36 CFD simulations produced one snapshot. In order to recast the unstructured CFD data in a structured order, the two-dimensional wind fields are extracted from a uniformly distributed grid with a resolution of 1 m × 1 m at an altitude of $z = 2$ m. Each column of the snapshot matrix S represents one wind direction and is built with the velocity components u_x and u_y as $(u_{x,1}, u_{x,2} ... u_{x,n}, u_{y,1}, u_{y,2} ... u_{y,n})^T$, where each component consists of n grid points (n ≈ 750000). Thus, the ROM is reduced into a two-dimensional problem. The idea of the POD is to decompose the wind vector fields $u(\mathrm{x}, \varphi)$ in spatial POD modes Ψ and their expansion coefficients a_k as defined in Eq. 1, where x has two-dimensions, $\mathrm{x} = (x, y)$.

$$u(\mathrm{x}, \varphi) = \sum_{k=1}^{36} a_k(\varphi) \Psi_k(\mathrm{x}) \tag{1}$$

A detailed description of the POD method can be found in Weiss [20] and in Ebert et al. [19]. The POD is performed by using the singular value decomposition (SVD) that breaks down the snapshot basis S into the POD modes U, the singular values Σ and the singular vectors V. The SVD is defined as follows:

$$S = U\Sigma V^T \approx \tilde{U}\tilde{\Sigma}\tilde{V}^T, \tag{2}$$

where $\tilde{U}$ covers the dominant POD modes of the velocity components Ψ_{u_x} and Ψ_{u_y} of the system with a truncation order r:

$$\tilde{U} = \Psi = \begin{bmatrix} \Psi_{u_x} \\ \Psi_{u_y} \end{bmatrix} = \begin{bmatrix} \psi_{1,1} & \psi_{1,2} & \cdots & \psi_{1,r} \\ \psi_{2,1} & \psi_{2,2} & \cdots & \psi_{2,r} \\ \vdots & \vdots & & \vdots \\ \psi_{2n,1} & \psi_{2n,2} & \cdots & \psi_{2n,r} \end{bmatrix}. \tag{3}$$

Sensor Placement Problem. The POD modes calculated by SVD are used for the data reconstruction by Gappy POD and to solve the sensor placement problem that identifies appropriate sensor locations for the wind field reconstruction. In the present work, two separate sensor placement algorithms are applied to design a sparse measurement matrix P_{mask} that will be used in the Gappy POD algorithm (see below). One method is based on the discrete empirical interpolation method (DEIM) [21] and one is based on the QR factorization with column pivoting [22]. In both methods, the sensor placements are tailored to a scalar POD basis Ξ due to the fact that the components u_x and u_y need to be captured at the same spatial locations by the wind sensors. The so-called sensor basis is defined by:

$$\Xi = \Psi_{u_x} + \Psi_{u_y}. \tag{4}$$

Detailed information of the use of the sensor basis and the definition of the DEIM is given in Ebert et al. [19]. The rank-revealing QR factorization is a column-wise pivoting of Ξ and is defined by:

$$\Xi P = QR, \tag{5}$$

where Ξ is decomposed into an unitary/orthogonal matrix Q, an upper triangular matrix R and a column permutation matrix P such that the diagonal of R is non-increasing. Both the DEIM and the QR factorization search one sensor location on each column of Ξ so that the number of sensors is equal to the rank of the sensor basis Ξ as well as the number of chosen POD modes.

4 Wind Field Reconstruction by Gappy POD

The data reconstruction of an incomplete wind field into a high resolution wind field is performed by Gappy POD on sparse velocity measurements. Specifically, the incomplete sparse wind field $\tilde{s}$ is calculated by a measurement matrix P_{mask} that includes the sensor locations times the velocity data from a simulation test case s_{trial} (Eq. 6). An approximation of the incomplete wind field is given by the standard POD projection as defined in Eq. 7. The idea of the Gappy POD is to compute the expansion coefficient $\tilde{a}$ by solving a linear system with a set of POD modes and the sparse measurement vector.

$$\tilde{s} = P_{mask} s_{trail} \tag{6}$$

$$\tilde{s} \approx P_{mask} \sum_{k=1}^{r} \tilde{a}_k(\varphi)\psi_k \tag{7}$$

$$s(x, y) \approx \psi\tilde{a} \tag{8}$$

The expansion coefficient and the POD modes are then used to perform the reconstruction of the complete wind field vector $s(x, y)$ (Eq. 8). A detailed description of Gappy POD is given in Ebert et al. [19] and is based on the original method of Everson and Sirovich [12]. The reconstruction of the velocity components u_x and u_y was performed on a test case that was not included in the snapshot basis. The wind field magnitude was then calculated by $\sqrt{u_x^2 + u_y^2}$. The wind field reconstructed by using a ROM with 10 sensor measurements and 10 POD modes is called ROM_{10} and is shown for the DEIM sensor locations in Fig. 3(a) and for the QR factorization sensor locations in Fig. 3(b). In this test case the wind is coming from the east direction ($\varphi = 77°$) and the sensor locations are illustrated by the colored dots. The contour plot shows the wind field magnitude and the arrows show the wind angle. The original FOM from the CFD simulation is shown in Fig. 3(c). It can be seen that the ROM_{10} predicts the overall flow pattern of the wind field quite well by using 10 sensor measurements for both sensor placement methods (DEIM and QR factorization). The least-squares reconstruction error in Fig. 3(d) shows a quantitative assessment of the wind field reconstruction depending on the number of POD modes that, by construction, are equal to the number of sensors. It can be seen that the error of the wind field reconstruction behaves similarly for both sensor placement methods. The black line shows the wind field reconstruction error that depends only on the truncation of the POD basis when all grid points are used for the reconstruction and a number of POD modes respectively. Three types of reconstruction errors can be defined. An interpolation error that occurs because the test data of $\varphi = 77°$ is not included in the snapshot basis. It is obtained when using all data points and the full-rank of the POD basis with 36 modes, and is approximately 20% in the current test case. A truncation error that occurs because only a few modes are used for the reconstruction with all data points. Taking a truncation of $r = 10$ POD modes, the truncation error is approx. 27%–20% = 7%. And a sparsity error that is caused by the fact that a limited number of sensors and a few POD modes are used for the reconstruction (DEIM/QR factorization). The sparsity error with 10 sensors and 10 POD modes is 32%–27% = 5% for the DEIM and 30%–27% = 3% for the QR factorization. As can be seen in Fig. 4(d), both the sparsity and truncation error are relatively small in comparison with the interpolation error for $r > 5$. Compared to the results of the

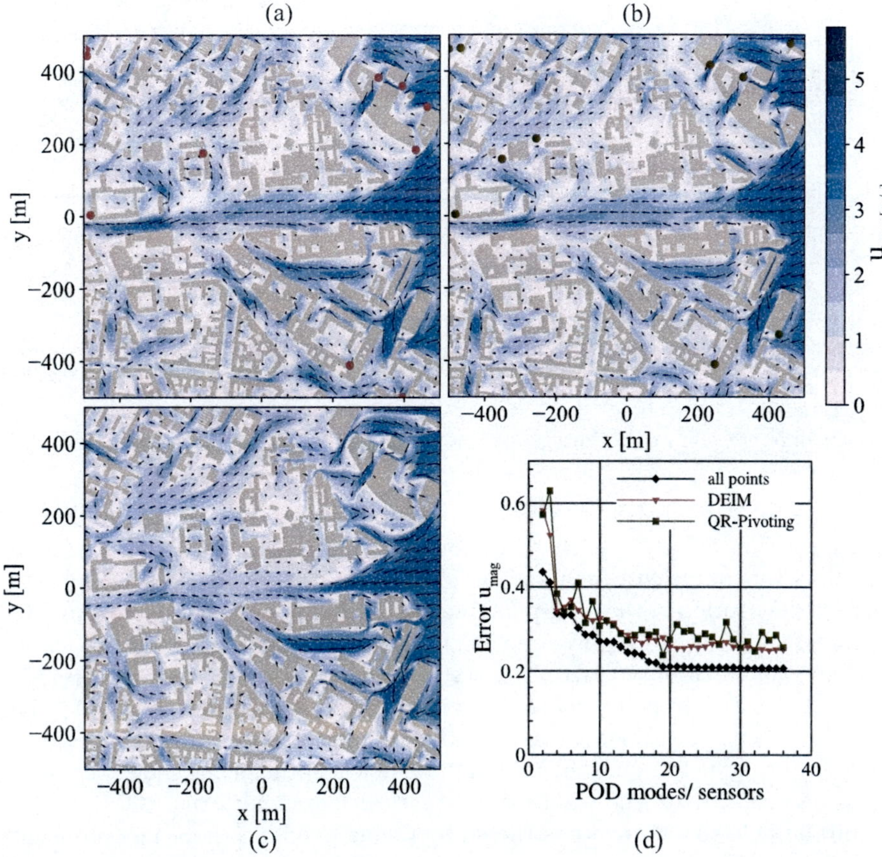

Fig. 3. Gappy reconstructed wind fields with (a) $\mathrm{ROM}_{10,\mathrm{DEIM}}$ and (b) $\mathrm{ROM}_{10,\mathrm{QR}}$ with sensor locations and (c) original from FOM at $\varphi = 77°$. (d) Least-squares reconstruction error.

wind field reconstruction of a small urban area with two buildings [19], where the truncation error was almost zero by using the full-rank of the POD basis, the truncation error of the wind field reconstruction of a large area in this study is much higher at approximately 20%. **We suspect that the differences in the** accuracy of the data reconstruction depend partly on the number of snapshots and partly on the complexity of the system. The larger area implicates a high level of complexity of the flow structures due to the flow around the buildings and a non-linear correlation between the formation of the coherent structures and the wind angle.

5 Wind Field for Wind-Based Path Planning

The reconstructed wind fields are used for a wind-based path planning of an exemplary UAV flight operation to find a minimum-energy path. The comparison between the path that is computed with the wind field of the FOM and the

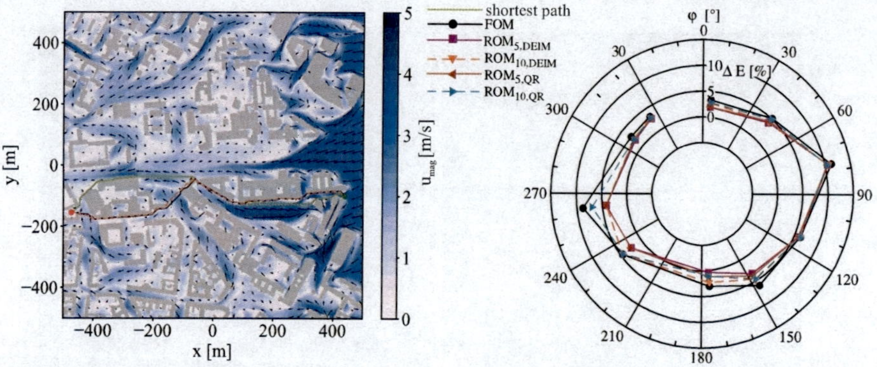

Fig. 4. Left: Path planning - shortest path and wind-based path. Right: Energy reduction for different wind directions φ.

one of the ROM is used to evaluate the accuracy of the ROMs. The path planning algorithm is based on the traditional A* path planner and was extended to include an energy cost function considering the prevailing wind conditions. A detailed description of the path planning algorithm and the energy cost function is given in Ebert et al. [19]. Figure 4(left) shows the path from a arbitrary chosen start location at $[x, y] = [-470, -155]$ m (red point) to a final location at $[x, y] = [440, -100]$ m (green point) when the wind is coming from east ($\varphi = 77°$). The green line shows the shortest path, the black line shows the wind-based path by taking the wind field from the FOM and the dashed orange line shows the wind-based path that takes the wind information calculated with the $ROM_{10,DEIM}$, i.e. a wind field that was reconstructed by taking only ten sensor measurements obtained by DEIM. The path that was calculated with the $ROM_{10,QR}$ is equal to the $ROM_{10,DEIM}$. It can be seen that the wind-based path is adjusted significantly in the area of $x = [-400, -50]$ m in order to avoid strong head wind to reduce the energy consumption of the UAV. Crucially, the wind-based path that was calculated on the $ROM_{10,DEIM}$ matches the path from the FOM very well although only 10 sensors were used for the reconstruction. The energy consumption measures E = 37092.7 J on the shortest path, E = 32918.5 J on the wind-based path from FOM and E = 33155.0 J on the wind-based path from $ROM_{10,DEIM}$. The percentage of the energy consumption that can be reduced by considering the wind-based path instead of taking the shortest one is shown in Fig. 4(right) for different wind directions. The black line shows the energy reduction that was calculated by taking the wind field from the FOM and the colored lines show the energy reduction that was obtained by the ROMs for different numbers of sensors and both sensor placement methods. Regarding the FOM, it can be seen that the energy reduction is highest with $\Delta E = 11.2\%$ when the wind is coming from east. The result of both ROM_{10} matches the result of the FOM very well ($\Delta\ E_{10,DEIM} = 10.6\% / \Delta\ E_{10,QR} = 10.4\%$). Small deviations occur with the ROM_5 when the wind is coming from south-west. However, a positive energy-reduction can still be achieved. Thus, the ROMs are capable of predicting a wind-based energy-reduced path from the chosen start to the final location by using a limited number of sensor measurements. Averaging over all wind directions, the overall mean energy reduction is 5.5% for the FOM, 3.8% for the $\mathrm{ROM}_{5,DEIM}$, 3.9% for the $\mathrm{ROM}_{5,QR}$, 4.5% for the $\mathrm{ROM}_{10,DEIM}$ and 4.7% for the $\mathrm{ROM}_{10,QR}$. Reconsidering that the wind field least-squares reconstruction error for the test case of $\varphi = 77°$ was 32% by using the $\mathrm{ROM}_{10,DEIM}$ (Fig. 3(d)), the calculated wind-based path from the $\mathrm{ROM}_{10,DEIM}$ matches the path of the FOM very well, with a deviation in the energy consumption along the path of just 0.6%. This is because the reconstruction error represents the mean integrated squared error of the total wind field, where the local deviations between the ROMs and the FOM are relatively small. Thus, the wind-based path finding algorithm is able to calculate a reliable path on a wind field that was reconstructed on just a few sensor measurements.

6 Conclusion

In the present work, wind fields of a large urban area were simulated with the wind angle as a parameter that builds the data basis for a model reduction by POD. The POD modes were then applied to reconstruct a high-resolution wind field from sparse sensor measurements by using the Gappy POD. The sensor measurements are obtained from a CFD test data. Two different sensor placement methods (DEIM/QR factorization) were investigated that calculate appropriate sensor locations. Three different error types were defined. The wind field reconstruction of the test data ($\varphi = 77°$) shows an interpolation error of 20%, a truncation error of 7% by using 10 POD modes and a sparsity error of 5% for the DEIM and 3% for the QR factorization by taking 10 POD modes and 10 sensor measurements. The reconstructed wind fields were then applied on an exemplary UAV flight mission to plan the minimum-energy path from a start to a final location. Compared to the shortest path, the wind-based path was adjusted to avoid strong head-wind which resulted in a maximum energy reduction of $\Delta E = 11.2\%$ when the wind was coming from east. The latter value was obtained by taking the complete full-order wind field. However, by using the wind field from ROM_5, for example, the energy reduction was shown to be almost as large with $\Delta E = 10.4\%$. An overall mean energy reduction was calculated of 3.8% by using the wind field from the $\mathrm{ROM}_{5,DEIM}$ and of 3.9% by using the $\mathrm{ROM}_{5,QR}$.

References

1. Wang, B.H., Wang, D.B., Ali, Z.A., Ting Ting, B., Wang, H.: An overview of various kinds of wind effects on unmanned aerial vehicle. Measur. Control **52**(7–8), 731–739 (2019)
2. Ware, J., Roy, N.: An analysis of wind field estimation and exploitation for quadrotor flight in the urban canopy layer. In: EIEE International Conference on Unmanned Aircraft Systems (ICUAS) (2016)

3. Kothari, M., Postlethwaite, I., Gu, D.W.: UAV path following in windy urban environments. J. Intell. Robot. Syst. **74**(3–4), 1013–1028 (2014)
4. Blocken, B., Stathopoulos, T., van Beeck, J.: Pedestrian-level wind conditions around buildings: review of wind-tunnel and CFD techniques and their accuracy for wind comfort assessment. Build. Environ. **100**(11), 50–81 (2016)
5. Mortezazadeh, M., Wang, L.L., Albettar, M., Yang, S.: CityFFD - city fast fluid dynamics for urban microclimate simulations on graphics processing units. Urban Clim. **41**, 101063 (2022)
6. Kastner, P., Dogan, T.: A cylindrical meshing methodology for annual urban computational fluid dynamics simulations. J. Build. Perform. Simul. **13**, 59–68 (2020)
7. Tong, Z., Li, Y.: Real-time reconstruction of contaminant dispersion from sparse sensor observations with gappy POD method. Energies **13**, 1956 (2020)
8. Bui-Thanh, T., Damodaran, M., Willcox, K.: Aerodynamic data reconstruction and inverse design using proper orthogonal decomposition. AIAA J. **42**, 1505–1516 (2004)
9. Willcox, K.: Unsteady flow sensing and estimation via the gappy proper orthogonal decomposition. Comput. Fluids **35**(2), 208–226 (2006)
10. Callaham, J.L., Maeda, K., Brunton, S.L.: Robust flow reconstruction from limited measurements via sparse representation. Phys. Rev. Fluids **4**(10), 267 (2019)
11. Erichson, N.B., Mathelin, L., Yao, Z., Brunton, S.L., Mahoney, M.W., Kutz, J.N.: Shallow neural networks for fluid flow reconstruction with limited sensors. Proc. Math. Phys. Eng. Sci. (2020)
12. Everson, R., Sirovich, L.: Karhunen-Loeve procedure for gappy data. J. Opt. Soc. Am. A **12**, 1657–1664 (1995)
13. Manohar, K., Brunton, B.W., Kutz, J.N., Brunton, S.L.: Data-driven sparse sensor placement for reconstruction. IEEE Control Syst. **38**, 63–86 (2018)
14. Clark, E., Askham, T., Brunton, S.L., Nathan Kutz, J.: Greedy sensor placement with cost constraints. IEEE Sens. J. **19**(7), 2642–2656 (2019)
15. Jayaraman, B., Al Mamun, S.M.A., Lu, C.: Interplay of sensor quantity, placement and system dimension in pod-based sparse reconstruction of fluid flows. Fluids **4**, 109 (2019)
16. Yildirim, B., Chryssostomidis, C., Karniadakis, G.E.: Efficient sensor placement for ocean measurements using low-dimensional concepts. Ocean Model. **27**, 160–173 (2009)
17. Hågbo, T.O., Giljarhus, K.E.T.: Pedestrian wind comfort assessment using computational fluid dynamics simulations with varying number of wind directions. Front. Built Environ. **8**, 69 (2022)
18. Franke, J., Baklanov, A.: Best practice guideline for the CFD simulation of flows in the urban environment: cost action 732 quality assurance and improvement of microscale meteorological models. University of Hamburg, Germany (2007)
19. Ebert, C., Weiss, J., Uijt De Haag, M., Ruwisch, M., Silvestre, F.: Trajectory planning in windy urban environment – a gappy POD approach for wind field estimates with sparse sensors. AIAA Aviation Forum (2022)
20. Weiss, J.: A tutorial on the proper orthogonal decomposition. AIAA Aviation Forum (2019)
21. Chaturantabut, S., Sorensen, D.C.: Nonlinear model reduction via discrete empirical interpolation. SIAM J. Sci. Comput. **32**(5), 2737–2764 (2010)
22. Drmac, Z., Gugercin, S.: A new selection operator for the discrete empirical interpolation method - improved a priori error bound and extensions. SIAM J. Sci. Comput. **38**, A631–A648 (2016)

Towards Indirect Assessment of Surface Anomalies on Wind Turbine Rotor Blades

Daniel Feldmann[1(✉)], Felix Oehme[2], Lennart von Germersheim[1], Rubén López Parras[1], Andreas Fischer[2], and Marc Avila[1]

[1] Universität Bremen, Center of Applied Space Technology and Microgravity, Am Fallturm 2, 28359 Bremen, Germany
daniel.feldmann@zarm.uni-bremen.de

[2] Universität Bremen, Institute for Metrology, Automation and Quality Science, Linzer Straße 13, 28359 Bremen, Germany

Abstract. We present results from novel field, lab and computer studies, that pave the way towards non-invasive classification of localised surface defects on running wind turbine rotors using infrared thermography (IRT). In particular, we first parametrise the problem from a fluid dynamical point of view using the roughness Reynolds number (Re_k) and demonstrate how the parameter regime relevant for modern wind turbines translate to parameter values that are currently feasible in typical wind tunnel and computer experiments. Second, we discuss preparatory wind tunnel and field measurements, that demonstrate a promising degree of sensitivity of the recorded IRT data w. r. t. the key control parameter (Re_k), which is a minimum requirement for the proposed classification technique to work. Third, we introduce and validate a local domain ansatz for future computer experiments, that enables well-resolved Navier–Stokes simulations for the target parameter regime at reasonable computational costs.

Keywords: Wind turbines · health monitoring · laminar-turbulent transition · infrared thermography (IRT) · direct numerical simulations (DNS)

1 Background and Motivation

With 132 TW h electricity fed into the grid in 2020, wind power has become Germany's leading source of energy (27 %) outperforming any fossil source [1]. However, cost reduction is indispensable to further strengthen the competitiveness of sustainable energy production beyond initial stages of massive subsidisation and political control. Rotor blade inspection of ageing wind turbines, for example, is elaborate and time-consuming and thus offers high potential to further reduce energy production costs through more efficient maintenance procedures [2,3].

Health monitoring of wind turbine blades usually employs visual inspection (from the ground or using drones and climbers), in order to assess damage

A. Dillmann et al. (Eds.): STAB/DGLR Symposium 2022, NNFM 154, pp. 529–538, 2024.
https://doi.org/10.1007/978-3-031-40482-5_50

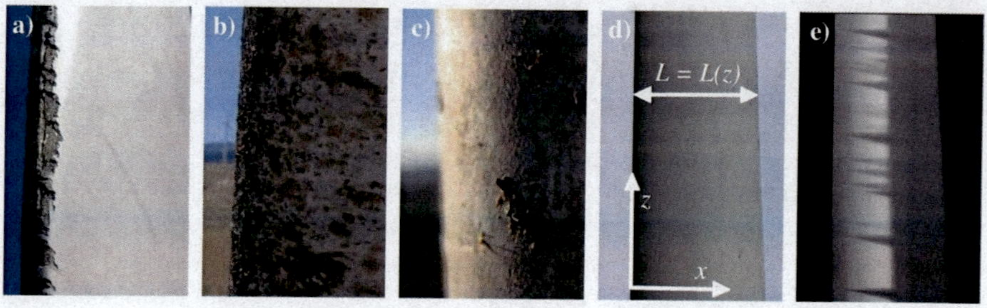

Fig. 1. Problem description. a–c): Wind turbine rotor blades with typical surface defects as they occur during operation [4]. a): Heavy erosion of the blade's leading edge due to impact (e. g., sand, rain, hail). b, c): Distributed and localised contamination with insect debris. d): Operating wind turbine in our preliminary field measurements in Thedinghausen. The blade suffers from naturally occurring surface anomalies, that are not visible from the ground (>100 m). The streamwise direction is approximately in x-direction and the blade's chord length ($L(z) \approx 5\,\mathrm{m}$) varies from root to tip. e): Thermographic image of the same blade (as in d), showing regions with higher (light) and lower (dark) surface temperature (black is blue sky compensation). Low temperature wedges indicate locally turbulent flow in the downstream wake of small localised surface anomalies in contrast to the otherwise laminar flow (high temperature).

and contamination level of the rotor blade's surface (Fig. 1a–c). These methods require a shutdown of the turbine and/or suffer from poor spatial resolution [5,6]. Common alternatives (e. g. strain-gauges, optical and acoustical sensors) require elaborate instrumentation and suffer from costly recertification [2]. Additionally, some of these techniques heavily impair the aerodynamic performance of the rotor blades. More advanced techniques (e. g. ultrasonic, radiographic and electromagnetic testing) are sometimes used in well-controlled lab environments, but are essentially impossible to transfer to running wind turbines in the field.

Infrared thermography (IRT) is an established diagnostic tool for indirect flow visualisation in aerofoil boundary layers [7], that overcomes the aforementioned drawbacks. It is based on the fact, that the convective heat flux between the fluid and the surface strongly depends on the local flow state (e. g. laminar or turbulent). For a given temperature difference between the incoming fluid and the aerofoil, flow state dependent spatial temperature gradients develop on the surface. This gradients can be measured by detecting the temperature dependent radiation from the surface with an infrared camera. For example, IRT has been used in well-controlled wind tunnel environments to detect laminar-turbulent transition and flow separation [8–10]. In these type of studies, the aerofoils are actively heated with $\mathcal{O}(10^3)\,\mathrm{W/m^2}$, to increase the convective heat flux density and thereby improving the thermal contrast (TC) between regions of different flow states. In contrast, applying IRT to operating wind turbines in the field is much more challenging because of several reasons [9]. First, active rotor blade heating is undesired. This heavily limits the maximum achievable TC, which now depends on local weather conditions and absorbed solar irradiation (i. e.

$\mathcal{O}(10^2)$ W/m^2 during clear sky). Second, measuring from remote ground locations (> 100 m) causes large transmission losses and low contrast to noise ratios (CNR). Third, the camera exposure time must not exceed $\mathcal{O}(10^{-4})$ s, in order to prevent blur due to rotor blade motion. Despite these constraints, IRT has been recently used to successfully detect premature laminar-turbulent flow transition due to surface anomalies in operating wind turbines [11–14]. These achievements open new avenues to monitor damage and contamination level of wind turbine blades (Fig. 1a–c) via IRT of the blade's surface (Fig. 1e) without shutting the turbine down for inspection nor modifying the rotor with additional sensors. For the application, however, it is not sufficient to simply detect the presence of an anomaly via IRT. Instead, it would be necessary to additionally infer exact position, size and shape of geometrical defects in order to enable remote classification of the surface anomaly; e. g. distinguish between erosion and deposition (Fig. 1a,b) or distinguish between different levels of contamination (Fig. 1b,c). This poses many methodological challenges and it is currently unclear, to what extent such a classification can be inferred faithfully from IRT images and whether physically informed modelling approaches might improve this endeavour.

Here we present results from our preliminary studies paving the way towards indirect assessment of surface anomalies on running wind turbine rotor blades. In Sect. 2, we briefly summarise the state of the art in early laminar-turbulent transition in flat plate (Blasius) boundary layers and establish terminology necessary for the rest of the paper. In Sect. 3, we introduce a parametrisation of our problem in terms of the roughness Reynolds number (Re_k), which is commonly used to characterise roughness induced transition in Blasius boundary layers. In Sect. 4, we discuss new results from our wind tunnel and field measurements, that demonstrate a promising degree of sensitivity of the recorded IRT data with respect to Re_k. In Sect. 5, we introduce and validate a local domain ansatz for future computer simulations, that enables well-resolved 3d DNS for the target parameter regime at reasonable computational costs. In Sect. 6, we briefly propose future directions.

2 Turbulence Transition in Blasius' Boundary Layer Flow

Probably the best understood example of boundary layer (BL) transition is a flat plate BL flow. In this canonical benchmark system, the streamwise velocity profile, $u_x = u_x(x, y)$, and thus the thickness of the BL, $\delta_{99}(x)$, are both very well described by the self-similar Blasius solution [e. g. 15]. The Blasius profile only depends on the wall-normal coordinate (y) and can be re-scaled for arbitrary streamwise locations ($x > 0$) just by fixing the Reynolds number, $Re = {}^{x\,u_\infty}/_{\nu_\infty}$, where u_∞ and ν_∞ are the velocity and the viscosity of the fluid in the far field.

The Blasius BL becomes linearly unstable at $Re = \mathcal{O}(10^5)$. In experiments, however, this instability is bypassed and laminar-turbulent transition can occur at lower Re due to finite-amplitude perturbations, which are usually present in the free-stream [16]. Especially relevant to our present work is the transition due to single (isolated) geometric perturbation elements on the plates surface, which perturb the velocity profile locally. Following a number of studies chiefly employing flow visualisations [17,18], Klebanoff et al. [19] provided an in depth, quantitative analysis of the transition to turbulence caused by a semisphere glued to a flat plate's surface. They showed that the key parameter governing the transition is the roughness Reynolds number, $Re_k = {u_k k}/{\nu_\infty}$, where k is the height of the roughness element (e.g. radius of the semisphere) and $u_k = u_x(x = x_k, y = k)$ is the undisturbed streamwise velocity component at that height ($y = k$) and streamwise location ($x = x_k$). They placed semispheres with different k at several x_k and found a critical Reynolds number of $Re_k \approx 325$, that is independent of x_k. They also studied a cylinder of unit aspect ratio ($\Gamma_k = {k}/{d} = 1$, where d is the cylinders diameter) and found a critical Reynolds number of $Re_k \approx 450$, which was lower than in most previous studies [18]. Since then, many experimental [15,20,21] and numerical [22–26] studies have been carried out, and it has been repeatedly highlighted that Re_k is the main parameter governing transition, albeit the geometry of the element (including Γ_k) plays an important role as well. For the specific case of cylinders, Puckert & Rist [15] demonstrated a changeover from purely convective to global instability for transcritical Re_k and an additional competition between varicose and sinuous instabilities, when the element is especially thin (i.e. large Γ_k). This indicates that the perturbation element solely acts as an amplifier in the subcritical regime (leading to symmetrical wake structures) and, additionally, as a wavemaker in the supercritical regime (leading to wake vortex shedding). We expect, that already these subtle differences, for example, will be distinguishable in the instantaneous wall-shear stress and thus in the plate's surface temperature distribution as well, assuming a unit Prandtl number and Reynolds' analogy to hold.

For wind turbines, however, we expect the curvature of the blade's surface to play an important role for this specific transition scenario. First, because the curved geometry results in streamwise pressure gradients and velocity profiles that can deviate notably from the Blasius solution. Second, because the velocity profile is not self-similar, implying that a particular Re_k can result from different combinations of (x_k, k), as discussed in Sect. 3. It is currently unclear to what extent these conditions change the underlying mechanisms (compared to Blasius' scenario) and whether these changes might help or hinder our endeavour.

3 Relevant Parameter Space (Re-Re_k)

Modern wind turbines in the multi-megawatt class operate at Reynolds numbers up to $Re = 10^7$, where x is taken to be L (i.e. the local chord length of the rotor blade), and u_∞ is taken to be the relative wind speed at a particular radial location (z) of the rotating blade (Fig. 1e) [27]. Modern wind turbines also employ rather thick rotor blade profiles (Fig. 2a) [28], so that even for large angle of attack (α), the most prone location for surface anomalies due to foreign impact remains the vicinity of the blade's leading edge (${x}/{L} \lesssim 0.2$). In this region, the BL is typically laminar, since natural transition in wind turbine rotor blades usually occurs later (${x}/{L} \gtrsim 0.3$). Now, we here consider generic geometric perturbation

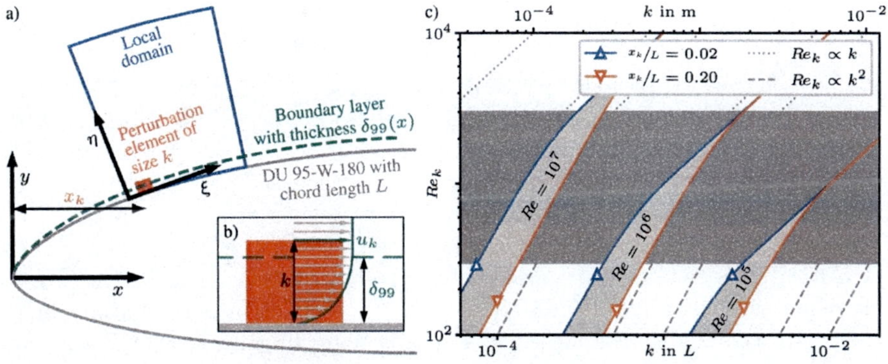

Fig. 2. Parametrisation of the problem. a): Sketch of a typical wind turbine rotor blade profile (DU 95-W-180 [28]) with its leading edge at $(x = 0, y = 0)$ in the global coordinate system. A laminar boundary layer (broken line) is growing from its leading edge and a generic perturbation element is located at $x_k/L \approx 1/10$. A local computational domain is placed around the perturbation element. b): Zoom of a. c): Relevant parameter space in terms of element size (k), position (x_k), chord (Re) and roughness (Re_k) Reynolds number, based on the crude assumption of a Blasius BL to estimate the local BL thickness, $\delta_{99} = \delta_{99}(x_k)$, and the relevant velocity scale, $u_k = u_k(x_k, k)$.

elements with a characteristic size ranging from $k = \mathcal{O}(10^{-5})$ m (reflecting e. g. small sand grain) up to $k = \mathcal{O}(10^{-2})$ m (reflecting e. g. insect debris) in order to model highly localised deposition on operating wind turbine rotor blades (Fig. 1c). Note, that more severe (distributed) contamination (Fig. 1b) and surface ablations (Fig. 1a) will be considered elsewhere. Furthermore, we begin by making the crude assumption, that the laminar BL in this region can be approximated by Blasius' solution. This allows us, for the first time, to roughly quantify the relevant range of Re_k for our problem by estimating the a priori unknown u_k from the given parameter values (Re, x_k, k) discussed above.

The results are summarised in Fig. 2c. The dark grey region denotes the range of Re_k realised in our preliminary wind tunnel campaign (at $Re = 10^5$, see Sect. 4), covering one order of magnitude from the sub-critical ($Re_k \approx 300$) up to the fully turbulent ($Re_k \approx 3000$) regime. Figure 2c also indicates, how these values translate to other situations; e. g., other combinations of (x_k, k) or higher Re. In this context, it is important to note, that for $k \ll \delta_{99}$ (i. e. element fully embedded in the linear part of the Blasius BL), Re_k scales quadratic with k for a given location (x_k), whereas for $k \gg \delta_{99}$ (i. e. element penetrates the flow outside the Blasius BL, as in Fig. 2b), Re_k scales only linear with k and is thus basically independent of x_k. Where exactly this transition from the quadratic to the linear scaling takes place, depends on Re. In any case, relevant locations for generic perturbation elements would be $0.02 < x_k/L < 0.2$, according to the discussion above. For a thick rotor blade profile (in contrast to the crude Blasius assumption here), the scaling in this two regimes will not be exactly quadratic and linear, due to curvature effects, but we expect the general behaviour to be qualitatively similar.

4 Infrared Thermography (IRT)

We have performed preparatory IRT measurements on a running 1.5 MW-wind turbine in Thedinghausen, Germany, and on a down-scaled wind tunnel model using the Deutsche WindGuard facilities near Bremerhaven. Our results show, that it is feasible to distinguish between different localised perturbation elements in terms of their size (k) and location (x_k) by means of an IRT image of the blades' surface temperature.

Already our field measurement at $Re = \mathcal{O}(10^6)$ demonstrates a very promising degree of sensitivity of the IRT method w. r. t. different perturbations in the vicinity of the blade's leading edge (Fig. 1e). Note, that here natural perturbations on the rotor blade's surface (as they occur during operation) generate a variety of distinguishable thermal wake patterns in the IRT image. In particular, the different shades of dark grey wedges indicate regions of different temperatures at different locations on the blade's surface that are presumably generated by surface anomalies of different size and location.

Our wind tunnel measurements at $Re = \mathcal{O}(10^5)$ are a first successful effort to better quantify the underlying relationships between surface anomalies (cause) and their footprint in the surface temperature (effect). Here, we equipped the suction side of a typical rotor blade profile (DU 97-W-250 [28], $L = 30\,\text{cm}$, $\alpha = 2°$) with ten cylindrical perturbation elements of varying size ($k \in [0.5, 1.6]$mm, constant $\Gamma_k = 0.3$) and repeated the IRT measurements for three different element locations ($^{x_k}/_L \in \{0.03, 0.1, 0.2\}$). Our results are summarised in Fig. 3. Although Re_k is the only control parameter in the flat plate transition scenario, it is apparent from Fig. 2, that on an aerofoil, one particular Re_k can be realised with different combinations of (x_k, k) for given free-stream conditions (Re). Thus, we first fix $^{x_k}/_L = 0.2$ to isolate the effect of varying k (Fig. 3a). In this particular scenario, IRT is remarkably sensitive with respect to a varying k, which here corresponds to $Re_k \in [270, 800]$. From element #3 on (i. e. $Re_k > 500$), a thermal signature is clearly visible in the downstream wake (Fig. 3a). This indicates that IRT is indeed capable of detecting weak transitional wakes at low trans-critical Re_k. If we now translate this first rough estimate of a minimum threshold for detectability ($Re_k \approx 500$) to a real-word wind turbine scenario by making use of Fig. 2b, we can conclude that our proposed measurement approach is suitable for detecting rotor blade defects as small as roughly 0.1 mm, with the exact threshold depending on both, $^{x_k}/_L$ and Re. Additionally, the size of the wake pattern as well as its CNR w. r. t. the undisturbed (i. e. warmer) surrounding increase with Re_k (Fig. 4a), thereby demonstrating a continuous sensitivity of our measurement approach.

Comparing Figs. 3a, b, and c additionally demonstrates that the IRT method is clearly capable of distinguishing different x_k in the entire range of relevant distances from the blade's leading edge ($^{x_k}/_L \in [0.03, 0.2]$). However, for elements at $^{x_k}/_L = 0.03$, the wake pattern in the thermogram is less distinct when compared to the wake patterns originating from elements placed farther downstream (e. g. $^{x_k}/_L = 0.1$). In particular, the CNR of the thermal wake patterns increases twofold, when the perturbation elements are moved farther downstream (i. e.

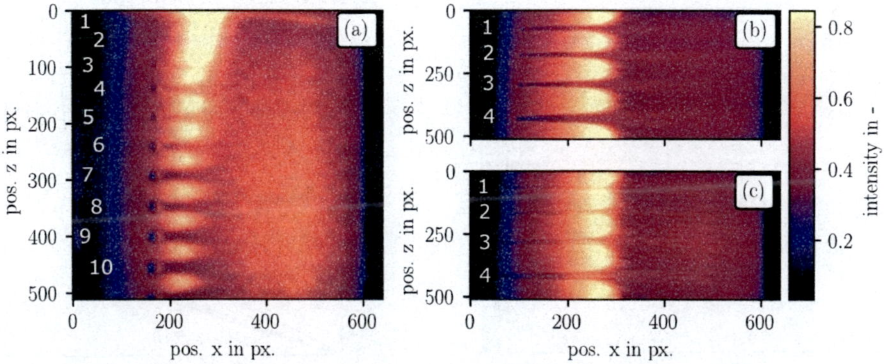

Fig. 3. Sensitivity analysis in the wind tunnel. Shown are IRT images of the blade's surface for fixed ($Re = 10^5, \alpha = 2°$) and varying (Re_k, x_k), where higher/lower intensity reflects a warmer/cooler surface. The blade geometry is a DU 97-W-250 profile [28] with a constant chord length of $L = 30$ cm. a): Sensitivity of the thermogram for fixed ($^{x_k}/_L = 0.2$) and varying $Re_k \in [270, 800]$ in the transcritical regime, where we expect the onset of unsteady flow. The cylindrical perturbation elements are numbered from one to ten with increasing Re_k. b): Same as in a, but for ($^{x_k}/_L = 0.1$). The four elements are numbered with increasing $Re_k \in [800, 3100]$. c): Same as in b, but for ($^{x_k}/_L = 0.03$).

from $^{x_k}/_L = 0.03$ to $^{x_k}/_L = 0.1$). We conclude, that the position of a surface defect directly affects the detectability of its thermal wake pattern. Note, that for the four elements shown in Figs. 3b and c, $Re_k \in [800, 3100]$, which is considerably higher than the roughness Reynolds numbers shown in Fig. 3a.

5 Direct Numerical Simulations (DNS)

We have performed preparatory DNS of the incompressible Navier–Stokes equations using the spectral-element framework `nektar++`, which is purposely designed for high-fidelity flow simulations on unstructured grids in moderately complex domains. In order to enable well-resolved 3d DNS of laminar-turbulent transition in a rotor blade boundary layer with geometric perturbation elements at reasonable computational costs, we introduce and validate a *local domain ansatz* in massively reduced computational domains. We stress, that otherwise an extensive variation study of the relevant control parameters would be prohibitively expansive for the intended range of parameter values ($Re > 10^5$, $Re_k \gg 300$).

The general idea of our local ansatz (Fig. 2a) goes as follows. First, we perform a 2d simulation of a typical wind turbine rotor blade (DU 95-W-180 [28]) in a large (global) computational domain without any perturbation element. Note, that we here use a sufficiently large domain (C-type, dimensions $15L \times 12L$), in order to generate tailored inflow and pressure boundary conditions (BC) for much smaller local domains, that are located around the desired x_k. Recall, that

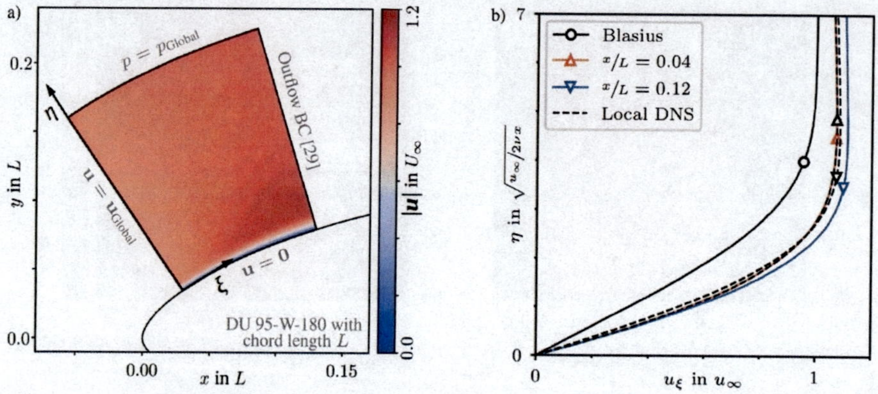

Fig. 4. Local domain ansatz. a): Velocity field ($|\boldsymbol{u}| = \sqrt{u_x^2 + u_y^2}$) generated in a local 2d simulation ($Re = 10^4$, domain at $0.03 \leq x/L \leq 0.13$) of a clean rotor blade (DU 95-W-180 [28], $\alpha = 2°$, w/o perturbation) using Dirichlet BC for $\boldsymbol{u}$ and p as indicated and a robust outflow BC [29] at the right. b): Comparison of the velocity profile, $u_\xi(\eta)$, in the boundary layer at two different streamwise locations, x/L, as predicted in local (broken lines) and global (full lines) DNS. The Blasius solution is given as reference.

the flow in the region of interest (i. e. $x/L < 0.2$) is always laminar, justifying our 2d approach. In a second step, we perform well-resolved 3d DNS in a local domain capturing only a small part of the curved rotor blade surface. In the future, this will allow us to include different isolated perturbation elements and, for the first time, investigate the transitional/turbulent flow in their downstream wake and the resulting local heat transfer between the fluid and the curved rotor blade surface in unprecedented detail.

Here we demonstrate the general viability of our ansatz (Fig. 4). For example, Fig. 4b compares velocity profiles from a global and a local DNS run (here both 2d, $Re = 10^4$) and shows two things. First, the velocity profiles in the rotor blade boundary layer are, as expected, substantially different from the Blasius solution. Second, the results obtained from local DNS runs agree very well with the global DNS results. In particular, close to the inflow BC of the local domain ($x/L = 0.04$, where later perturbation elements will be placed), the velocity profiles are identical (judged by the given line width of the plot). Close to the outflow BC of the local domain ($x/L = 0.12$), the BL profiles have slightly diverged ($||u_{\zeta,\text{local}} - u_{\zeta,\text{global}}||_\infty \leq 10^{-2}$), but still show very good qualitative agreement. Note, that these results represent a worst-case scenario, since the local domain starts very close to the blade's leading edge, where surface curvature has the strongest effect. For local domains that are located farther downstream (e. g. starting at $x/L = 0.1$ in order to capture perturbation elements farther away from the leading edge), we find that the deviations between local and global DNS results are even smaller (not shown). Note, that we have tested and analysed many different combinations of BC for the local DNS ansatz. In all cases, we have used Dirichlet and Neumann BC for u and p, respectively, at the inflow (left) and the blades surface (bottom), where the values for the Dirichlet BC were extracted from the global run. The results presented in Fig. 4, we obtained using pressure Dirichlet BC at the top and the robust outflow BC of Dong et al. [29] at the right. Other combinations of BC for the top/right boundaries yield very similar results with locally slightly better agreement, depending on the region of interest (i. e. either inside or outside the BL).

6 Perspective

Our preliminary studies show two things. First, IRT is sufficiently sensitive to distinguish between different size and location of highly localised perturbation elements in typical wind turbine rotor blade boundary layers. Second, the proposed local domain ansatz is feasible to approximate the boundary layer flow on a typical rotor blade profile sufficiently well, allowing for well resolved Navier–Stokes simulations of the underlying problem at reasonable computational costs.

Future works should aim at generating a comprehensive surface-state-flow-pattern catalogue and a deeper physical understanding of how surface irregularities in the vicinity of the blade's leading edge (cause) affect the downstream surface temperature distribution (effect). Here, it is particularly interesting to clarify, how much information about the three-dimensional flow in the wake is actually contained in its two-dimensional surface temperature projection, since the inverse problem (effect to cause) is not bijective. To this end, extensive wind tunnel and computer experiments could be combined in order to systematically characterise the effect of different generic surface anomalies on early transition and the resulting (turbulent) wake in a rotor blade boundary layer and the resulting temperature footprint. Hopefully, this will lay the foundation for the development of a data-trained physics-informed model to reliably relate individual thermographic images to particular rotor blade conditions based on image recognition and machine learning techniques.

Acknowledgement. We gratefully acknowledge resources and support provided by Nicolas Balaresque (Deutsche WindGuard) and the North German Supercomputing Alliance (HLRN).

References

1. Strom-Report. Germany's Power Mix 2020—Data, Charts & Key Findings (2021). https://strom-report.de/germany-power-generation-2020/
2. Li, D., Ho, S.C.M., Song, G., Ren, L., Li, H.: Smart Mater. Struct. **24**(3) (2015)
3. García Márquez, F.P., Peco Chacón, A.M.: Renew. Energy **161**, 998 (2020)
4. Ehrmann, R.S., Wilcox, B., White, E.B., Maniaci, D.C.: Effect of surface roughness on wind turbine performance. Technical report 10, Sandia National Laboratories, Albuquerque (2017)
5. Kim, D.Y., Kim, H.B., Jung, W.S., Lim, S., Hwang, J.H., Park, C.W.: 44th International Symposium on Robotics, pp. 1–5. IEEE (2013)

6. Shafiee, M., Zhou, Z., Mei, L., Dinmohammadi, F., Karama, J., Flynn, D.: Robotics **10**(1), 1 (2021)
7. Gartenberg, E., Roberts, A.S.: J. Aircr. **29**(2), 161 (1992)
8. Montelpare, S., Ricci, R.: Int. J. Therm. Sci. **43**(3), 315 (2004)
9. Dollinger, C., Sorg, M., Balaresque, N., Fischer, A.: Exp. Thermal Fluid Sci. **97**, 279 (2018)
10. Dollinger, C., Balaresque, N., Gaudern, N., Gleichauf, D., Sorg, M., Fischer, A.: Renew. Energy **138**, 709 (2019)
11. Traphan, D., Herraéz, I., Meinlschmidt, P., Schlüter, F., Peinke, J., Gülker, G.: Wind Energy Sci. **3**(2), 639 (2018)
12. Reichstein, T., et al.: Energies **12**(11), 2102 (2019)
13. Gleichauf, D., Sorg, M., Fischer, A.: Appl. Sci. **10**(18) (2020)
14. Parrey, A.M., Gleichauf, D., Sorg, M., Fischer, A.: Appl. Sci. **11**(18), 8700 (2021)
15. Puckert, D.K., Rist, U.: J. Fluid Mech. **844**, 878 (2018)
16. Brandt, L., Schlatter, P., Henningson, D.S.: J. Fluid Mech. **517**, 167 (2004)
17. Klanfer, L., Owen, P.R.: The effect of isolated roughness on boundary layer transition. Technical report, Royal Aircraft Establishment (1953)
18. Smith, A.M.O., Clutter, D.W., Aeros, J.: Sciences **26**(4), 229 (1959)
19. Klebanoff, P.S., Cleveland, W.G., Tidstrom, K.D.: J. Fluid Mech. **237**(101), 101 (1992)
20. Ye, Q., Schrijer, F.F., Scarano, F.: Int. J. Heat Fluid Flow **61**, 31 (2016)
21. Ye, Q., Schrijer, F.F., Scarano, F.: J. Fluid Mech. **837**, 597 (2018)
22. Masad, J.A., Iyer, V.: Phys. Fluids **6**(1), 313 (1994)
23. Fransson, J.H., Brandt, L., Talamelli, A., Cossu, C.: Phys. Fluids **17**(5), 1 (2005)
24. Loiseau, J.C., Robinet, J.C., Cherubini, S., Leriche, E.: J. Fluid Mech. **760**, 175 (2014)
25. De Grazia, D., Moxey, D., Sherwin, S.J., Kravtsova, M.A., Ruban, A.I.: Phys. Rev. Fluids **3**(2), 1 (2018)
26. Casacuberta, J., Groot, K.J., Ye, Q., Hickel, S.: Flow Turb. Combust. **104**(2–3), 533 (2020)
27. Ge, M., Tian, D., Deng, Y.: J. Energy Eng. **142**(1), 04014056 (2014)
28. Timmer, W.A., van Rooij, R.P.J.O.M.: J. Solar Energy Engin. **125**(4), 488 (2003)
29. Dong, S., Karniadakis, G.E., Chryssostomidis, C.: J. Comput. Phys. **261**, 83 (2014)

Development of a Low-Cost Thermal Human Manikin with Seven Zones for Comfort Assessment in Passenger Cabins

Daniel Schiepel(✉), Felix Werner, and Daniel Schmeling

German Aerospace Center (DLR), Institute of Aerodynamics and Flow Technology, Göttingen, Germany
daniel.schiepel@dlr.de
http://www.dlr.de

Abstract. A low-cost thermal human manikin with seven zones has been developed for comfort assessment in passenger cabins. The total costs for tor the manikin amount to less than 1, 000 € and it allows for temperature measurements in the individual zones, where the small differences in the resistance depending on the temperature are recorded. Based on this dependence both, the surface temperature and the equivalent temperature, can be determined on a zonal basis. The latter allows for a comfort assessment of the individual zones in accordance with common standards.

Keywords: thermal manikin · zonal heat release · comfort assessment · equivalent temperature

1 Introduction

According to a recent study of Kim et al. (2019) people spend a considerable amount of their time in vehicles (up to 51 min per day) [1]. Hence, for the manufacturers thermal comfort in cars is a key argument in terms of the competitive edge. Thermal comfort is generally influenced by temperature, radiation, airflow properties of the air-conditioning system and individual sensation. Today, standard tests to characterize thermal comfort are based on e.g. the concept of equivalent temperature (T_{EQ}) as defined in the German industrial standard [2]. There are various approaches to determine T_{EQ} using sensors or infrared thermometry, see [3]. However thermal manikins are also suitable for this task. Various models are already available for this purpose, but they cost more than 100, 000 US dollars. Furthermore, these systems are often bulky and heavy. Consequently, they are only used sporadically and only at predefined positions. This means that the comfort information is only available at one seat and not at other relevant places. Accordingly, data acquisition at different locations is only possible in successive measurements. This results in time-consuming and cost-intensive planning and

A. Dillmann et al. (Eds.): STAB/DGLR Symposium 2022, NNFM 154, pp. 539–548, 2024.
https://doi.org/10.1007/978-3-031-40482-5_51

execution. Therefore, in experimental studies with several considered positions, the comfort information is often omitted and generic heat loads, e.g., heating mats, are used instead. For more detailed investigations, these generic loads are substituted by heated thermal passenger models, which simulate not only a realistic heat dissipation but also the geometrical obstruction of the passengers. For these investigations, independent measurement systems are required to record and evaluate the comfort-specific variables. However, since the thermal sensation of people is influenced by the interaction of various comfort-relevant flow variables, such as air temperature, air velocity and radiation temperature, the sole and selective measurement of these variables is often insufficient for a holistic and robust evaluation of thermal comfort and can lead to an inadequate assessment. For this reason, the already established measuring device "thermal human manikin" [4], which allows to record the equivalent temperature in accordance with EN ISO 14505-2 [2], was further enhanced. This measuring device is a heatable human model, which simulates a realistic heat emission of a human being as well as its obstruction and offers the possibility to simultaneously determine the equivalent temperature by means of infrared thermography. However, the human model can only be operated with a predefined heating power distribution. Yet, there are a number of applications where zonal control of the heating power of individual body parts is necessary, e.g., to simulate the human metabolism for different body parts. Moreover, thanks to the zonal implementation the surface temperature can be determined without the use of infrared cameras, ultimately allowing zonal comfort to be determined. This acquisition of the comfort variables is important because of typical asymmetries within cabins, e.g., a colder arm at the side of the window during the winter. In this paper, the development of a light-weight low-cost thermal human model with seven zones for measuring thermal comfort in passenger cabins is presented. The model can acquire the zonal T_{EQ} and the associated heating power required for determining thermal comfort.

2 Experimental Methods

A prototype of a heatable thermal manikin was developed which allows to control seven body zones individually by means of a defined heat flow. The basic body for the thermal human model is a male, fully-mobile manikin (Livelymen Volo by lideimmagine) [5]. The body consists of expanded polypropylene (EPP) and is wrapped with heating wires to allow for the heat generation. The wires are covered with a thin black anodized aluminium layer thus distributing the heat more efficiently and generating more homogeneous surface temperatures. The prototype has seven independently controllable zones, which are indicated in Fig. 1a). The assembled human model with black aluminium cover is shown in Fig. 1b).

Basic tests on heating power, wiring and durability are performed for the assembled model. In this validation run the model could not be heated and measured with high precision at the same time as the deduced values are only taken

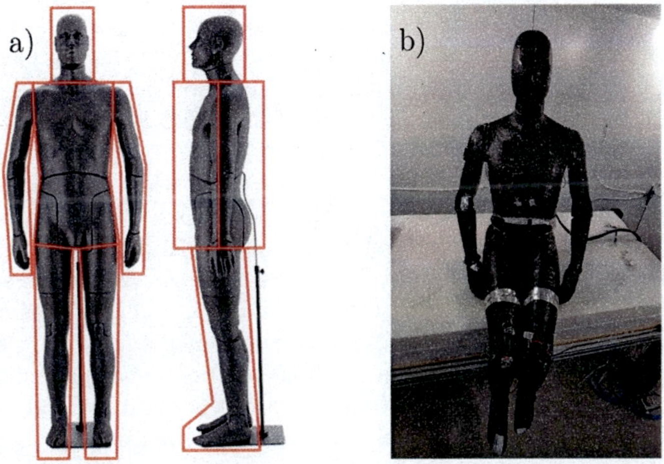

Fig. 1. a) Illustration of the seven individual zones of the thermal manikin based on [5]. b) Assembled thermal manikin with heating wire and outer aluminium cover.

from seven small power supply units (PSU). For a simplified assumption on the heat distribution of the individual body segments the power fraction is chosen as summarized in Table 1. Moreover, for a total heat output of 100 W the measured zonal resistance, the selected powers, the resulting voltages and the current are reported. As these values depend on the surrounding temperatures, an approximate room temperature of 21 °C was prescribed during the tests. Based on the measured values, the electronics can be chosen to withstand currents exceeding 2.7 A occurring for the head region and to operate with voltages between 5.6 V for the arms and 10.4 V for the legs. This reference heat output of 100 W corresponds to a human at light activity level whereas a reduced power of 75 W is released by a sitting human.

Table 1. Characterization of the seven zones of the zonal thermal manikin for a total power of 100 W. For each zone the measured resistance, power and power fraction, as well as current and voltage are indicated.

Body part	**Resistance in Ω**	Power in W	Fraction in %	Current in A	Voltage in V
Head	2.25	16.7	17	2.7	6.1
Front	4.35	16.7	17	2.0	8.5
Back	4.75	16.7	17	1.9	8.9
Left Arm	3.8	8.3	8	1.5	5.6
Right Arm	3.8	8.3	8	1.5	5.6
Left Leg	6.5	16.7	17	1.6	10.4
Right Leg	6.5	16.7	17	1.6	10.4

The electronic equipment was chosen based on the information of the pre-tests and the control of the individual zones was realized by means of a server-like structure. There is one control module and four power modules for two channels each. The control module contains an Adafruit FTDI Ft232h and a SparkFun 8-channel Qwiic Mux Breakout (TCA9548A) for the communication with the components. The power is specified using a pulse width modulation module (Adafruit 16-channel 12-bit PWM module), which is the interface between the two module types. The power modules use Cytron 10 A 5–30 V dual-channel DC motor drivers for the power supply. These motor drivers receive the power input via the PWM signal. A power measurement on the output side was implemented to exclude any power losses on the upstream components. The provided power is measured channel-wise by means of a voltage and current measurement module (Adafruit INA260), which transmits the data to the control module. Since each power module has its own supply input, the body zones can be grouped by their supply voltage as shown in Table 1. Therefore, it is possible to use optimally sized power supplies. However, any power supply with an output voltage of up to 40 V and sufficient current supply can be used. A photograph of the assembled control unit is presented in Fig. 2. In a) a power supply blade is shown with the main supply at the top, the wired electronics in the middle housed in a 3D printed enclosure and two supply outputs on the bottom of the image which lead to the thermal manikin. In Fig. 2b) the fully assembled control unit within its 3D printed enclosure is shown. The power supplies for the seven zones, i.e. four main supply inputs, and the USB connector can be seen.

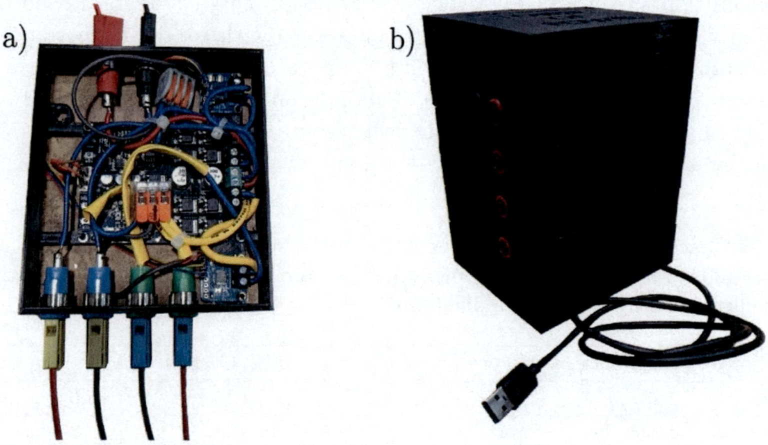

Fig. 2. Photograph of the assembled control unit. a) one power supply blade and b) the fully assembled control unit.

The control flow for the thermal manikin is illustrated in Fig. 3. The software for controlling the power distribution and data acquisition is implemented in Python3 on a standard laptop computer. Using the Adafruit FTDI Ft232h module the computer communicates with the Adafruit PWM module. This provides seven individual PWM outputs for the segment PSUs realized by DC motor drivers which are powered by one main power supply unit. The resulting power, current and voltage per segment are measured by individual power meters (Adafruit INA260) while the power is being provided to the body parts of the manikin. The data of the power meters is returned to the laptop computer. This allows for adjustments of the PWM duty cycle and thus to control the heating power of the body parts.

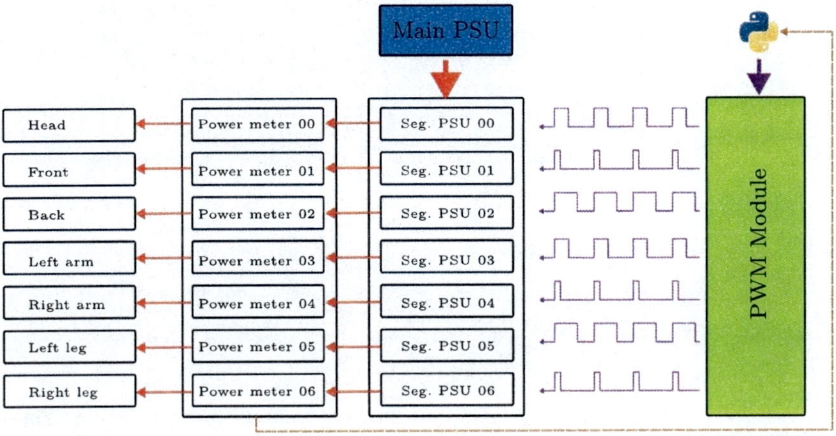

Fig. 3. Illustration of control (purple), data (orange) and power (red) flow of the thermal human manikin.

The seven-zone human model only weights 15 kg and can therefore be transported easily. The aim of designing a low-cost thermal manikin was achieved with a total costs for control and regulation electronics amounting to 150 € and 600 € for the basic body.

3 Results

In order to determine the thermal properties of the thermal manikin, it was put into a calibration chamber with temperature-controlled walls, see [4] for details. The inner walls are made of black aluminium with capillary tubes, which are positioned on their outer sides and connected to a temperature control device (Lauda Variocool VC 5000) with a precision of ± 0.1 K. The tubes are covered with styrodur insulation to reduce heat conduction. Using wires the thermal human manikin was hovering contact-free in the chamber to avoid undesired heat conduction between the model and the iso-thermal walls.

Before continuing with the calibration process of the thermal manikin it had to be validated that the manikin and the calibration chamber temperature were in agreement for the relevant range between 25 and 45 °C. Thus, in Fig. 4 the difference between the two temperatures is presented. It can be seen that the temperature differences are generally ranging from −0.1 to 0.4 K with slight increases to 0.6 K for a chamber temperature of 45 °C for the front and the back region. The temperature of the calibration chamber is slightly higher compared to the manikin. However, a sufficiently good agreement is achieved and the calibration procedure can be continued.

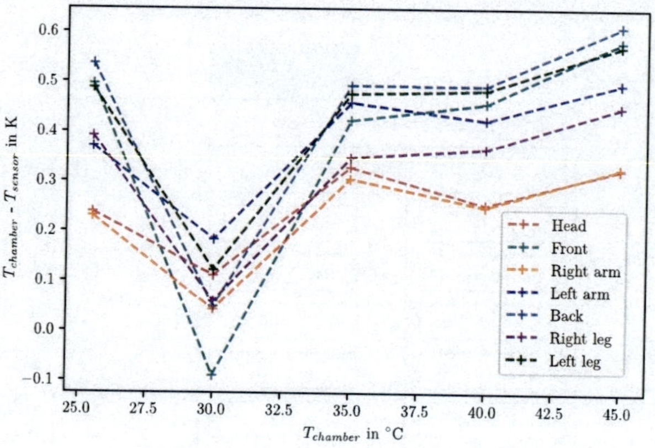

Fig. 4. Difference between the calibration chamber temperature and the manikin temperature depending on the chamber temperature. The thermal manikin released no heat.

According to the patent claim [6] the temperature-dependent resistance of the individual zones is of utmost importance for the acquisition of surface and equivalent temperatures. Here, the achieved accuracy of the electronics is put to the test. A temperature range of the calibration chamber between 5 and 45 °C is tested. The thermal manikin is operated in non-powered mode and - using several temperature sensors - it is guaranteed that the model and the chamber achieve the same temperature which typically takes up to six hours to reach equilibrium. This resistance is determined during a short, two-second current pulse to prevent the manikin's temperature from deviating from the chamber temperature. The resulting body-segment-wise resistance depending on the calibration chamber temperature is presented in Fig. 5. A clear linear dependence is apparent. The head with the shortest wires reflects the lowest resistance, whereas the highest resistance is determined for the legs. The body parts occurring twice (legs and arms) show the same values and dependence per pair. Moreover, all values are in good agreement with the observations presented in Table 1. This proves that the manikin and electronics allow for the acquisition of temperature-dependent resistances for the seven body segments.

The manikin temperature is in agreement with the calibration chamber temperature and can therefore be interpreted as the surface (skin) temperature of

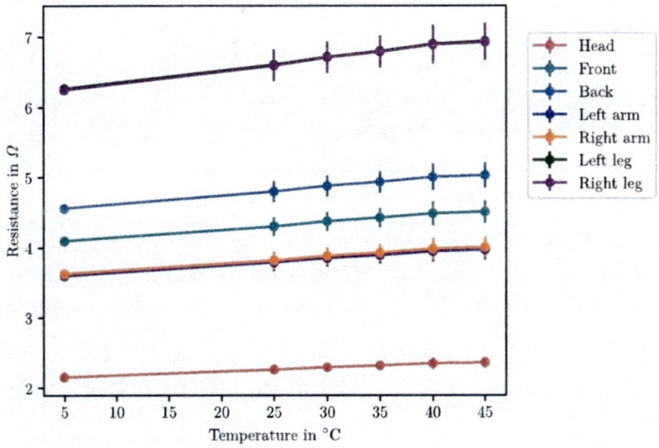

Fig. 5. Body-segment-wise resistance depending on the calibration chamber temperature. The thermal manikin released no heat.

the manikin, which is an important quantity for the comfort sensation. A linear curve fit to the data is performed allowing to calculate the surface temperature T_S from the resistance R:

$$T_S = m \cdot R + b \tag{1}$$

As an example, Fig. 6 shows the data for the head zone in red and the resulting regression curve is given in black. The estimated fitting parameters are indicated in the lower right corner. A good agreement can be seen, thus making the approach trustworthy.

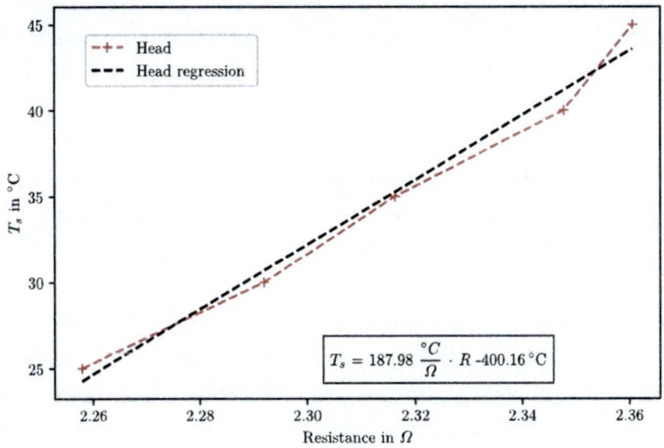

Fig. 6. Body-segment-wise resistance depending on the calibration chamber temperature for the head (red). A linear curve fit is performed and the resulting regression curve is given in black. The thermal manikin released no heat.

In the next step a sitting human is simulated by means of an accumulated heating power of 75 W distributed in fractions (indicated in Table 1) over the individual body parts. The manikin is placed in the calibration chamber at different chamber temperatures between 15 and 45 °C. The chamber temperature now corresponds to the temperature sensation, the equivalent temperature T_{EQ} (for details see [4]). This directly allows for a comfort assessment according to [2]. The determined dependence of T_{EQ} on the zonal resistance R is presented in Fig. 7. The clear trend and linear dependence of both quantities demonstrate the capability of the manikin to assess the thermal comfort reflected by T_{EQ}.

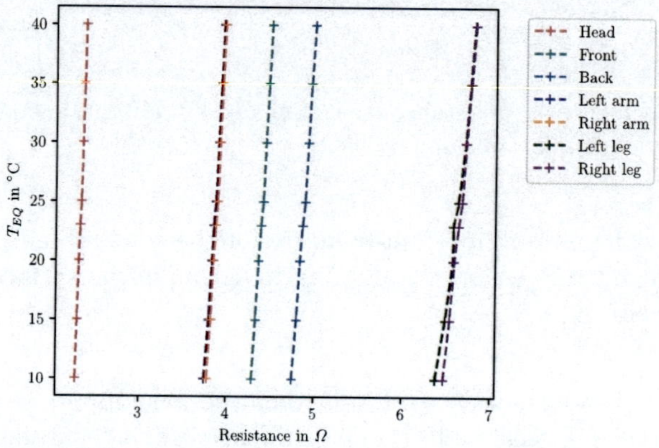

Fig. 7. Dependence of the equivalent temperature T_{EQ} on the zonal resistance R for a constant heat release of 75 W.

Furthermore, the surface temperature has often been used to determine the comfort sensation defined by T_{EQ} by means of infrared thermometry. However, all body parts have to be visible for the camera, which can be difficult - especially for the back of the manikin. The presented method allows for the acquisition of T_S and T_{EQ} for each body part without suffering from such limitations. In Fig. 8 the dependence of T_S on T_{EQ} is presented. This reveals that the heating power of the thermal manikin generally yields a higher temperature for T_S compared to T_{EQ}.

Although the dependence of T_S on the resistance is inherent in the calibration procedure, it is important to demonstrate the stability of the system for a long operating time of the heat-releasing manikin (more than seven days). In Fig. 9 the dependence of T_S on R for the heated manikin is presented. The curves reflect a linear dependence as the surrounding temperature is increased (which yields the presented increase in resistance). A drift in acquisition accuracy or manikin stability would be reflected by a deviation from this linear slope. As this is not observed, the system is stable for this long operating time. Moreover, the dependence $T_S(R)$ is of utmost importance if not only a comfort assessment by means of the equivalent temperature is desired but rather a full simulation

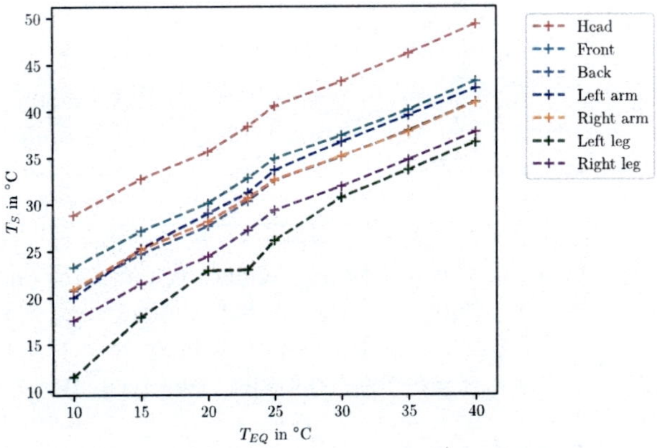

Fig. 8. Dependence of the surface temperature T_S on the equivalent temperature T_{EQ} for a constant heat release of 75 W.

of the human metabolism [7]. The latter requires the knowledge of the heating power for each body part and its surface temperature to determine the thermal comfort and to adjust the heating power. Both quantities are accessible with this new zonal thermal manikin.

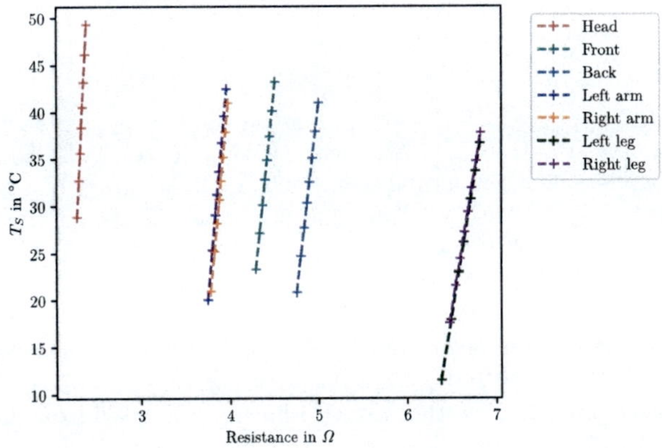

Fig. 9. Dependence of the surface temperature T_S on the zonal resistance R for a constant heat release of 75 W.

The power meter, INA260, exhibits an uncertainty of 25 mA. For a typical voltage of 6.65 V and a current of 3.2 A an uncertainty of $\sigma_R = 0.01\,\Omega$ is reported. Concerning the uncertainty of the surface temperature $\sigma_{T_S} = 1.4\,\text{K}$ is deduced by uncertainty propagation. For a investigated surface temperature range of 40 K this results in a dynamic range of 28 levels.

4 Conclusion and Outlook

A low-cost thermal human manikin with seven zones has been developed for the comfort assessment in passenger cabins using a fully mobile manikin made of expanded polypropylene. The total cost for the components amount to less than 1000 €. The manikin was calibrated in a temperature calibration chamber to allow for the temperature measurements of the individual zones. To do so, the small differences in the resistance depending on the temperature of the individual zones were determined. Based on this dependence, both the surface temperature and the equivalent temperature can be evaluated on a zonal basis. The equivalent temperature allows for a comfort assessment for the individual zones in accordance with common standards. In terms of the uncertainty of the surface temperature measurements, a high dynamic range of 28 is reported. In the future, the resistance-dependent surface temperature acquisition in combination with the measured heating power can be used to interface this zonal manikin to a full metabolism simulation as a hardware in the loop implementation. Here, a potentially applicable comfort simulation is the 33 node comfort model, NOODEL [7,8].

Acknowledgement. The presented work has been supported by the grant “KliMat” of the European Regional Development Fund (ERDF) and the federal state of Lower Saxony.

The authors like to thank Annika Köhne for proof-reading this work.

References

1. Kim, W., Anorve, V., Trefft, B.: American Driving Survey, 2014–2017 (2019)
2. Beuth: Ergonomics of the thermal environment - Evaluation of thermal environments in vehicles - Part 2: Determination of equivalent temperature (2006). DIN EN ISO 14505-2:2007-04. https://doi.org/10.31030/9765403
3. Schiepel, D., Westhoff, A.: Study on the influence of turbulence on thermal comfort for draft air. In: Dillmann, A., Heller, G., Krämer, E., Wagner, C. (eds.) STAB/DGLR Symposium 2020. NNFMMD, vol. 151, pp. 494–503. Springer, Cham (2021). https://doi.org/10.1007/978-3-030-79561-0_47
4. Lange, P., Schmeling, D., Westhoff, A., Bosbach, J.: Cost-effective human comfort manikin with realistic thermal load for studies of convection-driven ventilation systems. In: Roomvent & Ventilation Conference, pp. 367–372 (2018)
5. Lideimmagine, Livelymen (2020). https://www.lideimmagine.com/wp-content/uploads/2019/10/cod.00001B.jpg. Accessed 20 June 2022
6. Schmeling, D., Bosbach, J., Lange, P.: Method of operating a thermal human model, electronic assembly and thermal simulator assembly. In: Offenlegungsschrift (DE 10 2018 108 573 A1), Patent Claim (2019)
7. Streblow, R.: Thermal sensation and comfort model for inhomogeneous indoor environments = Thermisches Komfortmodell für inhomogene Umgebungsbedingungen, thesis (2011)
8. Streblow, R., Müller, D.: NOODEL - 33 Knoten Komfortmodell. In: GI - Gebäudetechnik in Wissenschaft & Praxis, under review

Numerical Study of Aerosol Dispersion in the Aircraft Cabin

Andrei Shishkin(✉), Daniel Schiepel, and Daniel Schmeling

German Aerospace Center, Institute of Aerodynamics and Flow Technology, Bunsenstraße 10, 37073 Göttingen, Germany
Andrei.Shishkin@dlr.de

Abstract. We report the results of numerical simulations of aerosol formation and dispersion in the aircraft cabin. The simulation was performed in a full-scale model of the Do728 aircraft cabin using the open-source software OpenFOAM. Aerosol cloud formation was studied during normal human breathing, with particles modeled as a composition of solid nuclei (1% of total mass) and a solution of NaCl (1%) and water (98%). The time of transition to the equilibrium state as a function of particle size, as well as the spatial position and size distribution of the formed aerosol cloud were investigated. Further, dispersion of aerosol particles in the cabin was simulated and compared for two ventilation configurations: Mixing Ventilation (MV) and Cabin Displacement Ventilation (CDV). The comparison of particle residence time in air, the fraction of particles deposited, and the concentration of particles in the breathing area revealed certain advantages of the CDV configuration.

Keywords: Eulerian-Lagrangian simulation · aerosol dispersion · aircraft cabin · mixing ventilation · cabin displacement ventilation

1 Introduction

The cabin ventilation as part of the Environmental Control System (ECS) of a passenger aircraft is responsible for a comfortable, energy efficient and healthy ventilation of the passenger area. Fresh air from outside is pressurized, tempered and mixed with high-efficiency particulate air (HEPA) filtered recirculated air before being guided to the passenger compartments via cabin air outlets in the upper part of the cabin. High volume flow rates of about 10 l/s per passenger are required to provide homogeneous temperature distribution and to ensure sufficient removal of the passengers' heat release. These high flow rates result in nominal air changes per hour of up to 20, i.e., nominally, the air in the cabin is exchanged every three minutes. However, a high degree of air mixing in the cabin and local differences due to geometrical obstructions cause, that potentially released contaminants are not fully removed after this time period. Thereby, the contaminant removal effectiveness is an evaluation parameter for the air quality, besides e.g. local age of air, which gained importance in the last years driven by

A. Dillmann et al. (Eds.): STAB/DGLR Symposium 2022, NNFM 154, pp. 549–558, 2024.
https://doi.org/10.1007/978-3-031-40482-5_52

the SARS-CoV-2 pandemic. Many recent studies [5,6,9,12] address this point by investigating the spreading of aerosol particles in generic as well as realistic passenger aircraft cabin environments by means of experimental [9,12] and numerical approaches [5,6,13]. Even if differing on detail level, the main finding of most studies is, that the spreading of aerosol particles on higher concentrations, i.e., potentially higher amount of inhaled particles by other passengers, is confined to a few seat rows around the aerosol source. On neighboring seats increased local concentrations of the aerosol particles or other tracer gases are found. Under the assumption of different viral loads per exhaled fluid volume, e.g., for normal infected persons and so-called 'super-spreader', mean infection risks maps were calculated, considering different mean flight times for short and long-haul flights. Besides the pressure-driven forced airflow between supply and exhaust openings, see Fig. 1 (left), thermal convection is induced by the heat release of the passengers, generating an upward-oriented airflow above the occupied seat benches. This so-called mixing ventilation (MV) is state-of-the-art in all modern passenger aircraft [14], both, single and twin-aisle. The interaction of the high-momentum jets and the thermal driven upward flow results in a high degree of mixing between new fresh air and 'old' air within the cabin, explaining the name of the ventilation concept. The biggest advantages of this concept are the high temperature homogeneity in the cabin and the robustness regarding (smaller) changes of the thermodynamic boundary conditions and inflow parameters. With regard to aerosol spreading, however, the higher mixing is prone to increase the dwell-times of the particles and the spreading in the near field.

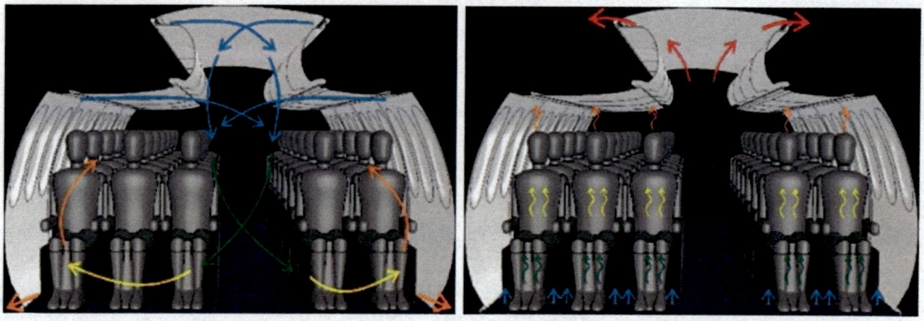

Fig. 1. Schematic view of the ventilation concepts. State-of-the-art Mixing Ventilation (MV) (left) and the novel Cabin Displacement Ventilation (CDV) (right)

Alternative ventilation concepts are within the scope of researchers since many years. Low-momentum trickle ceilings [7], personalized ventilation concepts [14] as well as cabin displacement ventilation (CDV) [2,14] are some examples. Thereby, the latter promises increased heat and contaminant removal efficiencies and thus is highly interesting in terms of energy efficiency and healthier cabin air. The CDV concept is based on a low-momentum air supply on floor level, e.g., through dense membranes below the seats, and the positioning of the exhaust openings in the upper cabin part, see Fig. 1 (right). The fresh air is forming a 'lake' on floor level and the upward motion of the air is driven by the thermal convection included by the heat release of the passengers. Consequently, airflow is generated only around the heat sources and removes the heat very efficiently. However, this ventilation concept increases the risk of uncomfortable temperature stratification since temperature increases from feet to head level. Further, an integration of this concept would require a completely new design of the whole ventilation pipe system in the aircraft and thus is not suitable for aircraft refits. Nevertheless, with the upcoming importance of the energy efficiency of all aircraft sub-systems following the greener aircraft, or the zero-emission aircraft approaches and the increased importance to the healthier cabin, the CDV concept and other ideas regain scientific interest.

In this respect, the present work aims to evaluate the aerosol spreading for classical MV and for CDV as an alternative ventilation concept. Different source locations are investigated and a particle diameter selective analysis of the average particle lifetime is performed.

2 Numerical Model

2.1 Computational Domain and Mesh

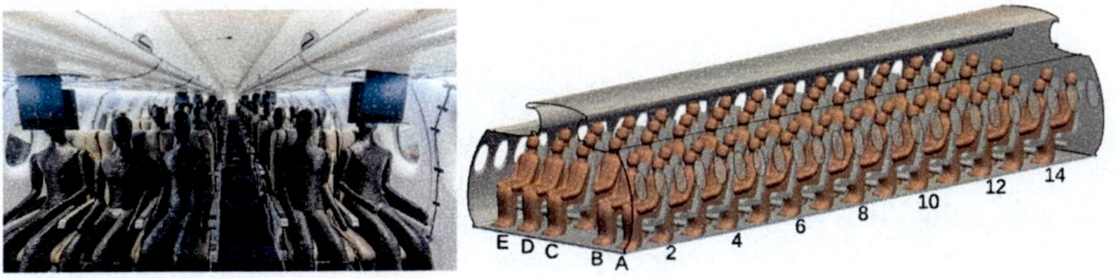

Fig. 2. The Do728 research facility (left) and computational domain (right).

The geometry of the computational domain (Fig. 2, right) corresponds to the Dornier 728 (Do 728) aircraft cabin test facility of the German Aerospace Center in Göttingen (Fig. 2, left). The cabin dimensions are 3.25 m × 2.14 m × 14.5 m (W × H × L) and it contains 14 five-seat rows. The configuration of the seats and the row numbering used here are shown in Fig. 2. The computational mesh is built using the standard OpenFOAM mesh generation utility. The base cell size of the mesh is equal to 2 cm, and the resolution becomes finer when approaching the walls. The thickness of the wall cell layer varies, and the smallest one is about 0.5 mm. The surface mesh cell size is ≈0.5 cm. The selected resolution yields approximately $1.2 \cdot 10^8$ cells.

Both MV and CDV ventilation configurations are modeled. The internal mesh used for both configurations is the same, only the corresponding inlet and outlet patches are defined depending on the simulated configuration.

2.2 Numerical Procedures

Particle flow simulation requires solving two problems: Navier-Stokes equations for ambient air and calculation of particle motion. The former is solved here as Reynolds-averaged Navier-Stokes (RANS) equations with the $k - \omega\, SST$ turbulence model. The latter is computed using the Lagrangian approach. The particles move in the air under the action of drag $\mathbf{F}_D$ and buoyancy $\mathbf{F}_B$ forces

$$\mathbf{F}_p = \mathbf{F}_D + \mathbf{F}_B. \tag{1}$$

It is assumed that the aerosol particle has a spherical shape. Then, as specified in [11],

$$\mathbf{F}_D = -C_D \frac{\pi D_p^2}{8} \rho |\mathbf{U}_{rel}| \mathbf{U}_{rel}, \tag{2}$$

with

$$C_D = \begin{cases} \frac{24}{Re_p}(1 + \frac{1}{6} Re_p^{2/3}) & \text{if } Re_p \leq 1000 \\ 0.424 & \text{if } Re_p > 1000, \end{cases}$$

where Reynolds number Re_p is based on the particle diameter D_p, the kinematic viscosity of the surrounding air, and the relative particle velocity $\mathbf{U}_{rel} = \mathbf{U}_p - \mathbf{U}$ of the particle velocity $\mathbf{U}_p$ with respect to the airflow velocity $\mathbf{U}$. The buoyancy force $\mathbf{F}_B$ is given by the formula

$$\mathbf{F}_B = m_p \mathbf{g} \left(1 - \frac{\rho}{\rho_p}\right), \tag{3}$$

where m_p is the mass of the particle, ρ and ρ_p are the densities of air and the particle, respectively, and $\mathbf{g}$ is the acceleration of gravity.

The model also takes into account the interaction of the droplet with the turbulent eddy. It is assumed that this interaction occurs during a time interval that is the minimum of two time scales, one of which is the typical lifetime of a turbulent eddy and the residence time of a droplet in the eddy [4]. In this case, the air velocity is represented as a sum $\mathbf{U} = \bar{\mathbf{U}} + \mathbf{U}'$ of the mean air velocity $\bar{\mathbf{U}}$ and the fluctuating eddy velocity $\mathbf{U}'$. The latter is modeled stochastically. In the model implemented in OpenFOAM,

$$\mathbf{U}' = \left(\frac{2k}{3}\right)^{1/2} \cdot N(0,1) \cdot \mathbf{e}_{dir}, \tag{4}$$

where k is the turbulent kinetic energy, $N(0,1)$ is a Gaussian normal random number with mean $a = 0$ and variance $\sigma = 1$, $\mathbf{e}_{dir}$ is a random unit vector.

A numerical study of both the formation of an aerosol cloud during respiration and its propagation in the aircraft cabin requires developed initial field data for Lagrangian modeling. To do this, steady-state RANS calculations are carried out. The MV and CDV ventilation configurations differ in their boundary conditions, namely the inlet volume rate and temperature of the supplied air. In MV configuration the inlet volume rate V_{in} is 700 l/s and the temperature T_{in} is 16 °C, in CDV configuration $V_{in} = 900$ l/s and $T_{in} = 20$ °C.

Further, the results of steady state simulation of the MV configuration are used as initial conditions for the study of aerosol cloud generation, and the results of both preliminary simulations are applied to the simulation of aerosol propagation.

3 Results and Discussion

3.1 Generation of Aerosol Cloud

In order to simulate the generation of the aerosol cloud we use the reactingParcelFoam solver which combines a compressible transient solver (URANS) for the air phase and a Lagrangian algorithm for tracking the particles as discussed in Sect. 2.2.

The sizes of exhaled droplets and their composition have been studied in the literature. Particularly, the summary can be found in Bagheri et al. [1], which also contains original experimental results. The relative humidity and temperature of exhaled air are discussed in [8], where the influence of various factors is experimentally investigated. Based on the literature we developed the initial conditions for modeling aerosol formation during respiration.

The exhaled air has a temperature of 35 °C and relative humidity of 80%, while the relative humidity of the air in the cabin is 20%. The diameters of particles are distributed in a range of 0.8 μm to 260 μm according to log-normal distribution with a median of 16 μm, see Fig. 4 (right). The droplet is composed of solid content (C) (1% of total mass) and of a solution of NaCl (1%) and water (98%).

The particle injector is the "mouth" of a dummy characterized by an opening area of 1.45 cm^2. It is supposed that the exhale-inhale cycle takes 3.75 s, and the air velocity U_{mouth} is defined by the equation

$$U_{mouth}(t) = A \cdot \sin \frac{2\pi}{T} t, \tag{5}$$

where $A = 2.8$ m/s and $T = 3.75$ s. The total volume of exhaled air for one cycle of breathing is equal to approximately 0.5 liters. The volume fraction of particles in the exhaled air is $\approx 1.5 \cdot 10^{-6}$. Therefore, following Elgobashi [3], we apply two-way coupling which takes into account the impact of particles on the environment but neglects the interaction of particles with each other. Figure 3 shows the process of the aerosol cloud formation as obtained in the simulation. It is found that in a short period of time (Fig. 3, $t = 1$ s and $t = 2$ s) the particles form significantly different sets: light dry particles at the top and heavy wet particles at the bottom of the evolving particle cloud. Later ($t = 3$ s and $t = 4$ s), young moist particles lose excess moisture, while heavy ones sink down and settle. Finally, light particles come into equilibrium with the environment and form a long-lived aerosol cloud.

The course of the evaporation process for droplets of different sizes is presented in Fig. 4. The graph of the dependence of the average water mass fraction

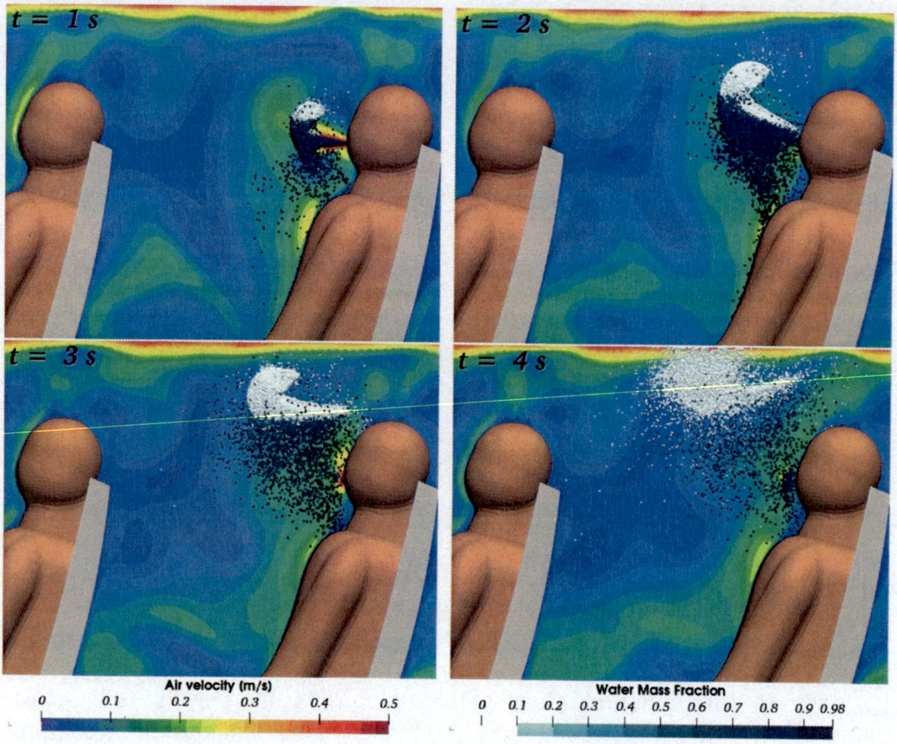

Fig. 3. Snapshots of developing the aerosol cloud formation process for t = 1 s to t = 4 s, obtained using URANS simulation. The velocity magnitude is colour coded as the background and the water-mass fraction of the particles is used as colour bar of the particles.

of a particle on its age (black dashed curve in Fig. 4, left) shows that within 4 s after release, the particle loses an average of about 99.7% of water as the average water mass fraction decreases from 0.98 to about 0.1. It is shown that the evaporation rate depends essentially on the particle size. Generally, the smaller the particle, the faster it loses excess water. A particle of less than 5 µm in size comes to an equilibrium state in an average time of about 0.5 s. For the particles with a diameter smaller than 10 µm it takes approximately 1.3 s to come into a balance with the surrounding air. As a result of evaporation, the particle size distribution changes significantly. Figure 4 (right) shows the transformation of the particle size distribution density over time from $t = 0$ s to $t = 3.7$ s. The number of particles with $D_p > 10$ is strongly reduced, and the median of the distribution is moved from 16 µm at the beginning to 3 µm at the end of the respiratory cycle. In addition to evaporation, particle scattering plays an important role in cloud formation, which is expressed in a decrease in particle concentration over time. The particle concentration is evaluated here by the count of particles in the ball of radius R = 1 cm. The particle cloud is defined as a connected region

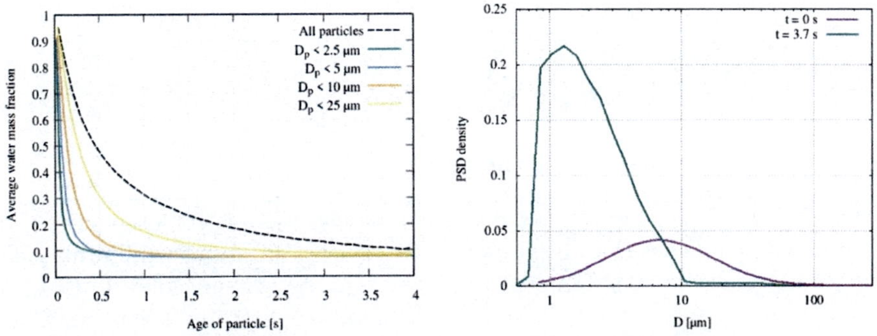

Fig. 4. Reduction of the water mass fraction of the particles depending on the age and diameter of the particles (left), and the density of particle size distribution (PSD) at t = 0 s and t = 3.7 s (right).

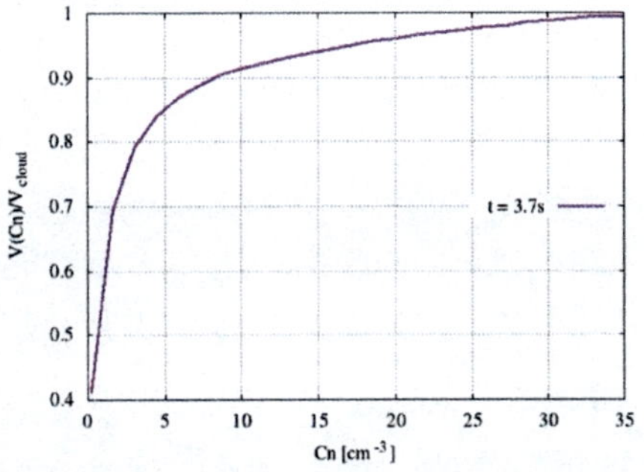

Fig. 5. Distribution of the volume of the particle cloud over the particle concentration at t = 3.7 s (left). Here $V(x)/V_{cloud}$ means the ratio of the cloud volume with particle concentration $Cn < x$ to the total cloud volume V.

of space with a concentration of particles $Cn > 0.24$ cm^{-3}, that is, within any sphere of radius $R = 1$ cm the cloud contains at least one particle. The initial concentration of exhaled particles is equal to 48.7 cm^{-3}, then it decreases due to dispersion. At the end of the breathing cycle at t = 3.7 s, a cloud with a volume of about 5.4 l is formed. The cloud contains about 78% of the ejected particles, the remaining 22% have settled, left the cabin, or dispersed outside the cloud with negligible concentration. Thus, the average concentration of particles in the cloud is estimated at 3.6 cm^{-3}. However, the particles are non-uniformly distributed over the cloud volume, as shown in Fig. 5. The concentration of particles in 90% of the cloud does not exceed 8.8 cm^{-3}, for 80% of the volume $Cn < 3$ cm^{-3}, and the minimum concentration $Cn \approx 0.24$ cm^{-3} accounts for more than 40% of the volume.

In general, as a result of evaporation and dispersion of particles in the process of aerosol formation during one respiratory cycle, the volume fraction of particles decreases from $1.5 \cdot 10^{-6}$ to $2.5 \cdot 10^{-9}$.

3.2 Aerosol Dispersion in the Cabin

After the cloud has formed, its further propagation runs under physical conditions different from the initial ones. All particles are in an equilibrium state and do not affect the environment due to the sparsity of the cloud. In this case, we solve the problem of kinematic particle cloud propagation using the velocity and turbulent fields obtained in previously performed RANS simulations. We use a transient solver uncoupledKinematicParcelFoam for the passive transport of particles [10]. The particle sources are located in the breathing zone at each position A–E in the eighth row. The initial clouds contain $8 \cdot 10^4$ particles with diameters uniformly distributed between 0.5 μm and 50 μm. The simulations cover 600 s of aerosol propagation. After this time, the number of active particles becomes negligible as the particles either settle or leave the cabin.

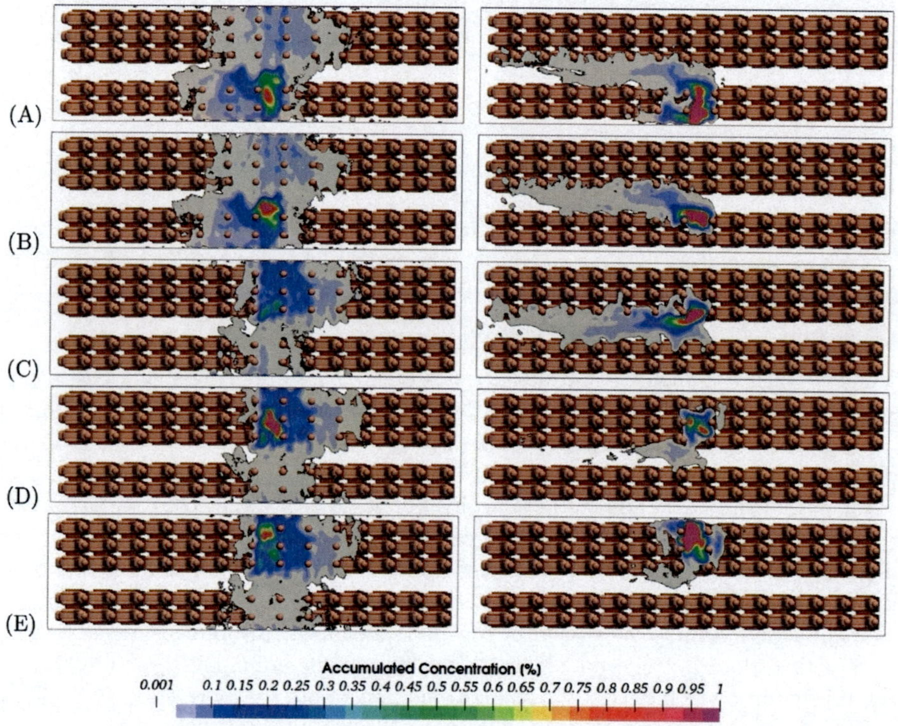

Fig. 6. Normalized accumulated concentration of the particles with $D_p < 5$ μm in the breathing zone at a height of 1.2 m, obtained in Lagrangian simulations of the configuration MV (left) and CDV (right) for the sources 8A–8E.

Considering the particle source as permanent with a constant particle production rate, we evaluate an accumulated particle concentration by integrating the number of particles per unit volume over time. The normalized accumulated concentration in the breathing zone is calculated as the ratio of the number of particles in the volume of air inhaled by a person to the number of particles produced by the source for the same time interval. For ease of calculation, it is assumed that the initial cloud is reproduced every second and that a person inhales an average of 0.1 liters per second. The results obtained for small particles ($D < 5$ μm) are presented in Fig. 6. The comparison reveals a certain advantage of the CDV configuration since it has a smaller area of noticeable concentration in all cases. The average lifetime and the fraction of settled particles are evaluated. The results are shown in Fig. 7. The lifetime of MV particles decreases monotonically with increasing size, while medium-sized particles (15 μm–20 μm) have the longest activity time in the CDV case. The CDV has a higher portion of exiting particles in almost the entire range of values except the largest ones (Fig. 7).

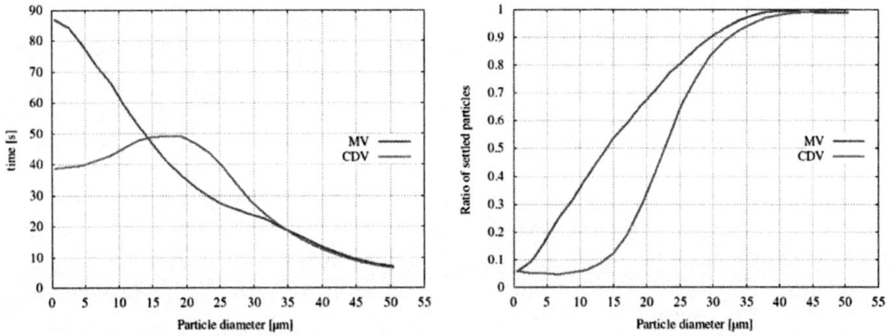

Fig. 7. Average lifetime of a particle depending on its diameter(left), and the average ratio of settled particles depending on diameter(right). The results obtained in Lagrangian simulations of MV and CDV configurations are averaged over all particle sources 8A–8E.

4 Conclusion

Numerical simulations of human breathing in the cabin of a 14-row Do 728 passenger aircraft have shown that during one breathing cycle, exhaled particles lose about 99.7% of moisture, come into equilibrium with the environment, and form a sparse particle cloud. The volume fraction drops by a factor of $\approx 10^3$ from the initial one due to dispersion and particle deposition. Further simulation of aerosol cloud propagation in the cabin for MV and CDV configurations revealed some advantages of the latter. The average residence time of fine particles in the air is shorter, and the fraction of particles exhausted by ventilation is higher in the case of CDV. Calculation of particle concentrations in the breathing zone in cases where the particle source is located in the central part of the cabin shows that the CDV configuration provides greater safety, since the area of significant particle concentration is smaller than in the MV case. The availability of the Do 728 cabin test facility at the German Aerospace Center in Göttingen allows a detailed comparison of experimental and calculated results in the future.

References

1. Bagheri, G., et al.: Exhaled particles from nanometer to millimeter and their origin in the human respiratory tract. Preprint at medRxiv (2021). https://doi.org/10.1101/2021.10.01.21264333
2. Bosbach, J., Lange, S., Dehne, T., Lauenroth, G., Hesselbach, F., Allzeit, M.: Alternative ventilation concepts for aircraft cabins. CEAS Aeronaut. J. **4**(3), 301–313 (2013). https://doi.org/10.1007/s13272-013-0074-z
3. Elghobashi, S.: On predicting particle-laden turbulent flows. Appl. Sci. Res. **52**(4), 309–329 (1994). https://doi.org/10.1007/BF00936835
4. Gosman, A.D., Ioannides, E.: Aspects of computer simulation of liquid-fuelled combustors (1981). https://doi.org/10.2514/6.1981-323
5. Gupta, J.K., Lin, C.-H., Chen, Q.: Transport of expiratory droplets in an aircraft cabin. Indoor Air **21**(1), 3–11 (2011). https://doi.org/10.1111/j.1600-0668.2010.00676.x
6. Gupta, J.K., Lin, C.-H., Chen, Q.: Inhalation of expiratory droplets in aircraft cabins. Indoor Air **21**(4), 341–350 (2011). https://doi.org/10.1111/j.1600-0668.2011.00709.x
7. Lange, P., Dehne, T., Schmeling, D., Dannhauer, A., Gores, I.: Realistic flight conditions on ground: new research facility for cabin ventilation. CEAS Aeronaut. J. **13**, 719–738 (2022). https://doi.org/10.1007/s13272-022-00594-2
8. Mansour, E., et al.: Measurement of temperature and relative humidity in exhaled breath. Sens. Actuators B Chem. **304**, 127371 (2020)
9. Netherland Aerospace Center (NLR) and National Institute for Public Health and the Environment (RIVM): CORSICA final report. NLR-CR-2021-232 (2021)
10. OpenFOAM. https://www.openfoam.com
11. Putnam, A.: Integratable form of droplet drag coefficient. ARS J. **31**, 1467–68 (1961)
12. Silcott, D., et al.: TRANSCOM/AMC Commercial Aircraft Cabin Aerosol Dispersion Tests (2020). https://www.ustranscom.mil/cmd/docs/TRANSCOM%20Report%20Final.pdf. Accessed Aug 2021
13. Talaat, K., Abuhegazy, M., Mahfoze, O.A., Anderoglu, O., Poroseva, S.V.: Simulation of aerosol transmission on a Boeing 737 airplane with intervention measures for COVID-19 mitigation. Phys. Fluids **33**(3), 033312 (2021). https://doi.org/10.1063/5.0044720
14. You, R., Lin, C.-H., Wei, D., Chen, Q.: Evaluating the commercial airliner cabin environment with different air distribution systems. Indoor Air **29**, 840–853 (2019). https://doi.org/10.1111/ina.12578

Check for updates

Generation and Distribution of Surgical Smoke During High Frequency Electrocauterization

Vera Stelzer[1,2(✉)], Sandra Melina Tauwald[1,2], Veronika Vielsmeier[3], Fabian Cieplik[3], Arne Kandulski[3], Wulf Schneider-Brachert[3], Olaf Wünsch[4], Markus Rütten[5], and Lars Krenkel[1,2]

[1] Ostbayerische Technische Hochschule Regensburg, Biofluid Mechanics, Galgenbergstraße 30, 93053 Regensburg, Germany
vera.stelzer@oth-r.de

[2] Regensburg Center of Biomedical Engineering (RCBE), OTH and University of Regensburg, and Regensburg Center of Health Sciences and Technology (RCHST), Galgenbergstraße 30, 93053 Regensburg, Germany

[3] University Hospital Regensburg (UKR), Franz-Josef-Strauß-Allee 11, 93053 Regensburg, Germany

[4] Institute of Mechanics, Chair of Fluid Mechanics, University of Kassel, Mönchebergstr. 7, 34125 Kassel, Germany

[5] German Aerospace Center, Institute of Aerodynamics and Flow Technology, Bunsenstr. 10, 37073 Göttingen, Germany
https://bfm.rcbe.de/

Abstract. Surgical Smoke is generated during the cauterization of tissue with high-frequency (HF) devices and consists of 95% water vapor and 5% cellular debris. When the coagulation tweezers, which are supplied with HF voltage by the HF device, touch tissue, the electric circuit is closed, and smoke is generated by the heat. In-vivo investigations are performed during tracheotomies where surgical smoke is produced during coagulation of tissue. Furthermore, in-vitro parametric studies to investigate the particle number and size distribution and the spatial distribution of surgical smoke with laser light sheet technique are conducted. With higher power of the HF device, the particles generated are larger in size and the total number of particles generated is also higher. Adding artificial saliva to the tissue shows even higher particle counts. The study by laser light sheet also confirms this. The resulting characteristic size distribution, which may include viruses and bacterial components, confirms considering the risk arising from surgical smoke. Furthermore, the experiments will provide the database for further numerical investigations.

Keywords: Surgical smoke · Fluid mechanics · Aerosols · Tracheotomies

1 Introduction

Surgical smoke lacks attention in the context of the current pandemic. Surgical smoke is generated during dissection or cauterization of tissue with

A. Dillmann et al. (Eds.): STAB/DGLR Symposium 2022, NNFM 154, pp. 559–568, 2024.
https://doi.org/10.1007/978-3-031-40482-5_53

heat-generating devices [11,14,15]. It consists of 95% water vapor and cellular debris in form of particulate matter (5%) [14–16]. The particulate matter consists of particles from combustion and chemicals like hydrocarbons, acrylonitrile, fatty acids, and phenols [4,8], but can also contain parts of viruses and bacteria [10,14]. In 1988, Garden et al. [6] found intact bovine and human papillomavirus (HPV) deoxyribonucleic acid (DNA) from the vapor of laser-treated verrucae. In 2002, they furthermore showed the possible disease transmission through surgical smoke from infected tissue that was treated with a carbon dioxide laser [7]. Viable oral polio virus was also found in smoke produced by excimer laser systems [12]. Human immunodeficiency virus DNA (HIV) also occurred in surgical smoke [1]. Furthermore, Hepatitis B virus (HBV) was identified in 10 of 11 samples from surgical smoke generated during laparoscopic abdominal surgery [9,16]. Besides viruses viable bacteria were also isolated in surgical smoke [5,16]. This possibility of disease transmission via surgical smoke, particularly in times of a global pandemic, makes further investigations regarding contagiousness essential. Electrocautery and laser applications, ultrasonic scalpels, and other tissue- and vessel-coagulating techniques are commonly used for dissection, hemostasis and tissue mobilization [16]. Electrocauterization creates particles in ranges of 10 nm up to 1 μm [4]. Due to the fact, that SARS-CoV-2 is 0.05 μm–0.25 μm in diameter [16], it would be possible to carry it within the produced aerosols. To be able to estimate the risk for surgical personnel, an exact number and size distribution must be determined, e.g., depending on the device used and its application under certain conditions. Bipolar electrocoagulation is used in otorhinolaryngology, among other technologies, for hemostasis and has been identified as an aerosol-generating procedure (AGP) during tracheotomies by our group. Berges et al. [3] also confirmed this. Tracheotomies are also ranked as the second highest risk AGP [3,13]. The generation of potentially pathogenic particles during electrocautery, particularly during tracheotomies, makes further investigation essential to protect clinical personnel, especially in the context of a pandemic. For this purpose, it is necessary to characterize the particles produced during electrocautery under different conditions in terms of their number and size. In addition, knowledge of their distribution in space becomes necessary for risk assessment. This study aims to determine the spatial distribution of surgical smoke, taking into account the number and size distribution of particles. In addition, the experiments will provide the database for further numerical investigations.

2 Materials and Methods

To determine particle number and size distribution during tracheotomies, experiments are performed in-vivo in an operating room (OR) and in the intensive care unit. Furthermore, in-vitro investigations were conducted to characterize essential parameters of the smoke and the spatial distribution using a laser light sheet technique. Figure 1 shows the setups of the in-vitro investigations performed. The methods are described in detail in the following sections.

2.1 Experimental Setup

Cauterization of tissue with high-frequency-devices produces aerosols. Optical aerosol spectrometers determine their size and number.

Optical Aerosol Spectrometers. Optical aerosol spectrometers (OAS) (Fidas Frog®, Palas GmbH, Karlsruhe, Germany) according to ISO 21501-1 and VDI 3867-4 record particle numbers and size distribution per second, within a measurement range of 0.2 µm–18 µm (256 raw data channels). The OAS are calibrated with respect to their volume flow rate using external volume flow rate sensors (D6F-02A1-110, Omron Corporation, Kyōto, Japan), prior to each measurement. Certified monodisperse test dust (MonoDust 1500, Palas GmbH) calibrates aerosol size. The resulting data are processed and evaluated using Matlab® R2021b (Mathworks, Inc., Natick, MA, US).

High Frequency Electrocautery. For electrocauterization, a high-frequency (HF)-device generates HF alternating current. This device supplies the connected tweezers with HF voltage. As soon as the tweezers contact biological tissue, the current circuit closes due to the tissue's water content and the resulting conductivity. Thus temperature increases, leading to coagulation or cutting of the intervened tissue. Interdisciplinary pre-studies of the authors identified electrocauterization as an aerosol-generating event during tracheotomies. To investigate the same settings used during medical events, power supplies were set to 20 W respectively 40 W.

2.2 In-Vivo Investigation of Aerosol Generation During Tracheotomies

To analyze particle number and size distribution in-vivo, the OAS are placed at dedicated positions next to the neck of the patients. While this, possible aerosol generating events are logged. The recorded particle distribution can be transferred to the respective situation during the medical intervention.

2.3 In-Vitro Parametric Study of Surgical Smoke in Terms of Particle Size and Number

A parametric study evaluates the parameters, influencing aerosol generation. Therefore, pig tissue samples with a diameter of 28 mm and a thickness of 5 mm are placed in an antistatic box. The resulting particle number and size distribution during bipolar electrocautery is determined. To quantify the influences of the tissue sample, one sample is used several times. 4 pig tissue samples (each investigated 4 times) result in 16 measurements per parameter. Sample preparation includes placing samples in a beaker filled with 0.9% NaCl solution and preheating them to the desired temperature of 37 °C by an incubator. A specific number of tissue specimens (32) are examined after applying a thin wetting layer of artificial saliva (AS) (Glandosane®, STADA AG, Bad Vilbel, Germany). Further parameters are power of the HF-device (20 W and 40 W) and time (cauterization for 5 s and 10 s). In the following, the abbreviations N for native and AS for artificial saliva are appended to the respective HF-device power and time used. To imitate real scenarios, the tweezers are inserted at a fixed penetration depth of 1 mm angled 45° to the specimen. Prior to the measurement, the box was prepared to be particle-free with a maximum accepted concentration of 0.1 particles per cm^3. HEPA filters connect the box to the environment (Fig. 1a). Each measurement is performed for 45 min. After one minute each, cauterization is applied. Calculating the arithmetic mean of all 16 measurements per parameter enables the averaging of the particle counts versus particle size in the respective channels. Summing up all the resulting averaged particle counts (APC) leads to the averaged total particle counts (AtPC). To investigate the particle counts over time both, overall and for dedicated channels, dynamic time warping (DTW) [2] is applied.

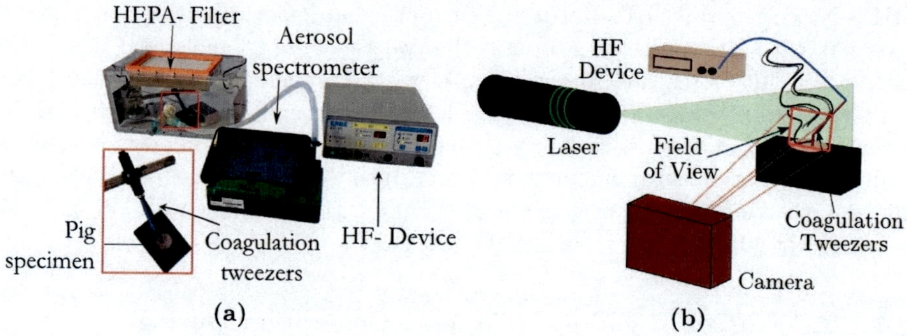

Fig. 1. In-vitro setups. **a** Setup for the parametric study in terms of particle number and size. **b** Setup for determining the spatial distribution of smoke using laser light sheet technique

2.4 In-Vitro Parametric Study of the Spatial Distribution

To investigate the spatial distribution of surgical smoke at different parameters, further in-vitro measurements are conducted. Therefore, the previously described setup (Sect. 2.3) is extended using laser light sheet technique. The surgical smoke is illuminated by a laser light sheet (DPSS laser RayPower 450, Dantec Dynamics, Skovlunde, Denmark) equipped with light sheet optics (series 80 × 80, Dantec Dynamics). The laser power is set to 450 mW. A camera, rectangular placed to the laser captures illuminated particles (Flare 48M30MCX, IO Industries, Inc., Ontario, Canada) (Fig. 1b). Recording properties are set to an exposure time of 300 µs, and a frame rate of 61.3 fps. Specimens with a size of 28 mm in diameter and thickness of 5 mm are investigated. The tweezers penetrate the tissue within a depth of 1 mm. The total measurement takes 10 s, after 2 s the electrocautery is applied for 5 s. The smoke propagation is examined at 20_{N} W, 20_{AS} W, 40_{N} W and 40_{AS} W. For visualizations, the images are post-processed using Matlab®. Background noise is removed by setting pixel values lower than 2369 to zero and applying a median filter. For better visualization of the smoke besides overexposed pixels, the pixel values exceeding 1×10^4 are set to 0. All 16-bit images are color-coded using global normalization. The data is normalized to a range between 0 and 1 taking into account the subtracted background and the maximum value.

3 Results

The following section presents the results of the different setups.

3.1 In-Vivo Investigation of Aerosol Generation During Tracheotomies

Figure 2 shows the measurement of particle counts during a tracheotomy. The increase after electrocauterization occurs with a time delay of a few seconds to the incidence red (Fig. 2a). Peaks without electrocautery, especially at the beginning, are due to the large number of people in the room and opening doors during the operation. Due to the size of the OR and the distance of the measuring device to the coagulation tweezers, the measured particle count can only be evaluated qualitatively. The following parametric study (Sect. 3.2) allows quantitative investigations. Nevertheless, a characteristic profile of the total resulting particle numbers divided among the occurring sizes is recognizable (Fig. 2b). The

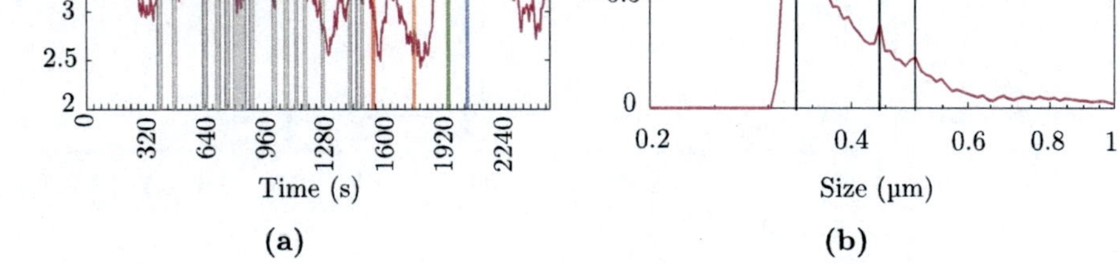

Fig. 2. Resulting particle counts over time and size during a tracheotomy. Areas marked in gray indicate when cauterization was applied. The red area marks the opening of the trachea, the orange one the spraying of AS, the green one the breathing stopped, and the blue one the placing of the tube. **a** Counts over time. **b** Counts over size. Black lines mark the occurrence of characteristic peaks

highest particle count occurs at a size of 0.33 µm, a second peak at 0.44 µm, and a third peak at 0.50 µm is visible (black lines Fig. 2b). The highest particle count at 0.33 µm, as well as the second and third peak could also be shown in the in-vitro parametric evaluations (Sect. 3.2). This distribution represents a characteristic size distribution of the particles generated due to electrocautery. Opening the trachea as well as placing the tube results in noticeable particle increases, which would confirm the classification of tracheotomies as a high-risk AGP. The highest particle count is led to spraying AS.

3.2 In-Vitro Parametric Study of Surgical Smoke in Terms of Particle Size and Number

Figure 3 shows the APC versus particle size (Fig. 3a) as well as 16 individual tests at 20_N W, 10 s and the resulting arithmetic mean (Fig. 3b). The total numbers of particles generated in the individual tests (V1-V16) show a large scattering (Fig. 3b), but show good agreement in terms of the size distribution. Consequently, the calculated arithmetic mean values agree well and characteristic curves are obtained for all parameters. The influence of the meat sample became apparent. When using the same sample, fewer deviations in the number occur than when comparing different samples. At both 20_N W and 40_N W, the curves show nearly the same shape at 5 s and 10 s cautery. The highest particle concentration occurs directly after the start of electrocautery after the current circuit is closed. This leads to immediate denaturation of the intervening tissue and an increase in resistance, which inhibits the cauterization process. Therefore, the difference in the cauterization times in our study has nearly no influence (Fig. 4a). Based on these findings, the test with AS was carried out only at 5 s

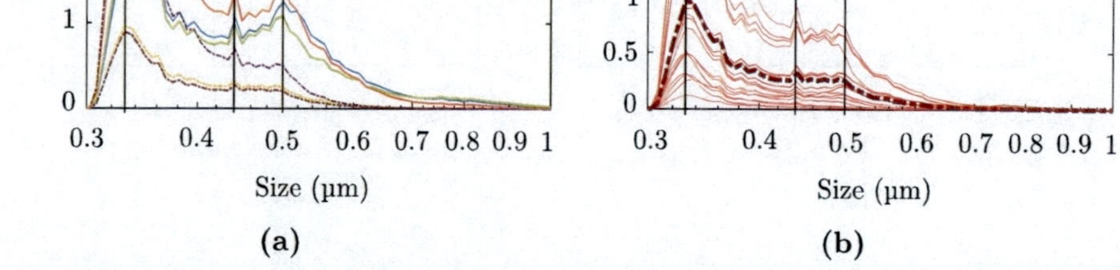

Fig. 3. Calculated arithmetic means of APC over size and individual tests at 20_N W, 10 s. **a** Size distribution results from the arithmetic mean of the individual tests per varied parameter. The black lines mark the occurrence of characteristic peaks. **b** Tests V1-V16 at 20_N W, 10 s and calculated mean (*dashed line*)

cautery. The parameter time is therefore neglected in the following considerations. The maximum APCs occur at a size of 0.33 µm for all measurements at 20_N W and 20_{AS} W. At 40_N W the maximum particle number occurs at a size of 0.34 µm. Both the measurements at 20_N W and 40_N W show a characteristic profile. At 20_N W, there is a second peak at a size of 0.44 µm. This peak is visible in all measurements at 40_N W. At 40_N W, a third peak at a size of 0.50µm is visible. Missing peaks at 20_N W and 20_{AS} W suggest higher applying power results in larger particles. The particle count at 40_N W at 0.33 µm is also higher than at 20_N W. This shows that a higher HF-device power increases the number of particles also in the smaller sizes, but by a smaller factor. The time pattern of the total particle count (Fig. 4a) again proves that particles produced

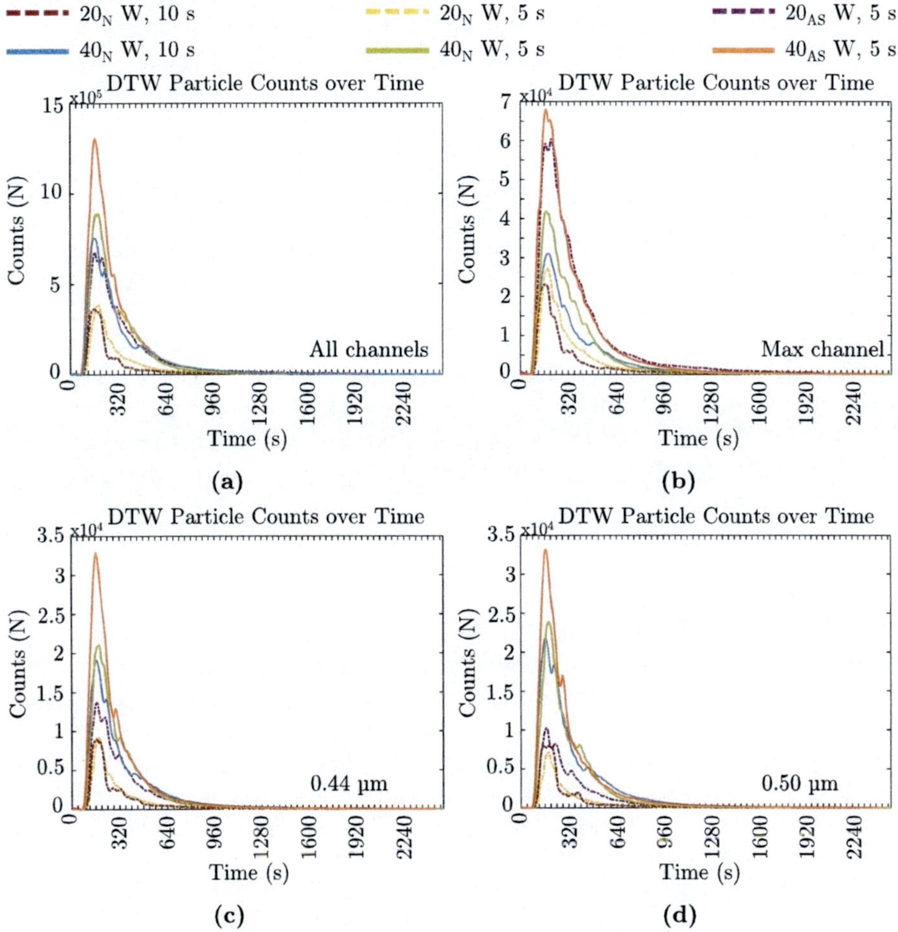

Fig. 4. DTW particle counts in all and in dedicated channels [2]. **a** Particle count of all channels. **b** Channel of maximum particle count. **c** 0.44 µm. **d** 0.50 µm

at 40_N W (4.01×10^7 at 5 s and 4.34×10^7 at 10 s integrated over time) are higher in number than those at 20_N W (1.12×10^7 at 5 s and 1.30×10^7 at 10 s integrated over time). Figure 4 shows the time pattern of the particle counts of the respective size at which the previously described peaks occur (black lines Fig. 3a). Consequently, selected sizes are the raw data channel where the maximum particle count occurs (0.33 µm or 0.34 µm) (Fig. 4b), 0.44 µm (Fig. 4c) where a second particle peak is evident at 20_N W, 20_{AS} W, 40_N W and 40_{AS} W and 0.50 µmm (Fig. 4d) where a third particle peak is evident at 40_N W and 40_{AS} W. The graphs with AS show that the wetting process produces noticeably more particles (3.81×10^7 at 20_{AS} W and 5.70×10^7 at 40_{AS} W). In the channel with the maximum individual particle counts, the maximum value of counts over time at 20_{AS} W is higher than the maximum value of all measurements at 40_N W. In the raw data channels of larger size, it is below those at 40_N W. The larger the particle size, the more pronounced this effect becomes. The particle count in total is, therefore, smaller as in all measurements at 40 W. This shows that the particles produced at 20_{AS} W are mainly in the size range of 0.33 µmm. Comparing 40_{AS} W and 40_N W, a higher particle count at 0.33 µmm and 0.44 µmm occurs at 40_{AS} W. This could suggest that at 40_{AS} W there is a higher increase in the smaller sizes than in the larger channels. Hence, it can be concluded that the formation of particles at channel size 0.33 µmm is increased by the cauterization of moistened tissue. However, when assessing the generated particles at 40_{AS} W, it must be taken into account that the maximum of the generated particles may have exceeded the maximum measurable concentration of the OAS at 2×10^4. When more than one particle occurs in the measuring section, the particle counter can no longer count or distinguish between them individually and coincidence occurs. This leads to statistical errors resulting in an underestimation of the actual number of particles. Thus, it can be assumed that the maximum peak at 40_{AS} W underestimates the real occurring value.

3.3 In-Vitro Parametric Study of the Spatial Distribution

Figure 5 shows the resulting distribution of surgical smoke at 20_N W (Fig. 5a), 20_{AS} W (Fig. 5b), 40_5N W (Fig. 5c), and 40_{AS} W (Fig. 5d). With AS more particles are generated. The least dense smoke is visible at 20_N W. At 40_{AS} W, the densest smoke is visible. Therefore, it can be concluded that there is a increased number in the total particle count at 40_{AS} W compared to 20_N W and 20_{AS} W. At 40 W, the smoke spread further into the room. At lower power, the spread dominates in the ascending flow direction. Thus, if medically possible, cauterization of tissue should be done at lower powers. This conclusions are in agreement with the previously shown results of the parameter study for the determination of the generated particle number and size distribution. To assess these effects, especially considering multiple slices in both horizontal and vertical directions, further experiments will be conducted.

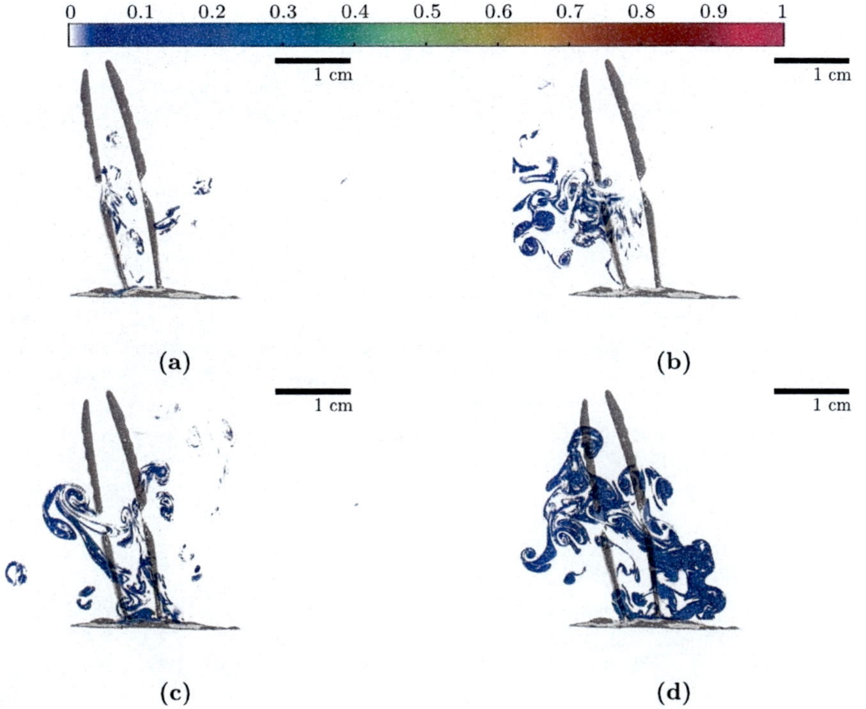

Fig. 5. Visualization of the resulting surgical smoke by laser light sheet technique. Times of identical smoke propagation are displayed. All 16-bit images are color coded and globally normalized to a range between 0 and 1. **a** 20_{N} W. **b** 20_{AS} W. **c** 40_{N} W. **d** 40_{AS} W

4 Conclusions

Finally, it can be concluded that the temporal duration of the cauterization in our study does not have a major influence. When ignited once, the highest number of particles occurs immediately after ignition and then steadily decreases. Furthermore, the meat sample influences the number of particles generated. Thus, fiber direction, density, as well as the muscular composition should be considered in further experiments. The higher the power of the HF-device, the higher the particle count and the larger the particles become. After applying AS, the number of particles increases especially in the small-size channels. Due to the size of the particles produced, which may well contain viruses, the risk of surgical smoke for clinical staff should not be neglected. Its contagiousness still needs to be investigated in further tests.

Acknowledgements. This work is supported by the RCHST and RCBE. Parts of this project were funded by the Bavarian State Ministry of Science and the Art's program to support Corona research projects. Special thanks also to Florian Erzinger for his support. Furthermore, special thanks to the clinical partners of the UKR for their cooperation.

References

1. Baggish, M.S., Polesz, B.J., Joret, D., Williamson, P., Refai, A.: Presence of human immunodeficiency virus DNA in laser smoke. Lasers Surg. Med. **11**, 197–203 (1991)
2. Bender, A., Bergmann, G.: Determination of typical patterns from strongly varying signals. Comput. Methods Biomech. Biomed. Eng. **15**(7), 761–769 (2012)
3. Berges, A.J., et al.: Quantifying viral particle aerosolization risk during tracheostomy surgery and tracheostomy care. JAMA Otolaryngol. Head Neck Surg. **147**(9), 797–803 (2021)
4. Brüske-Hohlfeld, I., et al.: Surgical smoke and ultrafine particles. J. Occup. Med. Toxicol. **3**, 31 (2008)
5. Capizzi, P.J., Clay, R.P., Battey, M.J.: Microbiologic activity in laser resurfacing plume and debris. Lasers Surg. Med. **23**(3), 172–174 (1998)
6. Garden, J.M.: Papillomavirus in the vapor of carbon dioxide laser-treated verrucae. JAMA **259**(8), 1199 (1988)
7. Garden, J.M., O'Banion, M.K., Bakus, A.D., Olson, C.: Viral disease transmitted by laser-generated plume (aerosol). Arch. Dermatol. **138**(10), 1303–1307 (2002)
8. Hensman, C., Baty, D., Willis, R.G., Cuschieri, A.: Chemical composition of smoke produced by high-frequency electrosurgery in a closed gaseous environment. Surg. Endosc. **12**(8), 1017–1019 (1998). https://doi.org/10.1007/s004649900771
9. Kwak, H.D., Kim, S.H., Seo, Y.S., Song, K.J.: Detecting hepatitis b virus in surgical smoke emitted during laparoscopic surgery. Occup. Environ. Med. **73**(12), 857–863 (2016)
10. Liu, Y., Song, Y., Hu, X., Yan, L., Zhu, X.: Awareness of surgical smoke hazards and enhancement of surgical smoke prevention among the gynecologists. J. Cancer **10**(12), 2788–2799 (2019)
11. Sanderson, C.: Surgical smoke. J. Perioper. Pract. **22**, 122–128 (2012)
12. Taravella, M.J., Weinberg, A., May, M., Stepp, P.: Live virus survives excimer laser ablation. Ophthalmology **106**(8), 1498–1499 (1999)
13. Tran, K., Cimon, K., Severn, M., Pessoa-Silva, C.L., Conly, J.: Aerosol generating procedures and risk of transmission of acute respiratory infections to healthcare workers: a systematic review. PLoS ONE **7**(4), e35797 (2012)
14. Ulmer, B.C.: The hazards of surgical smoke. AORN J. **87**(4), 721–738 (2008)
15. Vourtzoumis, P., Alkhamesi, N., Elnahas, A., Hawel, J.E., Schlachta, C.: Operating during Covid-19: is there a risk of viral transmission from surgical smoke during surgery? Can. J. Surg. **63**(3), E299–E301 (2020)
16. Zakka, K., et al.: Electrocautery, diathermy, and surgical energy devices: are surgical teams at risk during the Covid-19 pandemic? Ann. Surg. **272**(3), e257–e262 (2020)

High Spatial Resolution Tomo-PIV of the Trachea Focussing on the Physiological Breathing Cycle

Sandra Melina Tauwald[1,2(✉)], Maurizio Quadrio[3], Markus Rütten[4], Christian Stemmer[5], and Lars Krenkel[1,2]

[1] Regensburg Center of Biomedical Engineering (RCBE), OTH and University of Regensburg, and Regensburg Center of Health Sciences and Technology (RCHST), Galgenbergstraße 30, 93053 Regensburg, Germany
melina.tauwald@oth-regensburg.de

[2] Department of Biofluid Mechanics, Technical University of Applied Sciences (OTH) Regensburg, Galgenbergstr. 30, 93053 Regensburg, Germany

[3] Department Aerospace Science and Technology, Politecnico Milano, Via Privata Giuseppe La Masa 34, 20156 Milan, Italy

[4] German Aerospace Center (DLR), Bunsenstr. 10, 37073 Göttingen, Germany

[5] Technical University of Munich (TUM), Chair of Aerodynamics and Fluid Mechanics, Boltzmannstr. 15, 85748 Munich, Germany
http://bfm.rcbe.de

Abstract. Investigations of complex patient-specific flow in the nasopharynx requires high resolution numerical calculations validated by reliable experiments. When building the validation base and the benchmark of computational fluid dynamics, an experimental setup of the nasal airways was developed. The applied optical measurement technique of tomo-PIV supplies information on the governing flow field in three dimensions. This paper presents tomo-PIV measurements of the highly complex patient-specific geometry of the human trachea. A computer-tomographic scan of a person's head builds the basis of the experimental silicone model of the nasal airways. An optimised approach for precise refractive index matching avoids optical distortions even in highly complex non-free-of-sight 3D geometries. A linear-motor-driven pump generates breathing scenarios, based on measured breathing cycles. Adjusting **of the CCD cameras' double-frame-rate PIV-Δ t enables the detailed analysis of flow structures during different cycle phases.** Merging regions of interest enables high spatial resolution acquisition of the flow field.

Keywords: Tomographic PIV · Flow visualisation · Breathing cycle · Nasal airflow

1 Introduction

The main component of the upper airways, the nasopharynx, is a key element in the breathing apparatus [1]. While its major function is to condition the inhaled

A. Dillmann et al. (Eds.): STAB/DGLR Symposium 2022, NNFM 154, pp. 569–578, 2024.
https://doi.org/10.1007/978-3-031-40482-5_54

air, the nasopharynx influences the whole breathing cycle [2,3]. According to various medical studies, nasal breathing difficulties affect tens of millions of people, resulting in public health costs of up to a billion euros annually [4]. To reduce the booming costs and maximise the rate of satisfactory results following surgery, surgeons require a system of indicators to support them in selecting the appropriate treatment [5]. With regard to respiration, computational methods offer great promise [6,7]. Examining the anatomical shape and physiological function of the upper airways from a fluid mechanical point of view considerably improve diagnostic precision while also minimising the number of surgery failures [4,7,8]. Such investigations are challenging, largely due to the complex flow characteristics created by the manifold structures of the inter- and intra-individual variations in geometry [6,9,10].The literature contains extensive information regarding ways of predicting nasal flow patterns by Computational Fluid Dynamics (CFD) and optical measurements [11]. CFD enables the analysis of patient-specific geometries based on CT scans. However, the lack of information complicates the choice of a suitable calculation model. Experimental studies are therefore indispensable for assessing a calculation's meaningfulness. Studies assessing experimental investigations are mainly based on the optical method of Particle Image Velocimetry (PIV) [4,12,13]. These investigations were primarily conducted on nose similar geometries and/or analysed with planar or stereoscopic (stereo)-PIV. One disadvantage of these methods is a loss accuracy, especially for thin volumes and turbulent flows [14]. Data on realistic nose models, investigated by tomographic (tomo)-PIV, remains very rare - mainly due to the complexity of the geometry and the challenging experimental setup. Since tomo-PIV considers voxels (vox) as uniform cubes in all dimensions, in-plane and cross-plane properties can be analysed. Tomo-PIV enables all three components of velocity to be analysed. A cyclic respiratory flow and a physiological geometry of the upper airways is assumed in experimental investigations in order to create the appropriate physiological conditions. Since tomo-PIV calculates the velocity field based on the particle shift between consecutive images, the timing of PIV components needs to be adjusted based on the breathing cycle phase. The presend paper uses tomo-PIV to present a method of analysing the three-dimensional velocity vectors and the vorticity in a patient-specific model of the nasopharynx while setting a boundary condition based on the physiological breathing cycle [5,11].

2 Methods

2.1 Development of the Phantom Model

The basis of the workflow was a clinical computer-tomographic (CT) scan of a person's head presenting a normal sino- nasal anatomy (approved by the Institutional Review Board of the San Paolo Hospital) (Fig. 1a). The CT scan was used to build a rapid-prototyped core representing the airway-part of the phantom model. To enable investigation of the smallest cavities of the upper airways in a sufficient resolution it was necessary to upscale (magnification factor M = 2)

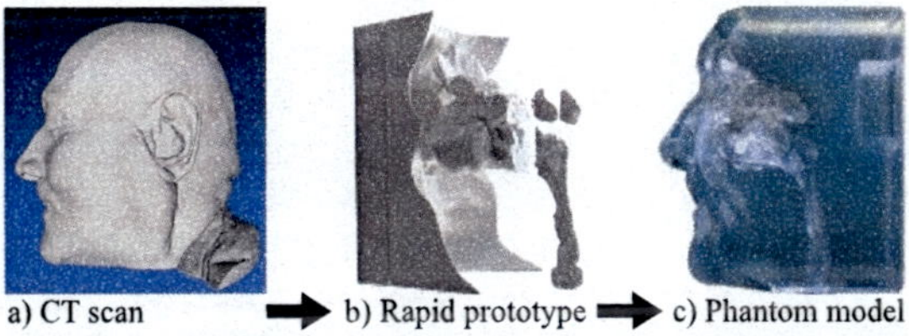

Fig. 1. Workflow of manufacturing the phantom model a) the initial CT scan, b) the scan-based water soluble rapid prototype and c) the final phantom model.

of the geometry. The rapid-prototyped airway model was made from calcium sulphate and a water-soluble binder (zp 131 powder/ zb 60 binder, Z Corporation, Burlington, USA) printed with ZPrinter 660Pro (3D Systems, South Carolina, USA) (Fig. 1b). **Since rapid-prototyped models suffer from aliasing-effects, the 3D print was coated with 20 layers (final thickness of 0.15 mm)** of a 2.5 wt/vol % polyvinylalcohol-distilled water solution (VWR chemicals, Pennsylvania, USA) for smoothing the surface. Fluorescence particles (Polysterol, Diameter 793 µm, ex/em = 530/607 nm, microparticles GmbH, Berlin, Germany) were then attached at defined locations of the model. These marker particles allow for precise local matching of the (subsequently calculated) velocity field and the stereolithography (STL) data of the rapid prototype using the software CADfix® (ITI, Milford, OH, US). For this alignment, a micro-CT scan of the rapid prototype was prepared and post-processed using the software Volume Graphics Studio (Volume Graphics, Heidelberg, Germany) and Netfabb® (Autodesk, Inc., San Rafael, CA, US). Then, the airway model was placed in an outer cast. As a final step, the mold was slowly filled with PDMS (Sylgard 184, Dow Corning, Michigan, USA). The model was cured under 2.5 bar pressure and 60° C in an autoclave. Finally, the airway-core was dissolved by repeated rinsing with water and subsequently cleaning with citric acid (Fig. 1c).

2.2 Experimental Setup

Tomo-PIV. Figure 2 shows the experimental tomo-PIV setup. It consists of a tomo-PIV system comprising three cameras, each fitted with CCD sensors (Imager Pro X 10, 1600 × 1200 pixel, LaVision GmbH, Göttingen, Germany), which is controlled centrally by a programmable timing unit (PTU). The PTU receives a TTL signal from the control unit of the linear-motor-driven piston pump (LinMot GmbH, Essen, Germany). The cameras were fitted with macro **objectives (ZEISS, Milvus 2/135), Scheimpflug adapters, and cut-off filters.** The light source used was a double pulsed Nd:YAG laser (EverGreen200 70–200 mJ @ 532 nm, 15 Hz, Lumibird, France) equipped with optical components. Knife edges were used to cut the diverging volume illumination into a RoI (region of interest) fitting volume. The RoI's volume was defined by the initial scale of the cameras' field of views (FoV)s and the depth of field (DoF) (50.5 mm × 34 mm × 17 mm). The geometry was mounted on a 3-axis traversing unit (isel Germany AG,

Eichenzell, Germany) and placed in a 800 L fish tank containing three portholes for the cameras at angles of 60° to each other. The fish tank was then filled with the working fluid (WF) (water-glycerol solution). The flow is induced at the trachea and exits the geometry at the nostrils. The tracer particles (PMMA-RhB, 20–50 µm, abs/emm = 560/584 nm, LaVision GmbH) exhibit a density ρ_{TP} of 1.19 g/ccm. The density of the WF $\rho_{WF} = 1.152\,\mathrm{g/cm^3}$ was analysed by a pycnometer (DMA 35, Anton Paar GmbH, Osftildern, Germany). Particles with WF were premixed and injected into the flow with a syringe. The dynamic viscosity of the WF μ_{WF} was investigated using a rotational rheometer with a cylinder geometry (Discovery HR20, TA Instruments, New Castle, Delaware, USA). μ_{WF} was measured by conducting a flow-sweep test within shear rates of between 10^{-1} to 10^2 1/s.

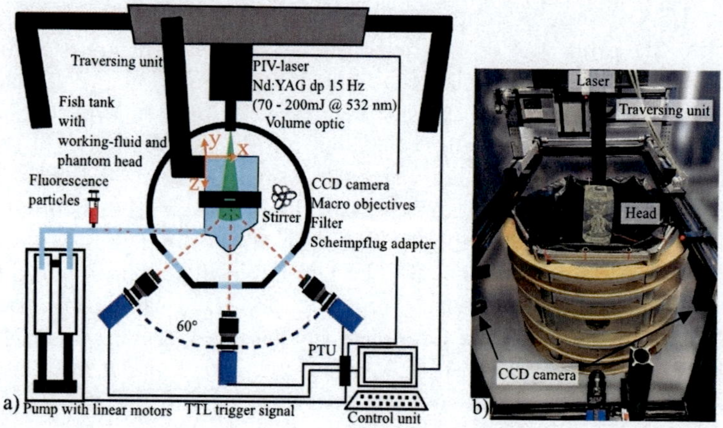

Fig. 2. Setup showing the main tomo-PIV components. Fig. 2a) shows a schematic view comprising three CCD cameras equipped with Scheimpflug adapters, macro objectives and filters positioned at 60° angles. The phantom head is placed in a fish tank filled with WF by a traversing unit. The flow is induced by a linear-motor-driven piston pump. Induced tracer particles are illuminated by a laser. All components are controlled centrally by a PTU. Fig. 2b) shows a photography of the setup.

Refractive Index Matching (RIM). To achieve an optimised match of the refractive indices (RI) between the WF (water-glycerol-mixture) and the PDMS head, an in-site method was developed. The method enables avoiding inaccuracies introduced by the refractometer measuring RI of solids (ORM 1RS, KERN & SOHN GmbH, Balingen, Germany). A light sheet from a continuous line laser (Ray Power 450 DPSS, Dantec Dynamics GmbH, Ulm, Germany) enters the fish tank through one of the portholes, is deflected by a mirror, and detected on it exiting the fish tank through the 120° angled porthole. The light sheet exiting the fish tank is visualised on scaled paper and captured by a camera (D3400, Nikon Corporation, Tokyo, Japan) fitted with a magnifying objective (AF-P DX

NIKKOR, 22/18, Nikon Corporation, Tokyo, Japan). A reference image of the light sheet was captured without crossing the model and images were captured while the phantom head was placed in the laser-line. The RI of the WF was adopted while adding specific amounts of water/glycerol and measured by a digital refractometer.

Linear-Motor-Driven Piston Pump. Flow rates were induced by a pump containing two linear motors and two pistons (diameter $ø_{\text{pump}} = 50\,\text{mm}$, driving length $l_{\text{pump}} = 170\,\text{mm}$), driven by a programmable logic controller (PLC) and controlled by Twincat software (Beckhoff Automation GmbH, Verl, Germany). To investigate the flow field by a constant speed of the pump's pistons, the speed of a single piston was set to 0.03 m/s forwards/backwards (cycle length 9.2 s), resulting in a flow rate of 130 ml/s with both pistons. To induce a physiological flow cycle, 30 cycles of a healthy subject were captured by a spirometer (Pneumotrac^{TM}, Vitalograph$^{®}$ GmbH, Hamburg, Germany) and processed by the proprietary Spirotrac 6 software. The cycles were processed using Matlab software (Matlab$^{®}$ R2020b, Mathworks Inc., MA, US) applying dynamic time-warping and fitted by a 7th degree polynomial function. The captured flow rate Q_{vivo} was adjusted to the experimental conditions Q_{vitro} using Reynolds' similarity [4].

$$Q_{\text{vitro}} = Q_{\text{vivo}} * M * \frac{\nu_{\text{vitro}}}{\nu_{\text{vivo}}} \tag{1}$$

2.3 Tomo-PIV Measurements and Timing

All components of the tomo-PIV setup were controlled by DaVis software. The tomo-PIV's capturing was triggered by starting the pump cycle. The PIV's intra-doubleframe-rate (PIV-Δ t) was set to 700 µs during a constant flow rate for a particle shift of between 5 and 8 pixels (pix). During breathing flow rates, loops of PIV-Δ t ranging from 120 µs to 2120 µs with an increment step of 400 µs were analysed. The images with a suitable particle shift were then post-processed. The cameras were precalibrated by a plate calibration target and following by Volume Self Calibration (VSC) based on the post-processed particle images. For PIV-algorithm, the final window size was set to 64 vox, with a maximum expected displacement to 16 vox. Four different FoVs were captured by moving the phantom head through a distance of 10 mm on the z-axis and 20 mm on the x-axis. The flow fields were merged by calculating the arithmetic mean of the overlapping areas using Matlab.

3 Results

3.1 Refractive Index Matching

The components of the WF were water (RI water $\eta_{H_2O} = 1.333$) and glycerol (RI glycerol $\eta_{\text{Glyc}} = 1.473$). The light sheet's reference projection, without crossing the phantom head, displays a straight, uninterrupted line, as shown in Fig. 3, while the projection of the line crossing the phantom head at an RI of $\eta_{WF} = 1.4142$ displays kinks. Adding water down to $\eta_{WF} = 1.4138$, again results in a continuous projection of the line.

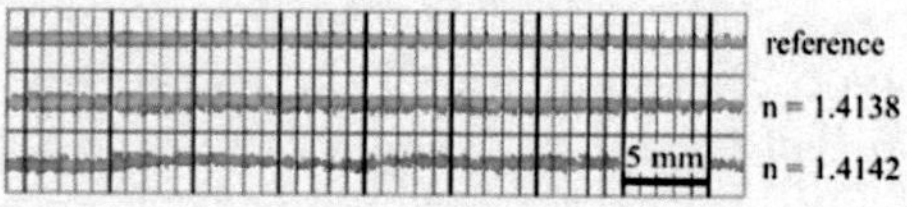

Fig. 3. Results of RIM of the WF and phantom head, reference image of the projected light sheet without passing the geometry (top), best matching RI (middle) and kinked light sheet caused by different RIs (bottom).

3.2 Positioning and Merging of RoIs

The final resolution achieved was 32 pix/mm. The DoF of the cameras was 17 mm. The reconstruction and vector calculation were performed for the whole volume of the FoV, while the vector fields were subsequently cropped to a volume of 45 mm × 33 mm × 17 mm. To increase the RoI, vector fields are measured in four different positions and were merged afterwards. Moving the traversing unit trough 10 mm in the z-dimension led to an overlap of 42 % of the RoIs, while a movement through 20 mm in the x-dimension led to an overlap of 28 %. Marker particles were used to combine the vector field's coordinate system with that of the scan. Fig. 4 shows the merging process of the four measurement positions (RoI1-4, Fig. 4 corners). For illustrating the high-resolution within the boundary layer regions, isosurfaces of each RoI with a low velocity magnitude ($|V| = 0.124$ m/s) are shown. RoI1 indicates the reference position, while RoI2 was moved along the z-axis, RoI3 along the z-axis and x-axis and RoI4 along the x-axis. Merging RoI1-4 resulted in a RoI's volume of 70.18 mm × 32.79 mm × 26.35 mm (Fig. 4 centre). The smallest cavity of the geometry, located at the ethmoidal sinuses, comprises a cross sectional area of 4 mm^2, which can be investigated with the spatial resolution of 32 pix/mm thus resulting in $4 * 10^3$ pix for this area. The investigated RoI reaches 60 cm^3 and therefore $2 * 10^9$ vox.

3.3 Measurements of Cyclic Flow Rates

The overall cycle had a duration of 3.59 s. The transition point from inhalation to exhalation was located at 1.97 s. The measurements of the dynamic viscosity of the WF resulted in $\mu_{WF} = 9.27$ mPas, the density measurements of the WF yielded a value of $\rho_{WF} = 1.152\,g/cm^3$ and thus brings with it a difference to the density of the tracer particles $\rho_{TP} - \rho_{WF}$ of $0.04\,g/cm^3$. Including the enlargement ($M = 2$), the ratio of flow rates results a value of $Q_{vitro}/Q_{vivo} = 0.94$. Since cyclic breathing results in time varying flow rates, an adaption of the

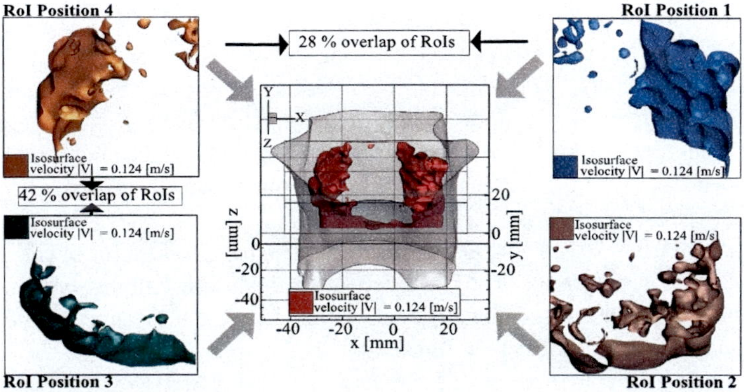

Fig. 4. Visualisation of the combining process of four RoIs. Isosurfaces at a velocity magnitude of $|V| = 0.124\,\text{m/s}$ of the ROIs for positions 1–4 are shown at the corners, coded in different colours. The centre shows the isosurfaces for the combination of the four RoIs after calculation of the arithmetic mean in the overlapping areas at the same velocity magnitude.

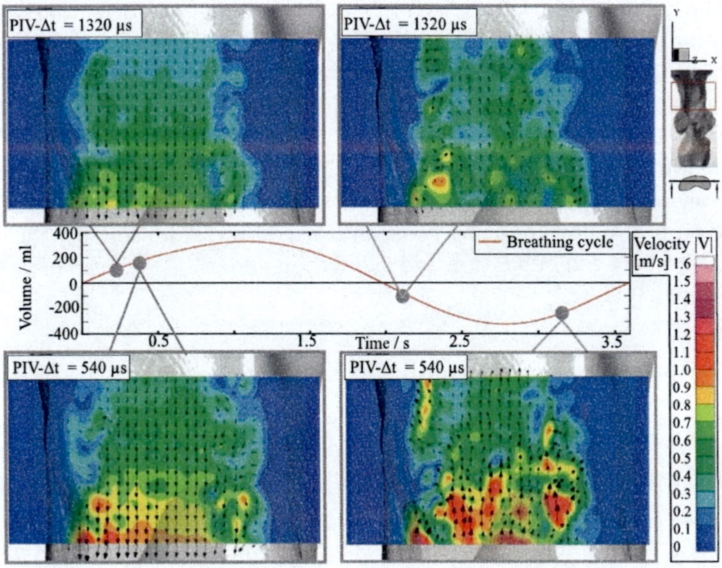

Fig. 5. Snapshots of four breathing cycle phases. Representative velocity profiles at cycle phase 0.045, 0.29, 2.21 and 3.17 s are shown colour-coded. The lengths of the projected vectors is relative to their velocity magnitude.

PIV-Δ t between different phases was necessary. This is shown on exemplary cycle phases at 0.29 s and 2.21 s for a PIV-Δ t of 1320 μs, cycle phases at 0.045 and 3.17 s showed a suitable particle shift for PIV-Δ t of 520 μs. Figure 5 shows four snapshots of the flow. An example slice is visible at z-position –

0.49 mm, where the volume consists of RoI1 and RoI2 with an overlap of 28 %. The velocity magnitude is shown colour-coded and velocity components are visualised by vectors. The lengths of the vectors are relative to their velocity magnitude. A decreasing velocity while reaching the boundaries of the geometry can be seen in Fig. 5. The vector projections indicate a main flow direction that depends on the inhalation or exhalation phase within the breathing cycle. The cycle phase at 0.29 s with a pump flow rate of 130 ml/s shows a velocity magnitude of up to 0.7 ms in the region where the inner diameter of the geometry decreases. Velocities range from 0 to 1.6 m/s, depending on the cycle phase. As shown in Fig. 5, the inhalation velocity fields display smoother gradients than the exhalation fields. The cycle phase at 2.21 s displays a peak in the lower left corner, while the cycle phase at 3.17 displays several peaks spread throughout the entire investigated volume. A difference in speed at the upper and lower edges of the RoI is noticeable. This can be explained by the reducing cross-sectional area of the trachea at RoI's location. A difference in speed at the upper and lower edges of the RoI is noticeable. This can be explained by the reducing cross-sectional area of the trachea at RoI's location. A difference between inhalation and exhalation flow profiles in Fig. 5. is visible. The smoothness of the flow profile during the inhalation in comparison to the exhalation may be due to the position of the pump connector as well as the geometry itself. The flow is conditioned by almost the entire geometry during inhalation at the investigated RoI in the trachea. Whereas in case of the exhalation, the flow is analysed right after exiting the connector of the pump after passing the complex geometrical part of the pharyngeal tonsils. The tonsils show two significant protrusions directly in front of the RoI. In order to make a statement about the flow profiles, optimisation with regard to the flow inducing position is necessary. Additionally, the different flow patterns between inhalation and exhalation underline the need of an analysis of the entire geometry.

3.4 Comparison Constant and Cyclic Flow Rate

Since a flow rate of 130 ml/s was induced during investigations of the constant flow cycle and breathing cycle phase at 0.045 s, the velocity profiles were compared. Figure 6 shows isosurfaces (Fig. 6 left) for the velocity magnitudes $|V| = 0.56$ m/s and 0.11 m/s. The right-hand side of Fig. 6 shows the velocity magnitude of a representative z-slice, colour-coded. The length of the visualised vectors is relative to their velocity magnitude. The upper part shows the results for constant flow, the lower part for the breathing cycle. While the isosurfaces are similar in the x-dimension at a velocity magnitude of 0.11 m/s, the isosurfaces for a velocity magnitude of 0.56 m/s occupy a larger area for the constant flow. The same applies to the visualisation on the representative z-slice. Vector projections show for both investigations a main direction in decreasing y-dimension. While comparing the flow profile of the constant flow rate and the cyclic flow rate, similarities are noticeable with regard to the direction of the flow, but there are differences in the velocity magnitude. Cyclic flow rates can induce additional transient effects, that might not occur in a steady flow rate boundary condition.

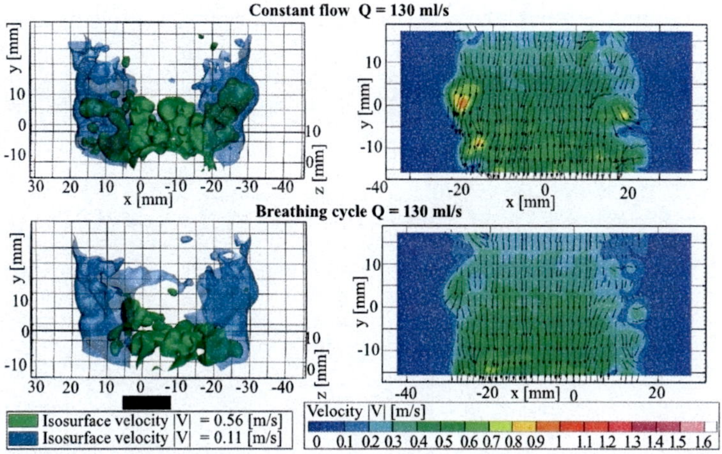

Fig. 6. Comparison of the constant flow rate Q = 130 ml/s and the same flow rate during the breathing cycle.

Since flow cycles in this study are not adapted by Womersley similarity, the high density of the working fluid in comparison to air, can enhance the differences due to the fluids intertia.

4 Conclusion

The resolution of 32 pix/mm and the two-fold magnification of the geometry enable an accurate analysis of smaller flow structures. The new RIM method, integrated in the setup, minimizes the errors caused by optical distortion, leading to a satisfactory VSC. The linear-motor-driven pump provides a suitable method for inducing a physiological breathing cycle. Nevertheless, since the present investigation is based on data from a single test-person, larger studies with several test-persons are essential in order to obtain a representative breathing cycle. The output trigger of the pump's control unit proved to be a suitable signal for activating image capturing in an investigation of several cycle phases. The micro-CT scan of the rapid prototype enables precise realignment of the experimental data, since marker particles are visible both in the final data and in the captured images. The traversing unit has the advantage of creating traceable locations for the experimental results relative to the geometry. Results can be merged by calculating an arithmetic mean for the overlapping areas. However, since the arithmetic mean was investigated on single snapshots of the flow, statistical surveys to determine standard deviations of the velocity profiles in the overlapping areas are needed. As shown in Fig. 5, adjusting the PIV-Δ t enables the investigation of velocity fields with different velocity magnitudes, caused by the breathing-cycle-dependent flow rate. To prove this requires multiple repeat measurements. The described methods were applied to an example position of the phantom head with a flow rate derived from a representative breathing cycle. Investigations on the whole flow field as a whole requires parametric studies to determine a suitable overlap of RoIs and averaging of overlapping flow parts. Additionally, since breathing is a repeating cycle, measurements under identical conditions over several cycles are necessary to analyse the development of flow over time.

Acknowledgements. This work is supported by the RCHST, the RCBE, and the Bavarian Academic Forum (BayWISS). Special thanks to Florian Erzinger for his support in the implementation of the experiments. Further thanks go to our colleagues of the project OpenNOSE for the intensive interdisciplinary exchange and the active support.

References

1. Morris, R.: Functional anatomy of the upper airway. Emerg. Med. Clin. North. Am. **6**, 639–669 (1988)
2. Kanda, A., Kobayashi, Y., Asako, M., Tomoda, K., Kawauchi, H., Iwai, H.: Regulation of interaction between the upper and lower airways in united airway disease. Med. Sci. (Basel) **7**(2), 27 (2019)
3. Licari, A., Castagnoli, R., Denicolo, C.F., Rossini, L., Marseglia, A., Marseglia, G.: The nose and the lung: united airway disease? Front. Pediatr. **5**, 44 (2017)
4. Cozzi, F., Felisati, G., Quadrio, M.: Velocity measurements in nasal cavities by means of stereoscopic piv - preliminary tests. In: Journal of Physics: Conference Series, vol. 882, p. 012010 (2017)
5. Quadrio, M., et al.: Review of computational fluid dynamics in the assessment of nasal air flow and analysis of its limitations. Eur. Arch. Othrhinolaryngol. **271**, 2349–2354 (2014)
6. Lynch, K.P., Scarano, F.: Experimental determination of thomographic piv accuracy by a 12-camera system. Meas. Sci. Technol. **25**, 1–10 (2014)
7. Lintermann, A., Schröder, W.: A hierarchical numerical journey trough the nasal cavity: from nose-like models to real anatomies. Flow Turbul. Combust. **102**, 89–116 (2019)
8. Naftali, S., Rosenfeld, M., Wold, M., Elad, D.: The air-conditioning capacity of the human nose. Ann. Biomed. Eng. **33**, 545–553 (2005)
9. Gambaruto, A.M., Taylor, D.J., Doorly, D.J.: Decompostition and description of the nasal cavity form. Ann. Biomed. Eng. **40**, 1142–1159 (2012)
10. Elad, D., Naftali, S., Rosenfeld, M., Wolf, M.: Physical stresses at the air-wall interface of the human nasal cavity during breathing. J. Appl. Physiol. **100**, 1003–1010 (2005)
11. Hörschler, I., Schröder, W., Meinke, M.: On the assumption of steadiness of nasal cavity flow. J. Biomech. **43**, 1081–1085 (2010)
12. Hopkins, L.M., Kelly, J.T., Wexler, A.S., Prasad, A.K.: Particle image velocimetry measurements in complex geometries. Exp. Fluids **29**, 91–95 (2000)
13. Spence, C.J.T., Buchmann, N.A., Jermy, M.C.: Unsteady flow in the nasal cavity with high flow therapy measured by steroscopic piv. Exp. Fluids **52**, 569–579 (2012)
14. Michaelis, D., Wieneke, B.: Comparison between tomographic piv and stereo piv. In: International Symposium of Laser Techniques to Fluid Mechanics (2008)

An Approach to Automated Detection of Sessile Droplets in Mixed Convection

Marie-Christine Volk[1(✉)], Konstantin Niehaus[2,3], Andreas Westhoff[2], and Claus Wagner[2,3]

[1] Faculty of Physics, Georg-August-Universität Göttingen, Friedrich-Hund-Platz 1, 37077 Göttingen, Germany
marie-christine.volk@dlr.de
[2] Aerodynamics and Flow Technology, German Aerospace Center, Bunsenstraße 10, 37073 Göttingen, Germany
[3] Institute of Thermodynamics and Fluid Mechanics, Technische Universität Ilmenau, Helmholtzring 1, 98684 Ilmenau, Germany

Abstract. A method for the measurement of position and size of sessile water droplets is presented. Droplets originate from condensation on a plane vertical surface in a vented cuboidal cavity with a mixed convective flow with humid air as the working fluid. Condensation is observed through a transparent cooling device coated with polyvinyl chloride with an average contact angle of 80.0(3)°. The implemented detection algorithm is based on the circle Hough Transform together with sophisticated pre- and post-processing steps, which are detailed in this work. Validation experiments yield a detection of over 97% of the area covered by droplets by detecting a minimal radius of 13.8 µm. Additionally, first experimental results of droplet size distributions are presented.

Keywords: dropwise condensation · evaporation · droplet growth · image recognition

1 Introduction

When humid air interacts with a surface below dew-point temperature, condensation occurs. The condensation rate on a surface depends on the surface wettability and forms either a continuous film or a multitude of droplets. Schmidt et al. [8] were the first who reported an increased heat transfer rate of dropwise condensation compared to filmwise condensation. This is used, for example, in polymer heat-exchangers which are less affected by corrosion compared to their metal counterparts [12]. Dropwise condensation also changes the transmittance of normally incident light through transparent materials from about 90% to almost 50% [2]. This phenomenon is mostly unwanted and can even pose a safety risk, considering windows, camera lenses, automotive windshields or headlights. Mixed convection occurs when natural and forced convection have similar impact on a fluid flow. If phase transition occurs in the system, the heat

A. Dillmann et al. (Eds.): STAB/DGLR Symposium 2022, NNFM 154, pp. 579–589, 2024.
https://doi.org/10.1007/978-3-031-40482-5_55

transfer is not only a function of the temperature and velocity distribution but also depends on the concentration field. The vast majority of publications investigate phase transition of water vapor in the absence or with small fractions of non-condensable gas (e.g. [3,11]), while the aforementioned day-to-day examples are exposed to low mass fractions of water vapor. More recently, Westhoff and Wagner [13] studied condensation of moist air in a channel flow at Reynolds number Re = 3415 using the change of humidity before and after the condensing surface to quantify the mass-transfer rate. Zheng et al. [14] recorded microscope images of condensation of moist air (at relative humidities 94% and 80%) in a channel flow on a hydrophobic polyethylene coated surface and used the results as validation for their numerical algorithm of droplet dynamics. Their droplet detection-algorithm detects a minimum droplet radius of 13.5 µm. Danilo et al. [5] studied condensation of humid airflow in a channel on a horizontal flat plate. They recorded microscope images to observe the droplet populations without explicitly measuring it. Tancon et al. [10] studied condensation of humid airflow on a vertical aluminum surface evaluating the drop-size distribution up to a minimum droplet radius of 3 µm, but did not provide details on the accuracy of their detection results.

This work presents the development and validation of an automated droplet detection-algorithm suited to investigate the time-resolved development of droplet-size distribution. The contact lines of sessile droplets are observed through a transparent cooling device. After brief description of the experimental set-up, the image pre- and post-processing is detailed together with the detection algorithm employing the circle Hough Transform (CHT) [6]. This is followed by the validation against a manually labeled set of reference images and first results showing the temporal evolution of droplet-size distribution during condensation at Re = 428. Closing remarks summarize this work and give a brief research outlook.

2 Experimental Set-Up

The experimental set-up is depicted schematically in Fig. 1. The rectangular container with length $L = 0.5\,\text{m}$, height $H = 0.25\,\text{m}$, and width $W = 0.25\,\text{m}$ is isothermally cooled at the front to $T_c = 9°\text{C}$ (blue) and heated at the vertical back wall to $T_h = 25°\text{C}$ (red). Air enters the cell through the left inlet with defined dew-point temperature Θ_{in}, mass flow rate $\dot{m}$, and fluid temperature T_{in}. The air exits through the outlet slot on the right. The ventilation openings extend of the entire cell height and have an aspect ratio of 10. To minimize the heat transfer through the left, right, top, and bottom sidewall the container is placed in a climate chamber to obtain almost adiabatic boundary conditions. A transparent cooling liquid (water-glycol mixture with ratio 1:1) flows inside the cooling plate which consists of two parallel glass walls. Inside the convection cell, the cooling plate is coated with a polyvinyl chloride (PVC) sheet with a thickness of $d_{\text{pvc}} = 60$. The droplet contact-angle on the PVC sheet was determined using

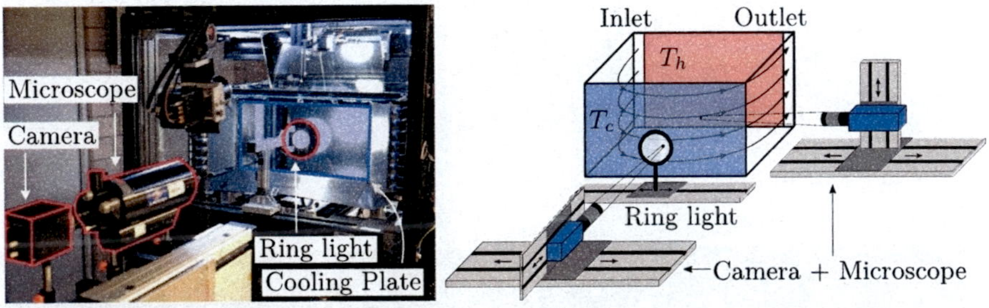

Fig. 1. Experimental set-up: (left) Photograph of sample with the cooling plate (blue) and the camera with the microscope and the ring light (red). (right) Schematic set-up with the heating plate at T_h (red), the cooling plate at T_c (blue), two ventilation openings on the rear, and the contact line and droplet profile camera (blue) on the front and side. (Color figure online)

a contact-angle measurement-device[1] and equals 80.0(3)°. To characterize the flow, the Reynolds number $\mathrm{Re} = \dot{m}/H\mu$ with μ the dynamic viscosity of the humid air at inlet temperature and two Grashof numbers $\mathrm{Gr}_{hc} = g\beta_T(T_h - T_c)H^3/\nu^2$ and $\mathrm{Gr}_{\mathrm{in}} = g\beta_T(T_{\mathrm{in}} - T_c)H^3/\nu^2$ with ν the kinematic viscosity, β_T the thermal expansion coefficient of humid air and g the gravitational acceleration are defined. A detailed description of the set-up and its boundary conditions can be found in Niehaus et al. [7]. To study the process of condensation on the cooled surface inside the sample with a non-invasive method, a camera[2] equipped with a far-field microscope[3] is installed at a working distance of about 1.20 m with an exposure time of 200 ms and a frame rate of 0.5 Hz. The resolution yields 2560 × 2160 pixels, resulting in a field of view of 4.4 mm by 3.7 mm. For the purpose of measuring the droplets-contact area, a diffuse, angled incident LED ring-light[4] is used. Since there is no back-reflecting surface inside the sample, the resulting images of the droplets originate from total reflections inside the droplets. In addition, a droplet profile camera is installed with a microscope to investigate the contact angle of the droplets inside the sample.

The average measurement time is two hours, starting with no droplets on the cooled surface at the beginning. $45\,\mathrm{min} < t < 120\,\mathrm{min}$ after condensation starts, the evaporation ($\Theta_{\mathrm{in}} = 0\,°\mathrm{C}$) is initiated. This time period depends on the droplet size to prevent the droplets from sliding down the foil. The measurement ends when all droplets have evaporated.

3 Image-Processing Method

In the following, the droplet-detection algorithm is explained. First the image preprocessing methods are motivated, then the algorithm itself is presented. The

[1] Krüss MobileDrop GH11.
[2] PCO Edge 5.5 camera (Camlink, rolling shutter).
[3] Long Distance Microscope QM1 by LaVision.
[4] Falcon FLKR-Si100 (white) with 6.72 W.

image post-processing, where corrections to the detection results are described in detail, concludes this section.

3.1 Preprocessing

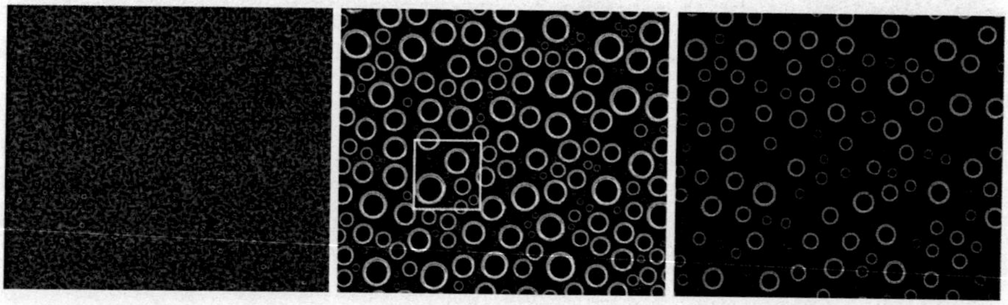

Fig. 2. Raw images showing different radii distributions of droplet contact lines: (left) almost uniform distribution during the onset of condensation, (middle) wide range of droplet sizes after first coalescences, (right) a distribution without small droplets during evaporation.

To extract the geometric data of the droplets from an image, a series of preprocessing steps is required. Raw gray-scale images from the camera (2560×2160 pixels, 16-bit scale) are shown in Fig. 2. To emphasize relevant structures, the image is corrected by a minimum image and the contrast is stretched to use the full 16-bit range. Thereafter, a Canny edge detection-algorithm is applied [4]. The algorithm first reduces noise by convolving a $(2k+1) \times (2k+1)$ $(k \in \mathbb{N})$ Gaussian filter with the image. An edge in a gray-scale image corresponds to a change in the pixels intensity. Hence, the edges can be found with a discrete differentiation operator, in the present case the Sobel operator [9] in an implementation of OpenCV [1] is used. This is followed by a non-maximum suppression and is applied to the resulting gradient image to thin out the edges and to filter

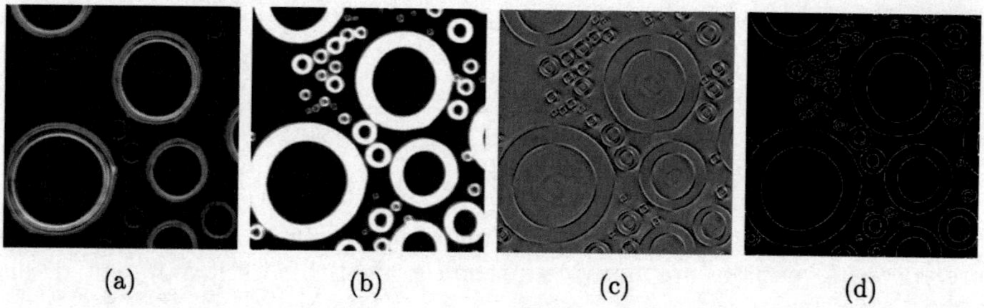

Fig. 3. Image preprocessing using the white area Fig. 2: (a) raw image, (b) enhanced contrast and blurred with 3×3 Gaussian filter (c) magnitude of intensity gradient using Sobel filter, and (d) binary edge detection result

out unwanted edges. It finds the pixel of an edge with the maximum value in gradient direction. All non-maximum values are set to zero. Hereby, for greater accuracy, the pixels are interpolated in the according gradient directions. All maximum values beyond an upper threshold are considered strong edges. Values between a lower and upper threshold are considered weak edges. These weak edges are only allowed to remain when connected to a strong edge. The result is a binary image. Figure 3 shows this as an example, for the image processing described above.

3.2 Droplet Detection

The goal of the droplet detection is to locate the droplet position and to parameterize its contact line. Analog to [15] the contact line is assumed circular and hence the contact line of a single droplet is determined by the triple (x, y, r), where x and y indicate the centroid position and r the radius. A well-documented method to detect circles in images is the CHT [6]. The CHT places circles of a certain predefined radius on the detected edges. The area with the most circle intersections is then assumed to be the center of a droplet with the selected radius. To reduce the computational cost, predefined radius ranges are defined. The smallest droplet radius to be detected is set to three pixels ($\approx$5 µm), and the maximum radius is set according to the maximum radius in the specific time series. Due to our condensation time, the maximum droplet radius reaches up to approximately 250 pixels ($\approx$430 µm). If the same Gaussian kernel size was used throughout all droplet sizes, small droplets would be smoothed out completely. This is avoided by segmenting the radius range into four sub-ranges with manually selected parameter sets (see Table 1 for designations), with small droplet detection using smaller kernel sizes but increased contrast compared to large droplet detection. Since the measurement starts with a dry surface, the droplet detection during the initial condensation period is only performed for smallest and small droplet sizes. When large droplets are present, they cast shadows upon the smaller droplets. This has to be compensated by enhancing local contrasts. When evaporation begins, the smallest and small droplets disappear first. Once these droplet sizes have disappeared, they are no longer observed in the process of evaporation. Hence, after a specified time period, these droplet sizes are excluded from the detection. The detection parameters are in summary the radius range, the parameter for the brightness and contrast adjustment, the blur kernel, a threshold for the edge detector, a threshold for the CHT and a parameter for the allowed overlap of the circles.

Table 1. Subdivision of the droplets into 4 manually defined size ranges.

Group	smallest	small	medium	large
Max. r/px	10	24	50	400
Max. r/µm	17	41	86	688

3.3 Post-processing

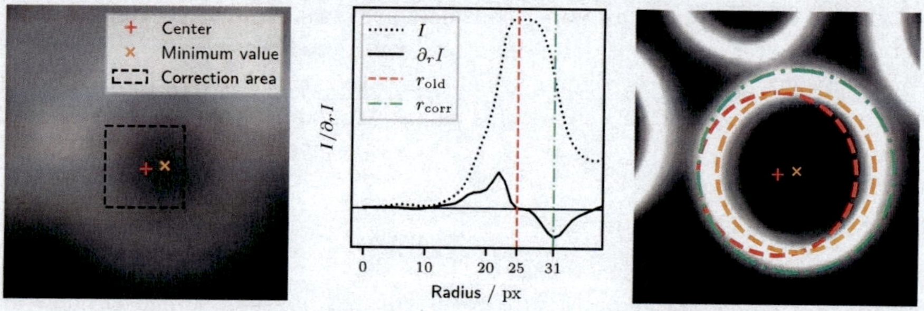

(a) (b) (c)

Fig. 4. Center and radius correction: (a) Droplet blurred with Gaussian kernel $r/2$. (b) Radial brightness profile I and its derivative $\partial_r I$. (c) Corrected center (orange x) and radius (green dot-dashed line). (Color figure online)

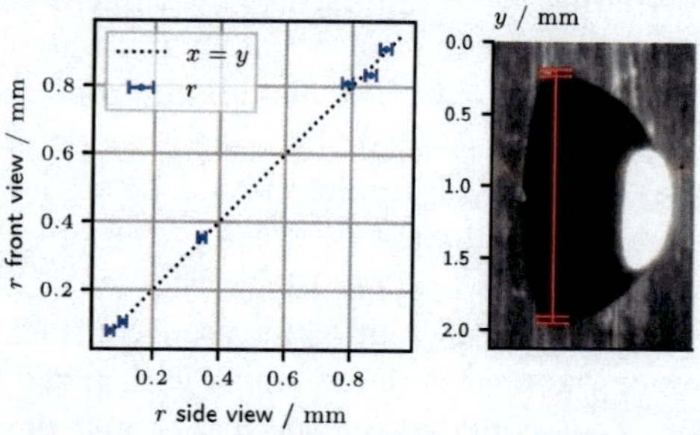

Fig. 5. Validation of the radius-correction method: (left) Comparison of the radius manually measured from the profile images and the radius obtained through the droplet detection-algorithm. (right) Measurement of the diameter of the droplet (red) through the profile-view. (Color figure online)

By inspecting the detected edges in Fig. 4c, it is found that every droplet consists of an inner and outer edge. Frayed edges become an increasing problem with decreasing droplet sizes. This increases the risk of detecting multiple droplets inside the outer droplet perimeter. Detection results are corrected by deleting overlapping circles with an area of intersection greater than half of their total area. In the next step the centroid position of the remaining droplets is corrected. Figure 4 illustrates the approach. The droplet is first blurred with a Gaussian kernel of $r/2$. This results in the droplet perimeter being blurred into its center. Figure 4a shows the blurred droplet and detected center position (red). The minimum value in the square area with edge length r around the droplet center is consequently the corrected center (orange). Figure 4c presents an image of the droplet with its first detected circle (red) and the center-corrected circle (orange). In order to correct the detected radius, a profile of the mean radial brightness around the corrected center is calculated. The radial profile is displayed by the black dotted line in Fig. 4b and the detected radius r_{old} is indicated by a vertical red line. To avoid the radial correction leading to errors, limits in defining the new radius r_{corr} are set according to the old radius $r_{\mathrm{corr}} = 0.75\,r \leq r \leq r_{c,\max}$. The maximum correction radius $r_{c,\max}$ is set to $1.4\,r$, except for small and smallest droplets, where it is set to $2\,r$ due to low initial detection accuracy. The corrected radius is finally defined by the minimum of the derivative of the radial brightness profile $I(r)$ (black solid line in 4b) leading to the final result of the center and radius correction presented in Fig. 4c. After the radius correction, the detected droplets may overlap again. To delete only the droplets that were likely detected incorrectly, all droplets that overlap with three or more others are deleted, regardless of their size. To validate the results, droplets placed on the cooling plate using a syringe, were photographed using the droplet profile and contact line camera. Consecutively, the radii obtained from the profile view are compared with the detection results. Figure 5 shows the scaling of the corrected radius and the radius measured by hand from the profile-images as well as an example of a droplet recorded with the profile camera.

4 Validation of the Image Recognition Method

Detection results are susceptible to two types of errors. The first error class includes droplets being detected although no droplet is physically present (invalid), while in the second class droplets are physically present but have not been detected (missed). To validate the present results, reference data was generated by labeling $N = 9$ images during condensation and $M = 3$ during evaporation by hand. A valid detection requires the correct determination of the position and size of a droplet. This is expressed by the condition

$$\|\vec{x}_{d,d} - \vec{x}_{d,h}\| < 0.75\,r_{d,h} \quad \text{and} \quad |r_{d,h} - r_{d,d}| < \kappa r_{d,h}, \tag{1}$$

with $\vec{x}_{d,d}$, $\vec{x}_{d,h}$, $r_{d,h}$, $r_{d,d}$, the location vector of the detection result, the location vector of the hand-labeled result, the hand-labeled radius, and the detected radius, respectively. κ signifies the maximum deviation between both radii, which is 0.1 for medium to large droplets and 0.33 for the rest. These thresholds are determined empirical and include increasing uncertainties of the hand-labeling process with decreasing $r_{d,h}$. The threshold for the positional deviation assures that the detected center is not situated on the bright ring and is uniform for all droplet sizes. All droplets that do not fulfill the condition Eq. 1

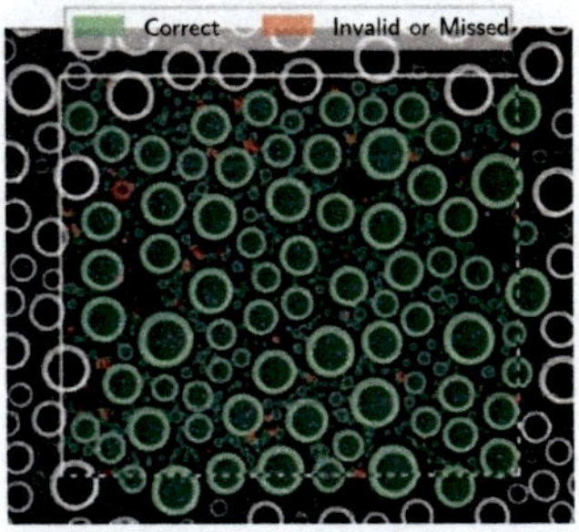

Fig. 6. Exemplary the detection result for Re = 448, $\mathrm{Gr}_{hc} = \mathrm{Gr}_{\mathrm{in}} = 2.57 \times 10^{8}$ and $\Theta_{\mathrm{in}} = 15.59\,°\mathrm{C}$ at $t = 101\,\mathrm{min}$.

but have an overlap with a hand-labeled droplet are considered invalid. Representative validation results are shown in Fig. 6 with droplets being detected correctly (green) or invalid and missed (red). The white frame indicates the area where hand-labeling has been performed. Droplets touching the left and top border are excluded, whereas the right and bottom border allow an overlap. To identify a minimum valid radius r_{min} above which detection results are significant, the number of missed and incorrectly detected droplets is displayed against r_{min} in Fig. 7a. The number of missed and incorrectly detected droplets is minimized at radii $r \geq 8\,pixel$ ($\approx$13.8 µm). Although detection is initially performed with a minimum radius of three pixels (($\approx$5.2 µm)), the validation results indicate, that most circles of this size are detected incorrectly or missed. Using this threshold, approximately 90% of the droplets are detected correctly, which corresponds to a detected covered area of 97%. Figure 7b shows the covered area fraction of the missed (black), invalid (red), and correctly detected (green) droplets at different times. With the threshold r_{min} the total number of missed droplets increases, while in turn the number of invalid droplets is minimized. With advanced condensation and evaporation, even 99% of the covered area is detected correctly.

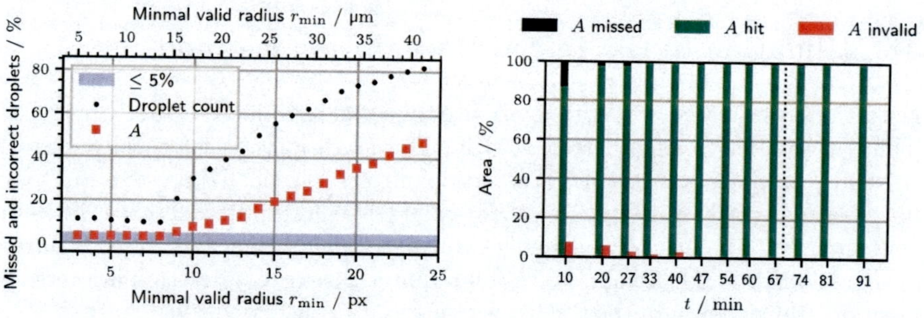

(a) (b)

Fig. 7. Validation of the droplet detection using data of the case Re = 428, $\text{Gr}_{hc} = \text{Gr}_{\text{in}} = 2.71 \times 10^8$ and $\Theta_{\text{in,cond.}} = 11.68\,°\text{C}$ and $\Theta_{\text{in,evap.}} = 0.0\,°\text{C}$: (a) The percentage of the total missed or incorrectly detected droplets against the detection threshold r_{min}. Black dots represent the droplet amount, red squares the associated covered area A. (b) Total covered area fraction of correctly (green) detected, missed (black) and invalid droplets (red) during a condensation and an evaporation period. The vertical black dotted line indicates the start of evaporation. (Color figure online)

5 Evolution of Droplet-Size Distribution

The described algorithm is used to analyze the droplet-size distribution during a measurement run exhibiting condensation and evaporation on the cooling plate surface. A sample distribution 60 min after the start of condensation is shown in Fig. 8a. The droplet distribution is approximated using the superposition of two

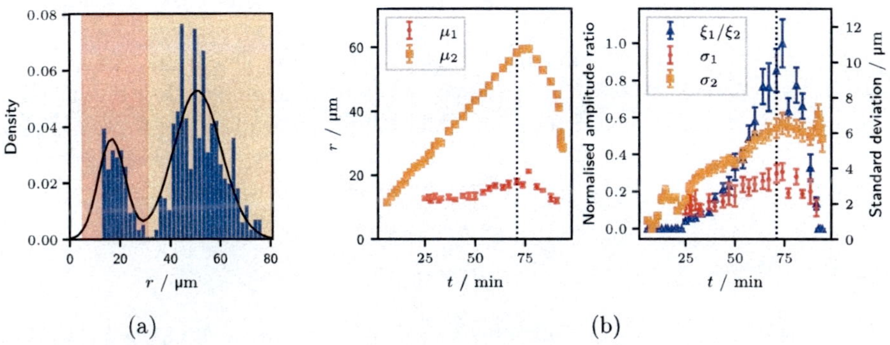

Fig. 8. Droplet distribution at Re = 428, $Gr_{hc} = Gr_{in} = 2.71 \times 10^8$ **and** $\Theta_{in,cond.} = 11.68\,°C$ **and** $\Theta_{in,evap.} = 0.0\,°C$: (a) Droplet distribution at $t = 60$ min. (b) The mean radii μ_i, the ratio of the amplitudes ξ_1/ξ_2 and the standard deviations σ_i of the Gaussian fits over the whole condensation process. The vertical black dotted line indicates the start of evaporation. (Color figure online)

Gaussian curves g_1 and g_2 with $g_i(x) = \xi_i \exp\left(-1/2\left(x-\mu_i/\sigma_i\right)^2\right)$, one for smaller droplets ($r < 33\,\mu m$) (red) and one for bigger droplets (yellow). The regression results during the condensation process are plotted in Fig. 8b. First, the Gaussian associated to the smaller droplets dominates the distribution, while for greater t a division into two curves is observed. The resulting gap in the distribution is associated to droplet coalescence. Due to the steady growth of the droplets, the mean values μ_1 and μ_2 of the two distributions shift towards larger radii until evaporation. The amplitude ξ_2 decreases during condensation as coalescence reduces the number of large droplets and therefore the distributions range expands (increasing σ_2), whereas the amplitude ξ_1 increases until evaporation starts.

6 Conclusion

The detection of sessile droplets on a subcooled, transparent surface in mixed convection in a humid airflow has been presented. With the present experimental set-up and the implementation of a detection algorithm based on the CHT, it is possible to obtain droplet center positions and radii while observing their change of size due to condensation or evaporation. Experimental validation results are presented showing that for radii greater than 8 pixel ($\approx 13.8\,\mu m$) more than 97% of the surface covered with droplets is detected correctly. Further, first results of the temporal evolution of radii distributions were presented. We observe a division into two droplet radii distributions. We assume, that for even larger t, the coalescence of droplets frees up space for a new generation of small droplets, and thus new small-droplet distributions arise. We were unable to confirm this assumption due to insufficiently long measurement periods.

We initially focused on condensation, since a constant contact angle can be assigned to the droplets. First investigations of the droplet profile camera show that this does not apply to evaporation. Therefore, modeling the contact angle of shrinking droplets at different boundary conditions will be the focus of future studies. Additionally, results will be presented in the future detailing the lateral radii distribution recorded using a far-field microscope on a linear motion system.

References

1. OpenCV - Open Source Computer Vision Reference Manual. OpenCV, 4th Edition(4.5.5.64), December 2021
2. Briscoe, B.J., Galvin, K.P.: The effect of surface fog on the transmittance of light. Sol. Energy **46**(4), 191–197 (1991)
3. Bum-Jin, C., Sin, K., Min Chan, K., Ahmadinejad, M.: Experimental comparison of film-wise and drop-wise condensations of steam on vertical flat plates with the presence of air. Int. Commun. Heat Mass Transf. **31**(8), 1067–1074 (2004)
4. Canny, J.: A computational approach to edge detection. IEEE Trans. Pattern Anal. Mach. Intell. **PAMI-8**(6), 679–698 (1986). Conference Name: IEEE Trans. Pattern Anal. Mach. Intell
5. Danilo, S., Dominique, C., Frédéric, P.: Experimental dropwise condensation of unsaturated humid air - influence of humidity level on latent and convective heat transfer for fully developed turbulent flow. Int. J. Heat Mass Transf. **102**, 846–855 (2016)
6. Davies, E.R.: A modified Hough scheme for general circle location. Pattern Recognit. Lett. **7**(1), 37–43 (1988)
7. Niehaus, K.A., Westhoff, A., Wagner, C.: Characterization of a mixed convection cell designed for phase transition studies in moist air. In: Dillmann, A., Heller, G., Krämer, E., Wagner, C. (eds.) STAB/DGLR Symposium 2020. NNFMMD, vol. 151, pp. 483–493. Springer, Cham (2021). https://doi.org/10.1007/978-3-030-79561-0_46
8. Schmidt, E., Schurig, W., Sellschopp, W.: Versuche über die Kondensation von Wasserdampf in Film- und Tropfenform. Technische Mechanik und Thermodynamik **1**(2), 53–63 (1930)
9. Sobel, I., Feldman, G.: A 3 × 3 isotropic gradient operator for image processing. Pattern Classif. Scene Anal. 271–272 (1973)
10. Tancon, M., et al.: Simultaneous measurement of heat flux and droplet population during dropwise condensation from humid air flowing on a vertical surface. Exp. Therm. Fluid Sci. **136**, 110677 (2022)
11. Tanner, D.W., Potter, C.J., Pope, D., West, D.: Heat transfer in dropwise condensation-Part I the effects of heat flux, steam velocity and non-condensable gas concentration. Int. J. Heat Mass Transf. **8**(3), 419–426 (1965)
12. T'Joen, C., Park, Y., Wang, Q., Sommers, A., Han, X., Jacobi, A.: A review on polymer heat exchangers for HVAC&R applications. Int. J. Refrig. **32**(5), 763–779 (2009)
13. Westhoff, A., Wagner, C.: Experimental study of moist air flow in the gap between the aircraft's fuselage and its cabin wall. CEAS Aeronaut. J. **11**(3), 591–607 (2020)

14. Zheng, S., Eimann, F., Philipp, C., Fieback, T., Gross, U.: Modeling of heat and mass transfer for dropwise condensation of moist air and the experimental validation. Int. J. Heat Mass Transf. **120**, 879–894 (2018)
15. Zheng, S., Eimann, F., Philipp, C., Fieback, T., Gross, U.: Dropwise condensation in the presence of non-condensable gas: interaction effects of the droplet array using the distributed point sink method. Int. J. Heat Mass Transf. **141**, 34–47 (2019)

Aerosol Spread in a Generic Train Entrance: Comparison Between Experiment and Numerical Simulation

Florian Webner[1(✉)], Andreas Kohl[1], Daniel Schmeling[1], and Claus Wagner[1,2]

[1] Institute of Aerodynamics and Flow Technology, German Aerospace Center (DLR), 37073 Göttingen, Germany
{Florian.Webner,Andreas.Kohl,Daniel.Schmeling,Claus.Wagner}@dlr.de
[2] Institute of Thermodynamics and Fluid Mechanics, Technische Universität Ilmenau, 98693 Ilmenau, Germany

Abstract. The global COVID-19 outbreak in 2020 has made understanding pathogen-laden aerosol transport and the associated transmission routes more relevant than ever. To determine how aerosol particles generated by continuous breathing accumulate in confined spaces, the particle concentrations in a small room resembling a train entrance are investigated. The room is ventilated and equipped with two heated manikins, one of which is continuously exhaling aerosol through the mouth for 30 min. For this setup we conducted local particle measurements in the center plane and a RANS simulation including the prediction of the transient particle transport. It is shown that the particle concentration increases logarithmically and attains a nearly steady state. The resulting local particle concentrations normalized to the source concentrations are subsequently compared. We find good agreement with the experiment in the exhalation zone of the breathing manikin and larger differences for the sensor positions beneath the ventilation inlet.

Keywords: aerosol spread · particle concentration · RANS · experiment

1 Introduction

Understanding the transport of pathogen-laden aerosols has been relevant for a long time since many diseases like the measles or the flu can be transmitted via the airborne route. The global outbreak of COVID-19 in 2020 further increased the interest in aerosol science in general. In this respect, public transport and indoor rooms have received extra attention since many people stay together in closed rooms for long periods of time. Hence, many researchers investigated aerosol transport in confined spaces by means of numerical simulation or experiments. Shang et al. (2022) [11] performed a numerical simulation solving the unsteady Reynolds-average Navier-Stokes (URANS) equation predicting the flow induced by a person coughing in a ventilated room for different ventilation scenarios. They showed

A. Dillmann et al. (Eds.): STAB/DGLR Symposium 2022, NNFM 154, pp. 590–600, 2024.
https://doi.org/10.1007/978-3-031-40482-5_56

that the thermal plume rising at the person entrains some of the particles so that these particles rise to a zone above the emitter's head instead of following the puff produced by coughing. Likewise, Wan et al. (2009) [14] investigated how aerosols emitted during coughing spread in an aircraft cabin performing URANS simulations for different ventilation settings. Furthermore, Gupta et al. (2012) [3] performed a URANS simulation of a coughing event in an aircraft cabin and estimated the infection risk. Hanna et al. (2021) [4] conducted experiments with inert tracer particles emitted in an aircraft cabin during one minute of simulated breathing. Other studies concentrate on the flow and spread of particles in the compartment of cars and trains, e. g., Mathai et al. (2020) [8]. Similar problems focusing on the flow in compartments were already of interest before the pandemic. For example, Konstantinov et al. (2015) [6] predicted the flow field and the thermal comfort in train compartments to investigate the influence of the heat released by the human metabolism on the flow field, which was found to be significant. Later, Konstantinov et al. (2022) [5] performed a URANS simulation of the flow field in a ventilated train compartment in combination with particle spreading induced by coughing, talking, or breathing.

While there is plenty of research on coughing events, literature on sustained emissions by breathing is scarce. Schmeling et al. [10] measured the concentrations of artificially produced aerosol over an exhalation period of 20 min on the seats in a train carriage. However, no CFD simulation was carried out, to support the scarce spatial resolution in the proximity of the aerosol source, where the highest concentrations were measured. Compared to the violent emission induced by coughing, breathing velocities are much lower. Furthermore, breathing is a continuous process, which means the emitted particles may accumulate in the environment, unlike particles emitted by singular short term events like coughing.

This leads to the question how particles accumulate in a confined ventilated space with continuous aerosol emission over a long period of time and how other passengers are exposed to the aerosol particles over time. To answer this question, we investigate the aerosol spread in a domain resembling a train entrance area over a time period of 1800 s by means of a RANS simulation and compare selected results to those obtained in an experiment. In our generic train entrance area, two standing heated manikins face each other and one of them is constantly emitting aerosol particles. An HVAC system supplies cold air from the ceiling while warm air leaves the room near the floor. The particle concentrations are measured over time at different positions in the experimental setup and all particles are tracked in the whole domain in the RANS simulation.

2 Methodology

2.1 Generic Train Entrance

The geometry of the considered generic train entrance with the two standing thermal manikins, TM1 and TM2, facing each other is shown in Fig. 1a. It is a cuboid room with a height of 2.4 m, a width of 3 m and a depth of 1.5 m resulting in a total volume of approximately 11 m^3. The entering air is prescribed at a

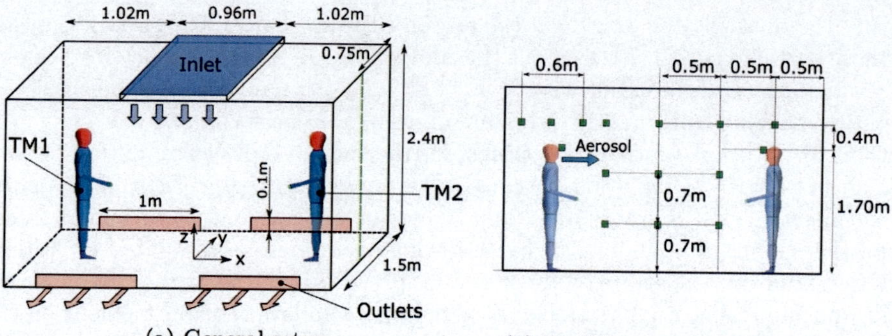

(a) General setup

(b) Particle sensor positions at y = 0

Fig. 1. a) Sketch of the generic train entrance with two standing thermal manikins. The air is supplied homogeneously through the 0.96 m × 1.46 m air inlet in the ceiling of the generic train entrance. Four passive air outlets are located close to the floor symmetrically on both sides of the setup. b) Positions of local particle concentration measurements in the center plane at y = 0 m.

rectangular inlet surface by setting constant air velocities over the whole inlet area which has a size of 0.96 m × 1.46 m. It is positioned 5.5 cm above the ceiling. An inlet flow with a volume flow rate of $\dot{V}_{inlet} = 25\,\mathrm{l/s}$ is prescribed, which is more than twice the flow rate provided by an HVAC system if two passengers are in a train compartment according to the norm DIN EN 13129. This results in a total of 8.2 air changes per hour (ACH), which is comparable with the ACH $= 8\,\mathrm{h}^{-1}$ of fresh air in a German ICE train carriage [10]. Besides, four air outlets, each with an outlet area of 0.1 m × 1 m, are positioned near the floor. For reasons of simplicity the walls are specified to be adiabatic. Each of the two thermal manikins (TMs) generates a constant heat flux of 100 W, which almost corresponds to the heat emitted by a human while standing. In our setup, TM1 (see Fig. 1a) exhales aerosol with a constant volume flow $\dot{V}_{in,aerosol} = 0.2\,\mathrm{l/s}$ for 30 min. This is twice the volume flow of an average female subject [2] mimicking the average exhalation momentum in a breathing cycle since the used TM does not inhale. A tube inlet with an inner diameter of 12 mm simulates the mouth opening with an area of 1.13 cm^2, which resembles the opening area of a female subject breathing through her mouth [2]. The resulting Reynolds number based on the jet diameter is $Re_{Jet} = \frac{U_0 D}{\nu} = 1385$ with U_0 being the mean jet velocity, $D = 12\,\mathrm{mm}$ the jet diameter at the orifice and ν the kinematic viscosity of the fluid [7].

2.2 Experimental Setup

In the experiment the local particle concentration C_{exp} in the room is measured using the particulate matter sensors SPS30 by Sensirion shown in Fig. 1b and 2a. This sensor is able to measure the concentration of particles with a diameter of 0.3 to 10 µm in 5 different size intervals (measurement bins). The data

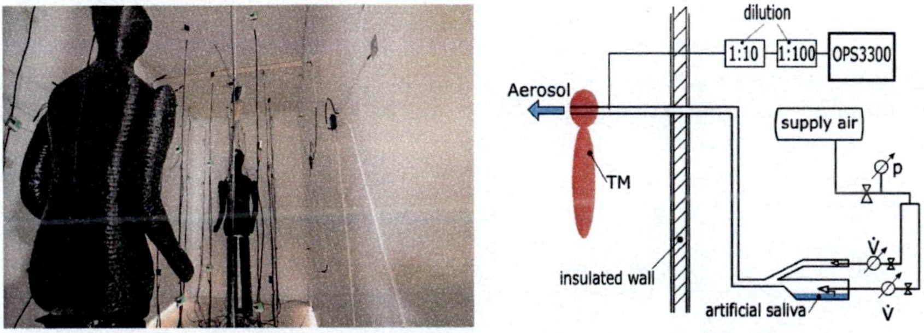

(a) (b)

Fig. 2. a) Image of the experimental setup with standing TMs and SPS30 particle sensors mounted on thin lines from the ceiling. b) Sketch of the aerosol source with air supply, measurement of pressure, volume flow input, and particle concentration of resulting aerosol with the optical particle sizer OPS3300 [12].

acquisition for all sensors is conducted with an inhouse, open source measurement system developed by Niehaus et al. [13]. For the comparison presented below, only particle sizes between 0.3 and 2.5 m are evaluated since they lead to the highest number concentrations during human tidal breathing [9] and are also associated with the highest concentration the aerosol source produces in the experiment (Table 1). The schematics of the aerosol source are shown in Fig. 2b and further information is provided in [10]. The produced aerosol particle concentration is continuously monitored by an optical particle sizer (OPS3300). The upper and lower boundaries (UB and LB) of the measured bins of particle sizes and the number concentration of the aerosol source $C_{s,exp}$ are summarized in Table 1. Samples are taken 200 mm upstream of the generic mouth at 1 Hz. The total volume fraction of all bins is $V_F \approx 1 \times 10^{-7}$ which is one magnitude below the critical volume fraction of 1×10^{6} mentioned by [1], which means that the **particles are not affecting the airflow.**

The supply air temperature is measured directly at the inlet before a dense membrane enforces a homogeneous air supply over the whole inlet area. The inlet air, coming from an HVAC system, also passes a HEPA filter system (H13 filter class) filtering more than 99.95% of the airborne particles in our measurement range. The TMs (see Fig. 2a) are heated with 100 W and have been used successfully, e. g., for measurements in a generic train compartment in [15]. In the experiment, the inner and outer wall temperature is measured at 8 positions returning an average heat flux of approximately 10 W through the walls.

2.3 Numerical Setup

The particle transport is simulated in two main steps: First, a RANS simulation is performed to produce the flow field. Then, a transient solver predicts the particle movement in the flow field. In the following, the RANS simulation and the transient particle tracking are described in more detail.

Table 1. Measured particle concentration inside the aerosol source tube, 200 mm upstream of the mouth opening of TM1. LB and UB are the lower and upper boundaries of the bin size of the OPS3300. $C_{s,exp}$ is the mean concentration in the interval of 30 min of aerosol production and emission. CV is the coefficient of variation–the ratio of the standard deviation σ to the mean $C_{s,exp}$–for the measurement bin and time interval. D_i is the midpoint diameter of the respective bin. V_F is the particle volume fraction calculated with D_i, assuming spherical particles.

LB [μm]	0.30	0.37	0.47	0.58	0.72	0.90	1.12	1.39	1.73	2.16
UB [μm]	0.37	0.47	0.58	0.72	0.90	1.12	1.39	1.73	2.16	2.69
$C_{s,exp}$ [10^3/cm^3]	131.6	109.2	70.4	38.8	23.0	16.6	9.2	13.4	4.6	0.8
CV [%]	6.1	5.9	6.1	5.4	5.4	5.5	5.1	4.8	5.4	6.1
D_i [μm]	0.34	0.42	0.52	0.65	0.81	1.00	1.25	1.56	1.94	2.42
V_F [10^{-9}]	2.6	4.2	5.2	5.6	6.4	8.9	9.5	26.7	17.7	6.2

The RANS simulation is performed using the k-ω shear stress transport turbulence model provided in the OpenFOAM v2112 toolbox. The computational domain is discretized using 5.77 million hexahedral cells in total, where the largest cells with a mean width of 22 mm are located in the center and the grid size is refined near the walls, around the thermal manikins, as well as in the regions where the thermal plumes and breathing jet are expected.

The dimensions and boundary conditions are chosen to mimic the generic train entrance with adiabatic, impermeable, and no-slip walls, a volumetric flow rate of 25 l/s with a temperature of 17 °C at the inlet, and a volumetric flow rate of 0.2 l/s with a temperature of 26 °C exhaled by TM1. The mouth is modelled as a pipe segment intruding 3.05 cm into the head with a diameter of 12 mm and the air is supplied from the inner base surface of this pipe segment. Finally, the volumetric flow rate at the outlets is computed by enforcing a Neumann boundary condition in streamwise direction. The surface temperatures of the TMs are prescribed in accordance with the temperatures measured in the experiment with an infrared camera. Three areas with different temperatures are distinguished: head (35.5 °C), torso (32 °C), and legs (30 °C).

Besides, the particle transport is computed using a stochastic dispersion model, which calculates forces acting on particles, such as gravity, fluid drag and inertial forces as well as the resulting acceleration of the particles. Additionally, the direction and amplitude of the associated velocities are perturbed by a random Gaussian number generator to model turbulent dispersion. In total, 2000 particles per second are seeded into the domain from the mouth of TM1. Ten different particle diameters D_i are considered (see Table 1), resulting in a seeding rate of 200 particles per second for each diameter. Assuming that particles do not influence each other, any arbitrary particle size distribution can be constructed by combining emitted particles of different sizes in the post-processing as previously demonstrated by [14].

By definition, the particles follow the considered steady flow without affecting it, in the sense of one-way coupling. This assumption is verified above in Sect. 2.2.

In the simulation, particles cannot rebound off the walls but they can stick to the walls on the first impact. The validity of the latter assumption is supported by [14], who calculated the force balance between the detachment forces to resuspend a particle and the adhesive forces between the particles and the wall. Higher particle concentrations are to be expected in the experiment than in the simulation since the aerosol source of the experiment emits orders of magnitude more particles. Therefore, to compare the predicted particle concentration C_{sim} [1/cm^3] with concentration C_{exp} measured in the experiment, both concentrations are normalized with the corresponding particle seeding rate s_r. First, to determine the predicted particle concentration at any position $\vec{r}$, all particles at position $\vec{p}$, for which $|\vec{r}-\vec{p}| < 7.5\,\text{cm}$ holds true, are counted. This yields $n_{r,7.5}$, i. e., number of particles in a sphere with a radius of 7.5 cm around $\vec{r}$. Then, $n_{r,7.5}$ is divided by the volume of a sphere with a radius of 7.5 cm to determine the particle concentration C_{sim}. Finally, the particle concentration is normalized by the seeding rate – the product of the particle concentration of the source C_s and the volumetric flow rate of the source $\dot{V}_s$. Since the volumetric flow rates in the experiment and the simulation are equal, normalizing by the seeding rate is equivalent to normalizing by the concentration of the source as shown in Eq. (1):

$$\frac{C_{exp}}{s_{r,exp}} = \frac{C_{sim}}{s_{r,sim}} \Leftrightarrow \frac{C_{exp}}{\dot{V}_s C_{s,exp}} = \frac{C_{sim}}{\dot{V}_s C_{s,sim}} \Rightarrow \frac{C_{exp}}{C_{s,exp}} = \frac{C_{sim}}{C_{s,sim}}, \tag{1}$$

where $C_{s,exp}$ and $C_{s,sim}$ denote the particle concentration at the source for the experiment and the simulation, respectively.

Thus, Eq. (1) demonstrates that the normalized particle concentration C_n [%] (particle concentration at any location divided by the particle concentration at the source) of the simulation and the experiment is equal, if the particle transport is equal. C_n is then calculated for 57,590 positions on the central xz-plane of the domain, covering a grid with 1 cm distance between the nodes and 7.5 cm distance to the walls. In the region closer than 7.5 cm to the manikins, a hemisphere with the flat side facing the manikin is used to calculate C_n. This is necessary since the spheres used for particle counting intercept with the wall, if they are located too close to the wall.

3 Results

3.1 Normalized Particle Concentrations on the Center Plane

The particle concentration C_n predicted in the flow simulation determined by time averaging from 1780 s to 1800 s is visualized in the center plane in Fig. 3. For comparison, the black circles include the concentrations measured at the sensor locations in this plane. The numbers in the circles indicate the labels of the corresponding sensor positions P0 - P14. The colorbar is capped at 1.8% to highlight the lower concentration values.

Near the walls and the floor, the predicted particle concentration C_n varies between 0.4 and 0.7%. Much higher C_n are obtained in front of the TMs, with values of up to 1.2% in the vicinity of TM2 and up to 4.7% (not visible in Fig. 3 due to the capped colorbar) in the breathing jet of TM1.

The predicted C_n agree fairly well with the measured particle concentrations ranging from 0.73% (P5) to 1.67% (P3). The experiment yields the highest concentrations at P2 and P3 with 1.54% and 1.67%, respectively, while the simulation predicts similar values at these positions with 1.62% and 1.59%, respectively. A general observation for all positions–except P2 and P3–is that the range of the values observed in the experiment does not vary as much between the different positions as the predicted concentration does. The largest differences between simulation and experiment are obtained just underneath the inlet at P4, where a low value of 0.19% is predicted in the simulation in comparison with the C_n of 0.9% measured in the experiment. The mean C_n obtained by averaging over all sensor positions is 0.85% for the simulation and 0.92% for the experiment, with an average absolute deviation of 0.24%.

3.2 Normalized Particle Concentration over Time

The normalized particle concentrations C_n over time predicted and measured at four selected positions, i. e., near the bottom (P12), near the top (P5), in the center below the inlet (P10), and in/near the exhalation jet (P2), are visualized in Fig. 4.

The markers on the black solid line indicate the mean value obtained by time-averaging ten values in a time interval of 2 s lying in the past and the future of the corresponding time. Furthermore, the black dashed line labeled 'scaled sim.' represents the value predicted in the RANS simulation normalized with the maximum of the predicted time series and multiplied by the mean measured value representing the time average of the interval between 1600 s and 1800 s. Thus, this scaling of the time series allows to compare the growth in time (i. e., doubling in 200 s) since the scaled time series of the simulation ends with a mean value similar to that of the measured time series.

Figure 4 reveals strong fluctuations in the particle concentrations at P2 and P10. These are positions where the simulation predicted the highest C_n. In contrast, the fluctuations of C_n at P5 and P12, which are further away from the aerosol source, are significantly lower. Further, the scaled time series corresponds quite well to the time series of the measurement, which demonstrates that the growth is comparable, although the C_n values at corresponding positions differ to a certain extent. P5 is an exception, where the measured C_n grows faster in the first 500 s.

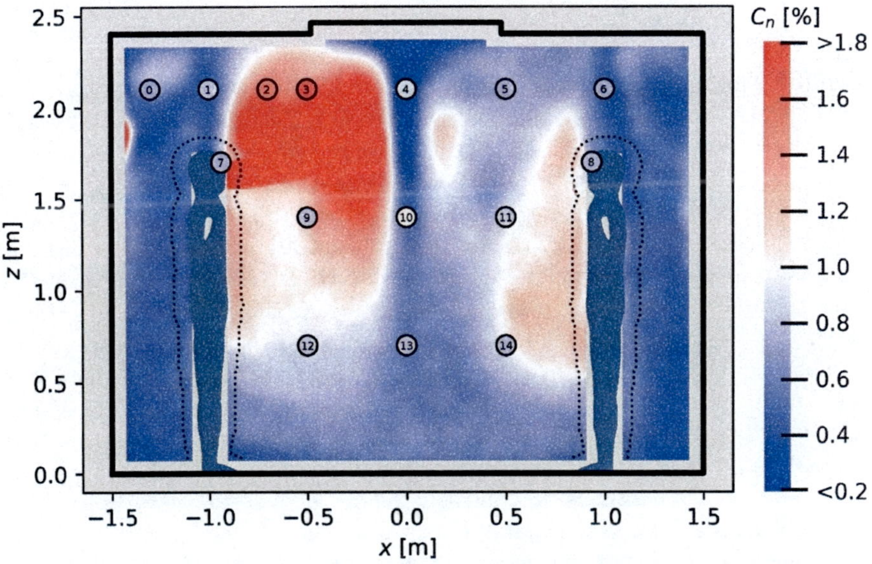

Fig. 3. Comparison of the distribution of the particle concentration C_n in the center plane with the particle concentration measured at the sensor positions in this plane highlighted by circles. The numbers 0–14 in the circles indicate the position of the sensors and the black dotted lines mark the distance to the layer around the manikin in which the particle concentrations C_n are determined in a hemisphere instead of the sphere used in the bulk due to the limited distance to the manikin.

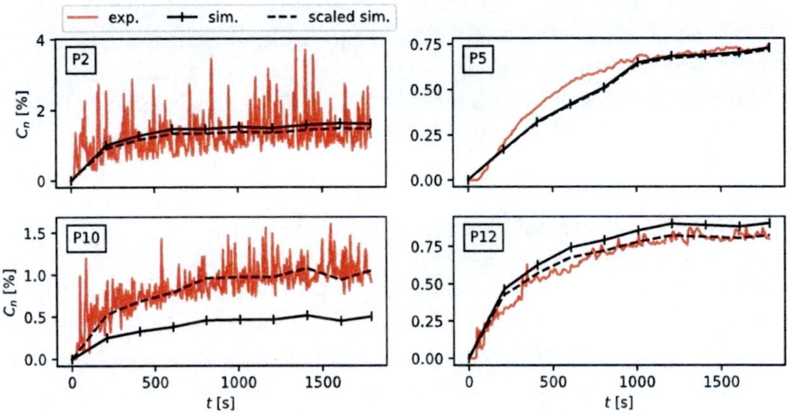

Fig. 4. Comparison of predicted and measured time series of C_n.

4 Conclusion

Figure 3 shows that the predicted and measured particle concentrations C_n are in the same order of magnitude, with values ranging from 0.5% to 1.6% at the considered sensor positions. Therefore, a person anywhere in the room inhales air with an aerosol particle concentration of 0.5 % to 1.6 % of the particle concentration of the source for the given exhalation rate of $\dot{V}_{in,aerosol} = 0.2\,\mathrm{l/s}$. The predicted distribution of the particle concentration visualized in Fig. 3 also reveals regions where particles seem to be 'locked in', e. g., in front of the TMs. This results in an elevated particle concentrations throughout 30 min. In contrast, particle concentrations measured at the sensor positions indicate a more homogeneous distribution in the experimental setup. Under the inlet, at P5, the numerical simulation yields very low particle concentrations and the particle transport between the manikins is low. This is not surprising since the RANS simulation predicts a continuous airflow from the inlet at the ceiling down to the floor between the manikins. In contrast, the descending flow is turbulent in the experiment, which leads to enhanced horizontal particle transport in this region. Consequently, the lower concentration below the inlet and the generally more heterogeneous particle distribution of the numerical simulation, suggest that our simulation underestimates the turbulence intensity and thus, the particle dispersion.

When comparing the accumulation of particles over time shown in Fig. 4, it is evident that the RANS simulation agrees remarkably well with the experiment. Both approaches suggest a strong increase of particle concentration in the first 500 s to 1000 s and a nearly steady state at 1800 s. Therefore, a person entering the room 1000 s after the particle emission has started and then staying in the room for 5 min is exposed to a much higher particle concentration than a person already present in the room from the start of particle emission and then staying for 5 min. At positions close to the aerosol source, e. g., at P2 and P10, the measured time series of C_n shows strong fluctuations over time. These are positions near an aerosol cloud with particularly high particle concentrations observed in the RANS simulation. Therefore, the unsteady flow field of the experiment would shift the edges of the aerosol cloud intermittently and cause strong fluctuations in the particle sensors nearby, as observed at P2 and P10.

In addition, the evaluation of the time series reveals that further away from the aerosol source, e. g., at P5, C_n of the experiment increases faster as than the C_n the RANS simulation, even though the steady state concentrations are nearly equal. This underlines that the particle dispersion is underestimated in the RANS simulation.

In conclusion, both the experiment and the flow simulation predict a normalized particle concentration of around $1 \pm 0.5\%$ of the concentration emitted by the aerosol source. Both approaches show a nearly steady state after 1800 s and a strong increase of aerosol concentration in the first 500 to 1000 s. Furthermore, the RANS-based flow simulation seems to underestimate the particle dispersion with the presented approach of long-term transient particle seeding into a steady state flow field.

We recommend further investigation of the factors influencing particle dispersion, e. g., the turbulence and the particle dispersion model.

Acknowledgments. This work was funded by the DLR project GANDALF. Special thanks to André Volkmann and Felix Werner (preparation of experiments), Andrey Shishkin (support for the CFD modeling), and Annika Köhne (proofreading).

References

1. Elghobashi, S.: On predicting particle-laden turbulent flows. Appl. Sci. Res. **52**(4), 309–329 (1994). https://doi.org/10.1007/bf00936835
2. Gupta, J.K., Lin, C.H., Chen, Q.: Characterizing exhaled airflow from breathing and talking. Indoor Air **20**(1), 31–39 (2010). https://doi.org/10.1111/j.1600-0668.2009.00623.x
3. Gupta, J.K., Lin, C.H., Chen, Q.: Risk assessment of airborne infectious diseases in aircraft cabins. Indoor Air **22**(5), 388–395 (2012). https://doi.org/10.1111/j.1600-0668.2012.00773.x
4. Hanna, S.: Transport and dispersion of tracer simulating COVID-19 aerosols in passenger aircraft. Indoor Air **32**(1), e12974 (2021). https://doi.org/10.1111/ina.12974
5. Konstantinov, M., Schmeling, D., Wagner, C.: Numerical simulation of the aerosol formation and spreading in a train cabin. J. Aerosol Sci. **170**, 106139 (2023). https://doi.org/10.1016/j.jaerosci.2023.106139
6. Konstantinov, M., Wagner, C.: Numerical simulation of the thermal comfort in a train cabin. Int. J. Railw. Technol. **4**(3), 69–88 (2015). https://doi.org/10.4203/ijrt.4.3.3
7. Lee, J.H.W., Chu, V.H.: Turbulent Jets and Plumes: A Lagrangian Approach. Springer, Boston(2013). https://doi.org/10.1007/978-1-4615-0407-8, ISBN: 978-1-4615-0407-8
8. Mathai, V., Das, A., Bailey, J.A., Breuer, K.: Airflows inside passenger cars and implications for airborne disease transmission. Sci. Adv. **7**(1), eabe0166 (2021). https://doi.org/10.1126/sciadv.abe0166
9. Morawska, L., Johnson, G.R., Ristovski, Z.D., Hargreaves, M., et al.: Size distribution and sites of origin of droplets expelled from the human respiratory tract during expiratory activities. J. Aerosol Sci. **40**(3), 256–269 (2009). https://doi.org/10.1016/j.jaerosci.2008.11.002
10. Schmeling, D., Kühn, M., Schiepel, D., Dannhauer, A., et al.: Analysis of aerosol spreading in a German Inter City Express (ICE) train carriage. Build. Environ. **222**, 109363 (2022). https://doi.org/10.1016/j.buildenv.2022.109363
11. Shang, Y., Dong, J., Tian, L., He, F., et al.: An improved numerical model for epidemic transmission and infection risks assessment in indoor environment. J. Aerosol Sci. **162**, 105943 (2022). https://doi.org/10.1016/j.jaerosci.2021.105943
12. TSI. Optical-particle-size (OPS)-3330. https://tsi.com/products/particle-sizers/supermicron-capable-particle-sizer-spectrometers/optical-particle-sizer-(ops)-3330/
13. Niehaus, K., Westhoff, A.: An Open-Source Data Acquisition System for Laboratory and Industrial Scale Applications. Meas. Sci. Technol. **24**(2) (2022). https://doi.org/10.1088/1361-6501/ac9994

14. Wan, M.P., To, G.N.S., Chao, C.Y.H., Fang, L., et al.: Modeling the fate of expiratory aerosols and the associated infection risk in an aircraft cabin environment. Aerosol Sci. Technol. **43**(3), 322–343 (2009). https://doi.org/10.1080/02786820802641461
15. Winter, S., Meyenberg, M., Schmeling, D.: A moving thermal manikin for the simulation of walking passengers in aircrafts or trains. In: Roomvent & Ventilation Conference 2018, pp. 1097–1102 (2018)

Rotorcraft

Dynamic Stall Computations of a Double-Swept Rotor Blade with Rigid and Elastic Modelling

Georg Babij(✉) and Martin Michael Müller

German Aerospace Center (DLR), Institute for Aeroelasticity, Bunsenstraße 10, 37037 Göttingen, Germany
georg.babij@dlr.de
https://www.dlr.de/ae/en

Abstract. Innovative helicopter rotor blades with a combined forward-backward double-sweep at the outer part of the blade enable a reduction in noise emission and enhance the overall performance of a rotor. In this context, the influence of the aeroelastic behaviour in connection with the dynamic stall phenomenon is of great importance. It is accompanied by large aerodynamic load peaks, primarily seen in the lift and the pitching moment, impacting the structural integrity of the blades and adjacent control components. Double-swept model rotor blades were developed and investigated experimentally at the German Aerospace Center (DLR) in Göttingen regarding the dynamic stall behaviour in a four-bladed rotor configuration at the Rotor Test Facility Göttingen. Due to an axial inflow to the rotor disc a sinusoidal variation in pitch angle is introduced to trigger the dynamic stall phenomenon once per revolution. The numerical study simulates the conducted experiments utilizing two different blade modelling approaches: elastic and rigid bodies. The corresponding computations are carried out with the use of computational fluid dynamics (CFD) and a multibody system (MBS). With the inclusion of blade elasticity both domains are connected together by using a strong aeroelastic coupling scheme. Three test cases with a rotor speed of 23.6 Hz will be presented comprising two test cases with fixed collective pitch angle and one with a superposed cyclic variation in pitch angle in order to introduce the dynamic stall phenomenon. Finally, a comparison is carried out with experimental data including the measured rotor thrust as well as the displacements at the blade tips. They show good agreement with the numerical results in both of the considered fields comprising the structural behaviour of the blades as well as the surrounding fluid flow.

Keywords: Dynamic Stall · Flexible Multibody System · Computational Fluid Dynamics · Rotor Aeroelasticity · Rotor Systems

1 Introduction

Innovative helicopter rotor blades with a combined forward-backward sweep at the outer part of the blade enable a reduction in noise emission and enhance the

A. Dillmann et al. (Eds.): STAB/DGLR Symposium 2022, NNFM 154, pp. 603–612, 2024.
https://doi.org/10.1007/978-3-031-40482-5_57

overall performance of a rotor [11]. In this context, the influence of the aeroelastic behaviour in connection with the dynamic stall phenomenon is of great importance. It is accompanied by large aerodynamic load peaks, primarily seen in the lift and the pitching moment, impacting the structural integrity of the blades and adjacent control components [6]. Double-swept model rotor blades were developed [9] and investigated experimentally in a four-bladed rotor configuration at the Rotor Test Facility Göttingen (RTG) [10] in the German Aerospace Center (DLR), see Fig. 1. They are built from multiple layers of carbon fibre reinforced polymer with an integrated foam core and measure 541 mm in blade length as well as 72 mm in chord length at the root. The rotor radius amounts to 652 mm. Due to an axial inflow to the rotor a sinusoidal variation in pitch angle Θ is introduced to bring about the dynamic stall phenomenon once per revolution. The following numerical study simulates the conducted experiments at a rotor speed $\Omega = 23.6$ Hz with two different modelling approaches considering the rotor blades as either elastic or rigid bodies. The corresponding computations are performed with the use of computational fluid dynamics (CFD) and a multibody system (MBS) which prescribes the rigid body motions for both blade modelling approaches, i.e. rigid and elastic blade modelling. With the inclusion of blade elasticity both domains are connected together by a strong aeroelastic coupling scheme. As a first step towards validation, a comparison with experimental data is carried out in terms of aerodynamic forces, namely the integral rotor thrust, and structural displacements measured at the blade tips for selected operating points.

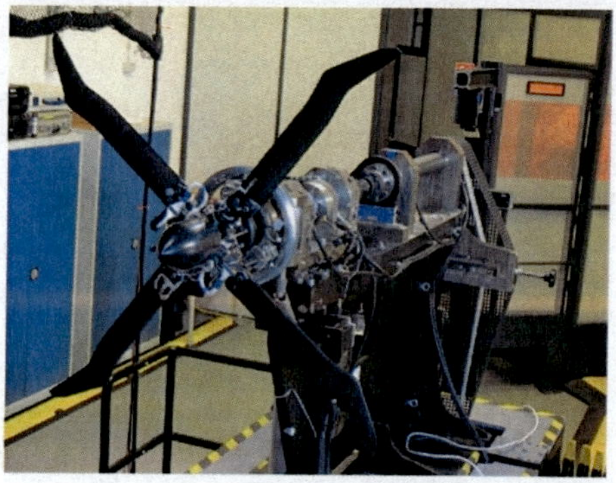

Fig. 1. Rotor Test Facility Göttingen with four-bladed rotor

2 Numerical Modelling

2.1 Structural Dynamics

Simpack is used to describe the structural dynamics of the entire rotor by means of a multibody system. For the approach with elastic modelled blades, the structural displacements are described in a modal representation with an underlying linear elastic material behaviour. 23 modal degrees of freedom are considered compromising the first 20 natural normal

Table 1. Eigenfrequencies of one clamped double-swept blade with 0° pitch angle for rotor speeds $\Omega = 0$ Hz [1] and $\Omega = 23.6$ Hz

Mode No.	$\Omega = 0$ Hz	$\Omega = 23.6$ Hz	Characterisation of normal mode
1	43.81 Hz	51.30 Hz	**1st flap**
2	177.14 Hz	185.45 Hz	**2nd flap**
3	242.08 Hz	242.95 Hz	1st lag
4	316.26 Hz	319.91 Hz	1st torsion
5	455.85 Hz	461.27 Hz	**3rd flap**

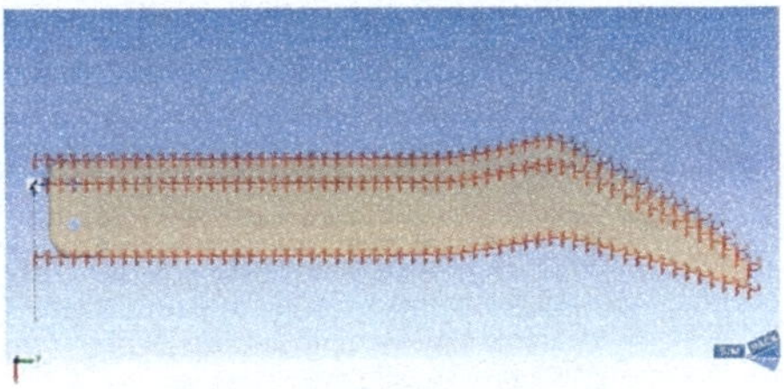

(a) Distribution of markers (red) on one blade

(b) Four-bladed rotor configuration

Fig. 2. Modelling of structural dynamics in Simpack (MBS)

modes of the blade up to a corresponding eigenfrequency of 3129.90 Hz and 3 further inertia relief modes as interface modes. The huge centrifugal forces on the rotor blades are taken into account by adding a geometric stiffness matrix. The first 5 eigenfrequencies in clamped boundary condition with their corresponding mode characterization can be found in Table 1 for the non-rotating system, i.e. $\Omega = 0$ Hz, and the considered operating point with $\Omega = 23.6$ Hz. 174 markers are distributed on 58 aerofoil cut sections which are equally spaced across the span. On each cut section, a marker is placed at the leading and trailing edge as well as at quarter chord position, see Fig. 2a. They serve as points to introduce aerodynamic forces on one hand and to readout structural displacements on the other. The entire model of the rotor configuration is depicted in Fig. 2b.

2.2 Fluid Dynamics

TAU utilizes the Reynolds-Averaged-Navier-Stokes equations to model fluid flow. Due to the unknown Reynolds stresses turbulence models are applied, such as the SST-k-ω-model according to Menter in this study. Furthermore, a finite volume method implemented for unstructured grids is utilized to obtain a second-order accurate spatial discretization. Against this backdrop, an approximate solution is sought on a dual grid being computationally created from the user-defined primary grid. Taking into account grid motion and deformation, the Chimera technique was exploited in connection with the geometric conservation law. The primary grid contains approximately 13.555 million nodes and it consists of a background grid as well as four Chimera grids representing the blades. A corresponding slice obtained by cutting through the middle of the domain is shown in Fig. 3a. Figure 3b provides a zoom view of the primary grid nearby the blades. The time integration is performed by a dual-time stepping approach with 500 iterations per physical time step Δt relying upon a two-step backward differentiation formula.

2.3 Strong Aeroelastic Coupling Scheme

The aeroelastic coupling is based on a sequential exchange of data in each physical time step Δt between both solvers. Therefore, it represents in the sense of

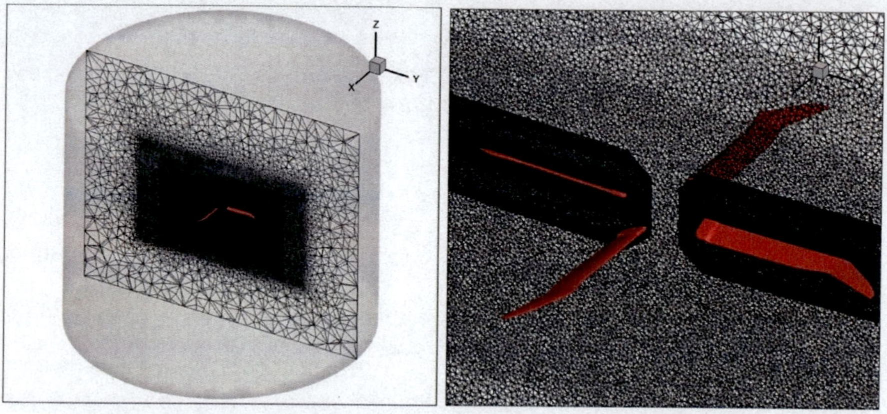

(a) Primary grid for the fluid domain (b) Primary grid nearby the rotor blades

Fig. 3. Modelling of fluid dynamics in TAU (CFD)

the terminology used within the rotorcraft community a strong coupling scheme. In detail, aerodynamic forces being calculated by TAU are transferred to Simpack and, vice versa, structural displacements together with rigid body motions are transferred from Simpack back to TAU [2,12] In order to exchange this data between both differently discretized domains, i.e. forces and displacements, scattered data interpolation methods based on radial basis functions are exploited. The entire procedure is illustrated by means of an eXtended Structure Design Matrix (XDSM) in Fig. 4. The associated software is implemented in Python as well as C and mostly relies on DLR's framework FlowSimulator. The physical time step in this study is chosen according to an azimuthal discretization $\Delta\Psi = 1°$, which results in approximately $\Delta t \approx 1.177 \cdot 10^{-4}$ s.

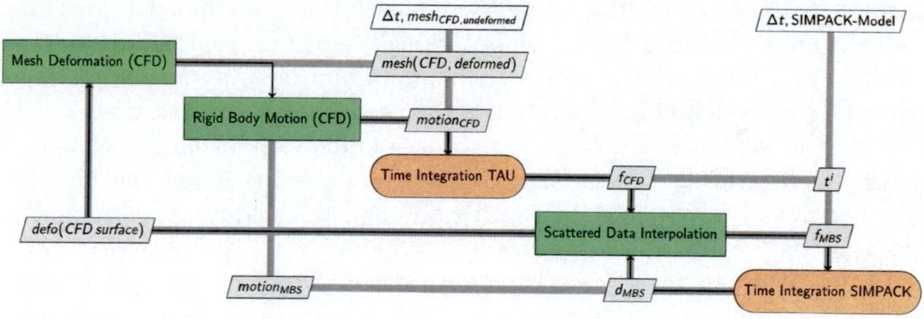

Fig. 4. Strong coupling scheme illustrated as eXtended Design Structure Matrix

3 Numerical Investigations and Comparison to Experimental Data

Three test cases (TC) with a rotor speed $\Omega = 23.6\,\text{Hz}$ will be presented comprising two operating conditions with a fixed collective pitch angle $\Theta_{coll} \in \{24°, 27°\}$ and one with an additionally superposed cyclic pitch angle variation $\Theta_{coll} \pm \Theta_{cyc} = 27° \pm 6°$. The latter serves to bring about the dynamic stall phenomenon. Their collective and cyclic pitch angles are listed as an overview in Table 2 together with their predominant flow characteristics at the blades. All test cases are performed for two different modelling approaches: One that incorporates aeroelasticity connected with elastic rotor blades as well as one that considers the blades as rigid bodies. Similar to purely aerodynamic numerical studies in which blade elasticity is neglected, cf. Kaufmann et al. [4,5] and Letzgus et al. [7,8] each test case is performed in such a manner that transients have subsided and a steady state has been reached for evaluation. Therefore, a vortex system is established in the computational domain that continuously exits at its boundaries. As a result, the differences between both approaches are elaborated in order to draw conclusions under consideration of blade elasticity. For this purpose, the integral force F_n acting normal to the planform of a blade and, additionally in case of fixed collective pitch angle, the pitch moment M_y relative to an axis passing through the quarter chord point ($x = \frac{c}{4}$) at the blade root are monitored. Finally, a comparison with experimental data is carried out for cases with fixed collective pitch angle. Therefore, the rotor thrust is extracted from the simulations in addition to the displacements of the blade tips.

Table 2. Overview of the presented test cases (TC)

TC No.	Θ_{coll}	Θ_{cyc}	predominant characteristics of fluid flow
1	24°	± 0°	mostly attached flow conditions
2	27°	± 0°	stall conditions
3	27°	± 6°	dynamic stall conditions

3.1 Test Cases with Fixed Collective Pitch Angle (TC 1 and 2)

TC 1 with $\Theta_{coll} = 24°$ and Mostly Attached Flow Conditions: Despite the occurrence of unsteady vortex shedding due to flow separation in the boundary layer at the root of the rotor blades, attached flow conditions at the blade surfaces as well as distinctive vortices trailing the tips are characteristic for this test case with both blade modelling approaches. Although no significant lift forces are induced at the root area, the vortex shedding leads to slight fluctuations in the aerodynamic loads on the blades. For an approach with rigid blade modelling, each of the four blades experiences a mean normal force $F_n \approx 61.5$ N and a mean pitch moment $M_y \approx -0.631$ Nm. In contrast to this, an approach with elastic blade modelling shows reduced mean blade normal forces $F_n \approx 60.7$ N which is due to a twisting of the blade in nose-down direction. The twist constantly increases in spanwise direction and reaches its maximum at the blade tip. Since the centre of pressure moves towards the trailing edge with increased twist towards nose-down direction, the magnitude of the pitch moment acting in nose-down direction is also increased. Therefore, it accounts $M_y \approx -0.731$ Nm when considering blade elasticity. The blade tips bend up with a normal deflection $w_n \approx 11.9$ mm. An illustration of the bending behaviour is shown by means of the displacements in blade normal direction in Fig. 5. In addition, the vortex system is illustrated by means of the λ_2-criterion [3] to give insight into the flow field.

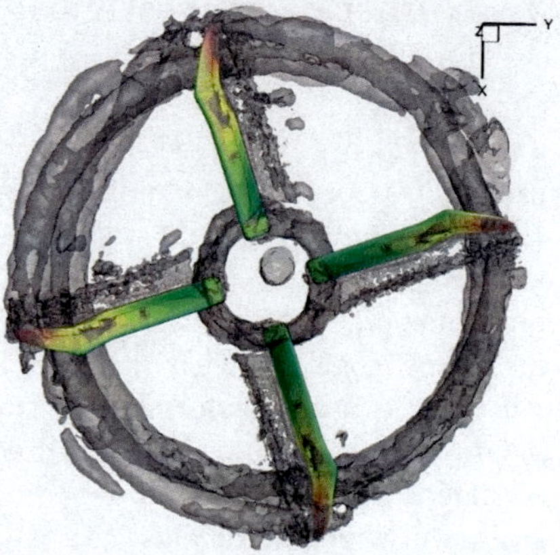

Fig. 5. TC 1 (Θ = 24° ± 0°) with elastic blade modelling: Vortex system (λ_2-criterion in grey), Displacement in blade normal direction (coloured) (Colour figure online)

TC 2 with $\boldsymbol{\Theta_{coll}}$ = 27° and Stall Conditions: Starting from TC 1, the collective pitch angle is raised so that now additional vortex shedding due to flow separation can be observed on the backward swept part of the rotor blade in contrast to TC 1. These shed vortices immediately interact with the trailing vortex at the blade tip and unite to form a larger vortex. Furthermore, there are still vortices shed at the blade roots. It is therefore characterized by stall conditions. Due to the strong flow separation, the values for the aerodynamic loads also fluctuate strongly. For an approach with rigid blade modelling, each of the four blades experiences a mean normal force $F_n \approx 72.0$ N and a mean pitch moment $M_y \approx -0.823$ Nm. Due to the aforementioned change in twist with the consideration of blade elasticity, the normal force on one blade decreases to $F_n \approx 71.5$ N and the pitch moment changes to $M_y \approx -0.834$ Nm. However, the differences in the aerodynamic loads between both modelling approaches are not as large as in TC 1 which can be attributed to the present flow separation on the backward swept part of the rotor blade. As a result, the centre of pressure only moves slightly to the trailing edge, so that there is also only a slight change in the pitch moment. The displacement of the blade tip amounts for this test case $w_n \approx 13.9$ mm.

Comparison to Experimental Data for TC 1 and 2: In order to account for varying ambient air conditions between each measuring point and the numerical simulations, the rotor thrust T is converted via the thrust coefficient $C_T = T/(\rho_\infty A_{Rotor}(2\pi\Omega \cdot R_{Rotor})^2)$ and the rotor solidity $\sigma = 4 \cdot A_{Blade}/A_{Rotor} = 0.0977$ to a specific blade loading C_T/σ. Apart from this, no other methods of data correction were used in this study. Together with data from the conducted

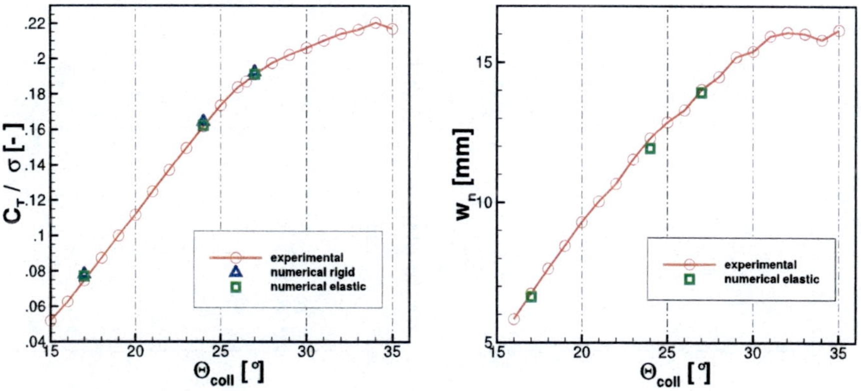

(a) Specific blade loading C_T/σ vs. Θ_{coll} (b) Blade tip displacement w_n vs. Θ_{coll}

Fig. 6. Comparison of experimental data and numerical results

experiments and an additional simulation point at $\Theta_{coll} = 17°$, C_T/σ is extracted and plotted for TC 1 & 2 in Fig. 6a as static polar. The slope of the curve decreases in the region of $\Theta_{coll} \approx 25°$ due to onset of flow separation. This curve bending is also predicted by the numerical simulations, cf. TC 1 & 2. The integral aerodynamic force as well as the tip displacement on one blade are predicted quite well, where in the case of elastic blade modelling the C_T/σ-values are closer to the experiment.

3.2 Test Case with Superposed Cyclic Pitch Angle Variation (TC 3)

TC 3 with $\Theta = 27° \pm 6°$ Inducing Dynamic Stall: Starting from TC 2, the dynamic stall phenomenon is induced by an superposed cyclic pitch angle variation $\Theta_{cyc} = 6°$ once per revolution. Additional large vortex structures are shed during the upstroke, even at the forward swept part of the blade which results in weakened blade tip vortices. This periodically driven flow separation behaviour is also reflected in a hysteresis loop of the aerodynamic loads as well as structural displacements by plotting them for one revolution against pitch angle Θ. It is characterized by different curves during up- and downstroke, see Fig. 7a for C_N/σ and 8a for w_n. Analogous to the procedure in TC 1 & 2, varying ambient air conditions between experiment and simulation are taken into account by calculating a C_N/σ-value w.r.t. $C_N = 4 \cdot F_n/(\rho_\infty A_{Rotor}(2\pi\Omega \cdot R_{Rotor})^2)$ and the blade normal force F_n.

A breakdown of C_n/σ in the vicinity of maximum pitch angle $\Theta_{max} = 33°$ is apparent which is larger in case of rigid blade modelling. Due to the fact that the hysteresis loop for rigid blade modelling almost envelops the one for an elastic modelling, the differences between up- and downstroke are greater in case of rigid blade modelling. This also means that both curves deviate strongly from each other. A hysteresis behaviour can also be seen for the blade tip displacement

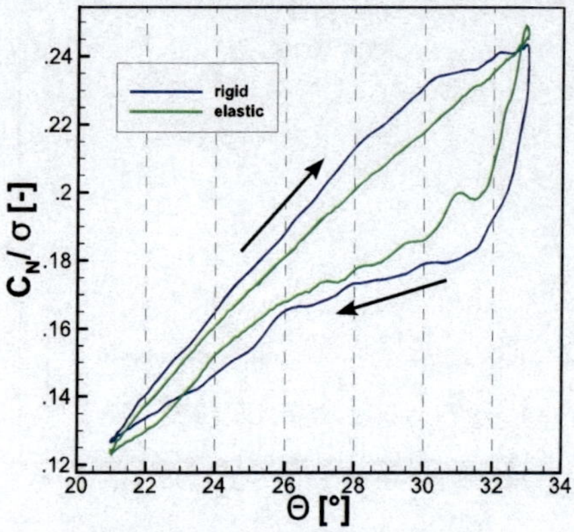

(a) Hysteresis loop of $C_N/\sigma(\Theta)$, course of time indicated by arrows ↑

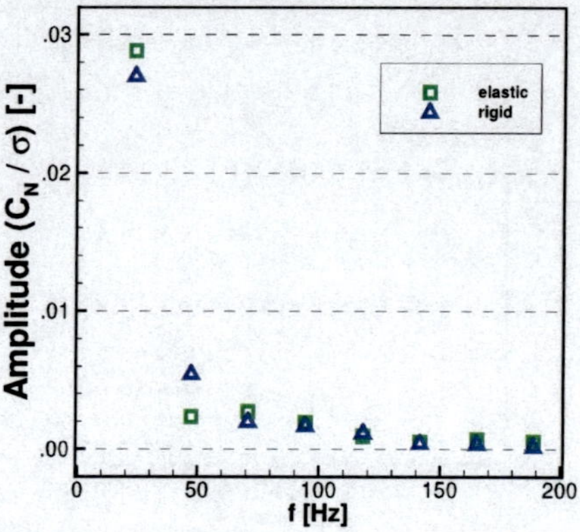

(b) Rotor harmonics by Fourier analysis

Fig. 7. C_N/σ-value for TC 3 ($\Theta = 27° \pm 6°$)

w_n, whereby an upward motion occurs during the downstroke respectively a downward motion during the upstroke for pitch angles $21° \leq \Theta \leq 25°$, see Fig. 8a. It means that the rotor blade bends up again during downstroke and vice versa.

By casting the time signals of F_n and w_n into frequency domain the influence of higher rotor harmonics becomes clear, see Fig. 7b and 8b. In case of the blade normal force F_n, the Fourier analysis reveals significant participation of rotor harmonics up to the fifth order f_5 = 118.0 Hz, whereby the first three orders f_1 = 23.6 Hz, f_2 = 47.2 Hz and third f_3 = 70.8 Hz deviate from each other

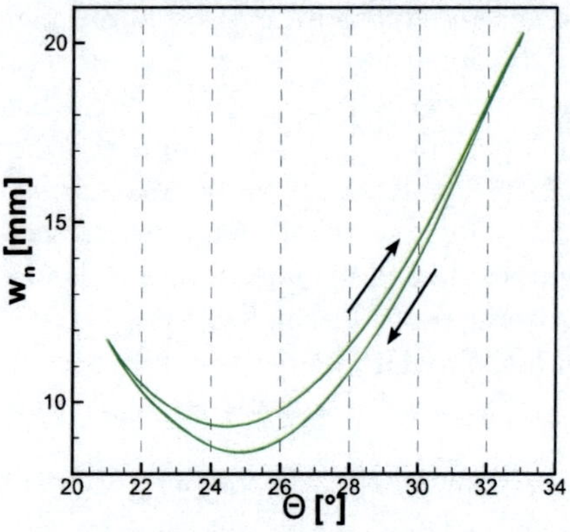

(a) Hysteresis loop of $w_n(\Theta)$, course of time indicated by arrows ↑

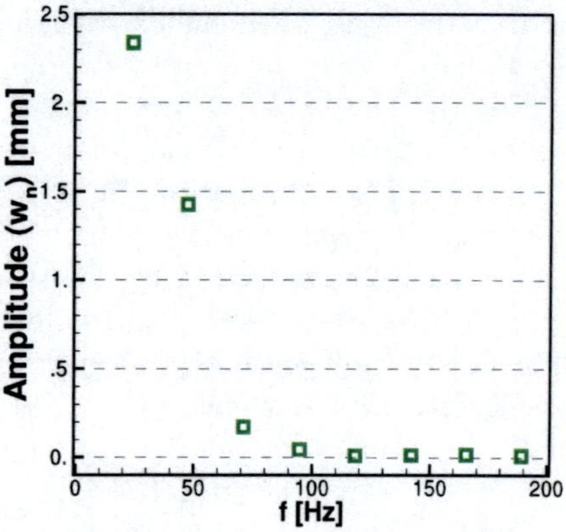

(b) Rotor harmonics by Fourier analysis

Fig. 8. Blade tip displacement w_n for TC 3 ($\Theta = 27° \pm 6°$)

for a rigid and elastic blade modelling. They contain the essential information upon the hysteresis curve compromising the slope during the upstroke followed by the breakdown behaviour. Higher orders of rotor harmonics mainly represent superposed high-frequency oscillations being observable during the up- and downstroke. Looking at the blade tip displacement w_n which is mainly excited by normal blade forces, harmonics up to the third order play a significant role.

4 Summary and Conclusion

Numerical simulations of a four-bladed model rotor with double-swept blades and a rotor speed $\Omega = 23.6$ Hz were carried out using DLR's CFD solver TAU and Simpack as a multibody system solver. Furthermore, the rotor blades were modelled with two different approaches: as rigid and elastic bodies. Afterwards, a comparison was made to experimental data. Three investigated test cases are shown comprising two operating conditions with a fixed collective pitch $\Theta_{coll} \in \{24°, 27°\}$ and one with a superposed cyclic variation in pitch angle $\Theta = 27° \pm 6°$ in which the dynamic stall phenomenon is induced.

Looking at test cases with fixed collective pitch angle, the integral normal blade force F_n is slightly reduced ($\Delta F_n \leq 0.7$ N) whereby the magnitude of the pitch moment increases acting minimally more in nose-down direction ($\Delta M_y \leq 0.1$ Nm). The increased magnitude of the pitch moment leads to higher forces in the pitch link rods which have not been included in the Simpack model. A comparison with experimental data shows good agreement for the blade tip displacements and aerodynamic blade loads in case of the integral normal blade forces. In consideration of blade elasticity, the reduction in the integral normal blade forces improves the numerical results moving marginally closer to the experiment. Furthermore, the magnitude of the pitch moment increases by considering blade elasticity which results in higher loads at the pitch link rods.

For test cases with additional cyclic pitch angle and occurring dynamic stall phenomenon, the hysteresis loops can be seen which differ for a rigid and elastic blade modelling. The breakdown of the normal blade force is larger in case of rigid blade modelling. Furthermore, the Fourier analysis shows the participation of rotor harmonics up to the fifth order in case of the normal blade force F_n and up to the third order in case of the blade tip displacement w_n. When comparing the rotor harmonics for both blade modelling approaches, it can be stated that the first rotor harmonic is increased whereas the second one is decreased when coming from a rigid to an elastic blade modelling approach. The changes in the rotor harmonics are due to blade bending vibrations which are captured by the blade tip displacement. They are excited by normal forces acting on the blades and mainly include a participation of the first, second and third rotor harmonic. However, for this specific rotor blade the second rotor harmonic is easily made to resonate, since it lies close to the first natural bending frequency of the blade. As a final conclusion, the inclusion of blade elasticity influences the aerodynamic blade loads in both cases: with fixed collective pitch angle and a superposed cyclic pitch angle. They increase with occurring non-linearities such as the onset of dynamic stall in this study.

5 Outlook

In order to simulate forward flight conditions in a wind tunnel experiment, further investigations will focus on a rotor configuration with a radial inflow from the front at 180° azimuth. In this case, the present model rotor blades are connected via flexible adapters to the rotor hub enabling a flapping as well as a lead-lag motion of the blades at its roots. Consequently, the rotor dynamics of the rotor will change substantially and, furthermore, they come closer to those of a flying helicopter rotor.

References

1. Arnold, J., Waitz, S.: Using multibody dynamics for the stability assessment of a new double-swept rotor blade setup. European Rotorcraft Forum, Delft (2018)
2. Babij, G.: Aeroelastic dynamic stall computations of a double-swept blade in a four-bladed rotor configuration. In: 48th European Rotorcraft Forum (2022)
3. Jeong, J., Hussain, F.: On the identification of a vortex. J. Fluid Mech. **285**, 69–94 (1995). https://doi.org/10.1017/S0022112095000462
4. Kaufmann, K., Costes, M., Richez, F., Gardner, A.D., Pape, A.L.: Numerical investigation of three-dimensional dynamic stall on an oscillating finite wing. In: AHS 70th Annual Forum and Technology Display, pp. 1–12, no. 126 in Conference Proceedings on CD-ROM (2014). https://elib.dlr.de/85019/
5. Kaufmann, K., Müller, M.M., Gardner, A.D.: Dynamic stall computations of double-swept rotor blades. In: Dillmann, A., Heller, G., Krämer, E., Wagner, C., Tropea, C., Jakirlić, S. (eds.) DGLR 2018. NNFMMD, vol. 142, pp. 351–361. Springer, Cham (2020). https://doi.org/10.1007/978-3-030-25253-3_34
6. Leishman, J.G.: Principles of Helicopter Aerodynamics, 2nd edn. Cambridge University Press, Cambridge (2006)
7. Letzgus, J., Gardner, A., Schwermer, T., Keßler, M., Krämer, E.: Numerical investigations of dynamic stall on a rotor with cyclic pitch control. J. Am. Helicopter Soc. **64**, 1–14 (2019)
8. Letzgus, J., Keßler, M., Krämer, E.: CFD-Simulation of three-dimesional dynamic stall on a rotor with cyclic pitch control. European Rotorcraft Forum, Delft (2015)
9. Müller, M.M., Schwermer, T., Mai, H., Stieg, C.: Development of an innovative double-swept rotor blade tip for the rotor test facility Goettingen. In: DLRK 2018 - Deutscher Luft- und Raumfahrtkongress (2018). https://elib.dlr.de/122702/
10. Schwermer, T., Richter, K., Raffel, M.: Development of a rotor test facility for the investigation of dynamic stall. In: Dillmann, A., Heller, G., Krämer, E., Wagner, C., Breitsamter, C. (eds.) New Results in Numerical and Experimental Fluid Mechanics X. NNFMMD, vol. 132, pp. 663–673. Springer, Cham (2016). https://doi.org/10.1007/978-3-319-27279-5_58, https://elib.dlr.de/13217/
11. Splettstößer, W., et al.: The ERATO programme: wind tunnel results and proof of design for an aeroacoustically ouptimized rotor. In: 25th European Rotorcraft Forum (1999). https://elib.dlr.de/13217/
12. Surrey, S., Ortun, B., Truong, K.V., Wienke, F.: Investigation of the structural blade dynamics and aeroelastic behavior of the 7A rotor. In: American Helicopter Society 72nd Annual Forum, West Palm Beach (2016)

Wind Tunnel Analysis of Helicopter Aerodynamics with Varying Rotor Head Blade Stub Lengths

Ulrich Hartmann(✉), Johannes Ruhland, and Christian Breitsamter

Chair of Aerodynamics and Fluid Mechanics, Technical University of Munich, Boltzmannstr. 15, 85748 Garching, Germany
ulrich.hartmann@tum.de

Abstract. Experimental investigations are performed on a helicopter wind tunnel model for three rotor heads with different blade stub lengths. This investigation is intended to judge the influence of the blade stub spanwise dimension on the drag, lift, and especially pitching moment characteristics. Aerodynamic forces, moments, and the wake flow fields are analyzed. As expected, the longest blade stubs provide the highest drag and lift coefficient along the angle of attack polar, while both shorter blade stubs show a similar force course. The flow field of the largest rotor head exhibits pronounced blade tip vortices, which influence the pitching moment polar at high angles of attack. The medium and long stubs show a similar flow field, while the hubcap vortex dominates the smallest blade stub configuration.

Keywords: helicopter · rotor wake · blade stubs

1 Introduction

Wind tunnel experiments play an important role in investigating the complex flow phenomena occurring around helicopters. The rotating rotor head generates noise and vibrations, while the turbulent wake of the rotor head can interact with the tail boom and the helicopter's vertical and horizontal stabilizers [3]. However, helicopter experiments are challenging, as scaled models are typically used. Nevertheless, most scaled rotor blades are still too big for standard wind tunnel facilities, and the flow phenomena occurring on scaled rotor blades may not match the ones of the full-scale variant. Consequently, only the rotor head is considered, truncating the blade attachments close to the first aerodynamic section. Numerical simulations then obtain the main rotor aerodynamic data at full-scale conditions while characteristics of the rotor head are taken from the wind tunnel tests. The data sets are then combined, for example, in a flight mechanics simulation tool. There are no clear guidelines for the design and length of the rotor blade stubs, and different variants can be found. A specific investigation of different rotor blade stub designs on a scaled wind tunnel model is needed to derive strategies and methods for future helicopter models.

A. Dillmann et al. (Eds.): STAB/DGLR Symposium 2022, NNFM 154, pp. 613–622, 2024.
https://doi.org/10.1007/978-3-031-40482-5_58

Previous experimental and numerical investigations were conducted at the Technical University of Munich [2,8]. A five-bladed rotor head was applied, and all blades were cut at the first aerodynamic section. A significant contribution of the blade stubs to the overall drag was shown. Furthermore, hub cap vortex pairs and blade stub tip vortices were detected. Some experiments used one-third radius blade stubs [1], while for investigating the hub wake, often no blades or small blade shanks were used [4,5]. In numerical analysis, the blades were cut at the root for a drag breakdown analysis [6], or a full-scale helicopter was investigated numerically [7].

All previously mentioned investigations had cut their blades at different positions. For each analysis, experimental and numerical data agreed well, and the wake propagating downstream was analyzed. However, the influence of the blade stub length has not been systematically investigated yet. This study presents wind tunnel results for three different blade stubs. The resulting aerodynamic forces and moments, along with the results of Particle Image Velocimetry measurements, are shown.

2 Experimental Setup

2.1 Aerodynamic Forces and Moments Measurements

The experiments are performed in the Wind Tunnel A of the Chair of Aerodynamics and Fluid Mechanics at the Technical University of Munich. The wind tunnel has an open test section with a height of 1.80 m, a width of 2.40 m, and a length of 4.80 m. The turbulence level of the freestream is below 0.4 %. The setup is shown in Fig. 1. The experiments are performed on a 1:7.3333 twin-engine light (TEL) class helicopter model, which consists of a fuselage, skid landing gear, tail boom, and horizontal and vertical stabilizer. Furthermore, a five-bladed rotor

Fig. 1. Wind tunnel model of the helicopter with rotor head *R1*.

head is mounted on top of the helicopter. The whole model is connected to a tail sting, including a motor unit, allowing the helicopter to rotate around the roll axis. This system is mounted on a support, which connects the whole model with an under-floor balance featuring a turn table. This six-component system has an accuracy based on maximum loads of 0.025 % for the aerodynamic force and moment components. The resulting maximum error for the drag coefficient is $\Delta C_D = 1.6 \times 10^{-2}$, for the lift coefficient $\Delta C_L = 3.3 \times 10^{-2}$, and for the pitching moment coefficient $\Delta C_m = 1.8 \times 10^{-2}$. The turn table allows simulating different inflow conditions with varying angles of attack α and angles of side slip β from $-30°$ to $30°$.

CAD model-based figures of the rotor head configurations are displayed in Fig. 2, consisting of a hubcap, one steering rod, and one blade. On the left-hand side, rotor head variant *R1* is shown. Here, the stub extends to the first aerodynamic blade section, corresponding to a ratio relative to the hubcap diameter of $\frac{l_1}{d_H} = 2.04$. The middle blade (*R2*) is shortened by nearly 50 % of the length of *R1*, resulting in a ratio of $\frac{l_2}{d_H} = 1.00$. Only the most-inner blade section with the pitch horn is modeled for the smallest stubs (*R3*). This length corresponds to a ratio of $\frac{l_3}{d_H} = 0.43$.

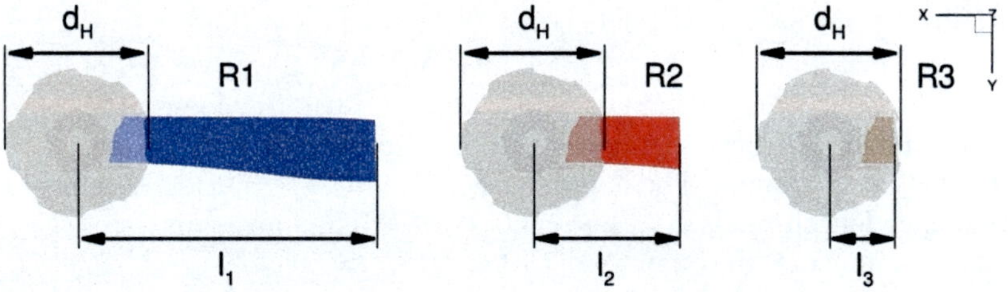

Fig. 2. CAD model of rotor head variant *R1* (left), *R2* (middle) and *R3* (right).

Figure 3 shows the rotor heads from Fig. 2 mounted on the helicopter model. Each rotor head configuration is orientated at a rotor position $\psi = 180\,°$, meaning one blade faces straight in the forward direction. On the left-hand side, the helicopter with variant *R1* is shown, while the middle picture displays *R2* and the right one *R3*.

Initially, only the fuselage and the skid landing gear with the rotor heads were investigated. Subsequently, the tail boom and vertical fin were added. The final measurements were performed with the full helicopter model, consisting of the fuselage, skid landing gear, rotor head, tail boom, and T-tail. This approach allows for distinguishing the aerodynamic force and moment components between each configuration.

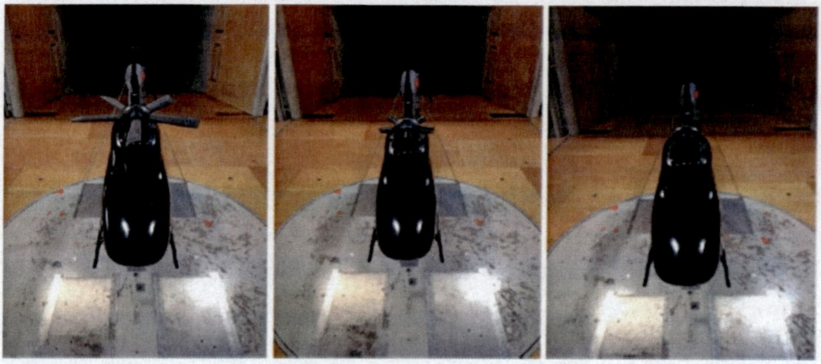

Fig. 3. Rotor head variants *R1* (left), *R2* (middle) and *R3* (right) mounted on the helicopter model.

2.2 Particle Image Velocimetry Measurements

The wake of the whole helicopter model is investigated using Stereo Particle Image Velocimetry (PIV) measurements (see Fig. 4 on the left-hand side). The PIV system is mounted on a three-axis traversing system, which allows to measure multiple flow field planes downstream of the rotor. A double-pulsed Nd:Yag laser with a maximum power of 325 mJ per pulse and a wavelength of 532 nm is used. Seeding particles (DEHS) of approximately 1 µm diameter circulate inside the wind tunnel. The flow field is captured by two sCMOS cameras with a resolution of 2560 × 2160 pixel. The nondimensional spatial resolution with respect to the reference length reads $\Delta d/l_{ref} = 3.0 \times 10^{-3}$. The uncertainties of the mean velocity components are quantified to $|u_{err}/U_\infty| = |v_{err}/U_\infty| = |w_{err}/U_\infty| < 0.015$. Each camera captures 400 images. The initial interrogation window is set to 48 × 48 pixel with 50 % overlap, the final to 32 × 32 pixel and 75 % overlap by four passes. The sampling frequency is adapted to the rotational speed of the rotor head, which allows recording pictures at a rotor position of $\psi = 180\,°$.

On the right-hand side of Fig. 4, the position of the PIV planes is shown. The first plane (*P1*) is located at the beginning of the tail boom, corresponding to a distance from the rotor hub center of $x_{P1} = 3.89 \times \frac{l_1}{d_h}$. The second plane is located further downstream at $x_{P2} = 6.23 \times \frac{l_1}{d_h}$. This placement allows to investigate the development of the wake structures from the rotor head. Furthermore, the size and strength of the vortical flow structures reaching the vertical and horizontal tail can be analyzed.

3 Results and Discussion

The following sections present the experimental results of the different rotor head configurations. Section 3.1 shows the measured forces and moments for the configurations as function of angle of attack α. Then, the influence of the rotor heads on the vertical and horizontal tail are presented. The results from the

Fig. 4. PIV system installed in the wind tunnel (left) and wind tunnel model with rotor head *R1* and axial velocity flow field data at plane *P1* and *P2* (right).

PIV measurements are presented in Sect. 3.3. All measurements are performed with an advance ratio of $\mu = \frac{U_\infty}{U_{TIP}} = 0.354$ and a Reynolds number of about 0.5×10^6. The inflow velocity is defined by U_∞, and U_{TIP} is calculated by the scaled rotor blade. Furthermore, a constant collective pitch angle is set. For all calculations, the same reference area S_{ref} and reference length l_{ref} are used. The drag coefficient is calculated by the formula $C_D = \frac{D}{q_\infty S_{ref}}$, lift coefficient $C_L = \frac{L}{q_\infty S_{ref}}$ and pitching moment coefficient $C_m = \frac{M}{q_\infty S_{ref} l_{ref}}$, where the dynamic pressure corresponds to $q_\infty = \frac{\rho}{2} U_\infty^2$, with density ρ.

3.1 Forces and Moments of the Rotor Head Configurations

This section presents a force and moment coefficient comparison between the different blade stubs. Figure 5 shows the α polar for $\beta = 0\,°$ of the drag coefficient (left), lift coefficient (middle), and pitching moment coefficient (right) for the helicopter configurations with long (*R1*), medium (*R2*), and short blade stubs (*R3*). All drag graphs show a parabolic shape. With a minimum at approximately $\alpha = 3\,°$, the drag increases with increasing and decreasing angles of attack. The longest stubs generate more drag than the medium and short ones, as expected. Especially at high angles of attack, the differences between *R1* and *R2* or *R3* increase. In contrast, the medium stubs have a slightly higher C_D value than *R3*, and the offset between both stubs remains similar over the whole α polar.

The middle diagram of Fig. 5 shows the lift coefficients of the configurations. Compared to *R2* and *R3*, the longest stubs have an aerodynamically shaped blade design, leading to a higher C_L and a higher C_L gradient. However, for low angles of attack ($\alpha \leq -12\,°$), where the influence of the horizontal tail dominates, the *R1* blades generate the lowest lift. The shorter blade stubs (*R2* and *R3*) show a similar behavior along the whole polar. While having a higher lift for low angles of attack, the lift gradient intersects the *R1* lift gradient by $\alpha = -10\,°$ and leads to a lower overall lift coefficient for higher angles of attack. At $\alpha = \beta = 0\,°$, the lift coefficient of *R2* and *R3* are almost zero. However, blade stub *R1* has a positive C_L value.

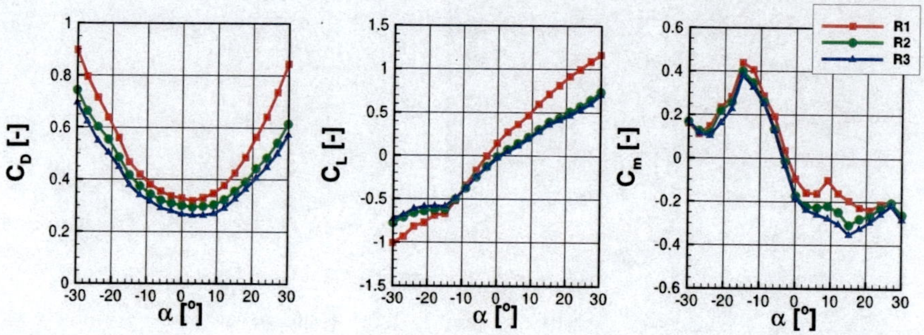

Fig. 5. Drag coefficient (left), lift coefficient (middle) and pitching moment coefficient (right) α polar for $\beta = 0\,°$ for the long (*R1*), medium (*R2*) and short (*R3*) blade stubs.

The pitching moments of the configurations are shown on the right-hand side of Fig. 5. For high negative α values, C_m is positive, increases with increasing angles of attack, and reaches maximum values at $\alpha = -15\,°$ for all configurations. However, *R3* provides the lowest maximum. With increasing angles of attack, the influence of the horizontal tail becomes more effective and decreases the pitching moment for all three rotor head configurations with a similar gradient. From $\alpha = -3\,°$, C_m becomes negative for all rotor heads. After passing $\alpha = 0\,°$, the gradients of all pitching moments increase. While C_m of *R1* reaches a local maximum at $\alpha = 10\,°$, C_m of rotor head variants *R2* and *R3* further decreases. At around $\alpha = 15\,°$, the medium and short rotor head configurations reach their global C_m minimum while the course of *R1* further decreases. For higher angles of attack ($\alpha \geq 27\,°$), the pitching moments of all rotor head configurations obtain similar values.

3.2 Influence of the Rotor Head Configurations on the T-Tail

During the wind tunnel measurements, the isolated fuselage with the rotor heads was also investigated. By subtracting the values of the isolated fuselage-rotor configuration from the full helicopter model, the resulting forces acting on the tail boom and horizontal and vertical tail can be calculated.

For a side slip angle of $\beta = 0\,°$, the resulting drag (left), lift (middle), and pitching moment (right) difference is shown in Fig. 6. Similar to the C_D polar plot of the complete helicopter configuration (see Fig. 5), a parabolic course is observed for all rotor heads. The influence of the rotor heads on the T-tail is stronger for high angles of attack. Rotor head *R1* has a slightly higher drag value in the range of $-30\,° \leq \alpha \leq 10\,°$ than *R2* and *R3*, while it becomes lower for $11\,° \leq \alpha \leq 27\,°$. The drag behavior of *R2* and *R3* are similar along the whole α polar. The lift values are shown in the middle diagram. For negative angles of attack, the lift difference is negative. This means, the configurations with vertical and horizontal tail generate less lift than the isolated fuselage configurations. Especially for *R1*, the lift difference is higher. At $\alpha = 0\,°$, the lift coefficient of

R1 drops significantly, while C_L of *R2* and *R3* further increases. For even higher angles of attack ($\alpha \geq 20\,°$), the lift difference of *R1* further increases, while the differences decrease for *R2* and *R3*. On the right-hand side of Fig. 6, the pitching moment is given for all three rotor configurations. Along the whole α polar, the three rotor heads have a similar course. Except in the range from $0\,°$ to $20\,°$, *R1* produces a higher pitching moment difference associated with a change in the gradient $dC_m/d\alpha$. This change was also visible in the pitching coefficient polar from Fig. 5.

3.3 Results of the Particle Image Velocimetry Measurements

This section highlights the flow fields of the different rotor head configurations for two cross flow planes located downstream of the rotor wake. The exact position of the planes was given in Sect. 2.2. Figure 7 displays the resulting flow fields of plane *P1* for rotor head *R1* on the first, *R2* on the second, and *R3* on the third row. The normalized axial velocity is displayed on the left-hand side, while the middle pictures show the vertical velocity. On the right-hand side, the non-dimensional axial vorticity fields ($\omega_x\, l_{ref}/U_\infty$) are displayed.

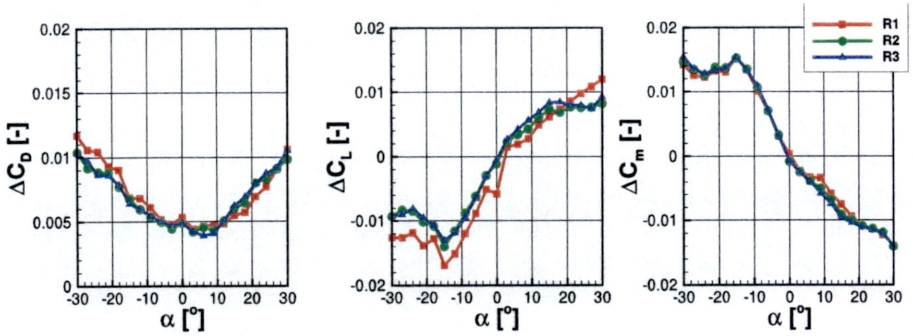

Fig. 6. **Influence of rotor heads on the T-tail: Drag difference (left), lift difference (middle) and pitching moment difference (right).**

Two main regions of velocity deficit can be detected for the axial velocity of *R1*. The size of the right area, which corresponds to the velocity deficit of the advancing blade (B), is slightly larger than the left one, linked to the retreating blade (A). Additionally, a smaller spot is separated from the retreating blade velocity area. Further right, two small velocity drop points can be detected, which correspond to blade tip vortices of the profiled blades (C). For the medium blade stubs (*R2*), the axial velocity flow field in the center looks similar to the long blade stubs. However, the areas are more centered, and no pronounced blade tip vortex is detected. Again, the region of the advancing blade (B) is smaller than the retreating blade's area (A). The short blade stubs (bottom row) show the smallest velocity deficit area, and their shape is curved to the top. Furthermore,

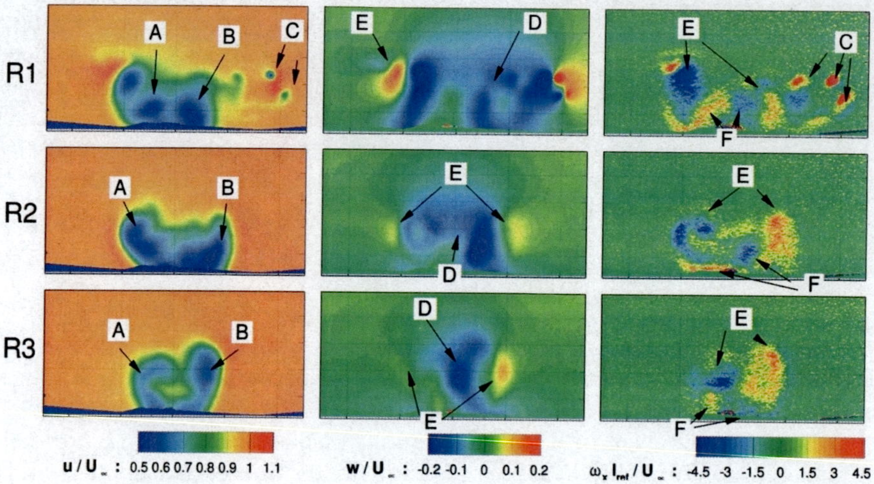

Fig. 7. Distribution of axial velocity (left), vertical velocity (middle) and vorticity (right) for rotor head *R1* (top line), *R2* (middle line) and *R3* (bottom line) at plane *P1*.

the velocity deficit in the center is higher compared to the edge region. In contrast to *R1* and *R2*, the velocity decrease area of the advancing blade (B) is more extended than that of the retreating blade (A).

The middle column shows the vertical velocity component of plane *P1*. The velocity deficit areas in the center (D) are connected to the lift generation of the hubcap. Especially at the advancing blade, the size of negative vertical velocity is more significant for *R1* than for the other blade stubs. This effect is due to the lift generation of the profiled blade (*R1*). Furthermore, the area of negative vertical velocity located behind the retreating blade is minor, as the rotor rotates with constant pitch angles. The downwash of the retreating medium blade stubs is also detected for *R2*, while it is less present for the *R3* rotor head. The positive vertical velocity areas next to the downwash areas are linked to the hubcap vortices (E). For rotor head *R1*, it is detected on the retreating blade side, and for *R2* and *R3* on the advancing and retreating blade. On the right-hand side of Fig. 7, the vorticity fields of the different blade stubs are displayed. Hubcap vortex pairs can be detected above the tail boom (E) and small parts of the mast fairing vortex pair (F) below them. Additionally, the red point-like areas at the top-right pictures correspond to the blade tip vortices (C) of *R1*.

Further downstream on the tail boom at plane *P2*, the resulting flow fields are given in Fig. 8. The same layout, as shown previously for plane *P1* in Fig. 7, is applied. At this plane position, high light sheet reflections occurred at the bottom-left part of the plane. Therefore, this area was blanked out in the analysis process.

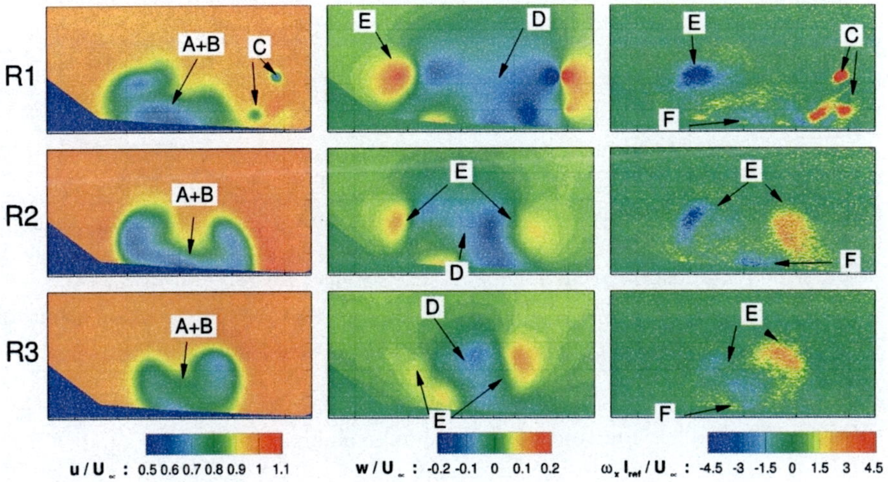

Fig. 8. Distribution of axial velocity (left), vertical velocity (middle) and vorticity (right) for rotor head *R1* (top line), *R2* (middle line) and *R3* (bottom line) at plane *P2*.

For the axial velocity, the two areas from Fig. 7 are merged into one velocity deficit area (A+B) for all rotor head variants. However, an additional velocity deficit area is visible at the advancing blade side for *R3*. Furthermore, the axial velocity magnitude is increased compared to *P1*. Again, *R1* and *R2* have a larger velocity deficit area for the retreating blade compared to the shortest blade stubs. For *R1*, two blade tip vortex areas (C) are shown, which have moved slightly to the bottom right compared to plane *P1*. The area with negative vertical velocity deficit (D) above the tail boom is smaller, while the magnitude is bigger for all blade stubs. Furthermore, the upwind part of the hubcap vortex pair (E) can be detected for all rotor heads. Again, the hubcap vortex pair (E) and the mast fairing vortex pair (F) are recognized in the vorticity flow field. In addition, the blade tip vortices (C) are displayed for *R1*.

4 Conclusion

A detailed analysis of the flow around a rotating rotor head with three different blade stub extensions has been carried out based on experimental methods. Aerodynamic forces and moments for the different rotor heads are compared, as well as the wake flow field for two downstream positions. Significant differences between the variants are shown. The longest blade stubs generate more drag and lift, while both smaller rotor head variants show a similar behavior along the whole α polar. In contrast, high angles of attack influence the pitching moment coefficient significantly. Especially for *R1*, a change of sign in the gradient $dC_m/d\alpha$ is observed. In addition, a drop in the lift coefficient shows an influence of the largest rotor head on the T-tail. A comparison of the flow fields for all rotor head variants shows a similar wake development for both larger blade stubs. Pronounced blade tip vortices are detected for the largest stubs only, while the hubcap vortex structures dominate the flow field of the medium and small stubs.

The presence of the blade tip vortices of the large blade extensions can interact with the vertical and horizontal stabilizers. Especially for higher angles of attack, these interactions affect the wake flow phenomena and may result in force and moment characteristics, especially for the pitching moment development, which would not occur for a full-scale helicopter configuration.

Therefore, blade stubs with a smaller spanwise extent not substantially exceeding the diameter of the hubcap result in less pronounced stub tip vortices (not present for a real rotor head configuration), thus minimizing interactions with the tail section. The pitching moment derivative $dC_m/d\alpha$ stays negative within $-10°$ to $10°$ exhibiting no slope change as observed for the smaller blade stub variants in contrast to the large one.

Acknowledgement. The funding of this investigation within the LUFO project NANNY (FKZ: 20Y2109E) by the German Federal Ministry for Economic Affairs and Climate Action (BMWK) is gratefully acknowledged.

References

1. Felker, F. F.: An experimental investigation of hub drag on the XH-59A. In: AIAA 3rd Applied Aerodynamics Conference, AIAA Paper, p. 4065 (1985). https://doi.org/10.2514/6.1985-4065
2. Grawunder, M., Reß, R., Stein, V., Breitsamter, C., Adams, N.A.: Validation of a flow simulation for a helicopter fuselage including a rotating rotor head. In: Dillmann, A., Heller, G., Krämer, E., Wagner, C., Breitsamter, C. (eds.) New Results in Numerical and Experimental Fluid Mechanics X. NNFMMD, vol. 132, pp. 303–313. Springer, Cham (2016). https://doi.org/10.1007/978-3-319-27279-5_27
3. Leishman, J.G.: Principles of Helicopter Aerodynamics. Cambridge University Press, New York, NY (2000)
4. Petrin, C.E., Jayaraman, B., Elbing, B.R.: Characterization of a canonical helicopter hub wake. Exp. Fluids **60**, 9 (2019). https://doi.org/10.1007/s00348-018-2655-4
5. Reich, D., Willits, S., Schmitz, S.: Scaling and configuration effects on helicopter rotor hub interactional aerodynamics. J. Aircr. **54**(5), 1692–1704 (2017). https://doi.org/10.2514/1.C034250
6. Renaud, T., Le Pape, A., Péron, S.: Numerical analysis of hub and fuselage drag breakdown of a helicopter configuration. CEAS Aeronaut. J. **4**(4), 409–419 (2013). https://doi.org/10.1007/s13272-013-0081-0
7. Schäferlein, U., Keßler, M., Krämer, E.: Aeroelastic simulation of the tail shake phenomenon. J. Am. Helicopter Soc. **63**(3), 1–17 (2018). https://doi.org/10.4050/JAHS.63.032003
8. Stuhlpfarrer, M., Kümmel, A., Breitsamter, C.: Investigations of helicopter wake flow including rotor-head motion. J. Aircr. **55**(5), 2114–2126 (2018). https://doi.org/10.2514/1.C034689

Vortex-Induced Stall on an Actively Twisted Highly Loaded Model Rotor Blade

Gunther A. Wilke[1(✉)], Berend G. van der Wall[1], Yasutada Tanabe[2], Hideaki Sugawara[2], Do-Hyung Kim[3], Hee J. Kang[3], Sung N. Jung[4], Seong H. Hong[4], Joëlle Bailly[5], George Barakos[6], Rinaldo Steininger[6], and Joon W. Lim[7]

[1] German Aerospace Center, Cologne, Germany
gunther.wilke@dlr.de
[2] Japan Aerospace Exploration Agency, Chōfu, Japan
[3] Korea Aerospace Research Institute, Daejeon, Korea
[4] Konkuk University, Gwangjin-gu, Korea
[5] ONERA, The French Aerospace Lab, Meudon, France
[6] University of Glasgow, Glasgow, UK
[7] US Army, Ames Research Center, Mountain View, CA, USA

Abstract. In preparation for a future wind tunnel test of a rotor with twist-actuated blades (STAR II), numerical predictions of this test have been carried out by the contributing partners. In this paper, the simulated results of the vortex-induced stall operating condition, synonymous with a highly loaded flight condition, are presented. The current conclusion is that a noticeable spread of the results is given when searching for the maximum attainable thrust, depending on the onset of stall once perceived by the individual simulation methodology. Some general trends among the simulations could still be identified with respect to the actuation settings that reduce vibrations and the required power. Generally, most CFD-based results allowed to capture this physical phenomenon, but carefully tuned low-fidelity aerodynamic tools also managed to capture the same trends at a fraction of the computational cost.

Keywords: helicopter rotor · STAR II · active twist · dynamic stall · vortex-induced stall · blade vortex interaction · rotor vibration

1 Introduction

In order to properly design a helicopter rotor, many different flow phenomena need to be modelled correctly: On the advancing side, shock waves may occur, while on the retreating side reverse flow is encountered. On top of that, the

Distribution Statement A. Approved for public release: distribution unlimited.
23. DGLR-Fachsymposium der STAB, Berlin, Germany, November 9-10, 2022.

A. Dillmann et al. (Eds.): STAB/DGLR Symposium 2022, NNFM 154, pp. 623–633, 2024.
https://doi.org/10.1007/978-3-031-40482-5_59

released tip vortices remain in the rotor plane for some time and interact with the rotor blades. These blade-vortex interactions (BVI) give the helicopter the distinctive noise in descent flight, but may trigger stall in hover as well as in forward flight.

Since the rotor blade has to operate reliably in many flight conditions, any passive blade design will always be a compromise to fulfill the various requirements arising from them. For example, Wilke [1] shows that the blade designs may differ significantly depending on the operational state, on which the emphasis has been laid upon. Therefore, the idea of a smart rotor blade exists, where the rotor adapts to the current flight condition.

Without exploring every research direction followed, the avid reader is referred to the surveys of Chopra [2] and Friedmann [3]. Only a brief categorization is done here. The first option, besides the regular swashplate of the rotor that is utilized for steering, a higher harmonic controlled [4] swashplate or a multi-swashplate [5] allows to actuate the rotor blades with more harmonics than the simple swashplate. Depending on the aim, the power, noise, or vibration, and to a certain extend a mixture of these, may be reduced versus the non-actuated blade. Opposed to the swashplate approach, other means of actuation exist. For example pitch link actuation [6] or flaps [7] are means of active flow control for helicopter rotors. Another option is the application of actively twisting the blades. In the NASA/Army/MIT Active Twist Rotor (ATR) project [8], piezo-electric elements actively twisted the rotor blades during their operation. DLR initiated the Smart Twisting Active Rotor (STAR) project in 2007 [9]. It is currently in the second phase, where the blades have been manufactured, but not yet been tested in the wind tunnel. Partners from all over the world participate in the STAR II project, namely: DLR, JAXA, ONERA, KARI, Konkuk University, NASA, University of Glasgow, and the U.S. Army.

In order to assess the viability and identify interesting operating conditions and actuation strategies for the wind tunnel test of the STAR II rotor, a prediction team was formed to perform simulations prior to the wind tunnel test. While the STAR II blades are investigated in a variety of flight conditions as published in [10], this paper focuses purely on the activities of the vortex-induced stall of the highly loaded rotor. A more in-depth analysis of this flight condition than the previous publication is performed and the difficulties in the prediction of the rotor performance in this flight condition are elaborated.

2 Methodology

To accurately assess the merits of the STAR II blades along with their actuation, the rotor simulation needs to account for the aerodynamics, flight mechanics (trim control angles), and structural dynamics, namely the rigid body motion with the elastic deformation. The flight condition investigated here poses a major simulation challenge. On the retreating side, the upwash from a tip vortex of the preceding blade causes dynamic stall at roughly mid-span of the blade progressing towards both sides. To capture this phenomenon, the aerodynamics of the tip vortex, but also the stall needs to be modelled precisely.

In the rotorcraft community, comprehensive codes have been well established. They solve the flight mechanics of the rotor system along with the aero- and structural dynamics of the rotor. However, especially on the aerodynamic side, simplified approaches are applied, such as the blade element theory, where the inflow is semi-empirically modelled. In order to improve this, prescribed or free-wake models may be employed. To further enhance fidelity of the simulations, the comprehensive codes may be coupled with CFD solvers to more accurately capture the three-dimensional, unsteady aerodynamics of the rotor. In both scenarios, the elastics of the blade are represented by a beam model.

For the upcoming results, a common legend is applied, see Fig. 1. The simulations are roughly grouped into plain comprehensive analysis (CA) codes (dashed lines, details in Table 1) and CFD coupled comprehensive analysis (solid lines, details in Table 2). All CA codes use a wake model to improve the inflow modelling of the blade element theory. For a more in-depth explanation of the applied methods, please consult [10].

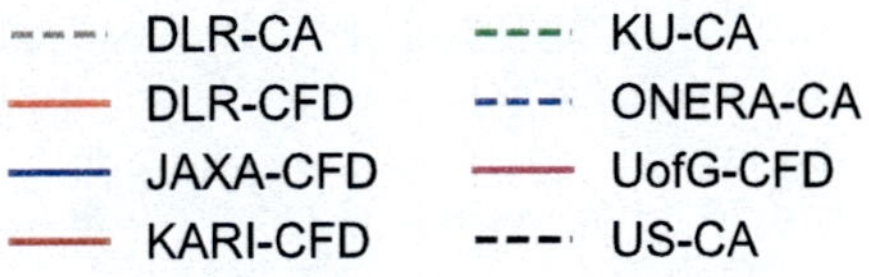

Fig. 1. General legend of the paper.

3 Predictions

The effort of the prediction team is to identify operational conditions worth to be considered during wind tunnel testing to focus resources on the most valuable ones. This includes testing of the unactuated blade, but also with the various actuation settings available. For the high load condition, the first step is to find the onset of stall and the maximum feasible thrust of the rotor, which is here specified as the operational condition, where zero roll and pitch moments of the rotor are still achieved. The second step investigates the actuation at a specified thrust, where most partners had a clearly visible stall, while still retaining a sufficient margin to the maximum thrust to allow for safe operations. The goal of this is to identify the setups of the actuation where required power and/or vibrations of the rotor can be reduced (through alleviation of dynamic stall).

The operational condition, key metrics of the rotor and settings of the actuation are summarized in Table 3. Note that when the activities of this prediction group started, the operational condition was similar to the current one, yet carried out at the nominal RPM (1041) of the rotor. However, foreseen limitations of the rotor balance and power of the model motor, required halving the RPM (521) of the rotor and the wind speed (to retain the advance ratio), while also too severe vibrations and structural stresses are avoided.

3.1 Step 1 - Thrust Sweep Without Actuation

In Fig. 2, trim controls, power and vibration for increasing thrust of the simulations are shown. It is observed that the pitch control angles at lower thrusts are close to each other, but the spread of the results grows with thrust. A similar tendency is found for the required power of the rotor, with even more diverse results for the Vibration Intrusion Index (VI) (defined as $\sum_{i=4,8} \sqrt{(0.5F^2_{x,i/rev}) + (0.67F^2_{y,i/rev}) + (F^2_{z,i/rev})}/W + \sqrt{M^2_{x,i/rev} + M^2_{y,i/rev}}/(RW)$ with $F_{j,i/rev}$ and $M_{j,j/rev}$ being amplitudes of the i/rev's component of the hub loads and moments in the respective coordinate direction and $W = 3600$ N the virtual model weight and R the rotor radius). Most of the spread at higher thrusts is attributed to the differently perceived onset of (deep) stall. For the required power as well as the trim control angles, most notably the collective pitch control angle θ_0, a change in the gradient is observed at this point. Looking at the maximum attainable thrust, the variation is large, with KU-CA arriving at $c_T/\sigma = 0.13$, and JAXA-CFD at 0.16 and ONERA-CA at 0.17. Regarding the VI, all codes except US-CA capture a growth with thrust, yet with different levels to begin with. The general expec-

Table 1. Summary of comprehensive code setups. box - modeling of the terminal box at the blade root, init. tip core - initial core radius of the rolled-up tip vortex model in $r_{c,0}/c$.

	DLR-CA	KU-CA	ONERA-CA	US-CA
box	no	no	no	no
fuselage	yes	no	no	no
init. tip core	0.033	0.5	0.3	0.8
blade elements	20	16	25	28
azimuth step size	2°	15°	5°	15°

Table 2. Summary of CFD setups. order - theoretical inviscid flux scheme order, transition - how laminar-turbulent transition is included, blade - resolution of the blade mesh in chordwise x normal x spanwise direction, background - spacing of the Cartesian grid area relative to the chord length.

	DLR-CFD	JAXA-CFD	KARI-CFD	UofG-CFD
box	yes	yes	yes	no
fuselage	yes	yes	yes	no
order	4	4	2	3
turbulence	SA-DDES-R	Menter SST	SA	Menter SST
transition	empirical	γRe_θ	no	no
blade resolution	$128 \times 64 \times 128$	$141 \times 81 \times 157$	-	$142 \times 64 \times 148$
background $\Delta x/c$	14%	10%	15%	15%
total # grid pts	15 M	88 M	29 M	35 M
step	0.25°	0.1°	1.0°	0.25°

Table 3. Key rotor metrics and reference values.

Radius [m]	2
chord [m]	0.121
RPM	521
advance ratio	0.3
shaft angle α_q	0°
density of air $[kg/m^3]$	1.226
temperature [K]	288.15
Step 1 c_T/σ	0.1 ... 0.17 in step of 0.01
Step 2 c_T/σ	0.13
Step 2 frequencies	1,2,3,4,5/rev
Step 2 phases	0 ... 330° in steps of 30°
Step 1 & 2 trim	zero hub moments

tation, a growth with thrust, is confirmed, but no distinct reason for the large variation could be identified so far.

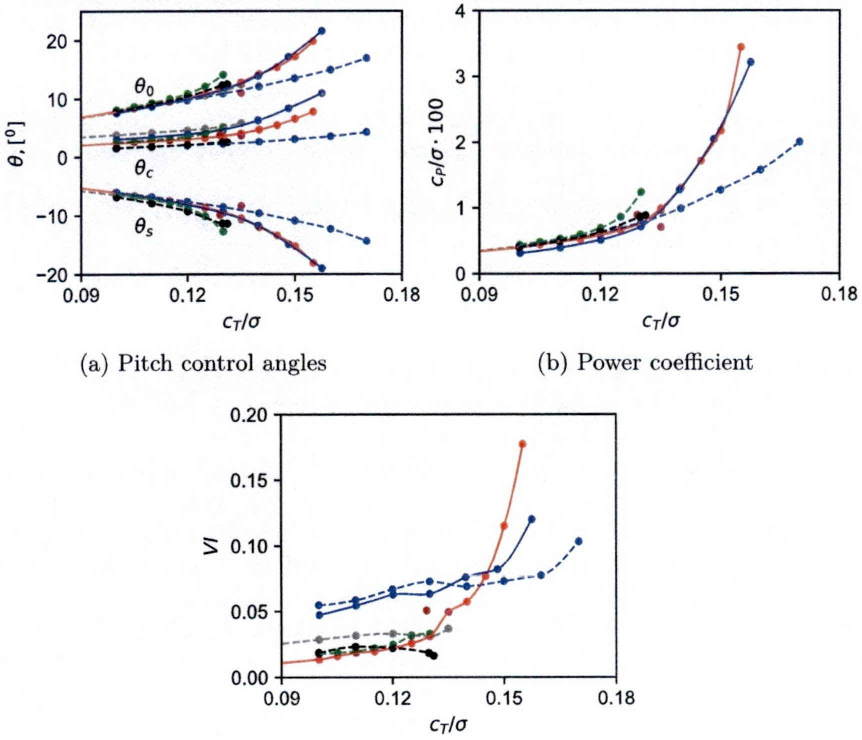

(a) Pitch control angles

(b) Power coefficient

(c) Vibration Intrusion Index VI

Fig. 2. Integral metrics over thrust variation.

To understand the physics better, the loads are compared at $c_T/\sigma = 0.10$, 0.13 and 0.15 (where results are available) in Fig. 3 to Fig. 5. The focus is laid upon the normal loads as well as the sectional pitching moments at the radial position $r/R = 0.67$, as it coincides with a section instrumented with pressure sensors of the wind tunnel model being closest to the occurring blade vortex interaction. The obvious general trend is that the local forces grow with increasing thrust, but also that the agreement among the partners is overall better at the lower thrust. The reason for this is that there is only very mild presence of pitching moment stall for DLR-CA, DLR-CFD and JAXA-CFD, with little feedback on the normal load and no stall for the other partners. The lower harmonic content of these curves therefore remains very similar, yet also looking at the advancing side, some blade vortex interaction (BVI) is also found, with DLR-CA having the largest ones. This is explained with the smallest tip core radius chosen, causing an inversely proportional stronger BVI event.

At the intermediate thrust, a larger spread of the results is observed, with the DLR- and JAXA-CFD based simulations clearly distinguishing themselves from the comprehensive analysis results in pitching moments, indicating a clear pitching moment stall, with also a clear normal force stall for DLR-CFD. Opposing this, US-CA is farthest away from these trends, explained with the greatest tip core radius used in the simulation coupled with a large time step, therefore capturing a too mild BVI. Interestingly enough, DLR- and KU-CA resolve a greater pitching moment stall than the other CA results and KARI- and UofG-CFD for this intermediate thrust. Moving onto the highest thrust, again, DLR- and JAXA-CFD results are relatively close together in behavior and magnitude of the normal loads and pitching moments, while ONERA-CA results show a strongly reduced stall behavior (Fig. 5).

For a final comprehension of the physics, the vorticity fields from the CFD-based simulations are plotted in Fig. 6. From these more qualitative impressions, multiple lessons are learned. The tip-vortex released from the preceding blade hits the following blade roughly at mid-span. The upwash side of this vortex therefore triggers the stall and thus aligns with other findings on this topic in the literature [11]. Due to the application of DDES for DLR-CFD, more secondary vortex structures are visible. However, when comparing the loads with JAXA-CFD, they are very similar except for some higher harmonic content for DLR-CFD. Moving onto KARI-CFD results, where the general vortex field is captured as well, the use of a second order scheme with classical URANS leads to a weaker vortex and therefore the pitching moment stall is not yet developed as strongly, but an onset of stall can be found. The UofG-CFD results resolve a similar vortex strength as DLR-CFD, yet the pitching moment stall is of similar magnitude to the one by KARI-CFD. While numerical order theoretically can be traded in for grid points, this case also demonstrates that the resolutions the partners achieved are at least an order of magnitude too small to be considered properly grid-converged and performing CFD simulations for rotor blades remains a challenging task in terms of computational resources.

3.2 Step 2 - Actuation Sweeps

In this phase, the potential benefits of an actuated blade are examined at a $c_T/\sigma = 0.13$. The actuation is imposed through a sinusoidal signal plus an offset. The amplitude of the actuation voltage is 500 V with an offset of 400 V. This has been simulated as a 2.08 Nm static nose-up pitching moment with 2.6 Nm dynamical varying pitching moment introduced over most of the airfoiled part of the blade ($r/R \approx 0.24 - 0.94$).

Due to the large design space for the actuation, not every partner could carry out the whole test matrix. DLR-CA, DLR-CFD and ONERA-CA were able to compute the whole range of data points. Their results are plotted as contour plots relative to their baseline results for the required power in Fig. 7, and VI index in Fig. 8. Note, the 0/rev results refer to the static actuation of the blade. While just a crude representation of the problem and with some discrepancies at the higher harmonics, with these plots is is quickly identified that the 2/rev actuation is likely to be the one with the greatest beneficial effect, and therefore the remaining partners were encouraged to compute the phase sweep at 2/rev, with the results given in Fig. 9. Similar to the baseline case, results are in better agreement for the required power than for the VI metric. Nevertheless, phases from $\phi \approx 0...90^o$ revealed improvements over the non-actuated rotor for most partners.

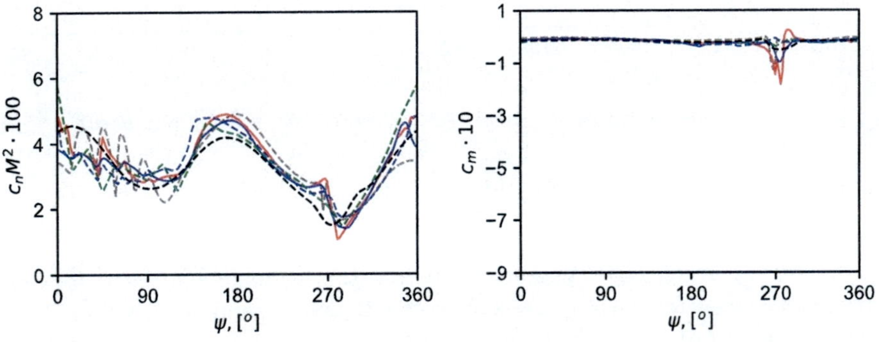

(a) $c_n M^2$ (b) $c_m(M = M_{tip}(\frac{r}{R} + \mu sin\psi))$

Fig. 3. Loads at $r/R = 0.67$ for $c_T/\sigma = 0.10$

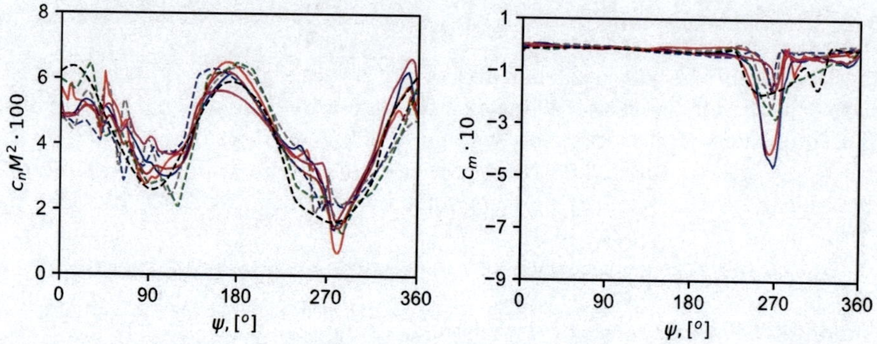

(a) c_nM^2 (b) $c_m(M = M_{tip}(\frac{r}{R} + \mu sin\psi))$

Fig. 4. Loads at $r/R = 0.67$ for $c_T/\sigma = 0.13$

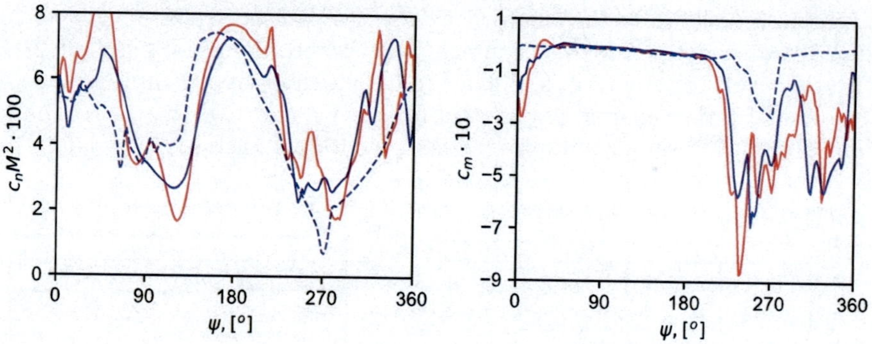

(a) c_nM^2 (b) $c_m(M = M_{tip}(\frac{r}{R} + \mu sin\psi))$

Fig. 5. Loads at $r/R = 0.67$ for $c_T/\sigma = 0.15$

In order to understand these merits better, the sectional loads at $r/R = 0.67$ are checked for cases with 2/rev actuation with $\phi = 0^o$ and $\phi = 240^o$ in Fig. 10 and Fig. 11, respectively. The observation here is that the pitching moment stall

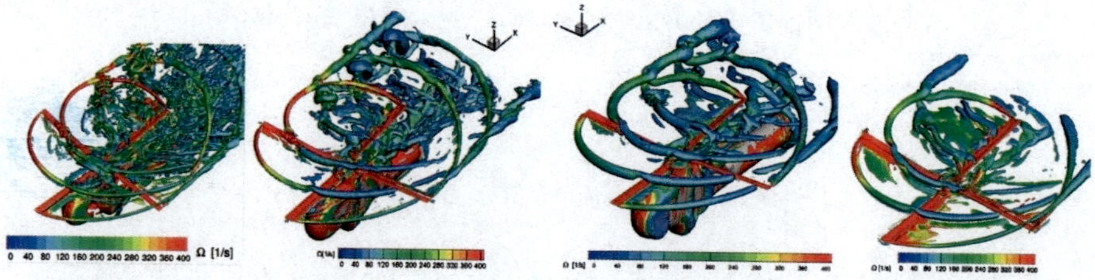

(a) DLR-CFD (b) JAXA-CFD (c) KARI-CFD (d) UofG-CFD

Fig. 6. Vorticity plots at $c_T/\sigma = 0.13$.

and its dynamic components are reduced for the $\phi = 0^o$ case relative to the baseline case, Fig. 2, but increased for the $\phi = 240^o$ case over the baseline results, with some impact on the normal loads.

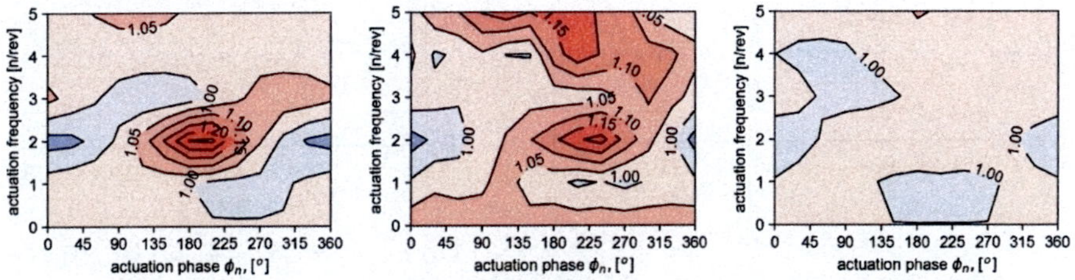

(a) DLR-CA (b) DLR-CFD (c) ONERA-CA

Fig. 7. Relative required power at $c_T/\sigma = 0.13$.

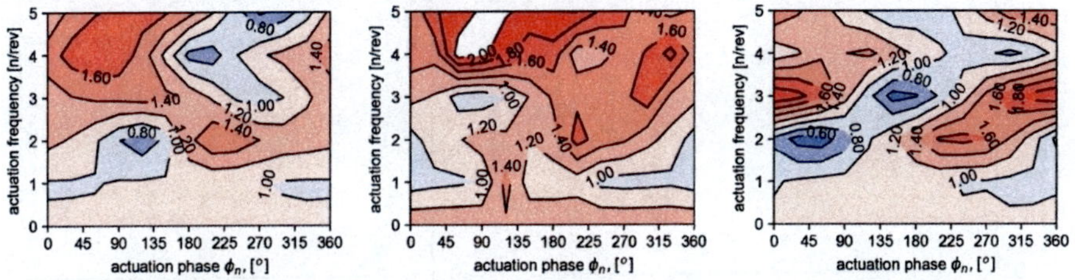

(a) DLR-CA (b) DLR-CFD (c) ONERA-CA

Fig. 8. Relative VI index at $c_T/\sigma = 0.13$.

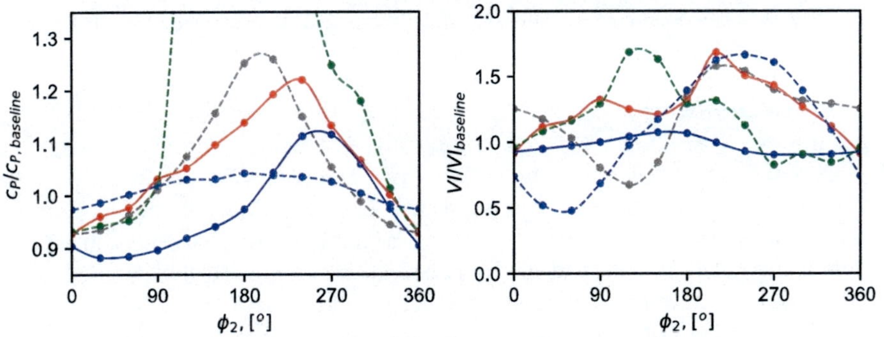

(a) power (KU results cropped) (b) VI index

Fig. 9. Integral results of phase sweep at 2/rev at $c_T/\sigma = 0.13$.

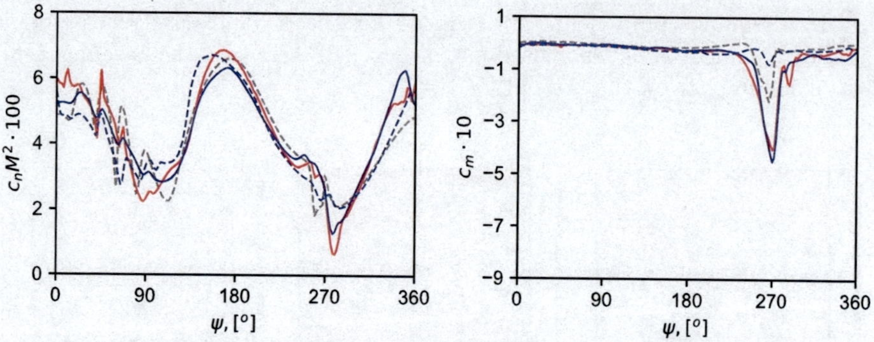

(a) $c_n M^2$

(b) $c_m(M = M_{tip}(\frac{r}{R} + \mu sin\psi))$

Fig. 10. Loads at $r/R = 0.67$ for the 2/rev $\phi = 0^o$ case at $c_T/\sigma = 0.13$.

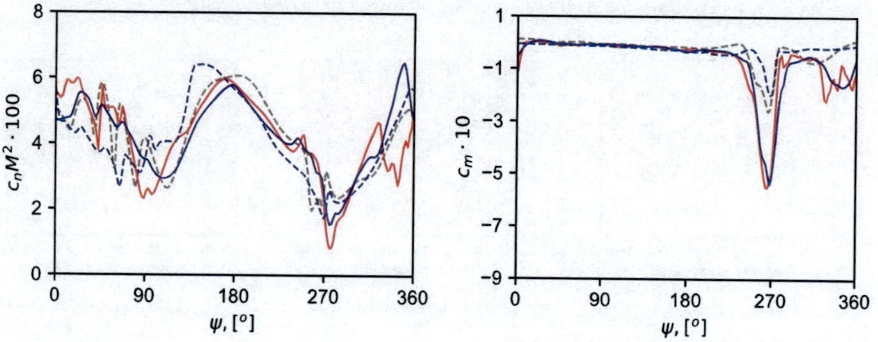

(a) $c_n M^2$

(b) $c_m(M = M_{tip}(\frac{r}{R} + \mu sin\psi))$

Fig. 11. Loads at $r/R = 0.67$ for the 2/rev $\phi = 240^o$ case at $c_T/\sigma = 0.13$.

4 Conclusions

In this paper, the prediction team activities for the vortex-induced highly loaded rotor condition of the Smart Twisting Active Rotor II (STAR II) are presented. These results pose the blind pre-test predictions prior to the wind tunnel test. While some noticeable spread in the potential maximum thrust of the unactuated blade is observed, the general tendencies and mechanisms were captured by most simulations. These results are still sufficient to narrow down the operating conditions for the wind tunnel test including the settings for the actuation that allow for a reduction in required power or vibrations. Yet, with the given computational resources, room for further improvement is given and more refined CFD simulations will likely show better agreement. Unfortunately, the current turnaround time for these have been in the magnitude of days to weeks, where low-fidelity, comprehensive code, simulations were in the magnitude of minutes.

References

1. Wilke, G.: Quieter and greener rotorcraft: concurrent aerodynamic and acoustic optimization. CEAS Aeronaut. J. **12**(1), 495–508 (2021). https://doi.org/10.1007/s13272-021-00513-x
2. Chopra, I.: Status of application of smart structures technology to rotorcraft systems. J. Am. Helicopter Soc. **45**(4), 228–252 (2000). https://doi.org/10.4050/JAHS.45.228
3. Friedmann, P.P.: On-blade control of rotor vibration, noise, and performance: just around the corner? J. Am. Helicopter Soc. **59**(4), 1–37 (2014). https://doi.org/10.4050/JAHS.59.041001
4. van der Wall, B.G., et al.: The HART II international workshop: an assessment of the state-of-the-art in comprehensive code prediction. CEAS Aeronaut. J. **4**(3), 223–252 (2013). https://doi.org/10.1007/s13272-013-0077-9
5. Küfmann, P., et al.: The first wind tunnel test of the dlr's multiple swashplate system: test procedure and preliminary results. In: AHS 72nd Annual Forum. West Palm Beach, Florida (2016)
6. Kretz, M.: Research in multicyclic and active control of rotary wings. Vertica **1**(1/2), 95–105 (1976)
7. Spangler, R.L., Hall, S.R.: Piezoelectric actuators for helicopter rotor control. In: Proceedings of the 31st AIAA/ASME/ASCE/AHS/ASC Structures, Structural Dynamics, and Materials Conference. Long Beach, CA (1990). https://doi.org/10.2514/6.1990-1076
8. Shin, S.J., Cesnik, C.E.S., Hall, S.R.: Closed-loop control test of the NASA/Army/MIT active twist rotor for vibration reduction. J. Am. Helicopter Soc. **50**(4), 178–194 (2005). https://doi.org/10.4050/1.3092854
9. Wierach, P., Riemenschneider, J., Opitz, S., Hoffmann, F.: Experimental investigation of an active twist model rotor blade under centrifugal loads. In: Wiedemann, M., Sinapius, M. (eds.) Adaptive, tolerant and efficient composite structures, pp. 391–407. Springer, Heidelberg (2013). https://doi.org/10.1007/978-3-642-29190-6_32
10. van der Wall, B.G., et al.:: Smart twisting active rotor (STAR) - Pre-Test Predictions. In: 48th European Rotorcraft Forum. Winterthur, Switzerland (2022)
11. Smith, M.J., Gardner, A.D., Jain, R., Peters, D., Richez, F.: Rotating wing dynamic stall: state of the art and future directions. In: 76th Annual Forum of the Vertical Flight Society. Virtual (2020). https://doi.org/10.4050/F-0076-2020-16472

High-Agility Configuration

Influence of Vertical Stabilizer on the Aerodynamic Stability of a Generic Multi Delta Wing Aircraft Configuration

Katharina Alt and Andreas Schütte(✉)

DLR - German Aerospace Center, 38108 Braunschweig, Germany
{katharina.alt,andreas.schuette}@dlr.de

Abstract. The current numerical investigations analyze the influence of the vertical stabilizer on the aerodynamics and aerodynamic stability, especially roll stability, of a multi delta wing configuration. Planform studies established to enhance the rolling moment behavior by means of strakes and wing leading edge slats are taken as a reference for the current vertical tail plane studies. The aim is to identify the differences between the configuration with and without vertical tails. In particular the influence of the vortical flow topology and interaction with the vertical tail planes on the aerodynamic behavior should be identified. This should lead to design recommendations for the current fighter design approaches within the DLR project Diabolo.

Keywords: computational fluid dynamics · vortex flow · vertical stabilizer · delta wing aerodynamics · aerodynamic stability

1 Introduction

Fig. 1. Flow topology of the DLR-FFD.

The aerodynamic requirements of modern fighter aircraft can be fulfilled by delta wings. The medium to high leading edge sweep angles of these configurations lead to a complex vortical flow field and challenging aerodynamic behavior. The aerodynamic requirements of such configurations include a high degree of agility, a high maneuverability, good supersonic flight characteristics, as well as low observability.

The DLR project Diabolo is looking at the design and assessment capabilities of modern fighter aircraft, to understanding the flow physics and aerodynamic behavior of fighter aircraft and to demonstrate the design and assessment methods by the development of a virtual DLR Future Fighter Demonstrator (DLR-FFD), see Fig. 1.

A. Dillmann et al. (Eds.): STAB/DGLR Symposium 2022, NNFM 154, pp. 637–646, 2024.
https://doi.org/10.1007/978-3-031-40482-5_60

Several experimental and computational investigations have been performed over the past years to predict the vortical flow aerodynamics of fighter type aircraft. The validation process for fighter type aircraft configurations have been progressing quite well and provide a sufficient confidence in the numerical methods to perform design studies and derive conclusions regarding the aerodynamic behavior and stability. For the DLR RANS method TAU several validation approaches have been performed. With this regard the investigations applying the F-16XL CAWAPI configuration by Fritz et al. [1] as well as Hitzel [2] should be mentioned. Investigations published by Schütte and Marini [3] and Werner et al. [4] look at a multi delta wing configurations within the NATO/STO AVT Task Group 316 on vortex-vortex interaction. Within these validation studies various physical models on different fidelity levels have been applied. The publications provide best practice for the grid applied physical model to predict the aerodynamics most accurately. These recommendations have been applied in roll stability studies within the DLR project Diabolo by Schütte and Hummel [5].

In the present work, the effect of vertical stabilizer on the aerodynamic stability of the generic multi delta wing aircraft configuration from [5] is analyzed to identify the influence on the flow physics and the aerodynamic behavior, with a focus on the rolling moment stability of the configuration.

2 Configuration

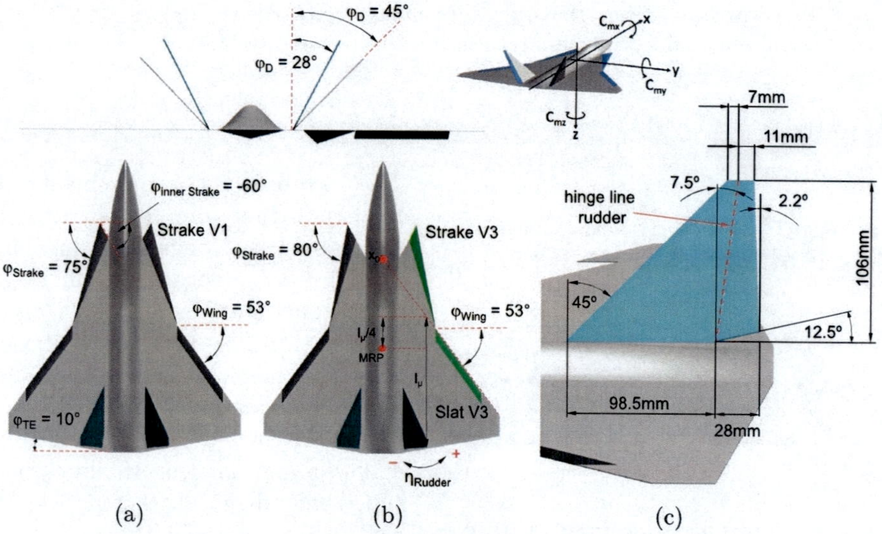

Fig. 2. Geometries: (a) planform 3, (b) planform 4 and (c) stabilizer setup.

Figure 2 shows the geometries of the multi delta wing configuration with stabilizer. Planform 3 and 4 without the vertical stabilizers were investigated by

Schütte and Hummel [5]. Both planforms are equipped with strake flaps and leading edge slats. The difference between planform 3 (Fig. 2a) and 4 (Fig. 2b) is the sweep angle of the strake with $\varphi_{Strake} = 75°$ for planform 3 and 80° for planform 4. The size of the strake flap and leading edge wing slats are the same and described in detail in [5]. The shape and position of the stabilizer have been defined within the DLR project Diabolo for the DLR-FFD. For both planforms the stabilizer with rudder are oriented parallel to the model x-axis with a dihedral angle of $\varphi_D = 28°$ for planform 3 and in addition with a dihedral angle of $\varphi_D = 45°$ for planform 4 only, see Fig. 2c. In the following a rudder deflection of $\eta_{Rudder} = 0°$ and 10° are considered. In addition, the body fixed coordinate system and sign conventions of the forces and moments are plotted in Fig. 2.

3 Computational Grids

For the grid generation, the software–system Centaur^{TM}, developed by Centaur-Soft [6], has been applied.

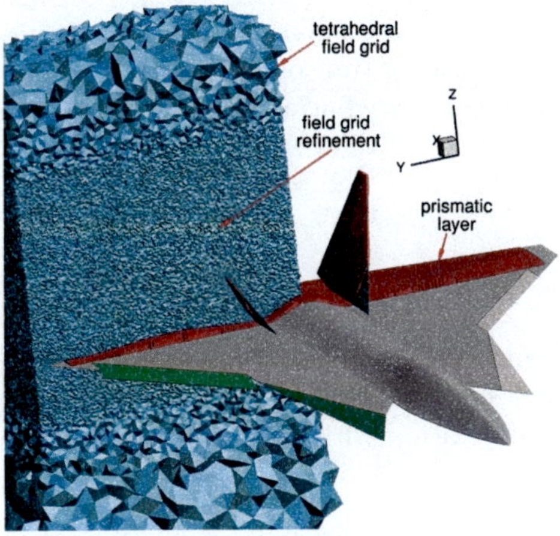

Fig. 3. Topology of the computational grid.

For numerical simulations hybrid meshes were generated. Figure 3 shows the global grid topology with the boundary layer resolving prism layer in red and the tetrahedral field grid colored in cyan. In addition, grid refinement to resolve the vortical over the configuration is marked.

For all meshes, the parameters of the grid generator were kept constant to achieve a comparable discretization for all configurations. All grids have 30 prismatic layers to resolve the boundary layer flow. The initial first wall distance is 0.002 mm or approx. $l_\mu \cdot 10^{-5}$. The leading edge thickness is 0.5 mm and resolved with 5 cells. The cell size of the surface triangulation on the main wing and fuselage as well as cell size of the tetrahedral within the source around the wing is 1.5 mm or $l_\mu \cdot 7 \cdot 10^{-3}$. The overall number of grid points is approx. $90 \cdot 10^6$. The best practice approach for the grid topology and discretization is validated in various publications by Schütte et al. [3,5].

4 CFD Approach

For the present work, the RANS equation flow solver DLR TAU is used. TAU is a CFD software developed by the DLR Institute of Aerodynamics and Flow Technology [7]. The flow solver TAU solves the compressible, three-dimensional, time-accurate Reynolds-Averaged Navier-Stokes equations using a finite volume formulation. The code is based on a hybrid unstructured grid approach to be able to handle structured and hybrid computational grids, which makes use of prismatic grid structures to resolve the viscous shear layer close to the wall. The current simulations were performed using the steady state and unsteady dual time-stepping approach. The dual time stepping approach was used in case no steady state solution was found. Therefore, the solution is provided by averaging over a time period of t = 0.5 s. The time period is estimated by the time a particle needs to pass forth and back through the entire grid. The unsteady physical time step applied is Δt = 0.0001 s and a Cauchy convergence criterion is applied using the lift and pitching moment coefficient (thresholds: $c_L = 1{\cdot}10^{-4}$ and $c_m = 1{\cdot}10^{-5}$). For the numerical simulations the one-equation Spalart-Allmaras turbulence model (SA) [8] in its negative formulation [9] and rotation correction (SARC) [10] is used. All computational simulations were performed on the DLR CARA high performance computing cluster located at the Technical University in Dresden. For each simulation 10 nodes with 64 processors each were used. The computational simulations were conducted at a Mach number of $M_\infty = 0.5$ and Reynolds numbers of $Re_{\infty,1m} = 9{\cdot}10^6$. The Mach number is chosen to provide a flight relevant and Mach number without taking transonic effect into account for the time being. All computations were conducted at asymmetric freestream conditions at an AoS (Angle of Sideslip) of $\beta = 5°$ and AoA (Angle of Attack) range from $\alpha = 15°$ to $31°$.

5 Computational Results

In the current section the influence of integrating the vertical stabilizer on the aerodynamic stability will be discussed. Figure 4a shows the rolling moment characteristic for planform 3 for the configuration with no strake flap and slat

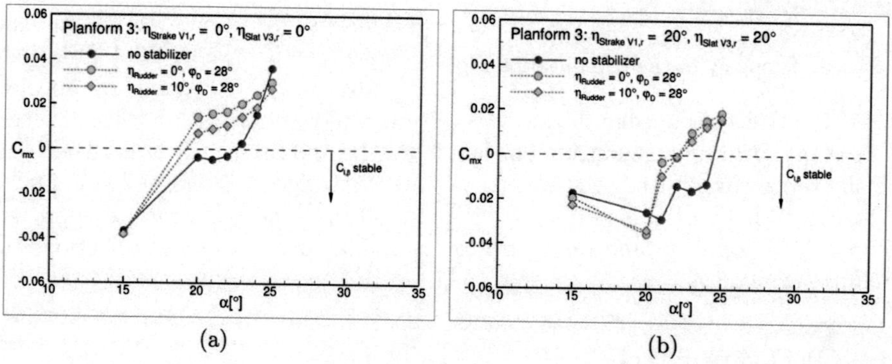

Fig. 4. Rolling moment characteristics planform 3: Comparison between stabilizer ($\varphi_D = 28°$) and without stabilizer configuration: (a) no strake and slat deflection and (b) with $\eta_{StrakeV1,r} = 20°$ and $\eta_{SlatV3,r} = 20°$ ($M_\infty = 0.5$, $\beta = 5°$).

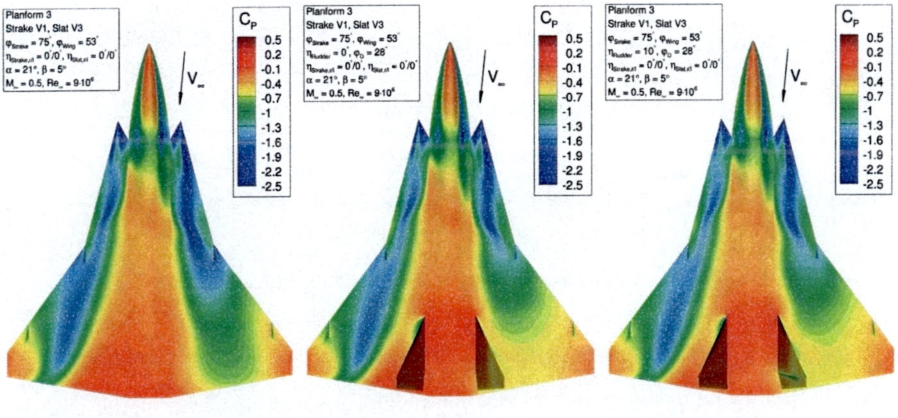

(a) (b) (c)

Fig. 5. Planform 3 with ($\varphi_D = 28°$) and without stabilizer: Surface pressure distribution for (a) no stabilizer, (b) $\eta_{Rudder} = 0°$ and (c) $\eta_{Rudder} = 10°$ at an AoA of $\alpha = 21°$ and no control surface deflection ($M_\infty = 0.5$, $\beta = 5°$).

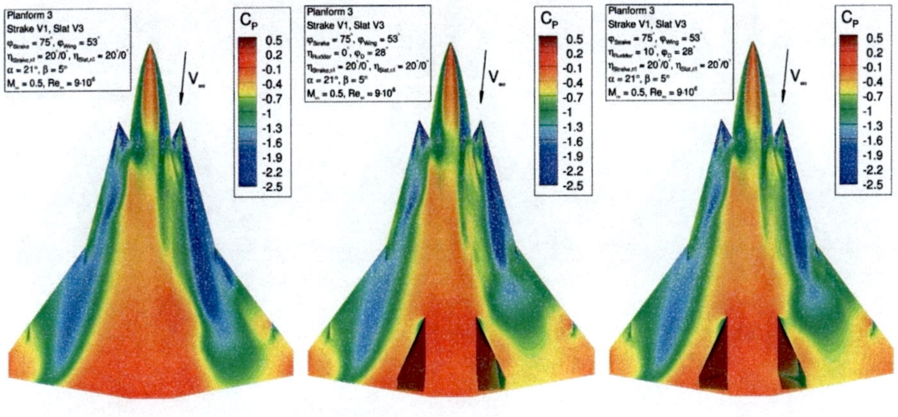

(a) (b) (c)

Fig. 6. Planform 3 with ($\varphi_D = 28°$) and without stabilizer: Surface pressure distribution for (a) no stabilizer, (b) $\eta_{Rudder} = 0°$ and (c) $\eta_{Rudder} = 10°$ at an AoA of $\alpha = 21°$ with $\eta_{StrakeV1,r} = 20°$ and $\eta_{SlatV3,r} = 20°$ ($M_\infty = 0.5$, $\beta = 5°$).

deflection and in Fig. 4b with $\eta_{StrakeV1,r}$ and $\eta_{SlatV3,r} = 20°$. It can be observed that due to the strake flap and leading edge slat deflection the roll instability is shifted to higher AoA from $\alpha = 23°$ to 25°. This is caused by a stronger strake and wing vortex on the windward side supported by the strake and slat deflection on the right hand side of the configuration, see [5]. By integrating the vertical stabilizer the roll stability margin is shifted to lower angle of attack. This can be explained by looking at the surface pressure distributions in Fig. 5 and

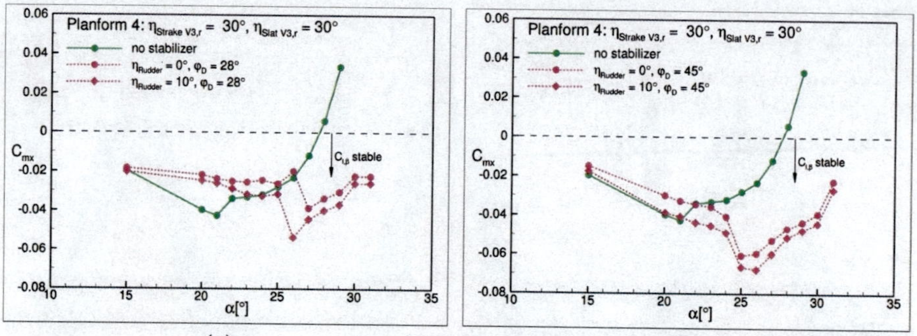

(a) (b)

Fig. 7. Rolling moment characteristics for planform 4: Comparison of the case with and without stabilizer and a dihedral angle of (a) $\varphi_D = 28°$ and (b) $\varphi_D = 45°$ ($M_\infty = 0.5$, $\beta = 5°$).

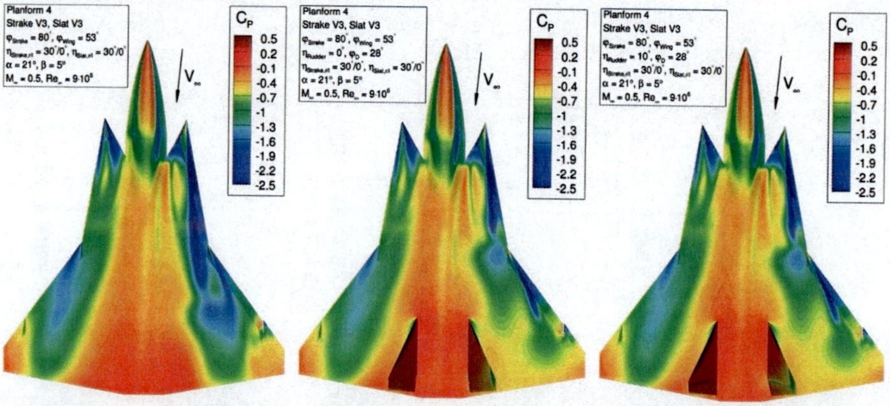

(a) (b) (c)

Fig. 8. Planform 4 with and without stabilizer: Surface pressure distribution for (a) no stabilizer, (b) $\eta_{Rudder} = 0°$ and (c) $\eta_{Rudder} = 10°$ at an AoA of $\alpha = 21°$ ($M_\infty = 0.5$, $\beta = 5°$).

Fig. 6. Depicted are the pressure distribution at an AoA of $\alpha = 21°$. For both configuration it can be seen that vortex breakdown of the wing and strake vortex are shifted upstream indicted by the suction pattern of both vortices on the windward side of the wing in comparison to the configuration without stabilizer, see Fig. 5b and c in comparison to Fig. 5a and Fig. 6 respectively. In addition the leeward side is much less effected by the stabilizer and thus the roll stability is shifted to lower AoA. Figure 7 shows the rolling moment characteristics for planform 4. Depicted are the characteristics for the configuration with stabilizer dihedral angle of $\varphi_D = 28°$ in Fig. 7a and for $\varphi_D = 45°$ in Fig. 7b. In comparison to planform 3 the roll stability margin is shifted to higher angle of attack due to a stronger strake vortex on the windward side provided by a higher leading edge sweep angle of the strake in comparison to planform 3 [5]. In contrast to planform 3 the roll stability for both setups of planform 4 remains stable along the entire AoA range. For the case $\varphi_D = 28°$ the roll stability is slightly reduced for smaller AoA. But for higher AoA the roll stability is increased significantly in comparison to the configuration without vertical stabilizer. In addition the roll stability increases by deflecting the rudder. Figure 8 shows the pressure distribution at an AoA of $\alpha = 21°$ for the configuration without stabilizer in Fig. 8a and with stabilizer ($\varphi_D = 28°$) for $\varphi_{Rudder} = 0°$ and $10°$ in Fig. 8b and c. As for planform 3 the integration of the stabilizer causes a shift of vortex breakdown upstream on the windward side of the wing. In contrast to planform 3 the suction pattern of the wing vortex remains but gets shifted outboard for the configurations with stabilizer. The more outboard located wing vortex trajectory seems to be less affected by the stabilizer. Similar to planform 3 the leeward side is less effected by the integration of the stabilizer. However, there is a significant increase in roll stability from $\alpha = 26°$ to $\alpha = 27°$ for the case with $\varphi_{Rudder} = 0°$ and from $\alpha = 25°$ to $26°$ AoA for the case with $\varphi_{Rudder} = 10°$ from Fig. 7a.

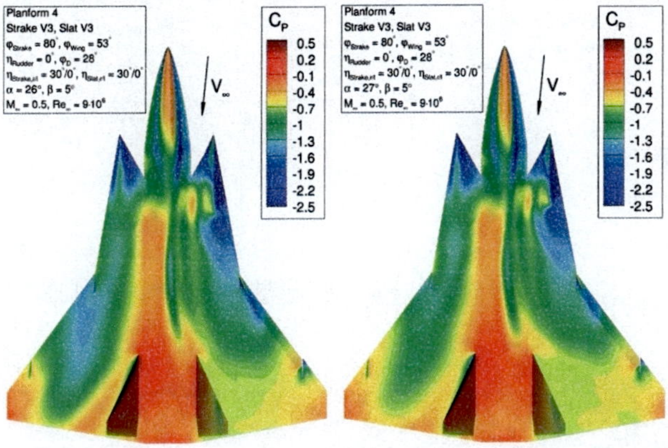

Fig. 9. Planform 4 with stabilizer $\eta_{Rudder} = 0°$: Surface pressure distribution at an AoA of $\alpha = 26°$ and $\alpha = 27°$ ($M_\infty = 0.5$, $\beta = 5°$).

This is actually an installation effect of the stabilizer regarding the leeward side. It can be observed that the suction of the merged strake and wing vortex on the leeward side for $\alpha = 26°$ in Fig. 9 is higher than for $\alpha = 27°$. Looking at the ω_x distribution in Fig. 10a it can be seen that for $\alpha = 27°$ the wing vortex and strake vortex are shifted away from the surface due to the higher AoA in comparison to $\alpha = 26°$. In addition, the vortex core of the strake vortex starts to burst causing the lower suction peak on the surface on the leeward side and thus a higher stable rolling moment.

Figure 10b is providing the rolling moment characteristics for the full configuration as well as the contributions by the stabilizer, forebody/strake and wing part of the configuration divided at the kink between strake and wing. It can be observed that for $\alpha = 26°$ and $27°$ the rolling moment contribution from the forebody/strake is not changing. A significant gain of roll stability is provided by the wing. This confirms the effect of vortex breakdown of the strake on the leeward side as the reason for the step in the rolling moment characteristics.

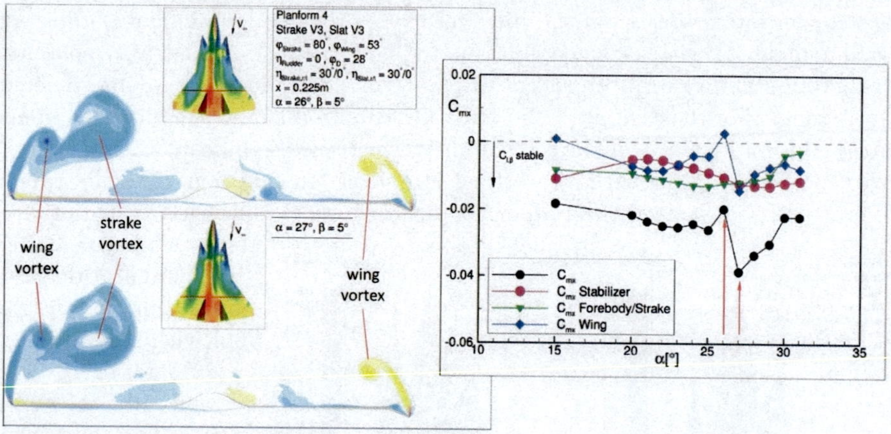

Fig. 10. Planform 4 with stabilizer $\eta_{Rudder} = 0°$: Left: $\omega_\mathbf{x}$ distribution at an AoA of $\alpha = 26°$ and 27° at a x = const. location, Right: Rolling moment characteristics versus AoA ($M_\infty = 0.5$, $\beta = 5°$).

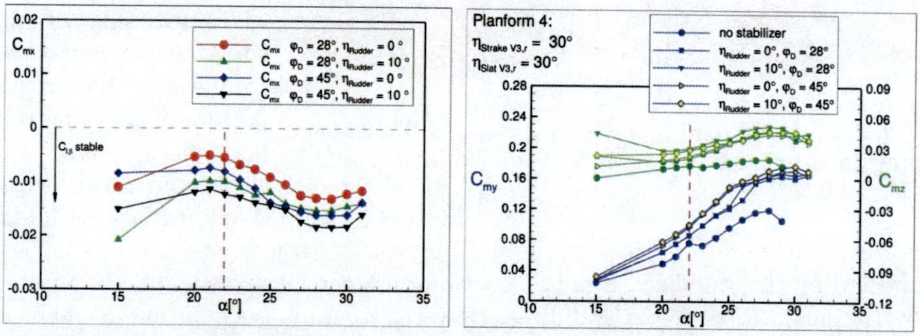

(a) (b)

Fig. 11. Planform 4: (a) Rolling moment characteristic of the stabilizers, (b) Pitch and yaw moment characteristics with and without stabilizer ($M_\infty = 0.5$, $\beta = 5°$).

In the following, the contribution of the stabilizer to the roll stability as well as to pitch and yaw moment characteristics will be discussed. Figure 11a shows the contribution to the rolling moment for the stabilizers of planform 4. Figure 11b shows the pitching and rolling moment characteristics for the configurations without stabilizer in comparison to the cases with stabilizer.

It can be observed that all stabilizer setups providing a stabilizing contribution to the rolling moment. In addition, there is a shift of the characteristics towards higher stability by deflecting the rudder from $\eta_{Rudder} = 0°$ to 10°. This applies for both cases $\varphi_D = 28°$ and 45°. Moreover, the case $\varphi_D = 45°$ is providing a higher roll stability for the same rudder deflections. However, for $\alpha = 15°$ the roll stability contribution for the case $\varphi_D = 45°$ is smaller than for $\varphi_D = 28°$. Finally, the stabilizer increases the pitch and yaw moment. Thus planform

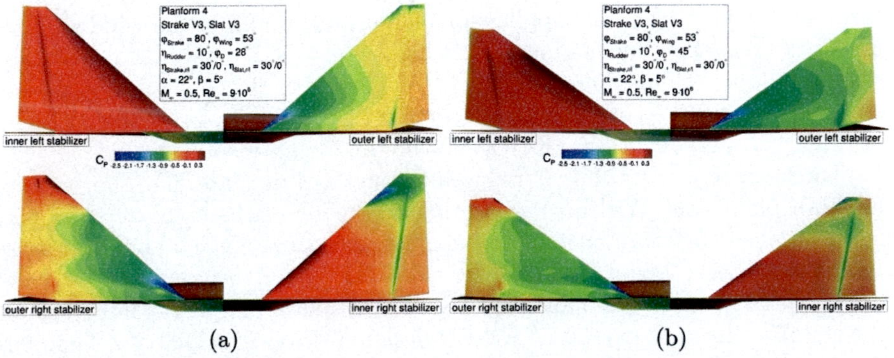

Fig. 12. Planform 4, Pressure distribution at the inner and outer surface of the stabilizer: (a) $\varphi_D = 28°$ and $\eta_{Rudder} = 10°$, (b) $\varphi_D = 45°$ and $\eta_{Rudder} = 10°$ at $\alpha = 22°$ ($M_\infty = 0.5$, $\beta = 5°$).

4 shows longitudinal instability and directional stability for both cases $\varphi_D = 28°$ and $\varphi_D = 45°$.

To understand the influence of the stabilizer in Fig. 12 the surface pressure distributions on the inner and outer surfaces of the vertical stabilizers are plotted at an AoA of $\alpha = 22°$ for the case $\varphi_D = 28°$ and $\eta_{Rudder} = 10°$ in Fig. 12a and $\varphi_D = 45°$ and $\eta_{Rudder} = 10°$ in Fig. 12b. It can be observed that on the outer left and right stabilizer for both cases a vortex occurs causing higher suction on the outer side of the stabilizer in comparison to the inner side. In addition, another vortex occurs at the top of the inner right stabilizer for both cases, indicated by a high suction pattern. The latter is supported by the interaction of the stabilizer with the windward side merged strake wing vortex. These vortex systems support a decrease in pitch stability but an increase in yaw stability.

6 Conclusion

The present work investigated the influence of the vertical stabilizer with rudder at asymmetric free stream conditions on the flow topology and aerodynamic stability of a multi delta wing aircraft configuration. The numerical investigations based on planform and control surface studies on the configurations without vertical stabilizers within the DLR project Diabolo.

It was found that in comparison to the configuration without stabilizer, the roll stability for planform 3 is reduced significantly. This applies for setups with and without control device deflection. This increased roll instability is caused by an upstream effect by the stabilizer which shifts vortex breakdown to lower angle of attack in comparison to the configuration without stabilizer in particular on the windward side of configuration. The deflection of the rudder is supporting the roll stability but the overall upstream effect on the vortex system increases.

For planform 4 with a higher leading edge sweep angle of the strake the roll stability remains in comparison to planform 3. Even though the upstream effect by the stabilizer also applies for planform 4, it effects the windward and leeward side in the same manner but at higher AoA. Even more the roll stability remains and no roll instability occurs in comparison to the configuration without stabilizer within the considered AoA range. The increase of the dihedral angle of the vertical stabilizers are increasing the roll stability additionally.

The longitudinal stability is reduced through the integration of the stabilizer for both planforms. With increasing dihedral angle of the stabilizer, the pitch moment is getting more unstable. In contrast, the directional stability increases due to the integration of the stabilizer.

The present results and analyses are the basis for an improved integration of the stabilizer into the overall DLR-FFD design developed in Diabolo. Therefore, in future investigations Mach number effects will be taken into account.

References

1. Fritz, W., Davis, M., Karman, S.J., Michal, T.: Reynolds-averaged Navier-Stokes solutions for the CAWAPI F-16XL using different hybrid grids. AIAA J. Aircr. **46**(2), 409–422 (2009). https://doi.org/10.2514/1.35106
2. Hitzel, S.: High-Angle-of-attack F-16XL flight simulations at sub- and transonic speeds. AIAA J. Aircr. **54**(6), 2014–2026 (2017). https://doi.org/10.2514/1.C034025
3. Schütte, A., Marini, R.N.: Computational aerodynamic sensitivity studies for generic delta wing planforms. In: Dillmann, A., Heller, G., Krämer, E., Wagner, C., Tropea, C., Jakirlić, S. (eds.) DGLR 2018. NNFMMD, vol. 142, pp. 338–348. Springer, Cham (2020). https://doi.org/10.1007/978-3-030-25253-3_33
4. Werner, M., Schütte, A., Weiss, S.: Turbulence model effects on the prediction of transonic vortex interaction on a multi-swept delta wing. In: AIAA Paper 2022-0295 (2022). https://doi.org/10.2514/6.2022-0295
5. Schütte, A., Hummel, D.: Numerical design studies on the roll-stability of a multi delta wing configuration. AIAA J. Aircr. **60**(3), 898–914 (2023). https://doi.org/10.2514/1.C037128
6. CentaurSoft: Centaur Tutorial Version 14.0 (2020). http://www.centaursoft.com
7. Schwamborn, D., Gerhold, T., Heinrich, R.: The DLR TAU-Code: Recent applications in research and industry. In: Proceedings of "European Conference on Computational Fluid Dynamics ECCOMAS CDF, Delft, The Netherland (2006). http://resolver.tudelft.nl/uuid:59162917-d24d-4f11-955f-feaeaa21e2c1
8. Spalart, P., Allmaras, S.: A one-equation turbulence model for aerodynamic-flows. In: AIAA Paper, pp. 92–0440 (1992). https://doi.org/10.2514/6.1992-439
9. Allmaras, S., Johnson, F., Spalart, P.: Modifications and clarifications for the implementation of the Spalart-Allmaras turbulence model. In: ICCFD7 Proceedings, ICCFD7-1902 (2012). https://www.iccfd.org/iccfd7/assets/pdf/papers/ICCFD7-1902_paper.pdf
10. Spalart, P., Shur, M.: On the sensitization of turbulence models to rotation and curvature. Aerosp. Sci. Technol. **1**(5), 297–302 (1997). https://doi.org/10.1016/S1270-9638(97)90051-1

Invariant Analysis of Vortical Delta Wing Flow Using the Extended Optimal Triple Tensor Decomposition

Markus Rütten[1(✉)], Michael Werner[1], and Jonas Zastrow[2]

[1] German Aerospace Center (DLR), Institute of Aerodynamics and Flow Technology, Bunsenstr. 10, 37073 Göttingen, Germany
markus.ruetten@dlr.de

[2] German Aerospace Center (DLR), Institute of Aeroelasticity, Bunsenstr. 10, 37073 Göttingen, Germany
http://www.dlr.de/as

Abstract. Vortices and vortex breakdown flow structures over a multi-delta wing configuration are often analyzed by applying flow field invariant analysis techniques from critical point theory and flow topology research. In particular the three invariants of the velocity gradient tensor are used to classify and analyze flow fields, mainly for the assignment of vortical flow regions and assessment of their stability. In this work, the extended optimal triple tensor decomposition is applied to the vortical flow field delivering new tensor fields which allow to recalculate the rate of strain and rotation tensors in an associated basic reference frame. These new tensor fields contain separated distinct flow properties. After retransformation into the original coordinate system, flow field invariant analysis techniques applied to these newly derived tensor fields provide further insights into the dynamics of vortical flows over delta wings.

Keywords: vortical flow field analysis · invariants · tensor decomposition

1 Introduction

Agility, maneuverability and speed performance of new combat aircraft are realized by low-aspect-ratio highly swept wings. Such aircraft designs have to meet challenging aerodynamic requirements, since they have to operate in subsonic, transonic and supersonic speeds. The associated flow topology features leading-edge flow separation throughout most flight regimes. This results in large-scale vortices over major parts of the aircraft suction side. There, pressure is lowered by high circumferential velocities surrounding dominant primary vortices, which, eventually, generates a great amount of locally concentrated lift. In order to further increase lift, modern combat aircraft planforms feature several leading edges and, hence, create multiple leading edge primary vortices. However, maneuvers lead to complicated vortex-vortex and, at transonic Mach numbers, vortex-shock

A. Dillmann et al. (Eds.): STAB/DGLR Symposium 2022, NNFM 154, pp. 647–657, 2024.
https://doi.org/10.1007/978-3-031-40482-5_61

interactions [6]. Below a planform-dependent angle of attack (AoA) the vortical flow system remains stable. However, as AoA increases, unsteady flow behavior and flow instabilities develop. Eventually, the so-called vortex breakdown phenomenon occurs and moves up-stream over the suction side [13] of the aircraft. Vortex breakdown can be described as a sudden widening of the vortex itself, and it is, in general, associated with a reversed flow and strong fluctuations within the vortex core. As a consequence, the vortex-induced lift fluctuates heavily and even subsides throughout the process of vortex breakdown. As this phenomenon has a major impact on the stability and control characteristics, its dynamics are of special interest in the current research. Thus, it is of crucial importance to understand the dynamics of vortical flow fields, beginning with the definition of a vortex and the physical conditions it has to fulfill. Additionally, the main dynamic flow characteristics, such as their genesis, growth and eventual breakdown, are considered. Therefore, this work briefly reminds some basic aspects of Vortex Dynamics: it begins with methods to define and identify vortices. Next, the definitions of the three basic invariants of the velocity gradient tensor and their physical meaning in regard to vortical flows are introduced. Thereafter, a focus will lay on the optimal triple tensor decomposition as a new vortex criterion, which also allows the formulation of new tensor invariants for the resulting reduced sub-tensor fields. After the flow physical part, a technical part will follow in which the implementation of the computational algorithms and the numerical simulation setup are explained. Thereafter, the new flow field analysis technique is applied to the numerical flow simulation data set, and the analysis results will be presented and discussed within the context of vortical flows.

2 Aspects of Vortex Dynamics

Although the term vortex is often used by most of the fluid dynamic scientists in a general sense, a precise mathematical description is difficult and not complete in the sense of describing all features of this kind of rotational fluid. Lugt [14] wrote about the dilemma of defining a vortex, and he proposes following definition of a vortex: *A vortex is the rotating motion of a magnitude of material particles around a common center.* This intuitive definition, which describes the visual observations in nature, is incomplete, since not being Galilean invariant to moving reference frames. The Lagrangian perspective is another difficulty to deal with. The approach followed here is to assign a flow region as vortex by using local pointwise (fluid-element wise) flow field criteria. In fact, the lack of a general vortex definition has led to the development of various vortex identification schemes which have been rather successful in practice, i.e. the Lambda2 definition [9], the invariant Q criterion and the associated discriminant condition, the kinematic vorticity number [16], the helicity model [15] and a lot more. No model proved to be perfect, and open questions still remain i.e. when considering the objectivity of a vortex definition. Hence, in the last years there was a fruitful discussion about local type of vortex definitions [3] and the inherent constraint of objectivity that a vortex definition hast to fulfill [7]. Nevertheless, the problem of defining a vortex is still not solved in detail, because essential flow features are not decomposed correctly, i.e. shearing flow over curved walls is incorrectly assigned to vortical flows. Here, the optimal triple tensor decomposition comes into play whose extension is proposed to analyze vortical flows.

2.1 The Extended Optimal Triple Tensor Decomposition

The optimal triple tensor decomposition (TDM) was originally developed by Kolár [11] in order to identify vortices in complicated flow fields. Now, the idea is to apply an extended version of it on the gradient of the velocity field, in order to recalculate the rate of strain and the rate of rotation tensor in a new associated basic reference frame (BRF), thereby fulfilling the physical constraint of objectivity. The proper re-orientation of those tensors in the new basic reference coordinate system allows distinguishing between rotational, effective shear and, finally, elongational fluid motion. Hence, this approach delivers new flow physical insights into the dynamics of the flow fields. The resulting tensor fields can then be used in the context of tensor invariant analysis so that their specific properties can be assigned to specific characteristics of vortical motion.

Kolář transforms the velocity gradient tensor into the BRF by iteratively applying rotation-tensor operations to the original coordinate system,

$$\nabla \mathbf{v}|_{BRF} = \mathbf{Q}\; grad\; \mathbf{v}|_{xyz}\; \mathbf{Q}^T, \tag{1}$$

until maximal effective shear is composed in $\nabla \mathbf{v}|_{BRF}$. Thereto, the components of the rotation tensor $\mathbf{Q}$ have to be determined by the following maximization:

$$\begin{aligned}&(\|D_{12}\Omega_{12}\| + \|D_{23}\Omega_{23}\| + \|D_{31}\Omega_{31}\|)_{BRF}\\ &= max\,(\|D_{12}\Omega_{12}\| + \|D_{23}\Omega_{23}\| + \|D_{31}\Omega_{31}\|)_{all}\,,\end{aligned} \tag{2}$$

in which the tensor components are associated to the tensors $\mathbf{D}$ and $\boldsymbol{\Omega}$, the symmetric and asymmetric parts of $\nabla \mathbf{v}$. Which algorithm is best to determine the final rotation tensor $\mathbf{Q}$ maximizing the shearing part of $\nabla \mathbf{v}|_{BRF}$ is point of discussion. We developed a novel approach which is explained in Sect. 2.3. Further hints to the geometrical interpretation can be found in [12]. The essential point is, that within this BRF it is possible to further decompose and reformulate the resulting rotation and strain parts into novel pure shear and elongational flow-depending straining and rotational terms. Thereto, the components of the transformed velocity gradient tensor are used to calculate the so-called residual tensor $\mathbf{RES}|_{BRF}$:

$$\mathbf{RES}|^{BRF} = \begin{pmatrix} u_x & (sign\; u_y)min(|u_y|,|v_x|) & (sign\; u_z)min(|u_z|,|w_x|)\\ (sign\; v_x)min(|u_y|,|v_x|) & v_y & (sign\; v_z)min(|v_z|,|w_y|)\\ (sign\; w_x)min(|u_z|,|w_x|) & (sign\; w_y)min(|v_z|,|w_y|) & w_z \end{pmatrix} \tag{3}$$

From the residual tensor a new tensor $\nabla \mathbf{v}^{SH}|_{BRF}$, containing only shear-rate related components (also possibly rotated), can be split off:

$$\nabla \mathbf{v}^{SH}|_{BRF} = \nabla \mathbf{v}|_{BRF} - \mathbf{RES}|_{BRF}. \tag{4}$$

The residual tensor can be further decomposed into a pure-rotation tensor and a strain tensor, in which only non-rotation related elongational respectively extensional rate of strain components occur:

$$\mathbf{RES}|_{BRF} = \frac{1}{2}\left[\mathbf{RES}|_{BRF} - \mathbf{RES}^{T}|_{BRF}\right] + \frac{1}{2}\left[\mathbf{RES}|_{BRF} + \mathbf{RES}^{T}|_{BRF}\right]. \quad (5)$$

This can be rewritten to:

$$\mathbf{RES}|^{BRF} = \mathbf{D}^{EL}|_{BRF} + \boldsymbol{\Omega}^{EL}|_{BRF}, \quad (6)$$

a decomposition, which Kolář utilized for identification of vortices, concentrating on the $\boldsymbol{\Omega}^{EL}|_{BRF}$ term only; this corresponds in his notation to $\boldsymbol{\Omega}^{vortex}|_{BRF}$ representing a rigid vortical rotation. For our interest, the tensor $\mathbf{D}^{EL}|_{BRF}$ in Eq. (6) is important since it contains the non-rotating elongation rates only.

The tensor $\nabla\mathbf{v}^{SH}|_{BRF}$ in Eq. (4) contains the effective shear rates but still includes rotational parts within its strain rate components. Thereto, a further Cauchy-Stokes decomposition is applied to this tensor, leading to an asymmetric rate of rotation tensor $\boldsymbol{\Omega}^{SH}|_{BRF}$ and a symmetric rate of shear tensor $\mathbf{D}^{SH}|_{BRF}$:

$$\nabla\mathbf{v}^{SH}|_{BRF} = \mathbf{D}^{SH}|_{BRF} + \boldsymbol{\Omega}^{SH}|_{BRF}. \quad (7)$$

Then, the extended optimal triple tensor decomposition takes the following form:

$$\nabla\mathbf{v}|_{BRF} = \mathbf{D}^{SH}|_{BRF} + \boldsymbol{\Omega}^{SH}|_{BRF} + \mathbf{D}^{EL}|_{BRF} + \boldsymbol{\Omega}^{EL}|_{BRF}. \quad (8)$$

Now, four tensors are available to feed the invariant analysis of the flow field for the multi delta-wing application. Eventually, the whole decomposition in Eq. (8) can be utilized to gain more information about the flow physical mechanisms i.e. of vortex interaction or of vortex breakdown by analyzing shearing and elongational straining of strain and vorticity in the BRF separately. Of course, this could be done in the BRF, but with regard to the aircraft flow system it is useful to transform the tensors in Eq. (8) back into the original coordinate system which facilitates interpretation and comparison of the results with other vortex describing methods.

2.2 Flow Field Invariants in Vortex Dynamics

One objective of the topological analysis in Fluid Dynamics is to gain information about the state and the dynamics of the flow system. The identification and classification of topological elements in the velocity field $\mathbf{v}$ and further associated flow fields deliver an overview of significant flow structures. In particular in Vortex Dynamics flow invariant analysis techniques from critical point theory and flow topology research are applied to in order to analyze the occurrence of vortices, their genesis and eventually their decay. Invariants of the velocity gradient tensor $grad\,\mathbf{v}$ play a major role in classification and analysis of flow fields. In particular, the three invariants P, Q and R of the associated field tensors can be linked to the tensor eigenvalues λ_i and associated eigenvectors, which describe rate and direction of information spreading. They are associated with physical properties of the flow field.

The first invariant, P, is determined as follows [4]:

$$P = Tr(\nabla \mathbf{v}) = \nabla \cdot \mathbf{v} = D_{ii} \, , \tag{9}$$

whereas D_{ii} are the diagonal elements of the rate of strain tensor $\mathbf{D}$. Following Einstein notation, indices are used as a shortage for the calculation of the sum of the diagonal elements. Since this sum represents the divergence of the velocity field, which vanishes for an incompressible flow, the invariant P delivers only for compressible flows further flow information.

The second invariant, Q, can be traced back to the rate of strain tensor, $D_{ij} = \frac{1}{2}\left(v_{i,j} + v_{j,i}\right)$, and the rate of rotation tensor, $\Omega_{ij} = \frac{1}{2}\left(v_{i,j} - v_{j,i}\right)$:

$$Q = \frac{1}{2}\left[P^2 - Tr\left[(\nabla \mathbf{v})^2\right]\right] = \frac{1}{2}\left(P^2 - D_{ij}D_{ji} - \Omega_{ij}\Omega_{ji}\right) \, . \tag{10}$$

Bearing in mind the problem to formulate a local fluid-element-based definition of a vortex, then the following interpretation of Eq. 10 is relevant: the invariant Q balances the rotational and shear part of the fluid element motion. Therefore, the following vortex definition can be proposed: The condition for a local fluid element belonging to a vortical region is that its local rotational motion exceeds its shearing motion, following i.e. Truesdell [16].

For the third invariant R the following holds:

$$\begin{aligned} R = Det(\nabla \mathbf{v}) &= \frac{1}{3}\left[-P^3 + 3PQ + tr\left[(\nabla \mathbf{v})^3\right]\right] \\ &= \frac{1}{3}\left(-P^3 + 3PQ + D_{ij}D_{jk}D_{ki} + 3\Omega_{ij}\Omega_{jk}D_{ki}\right) \, . \end{aligned} \tag{11}$$

Obviously, the third invariant features the most complex structure, consisting of, firstly, a cubed divergence term, secondly, a mixed term of first and second invariant, and, thirdly, a measure for the rate of strain and, at last, a measure of the rate of rotation. Both latter terms act as enforcement or weakening of the strain rate. They also indicate strain and vorticity stretching.

The splitting of the second and third invariant into terms, in which rate of shear and rotation tensors occur separately, enables a deeper flow analysis, when the results of the extended optimal triple decomposition are taken into account. It allows analyzing the impact of rate of shear and rotation tensors specifically by investigating:

$$Q^* = -\frac{1}{2}\left(D^*_{ij}D^*_{ji} + \Omega^*_{ij}\Omega^*_{ji}\right) \, , \quad R^*_{\mathbf{D}} = D^*_{ij}D^*_{jk}D^*_{ki} \, , \quad R^*_{\mathbf{\Omega}} = \Omega^*_{ij}\Omega^*_{jk}D^*_{ki} \, , \tag{12}$$

where the * indicates that the shear (SH) or elongation (EL) related tensors of Eq. (8), transformed back into the original coordination system, are used.

2.3 Implementation

In order to perform the TDM, the approach outlined in the previous section was implemented in Python. While the decomposition of the velocity gradient tensor itself is straightforward, the identification of the BRF, where the decomposition is actually valid, is computationally very expensive. For each point in the computational mesh Eq. (2) has to be satisfied. Based on symmetry considerations, each of the three Euler angles can vary between $0 \leq \theta < \pi$. Therefore, a brute-force approach would require $(\pi/\Delta\theta)^3$ rotations of the velocity gradient tensor per point to find the optimum, where $\Delta\theta$ is the step size of the Euler angles. As this would lead to very high computational requirements, Boukharfane et al. [1] proposed an approach, where the TDM is seen as an optimization problem. In this paper, a similar approach is used. However, as the condition outlined in Eq. (2) leads to an optimization problem with many local maxima, a two-step approach, outlined in algorithm 9, is used in the current work. In a first step, a limited number of iterations of a global optimization algorithm, the so-called *DIRECT* algorithm [10] is used to get an estimation of the location of the global maximum, see Algorithm 1. The choice of the number of iterations is a compromise between computational effort and accuracy of the estimation of the global maximum. For the current study, 3000 iterations were performed for each point. In a second step, the resulting estimation is used as a starting point for a gradient-based optimizer, the so-called *Nelder-Mead* algorithm [5], to further refine the estimation and find the global optimum. Both the *DIRECT* and the *Nelder-Mead* algorithm use the same objective function, which is shown in Algorithm 2. The result of the optimization process is a combination of Euler angles, $\theta_{1|BRF}, \theta_{2|BRF}$ and $\theta_{3|BRF}$, that are used to transform the velocity gradient tensor from the original reference frame into the BRF. As the determination of the optimal rotation angles for each point is independent of any other point in the mesh, the algorithm outlined above can easily be parallelized to further speed up the computations. Within the scope of the presented work this novel promising theoretical framework is applied to a flow simulation data set generated for a DLR multi delta-wing model.

Algorithm 1. Determination Basic Reference Frame

---

procedure FindBRF
    $N \leftarrow$ number of points in the mesh
    $\theta_1, \theta_2, \theta_3 \leftarrow$ Euler angles, bounded $[0, \pi)$
    while $n < N$ **do** ▷ loop over all points in the mesh
        $\nabla \mathbf{v}_n \leftarrow$ velocity gradient tensor at current point
        $\theta_1|_D, \theta_2|_D, \theta_3|_D \leftarrow$ DIRECT($\nabla \mathbf{v}_n$) ▷ get estimate of optimal angles via DIRECT algorithm
        $\theta_1|_{BRF}, \theta_2|_{BRF}, \theta_3|_{BRF} \leftarrow$ Nelder-Mead($\nabla \mathbf{v}_n$, $\theta_1|_D, \theta_2|_D, \theta_3|_D$)
    end while
end procedure

Algorithm 2. Definition of the objective function

function OBJFUN($\nabla\mathbf{v}|_{xyz}, \theta_1, \theta_2, \theta_3$)
  $\nabla\mathbf{v}|_{rot} \leftarrow \mathbf{Q}(\theta_1, \theta_2, \theta_3)\ \nabla\mathbf{v}|_{xyz}\mathbf{Q}(\theta_1, \theta_2, \theta_3)^T$ ▷ rotate gradient tensor
  $\mathbf{D} \leftarrow \frac{1}{2}(\nabla\mathbf{v}|_{rot} + \nabla\mathbf{v}^T|_{rot})$
  $\boldsymbol{\Omega} \leftarrow \frac{1}{2}(\nabla\mathbf{v}|_{rot} - \nabla\mathbf{v}^T|_{rot})$
return $(|D_{12}\Omega_{12}| + |D_{23}\Omega_{23}| + |D_{23}\Omega_{23}|)\,/\,\|\nabla\mathbf{v}|_{xyz}\|$
end function

3 Flow Field Data

The invariant analysis is applied to the vortical flow field around the DLR-F23 combat aircraft wind tunnel model, which is depicted in Fig. 1. A sketch of the geometry can be found in Fig. 1a. Its planform is derived from a 45°/75° double delta design, and includes a large diamond fillet, as well as a chined ogival forebody. With regard to the profiles, all sections are elliptic and merge either into a cropped trailing edge or a constant leading edge with a radius of 0.5 mm. With regard to the configuration and numerical set-up, it is very similar to and described in detail in [17]. One major difference is, that the grid design for this work was done with "Centaur" by CentaurSoft. In particular, the tetrahedral elements above the models suction side feature an edge length of 1.5 mm or 0.365% of MAC, which results in approximately 33×10^6 elements for the entire grid. The employed CFD solver is the DLR TAU-code in implicit hybrid-RANS-LES mode. The presented analysis method is first tested in a subsonic scenario. Thus, an inflow Mach number of 0.65 and an angle of attack of $\alpha = 18°$ are chosen. The time step size of $\Delta t = 1.25 \times 10^{-6}$ s relates to the tetrahedral size in the vortex cores, where the highest velocities and highest pressure fluctuations are present. These vortices can also be identified in the surface pressure distribution shown in Fig. 1b.

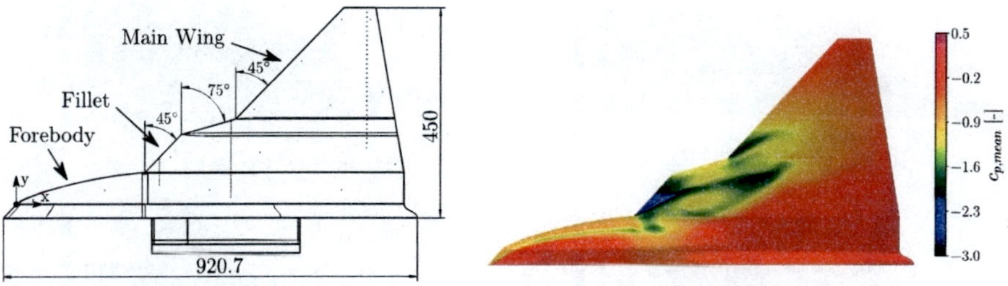

(a) Sketch of the DLR-F23 geometry (b) c_p distribution, $Ma = 0.65, \alpha = 18°$

Fig. 1. Sketch of the DLR-F23 model and surface pressure distribution

4 Vortical Flow Invariant Analysis

An impression of the vortical flow topology for the current test case is given in Fig. 2. The visualized structures always depend on the choice of the isosurface value, with higher values filtering out larger parts of the structures. As there is no objectively correct choice of the isosurface value, a value of 1 was used throughout the paper. In Fig. 2a the classical Q-criterion is visualized. It shows three distinct primary vortices, a forebody vortex, fillet vortex and main wing vortex. The forebody and fillet vortex are intertwining, while the main wing vortex is already beginning to break down. Additionally, an isosurface of the third invariant R is depicted in Fig. 2b, showing a much more complex structure, indicating strain and vortex stretching regions. In Fig. 3, the norms of the four different parts of the decomposed velocity gradient tensor, based on Eq. (8), are visualized using isosurfaces. As can be seen in Fig. 3a and 3b, the norms of the strain rate and vorticity tensors of $\nabla \mathbf{v}^{SH}$ are identical and closely resemble the structures shown

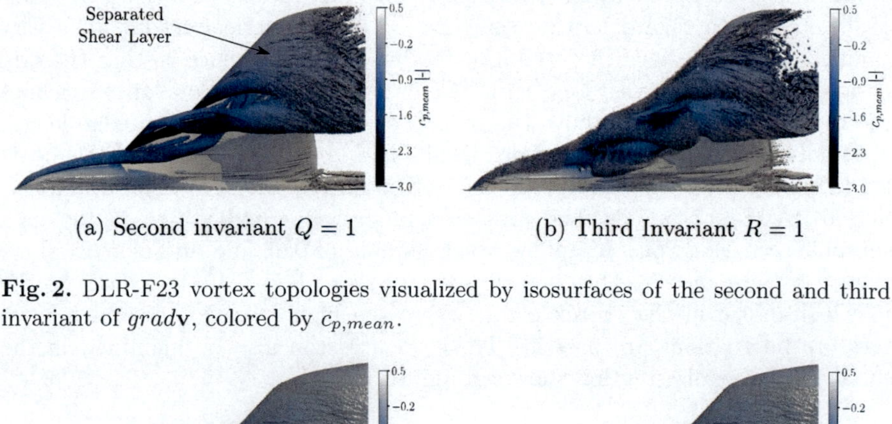

(a) Second invariant $Q = 1$ (b) Third Invariant $R = 1$

Fig. 2. DLR-F23 vortex topologies visualized by isosurfaces of the second and third invariant of $grad\mathbf{v}$, colored by $c_{p,mean}$.

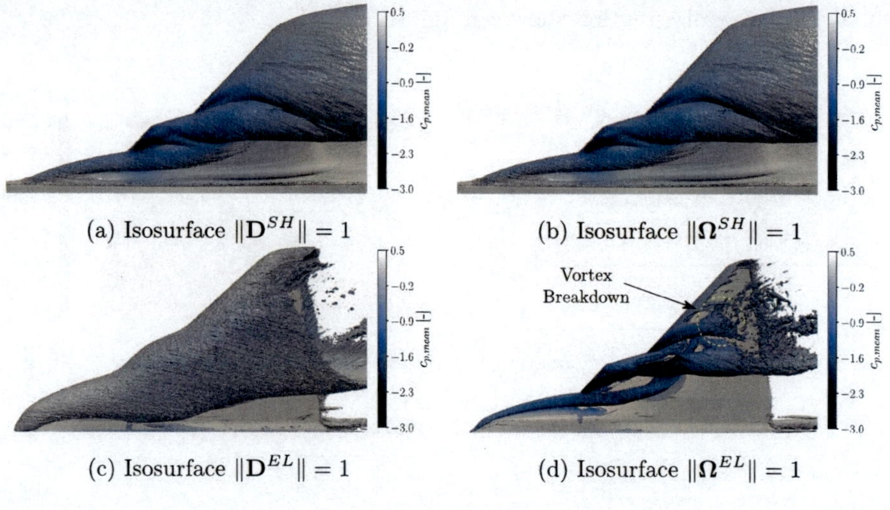

(a) Isosurface $\|\mathbf{D}^{SH}\| = 1$ (b) Isosurface $\|\mathbf{\Omega}^{SH}\| = 1$

(c) Isosurface $\|\mathbf{D}^{EL}\| = 1$ (d) Isosurface $\|\mathbf{\Omega}^{EL}\| = 1$

Fig. 3. Visualization of vortical flow structures above the DLR F23 delta-wing based on different parts of $\nabla \mathbf{v}$, colored by $c_{p,mean}$.

in Fig. 2a. In contrast, the structure of $\|\mathbf{D}^{EL}\|$, shown in Fig. 3c, completely envelops the vortices, with highest values found in the region surrounding the vortex cores, whereas in the core itself and far from it only low levels of $\|\mathbf{D}^{EL}\|$ are found. Finally, Fig. 3d shows $\|\boldsymbol{\Omega}^{EL}\|$, which corresponds to the rotational part of $\nabla\ \mathbf{v}^{EL}$. Now, nearly-rigid-body vortex regions are well separated from shear and elongational flow. Compared to the traditional vortex criterion shown in 2a, it offers a much more accurate representation of the vortex topologies, as non-rotational shear layers, for example at the main wing leading edge or downstream of the vortex breakdown location, are filtered out. To analyze the vortex topology further, it is possible to specifically look at the components of the second and third invariants of $\nabla\mathbf{v}^{SH}$ and $\nabla\mathbf{v}^{EL}$, which have been introduced in Eq. (12). For $\nabla\mathbf{v}_{SH}$ it is observed that $Q^{SH} = 0$, as $D_{ij}^{SH} D_{ji}^{SH} = -\left(\Omega_{ij}^{SH} \Omega_{ji}^{SH}\right)$ in the whole domain. The structure of Q^{EL} is not shown, as it is qualitatively very similar to the one shown in Fig. 3d, which confirms Q^{EL} as vortex criterion. Now shearing and elongational associated components of the third invariant are shown in Fig. 4. The shear related parts of the third invariant of $\nabla\mathbf{v}^{SH}$ are visualized in Fig. 4a and 4b, resembling the topology of $\|\mathbf{D}^{SH}\|$ and $\|\boldsymbol{\Omega}^{SH}\|$ shown in Fig. 3, their structures are focused on the region immediately around the vortex cores. There, shearing flow is intensified by shearing and vorticity respectively. In Fig. 4c and 4d isosurfaces of the third invariant of $\nabla\mathbf{v}^{EL}$ are shown. Here, $R_{\mathbf{D}}^{EL}$ is focused on the wider area around the vortices, depicting the self-intensifying elongation rates, whereas $R_{\boldsymbol{\Omega}}^{EL}$ is focused completely on vortex core regions reflecting the elongational flow structure intensification by vorticity.

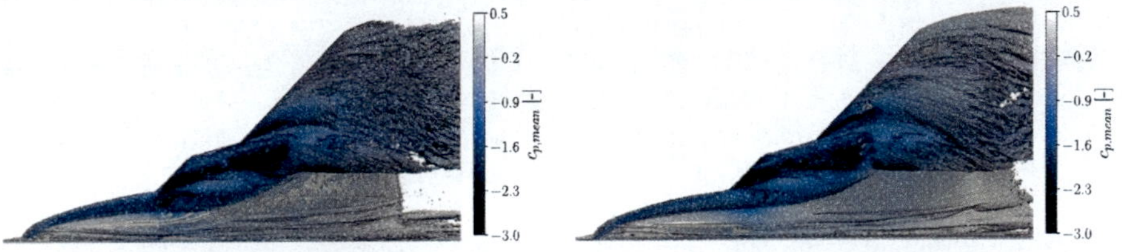

(a) Isosurface $R_{\mathbf{D}}^{SH} = 1$ (b) Isosurface $R_{\boldsymbol{\Omega}}^{SH} = 1$

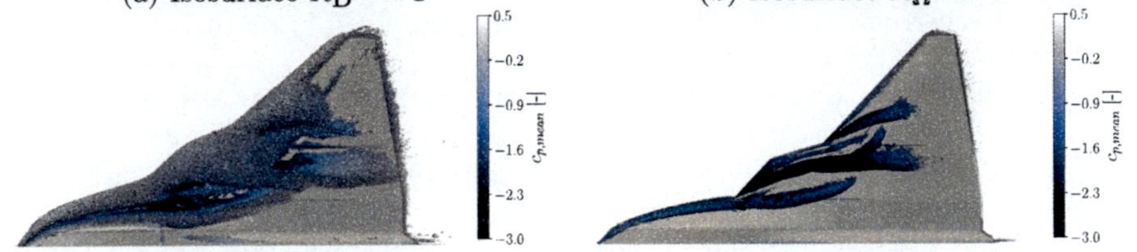

(c) Isosurface $R_{\mathbf{D}}^{EL} = 1$ (d) Isosurface $R_{\boldsymbol{\Omega}}^{EL} = 1$

Fig. 4. Visualization of vortical flow structures above the DLR F23 delta-wing based on the extended triple decomposition of *grad* $\mathbf{v}$, colored by $c_{p,mean}$.

5 Conclusion

In this work it has been demonstrated that the extended optimal triple decomposition applied to the complicated vortical flow fields over a multi-delta wing aircraft configuration appears to be a suitable vortex-identification method, allowing deeper flow analysis. Due to its ability to naturally decompose fluid motion into straining, shearing, and rigid-body rotation, it provides additional flow physical insights into the flow phenomena of developing and interacting vortices compared to the classical Q criterion. Previous implementations of the optimal triple decomposition suffered from very high computational effort to identify the basic reference frame needed for the decomposition. The current approach overcomes the difficulty by using a combination of global and local optimization algorithms. These well performing algorithms make the discussed novel invariant analysis possible, even for huge CFD data sets.

Acknowledgments. The authors gratefully acknowledge that this work could be conducted within the project "DIABOLO" funded by DLR.

References

1. Boukharfane, R., Er-raiy, A., Alzaben, L., Parsani, M.: Triple decomposition of velocity gradient tensor in compressible turbulence. Fluids **6**, 98 (2021)
2. Breitsamter, C.: Unsteady flow phenomena associated with leading-edge vortices. Progr. Aerosp. Sci. **44**, 48–65 (2008)
3. Chakraborty, P., Balachandar, S., Adrian, R.J.: On the relationships between local vortex identification schemes. J. Fluid Mech. **535**, 189–214 (2005)
4. Chong, M.S., Perry, A.E., Cantwell, B.J.: A general classification of three-dimensional flow fields. Phys. Fluids A **2**(5), 261–265 (1990)
5. Gao, F., Han, L.: Implementing the Nelder-Mead simplex algorithm with adaptive parameters. J. Comput. Optim. Appl. **51**, 259–277 (2012)
6. Gursul, I., Gordnier, R., Visbal, M.: Unsteady aerodynamics of nonslender delta wings. Progr. Aerosp. Sci. **41**, 515–557 (2005)
7. Haller, G., Hadjighasem, A., Farazmand, M., Huhn, F.: Defining coherent vortices objectively from the vorticity. J. Fluid Mech. **795**, 136–173 (2016)
8. Hunt, J.R.C., Wray, A.A., Moin, P.: Eddies, streams and convergence zones in turbulent flows. Center of Turbulence Research Report CTR-S88, pp. 193–208 (1988)
9. Jeong, J., Hussain, F.: On the identification of a vortex. J. Fluid Mech. **285**, 69–94 (1995)
10. Jones, D.R., Perttunen, C.D., Stuckman, B.E.: Lipschitzian optimization without the Lipschitz constant. J. Comput. Optim. Appl. **79**, 157–181 (1993)
11. Kolář, V.: Vortex identification: new requirements and limitations. Int. J. Heat Fluid Flow **28**, 638–652 (2007)
12. Kolář, V., Šístek, J., Fehmi Cirak, F., Moses, P.: Average corotation of line segments near a point and vortex identification. AIAA J. **51**(11), 2678–2694 (2013)
13. Lambourne, N.C., Bryer, D.W.: The bursting of leading-edge vortices - some observations and discussion of the phenomenon. ARC R & M, No. 3282 (1961)

14. Lugt, H.J.: The dilemma of defining a vortex. In: Müller, U., Roesner, K.G., Schmidt, B. (eds.) Theoretical and Experimental Fluid Mechanics, pp. 309–321. Springer, Heidelberg (1979). https://doi.org/10.1007/978-3-642-67220-0_32
15. Moffat, H.K., Tsinober, A. (ed.): Topological Fluid Mechanics. Proceedings of the IUTAM Symposium in Cambridge (1989)
16. Truesdell, C.: The Kinematics of Vorticity. Indiana University Press (1954)
17. Zastrow, J., Probst, A.: Vortex topology mode alternation in hybrid RANS-LES simulations of transonic flow over the DLR-F23 combat aircraft wind tunnel model. In: Stab Symposium (2022)

Numerical Investigation of the Unsteady Behaviour of Side Jet Induced Control Concepts

Christian Schnepf(✉)

Institute of Aerodynamics and Flow Technology, German Aerospace Center (DLR), Bunsenstraße 10, 37073 Göttingen, Germany
Christian.Schnepf@dlr.de

Abstract. In this study, the performance of side jet control for two different altitudes (10 km/30 km) was numerically investigated and compared with the effectiveness of secondary injection thrust vector control for a supersonic missile at M = 6.2 and 7.0. This comparison also considered the behavior of the transient force signal caused by the side jet thrust increasing with time, resulting in a time delay between reaching the maximum side jet thrust and reaching the maximum control force. It was shown that, in contrast to thrust vector control, the performance of side jet control decreases with increasing altitude as a result of a less effective shock wave boundary layer interaction. With respect to this interaction, it was found that the boundary layer state has a great influence on this interaction and affects the induced force to a certain extent.

Keywords: Missile control · side jet control · SWBLI · Transition · CFD

1 Introduction

Flight at high altitudes (upper tier) presents a particular challenge for conventional control concepts of high-agility missiles. The low air density in the upper tier and the associated low dynamic pressure have an unfavorable effect on the effectiveness of aerodynamic control surfaces. As a result, high agility missiles operating in this altitude range have specially adapted control methods, such as side jet control (SJC) [1,2] or thrust vector control [3]. The main interest of this study is SJC, which is investigated numerically. However, to evaluate its effectiveness, simulations are also performed for thrust vector control. For the latter, side jet induced thrust vector control (SITVC) is considered (Fig. 1). This concept works similarly to that of SJC. In both concepts, a (side) jet is injected transversely into a supersonic flow. The side jet acts as an obstacle to the supersonic flow and forces a change in flow direction which is accompanied by a bow shock. Near the wall, the bow shock is replaced by a shock system due to a shock wave boundary layer interaction (SWBLI). The flow topology can be similar to

A. Dillmann et al. (Eds.): STAB/DGLR Symposium 2022, NNFM 154, pp. 658–668, 2024.
https://doi.org/10.1007/978-3-031-40482-5_62

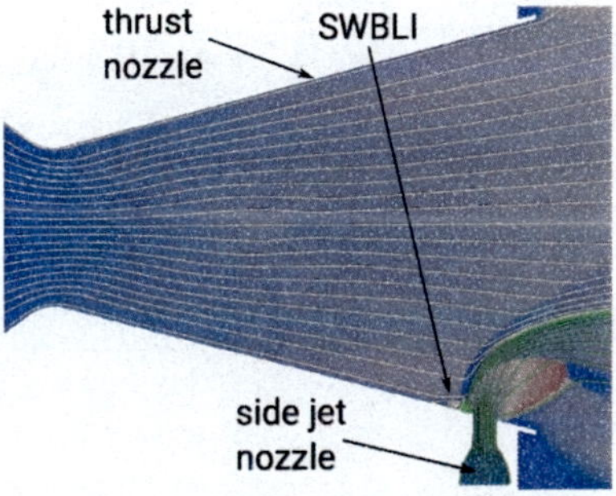

Fig. 1. Secondary injection thrust vector control (Flow in Symmetry plane).

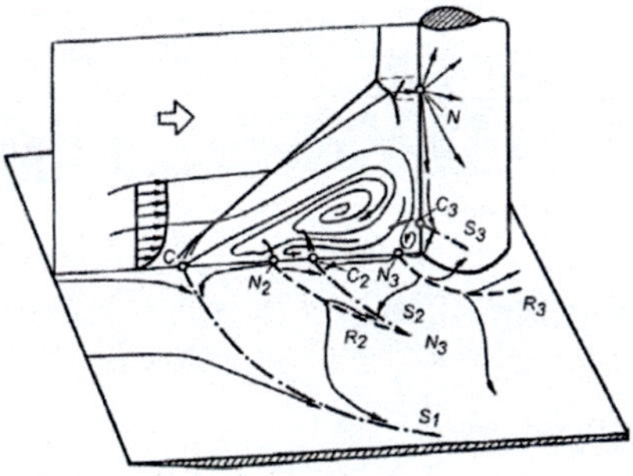

Fig. 2. Supersonic flow past a cylindrical obstacle on a flat plate [4].

the one of a 3D separation on a vertical cylinder (Fig. 2). Beneath the SWBLI, the surface plateau pressure [4] is significantly higher compared to the pressure of the undisturbed flow. Ideally, this will lead to an increase in the control force. In the worst case, however, this can also reduce the control force, as shown in [1].

In this study, the focus is not only on the effectiveness of the control methods, but also on the transient behavior of the control force when the side jet mass flow is ramped up from zero to the maximum mass flow, as it would occur in reality when the gas generator is started up. Exclusively for SJC, another focus under investigation is how the state of the boundary layer - laminar or turbulent - affects the force signal. Here it should be pointed out once again that the force signal is considerably influenced by the SWBLI and the SWBLI in turn is influenced by the state of the boundary layer [5]. The latter and the transient force behavior are important factors for the evaluation of the control concept with regard to the ability to initiate maneuvers rapidly. To the best of the author's knowledge, this has not yet been studied in the public literature. The investigation will be undertaken for a generic supersonic missile. The Simulations were performed using the DLR-TAU Code for two specific altitudes of 10 km and 30 km and a Mach number of $M = 7.0$ for the lower altitude and a Mach number of $M = 6.2$ for the higher altitude. These are representative flow conditions for such a missile and are intended to show how the flight altitude affects the control concepts.

2 Test Case and Numerical Setup

2.1 Test Case

The study was performed on a generic supersonic missile with three stages, only the first two having a thruster. Figure 3 shows the outer mold line of the missile designed in the German Aerospace Center's ITEM-FK project. The thrust nozzle of stage 1 is subject of this investigation for SITVC. Basic geometric data of the thrust nozzle and the altitude-independent flow conditions of the thrust nozzle flow are shown in Table 1. Figure 4 presents the geometry of the sonic side jet

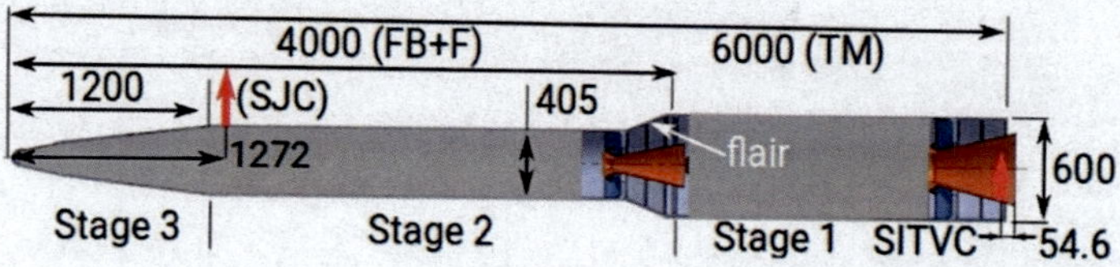

Fig. 3. Geometry of the generic three stage missile (Dimensions in mm).

Fig. 4. Geometry of the used Side-jet nozzle.

nozzle used for both control concepts. For SITVC actuation, such a nozzle is located 54.6 mm in the x-direction from the thrust nozzle exit. A geometrically identical nozzle is located at a distance of $\Delta x = 1272$ mm from the tip of the missile nose to study the effect of SJC. Reference values for the definition of the gas used in the side jet nozzles were taken from [2]. A common dimensionless number for characterising SJC is R_j. It is the ratio of the total pressure of the side jet to the static pressure of the ambient flow. Typically, R_j is in the three-digit range. This value was only varied in the SJC simulations for an altitude of 30 km. Table 2 shows the values for R_j and the properties of the side jet. SITVC and SJC are investigated for flow conditions presented in Table 3. The diameter $d_{Ref} = 405$ mm of the missile forebody was used to determine the Reynolds number and the reference area to define the aerodynamic coefficients ($A_{ref} = \pi\ d_{ref}^2/4$). The geometry from the nose to the flair of stage 2 is in the following referred to as FB, including flair as FB+F and the total missile as TM.

Table 1. Geometry definition and flow condition of the thrust nozzle (CC = Combustion Chamber, γ = heat capacity ratio, R = gas constant)

	Throat diameter	Exit diameter	Exit M	p_{0CC}	T_{0CC}	γ	R
	[m]	[m]	[–]	[bar]	[K]	[–]	[J/(kg K)]
thrust nozzle	0.158	0.410	2.82	40	3300	1.125	306

Table 2. Side-jet parameters at the inlet of the sonic nozzle

γ	R	T_s	p_{0CC}	$R_{j,10\,\mathrm{km}}$	$R_{j,30\,\mathrm{km}}$	ΔT_{QS}
[–]	[–]	[K]	[bar]	[–]	[–]	[s]
1.235	319	2151	100	377	8928	0.0050
1.235	319	2151	24	90	2130	0.0038
1.235	319	2151	4.2	16	377	0.0030

2.2 Numerical Setup

The DLR-TAU Code [6] with an AUSMDV flux vector splitting scheme was used for this study with the extension to model thermodynamic and transport properties for a non-reacting mixture of three thermally perfect gases (air, main thrust, side jet). TAU has already been used successfully in a study on the static behavior of SJC [7]. TAU itself is a hybrid structured/unstructured finite volume flow solver based on the compressible Reynolds-Averaged Navier-Stokes equations. A detailed description of TAU is given in [6]. The unsteady RANS simulations have been performed with the Spalart-Allmaras one-equation eddy viscosity model. Initially a fully turbulent boundary layer is assumed. Considering the results of [8], the assumption of a turbulent boundary layer within the thrust nozzle near the side jet nozzle is justified for the present combustor pressure and nozzle geometry. However, for the ogive nose of the missile, additional simulations were performed with a partial laminar boundary layer. Based on the data of [9,10], a laminar boundary layer may be present in the nose region at least for the higher altitude. In the partial laminar simulations, the boundary layer upstream from a position x_{TR} at the nose was defined as laminar and downstream as turbulent. The following transition positions x_{TR} were examined in this regard: 600 mm, 840 mm, 1015 mm and 1200 mm. Not for all x_{TR} a converged solution could be achieved. The convergence was depended on the altitude and on R_j. Figure 5 shows for which conditions converged solutions are available.

Table 3. Flow conditions of simulations

Altitude	p_∞	T_∞	M	Re
[km]	[Pa]	[K]	[-]	[-] (D_{Ref}=410 mm)
10	26500	223.25	7.0	26,404,992
30	1120	227.15	6.2	858,898

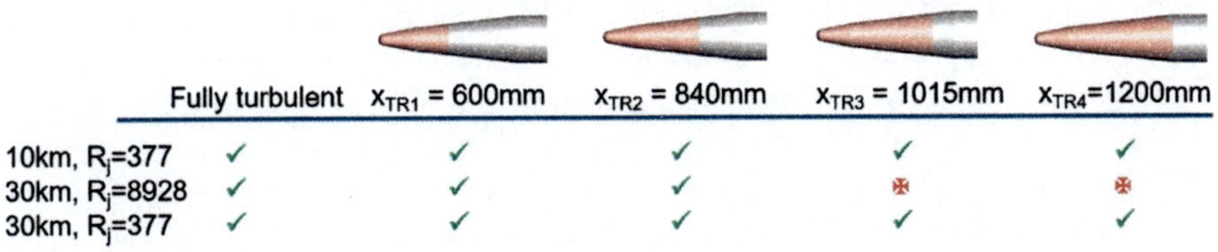

Fig. 5. Sensitivity of the simulation convergence to altitude and R_j. The pink color indicates where the boundary layer is defined laminar in the simulations.

The flows in the thrust nozzle and in the sonic side jet nozzle are initialized by an inflow boundary condition by specifying the total pressure p_{0CC} and the static temperature T_s. The simulation process is divided into two steps. First, a converged solution is generated without a side jet. Then, in the unsteady simulations, p_{0CC} is discretely increased from 0 to its maximum value within the time span ΔT_{QS} and kept there constant. With an increase of p_{0CC} the mass flow of the side jet also increases (Fig. 6). Different maximum values of p_{0CC} (R_j) were investigated for the higher altitude. A reduction of the maximum pressure leads to a decrease of ΔT_{QS} and R_j (see Fig. 6 and Table 2).

The initial physical time step of the unsteady RANS simulations was $\Delta t =$ 1e−5 s. A temporal convergence study was undertaken with additional time steps of 5e−6 s and 2.5e−6 s. The study showed that the characteristics of the solution were not affected by the time step. This is also true for the two fine meshes of a three stage mesh convergence study.

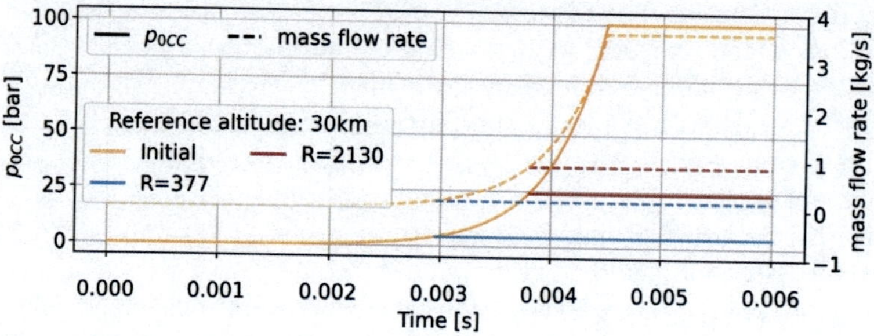

Fig. 6. Development of combustion chamber pressure and mass flow over time.

3 Results

This chapter presents the results of this study. First, the determined coefficients of the integral normal forces (control force) are compared. This is followed by the analysis of the flow field, in particular the analysis of SWBLI and its sensitivity on R_j, the flight altitude and the state of the boundary layer.

Figure 7 shows the normal force coefficient, which is composed of the side jet thrust and the pressure force induced by the SWBLI, for an altitude of 10 km. The pressure force coefficient is also shown in Fig. 7. A comparison of the force signals of the SJC for the forebody (SJC, FB) with respect to the influence of the boundary layer state on C_Z shows that the impact of transition position x_{TR} is rather small. The laminar simulations show only slightly smaller quasi-static pressure forces and thus lower total forces compared to the fully turbulent simulations. Although the difference for C_Z is small, the flow field induced by the SWBLI is significantly affected by transition, as shown below.

Regardless of the state of the boundary layer, a time delay exists between the time at which the maximum side jet mass flux is reached ($t = 0.0045$ s) and the time at which the final value of C_Z settles. The reason for this is that the SWBLI has not yet reached a quasi-steady state at $t = 0.0045$ s. For $t > 0.008$ s, however, the force signal exhibits only a small oscillation about a constant mean

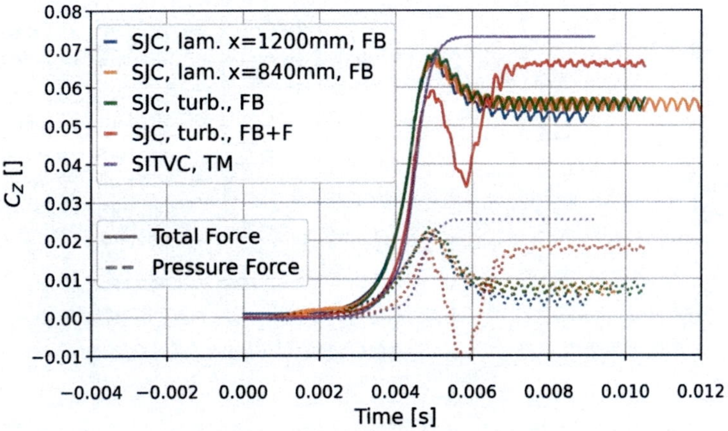

Fig. 7. Influence of the boundary layer condition, geometry and control system on the time-dependent normal force coefficient for an altitude of 10 km ($R_j = 377$).

value. The oscillation results from the unsteady characteristic of the SWBLI in its quasi-steady state, with a lower dominant frequency for the laminar data.

A comparison of C_Z of the forebody with (FB+F) and without flair (FB) for the turbulent case shows that the missile geometry downstream of the side jet can have a significant effect on the time history of C_Z and its final value. The SITVC also exhibits a delay between the time at which the final value of C_Z is reached and $t = 0.0045$ s. But the delay is smaller and the final static force is larger than in the case of SJC.

Figure 8 presents C_Z for an altitude of 30 km. At this altitude, the maxima of C_Z are much larger than for 10 km, but this is mainly an effect of the different dynamic pressures. For reference and to classify the values in comparison to Fig. 7, it is useful to know that the SITVC induces an identical force signal in Newtons at both altitudes. If the SITVC case is now considered as a reference, it becomes clear that SJC induces smaller total control forces at higher altitudes. This is equivalent to a drop in performance of the SJC with increasing altitude. It should be noted that the side jet for R = 8928 in Fig. 8 has the same properties as the one used for the simulations at 10 km altitude ($R_j = 377$). The reason for the drop in performance is that the contribution of the pressure force to the total force decreases significantly with increasing altitude. For an altitude of 30 km, an identical side jet induces a much larger separation than for 10 km (Fig. 9). As a result, the pressure distribution changes not only on the upper half of the missile on which the side jet emanates, but also significantly on the lower half of the missile which faces away from the side jet. This even leads to the fact that the c_p-distribution on the lower side more than compensates the favourable c_p-distribution on the upper side, resulting in an overall negative pressure force (Fig. 8). These adverse effects of SWBLI can be reduced by using a smaller R_j, which allows a positive pressure force to be generated even for 30 km. However,

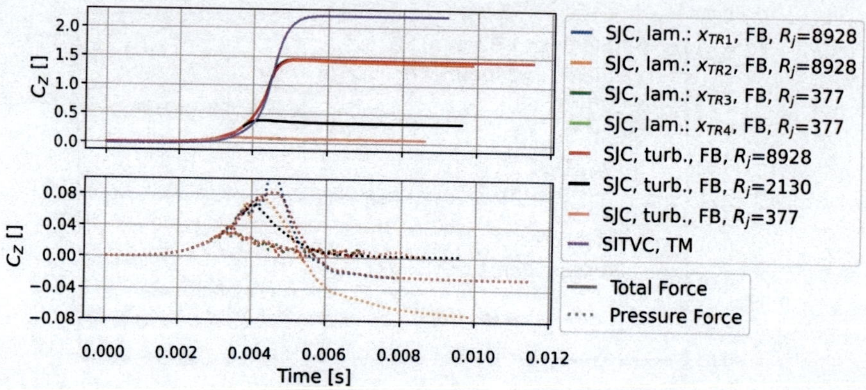

Fig. 8. Influence of the boundary layer condition, control system and R_j on the time-dependent normal force coefficient for 30 km.

the downside of a lower R_j is that the total control force decreases due to the lower thrust of the side jet itself (Fig. 8).

Another reason for the decreasing performance of SJC is the decreasing dynamic pressure with increasing altitude. This reduces the effectively achievable pressure force anyway. This is a similar effect to the performance drop of conventional aerodynamic control surfaces with increasing altitude.

Regarding the influence of boundary layer transition on the force signal, similar conclusions as for the 10 km case can be derived: the laminar simulations show a lower total force due to the unfavourable development of the SWBLI induced pressure force. Besides, and characteristic for 30 km, the pressure force signal of the laminar cases takes much longer to reach a quasi-static state.

Figure 9 shows the wall shear stress lines and the contour plot of the surface c_p-distribution in the area of the SWBLI. Analysing first the results of the turbulent simulations (upper three figures), the influence of the altitude and the ratio R_j becomes clearly visible. An increase in flight altitude with unchanged side jet properties leads to a significantly larger separation. Reducing the value of R_j to the same value as at 10 km altitude significantly reduces the size of the separation, but it remains larger than in the 10 km case[1]. It is noteworthy, however, that the overall flow topology is sustained and similar to that described in [4] and shown in Fig. 2. Regarding the influence of the boundary layer state on the flow topology, a comparison of plots 1 and 4 and 2 and 5 in Fig. 9 shows that the separation increases with increasing laminar region. This is a well-known correlation for general 2D cases, but has also been demonstrated for supersonic flows on a cone [5]. Furthermore, the complexity of the SWBLI flow topology increases when a laminar region is considered in the simulations.

Regarding the extent of the separation bubble normal to the wall, it can be seen in Fig. 10 that it is shallower in the laminar case than in the turbulent

[1] It should be noted that different Mach numbers were examined for different altitudes.

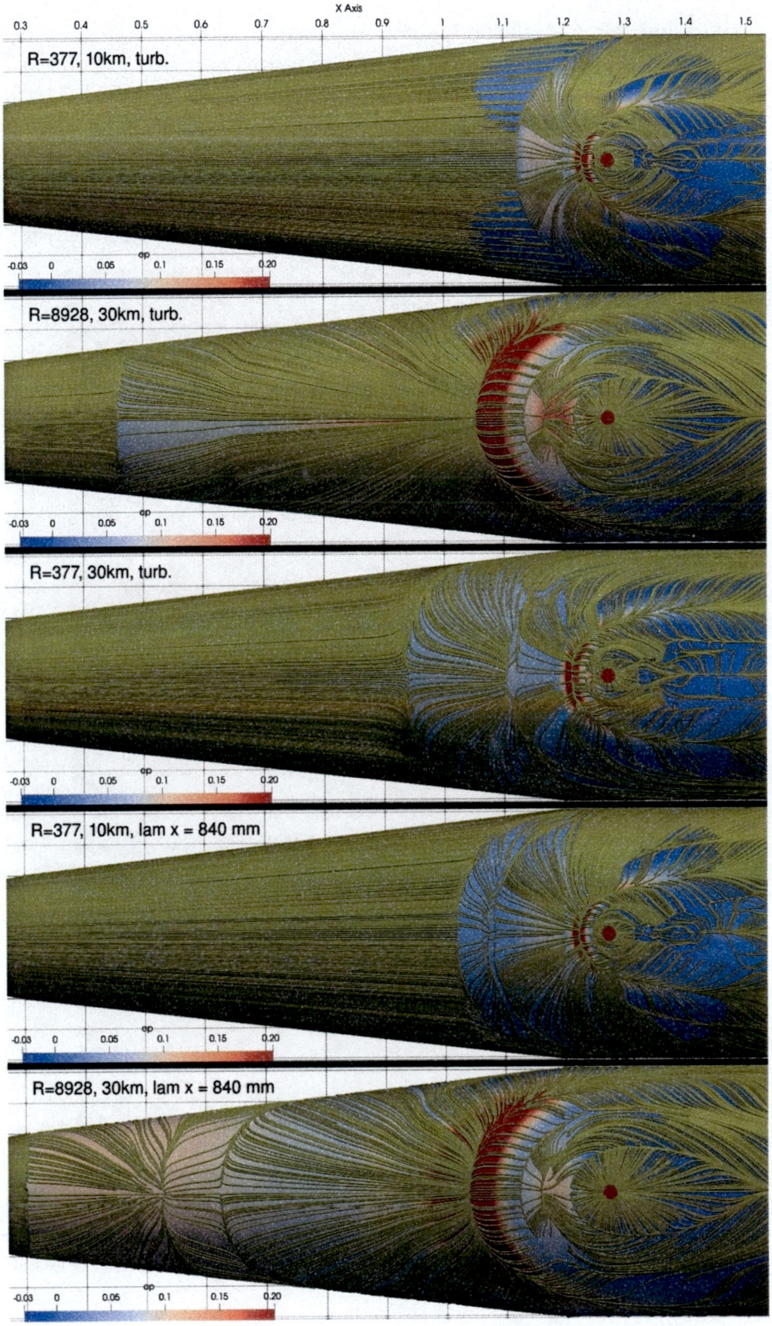

Fig. 9. Visualization of the flow on the surface with wall shear stress lines and c_p-distribution for quasi static conditions $t \gg \Delta T_{QS}$ (Scale in meter).

case. This results in less deflection of the supersonic flow and a weaker separation shock. Hence, when moving the transition further upstream, the plateau pressure beneath the separation alse decreases, compared to the fully turbulent case (see also [4,11]). This leads to an overall reduction in pressure force. A comparison of the two lower plots in Fig. 10 shows again that the SWBLI is significantly influenced by the altitude: Increasing size with increasing altitude.

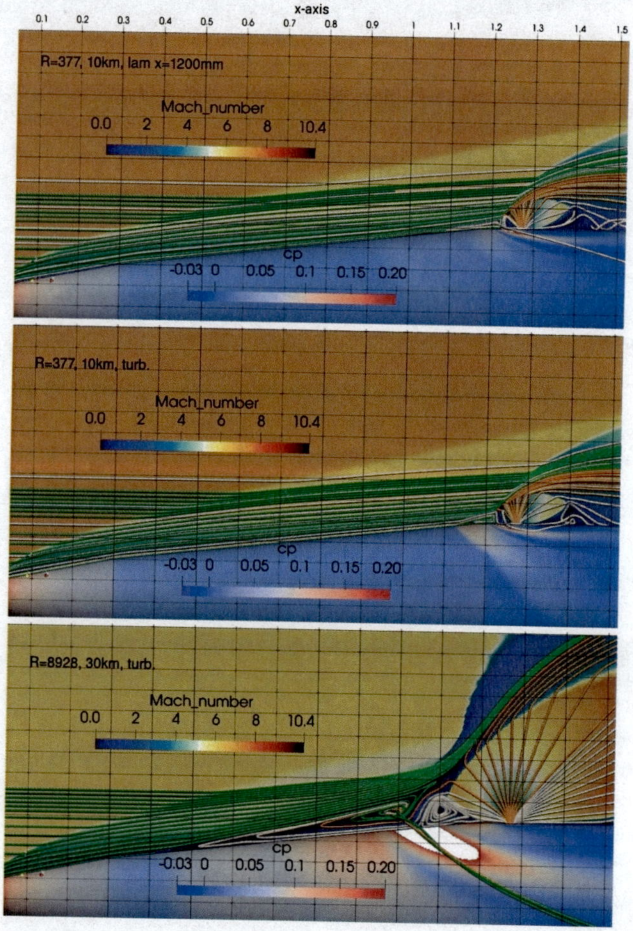

Fig. 10. Visualization of the SWBLI with streamlines, surface c_p-distribution, and a Mach contour plot in the symmetry plane for 10 km and 30 km for quasi static conditions $t \gg \Delta T_{QS}$ (Scale in meter).

4 Conclusion

In this study side jet control was numerically investigated for a supersonic missile for two altitudes (10 km/30 km) and altitude dependent Mach numbers of 7.0 and 6.2. In the study, the flow properties of the side jet and the state of the boundary layer (laminar/turbulent) at the nose of the missile were varied. Both parameters influence the shock wave boundary layer interaction, which occurs as a direct result of the side jet and affects the control force. Unsteady RANS simulations have been performed in which the mass flow of the side jet was ramped up from zero to a maximum value. The associated transient control force signal was investigated for its sensitivity to the boundary layer state, altitude and properties of the side jet. The performance of side jet control was compared with secondary injection thrust vector control. The main conclusions are:

- Using a side jet where the thrust increases steadily from zero to a maximum value, a discernible time delay exists between the time the side jet thrust becomes maximum and the time the induced force is quasi-static
- In contrast to the SWBLI and the accompanied separation, the control force is slightly affected by the boundary layer state
- A higher side jet thrust results in a higher control force, even though the side jet induced pressure force may decrease as a result
- In contrast to SITVC, SJC loses efficiency with increasing altitude, as the induced pressure force decreases.

References

1. DeSpirito, J.: Lateral reaction jet flow interaction effects on a generic fin-stabilized munition in supersonic crossflows. Technical report ARL-TR-6707, ARMY RESEARCH LABORATORY (2013)
2. Gnemmi, P., Gruhn, P., Leplat, M., Nottin, C., Wallin, S.: Computation validation on lateral jet interactions at supersonic speeds. Int. J. Eng. Syst. Modell. Simul. **5**(1–3), 68–83 (2013)
3. Deere, K.: Summary of fluidic thrust vectoring research at NASA Langley research center. In: 21st AIAA Applied Aerodynamics Conference, AIAA 2003-3800, Orlando, Florida (2003)
4. Babinsky, H., Harvey, J.K.: Shock Wave-Boundary-Layer Interactions, vol. 32. Cambridge University Press, Cambridge (2011)
5. Pandey, A., Casper, K.M., Spillers, R., Soehnel, M., Spitzer, S.: Hypersonic shock wave-boundary-layer interaction on the control surface of a slender cone. In: AIAA Scitech 2020 Forum, AIAA 2020-0815, Orlando, Florida (2020)
6. Schwamborn, D., Gerhold, T., Heinrich, R.: The DLR TAU-code: recent applications in research and industry. In: European Conference on Computational Fluid Dynamics, Delft, Netherland (2006)
7. Kovar, A., Schülein, E.: Effect of side jets in a supersonic flow measured and calculated on a flat plate and a generic missile configuration. RTO-MP AVT-135, German Aerospace Center (2013)

8. Smith, T.A.: Boundary layer development as a function of chamber pressure in the NASA Lewis 1030: 1 area ratio rocket nozzle. In: 24th Joint Propulsion Conference, AIAA-88-3301, Boston, Massachusetts (1988)
9. Schneider, S.P.: Flight data for boundary-layer transition at hypersonic and supersonic speeds. J. Spacecr. Rockets **36**(1), 8–20 (1999)
10. Berry, S.A., Kimmel, R., Reshotko, E.: Recommendations for hypersonic boundary layer transition flight testing. In: 41st AIAA Fluid Dynamics Conference and Exhibit, AIAA 2011-3415, Honolulu, Hawaii (2011)
11. Polivanov, P.A., Sidorenko, A.A., Maslov, A.A.: Transition effect on the shock wave/boundary layer interaction region and the wake at low supersonic Mach number. In: International Conference on the Methods of Aerophysical Research, Paper 1770 (2016)

Airplane Aerodynamics/Propulsion Integration

Aerodynamic Design of Shock Control Bumps Considering Structural Constraints

Andreas Goerttler[1(✉)], Sven Christian Künnecke[2], and Christian Sabater[3]

[1] Institute of Aerodynamics and Flow Technology, German Aerospace Center (DLR), Bunsenstraße 10, 37073 Göttingen, Germany
Andreas.Goerttler@dlr.de
[2] Institute of Lightweight Systems, DLR, Braunschweig, Germany
[3] Institute of Aerodynamics and Flow Technology, DLR, Braunschweig, Germany

Abstract. The design of realistic shock control bumps requires multiple disciplines. In this paper, several shapes of static shock control bumps are investigated numerically and evaluated under structural considerations due to constraints imposed by a realistic morphing spoiler structure.

The bumps are optimized to reduce mainly wave drag of two-dimensional airfoils at a given lift coefficient. The shapes are not only evaluated by their drag reduction potential but also by their structural feasibility. The aerodynamic optimization is done with surrogate based modeling with up to five parameters. It is shown that up to 30% of drag reduction is reached. The penalty of only allowing structurally feasible bump shapes is within an acceptable range of only few percentage points.

Keywords: shock control bumps · drag reduction · computational fluid dynamics · structural considerations

1 Introduction

During cruise flight, a shock emerges in the transonic flow field around an aircraft wing and leads to high wave drag, especially at high cruise speeds. Ashill et al. [1] proposed back in 1992 bumps in proximity to the shocks to control the shock strength of laminar airfoil sections. Since then, shock control bumps are known to be able to change the flow in the vicinity of the shock, changing the strong single shock into a weaker λ-shock characteristic, similar to the shock system displayed with green lines in Fig. 1. Either an oblique shock is created at the foot of the ramp of the bump or a smooth bump leads to multiple compression waves. Both shock systems lead to a reduction in wave drag compared to the original shock, and thus to a lower overall drag. More general information about shock control bumps can be found in the detailed review paper by Bruce & Colliss [2]. However, the reduction is very sensitive towards the shape of the bump. Especially the height and the position of the bump crest play a dominant role. A wrong combination of those parameters can lead to two shocks (see

A. Dillmann et al. (Eds.): STAB/DGLR Symposium 2022, NNFM 154, pp. 671–681, 2024.
https://doi.org/10.1007/978-3-031-40482-5_63

two single red shocks in Fig. 1), and thus to an increase in drag compared to a wing or airfoil without bump. To overcome this sensitivity towards bump parameters, either robust or adaptive shock control bump systems [5] need to be used. Robust means in this context acceptable off-design results of a static bump. A technology to fix the shock position is the concept of variable camber, which was investigated by Werner [15], who has already shown relevant drag reduction with a continuous bump on a three-dimensional wing. In contrast to that study, this analysis investigates a wing with a swept leading edge.

While aerodynamic optimization of the bump shape yields optimum performance, restrictions imposed due to the integration of the bump into the wing structure can limit the drag reduction significantly. Jinks et al. [4] optimize adaptive shock control bumps for a transonic airfoil considering structural constraints. They focused on an aero-structural optimization process with two actuation points deflecting the upper surface. In contrast to this study, the spoiler actuator shall be used for the bump deflection, thus only one single actuation point is necessary, see [6]. In this context, two-dimensional shock control bumps are designed in this study to robustly reduce drag within given structural constraints for a constant Mach and Reynolds number. This includes the constructional feasibility of integrating a retractable bump with a specific shape and height on the spoiler. The aerodynamic optimization of the shock control bump covers a variation of both the height and the crest position of the bump. Furthermore, a variation of the shape function is also investigated. The optimization goal is to diminish the overall drag at a constant lift coefficient. This optimization procedure is done for different lift coefficients to analyze the influence of different flow situations on the bump shape and the drag reduction potential. The optimized shape is then evaluated at off-design lift coefficients and modifications are made to increase the robustness.

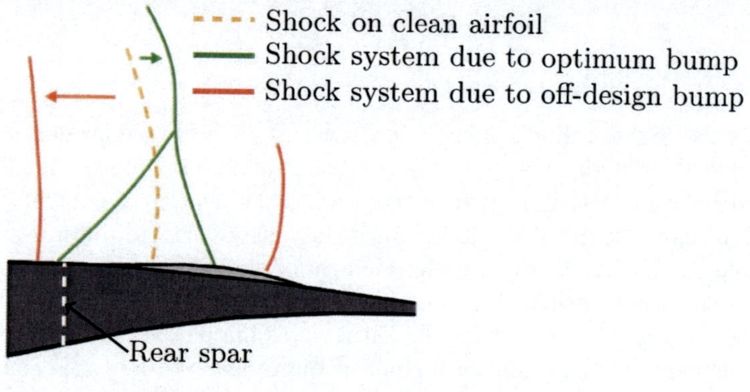

Fig. 1. Sketch of shock systems on airfoil.

2 Methods

The numerical computations are carried out with the DLR inhouse flow solver TAU-Code [12]. Mostly, the Spalart-Allmaras turbulence model was used for the stationary computations. The optimization is done with the Surrogate-Modelling for Aero-Data Toolbox (SMARTy) which is also developed by DLR. The toolbox is used for the sampling of the initial design of experiments with Sobol sequences and the creation of the Kriging surrogate [3]. This approach has already shown successful optimizations of shock control bumps under uncertainty by Sabater et al. [9].

The wing shape of the project HLFC-Win [13] is used as reference geometry. This aerodynamic shape consists of fuselage, wing, vertical tail plane, flap-track fairings, pylon and nacelle. Three-dimensional simulations were carried out as reference solutions. The HLFC (hybrid laminar flow control) suction panels were considered within the transition prediction process only. Transition prediction was performed using *coco* and *lilo* [10,11], with *coco* including the suction in the boundary layer computations and *lilo* performing the stability analysis. This semi-automatic process was done multiple times till sufficient convergence of the transition points was reached. The flow conditions (angle of attack α, transformation angle β and Mach number Ma_n) of a two-dimensional simulation of a normal cut were chosen to match the pressure distribution of the extracted three-dimensional line-in-flight cut, here, especially the shock position and strength. The two-dimensional geometry was created by transferring the extracted airfoil sections at a spanwise coordinate of $\eta = 57\%$ with the transformation angle β into the normal cut section. The pressure distributions were accordingly transferred. The transition position was taken from the corresponding three-dimensional simulation.

3 Investigation of Different Bump Shapes

A robust shock control bump has been optimized for the two-dimensional airfoil section at a spanwise coordinate of $\eta = 57\%$ of the GBD-DLR-2 [13] shape. In the optimization, the bump was considered only on top of an elongated spoiler in the rear part of the airfoil. In this context, the spoiler length was increased by moving the allowable starting point upstream towards the position of the rear spar (see sketch in Fig. 1) to facilitate structural feasibility. This section focuses mainly on two different bump shapes *LoadedBeam* and *StructConsid*, their drag reduction potential and a comparison with guidelines from literature.

3.1 Bump Shape Inspired by a Loaded Beam

The outcome of the optimization for three lift coefficients ($c_l = [0.58, 0.60, 0.62]$) is displayed in Fig. 2. The shown bump shape function *LoadedBeam* is inspired by a beam loaded with an individual force and is used as a reference shape. For all three optimized bumps the rear end of the bump is at the rearmost location of the spoiler. It can be seen that despite this geometrical constraint the height increases with increasing lift coefficient. Furthermore, the crest position of the bump shifts, according to the shock movement, in downstream direction with increasing lift coefficient.

The corresponding drag polars are shown in Fig. 3 to evaluate the bump performance at off-design conditions. Depending on whether the lift coefficient is higher or lower than the design lift coefficient of the bump (black border around colored circles), two different flow characteristics occur. Above the design lift coefficient, all three bump designs allow for higher lift at constant drag or lower drag at constant lift, respectively. At these conditions, the bump still induces a favorable shock system, similar to the green λ shock system in Fig. 1. This leads to lower drag values than the baseline case. However, at lift coefficients below the design point the drag values are significantly higher than the baseline case due to the appearance of two strong shocks, like the red shock system in Fig. 1. After the first shock, which occurs upstream of the bump, the flow accelerates due to the curvature of the bump, reaches supersonic speeds and generates a second strong shock. Those two shocks lead to a significant increase in wave drag. In order to avoid this additional drag, the bump shape must be retractable by an actuation mechanism for flow conditions below the design lift coefficient.

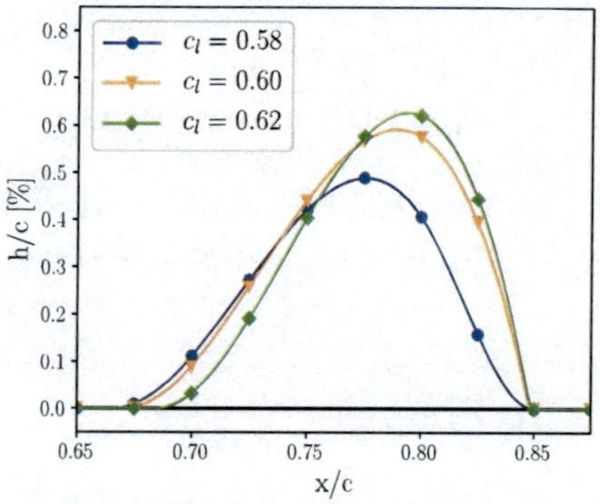

Fig. 2. Optimized *LoadedBeam* bump shapes for three different lift coefficients (not to scale).

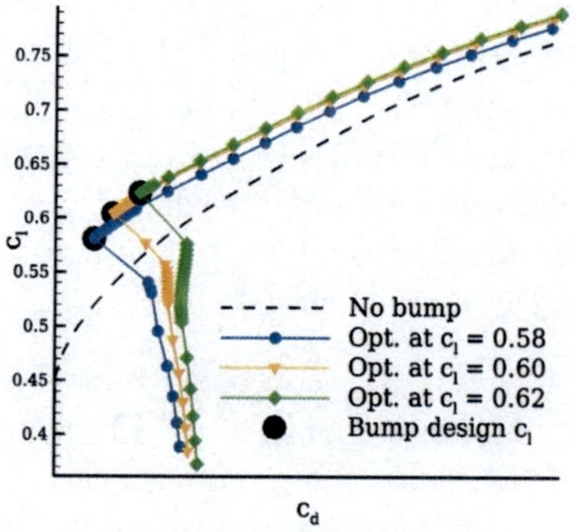

Fig. 3. Drag polars of two-dimensional airfoils with *LoadedBeam* bumps optimized at different lift coefficients.

3.2 Shapes Considering Structural Constraints

The bump shapes optimized with the above mentioned *LoadedBeam* approach have a continuous transition at both, the bump start and end points. However, since in this work a bump on a spoiler should be designed, the structural constraints must be considered. This means that not only the bump start and end points must coincide with the spoiler but also that the transition at the hinge points must be structurally feasible.

The goal was to define a twice continuously differentiable contour between the start and end point of the bump. Furthermore, the maximum value of the curvature should not exceed $7/m$ to ensure structural feasibility.

Figure 4 displays two optimized shapes of the *StructConsid* bump, which are now fixed to the elongated spoiler. They were optimized for $c_l = 0.55$ and $c_l = 0.60$. The shape optimized for the larger lift coefficient is higher and the crest of the bump is at a more downstream position. The shape starts with a kink and continues with a linear slope almost up to the crest. The rear bump edge is also linear. The crest top is defined with a spline of fifth order ensuring a smooth, curvature continuous, transition between the two linear parts. Those two parts can be interpreted as a rotation around the hinge point and are, therefore, structurally feasible. It was initially chosen since the spoiler was supposed to have two rigid parts with a flexible thin structure in between. The two rigid spoiler sections are realized by the linear front and rear edges of the bump shape and the flexible and morphing structure is implemented by using the fifth order spline. This spline around the bump crest was only implemented on a length of 1% of the chord length which corresponds to 5% of the spoiler length. Figure 5 displays the pressure coefficient around the shock position for the clean and bump case at $c_l = 0.60$. The strong shock of the reference case is split into two weaker shocks, with the second shock now being further downstream. With this bump, the total drag is reduced by 22%. This reduction can be attributed both directly to a lower wave drag and indirectly to a lower required angle of attack. Both aerodynamic shapes have been furthermore structurally optimized to design for example the laminate thickness and actuator position, see Künnecke et al. [6]. The structural investigations show, however, that the rear edge of the spoiler bump does not have to consist of a stiff structure but can also be flexible [6]. This would facilitate the structural realization even further and will be the next step in the iterative process between aerodynamic and structural simulations. Both, the structurally optimized bump shape and the simulated realistic clean shape have been aerodynamically investigated to quantify aerodynamic differences compared to the original shapes. This shape has also potential in delaying the buffet onset as shown by Müller et al. and Waldmann et al. [8, 14].

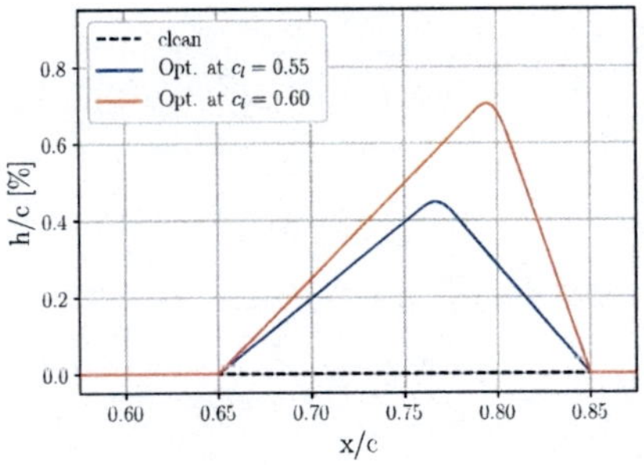

Fig. 4. Optimized *StructConsid* shapes for $c_l = 0.55$ and $c_l = 0.60$ (not to scale).

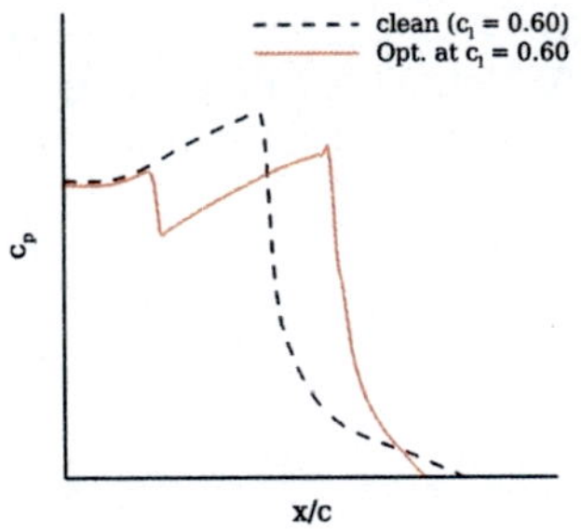

Fig. 5. Pressure coefficient around shock position for clean airfoil and *StructConsid* bump optimized at $c_l = 0.60$.

3.3 Comparison with Different Shapes

Figure 6 shows the drag reduction of the bump optimizations using three shape functions by displaying the normalized drag of the optimized 2D airfoil section for different lift coefficients. The *StructConsid* shape is fixed in its position, whereas the position of the other two shapes is described by the two additional optimization parameters start and length. The shape *LongQ* consists of a linear slope on the first half of the front flank and a cubic spline which describes the bump crest and the rear flank. The focus is not only on aerodynamic performance, but also on structural feasibility and the quantification and evaluation of the differences of the various bump shape functions. The structural constraints originate from a possible integration of a bump actuation mechanism in the wing, which is able to attain not only the clean airfoil and the optimized bump shape on the spoiler but also the airbrake functionality. All investigated shapes show almost similar drag reduction potentials in the analyzed lift coefficient range. The free optimized start points are always further downstream than the fixed start point of the StructConsid bump. This is an advantage at higher lift coefficients, as downstream bumps have lower drag values at these lift coefficients. However, for lower lift coefficients, the shape of the StructConsid bump seems to be advantageous due to the long straight front edge.

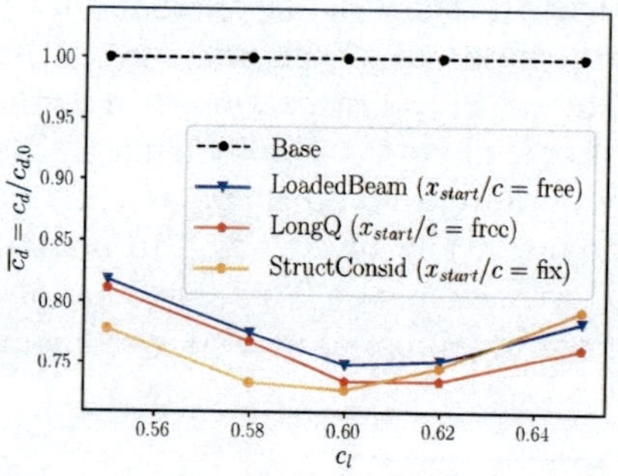

Fig. 6. Normalized drag values of two-dimensional airfoils optimized with different bump shapes over different lift coefficients.

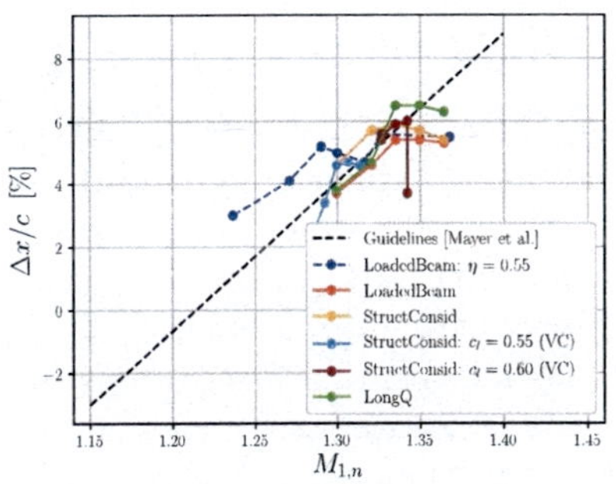

Fig. 7. **Difference between crest position** and clean shock position over the shock strength for multiple optimized bumps.

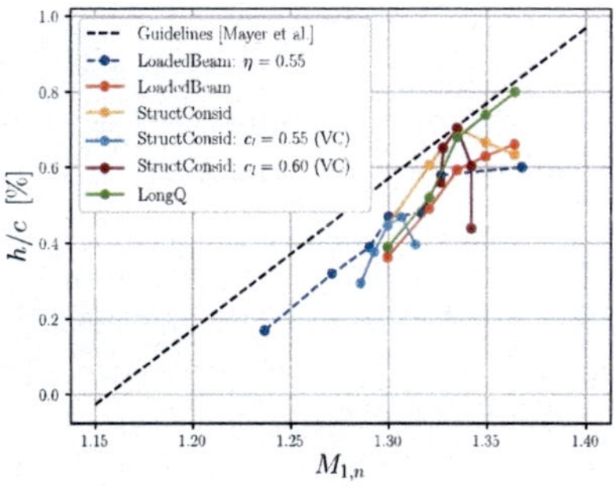

Fig. 8. Height of the optimized bumps over shock strength.

Figures 7 and 8 show a comparison of the analytic guidelines by Mayer et al. [7] and the results of multiple bump shapes obtained within this investigation. The guidelines have been derived with purely aerodynamic optimizations of the airfoiles NACA0012, OAT15A and OALT25. Figure 7 displays the distance between the crest of the bump and the shock position on the clean airfoil. The data are plotted over the shock strength, measured by the maximum pre-shock Mach number $M_{1,n}$ like Mayer et al. [7] proposed. It can be observed that the obtained bump crest positions are clustered around $\Delta x/c = 5\%$. The simulations agree well with the proposed guidelines. Figure 8 displays the crest height over $M_{1,n}$. It can be observed that the agreement is not as good as in Fig. 7. The bumps investigated in this study are smaller by roughly 0.1% to 0.2% chord length. The main reason of this almost constant offset are the structural constraints. Simulations without any structural limitations result in higher bumps with only a slightly greater drag reduction potential. The height increases almost linearly with the shock strength. On average, the bump height is between 0.4% and 0.7% chord length. Multiple simulations with two spanwise positions, different bump shapes and variations of starting point and variable camber (VC) have been performed. However, the differences are not significant and emphasize the robustness of the guidelines.

4 Influence of Different Parameters on *StructConsid* shock control bump

This section investigates the influence of specific parameters of the *StructConsid* shock control bump. First, modifications which trade peak performance against greater range of operation are investigated. The effect of suction rates is discussed and the influence of a deflected flap is analyzed.

4.1 Increase of Robustness

The optimized bumps are aerodynamically not robust in the sense that they result in poor performance in off-design conditions. To increase the aerodynamic robustness the simplest approach, reducing the height, is further investigated. This means that a static bump, without any height adaption, should cover a larger lift range. A multi-point design, however, results in a bump that corresponds to the bump optimized for the lowest lift coefficient, regardless of the weighting factors. This is due to the fact that a bump further back is always penalized by a high drag coefficient at lower lift coefficients. Figure 9 displays the drag polars and the increase of robustness due to a height reduction of 20%. The corresponding shapes are shown in Fig. 10 with the height reduced bumps in dashed lines. It can be seen, that the two original bumps form a sharp drag polar. Due to this sensitive reaction, the drag polar is resolved with up to 1/100° steps. The leftmost point in Fig. 9, the optimum with the lowest drag value and marked with a black border, is reached for a specific angle of attack. But even reducing the angle of attack by a hundredth of a degree causes this airfoil to lose almost all drag reduction. A smaller height, however, leads to a much smoother drag polar. Around the design point, only slightly higher drag values are obtained, but reducing the angle of attack still results in lower drag values than in the clean case. The bump is, therefore, more robust against a variation of the angle of attack with only a small penalty around the design point.

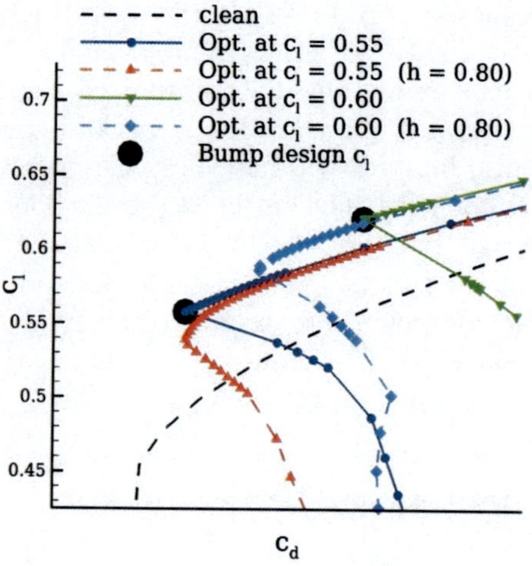

Fig. 9. Drag polars of optimized and height reduced *StructConsid* bumps.

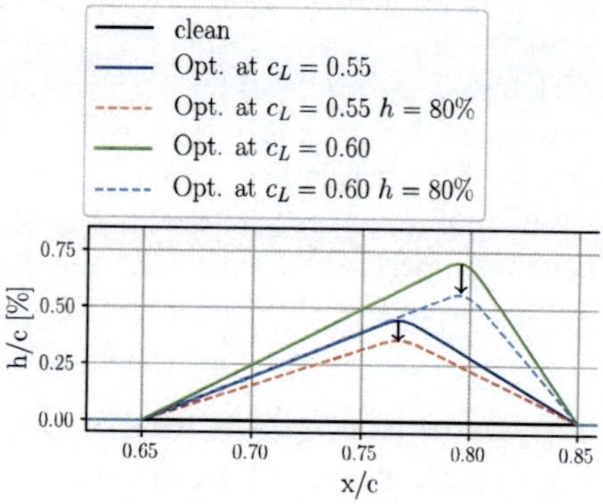

Fig. 10. Shapes of optimized and height reduced *StructConsid* bumps (not to scale).

4.2 Influence of Suction Rates and Transition Point

The influence of different suction rates has been studied with two-dimensional simulations also. However, it was found, that the suction rate does not have a direct, but only secondary effect on the drag reduction mechanism of the bump.

The suction rate particularly changes the location of the transition process from laminar to turbulent flow and thus, influences the shock position and strength. An optimized bump would, therefore, be different compared to a reference bump. However, the relation between specific shock strength and position and the optimized bump remains unchanged.

4.3 Influence of Variable Camber

The position of the shock depends on the flow conditions. The range of operation of a shock control bump can be increased by holding the shock on a desired position. A possible shock fixing mechanism can be used by deflecting a flap and, therefore, changing the camber of the airfoil.

First, the effect of variable camber is investigated for the airfoil without any bump. In this case, a downward deflection ($\delta > 0°$) leads to a downstream shock movement and vice versa for a constant lift coefficient. Therefore, in a second step, bump optimizations were performed for multiple flap deflections. It was investigated how the optimized bump shape differs for each flap deflection. Due to the changing shock position, different optimized bump shapes are expected. Figure 11 displays the drag polars of the corresponding optimized bump shapes in Fig. 12. It is revealed that the lowest drag values around the design point are reached with a flap setting of $\delta = 1°$. The most robust bump is, however, achieved with a flap setting of $\delta = 2°$. These results are based on the combination of shock position and allowed bump position. The flap changes the shock position and, therefore, influences the bump crest position accordingly. A downward deflected

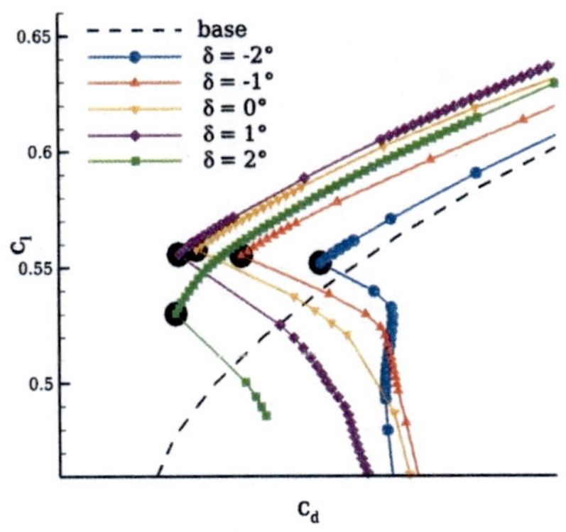

Fig. 11. Drag polars of *StructConsid* bumps optimized at $c_l = 0.55$ with different flap settings δ.

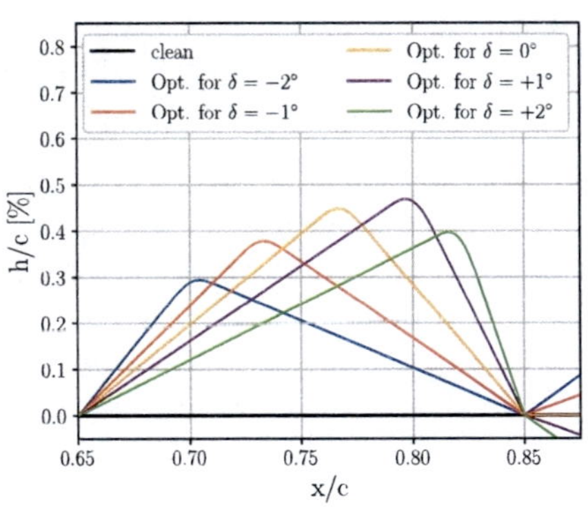

Fig. 12. Shapes of *StructConsid* bumps optimized at $c_l = 0.55$ with different flap settings δ (not to scale).

flap leads to a crest position further downstream. The bump shape optimized with $\delta = -2°$ leads to a crest position at around $x/c = 0.70$. Furthermore, it leads to the steepest front edge but smallest bump height. The shape obtained with a flap setting of $\delta = 1°$ has its crest position at around $x/c = 0.80$ which is roughly three quarters of the bump. This shape most resembles a bump optimized without any constraints. The most robust bump regarding lift changes is achieved by the shape with the shallowest front edge and the crest position furthest downstream.

5 Conclusion

It has been shown, that drag reduction bumps for transonic shock control are possible within structural constraints that occur with an implementation of the bump on a morphing spoiler of a modern transonic wing. Under these constraints, the allowed bump position dominates the possible drag reduction over the exact bump shape. Within the allowed bump position, the possible height and position of the crest define the drag reduction potential. The drag penalty compared to purely aerodynamically optimized bumps is negligible for airfoils with a shock occurring in proximity of the spoiler. Furthermore, some drag reduction potential needs to be traded for an increase in robustness and thus, range of operation of the bump. Boundary layer suction has only a secondary influence via the boundary layer transition and the resulting shock movement. A flap deflection influences the airfoil camber and has a major impact on the optimized bump shape as the shock moves accordingly. This shock fixing mechanism can, however, increase the range of operation of a shock control bump by holding the shock on the desired position. Furthermore, this allows to control the bump shape in case other optimization goals require specific positions or shapes. In future studies, the flap will be taken into account as an additional optimization parameter and the optimized bumps will be investigated on a three dimensional wing.

Acknowledgments. This research was funded in scope of the 5th Federal Aeronautical Research Program (LuFo) V-3 by the German Federal Ministry of Economic Affairs and Climate Action within the project Move-IntegR (FKZ: 20W1729C).

The used aeroshape GBD-DLR-2 was generously provided by the HLFC-Win consortium (Clean Sky 2 Joint Undertaking under the European Union's Horizon 2020 research and innovation programme (Grant agreements CS2-GAM-2018-LPA-AMD-807097-38 and CS2-GAM-2020-LPA-AMD-945583-11)).

References

1. Ashill, P.R., Fulker, J.L., Shires, A.: A novel technique for controlling shock strength of laminar-flow airfoil sections. In: Proceedings of the 1st European Forum on Laminar Flow Technology, Hamburg, 16-18 March 1992, pp. 175-183 (1992)
2. Bruce, P.J.K., Colliss, S.P.: Review of research into shock control bumps. Shock Waves **25**, 451–471 (2015)

3. Han, Z.-H., Görtz, S., Zimmermann, R.: Improving variable-fidelity surrogate modeling via gradient-enhanced kriging and a generalized hybrid bridge function. Aerosp. Sci. Technol. **25**, 177–189 (2013)
4. Jinks, E., Bruce, P., Santer, M.: Optimisation of adaptive shock control bumps with structural constraints. Aerosp. Sci. Technol. **77**, 332–343 (2018)
5. Künnecke, S.C., Vasista, S., Riemenschneider, J., Keimer, R., Kintscher, M.: Review of adaptive shock control systems. Appl. Sci. **11**(2), 817 (2021)
6. Künnecke, S.C., Goerttler, A., Vasista, S., Riemenschneider, J.: Strukturkonzept eines vorgebogenen morphenden Spoilers zur adaptiven transsonischen Stoßkontrolle. 4SmartS conference proceedings, Braunschweig (2022)
7. Mayer, R., Lutz, T., Krämer, E., Dandois, J.: Control of transonic buffet by shock control bumps on wing-body configuration. J. Aircr. **56**(2), 556–568 (2019)
8. Müller, J., Waldmann, A., Goerttler, A., Künnecke, S., Lutz, T.: Increased efficiency of shock control bumps for buffet onset delay due to variable camber. In: Dillmann, A., Heller, G., Krämer, E., Wagner, C., Weiss, J. (eds.) New Results in Numerical and Experimental Fluid Mechanics XIV. STAB/DGLR Symposium 2022. NNFM, vol. 154. Springer, Cham (2024). https://doi.org/10.1007/978-3-031-40482-5_64
9. Sabater, C., Bekemeyer, P., Görtz, S.: Efficient bilevel surrogate approach for optimization under uncertainty of shock control bumps. AIAA J. **58**, 5228–5242 (2020)
10. Schrauf, G.: COCO - A program to compute velocity and temperature profiles for local and nonlocal stability analysis of compressible, conical boundary layers with suction. ZARM Technik Report (1998)
11. Schrauf, G.: LILO 2.1 - User guide and tutorial. GSSC Technical report (2006)
12. Schwamborn, D., Gerhold, T., Heinrich, R.: The DLR TAU-code: recent applications in research and industry. In: ECCOMAS CFD 2006 Conference, 04-08 September 2006, Netherlands (2006)
13. Streit, T., Kruse, M., Kilian, T., Geyr, J.v., Petropoulos, I.: Aerodynamic design and analysis of HLFC wings within the European project HLFC-WIN. In: ICAS 2022 Stockholm
14. Waldmann, A., Müller, J., Goerttler, A., Lutz, T.: Shock control bumps for buffet onset delay on sections of a wing with hybrid laminar flow control. In: Dillmann, A., Heller, G., Krämer, E., Wagner, C., Weiss, J. (eds.) New Results in Numerical and Experimental Fluid Mechanics XIV. STAB/DGLR Symposium 2022. NNFM, vol 154. Springer, Cham (2024). https://doi.org/10.1007/978-3-031-40482-5_65
15. Werner, M.: Application of an adaptive shock control bump for drag reduction on a variable camber NLF wing. AIAA SciTech 2018. Kissimmee, USA (2018)

Increased Efficiency of Shock Control Bumps for Buffet Onset Delay Due to Variable Camber

Jens Müller[1(✉)], Andreas Waldmann[1], Andreas Goerttler[2], Sven Christian Künnecke[3], and Thorsten Lutz[1]

[1] University of Stuttgart, Faculty 6, Institute of Aerodynamics and Gas Dynamics, Pfaffenwaldring 21, 70569 Stuttgart, Germany
jens.mueller@iag.uni-stuttgart.de
[2] German Aerospace Center (DLR), Institute of Aerodynamics and Flow Technology, Bunsenstraße 10, 37073 Göttingen, Germany
[3] DLR, Institute of Composite Structures and Adaptive Systems, Lilienthalplatz 7, 38108 Braunschweig, Germany

Abstract. Within this work, 2D numerical simulations are used to assess the potential of variable camber (VC) to increase the effectiveness in terms of buffet onset delay of shock control bumps on the airfoil of a HLFC wing. Since the bumps are restricted to the spoiler, positioning of the bump relative to the shock is not ideal, which limits the potential in shifting the transonic buffet boundary to higher lift coefficients. VC is used to move the shock towards the leading edge. This increases the capabilities of shock control bumps in moving buffet onset to higher lift coefficients values over a wide range of bump crest positions and bump heights. In addition, VC enables the usage of a structurally designed bump on a morphing spoiler, which is optimized for wave drag reduction without VC and can be used for buffet onset delay with VC applied.

Keywords: transonic buffet · shock control bumps · variable camber

1 Introduction

On conventional transport aircraft, airspeed and angle of attack are limited, among other things, by the occurrence of unsteady shock oscillation, the so-called transonic buffet. In transonic flow, the shock wave/boundary layer interaction leads to shock induced separation. The mutual interaction between separation and shock wave leads to a periodic movement of the shock together with a combination of reattachment and separation of the flow downstream from the shock. Further details on transonic buffet and current research are provided by Giannelis et al. [5]. Due to the unsteady aerodynamics and load fluctuations resulting from the buffet, a sufficient distance from the buffet limit must be ensured during flight [12]. An extension of the flight envelope to higher Mach numbers and angles of attack can be achieved by actively influencing the shock using shock control bumps (SCBs) [10]. SCBs were first introduced by Ashill et al. [1] as an effective

A. Dillmann et al. (Eds.): STAB/DGLR Symposium 2022, NNFM 154, pp. 682–692, 2024.
https://doi.org/10.1007/978-3-031-40482-5_64

method for wave drag reduction in transonic flight. The bump's crest position should be located shortly downstream but close to the shock position to create a λ-shaped shock structure or, ideally, an isentropic compression. An overview of recent research regarding shock control bumps for wave drag reduction is provided in [3]. In terms of buffet onset delay, Birkemeyer et al. [2] found, that a bump located further downstream than for wave drag reduction is able to shift the buffet onset to higher lift coefficients c_L. Guidelines for bump design in terms of buffet onset delay are provided by Mayer et al. [10].

On today's commercial aircraft configurations, the available space for the integration of SCBs and their actuators is limited, which restricts the positioning of SCBs. Furthermore, the desirable use of a SCB for both drag reduction and buffet onset delay is very challenging as the design criteria regarding the position of bump and bump crest differ significantly. Both aspects can thus severely limit the effectiveness of SCBs in shifting the buffet boundary if the bump's crest position is too close to the shock location. If the position of the SCB is not optimal for buffet onset delay, there is a need to shift the location of the shock close to the buffet boundary to ensure an increase of the buffet c_L without violating geometric restrictions or reducing the effectiveness of the SCB to reduce wave drag in cruise flight. The shock position can be influenced, e.g. by static deflection of trailing edge flaps, which change the camber of the airfoil/wing (VC). Since VC has also an effect on the flow separation, it is possible to delay buffet onset as well. For supercritical airfoils, this is shown e.g. by Lee [9] and Despré et al. [4]. Werner [18] combines VC with the application of a SCB on a NLF wing in terms of an optimized wave drag reduction. It is shown that VC can be used to ensure an appropriate shock location relative to the SCB when the crest position of the SCB is limited to a small region of the wing. Richter et al. [13] demonstrate that when combining VC with a position fixed SCB drag reduction can be achieved over a broad range of Mach numbers. Hence, the question arises whether VC can also be used to increase the efficiency of SCBs in terms of buffet onset delay, when the SCB positioning is limited.

The current work is part of the LuFo project Move-IntegR, which focuses on the application of shock control bumps on a hybrid laminar flow control (HLFC) wing of a transonic airliner. It is based on the companion work by Waldmann et al. [17], where parameter studies of SCBs with a focus on buffet onset delay were performed. The aim of the present work is to increase the effectiveness of SCBs in terms of buffet onset delay by means of steady trailing edge flap deflections on a representative 2D airfoil of the HLFC wing. The SCBs are designed to fulfill the integration limitations by the aircraft design as well as considering structural feasibility of a morphing spoiler, which implements the bump shape. This limits the possible bump shapes and thus the SCB's effectiveness for buffet onset delay. The trailing edge flap deflection is used to change the camber of the airfoil in order to shift the shock to a desirable position in relation to the SCB. The impact of VC on the SCB performance is evaluated and the aerodynamic mechanisms are analyzed. Furthermore, it is examined whether VC enables the utilization of a structurally designed SCB on a morphing spoiler for buffet onset delay, which is without VC optimized for wave drag reduction in cruise flight.

2 Geometry and Simulation Methodology

The reference aircraft of the current work is the GBD-DLR-2 geometry provided by the DLR. This geometry is based on the Airbus XRF1 research aircraft, which was redesigned with HLFC within the HLFC-WIN project [16]. The aircraft contains fuselage, wing, nacelle, pylon, flap-track fairings, and vertical tail plane. From the engine mounting to the wing tip suction panels are installed in the leading edge area and the wing shape is modified to ensure laminar flow on the outer wing. The investigations within the work at hand are carried out on a 2D wing section extracted at $\eta = \frac{y}{0.5b} = 0.57$, which is located on spoiler 5. To account for the wing sweep of the 3D wing, the 2D airfoil geometry as well as the inflow conditions are transformed according to the cosine rule, see [12], with a sweep angle $\Lambda = 25°$. The reference cruise condition of the aircraft is $M_\infty = 0.85$ at $c_L = 0.5$. Details regarding the baseline flow and the 2D transformation are provided in [17]. Buffet onset for each bump/flap setting is determined from a lift polar with the $\Delta\alpha$ criterion described by Lawson et al. [8]. Details on the buffet determination and the corresponding linearization of the lift polar used within this work are provided in [17]. The lift polars are determined in stationary 2D RANS airfoil simulations utilizing the block structured, compressible, finite volume CFD solver FLOWer [14]. The Reynolds-Averaged Navier Stokes (RANS) equations are closed by the Spalart-Allmaras one-equation turbulence model. Each polar consists of 56 to 66 different angles of attack α with the inflow conditions equivalent to the 2D transformed cruise conditions of the reference aircraft. Due to the suction on the HLFC wing at this spanwise position, the laminar to turbulent transition is fixed at $x_{tr}/c = 48.4\%$ in all 2D simulations within this work. The simulation framework used in this work is described in the companion paper by Waldmann et al. [17]. In line with [17], the grid applied in the 2D CFD simulations is block structured and of C-H type. It is discretized with about 150 points located on the upper and lower side of the airfoil and point clustering in the vicinity of the shock and the spoiler. The grid topology is described by Mayer et al. [11].

Within the simulation framework, the bump shapes and the flap deflections are created by modifying the airfoil geometry prior to the mesh generation in Gridgen. The trailing edge flap of the GBD-DLR-2 geometry at $\eta = 57\%$ used for VC is located directly behind the spoiler at relative chord length $x/c > 85\%$. The flap deflection is realized by a rotation of the trailing edge around the hinge point M and an interpolation of this displacement vector to the rest of the flap depending on the x distance to the hinge point. The resulting flap deflection for $\delta = -2°$ and its orientation relative to a SCB is shown in Fig. 1.

To account for a robust parameter space, two different bump types, SCB1 and *StructConsid* (SC), are considered on the upper side of the airfoil based on the previous investigation by Waldmann et al. [17]. For each shape, bump crest position and bump height are varied while the geometrical extend is fixed to spoiler 5 ($70\% \leq x/c \leq 85\%$). This enables a structural integration via a morphing spoiler. The SCB1 shape is based on aerodynamic studies by Sommerer et al. [15]. It is designed for an optimized shock control but does not consider structural feasibil-

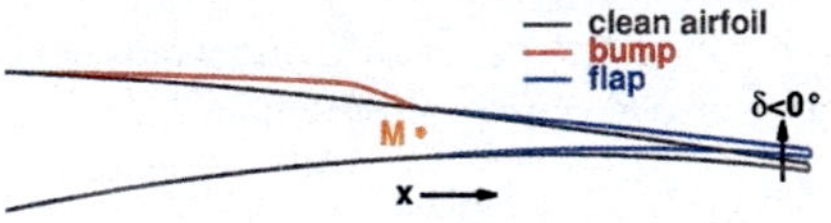

Fig. 1. Flap deflection

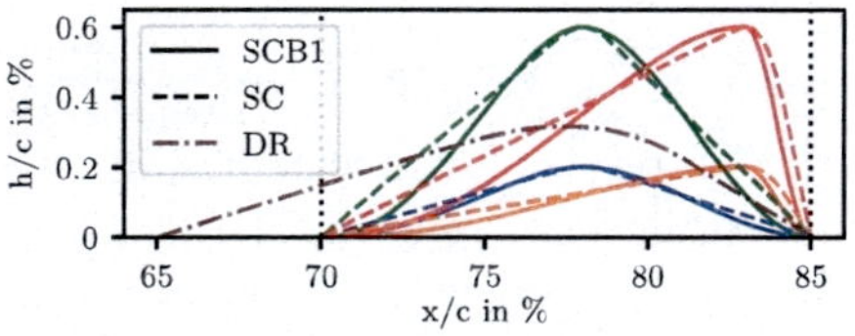

Fig. 2. Exemplary bump shape. Dotted vertical lines: spoiler start and end. Matching colors: similar x_{cr} and h_{cr}

ity. The SC bump shape, developed within the Move-IntegR project, takes geometric restrictions into account that make a structural realizability more likely. Within the current investigation, the maximum curvature of the SC bump is limited to 7/m to enable a feasible morphing spoiler structure. Details on the SC bump shape are provided by Goerttler et al. [6]. SCB1 and SC are exemplary shown in Fig. 2 for different crest positions x_{cr} and crest heights h_{cr}. Besides focusing on SCBs in combination with VC only in terms of shifting the buffet boundary to higher c_L values when the position and shape is restricted due to structural constraints, a combined SCB suitable for wave drag reduction and buffet onset delay is a major goal in SCB design. This approach of using a combined bump is motivated by the concept of using a single actuator to create a position-fixed SCB on a morphing spoiler as investigated e.g. in [7]. Therefore, a bump based on the SC shape but optimized for wave drag reduction by Goerttler et al. [6] is considered. Based on this aerodynamic bump shape and the corresponding aerodynamic forces, a structural design of an adaptive bump was conducted on a morphing spoiler with the process chain presented by Künnecke et al. [7]. The resulting bump shape of this structurally optimized morphing spoiler, further called *Drag Reduction* (DR), is used to investigate whether VC can be used to utilize a bump shape optimized for drag reduction also for buffet onset delay. Within the investigations of Goerttler et al. [6], the range of the spoiler was increased upstream to the position of the rear spar. Hence, the DR bump ranges from $65\% \leq x/c \leq 85\%$ with $x_{cr}/c = 77.5\%$ and $h_{cr}/c = 0.32\%$, see Fig. 2.

3 Buffet Onset Delay by Application of Variable Camber

The capabilities of VC to increase the efficiency of SCBs located on the spoiler of the HLFC wing section are presented in the following. First, the SCB performance without VC is briefly presented. It is followed by a detailed analysis of the impact of upward directed trailing edge deflections on the SCB performance in terms of buffet onset delay and the corresponding aerodynamic mechanisms.

3.1 SCB Performance Without VC

Waldmann et al. [17] performed an extensive study on different crest positions and bump heights of two selected SCB shapes for buffet onset delay at the same wing section of the XRF1-HLFC wing. They show, that the geometric restriction of the bump to be located on the spoiler geometry limits the capability of the SCB in terms of increasing the lift coefficient of buffet onset. Within the spatial constraints it is shown, that both, SCB1 and SC have the highest potential for buffet onset delay for crest positions $x_{cr}/c \approx 83.5\%$ to 84%, very close to the rear end of the spoiler, and high crest heigths of $h_{cr}/c \approx 7\%$. These high bumps with far downstream located crest positions are hard to realize on an aircraft structure due to the high curvatures and the necessary prebend of the spoiler to hold the spoiler's trailing edge against the aerodynamic loads down on the flap. Based on the findings of Waldmann et al. [17], the parameter range of SCB1 and SC bump shapes used for the subsequent VC investigations is selected. While the bump position is fixed to the spoiler ($70\% \leq x/c \leq 85\%$), the following crest positions and crest heights are considered: $x_{cr} = [78, 80, 82, 83]\%c$, $h_{cr} = [0.2, 0.4, 0.6]\%c$. The increase in lift coefficient c_{L_b} at which buffet occurs at the current configuration compared to the buffet $c_{L_{b,clean}}$ of the clean airfoil, $\Delta c_L = \frac{c_{L_b} - c_{L_{b,clean}}}{c_{L_{b,clean}}}$ is used as a measure of bump efficiency on buffet onset delay. Figure 3 displays Δc_L for the selected SCB1 and SC bumps as well as the DR bump without VC. A shift of c_{L_b} up to 9% is possible for the most downstream crest position but hard to realize from a structural point of view due to the aforementioned necessary prebend. Shifting the bumps crest position further upstream decreases Δc_L significantly with even a negative impact on the buffet c_L at $x_{cr}/c = 78\%$. The structurally designed drag reduction (DR) bump shows the lowest Δc_L and is accordingly not applicable to buffet onset delay. The unfavorable position of the bump crest compared to the shock at the clean airfoil leads to the poor buffet onset delay performance of the SCBs at the more upstream positions, see [17]. Hence, for both, structurally realizable SCB1/SC bump shapes and the DR bump applied for buffet onset delay, the position of the shock has to be shifted further upstream. In the following, this is realized by deflecting the trailing edge flap, which alters the camber of the airfoil.

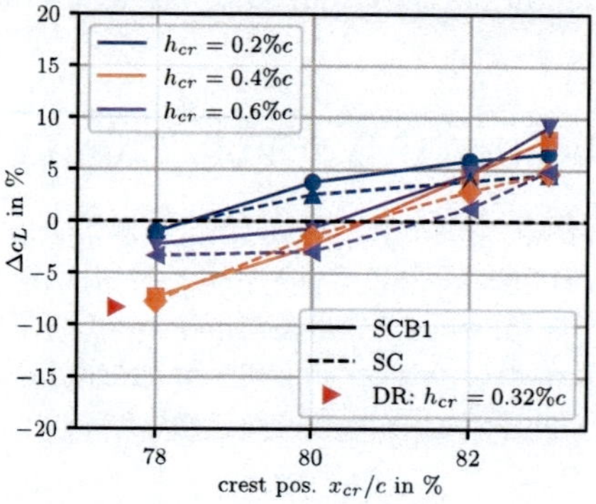

Fig. 3. Buffet onset delay of different bumps without VC

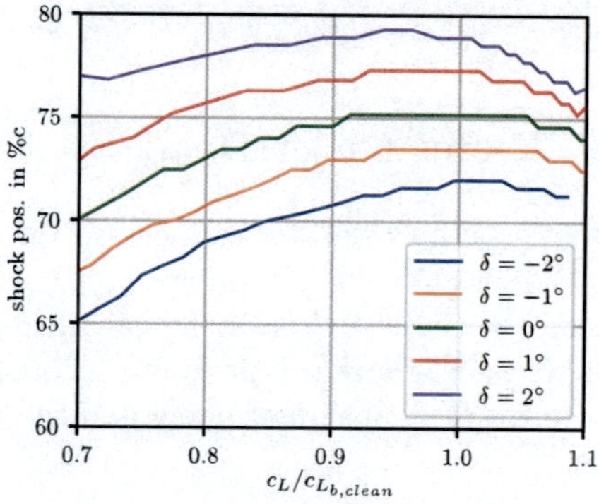

Fig. 4. Shock position for different flap angles (clean airfoil)

3.2 Effect of VC on the Efficiency of SCBs for Buffet Onset Delay

First, the influence of different flap deflections δ on the shock position on the upper side of the clean airfoil without SCBs is evaluated in Fig. 4 in the relevant lift region close to the buffet c_{L_b} of the clean airfoil. Consistent with the results of Werner on a NLF wing [18], a downward directed flap deflection ($\delta > 0°$) leads to a downstream movement of the shock, while an upward flap deflection ($\delta < 0°$) shifts the shock towards the leading edge. As pointed out in Sect. 3.1, the shock needs to be shifted further upstream due to the structural constraints. Hence, only upward directed flap deflections with $-2° \leq \delta < 0°$ are applied to the SCBs shown in Fig. 3 to increase the efficiency of the SCBs for buffet onset delay. This allows a reliable statement about the potential and limits of VC at an acceptable number of simulations. Figure 5 shows the impact of the selected flap angles on the buffet performance of the SCB1 (Fig. 5a) and SC/DR bumps (Fig. 5b). The effect of VC on the buffet c_L of the clean airfoil is plotted as reference. VC is capable of increasing c_{L_b} up to 4% on the investigated HLFC wing section even without a SCB. Combining VC with SCBs, the buffet boundary can be shifted to higher c_L values for a wide parameter range of x_{cr} and h_{cr} for both, SCB1 and SC bump shapes. As is the case without VC, the aerodynamically optimized bump shape SCB1 has a slightly higher potential in increasing c_{L_b} (up to $\Delta c_L = 19.5\%$) than the SC bump design (up to $\Delta c_L = 14.8\%$), which focuses on structural feasibility. The highest Δc_L is achieved with $x_{cr}/c = 83\%$ and $h_{cr}/c = 0.6\%$. The clean airfoil as well as some bump shapes (e.g. SC with $x_{cr}/c = 83\%$ and $h_{cr}/c = 0.2\%$) reveal, that the potential of VC is limited, which is discussed in detail in Sect. 3.3. Another key finding from Fig. 5b is, that VC enables the use of a SCB optimized for wave drag reduction, with a shape that is structurally sized, for buffet onset delay. Δc_L can be raised by 14.9% up to $\Delta c_L = 6.6\%$. The trend of the curve shows, that higher flap deflection might lead to even higher Δc_L.

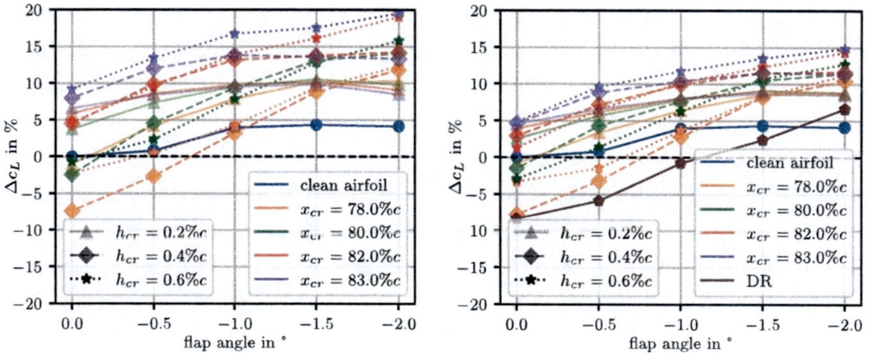

(a) SCB1 (b) SC and DR

Fig. 5. Buffet onset delay achievable with VC

3.3 Specific Analysis of Selected Bumps

Within this section, the aerodynamic effects leading to the increase in Δc_L when VC is applied are analyzed. Since the mechanisms are comparable for SCB1, SC and DR shape, only SC is considered within this section.

First, the SC bump with $x_{cr}/c = 78\%$ and $h_{cr}/c = 0.4\%$ at $\delta = 0°$ and $\delta = -2°$ is studied in more detail. Without VC, this bump has the most negative impact on the buffet c_L with $\Delta c_L = -7.8\%$, see Fig. 5b. With increasing upward directed flap deflection, c_{L_b} can be shifted up to $\Delta c_L = 11.6\%$. Figure 6 shows the density gradient for both cases at $c_L/c_{L_{b,clean}} = 90\%$ (Fig. 6a and 6b) and at $c_L = c_{L_{b,clean}}$ (Fig. 6c and 6d). Comparing the shock structures at $c_L/c_{L_{b,clean}} = 90\%$ it can be observed that without flap deflection (Fig. 6a) a λ-shaped shock structure develops. The first, weaker shock is fixed to the spoiler hinge due to the abrupt contour change. While there is no separation at $c_L/c_{L_{b,clean}} = 90\%$, shock induced separation extends over $\approx 10\%c$ at $c_{L_{b,clean}}$. Since the main shock is located at the bump crest, the bump is not capable of reattaching the separation within the extend of the bump, leading to an early shock oscillation as described in Sect. 1. With a flap deflection of $-2°$ at $c_L/c_{L_{b,clean}} = 90\%$ (Fig. 6b), there is only a single shock located at the spoiler hinge, which clearly shows, that VC is capable of shifting the shock position upstream even with a SCB. At higher c_L (Fig. 6d) a small λ-shaped shock structure develops but the main shock still remains further upstream than the bump crest. Due to the bump shape, the flow is accelerated between the shock and the bump crest, which effectively suppresses flow separation and accordingly delays buffet onset to higher c_L.

This effect can be seen in detail when looking at the pressure coefficient (c_p) and skin friction coefficient (c_f) distribution in the bump region for different bump heights at $c_{L_{b,clean}}$ (Fig. 7). The selected bump is the best performing SC bump ($x_{cr}/c = 83\%$, $h_{cr}/c = 0.6\%$) and its lower counterpart ($h_{cr}/c = 0.2\%$). Both are able to raise c_{L_b} with and without VC but the higher bump crest significantly increases the effectiveness of VC for buffet onset delay compared to the lower bump. The clean airfoil without bump and VC is plotted as reference. While the clean airfoil shows a significant shock induced separation region ($c_{f,x} < 0$), the $h_{cr}/c = 0.2\%$ bump without VC leads to a fast reattachment

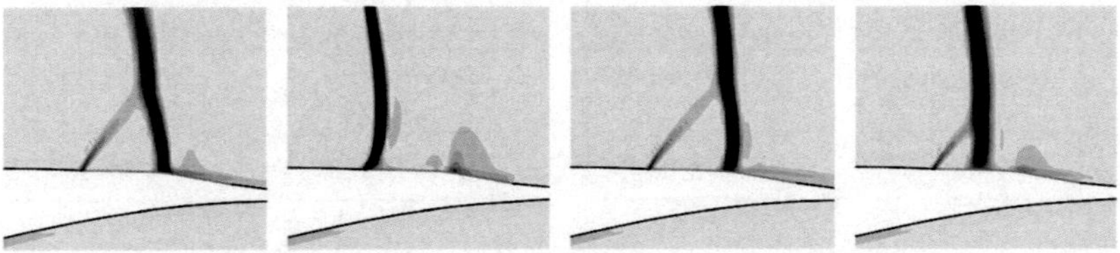

(a) $\delta = 0°$ and $c_L = 0.9c_{L_{b,clean}}$ (b) $\delta = -2°$ and $c_L = 0.9c_{L_{b,clean}}$ (c) $\delta = 0°$ and $c_L = c_{L_{b,clean}}$ (d) $\delta = -2°$ and $c_L = c_{L_{b,clean}}$

Fig. 6. Numerical Schlieren (SC: $h_{cr} = 0.4\%c$, $x_{cr} = 78\%c$), bump marked in red

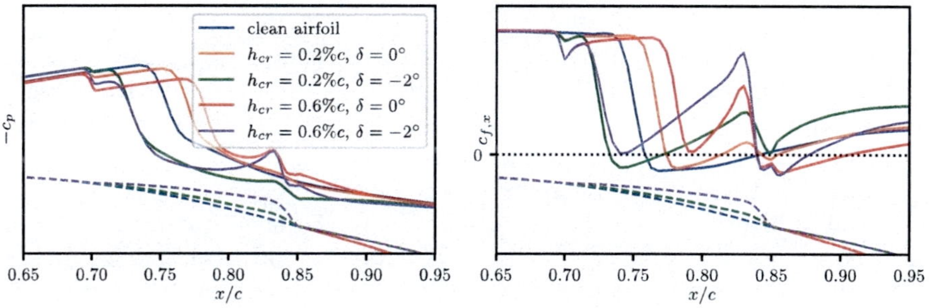

Fig. 7. c_p and $c_{f,x}$ on the suction side at $c_L = c_{L_b}$ **of the clean airfoil for different SC** bumps ($x_{cr} = 83\%c$) with and without VC. Dashed lines: Bump contour

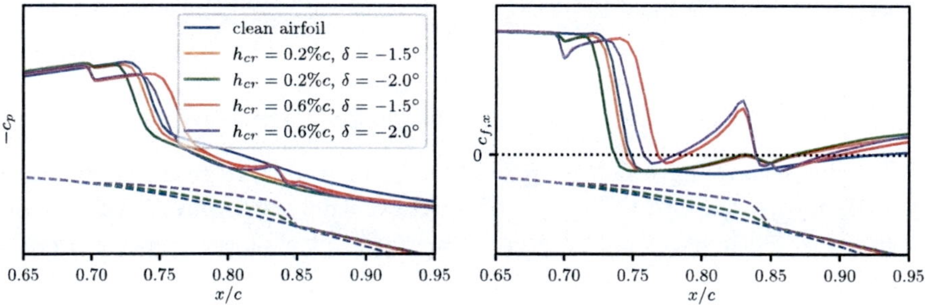

Fig. 8. c_p and $c_{f,x}$ on the suction side. $c_L = c_{L_b}$ of the SC bump with $h_{cr} = 0.2\%c$, $x_{cr} = 83\%c$, $\boldsymbol{\delta} = -1.5°$. All SC bumps $x_{cr} = 83\%c$. Dashed lines: Bump contour

of the flow on the uprising bump flank, which delays the shock oscillation. The $h_{cr}/c = 0.6\%$ bump is even capable of completely suppressing the shock induced separation but leads to a separation region downstream of the bump crest due to the high curvature. A flap angle of $-2°$ shifts the main shock further upstream. **This significantly increases the buffet onset delay. Even though the shock induced separation remains nearly unaffected for the shown** c_L, the extend of the separation region downstream of the bump crest is considerably reduced or completely suppressed by the flap deflection.

Further analysis on the mechanisms of VC can be done when comparing $\boldsymbol{\delta} = -1.5°$ with $\boldsymbol{\delta} = -2°$ for the aforementioned bumps at a higher c_L. Figure 8 shows c_p and c_f **distributions at the buffet** c_L of the $x_{cr}/c = 83\%$, $h_{cr}/c = 0.2\%$, $\boldsymbol{\delta} = -1.5°$ bump. While $\boldsymbol{\Delta c_L}$ of the SC bump with $h_{cr}/c = 0.6\%$ increases when the flap is further deflected upwards from $\boldsymbol{\delta} = -1.5°$ to $\boldsymbol{\delta} = -2°$, it leads to a slight reduction in $\boldsymbol{\Delta c_L}$ for $h_{cr}/c = 0.2\%$. For both bump heights, the shock is shifted upstream due to the increased flap deflection with a shock induced separation occurring at all cases. With $h_{cr}/c = 0.2\%c$, the higher flap deflection shifts the shock itself, and correspondingly the starting point of the shock induced separation to a more upstream position. This increases the total size of the separation region since the earlier reattachment of the flow downstream of the bump due to the higher flap deflection is less pronounced than the shift in shock position. The bump crest itself is not high enough to ensure a relevant reattachment of the flow on the bump. Hence, c_{L_b} cannot be shifted to a higher c_L when increasing the flap deflection from $\delta = -1.5°$ to $\delta = -2°$. A different situation occurs when the bump height is increased to $h_{cr}/c = 0.6\%$. Due to the high curvature at the bump crest, the flow is accelerated upstream from the crest, which leads to a fast reattachment of the flow upstream the bump crest position with a comparable chordwise extend of the shock induced separation region for both flap angles. The flow acceleration is also visible in the peak in the c_p distribution at the bump crest (see Fig. 8). The flow separates again downstream of the bump crest position, where the higher flap angle of $\delta = -2°$ leads to an earlier reattachment. Hence, for the higher bump crest, c_{L_b} can still be shifted to higher c_L when increasing the flap deflection from $\delta = -1.5°$ to $\delta = -2°$, which is not possible for the lower bump crest.

4 Conclusion

The analyses above have shown that VC can be used to increase the effectiveness of SCBs integrated into the spoiler on a HLFC wing section in terms of buffet onset delay. Without VC, the highest achievable Δc_L is 9% within the selected parameter range. However, this is only possible when the crest position is located close to the rear end of the bump, which impedes a structural realization. Hence, more upstream located crest positions are being pursued. Since this decreases Δc_L due to the unfavorable position of the bump crest relative to the shock, an upward directed flap deflection can be used to increase c_{L_b}. With moderate flap angles up to $\delta = -2°$, an increase in c_{L_b} up to 19.5% compared to the clean airfoil can be achieved. Even for the most upstream located bump crest positions, which fulfill the structural realization requirements, an increase in c_{L_b} of about 10% is possible. This demonstrates, that VC is applicable for different bump shapes over a wide range of bump crest positions and bump heights. In addition, VC enables the use of a common bump shape for wave drag reduction and buffet onset delay. This reduces the difficulty in designing multifunctional bumps, and simplifies system integration and complexity. VC is capable of shifting the shock position to more favorable positions relative to the bump. Hence, shock induced flow separation can be reduced, increasing the effectiveness of the SCB. Furthermore, the upward directed flap deflection diminishes the flow separation occurring downstream of the bump crest. Both effects shift buffet onset to higher c_L and extend the flight envelope to higher Mach numbers and angles of attack.

Acknowledgments. The funding of the activities through the Federal Ministry for Economic Affairs and Climate Action under contracts FKZ 20W1729E in the framework of the air transport research program is gratefully acknowledged. The used aeroshape GBD-DLR-2 has been generously provided by the HLFC-Win consortium (Clean Sky 2 Joint Undertaking under the European Union's Horizon 2020 research and innovation program (Grant agreements CS2-GAM-2018-LPA-AMD-807097-38 and CS2-GAM-2020-LPA-AMD-945583-11)).

References

1. Ashill, P., Fulker, J., Shires, A.: A novel technique for controlling shock strength of laminar flow airfoil sections. Technical report, DGLR, Part 6(92-01-022), pp. 175-183 (1992)
2. Birkemeyer, J., Rosemann, H., Stanewsky, E.: Shock control on a swept wing. Aerosp. Sci. Technol. **4**(3), 147–156 (2000). https://doi.org/10.1016/S1270-9638(00)00128-0
3. Bruce, P.J.K., Colliss, S.P.: Review of research into shock control bumps. Shock Waves **25**(5), 451–471 (2015). https://doi.org/10.1007/s00193-014-0533-4
4. Despré, C., Caruana, D., Mignosi, A., Reberga, O., Corrège, M.: Buffet active control-experimental and numerical results. Technical report, ONERA (FRANCE) (2001)
5. Giannelis, N.F., Vio, G.A., Levinski, O.: A review of recent developments in the understanding of transonic shock buffet. Prog. Aerosp. Sci. **92**, 39–84 (2017). https://doi.org/10.1016/j.paerosci.2017.05.004
6. Goerttler, A., Künnecke, S.C., Sabater, C.: Aerodynamic design of shock control bumps considering structural constraints. In: Dillmann, A., Heller, G., Krämer, E., Wagner, C., Weiss, J. (eds.) New Results in Numerical and Experimental Fluid Mechanics XIV. STAB/DGLR Symposium 2022. NNFM, vol. 154. Springer, Cham (2024). https://doi.org/10.1007/978-3-031-40482-5_63
7. Künnecke, S.C., Goerttler, A., Vasista, S., Riemenschneider, J.: Strukturkonzept eines vorgebogenen morphenden Spoilers zur adaptiven transsonischen Stosskontrolle. 4SmartS conference proceedings, Braunschweig, Germany (2022)
8. Lawson, S., Greenwell, D., Quinn, M.K.: Characterisation of buffet on a civil aircraft wing. In: 54th AIAA Aerospace Sciences Meeting. San Diego, USA (2016). https://doi.org/10.2514/6.2016-1309
9. Lee, B.H.K.: Effects of trailing-edge flap on buffet characteristics of a supercritical airfoil. J. Aircr. **29**(1), 93–100 (1992)
10. Mayer, R., Lutz, T., Krämer, E., Dandois, J.: Control of transonic buffet by shock control bumps on wing-body configuration. J. Aircr. **56**(2), 556–568 (2019). https://doi.org/10.2514/1.C034969
11. Mayer, R., Lutz, T., Krämer, E.: Numerical study on the ability of shock control bumps for buffet control. AIAA J. **56**(5), 1978–1987 (2018). https://doi.org/10.2514/1.J056737
12. Obert, E., Slingerland, R., Leusink, D.J.W., van den Berg, T., Koning, J.H., van Tooren, M.J.L.: Aerodynamic Design of Transport Aircraft. Ios Press, Amsterdam (2009)
13. Richter, K., Strohmeyer, D., Rosemann, H.: Numerische Untersuchungen an transsonischen Profilen mit flexibler Hinterkante und Konturbeule. Deutscher Luft- und Raumfahrtkongress 2000, Leipzig, Germany (2000)
14. Rossow, C.C., Kroll, N., Schwamborn, D.: The MEGAFLOW Project – Numerical Flow Simulation for Aircraft. In: Di Bucchianico, A., Mattheij, R., Peletier, M. (eds.) Progress in Industrial Mathematics at ECMI 2004. Mathematics in Industry, vol. 8, pp. 3–33. Springer, Berlin, Heidelberg (2006). https://doi.org/10.1007/3-540-28073-1_1
15. Sommerer, A., Lutz, T., Wagner, S.: Design of adaptive transonic airfoils by means of numerical optimisation. In: European Congress on Computational Methods in Applied Sciences and Engineering, Barcelona, Spain (2000)

16. Streit, T., Kruse, M., Kilian, T., v. Geyr, J., Petropoulos, I.: Aerodynamic design and analysis of HLFC wings within the European Project HLFC-WIN. In: 33rd Congress of the International Council of the Aeronautical Sciences (ICAS), Stockholm, Sweden (2022)
17. Waldmann, A., Müller, J., Goerttler, A., Lutz, T.: Shock control bumps for buffet onset delay on sections of a wing with hybrid laminar flow control. In: Dillmann, A., Heller, G., Krämer, E., Wagner, C., Weiss, J. (eds.) New Results in Numerical and Experimental Fluid Mechanics XIV. STAB/DGLR Symposium 2022. NNFM, vol. 154. Springer, Cham (2024). https://doi.org/10.1007/978-3-031-40482-5_65
18. Werner, M.: Application of an adaptive shock control bump for drag reduction on a variable camber NLF wing. In: 2018 AIAA Aerospace Sciences Meeting. American Institute of Aeronautics and Astronautics, Kissimmee, USA (2018). https://doi.org/10.2514/6.2018-0789

Shock Control Bumps for Buffet Onset Delay on Sections of a Wing with Hybrid Laminar Flow Control

Andreas Waldmann[1(✉)], Jens Müller[1], Andreas Goerttler[2], and Thorsten Lutz[1]

[1] University of Stuttgart, Pfaffenwaldring 21, 70569 Stuttgart, Germany
waldmann@iag.uni-stuttgart.de

[2] German Aerospace Center, Bunsenstr. 10, 37073 Göttingen, Germany

Abstract. This work describes parameter studies of shock control bumps focused on delay of buffet onset. A two-dimensional wing section simulation setup was derived, consistently with the 3D flow about the mid-wing of a swept wing aircraft equipped with a hybrid laminar flow control suction system. Two bump shapes respecting constraints of the existing spoiler were defined and analyzed for their ability to delay the onset of transonic buffet. Both types of geometry were found to be able to delay the occurrence of buffet at the given cruise conditions, with both cases requiring bumps whose crest is positioned far downstream. Both smooth curvature devices and structurally realizable bumps showed potential benefits, moving buffet onset to higher lift coefficients c_L.

Keywords: transonic buffet · shock control bumps

1 Introduction

Modern long range commercial aircraft cruise at transonic speeds and strive to balance economic operation with high travel speeds. Increased flight speeds, increased loading due to maneuvers or incoming gusts can move the operating point toward the edge of the flight envelope. The limit is imposed by the occurrence of buffet, which is a phenomenon involving shock/boundary layer interaction on the wing upper surface. This is an undesirable condition with oscillating loads, which may impede safe operation of the aircraft. Better understanding of this complex phenomenon has long been a goal of aerodynamics research, as improved knowledge may lead to reduced necessary safety margins in future aircraft designs.

Three-dimensional buffet involves spanwise propagation effects and broadband oscillation characteristics, whereas 2D buffet on airfoils is typically a single-frequency phenomenon. Giannelis et al. [4] provided a comprehensive review of buffet research, highlighting the differences. However, transonic buffet in both 3D and 2D buffet occurs at elevated Mach number M_∞ or increased angle of attack α, and is typically associated with an upstream displacement of the shock, its

A. Dillmann et al. (Eds.): STAB/DGLR Symposium 2022, NNFM 154, pp. 693–703, 2024.
https://doi.org/10.1007/978-3-031-40482-5_65

chordwise oscillation and a flow separation downstream from it. Bruce and Colliss [2] published a comprehensive overview of bump effects. While shock control bumps positioned at or near the shock can contribute to drag reduction, Birkemeyer et al. [1] observed that bumps located further downstream caused a delay of buffet onset. It has been established that this delay is caused by inhibited separation downstream from the shock. More complex configurations have been since described in literature (e.g. Colliss et al. [3]), such as 3D shaped bumps generating longitudinal wake vortices in order to control the boundary layer. Mayer et al. [9] described a comprehensive approach to the problem of designing shock control bumps aimed at delaying the onset of buffet. This includes both 2D and 3D cases, and the use of two-dimensional and three-dimensional bump shapes.

Shock control bumps have been historically investigated as a method to improve wave drag at increased speeds, enhancing efficiency at off-design conditions. An appropriately placed bump converts the compression via single shock into a λ-shaped two-shock structure or into isentropic compression waves. Both of these reduce total pressure loss, resulting in overall reduced wave drag. Similar bumps can also be used to delay the onset of transonic buffet oscillations, however the positioning and sizing requirements are different than those for wave drag reduction.

Practical application of laminar flow on wings is a promising avenue of research to achieve efficiency gains in commercial aviation. While such laminar flow may be achieved via pure shape optimization, this natural laminar flow approach is challenging to implement on swept wings at transonic speeds. The additional use of suction, termed hybrid laminar flow control (HLFC), can help achieve substantial regions of laminar flow at typical aircraft operating conditions.

The Move-IntegR project involves research on feasible designs of shock control bumps for application on a future transonic airliner wing equipped with hybrid laminar flow control. The baseline is the aircraft geometry GBD-DLR-2 [18] provided by DLR via the HLFC-Win project. This is a full wing-body configuration including flap track fairings, engine nacelles and a suction system on the wing upper side. The aim of the present work is an improved understanding of appropriate bump design characteristics with respect to buffet onset on such an aircraft wing within the constraints imposed by operation and wing systems design.

2 Simulation Methodology

The work was carried out using a framework based on automated dispatching and processing of 2D finite volume computational fluid dynamics (CFD) simulations. Buffet onset was determined using steady-state simulations of wing sections, avoiding more expensive time-resolved runs prioritizing a larger parameter space with given computational resources.

2.1 Flow Solver

All flow simulations employed the block-structured compressible flow solver FLOWer [15]. The solver ingests grids created by the Gridgen mesh generation software and uses a cell-centered approach. Such block-structured grids consist of about 150 points for upper and lower side, respectively, and are periodic in spanwise direction. Far field boundary conditions set at about 50 chord lengths from the airfoil surface determine the inflow. Steady state Reynolds-Averaged Navier Stokes (RANS) simulations were conducted throughout, using the Spalart-Allmaras one-equation turbulence model [17]. Each polar was computed by sampling at 56 incidences, with higher resolution in the upper half of the incidence range where buffet is expected.

2.2 Processing Framework

The aircraft working group of the Institute of Aerodynamics and Gas Dynamics (IAG) has been using a Python-based framework for simulation and optimization of shock control bumps over a period of several years, with an overview published by Mayer et al. [10]. The *SCBopt3* framework is visualized in Fig. 1.

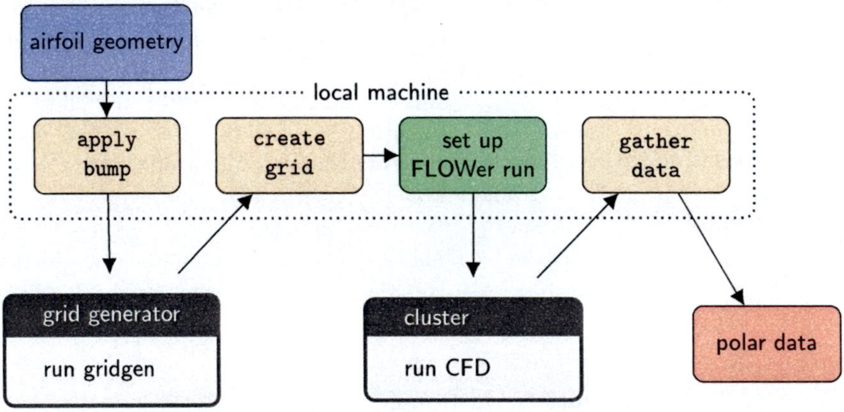

Fig. 1. Overview of the SCBopt simulation framework.

The approach enables consistent computational grids and simulation setups, ensuring that every case runs analogously. The framework includes automatic creation of C-H type grids using the Gridgen software, with the topology described by Mayer et al. [10]. Points are clustered near the datum shock position and the spoiler location on the upper surface. The grid generation approach is based on modifying the baseline airfoil shape by applying a displacement function normal to the baseline surface, which results from the bump parameterization. The resulting contour of the airfoil with applied bump is then input into the grid generator, avoiding grid morphing and ensuring consistent wall-normal size growth rates.

3 Baseline Flow

Exploration of the design space in 2D requires the replication of the flow condition encountered at the wing section with a two dimensional analogue. The present work focuses on the application of shock control bumps on spoiler 5 on the wing. A streamwise slice through the wing at $\eta = 57\%$, which is near the spanwise midpoint of the spoiler, serves as the two-dimensional geometry for the analysis below.

3.1 Flow Quantity Transformation

The present work is aimed at sampling a large part of the parameter space, which necessitates a large number of designs. Since each design implies the simulation of a sufficient α range on order to obtain a polar, the study focuses on 2D RANS simulations. The cruise condition was set to $M_\infty = 0.85$ and an aircraft reference lift coefficient of $C_L = 0.5$. A 3D RANS simulation result of the full aircraft at the aforementioned cruise conditions and incorporating controlled transition location is used to obtain a base pressure distribution representing cruise conditions at the spoiler.[1]

The appropriate conditions for the 2D simulation require accounting for the wing sweep, which is done using the cosine rule as described by Obert et al. [13]. The relevant quantities such as pressure and force coefficients, as well as wing section geometry, were transformed using the appropriate transformation equations. The suction results in a transition location of $x_{\mathrm{tr}}/c = 48.4\%$ at this spanwise position. No suction was modeled in the 2D simulations, with the effect accounted for by prescribed transition at the given x_{tr}/c.

The proposition by Mayer et al. [9] to use the sweep angle of the shock has been slightly altered. While the local shock sweep angle is 23.5°, an angle of 25° showed more consistent results with the 3D solution in terms of shock position and strength.

3.2 Buffet Determination via Polar

With buffet being an inherently unsteady phenomenon, indirect methods are required to detect its onset when only steady state data is available. Lawson et al. [7] described various methods for buffet detection, most of which rely on experimental or unsteady data. The present work employs a method based on lift polars acquired from steady state RANS data. The $\Delta\alpha$ criterion, as described by Lawson et al., detect the lift break in the polar as the angle of attack where the lift departs from linear behavior. This method involves estimation of the linear lift coefficient slope before the break. A parallel line shifted by $\Delta\alpha = 0.1°$ with respect to this linear region intersects the polar at a an angle of attack slightly above the lift break. The angle of attack α_{b} and the lift coefficient $c_{L,\mathrm{b}}$ of this

[1] This solution was provided to the authors by DLR in the context of the Move-IntegR project and was created using the TAU unstructured flow solver.

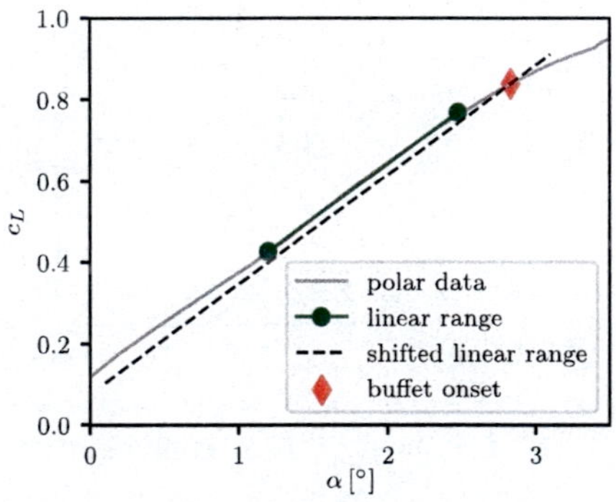

Fig. 2. Buffet determination on the baseline airfoil using the $\Delta\alpha$ criterion

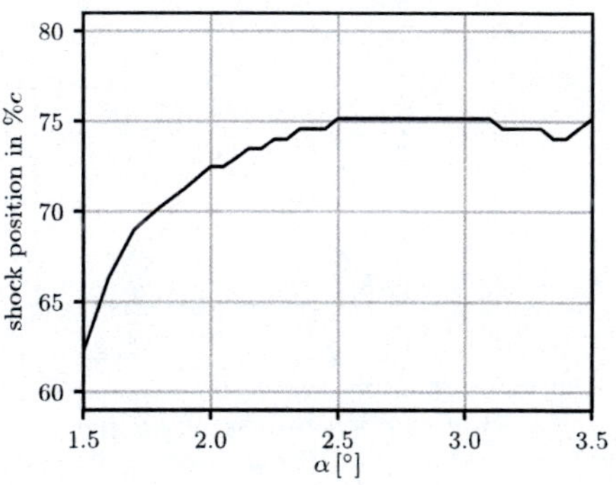

Fig. 3. Shock position on the baseline airfoil

intersection point are considered the angle of attack and the lift coefficient at which buffet onset occurs.

The lift break detection is employed for lack of better data, as buffet detection based on purely steady-state results is merely an approximation. The difficulty arises from the lift polar being not completely linear. Nevertheless, it provides a robust and quick method of approximating buffet onset. In the present work, the complex shape of the lift polar as shown in Fig. 2 was approximated using piecewise linear fitting. Using the pwlf package [6] for the python programming language, each polar was described using three line segments. The middle segment is used as a representation of the linear portion of the lift polar. The linear fitting parameters were specifically tuned for the airfoil type and provide reliable results in this particular case.

4 Shock Control Bumps on a Spoiler

Bump geometries are defined on the wing section upper surface in terms of displacement in normal direction from the baseline surface. In two dimensional shapes, the main parameters determining the buffet performance are the crest position x_{cr}/c and the crest height h_{cr}/c. c denotes the local chord length. The target parameter is $\Delta c_L = (c_{L,b,\text{bump}} - c_{L,b,\text{baseline}})/c_{L,b,\text{baseline}}$, i.e. the shift of the buffet boundary toward higher lift coefficient compared to the baseline geometry, expressed as a percentage.

4.1 Bump Types

Two different bump shape parameterizations are considered in the present work. SCB1 is the result of aerodynamic studies by Sommerer et al. [16] and includes a smooth transition from the baseline surface to the bump with tangent continuity. The second bump type *StructConsid* (SC) is a shape created within the Move-IntegR project aimed at obtaining a structurally realizable contour.

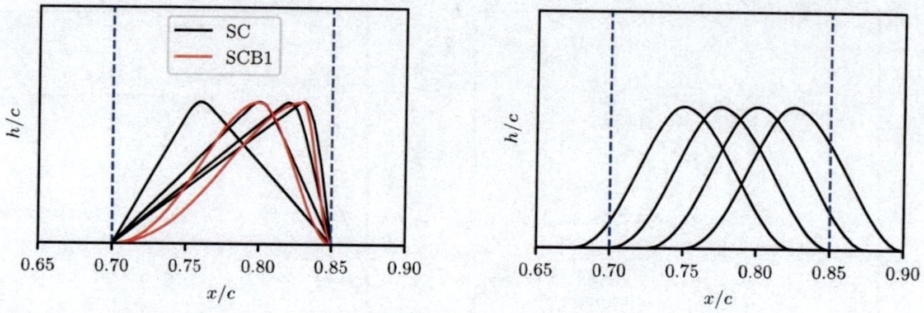

(a) Bumps constrained by spoiler (b) SCB1 without position constraint

Fig. 4. Shapes of bumps with a length of 15%c. Dashed vertical lines indicate spoiler start and end. Axes not to scale.

Smooth SCB1 type bumps are typically intended to create compression waves on the upstream flank, reducing the shock strength [2]. Bumps with a kink between the wing surface and the forward ramp of the bump cause a secondary oblique shock, resulting in a λ-shaped two-shock structure. The SC shape is intended to be realizable with a conventional hinge at the spoiler mounting point at $x/c = 70\%$. Its shape is intended to represent a rotation of the spoiler about its hinge, with a fifth order polynomial describing the portion around the crest. The rear flank is linear, and connects the crest portion with the wing surface at $x/c = 85\%$. More detail on the geometry and design intent can be found in the companion paper by Goerttler et al. [5]. Representative shapes of SCB1 and SC bumps with different crest positions are shown in Fig. 4a.

The HLFC wing configuration includes a spoiler on the upper side, with a hinge located at $x/c = 70\%$ and a chordwise extent of the spoiler of 15%. Ignoring realizability with respect to materials and actuator design, these constraints impose a spoiler location and maximum chordwise length. The buffet bump design guidelines by Mayer et al. [9] indicate that an appropriate bump crest location x_{cr}/c should be offset downstream by $\Delta x_{cr}/c \approx 5\%$ in the present case. This results from the calculated pre-shock Mach number $M_{1,n} = 1.33$ for the current geometry and flow conditions. More specifically, Mayer et al. [10] found an optimum crest location of 7% downstream from the shock for buffet onset delay.

The bump length realizable on the spoiler is shorter than the typical value of 25%c in bump literature [2], which presents a challenge. The shock position on the base geometry is shown in Fig. 3, showing that a bump designed according to the guidelines can be positioned on the spoiler, with a crest position near its downstream end.

4.2 Losses Due to Spatial Constraints

Spoiler-mounted bumps constrain the design space, since typical unconstrained bumps described above tend to be longer and can be positioned relative to the baseline shock position. The effect of the constraint on the achievable Δc_L is shown in Fig. 5. Rear-mounted bumps with crest positions $x_{cr}/c > 80\%$ are beneficial in all cases, albeit symmetric SCB1 bumps such as shown in Fig. 4b achieve higher Δc_L values (Fig. 5b) than those that are restricted to the spoiler's chordwise extent (Fig. 5a). While crest position is indeed the main parameter for the bump's function as a device for buffet onset delay, the steep rear flank necessitated by the restriction of the spoiler's extent decreases the absolute amount of buffet boundary shift.

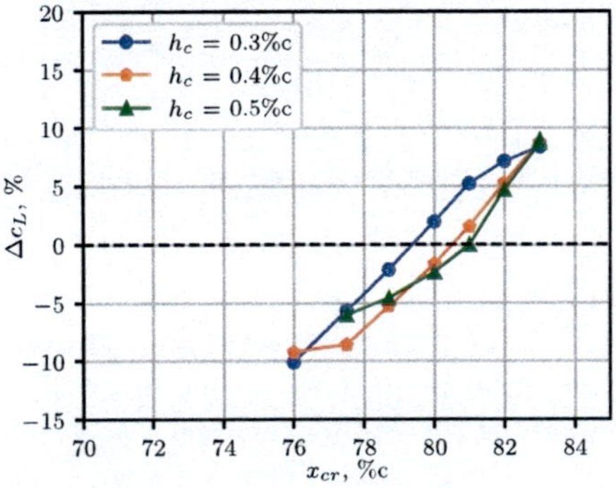

(a) Restricted to spoiler

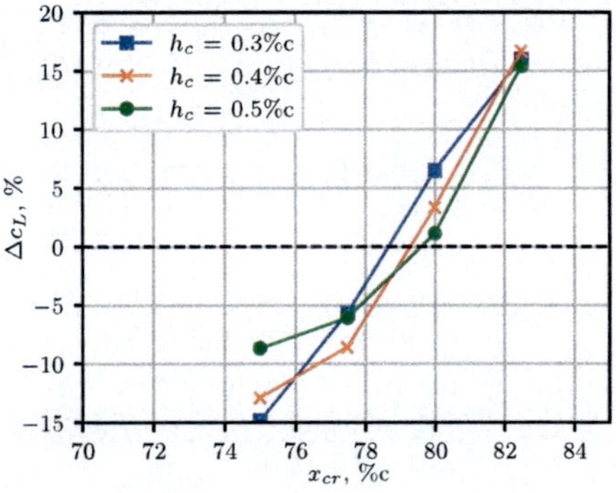

(b) No position constraint

Fig. 5. Buffet onset delay achievable by SCB1 bumps with varying x_{cr} and h_{cr}.

4.3 SCB1 and SC Parameter Variation

In order to gain an understanding of bump performance across the two-dimensional parameter space (x_{cr}, h_{cr}), a design of experiments based on a latin hypercube sampling (LHS) (McKay et al. [11]) approach is carried out. A Gaussian process regression (a.k.a. Kriging) is carried out based on sampled data points, which involves the use of the python package scikit-learn [14] to conduct hyperparameter tuning. The approach evaluates Δc_L for each sample and generates a response surface for the entire parameter space. For both bump types, 20 samples are used following Loeppky et al. [8].

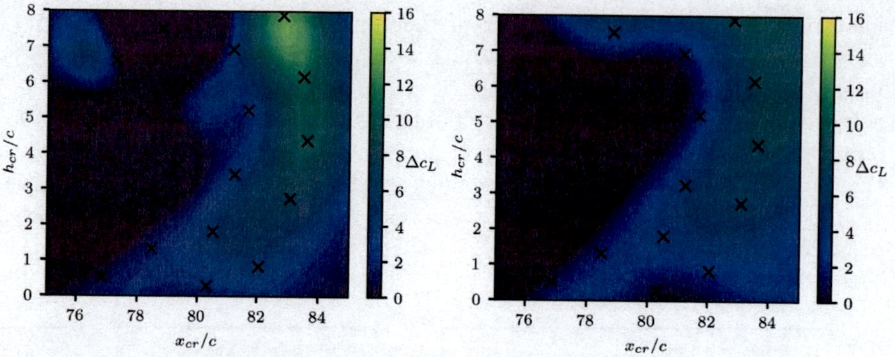

(a) SCB1 (b) SC

Fig. 6. Gaussian process regression response surfaces for SCB1 and SC bumps. Sampling points shown using black crosses.

Contour maps visualizing the resulting surrogate model are shown in Fig. 6. While small bumps achieve minor benefits with respect to buffet onset delay with x_{cr}/c positions near the chordwise center of the spoiler, worthwhile parameter combinations with significant Δc_L tend to be clustered around downstream extremes and greater bump crest heights. These observations remain valid irrespective of the actual bump shape. However, while bumps with extreme asymmetry and crest positions as x_{cr}/c near 83% perform well in terms of Δc_L, they are very difficult to realize using an actuator system and a structure which needs to enable both the baseline shape and the extended bump. The positioning of actuation devices within the available internal space is challenging, requiring a trade-off between realizability and potential aerodynamic benefit.

4.4 Parameter Impacts

The buffet onset, as detected via the polar, is associated with shock inversion which denotes beginning upstream motion of the shock. Delaying buffet onset is achieved by placing the bump in such a manner that the local surface curvature is reduced and the shock reaches further downstream. The upstream motion starts at higher levels of attack, effectively delaying the onset of buffet. The opposite is achieved by upstream placement of the crest, which causes a c_p increase at the upstream ramp of the bump followed by a second acceleration and a shock near the crest. This is visualized for both SCB1 and SC bumps at $x_{cr}/c = 76\%$ in Figs. 7a and 7b. The upstream pressure increase differs between the two bump shapes, with the smooth SCB1 transition creating a gradual increase, and the kink at the start of the SC bump causing a c_p jump consistent with an oblique shock. The main downstream shock is positioned directly on the crest on both bump shapes. Followed by the downward slope of the downstream ramp, this promotes early separation. In contrast, further downstream bump positions at $x_{cr} > 80\%$ permit the shock to move further downstream in chordwise direction.

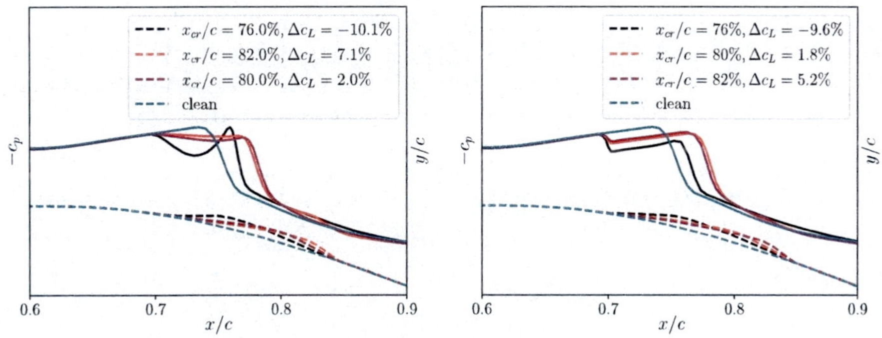

(a) SCB1 (b) SC

Fig. 7. Suction side pressure distributions, at constant $c_{L,b}$ of the baseline geometry. The c_p distributions (solid lines) are shown in the upper part, the bump contour (dashed lines) is in the lower part of the figures for reference.

At the $c_{L,b}$ value of the baseline geometry, it is still located on the upstream ramp, with the acceleration over the bump crest occurring only after the shock. This achieves the desired effect of shifting buffet onset to higher values of c_L.

5 Conclusion

The above analysis has shown that is is possible to find a shock control bump design capable of delaying buffet onset at typical cruise conditions within the constraints of the wing spoiler system. The buffet delay performance in terms of increase of $c_{L,b}$ trend in the same direction irrespective of the bump shape type. Higher bumps and far downstream crest positions are beneficial for buffet onset delay. Therefore, feasible bump shapes can be generated on the spoiler surface, although the crest position x_{cr}/c required for this is close to the spoiler trailing edge and therefore difficult to realize using an actuator system. A combined use of spoiler-mounted bumps and upward deflection of trailing edge devices can help alleviate this issue, as discussed in the companion work by Müller et al. [12]. Further analysis of the parameter space using infill methods in the upper right quadrant of the surrogate model representation may reveal optimum locations. In addition, forthcoming studies shall extend these findings to three-dimensional bumps.

The present investigation is based on steady-state CFD simulations, with the analysis methods for buffet detection tailored for the particular wing section geometry of the GBD-DLR-2. The analysis relies on a single method of buffet detection, which could be improved by incorporating additional methods described in Sect. 3.2. A specific validation using experimental data is currently not feasible due to lack of experimental data. Experimental studies of the buffet onset behavior of the GBD-DLR-2 wing would aid the validation of buffet analysis methods, and should be considered in the future.

Acknowledgement. The funding of the activities through the Federal Ministry for Economic Affairs and Climate Action under contracts FKZ 20W1729E in the framework of the air transport research program is gratefully acknowledged. The used aeroshape GBD-DLR-2 has been generously provided by the HLFC-Win consortium (Clean Sky 2 Joint Undertaking under the European Union's Horizon 2020 research and innovation program (Grant agreements CS2-GAM-2018-LPA-AMD-807097-38 and CS2-GAM-2020-LPA-AMD-945583-11)).

References

1. Birkemeyer, J., Rosemann, H., Stanewsky, E.: Shock control on a swept wing. Aerosp. Sci. Technol. **4**(3), 147–156 (2000). https://doi.org/10.1016/S1270-9638(00)00128-0
2. Bruce, P.J.K., Colliss, S.P.: Review of research into shock control bumps. Shock Waves **25**(5), 451–471 (2015). https://doi.org/10.1007/s00193-014-0533-4
3. Colliss, S.P., Babinsky, H., Nübler, K., Lutz, T.: Vortical structures on three-dimensional shock control bumps. AIAA J. **54**(8), 2338–2350 (2016). https://doi.org/10.2514/1.J054669
4. Giannelis, N.F., Vio, G.A., Levinski, O.: A review of recent developments in the understanding of transonic shock buffet. Prog. Aerosp. Sci. **92**, 39–84 (2017). https://doi.org/10.1016/j.paerosci.2017.05.004
5. Goerttler, A., Künnecke, S., Sabater, C.: Aerodynamic design of shock control bumps considering structural constraints. In: Dillmann, A., Heller, G., Krämer, E., Wagner, C., Weiss, J. (eds.) New Results in Numerical and Experimental Fluid Mechanics XIV. STAB/DGLR Symposium 2022. NNFM, vol. 154. Springer, Cham (2024). https://doi.org/10.1007/978-3-031-40482-5_63
6. Jekel, C.F., Venter, G.: pwlf: A Python Library for Fitting 1D Continuous Piecewise Linear Functions (2019). https://github.com/cjekel/piecewise_linear_fit_py
7. Lawson, S., Greenwell, D., Quinn, M.K.: Characterisation of buffet on a civil aircraft wing (2016). https://doi.org/10.2514/6.2016-1309
8. Loeppky, J.L., Sacks, J., Welch, W.J.: Choosing the sample size of a computer experiment: a practical guide. Technometrics **51**(4), 366–376 (2009). https://doi.org/10.1198/TECH.2009.08040
9. Mayer, R., Lutz, T., Krämer, E., Dandois, J.: Control of transonic buffet by shock control bumps on wing-body configuration. J. Aircr. **56**(2), 556–568 (2019). https://doi.org/10.2514/1.C034969
10. Mayer, R., Lutz, T., Krämer, E.: Numerical study on the ability of shock control bumps for buffet control. AIAA J. **56**(5), 1978–1987 (2018). https://doi.org/10.2514/1.J056737
11. McKay, M.D., Beckman, R.J., Conover, W.J.: A Comparison of three methods for selecting values of input variables in the analysis of output from a computer code. Technometrics **21**(2), 239–245 (1979). https://doi.org/10.2307/1268522. https://www.jstor.org/stable/1268522
12. Müller, J., Waldmann, A., Goerttler, A., Künnecke, S., Lutz, T.: Increased efficiency of shock control bumps for buffet onset delay due to variable camber. In: Dillmann, A., Heller, G., Krämer, E., Wagner, C., Weiss, J. (eds.) New Results in Numerical and Experimental Fluid Mechanics XIV. STAB/DGLR Symposium 2022. NNFM, vol. 154. Springer, Cham (2024). https://doi.org/10.1007/978-3-031-40482-5_64

13. Obert, E., et al.: Aerodynamic Design of Transport Aircraft. Ios Press, Amsterdam (2009)
14. Pedregosa, F., et al.: Scikit-learn: machine learning in python. J. Mach. Learn. Res. **12**, 2825–2830 (2011)
15. Rossow, C.C., Kroll, N., Schwamborn, D.: The MEGAFLOW project – numerical flow simulation for aircraft. In: Di Bucchianico, A., Mattheij, R., Peletier, M. (eds.) Progress in Industrial Mathematics at ECMI 2004. Mathematics in Industry, vol. 8, pp. 3–33. Springer, Heidelberg (2006). https://doi.org/10.1007/3-540-28073-1_1
16. Sommerer, A., Lutz, T., Wagner, S.: Design of adaptive transonic airfoils by means of numerical optimisation. In: European Congress on Computational Methods in Applied Sciences and Engineering (2000)
17. Spalart, P., Allmaras, S.: A one-equation turbulence model for aerodynamic flows. In: 30th Aerospace Sciences Meeting and Exhibit. American Institute of Aeronautics and Astronautics, Reno, NV, U.S.A. (1992). https://doi.org/10.2514/6.1992-439
18. Streit, T., Kruse, M., Kilian, T., Geyr, V.J., Petropoulos, I.: Aerodynamic design and analysis of HLFC wings within the European project HLFC-win. In: ICAS, Stockholm (2022)

Author Index

A

Alt, Katharina 637
Avila, Marc 529

B

Babij, Georg 603
Bagheri, Eman 329
Bahavar, Philipp 509
Bailly, Joëlle 623
Barakos, George 623
Basov, Leo 133
Batmaz, Ege 509
Becker, Stefan 329
Bell, James R. 27
Breitsamter, Christian 231, 613
Büchner, Adam 219
Buhr, Alexander 27
Bur, Reynald 361

C

Cieplik, Fabian 559

D

Danker, Marten 38
Davies, Kristopher 425

E

Ebert, Carola 519
Eck, Mario 48
Ehrenfried, Klaus 294
Ehrenfried, Klaus 58, 78, 88
Ehrmanntraut, Simon 219

F

Fechter, Stefan 305
Fehrs, Michael 3, 415, 425
Feldmann, Daniel 529
Fey, Uwe 68
Fischer, Andreas 529

G

Gageik, Manuel 68
Giehler, Julian 361
Goerttler, Andreas 671, 682, 693
Gottfried, Sebastian 219
Grabe, Martin 133
Grenson, Pierre 361
Gries, Torsten 68, 78
Grille Guerra, Adrián 3

H

Hannemann, Volker 143
Hartmann, Ulrich 613
Hein, Stefan 475
Heine, Daniela 58, 78
Heine, Gerrit 78
Helm, Sebastian 415, 425
Henning, Arne 27, 339
Hepperle, Martin 186
Hollmann, René 196
Holste, Kristof 153
Hong, Seong H. 623
Huismann, Immo 305
Hussong, Jeanette 496

J

Jakirlic, Suad 496
Jentys, Mauricio M. 231
Jung, Sung N. 623

K

Kandulski, Arne 559
Kang, Hee J. 623
Keßler, Manuel 349
Kilian, Thomas 382
Kim, Do-Hyung 623
Klar, Peter J. 153
Klein, Marten 435
Knopp, Tobias 465
Kohl, Andreas 590
Koppenwallner, Georg E. 177

A. Dillmann et al. (Eds.): STAB/DGLR Symposium 2022, NNFM 154, pp. 705–707, 2024.
https://doi.org/10.1007/978-3-031-40482-5

Kowalski, Thomas 68
Krämer, Ewald 349
Krengel, Markus Dino 186
Krenkel, Lars 559, 569
Kunhappan, Deepak 241
Künnecke, Sven Christian 671, 682

L
Langer, Stefan 252, 262
Lange-Schmuckall, Fabian 196
Leicht, Tobias 262
Lim, Joon W. 623
Löffler, Stephan 119
López Parras, Rubén 529
Löwe, Johannes 241
Lübon, Christian 445
Lüdeke, Heinrich 372
Lutz, Thorsten 682, 693

M
Maigler, Maximilian 38
Mertens, Christoph 3
Martinez Schramm, Jan 163
Meyer, Oliver 38
Müller, Jens 682, 693
Müller, Martin Michael 603
Müller, Max 78, 88
Müller, Tobias Simon 78, 88

N
Nabikhani, Amin 339
Nagel, Aaron 133
Niehaus, Konstantin 579

O
Oblapenko, Georgii 143
Oehme, Felix 529

P
Peitsch, Dieter 48
Probst, Axel 315

Q
Quadrio, Maurizio 569
Quero, David 14

R
Rädel, Martin 196
Ramaswamy, Deepak Prem 99
Rempke, Arne 273
Rist, Ulrich 392, 402, 455
Ritter, Markus R. 3
Rohlfs, Lennart 109
Römer, Tristan M. 455
Ruhland, Johannes 613
Rütten, Markus 559, 569, 647

S
Sabater, Christian 671
Schäfer, Clara 153
Schein, Jochen 38
Schiepel, Daniel 539, 549
Schmalz, Martin 14
Schmeling, Daniel 509, 539, 549, 590
Schmidt, Heiko 435
Schmidt, Leni 163
Schmidt, Oliver T. 392
Schneider-Brachert, Wulf 559
Schnepf, Christian 658
Schrauf, Geza 382
Schreyer, Anne-Marie 99
Schütte, Andreas 637
Schwöppe, Axel 262
Selent, Björn 392
Shishkin, Andrei 549
Siebert, Florian 206
Steininger, Rinaldo 623
Stelzer, Vera 559
Stemmer, Christian 569
Ströer, Philip 465
Stück, Arthur 219
Sugawara, Hideaki 623

T
Tanabe, Yasutada 623
Tauwald, Sandra Melina 559, 569
Theiss, Alexander 475
Thieme, Mathis 119
Togiti, Vamshi 284
Tsai, Pei-Yun 435
Tschüter, Ronny 305

V
van der Wall, Berend G. 623
van Oudheusden, Bas W. 3
Vielsmeier, Veronika 559
Volk, Marie-Christine 579

von Germersheim, Lennart 529
von Soldenhoff, Richard 486

W

Wachter, Florian 329
Wagner, Claus 88, 509, 579, 590
Waldmann, Andreas 682, 693
Wang, Xiaoyu 496
Webner, Florian 590
Weinman, Keith A. 294
Weiss, Julien 109, 119, 519
Wendler, Johannes 305
Wenzel, Christoph 402, 455
Werner, Felix 539
Werner, Michael 647
Westhoff, Andreas 579
Westphal, Christian 48
Wickersheim, Robin 349
Wilke, Gunther A. 623
Wu, Yongxiang 402
Wunderlich, Tobias Franziskus 206
Wünsch, Olaf 559

Z

Zakrzewski, Alexander 196
Zastrow, Jonas 315, 647
Zorn, Jana 153